Phosphate deposits of the world

VOLUME 1

Proterozoic and Cambrian phosphorites

EDITED BY
P.J. COOK AND
J.H. SHERGOLD
*Division of Continental Geology,
Bureau of Mineral Resources, Canberra, Australia*

International
Geological Correlation Programme
Project 156:
Phosphorites

IUGS UNESCO IGCP

The right of the University of Cambridge to print and sell all manner of books was granted by Henry VIII in 1534. The University has printed and published continuously since 1584.

CAMBRIDGE UNIVERSITY PRESS

*Cambridge
London New York New Rochelle
Melbourne Sydney*

PUBLISHED BY THE PRESS SYNDICATE OF THE UNIVERSITY OF CAMBRIDGE
The Pitt Building, Trumpington Street, Cambridge, United Kingdom

CAMBRIDGE UNIVERSITY PRESS
The Edinburgh Building, Cambridge CB2 2RU, UK
40 West 20th Street, New York NY 10011–4211, USA
477 Williamstown Road, Port Melbourne, VIC 3207, Australia
Ruiz de Alarcón 13, 28014 Madrid, Spain
Dock House, The Waterfront, Cape Town 8001, South Africa

http://www.cambridge.org

First published 1986
First paperback edition 2005

A catalogue record for this book is available from the British Library

Library of Congress catalogue card number: 84-21396

ISBN 0 521 25034 X hardback
ISBN 0 521 61921 1 paperback

Contents

Contributors

Dr D.M. Banerjee
Department of Geology, University of Delhi, Delhi 11007, India

Dr M.D. Brasier
Department of Geology, University of Hull, Hull, HU6 7RX, UK

Dr Z. Bjamba
CMEA International Geological Expedition, PO Box 324, Central Post Office, Ulan Bator, Mongolian People's Republic

Dr Chen Qiying
Institute of Geology, Academia Sinica, Beijing, China

Dr R. Choudhuri
Rajasthan State Mines and Minerals Ltd, 133 Saheli Marg, Udaipur 313001, India

Dr R.L. Christie
Institute of Sedimentary and Petroleum Geology, Geological Survey of Canada, 3303 33rd Street NW, Calgary, Alberta T2L 2A7 Canada

Dr P.J. Cook
Division of Continental Geology, Bureau of Mineral Resources, GPO Box 378, Canberra, ACT 2601, Australia

Dr M.A. Dardenne
Laboratoire de Géologie Appliquée, Tour 26–16, 5e Etage, Université Paris vi, 4, place Jussieu, 75230 Paris Cedex 05, France

Dr E.A. Eganov
Institute of Geology and Geophysics, USSR Academy of Sciences, Siberian Division, 630090 Novosibirsk, USSR

Ms A.S. Elgueta
Estrella Solitaria 4690, Nunoa, Santiago, Chile

Dr Guo Shizeng
Institute of Geology, Academia Sinica, Beijing, China

Dr Mirza Talib Hasan
Geological Survey of Pakistan, D66, Block F, North Nazimabad, Karachi, Pakistan

Professor P.F. Howard
School of Earth Sciences, Macquarie University, North Ryde, NSW 2113, Australia

Dr R.G. Humphreville
Amoco Production Company (International), PO Box 4381, Houston, Texas USA

Dr A.V. Ilyin
Institute of the Lithosphere, USSR Academy of Sciences, Staromonetny per. 22, 109180 Moscow, USSR

Dr J. Kirschvink
Division of Earth and Planetary Sciences, Caltech 170–25, Pasadena, California 91125, USA

Dr A. Kühn
Geological Institute, Rakowiecka 4, 00–975 Warsaw, Poland

Dr F. Leyva
Empresa Nacional Adaro de Investigaciones Mineras SA, Serrano 116, Madrid 16, Spain

Dr Li Yueyan
Department of Geology and Mineral Resources, Ministry of Geology, Beijing 100812, China

Professor J. Lucas
Institut de Géologie, Université Louis Pasteur, 67084 Strasbourg Cedex, France

Dr G.H. McClellan
International Fertilizer Development Center, PO Box 2040, Muscle Shoals, Alabama 35660, USA

Dr L.F. Magalhães
Metais de Goias S/A (METAGO), Rua 68, No. 727, Centro, Goiania, Goias 74000, Brazil

Nguyen Dang Khoa
General Department of Chemistry, Hanoi, Socialist Republic of Vietnam

Mr A.J.G. Notholt
12 Thornhill Road, Ickenham, Uxbridge UB10 8SF, UK

Dr J.T. Parrish
US Geological Survey, Denver Federal Center MS 971, PO Box 25046, Denver, Colorado 80225, USA

Dr E. Perconig
Minas de Almaden y Arrayanes, S.A., Paseo de la Castellana 18, 28046 Madrid, Spain

Professor S.R. Riggs
Department of Geology, East Carolina University, Greenville, North Carolina 27834, USA

Dr A.B. Roy
Department of Geology, State University of Rajasthan, Udaipur, India

Mr F.N. Saavedra
Instituto de Investigaciones Mineras, Avenida Liberador San Martín 1109 Oeste, 5400 San Juan, Argentina

Dr M.W. Sandstrom
Australian Institute of Marine Sciences, PMB 3, Townsville, QLD 4810, Australia

Dr C.R. Scotese
Institute of Geophysics, University of Texas, 4920 North IH 35, Austin, Texas 78751, USA

Dr R.P. Sheldon
3816 T Street NW, Washington, D.C. 20007, USA

Dr J.H. Shergold
Division of Continental Geology, Bureau of Mineral Resources, GPO Box 378, Canberra. ACT 2601, Australia

Dr M. Slansky
Bureau de Recherches géologiques et minières (BRGM), BP 6009, 45018 Orleans Cedex, France

Dr L.A. Soares
Metais de Goias S/A (METAGO), Rua 68, No. 727, Centro, Goiana, Goias 74000, Brazil

Dr P.N. Southgate
Bureau of Mineral Resources, GPO Box 378, Canberra, ACT 2601, Australia

Dr Yu.K. Sovetov
Institute of Geology and Geophysics, USSR Academy of Sciences, Siberian Division, 630090 Novosibirsk, USSR

Dr Sun Shu
Institute of Geology, Academia Sinica, Beijing, China

Tran Quoc An
General Department of Geology, 6 Pham Ngu Lao Street, Hanoi, Socialist Republic of Vietnam

Dr R. Trompette
Laboratoire Geologie Structurale, Université Aix-Marseille III, Centre de St Jérome, 13.397 Marseille Cedex 13, France

Dr F. Vázquez
Instituto Geológico y Minero de España, Ríos Rosas 23, Madrid 3, Spain

Dr F. Velando
Empresa Nacional Adaro de Investigaciones Mineras SA, Serrano 116, Madrid 16, Spain

Academician A.L. Yanshin
Institute of the Lithosphere, USSR Academy of Sciences, Staromonetny per. 22, 109180 Moscow, USSR

Professor Yeh Lientsun
Institute of Geology, Academia Sinica, Beijing, China

Dr N.A. Zaitsev
Geological Institute, USSR Academy of Sciences, Pizhevsky per. 7, 109017 Moscow, USSR

Professor A.M. Ziegler
Department of Geophysical Sciences, University of Chicago, 57345 Ellis Avenue, Chicago, Illinois 60637, USA

Preface

'*Phosphate deposits of the world*' is a series of multi-author volumes summarising the results of an international research programme on phosphate deposits (IGCP Project 156). The International Geological Correlation Programme (IGCP) was jointly established by UNESCO and the International Union for Geological Sciences (IUGS) in 1972 in order to facilitate the acquisition and distribution of knowledge of mankind's resources and environment. A project to study sedimentary phosphate deposits (phosphorites) was formally accepted as IGCP Project 156 in early 1977.

Phosphate rock is the basis of a major world industry for the manufacture of phosphate fertilizers and phosphorus-based chemicals. Since 1972, world production of phosphate rock has exceeded 100 million tonnes a year. Without this phosphate the spectre of starvation would loom much larger in the world today. As Isaac Asimov has put it: 'We can substitute atomic energy for oil and coal, but there is absolutely no substitute for phosphorus'. Although there is no shortage of phosphate rock in the world as a whole, with resources of many billions of tonnes of ore having already been identified, there is a rather inequitable distribution of those resources. Countries such as Morocco, the USA and the USSR are major producers of phosphate rock. However, many countries, especially in Asia, Africa and South and Central America, have no domestic production and there is much under-utilisation of phosphate-based fertilizers by developing countries in these regions. For this reason, the question of locating new sources of phosphate rock, particularly phosphorites (phosphatic sedimentary rocks), has received increased attention in recent years. Phosphate rock is also of interest because of the abundance of elements such as uranium, rare earths, vanadium, and fluorine, which may be extracted as by-products of fertilizer production. Moreover some phosphatic sedimentary rocks and their associated sediments are believed to be important petroleum source rocks. Phosphorus plays a crucial role in all life processes, and the distribution of phosphorus in the sedimentary record can be linked to biological evolution and to the evolution of the oceans. Phosphorites are also of great interest in their own right for their chemistry is complex, their microscopic structure is frequently puzzling, and their mode of formation a matter of considerable scientific debate. There are few other earth materials which combine such fundamental scientific significance with great socio-economic importance.

It was recognised early in the development of IGCP Project 156 that in order to be successful in addressing this wide range of questions it would be necessary to set up various working groups within the Project. A total of four such groups have been established, each of them co-ordinated by two Chairmen. Each of the volumes in this series 'Phosphate Deposits of the World' summarise the results of a working group, with the working group chairmen also acting as Editors.

Working Group 1 – Proterozoic–Cambrian Phosphorites (Co-Chairmen: P.J. Cook and J.H. Shergold) – provided the initial focus of the Project. At the start of the Project, deposits of this age were perhaps the least known group of phosphorites. Working Group 1 has greatly added to our knowledge of them, with a large number of deposits and occurrences now documented. Not only has this led to a far better understanding of this ancient period of phosphogenesis, it has also in part resulted in exploration efforts focusing on deposits of this age in various parts of the world. This in turn has led to at least one important discovery and the re-evaluation of poorly known occurrences.

Working Group 2 – International Phosphate Resource Data Base (Co-Chairmen: A.J.G. Notholt and R.P. Sheldon) – arose from the need to organise present geological data in order to assess undiscovered phosphate resources. Resources assessment is facilitated by ready access to the geological literature on phosphate deposits. Until the present activities of Project 156 were initiated, no systematic, comprehensive effort had been made to organise this information. The efforts of this working group in compiling a comprehensive data base are of great importance to a reliable evaluation of national, regional, and world phosphate resources.

Working Group 3 – Young Phosphogenic Systems (Co-Chairmen: W.C. Burnett and S.R. Riggs) – is based on one of the basic tenets in geology, that we can use present-day processes as a key to understanding how ancient rocks were formed. An understanding of present-day systems where phosphorites are forming will clearly be of very great relevance to understanding how ancient deposits formed. The interaction between land and sea geologists and geochemists stimulated by Working Group 3 has resulted in a more careful comparison of the differences and similarities between ancient and modern sedimentary phosphorite environments and has greatly facilitated our understanding of phosphogenesis. It has also generated a great deal of new information on phosphogenesis through the study of experimental systems.

Working Group 4 – Cretaceous–Eocene Phosphorites (Co-Chairmen: K. Al-Bassam and J. Lucas) – This group which was the last of the Working Groups to be established is concerned with what was probably the world's most important phosphogenic event. Deposits of Cretaceous to Eocene age are of enormous economic importance to the countries of the Middle East, North Africa, West Africa and the northern part of South America and of great importance to many other countries which rely on deposits of this age as their primary source of imported rock phosphate. Despite their importance there was no comprehensive study of deposits of Cretaceous–Eocene age until Working Group 4 was established. The aim of this Working Group has been to assemble a comprehensive data base on which to develop a far better understanding of this massive phosphogenic province.

In addition, other groups evolved as the Project progressed. The Committee on Rock Phosphate Standards (CORPS) was convened by Z.S. Altschuler, L. Prévôt and Y. Zanin because of a general feeling amongst project participants concerned with the geochemistry of phosphorites that existing rock phosphate standards were indequate for their purposes.

National working groups also evolved in part to arrange international meetings but more commonly as a mechanism to foster interchanges between phosphate geologists in various countries. Contacts with national working groups have been made formally through National Representatives nominated by the National IGCP Committees, or informally through a National Correspondent.

Links with all these groups and the several hundred individuals within the project have been maintained by letter, telegram and telephone. However, undoubtedly, one of the most important forms of communication for the project as a whole has been the Project Newsletter for which A.J.G. Notholt has acted as Editor for most of the life of the Project. All of these links, sometimes formal, but most commonly quite informal, have given everyone who wished to participate in the Project the opportunity to do so.

The Project has also aimed at being outward-looking. For example, it has co-operated with organisations such as the Resource Systems Institute of the East–West Center, Honolulu, and the Geological Society of London in convening joint meetings. A particularly important feature of Project's activities has been the holding of or participation in training courses on phosphate geology for geologists from developing countries. These have been held in Sydney (Convenor, P.F. Howard), Bangkok (Convenors, Z.S. Altschuler, W.C. Burnett, W. Hamilton, R.P. Sheldon), Dakar (Convenor, M. Luleu of the International Center for Training and Exchanges in Geosciences) and Greenville, N.C. (Convenors, W.C. Burnett, S.R. Riggs, R.P. Sheldon). Together, these various activities have, we believe, resulted in a stimulating, diverse and relevant Project.

However, through the life of the Project, the scientific aim of all participants has been to develop a better understanding of the distribution, nature, and origin of phosphate deposits. It is up to the reader to judge the extent to which this aim has been achieved. In our view, this volume, together with its forthcoming associated volumes, provides clear evidence that our knowledge and understanding of phosphate deposits has greatly advanced in recent years through the work of IGCP Project 156. We hope that they will also stimulate new ideas on the origin of phosphate deposits. It is also our hope that these volumes make not only an important contribution to basic scientific knowledge but also to the search for and exploitation of phosphate deposits in the many parts of the world that currently lack this vital resource.

Peter J. Cook John H. Shergold
(Co-Project Leaders 1977–1984)

Acknowledgements

Many people and organisations have given us assistance in organising IGCP Project 156 and particularly in visiting the Proterozoic and Cambrian phosphate deposits which form the basis of this book. This assistance, whether financial, logistical, organisational, or concerned with publication, is gratefully acknowledged here.

Organisations which have assisted us include the following:

Aberfoyle Minerals Pty Ltd, Australia
Academia Sinica Department of Earth Sciences. Beijing
Adelaide & Walleroo Fertilizers, Australia
Alexander von Humboldt Foundation, W. Germany
Australian Bureau of Mineral Resources, Canberra
Australian Department of Education, Canberra
Australian Department of Foreign Affairs, Canberra
Australian Department of Science & Technology, Canberra
Australian Development Assistance Bureau, Canberra
Australian Fertilizers Ltd
Australian National Commission for UNESCO, Canberra
Australian National University, Canberra
Baja California district of the Asociación de Ingenieros de Minas, Metalurgistas y Geólogos de México.
Beke Industries Fertilizers, USA
BH South Ltd (now part of the Western Mining Corporation), Australia
Centre national de la Recherche scientifique, Paris
China National Committee for IGCP, Beijing
Chinese Ministry of Chemical Industry, Beijing
Chinese Ministry of Geology and Mineral Resources, Beijing
Circum-Pacific Energy and Minerals Resources Council
Commission national du Sénégal pour l'UNESCO
Committee for the Coordination of Joint Prospecting for Mineral Resources in Asian off-shore areas
Committee for the Coordination of Joint Prospecting for Mineral Resources in South Pacific off-shore areas
Commonwealth Foundation, London
East Carolina University, Greenville
Florida State University, Tallahassee
Geological Bureau of Guizhou Province, Guiyang
Geological Bureau of Yunnan Province, Kunming
Geological Survey of Queensland, Brisbane

Geological Survey of India
Hindustan Zinc Ltd, Zawar
Ian Potter Foundation, Melbourne
ICI Australia, Melbourne
Indian Department of Science and Technology, Delhi
Indian National Committee for IGCP, Delhi
Institute de Géologie, Université Louis Pasteur, Strasbourg
Institut Mondial du Phosphate, Casablanca
Institute of the Lithosphere, USSR Academy of Sciences, Moscow
Institut sénégalais de Recherche agronomique, Dakar
Institute of Geological Sciences, London
International Fertilizer Industry Association, Paris
International Food Policy and Research Institute, Honolulu
International Geological Correlation Programme, Paris
International Institute for Applied System Analysis, Honolulu
International Mineral and Chemical Corporation, USA
International Union of Geological Sciences, Paris
Kazakhstan Academy of Sciences, Alma-Alta
Macquarie University, Sydney
Mineral Exploration Corporation Ltd, India
Mines Exploration Pty Ltd, Australia
Ministère de l'Education nationale, Senegal
Ministère de l'Energie et des Mines, Morocco
Ministère de la Recherche scientifique et de la Technologie, Senegal
Mongolian People's Republic Academy of Sciences
Montsanto Ltd, USA
Mount Isa Mines Ltd, Australia
Office Chérifien des Phosphates, Morocco
Office national de Recherches et de l'Exploitations petrolières, Morocco
Pyrites, Phosphates and Chemicals Ltd, India
Queensland Phosphate Pty Ltd, Australia
Rajasthan State Directorate of Mines and Geology, Udaipur
Rajasthan State Mines and Minerals Ltd, Udaipur
Research School of Earth Sciences, Canberra
Resource Systems Institute, East–West Center, Honolulu
Roca Fosforica Mexicana, La Paz, Mexico
State Institute for Chemical Raw Materials, Moscow
Société sénégalaise des Phosphates (de Thies et de Taiba), Dakar
UNESCO, Paris
Union Oil Corporation, USA
Universidad Autonoma de Baja California Sur, Mexico
Université de Dakar, Senegal
US Geological Survey
US National Science Foundation, Washington
USSR Academy of Sciences, Moscow
Utah Geological Association
Uttar Pradesh Mineral Development Corporation
Wyoming Geological Survey

Our thanks also go to very many people in all parts of the world for their assistance in organising field workshops and seminars, and for facilitating funding, transportation, translation etc. over the past seven years. We would especially like to thank the following:

Australia: J.N. Casey, J. Elsom, L. Fisher, N.H. Fisher, M.F. Glaessner, L.B. Gustafson, D. Hackett, J.F. Lovering, I. Newnham, P.W. Pritchard, J.K. Rogers, R.W.R. Rutland, P.T. Stoker.
China: Cheng Yuqi, Li Yueyan, Li Xiji, Luo Yiqing, Rui Wentong, Wang Daxion, Yan Shutan.
India: G.M. Banerjea, D.M. Banerjee, R. Choudhuri, B.N. Dhoundial, D.P. Dhoundial, T.N. Jaggi, D.R.S. Mehta, S.P. Nautiyal, A. Pant, H. Prasad, V.N. Sant, M.N. Sehgal, B.N. Sinha, R.N. Srivastava, J. Swami Nath, V.K.S. Varadan.
Mexico: F.J. Escandon, R. Jiminez, A. Martinez, V. Moyron.
MPR/USSR: Irina Beu (nee Lozovoya), Yu.V. Bufeev, B. Chinbat, B. Luvsandansan, S.D. Scotarenko, B. Shirendyb, S. Tsegmid, O. Tumurtogoo.
Morocco: M. Bensaid, M. El Kheir, M. Saadi
Senegal: B.M. Daffe, M.P. Hameh
USA: D. Blackstone, W.C. Burnett, L.G. Duncan, G.D. Emigh, G. Greer, A. Lee, K.C. Lohman, V.E. McKelvey, E. Maughan, D. Miller, A.R. Palmer, L. Raymond, H. Ritzma, R.P. Sheldon, Thon Gin, R. Willden, N. Bridges.
USSR: A.L. Yanshin, A.A. Abdullin, S.E. Chakabaev, E.A. Eganov, A.V. Ilyin, A.V. Nedogon, A.S. Sokolov, A.I. Timchenko.

A particular debt of gratitude is owed to people associated with the organisational side of IGCP and IUGS. We would like to especially thank Eric von Braun (IGCP) and Christian Weber (IUGS) for their help in many aspects of the Project organisation and funding.

We would like to also record our indebtedness to those who have helped us in the preparation of this book. For their typing we acknowedge Carol Britten, Lynne Fitzsimmons and Debbie Jones (BMR) and Joyce Barrett, Karen Gibbs and Alaine Moore (ANU); for reference collating John Laurie; and for redrawing the great majority of the figures Terry Kimber.

Finally, special thanks to our families, Norma, John and Julian Cook, Judy and Julia Shergold, for their tolerance whilst we have spent prolonged periods away from home looking at phosphate deposits in distant parts of the globe.

In addition to these general acknowledgements, authors of some chapters also wish to acknowledge assistance and constructive criticism from a number of other individuals and/or organisations:

Tran Quoc An and Nguyen Dang Khoa acknowledge the encouragement of Dr Tran Duc Luong, Director General of the General Department of Geology, Hanoi; and Drs Pham Quoc Tuong and Le Thac Xinh for assistance with their manuscript.

D.M. Banerjee acknowledges N.S. Bhalti and R. Bashyal for providing samples; P.F. Howard, J. Lucas, L. Prévôt, J. McArthur, G. Kelling and R.P. Sheldon for providing sample analyses; M. Schidlowski and J.D. Arneth for providing analytical facilities at the Max Planck Institute for Chemistry, Mainz; and the Alexander von Humboldt Foundation, Bonn, for financial support; and G. Saigal, M.W.Y. Khan, N. Srivastava (Saigal), and R. Choudhuri for assistance in various capacities.

R. Choudhuri and A.B. Roy are grateful to the Rajasthan State Mines and Minerals Ltd, Udaipur, Rajasthan, and the

Department of Geology, University of Rajasthan, Udaipur for providing facilities; help received from the State Directorate of Mines and Geology, Rajasthan is also gratefully acknowledged; and for assistance in the preparation of manuscript D.K. Thakur, S.K. Bhattacharya, R.S. Madan and S. Vijayan.

P.J. Cook and Sara Elgueta acknowledge Queensland Phosphate Ltd (now a part of the Western Mining Corporation) for allowing their study to be undertaken on the Lady Annie prospect, and for providing unpublished information; J.K. Rogers and M.C. Brown (then of BH South Ltd) assisted in all aspects of the work; and E. Kiss, P. Berrie and H. Cutten provided technical assistance.

M.A. Dardenne acknowledges the receipt of Brazilian CNPq grant 0401400/81; and with his co-authors thanks T. Fairchild who assisted with the manuscript.

Li Yueyan acknowledges the various organisations of the Ministry of Geology and Mineral Resources who provided information; Engineers Zhou Maoji, Xu Wuyang and others at the Geological Bureau of Guizhou and Sichuan; Engineers, Wang Chongwu, Ge Hongru and other scientists of the Geological Bureau of Yunnan, and Engineer Chen Congyun of the Shenyang Institute of Geology and Mineral Resources who provided much valuable data and the photographs used in the publication; and Chief Engineer Du Xianglin, No. 115 Geological Brigade of the Geological Bureau of Guizhou, for his great assistance in the compilation of the manuscript.

G.H. McClellan and F.N. Saavedra recognise the assistance of a number of colleagues in obtaining samples that were used to broaden the scope of this work, including G.J. Calder, W.F. Cannon, R. Choudhuri, R. Flicoteaux, J.R. Lehr, T.H. McCosker, R.P. Sheldon, M. Slansky, and E. Ullmer; and also thank Michelle Cooper for her valuable assistance in the final stages of this chapter preparation.

A.J.G. Notholt and M.D. Brasier are indebted to A.W.A. Rushton, British Geological Survey, London, for his helpful comments and critical reading of the manuscript.

Judith Parrish and co-authors would like to thank J.J. Sepkoski Jr, for assistance with Cambrian correlations; M.E. Taylor, J.H. Shergold, P.A. Jell, P.J. Cook, P.N. Southgate, A.R. Palmer and E.K. Maughan for discussion and comment; and Ms Linda Lashmet for drafting assistance.

E. Perconig and co-authors acknowledge the Director of the Geological & Mining Survey of Spain (IGME) for authorising publication; fellow workers from Empresa Nacional Adaro de Investigaciones Mineras, S.A. (ENADIMSA) who contributed to this study; and Jaime Garcia-Rodriquez for undertaking the difficult task of translating the manuscript from Spanish.

M. Sandstrom thanks P.J. Cook for comprehensive advice and encouragement; R.P. Philp for providing GC-MS facilities; Henry Pollach for stable carbon isotope analysis; E. Kiss, P. Berrie and J. Fildes for assistance in sample analysis; and A.J. Cook, R.J. Kantsler, A.C. Hutton, M. Glikson for assistance in organic petrography of the Cambrian sediments; samples were kindly provided by Bureau of Mineral Resources and Broken Hill South Ltd, and the work was supported by a postgraduate scholarship from the Australian National University.

J.H. Shergold and M.D. Brasier particularly acknowledge Zhang Wentang, Nanjing Institute of Geology and Palaeontology for providing palaeontological information on the Cambrian phosphorites of China; and A. Yu. Rozanov, Palaeontological Institute, Moscow for assisting with Russian information.

P.N. Southgate acknowledges R.E. Garrison, P.F. Howard, S.R. Riggs, R. Brown, P.G. Haslett, B. Logan, B.M. Radke, P.J. Cook and J.H. Shergold for constructive criticism, comments, and discussion; and financial support provided by the Commonwealth Postgraduate Research Award.

Yeh Lientsun and colleagues are indebted to Zhang Shishan for assistance during field work; and Wang Qingehen, Liu Tiebing, and Liu Kuiwu for assistance with the manuscript.

P.J. Cook, J.H. Shergold and P.N. Southgate publish with the approval of the Director, Bureau of Mineral Resources, Canberra, Australia.

1

Proterozoic and Cambrian phosphorites–an introduction

P.J. COOK AND J.H. SHERGOLD

The project

Proterozoic and Cambrian Phosphorites is the first of a multi-volume series 'Phosphate Deposits of the World' which together summarise the results of a major international research project on phosphate deposits. This work, conducted under the aegis of the International Geological Correlation Program (IGCP) Project 156 was first mooted by us in 1975, but it was not until 1976, and particularly during the International Geological Congress held in Sydney, that we were able to start to canvas international opinion. Encouraged by the overwhelmingly favourable response to the proposal, it was formally submitted to the IGCP Board in late 1976, and was officially accepted as Project 156 in early 1977.

Our initial idea was that such a project stood the greatest chance of scientific success if it focused clearly on a specific question rather than considering phosphate deposits in general. It was decided that this focus should be on Proterozoic and Cambrian phosphorites, a decision in part based on personal bias in that we both have had a longstanding interest in the Cambrian phosphate deposits of Australia and saw a need to test a hypothesis that there was indeed an Australian–Asian phosphogenic pr ›vince. In addition, it seemed to us that despite the apparent importance of phosphate deposits of this age, they were for the most part poorly documented. There was also a further consideration which was to greatly influence us: the countries of Asia were for the most part deficient in phosphate deposits, and it seemed to us that in many of these countries the greatest potential for finding phosphate deposits lay in rocks of Proterozoic or Cambrian age. Therefore, in determining our initial focus, we saw scientific interest and economic significance coming together in phosphate deposits of this age. It was to transpire as the Project progressed that not only did this scientific and economic relevance apply to Asia, it applied also to Africa, Europe and South America. So, the Project was inevitably required to take a world-wide view.

The scope of the Project was to be still further expanded by the great interest that it generated. Prior to the setting up of Project 156 there was no forum for bringing together the many phosphate researchers around the world. As a consequence, despite the original intention to limit the stratigraphic scope of the Project, it soon became clear that it was providing a much needed rallying point for scientists interested in phosphate deposits of all ages. In fact, phosphate deposits are not found 'of all ages'. They had been shown by Cook & McElhinny (1979) to occur in quite discrete pulses so that it was logical to set up Working Groups within the project, each dealing with an important phosphogenic episode. The Proterozoic–Cambrian (Working Group 1), Miocene–Recent (Working Group 3) and Cretaceous–Eocene (Working Group 4) periods were identified globally as perhaps the three most important phosphogenic episodes. Additionally, because of the great socio-economic significance of phosphate deposits, the need was recognised to form a group (Working Group 2) concerned with establishing a data base for world phosphate resources of all types and ages.

Together, the work of these various Working Groups is to be presented as a collection of four volumes (of which this volume is the first) in a series entitled 'Phosphate Deposits of the World'.

Since its small beginnings in 1977, Project 156 has grown to a point where there are now over 250 active geoscientists registered with it. One of the important ways of facilitating international collaboration is through meetings. Since 1978, Project 156 has organised international field workshops and associated seminars in which scientists from 51 countries have participated. To date the Project has examined Proterozoic phosphorites in India (1981) and China (1982), Proterozoic or Early Cambrian deposits in the Mongolian People's Republic (1980), Early Cambrian deposits in China (1982) and Kazakhstan (1984), Middle Cambrian deposits in Australia (1978), Late Cambrian potentially phosphatic sediments in the Great Basin of the western USA (1979), and phosphorites believed to be close to the Cambrian–Ordovician Boundary in the central Himalaya (1981). In addition, members of the Project have had the opportunity to compare late Proterozoic and Cambrian phosphorites with younger phosphorite deposits in the Permian Phosphoria Formation of western USA (1979), the Late Cretaceous and Early Eocene deposits of Morocco and Senegal (1983), the Oligocene deposits of Baja California, Mexico (1980), and the Neogene deposits of Southeastern USA (1985). As a result of these meetings we and many other geologists have been able to see at first hand many of the deposits dealt with in this book. We believe it is true to say that since 1977 there has been an enormous increase in our knowledge of Proterozoic and Cambrian phosphorites in particular and phosphorites in general.

The book

In this 'Introduction' we provide the reader with a brief overview of the early history of research into Proterozoic and Cambrian phosphorites in order to place this volume in an historical perspective. We also give an outline of the terminology that we have used in order to rationalise the multiplicity of terms in the phosphate literature. The main part of the volume is divided into three parts: the first, a series of regional chapters, the second, a group of chapters on specific deposits, and the third, a number of specialist and overview chapters.

In the regional chapters A.J.G. Notholt and R.P. Sheldon (Chapter 2) present a world review of phosphate resources of Proterozoic and Cambrian age pointing to their economic importance and setting the scene for the regional papers which follow. In Chapter 3, on Australian phosphorites, P.F. Howard reviews the occurrence of phosphorites ranging in age from early Proterozoic, or even Archaean, to Late Cambrian. However, it is the Cambrian deposits, and particularly those of the Georgina Basin, that are dealt with at length because of their potential economic importance. The Proterozoic and Cambrian deposits of Asia were first reviewed by Bushinsky (1966). A series of chapters helps to summarise much of the new information obtained on Asian deposits since then. With its large population and intense agricultural practices, phosphate deposits have a particular importance for China. Over the past 30 years a massive exploration and research programme on phosphorites has been undertaken and the results of some of this work can be seen in Chapter 4 by Li Yueyan. The deposits of southwestern China are especially important and are therefore considered in some detail, with the author highlighting the particular significance of palaeogeography in the development and localisation of deposits.

The deposits of Soviet Asia and the Mongolian People's Republic are reviewed by A.L. Yanshin (Chapter 5). This is a vast and geologically complex area where many important, but mostly small-scale deposits have been found, and where a vigorous search for more deposits is continuing. On the Indian subcontinent, where there is a trophic situation similar to that in China, the search for phosphate deposits has been accorded particular importance. Unfortunately, it has not so far resulted in the discovery of large deposits, although a number of discoveries are presently being exploited on a limited scale, notably the deposits of Rajasthan. D.M. Banerjee, in Chapter 6, has reviewed the very considerable amount of information that has accrued on the deposits of the Indian subcontinent as a result of the many exploration programmes. Much work still remains to be done and Banerjee points to a number of important geological questions which await resolution. Cambrian phosphorites were first discovered in Europe around 1875. Since that time a considerable amount of work has been done and many occurrences have been documented. Much detail is available on the European occurrences and this is summarised in Chapter 7 by A.J.G. Notholt and M.D. Brasier. Unfortunately, no major phosphorite deposits of Proterozoic or Cambrian age have been discovered to date in Europe. Nevertheless, documentation of the nature, age and origin of the known occurrences are very important to our understanding of phosphogenesis. The same is true of the Proterozoic–Cambrian phosphorites of North America where again no major deposits are known, but which provide an important part of the total world picture. Unlike many other parts of the world, there has been relatively little phosphate exploration to date on potential Proterozoic or Cambrian sequences of North America, in part because of the occurrence of vast deposits of Permian and Neogene age. However, R.L. Christie and R.P. Sheldon summarise the information presently available for North America in Chapter 8. Many of the deposits discussed in Chapters 7 and 8 were formed on the same Avalon Platform, a feature which predated Atlantic rifting. This provides a graphic example of the comparative irrelevance of present-day geography to the original disposition of Proterozoic or Cambrian phosphorites in many cases. With the notable exception of the north and west, large parts of Africa appear to be totally devoid of sedimentary phosphate deposits. In many of these areas the best prospects for phosphorites are in rocks of Proterozoic age. Minor occurrences of this age are known in East Africa (Somalia) but the only major deposits with a potential for exploitation are in West Africa. Little has been published in English on this region to date but in Chapter 9 M. Slansky summarises the regional geology and phosphate occurrences of this large area. Finally, like Africa, many parts of South America have considerable potential for the discovery of Proterozoic or Cambrian phosphate deposits. There are, for example, extensive Cambrian sediments in Argentina. However, to date, the only important phosphorites in South America are found in the Bambuí Group of Brazil. These are described by M.A. Dardenne, R. Trompette, L.F. Magalhães and L.A. Soares in Chapter 10. These authors comment on the similarities between the South American and West African deposits. The juxtaposition of the chapters dealing with these two regions may help the reader to reach his own conclusions about the similarities and differences. The only continent for which we have no regional chapter or, indeed, any report of Proterozoic or Cambrian phosphorites is Antarctica. Cambrian sediments similar to the Australian Cambrian (which is phosphatic in part) are known to occur but so far no phosphorites of Cambrian or Proterozoic age have been discovered.

The second section of the volume deals with individual deposits or groups of deposits. Here we have not aimed at covering all major phosphate deposits of Proterozoic or Cambrian age. This will be done in Volume 2 of the series. Rather, we have aimed to give the reader a cross-section of deposit types of both Proterozoic and Cambrian age and from most of the major phosphogenic regions. Some chapters provide the first details of deposits to be published in English. Like the regional chapters, the chapters concerned with individual deposits or groups of deposits deal first with Australian, then Asian, European and finally African deposits. In Chapter 11 P.J. Cook and S. Elgueta provide details of the Lady Annie deposit in northern Australia. They focus on the accurate reconstruction of a depositional model for this deposit and through this they are able to develop a clear picture of the depositional environment. Many of the details available for the Chinese deposits have already been provided in Chapter 4. However, additional information is provided by Yeh Lientsun, Sun Shu, Chen Qiying and Gui Shizung in Chapter 12, which describes the Kunyang deposit of Yunnan Province, one of China's most important phosphate deposits. A little to the south and west of Kunyang are the large phosphate deposits of Lao Cai. The stratigraphy and structure of these deposits is described, we believe for the first time in English, in Chapter 13 by Tran Quoc An and Nguyen Dang Khoa. Similarly for the Khubsugul deposits of the Mongolian People's Republic, where Soviet and Mongolian scientists have been jointly working for a number of years, only a limited amount of information has been published in English. In Chapter 14, A.V. Ilyin, N.S. Zaitsev and Z. Bjamba provide a clear picture of the nature and origin of this important group of deposits. The Karatau deposit of southern Kazakhstan was the first major Cambrian deposit to be discovered and still remains one of the world's largest phosphate deposits of this age. Little has been published on the deposit in English although a very considerable amount of research has been undertaken on it by Soviet scientists. In Chapter 15, E.A. Eganov, Yu.K. Sovetov and A.L. Yanshin provide details of this deposit, including important new information on its geology and origin. To the south, across the Parmir and Karakoram Ranges, Early Cambrian phosphorite also occurs in northern Pakistan. The Hazara deposit is described by Mirza Talib Hasan in Chapter 16. The deposit has not been extensively studied, and its importance rests in providing evidence that the Cambrian phosphogenic province which we see so clearly in China and parts of central Asia did indeed extend into the present Himalayan region. The Jhamarkotra deposit of Rajasthan is to date the most important phosphate deposit discovered in India. Because of its economic importance it has been extensively studied. However, the scientific importance of the deposit is much wider for it constitutes perhaps the world's best known stromatolitic phosphorite deposit. R. Choudhuri and A.B. Roy review the geology of this deposit and speculate on its origin in Chapter 17. The Fontanarejo deposit in Spain, like Hazara, has not been extensively studied so far. The information presently available is summarised in Chapter 18 by E. Perconig, F. Vazquez, F. Velando and F. Leyva. Again, the importance of this deposit rests not in its potential for economic exploitation but rather for the indication it provides that deposits of greater magnitude may be waiting to be found in the Proterozoic or Cambrian rocks of other parts of Europe. The deposits of the Volta Basin, whilst somewhat better known than the Spanish deposits, have yet to be investigated in detail. In Chapter 19, J. Lucas of France, A.V. Ilyin of the USSR and A. Kühn of Poland not only summarise the information available on the deposits of the Volta Basin in West Africa, they also provide a graphic illustration of the type of international co-operation that we have aimed at fostering through Project 156.

In the third major section we have brought together a number of specialist papers which we believe to have wide significance for Proterozoic and Cambrian deposits throughout the world and indeed, in many cases, for deposits of other age. In Chapter 20 G.H. McClellan and F.N. Saavedra summarise the chemistry and mineralogy of Proterozoic and Cambrian phosphate deposits from many parts of the world. They point to a number of common features in phosphorites of this age and provide compelling evidence of systematic compositional changes with increasing age. Many of their findings are of great relevance to the exploitation and beneficiation of ancient phosphate deposits. In his chapter on organic matter in phosphorites (Chapter 21), M.W. Sandstrom is forced to rely mainly on the results of his work on the Cambrian phosphorites of the Georgina Basin. Many of his conclusions are undoubtedly relevant to other Cambrian and probably also Proterozoic phosphorites. The difficulties are that the work calls not only for sophisticated laboratory techniques, but also for access to unweathered samples. We hope that Sandstrom's chapter will lead to the recognition that more organic geochemistry should be undertaken on ancient phosphorites, not only because of its intrinsic scientific interest, but also because some phosphatic sequences are believed to be important petroleum source rocks. The chapter by J.T. Parrish, A.M. Ziegler, C.R. Scotese, R.G. Humphreville and J.L. Kirschvink (Chapter 22) draws together a number of lines of geological and geophysical data to produce a global palaeogeographic setting for Cambrian phosphogenesis. In our view their attempt is to be applauded not only because it is done here for the first time on admittedly sparse data, nor because their model seems to work, but because it does provide us with a global model which can be tested as more information accumulates on both Cambrian and Proterozoic deposits. J.H. Shergold and M.D. Brasier present in Chapter 23 a definitive analysis of the biochronology of Proterozoic and Cambrian phosphatic sediments and deposits. Their analysis will be of major significance not only for Proterozoic and Cambrian palaeontology and phosphogenesis, but also to our understanding of the Precambrian–Cambrian boundary and the occurrence of phosphate deposits around that boundary. P.N. Southgate (Chapter 24) provides a very detailed analysis of facies relationships in phosphatic sequences of the Georgina Basin, focusing in particular on the occurrence of phosphatic hardgrounds and stromatolites. Drawing on his experience in the Georgina Basin, he is able to recognise similar sequences in other Proterozoic and Cambrian rocks in various parts of the world. S.R. Riggs (Chapter 25) is similarly able to draw on his detailed experience of a particular area (the Miocene phosphorites of the southeastern United States) as well as his experience in phosphate exploration to develop a number of models for phosphogenesis which are potentially of significance as exploration tools.

Finally, in Chapter 26, we have endeavoured to summarise what we see as the main points coming out of the previous 25 chapters. We have also attempted to produce a global model which we believe explains many of the features of Proterozoic (particularly Late Proterozoic) and Cambrian phosphorites and their clustering around the Precambrian–Cambrian boundary.

The volume is broad in scope, its contents ranging from general to highly detailed presentations. As such, there is a limited overlap, particularly in the degree of resolution between the three sections, but this does mean that maximum coverage, from different points of view, is achieved.

Historical review

The recognition of phosphatic sedimentary rocks was first made early in the nineteenth century, the work of Buckland (1829) being especially notable. Almost all of this early work was on Cretaceous phosphorites. A link between phosphorites and organic remains was recognised early and for this reason the search for phosphorites largely ignored the Precambrian and Cambrian. However, detailed work commencing in 1863 on the Cambrian of Wales by Davies (1875) proved that phosphorites do indeed occur in the Cambrian. Investigations by Hicks (1875) and Hudleston (1875) led them to conclude that Cambrian rocks were no less phosphatic than younger rocks. However it was work in eastern Canada by Dawson (1876) which finally led to the realisation that not only were phosphorites relatively abundant in the Cambrian, but they were also present in the Precambrian (Laurentian). No longer were Precambrian and Cambrian sediments to be regarded as non-phosphatic. This was reinforced by Andersson (1896) who found that phosphorites were also widespread in the Cambrian of Scandinavia. Jack (1919) was similarly to find phosphorites associated with sediments of Precambrian and Cambrian ages in the Flinders Ranges of South Australia. However, throughout this early work these ancient phosphorites were to come in for little attention when compared with younger ones. The reason for this was simply that phosphate deposits of Precambrian or Cambrian age were for the most part no more than occurrences. Consequently, all the important concepts on phosphogenesis were developed from observations on younger phosphorites, culminating perhaps in the work of Kazakov (1938) and his so-called upwelling hypothesis. It was really not until this time that major phosphate deposits were to be discovered in the Cambrian and Precambrian. The first to be found were the Cambrian deposits of the Karatau region (Mashkara, 1938), soon to be followed by the discovery of the important deposits of southern China (Cheng, 1939). Despite this, the idea lingered on, in the English scientific literature at least, that ancient rocks were devoid of phosphorites (Geijer, 1962) largely because these Precambrian and Cambrian phosphorites were described in comparatively little known Russian and Chinese publications, a feature commented on by Davidson (1963). Bushinsky (1938, 1963, 1966) was to focus the attention of geologists on the importance of phosphorites of this age, his observations leading him to conclude that contrary to the ideas developed by Kazakov (and extended to younger phosphorites such as the Phosphoria Formation by McKelvey *et al.*, 1953), these ancient phosphorites were formed under shallow-water conditions. Additionally, he seriously questioned the concept of upwelling ocean currents being the source of the phosphate.

Discoveries of phosphate deposits were to continue apace in Asia leading Yanshin (1964) to point to the great importance of the Cambrian for the formation of phosphate deposits. This importance was to grow as deposits were discovered in Mongolia in 1964 (Zaitsev *et al.*, 1970), and though somewhat older, in Rajasthan (1966). Still further afield, important Proterozoic–Cambrian deposits were to be discovered in West Africa in 1964 (Sougy, 1964) and South America in 1961 (Guimarães, 1967). It was not until 1966 (Russell, 1967; Howard, 1972), however, that major phosphate deposits were found in Australia – in the Georgina Basin. This was to lead Cook (in de Keyser & Cook, 1972) to propose that a major Cambrian phosphogenic province extended through the Asian and Australian region, comparing it to the Cretaceous–Eocene Tethyan province. It was this hypothesis that was to form the basis of the preliminary proposal for the establishment of IGCP Project 156. With the commencement of the Project in 1977, a series of important guidebooks has appeared as part of the programme of the Project. These include publications by Cook & Shergold (1979), Ilyin & Bjamba (1980), Indian National Committee for IGCP (1981a, b) and China National Committee for IGCP (1982, 1984). Increasingly, the data on these deposits have been used as a basis for developing new theories for the formation of phosphate deposits. Cook & McElhinny (1979) developed models to account for the spatial and temporal distribution of phosphorites drawing on experience of Cambrian deposits, and Cook & Shergold (1980) drew up a list of 'ingredients' for late Proterozoic and Cambrian deposits. During the period 1977 to present, there has been a quickening of interest in Proterozoic and Cambrian phosphorites to such an extent that it is difficult to summarise adequately the large number of papers that have appeared since 1978. However, as this volume is concerned primarily with the details of observation made during this period, we leave it to the subsequent chapters to do justice to the more recent work.

Terminology

Where possible, geological terminology used in this volume is based on the definitions given in the American Geological Institute's *Glossary of Geology* (Gary *et al.*, 1972). However, the classification of phosphorites poses particular problems. The first clear use of the term 'phosphorite' for a phosphatic sedimentary rock was by Daubeny & Widdrington (1845). Since that time, whilst the term has been used loosely, it has mostly been applied to sedimentary phosphate deposits. There is however no universally accepted definition of a phosphorite. Examples of definition include: 'Rocks composed essentially of carbonate apatites which are usually moderately high in fluorine and consequently are mineralogically francolite' (McConnell, 1973); 'A deposit of phosphate, directly or indirectly of sedimentary origin which is of economic interest' (Emigh, 1973); 'A sedimentary rock comprised principally of phosphate minerals (Ca, Al, Fe phosphate)' (Gary *et al.*, 1972); 'A sedimentary rock in which the mineral apatite is abundant' (Friedman & Sanders, 1978); 'I propose to take the limit between phosphorites (or phosphatite) and phosphatic rocks (such as phosphatic clay, phosphatic sandstone, phosphatic limestone) at 18% P_2O_5' (Slansky, 1979); '$>10\%$ of apatite grains would constitute a phosphorite... based on extensive grain counts which suggested that there was a distinct break in the rock types which contain greater than and less than 10% apatite grains' (Riggs, 1979a).

It is apparent from these definitions that, while most would agree on the restriction of the term to a sedimentary deposit, there is a marked divergence of opinion on just where the cut-off should be between a phosphorite and a phosphatic sediment. If there must be 50% or more of phosphate minerals in the rock, then only rocks with more than about 18–20% P_2O_5 (depending on the composition of the collophane) would be classified as phosphorites. This may be a satisfactory solution for sequences containing only high-grade phosphate deposits, but it would for instance exclude rocks making up many important sedimentary phosphate deposits. Elgueta (1981) found that for the Lady Annie phosphate deposits of the Georgina Basin there are two distinct phosphate populations with a cut-off between them at 15% P_2O_5, and used this figure to distinguish between a phosphorite and a phosphatic sedimentary rock. However, there is no inherent reason why the same cut-off figure of 15% P_2O_5 should apply to all other phosphorites. Additionally, it became clear to us during the compilation of this volume that many authors did not use a precise geochemical definition for a phosphorite. Therefore here, and indeed in most other publications on phosphorites, the term phosphorite is used no more precisely than meaning a sedimentary rock containing abundant phosphate, generally though not exclusively in the form of fluorapatite or carbonate fluorapatite.

A great many classificatory schemes have been proposed for phosphorites and for describing their constituent particles. Generally, a descriptor, such as pelletal, i.e. pelletal phosphorite, is used, thus denoting a phosphorite composed of rounded 'pellets'. However the use of pelletal in this way tends to epitomise the rather unsatisfactory nature of phosphorite nomenclature, since 'pelletal' in the carbonate literature denotes a particular type of particle. By contrast, in the phosphorite literature it denotes nothing more than a round particle of about sand- or coarse silt-size and the term is used whether that particle is an ooid, a faecal particle, a rounded phosphatic fossil fragment, a rounded intraclast, or a round particle of indeterminate origin. If pelletal indicates nothing more than rounded phosphate particles (of any origin) then perhaps a term which does not have a specific connotation would be more desirable?

A widely recognised distinction in phosphorites is based on grain size and specifically between phosphorites where the phosphate particles are of sand- or coarse silt-size and those which are clay- and fine silt-size. Various terms have been suggested for the fine-grained phosphorites including microsphorite (Riggs, 1979a,b) and mudstone phosphate (Russell & Trueman, 1971). In an attempt to rationalise this, Cook (1979) suggested the use of the terms phosphorudite, phosarenite, and phospholutite. More recently, Cook and his former students (S.A. Elgueta, M.W. Sandstrom and P.N. Southgate) have developed a scheme based on the carbonate classification of Dunham (1962) (see Table 1.1). This provides an easy-to-use systematic and non-genetic classification of phosphorites. It has the added advantage that the system is widely used for carbonate rocks and consequently many people are already familiar with the system. Even more important is the fact that the system works and phosphorites are easy to classify whether in hand specimen or thin-section. Under this system a 'pelletal' phosphorite generally becomes grainstone phosphorite and 'non-pelletal' phosphorite generally becomes mudstone phosphorite. There are also the intermediate types, packstone phosphorite and wackestone phosphorite, based in the carbonate literature on whether the rock is grain-supported or mud-supported. The term 'boundstone phosphorite' is also available although this is not a common type. The analogue of the crystalline carbonate is perhaps in part phoscrete, a term used by Cook (1972) for secondary apatite precipitates generally believed to form within the weathering profile and analogous to

Table 1.1 *Dunham's (1962) classification of carbonate rock modified for phosphorites*

Depositional texture recognisable					Depositional texture not recognisable
Original components not bound together during deposition				Original components were bound together during deposition... as shown by intergrown skeletal matter, lamination contrary to gravity, or sediment-floored cavities that are roofed over by organic or questionably organic matter and are too large to be interstices	Crystalline carbonate (subdivide according to classification designed to bear on physical texture or diagenesis)
Contains mud (particles of clay and fine silt size)			Lacks mud and is grain-supported		
Mud-supported		Grain-supported			
Less than 10% grains	More than 10% grains				
Mudstone	Wackestone	Packstone	Grainstone	Boundstone	
Classification of phosphorites					
'Non-pelletal phosphorites'		'Pelletal phosphorites'			
Mudstone phosphorite	Wackestone phosphorite	Packstone phosphorite	Grainstone phosphorite	Boundstone phosphorite	

After Cook, Elgueta, Sandstrom & Southgate (unpublished).

a calcrete. However elsewhere in this volume Southgate applies the term 'phoscrete profile' with greater precision and has further refined this general classificatory system (Southgate, 1983). For this publication the basic classification is regarded as adequate. Thus, we have made the editorial decision that where authors used the term 'pelletal phosphorite' this would in all cases be changed to 'grainstone phosphorite'. In some instances the term 'packstone phosphorite' may be more correct but seldom do authors who use the term 'pelletal phosphorite' provide details of whether the rock is mud-supported or grain-supported. Additionally, in our experience wackestone and packstone phosphorites are comparatively rare, and we suspect that our perhaps arbitrary translation of 'grainstone' for 'pelletal' is petrographically precise in almost all cases.

Another area requiring standardisation is in the nomenclature of phosphate particles. Perhaps the most precise and potentially valuable system is that proposed by Prévôt at recent Project 156 meetings. However, at the present time, the system of Mabie & Hess (1964) is probably the most widely used, and where possible, we have used their terminology here. In this scheme, a second descriptor is applied, e.g. oolitic grainstone phosphorite. In some cases terminology used in the carbonate literature is more applicable, e.g. skeletal grainstone phosphorite.

Stratigraphic terminology is as controversial as descriptive phosphate nomenclature. While most countries have a Stratigraphic Nomenclature Committee of some kind which is aligned with the International Stratigraphic Nomenclature Commission, and generally abides by the rulings of that group, terminology is not compatible throughout the world. Hence, the reader will find it necessary, especially in the European papers, to be aware of old established stratigraphic concepts. Thus, he or she will find a combination of biologically and lithologically named units, or others based on mineralogical criteria, e.g. Exporrecta Conglomerate, Alum Shale, etc. Alternatively, the reader may find an alpha-numeric reference used in some areas. For example, the Norwegian Cambrian–Silurian chronostratigraphic subdivisions have the Symbols 1, la, laα etc., equivalent to Series, Stages and Zones (for explanation of this and other Scandinavian terminology, see Henningsmoen, 1961). It has been necessary to accept this nomenclature. Similarly, we have accepted the translations of Soviet authorities from Russian concepts into western styles so that a yarus becomes a stage, and a gorizont a zone, even though there are admitted differences in philosophy (see Öpik, 1975, pp, 49–51 for discussion). It is only during the last five years that Chinese geologists have begun to distinguish terminology for lithostratigraphic and biochronologic units in the Cambrian. We have used the most up-to-date information available. Pinyin renderings of Chinese stratigraphic units have also been adopted.

For time scales, we have generally accepted the regional usage as presented by the authors concerned. Post-glacial late Proterozoic time is referred to as Vendian in the Soviet Union and Europe and this is equivalent to late Sinian in China, late

Table 1.2. *Proterozoic to Middle Cambrian terminology*

Million years	North America	United Kingdom	USSR	China	Australia	
525	Lairdian	Menevian	Mayan	Zhangxia	Boomerangian	Cambrian
					Undillan	
			Amgan	Xuzhuang	Floran	
	Waucoban	Solvan		Maozhuang	Templetonian	
540		Caerfai	Lenian	Longwangmiao	Ordian	
			Atdabanian	Canglangpu	No stages designated	
				Qiongzhusi		
			Tommotian	Meishucun		
590	Hadrynian	Vendian	Vendian/Yudoman	Sinian	Adelaidean	Proterozoic
680						
800		Riphean	Riphean	Qingbaikou		
1000	Helikian			Jixian		
1800 ↓					Carpentarian	

Adelaidean in Australia, and late Hadrynian in North America. Riphean is used for the pre-Vendian scale in the Soviet Union and Europe, and is largely equivalent to the Helikian and early Hadrynian of North America (see Table 1.2). We consider the proposal for an Ediacarian System (Cloud & Glaessner, 1982) has great merit. However we regretfully felt unable to use the term in this volume because of the imprecision of most of the Proterozoic radiometric and biochronologic data available to us.

There is no international agreement at the time of writing on the limits of the Cambrian System. It seems likely, however, that the International Union of Geological Sciences, Commission on Stratigraphy, Precambrian–Cambrian Boundary and Cambrian–Ordovician Boundary Working Groups will come to decisions in the near future. It is generally agreed within these bodies that the Precambrian–Cambrian Boundary should be placed within a fossiliferous sequence. For this purpose sections on the Aldan River at Ulakhan–Sulugur on the Siberian Platform, at the base of the Tommotian Stage, and at the base of the Dahai Member of the Yuhucun Formation, within the Meishucun Stage at Xiaowaitoushan, Kunming, Yunnan Province, China, are currently under international scrutiny. With respect to the upper boundary of the Cambrian System, there appears to be a general consensus in favour of regarding the Tremadoc Series as initial Ordovician and seeking a suitable horizon close to its base, datable by conodonts, graptoloids and trilobites which ensure intercontinental correlation. Because even the internal divisions of the Cambrian are frequently disputed, we have adopted the simplified correlations suggested by Shergold & Brasier (this volume), and accept the *status quo* on the correlation of regional biochronological scales.

Palaeontological dating is of necessity relative compared to geochronology, and the two methods do not always give compatible results. We follow Harland *et al.* (1982) in regarding Vendian time as commencing about 670 m.y. ago, and Cambrian time at between 590 and 570 m.y. depending at what level the boundary is finally chosen. Younger proposed dates for the beginning of the Cambrian, between 540 and 530 m.y., are considered here to reflect a thermal event between the Early and Middle Cambrian epochs. One striking anomaly between biochronological and geochronological dating is raised in this book regarding the dating of some of the Indian Proterozoic deposits in which stromatolite morphology appears to suggest a younger age than the generally accepted geochronology.

There are a number of important points that we should make here regarding this volume and our underlying philosophy in creating it. In most multi-author volumes chapters vary in quality and quantity. This is perhaps more so in this volume than in many others. We have where possible endeavoured to apply a degree of uniformity of quality to all the chapters. In this we were helped by many referees to whom we extend our thanks. However, in some cases research or exploration programmes are at a very early stage and the data are not available. In other cases authors are from developing countries and may not have access to the same sophisticated investigative techniques available in developed countries. If we were to wait until all investigations were at a uniformly high level then it probably would have been necessary to delay this volume until the next century! Additionally, it is important to point out that the authors of more than half the chapters have a language other than English as their first language. This has inevitably led to difficulties both for the authors and ourselves. There have been times we must admit when we have felt it might have been easier to write this book ourselves, but this would have resulted in a volume which was neither as definitive as we believe this to be, nor would it have been in the spirit of international co-operation which the Project has engendered. In conclusion then, we make no apology for any perceived variations in the quality or quantity of the information presented here, for this is an inevitable consequence of producing what we believe to be a definitive yet truly international text on a topic for which until relatively recently there was little or no information available. We judge that this book will be of interest to both the specialist and the general reader alike, and will prove to be a valuable reference source for some years to come on phosphorites in general, and Proterozoic and Cambrian phosphorites in particular.

References

Andersson, J.G. (1896). Uber cambrische und silurische phosphorit – fuhrende Gesteine aus Schweden. *Bulletin of the Geological Institutions of the University of Uppsala*, **2**, 133–238.

Buckland, W. (1829). On the discovery of coprolites or fossil faeces in the Lias at Lyme Regis and in other formations. *Geological Society of London, Transactions*, 2nd Series, **3(1)**, 223–36.

Bushinsky, G.I. (1938). Petrography and certain problems of the genesis of the Aktyubinsk phosphorites. *Byulleten Moskovakogo Obshchestva Ispytatelei Prirody, Geologicheskiy*, **16(4)** (in Russian).

Bushinsky, G.I. (1963). The shallow occurrence of phosphoritic sediments and their genesis. In *Deltaic and Shallow Marine Deposits*, ed. P.L. Bezrukov, M.S. Shretsov & V.S. Yablokov. Akademiya Nauk SSSR, Moscow (in Russian).

Bushinsky, G.I. (1966). Old phosphorites of Asia and their genesis. *Akademiya Nauk SSSR, Trudy Geologicheskogo Instituta*, **149**, 195pp. (published in English by Israel Programme for Scientific Translations, Jerusalem, 1969).

Cheng Yuqi (1939). A brief report on the geology of the phosphorite deposit at Zhongyicun–Waitaoshan district, Kunyang, Yunnan. *Geological Review* (*Dizhi Lunping*), **4**, 185–92 (in Chinese).

China National Committee for IGCP (1982). *Guidebook to Field Excursions*. Fifth International Field Workshop and Seminar on Phosphorite, Kunming, China, November 17–24, 1982, 50 pp.

China National Committee for IGCP (Editor) (1984). International Geological Correlation Programme Project 156–Phosphorite. Symposium of 5th International Field Workshop and Seminar on Phosphorite. Vol. 1, 389pp., Vol. 2, 503pp. Geological Publishing House, Beijing, China.

Cloud, P.E. & Glaessner, M.F. (1982). The Ediacarian Period and System: Metazoa inherit the Earth. *Science*, **217(4562)**, 783–92.

Cook, P.J. (1972). Petrology and geochemistry of the phosphate deposits of northwestern Queensland. *Economic Geology*, **67**, 1193–213.

Cook, P.J. & McElhinny, M.W. (1979). A reevaluation of the spatial and temporal distribution of sedimentary phosphate deposits in the light of plate tectonics. *Economic Geology*, **74**, 315–30.

Cook, P.J. & Shergold, J.H. (Eds) (1979). *Proterozoic–Cambrian Phosphorites*. 106pp. ANU Press, Canberra.

Cook, P.J. & Shergold, J.H. (1980). Proterozoic and Cambrian phosphorites of Asia and Australia – a progress report. In *Fertilizer Mineral Potential in Asia and the Pacific* ed. R.P. Sheldon & W.C. Burnett, pp. 207–23. East–West Resource Systems Institute, Honolulu.

Daubeny, C. & Widdrington, Capt. (1845). On the occurrence of phosphorite in Estremadura. *Quarterly Journal of the Geological Society of London*, **1**, 52–5.

Davies, D.C. (1875). The phosphorite deposits of North Wales. *Quarterly Journal of the Geological Society of London*, **31**, 357–67.

Davidson, C.F. (1963). Phosphate deposits of Precambrian age. *Mining Magazine of London*, **109(4)**, 205–8.

Dawson, J.W. (1876). Note on the phosphates of the Laurentian and Cambrian rocks of Canada. *Quarterly Journal of the Geological Society of London*, **32**, 285–91.

de Keyser, F. & Cook, P.J. (1972). Geology of the Middle Cambrian phosphorites and associated sediments of northwestern Queensland. *Bureau of Mineral Resources, Geology and Geophysics of Australia, Bulletin*, **138**, 79pp.

Dunham, R.J. (1962). Classification of carbonate rocks according to depositional texture. In *Classification of Carbonate Rocks*, ed. W.E. Ham, pp. 108–21. *American Association of Petroleum Geologists, Memoir*, **1**, 279pp.

Elgueta, S. (1981). Sedimentology of Lady Annie Phosphate Deposit. M.Sc. Thesis, Australian National University (unpublished).

Emigh, G.D. (1973). Economic phosphate deposits. In *Environmental Phosphorus Handbook*, pp. 97–116, John Wiley & Sons, New York.

Friedman, G.M. & Sanders, J.E. (1978). *Principles of Sedimentology*. 792pp. John Wiley & Sons, New York.

Gary, McAfee Jr, R. & Wolf, C.L. (Eds.) (1972). *Glossary of Geology*. 857pp. American Geological Institute, Washington, DC.

Geijer, P. (1962). Some aspects of phosphorus in Precambrian sedimentation. *Arkiv for Mineralogie och Geologi*, **3**, 165–86.

Guimarães, D. (1967). Ocorrencia de fosforita no municipio de Abaeke, Minas Gerais. *Departamento Nacional da Produção Mineral, Notas Preliminares e Estudios*, **144**, 18 pp.

Harland, W.B., Cox, A.V., Llewellyn, P.G., Pickton, C.A.G., Smith, A.G. & Walters, R. (1982). *A Geologic Time Scale*, 131pp. Cambridge University Press, Cambridge.

Henningsmoen, G. (1961). Code of stratigraphical nomenclature for Norway. *Norges Geologiske Undersøkelse*, **213**, 229–33.

Hicks, H. (1875). On the occurrence of phosphates in Cambrian rocks. *Quarterly Journal of the Geological Society of London*, **31**, 368–76.

Howard, P.F. (1972). Exploration for phosphorite in Australia – a case history. *Economic Geology*, **67**, 1180–92.

Hudleston, W.H. (1875). On the chemical analyses of the rocks. Appendix to H. Hicks: On the occurrence of phosphates in the Cambrian rocks. *Quarterly Journal of the Geological Society of London*, **31**, 376–83.

Ilyin, A.V., & Bjamba, Z. (1980) *Guidebook for Excursion, Phosphorites of the Khubsugul Basin, Mongolian People's Republic*. Field conference of IGCP Project 156. Geologicheskii Institut Akademiya Nauk SSSR, Moskow, 135pp. (in Russian and English).

Indian National Committee for IGCP (1981a). *Guidebook for Excursion, Aravalli Phosphorites around Udaipur, Rajasthan, India*. Fourth International Field Workshop and Seminar of IGCP Project 156–Phosphorite, 57pp. Geological Survey of India, Jaipur.

Indian National Committee for IGCP (1981b). *Excursion Guide Book on Dehradun–Mussoorie Area*. Fourth International Field Workshop and Seminar on Phosphorite. International Geological Correlation Programme Project 156 – Phosphorite, 53pp. Geological Survey of India, Northern Region.

Jack, R.L. (1919). Alunite deposits. Section A., Hundred of Ramsay. *South Australian Mining Review*, **28**, 51.

Kazakov, A.V. (1938). Phosphate facies. *Trudy Nauchnogo Instituta po Udobreniyu i Insektofungisidam*, **145**, 108pp. (in Russian).

McConnell, D. (1973). *Apatite*, 111pp. Springer-Verlag, New York.

McKelvey, V.E., Swanson, R.W. & Sheldon, R.P. (1953). The Permian phosphorite deposits of the western United States. *19th Session of the International Geological Congress, Algiers*, **11**, 45–65.

Mabie, C.P. & Hess, H.D. (1964). Petrographic study and classification of Western phosphate ores. *US Bureau of Mines, Report of Investigations*, **6468**, 95pp.

Mashkara, I.J. (1938). The stratigraphy and tectonics of the northeastern branch of the Karatau (Southern Kazakhstan). *Problemy Sovietskoy Geologii*, **8(1)**, 49–59.

Öpik, A.A. (1975). Cymbric Vale fauna of New South Wales and Early Cambrian biostratigraphy. *Bureau of Mineral Resources, Geology and Geophysics of Australia, Bulletin*, **159**, 78pp.

Riggs, S.R. (1979a). Petrology of the Tertiary phosphorite system in Florida. *Economic Geology*, **74(2)**, 195–220.

Riggs, S.R. (1979b). A petrographic classification of sedimentary phosphorites. In *Proterozoic–Cambrian Phosphorites*, ed. P.J. Cook & J.H. Shergold, pp. 63–4. ANU Press, Canberra.

Russell, R.T. (1967). Discovery of major phosphate deposits in northwest Queensland. *Queensland Government Mining Journal*, **68**, 153–7.

Russell, R.T. & Trueman, N.A. (1971). The geology of the Duchess phosphate deposits, northwest Queensland, Australia. *Economic Geology*, **66**, 1186–214.

Slansky, M. (1979). Proposals for nomenclature and classification of sedimentary phosphate rocks, In *Proterozoic–Cambrian Phosphorites*, ed. P.J. Cook & J.H. Shergold, pp. 60–3. ANU Press, Canberra.

Sougy, J. (1964). Les formations paléozoïques du Zemmour noir (Mauritanie septentrionale). *Annales de la Faculté des Sciences, Université de Dakar*, **15(12)**, 695pp.

Southgate, P.N. (1983). Middle Cambrian phosphatic and calcareous depositional environments, the Undilla region of the Georgina Basin, Australia. PhD Thesis, Australian National University (unpublished).

Yanshin, A.L. (1964). Perspectives and scientific problems in the search for potassium salts and phosphorites in Siberia. *Byulleten Moskovskoye Obshchestvo Ispytateley Prirody, Otdelenie Geologicheskiy*, **39(5)**, 1–17.

Zaitsev, N.S. & Ilyin, A.V. (1970), Khubsugul Phosphorite Basin. *Doklady Akademii Nauk SSSR*, **192(2)**, 391–4.

2

Proterozoic and Cambrian phosphorites – regional review: world resources

A.J.G. NOTHOLT AND R.P. SHELDON

Introduction

During the latter half of the 19th century and the early part of this century, major marine phosphorites were found and developed in the Middle East. North Africa, and south-eastern and western United States. In addition, many smaller phosphorite deposits were discovered in Europe. All of these deposits are of Ordovician and younger age. A concept that major phosphorite deposits were deposited late in geological history thus prevailed in at least western geological thinking, so that exploration tended to be restricted to the younger rocks, even though a few small occurrences, all non-economic, were known in older sediments. For example, phosphorites of Proterozoic and Cambrian age had been described from Cambrian strata in the United Kingdom and Sweden in 1875 and 1896 respectively. Similarly, phosphatic nodules in the Upper Riphean shales of the United Kingdom described in 1907 are the first to have been recorded from Proterozoic rocks. Cambrian phosphorites were also found in Canada during this period.

Major discoveries in Proterozoic and Cambrian rocks were not made until the late 1930s, initially in the USSR (Kazakhstan), Poland, Korea, China and northern Vietnam. In rapid succession, important discoveries were made in Mongolia, Australia, India, Brazil and western and central Africa, and smaller deposits in the Baltic Platform of Scandinavia, north-central United States, France, Iran, Spain, Pakistan and Nepal. The early discoveries were made in the course of regional geological mapping and exploration for metal deposits, but later deposits were found mainly using more direct exploration techniques.

It is now apparent that what is perhaps the greatest global phosphogenic episode in geological history occurred in Late Proterozoic and Cambrian times on all continents except, as far as we know, ice-covered Antarctica. A less important phosphogenic episode occurred in Early Proterozoic time, represented by deposits in Australia, India, Korea and North America. Although these Proterozoic and Cambrian phosphorites are commonly folded, faulted and even metamorphosed by later tectonic activity, so that their economic value has been lessened, many are now economic deposits. More than 90 individual or groups of occurrences are now known, divided about equally between those of Proterozoic and Cambrian age. Pelletal phosphorites are common in most Cambrian deposits, but phosphorites of undoubted Proterozoic age contain mudstone phosphorite (microphosphorite) and stromatolitic phosphorite. Many of these deposits occur within the Siberian and Chinese cratons (Bushinsky, 1969; Notholt, 1972, 1979; Lee, 1980) where there is one major, predominantly Cambrian, phosphogenic province, which includes important resources in the USSR, China, Mongolia and the Socialist Republic of Vietnam. Another major Cambrian phosphogenic province occurs on the Australian craton in Queensland and the Northern Territory, and the Indian craton of India and Pakistan. The extent to which these areas were palaeogeographically related is apparent from Parrish *et al* (this volume).

On the basis of published estimates total Proterozoic–Cambrian resources amount to over 14 000 million tonnes of phosphate rock containing more than 6% P_2O_5 (Table 2.1), located mainly in the USSR, China, Mongolia, Vietnam and Australia (Fig. 2.1). These resources are equivalent to about 10% of total world resources. In excess of 10 000 million tonnes averaging 14% P_2O_5 or more may be assigned to the Cambrian. Although only a small proportion of the total identified resources is likely to attract commercial interest, at least in terms of mineable thickness or quality, phosphorites of Proterozoic–Cambrian age have become sufficiently well known to provide more detailed insight into phosphogenic processes and, in terms of exploration potential, offer new depositional environments that are likely to be favourable to phosphorite formation.

North America

The phosphorite resources of North America are contained mainly in Lower Tertiary and Permian sediments, and only minor occurrences of Proterozoic or Cambrian have so far been recorded, in Tennessee and Montana in the USA and in western Canada (Sheldon & Christie, this volume). However, among the most recent and noteworthy discoveries in the USA are the black phosphatic nodules (Cannon & Klasner 1976) from the Middle Proterozoic Marquette Range Supergroup in the central part of the Upper (Keweenaw) Peninsula of

Table 2.1. *Major resources of Proterozoic–Cambrian phosphorite*

Age/location	Number (Fig. 2.1)	Resources (million tonnes)	Host rocks
PROTEROZOIC			
Brazil			
Patos de Minas, Minas Gerais	1	454	Bambui Group
India			
Udaipur, Rajasthan	9	64	Aravalli Series
Korea			
Singpung	10	90	Mach'onnyang Series
Upper Volta			
Kodjari–Arly	2	500	Pendjari Formation
USSR			
Gornaya Shoriya, Kemerovsk	4	274	Yenesei Suite
Ukha–Gol, Buryat	5	2600	Boxon Series
CAMBRIAN			
Australia			
Duchess–Lady Annie, Queensland	11	2000	Beetle Creek Formation
Alexandria–Wonarah, Northern Territory	12	1000	Beetle Creek Formation
China			
Kunming, Yunnan	7	4000	Meishucun Formation
USSR			
Karatau, Kazakhstan	3	1700	Chulaktau Suite
Mongolia			
Khubsugul	6	1000	Khubsugul Series
Vietnam			
Lao Cai	8	1000	Coc-San Series

Michigan. At one locality, a conglomeratic bed 15 m thick averages about 15% P_2O_5. Sampling of a 100 m thick sequence showed 5–10% P_2O_5. The formations in the Upper Peninsula are poorly exposed, but resources in this area could be considerable particularly if extensions are proved, as seems to be the case with the discovery, in 1978, of phosphorite of the same age in Carlton County, east-central Minnesota (Ullmer, 1981). Although thin, three phosphatic units with a maximum thickness of about 30 cm have been found in the Thomson Formation in association with graphitic phyllites and quartzites. Samples contain up to 21% P_2O_5. The Thomson Formation is part of the Animikie Group, a sequence consisting predominantly of clastic sediments and iron ores which occurs also in northern Wisconsin and the Upper Peninsula of Michigan.

South America

In addition to the well-known igneous deposits in Brazil, phosphorites were discovered in 1967 in the Late Proterozoic Bambuí Group of Minas Gerais. Deposits, situated about 55 km north-west of Patos de Minas (Cathcart, 1977), occur as irregular lenses up to 500 m long and 8 m thick within a NE–SW trending belt of isoclinally folded and much faulted claystone and siltstone. They have been traced along strike for about 3 km, with resources estimated at nearly 454 million tonnes averaging 10.6% P_2O_5. The Bambuí Group crops out over a very large area in the states of Minas Gerais, Goiás and Bahia. It therefore seems likely that further exploration work, particularly in Goiás to the north of Patos de Minas, will lead to the delineation of additional deposits, some of which may be large. The Bambuí or São Francisco Basin is the largest of the Proterozoic areas in Brazil, occupying an area of about 400 000 km² (see Dardenne *et al.*, this volume).

Africa

Phosphate rock occurs in Africa in two markedly contrasting geological environments, sedimentary and igneous, with resources of the former residing predominantly in rocks of Late Cretaceous and Eocene age. Minor occurrences of Proterozoic age have been reported from Morocco (Viland, 1977), Senegal (Slansky, 1980) and Somalia (Ilyin, 1978), and Cambrian phosphorite from the Taoudeni Basin of Mauritania (Sougy, 1964). However, by far the most important are the Late Proterozoic deposits discovered astride the Upper Volta, Benin and Niger borders (Trompette, 1978; Lucas, Prévôt &

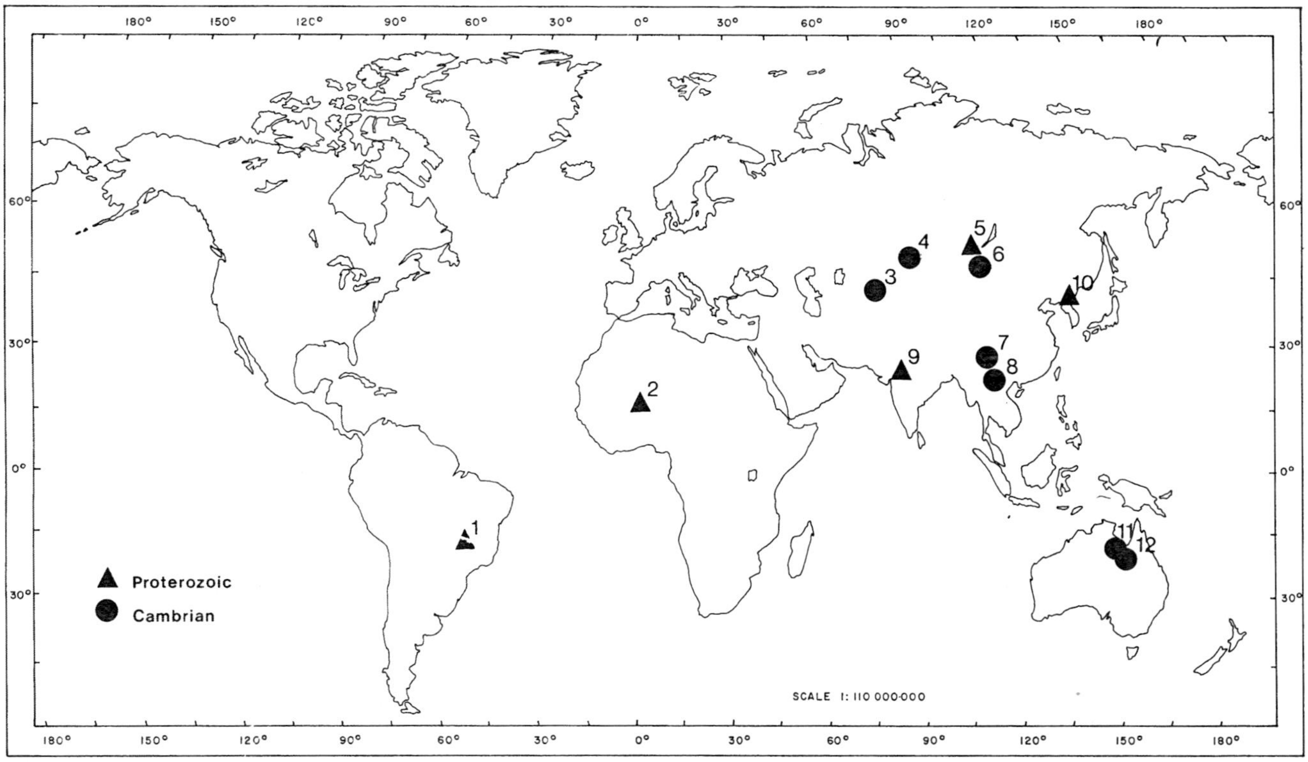

Fig. 2.1. Distribution of major Proterozoic–Cambrian phosphorite resources.
Key: 1, Patos de Minas, Minas Gerais, Brazil; 2, Upper Volta–Niger; 3, Karatau Basin, Kazakhstan, USSR; 4, Gornaya Shoriya, Kemerovsk, USSR; 5, Ukha–Gol, Buryat, USSR; 6, Khubsugul Basin, Mongolia; 7, Kunming area, Yunnan Province, China; 8, Lao Cai, Vietnam; 9, Udaipur, Rajasthan, India; 10, Singpung, Korea (DR); 11, Duchess–Lady Annie areas, Queensland, Australia; 12, Alexandria–Wonarah areas, Northern Territory, Australia.

Trompette, 1980). Their occurrence is an indication that considerable effort should be made to determine the location of other sedimentary basins in these Precambrian areas in order to determine their phosphate potential, for West Africa, like most of the African continent, is composed largely of Precambrian rocks. In Upper Volta, individual beds sometimes several metres thick with 15–32% P_2O_5 are reported to contain 500 million tonnes of phosphate rock, but this is likely to be a very conservative estimate. The beds are found in the upper part of the Pendjari Formation, a sequence comprising breccia, tillite, dolomitic limestone, bedded cherts and argillaceous siltstones. As at Patos de Minas in Brazil, weathering has produced a surface zone rich in wavellite. In Niger, a borehole drilled at Tapoa has revealed a main bed of phosphorite 60 m thick (Slansky, 1980) (see also Lucas *et al.*, this volume).

Europe

Although the presence of Proterozoic and Cambrian phosphorites in Europe has been known for many years, most of the occurrences are in the form of thin, reworked nodule beds, with the notable exception of relatively recent discoveries in the Cambrian strata of northern and southern France. Phosphogenic episodes are evidently more widespread on the Baltic Platform than elsewhere in Europe (Notholt & Brasier, this volume), with phosphorites of Early Cambrian age occurring, for example, at a relatively large number of localities along the eastern border of the Caledonides in Sweden. Individual beds are generally less than 1 m thick; nevertheless, their distribution is such that a Cambrian phosphogenic province may be delineated. Also, in the eastern part of the Baltic Platform are the uraniferous phosphorites in the Lower Proterozoic schist belts of central and northern Finland (Äikäs, 1980, 1981). These, and occurrences of comparable age in Sweden, give support to the view (Äikäs, 1980) that an Early Proterozoic phosphogenic province existed within the Baltic Platform. More extensive distribution of Late Proterozoic phosphorites than hitherto supposed is indicated by a further discovery in central Spain (Perconig *et al.*, this volume).

In Finland, the most important discoveries are in the marine Upper Jatulaina (Karelian) metasediments where a phosphate zone up to 50 m thick occurs in the Vihanti sulfide ore complex at Lampinsaari, south of Oulu, consisting of lenticular beds up to 0.5 m thick of apatite–dolomite, skarn and apatite–gneiss (Rehtijärvi, Äikäs & Makelä, 1979). Some examples show up to 25.5% P_2O_5, but the entire phosphate zone averages only 3–5% P_2O_5 and 0.03% U. A similar body, 3.5 km long and up to 90 m wide, occurs at Temo, about 20 km north-east of Kuopio. As far as is known, estimates of the phosphate resources potential of these Proterozoic rocks have not been made, but their occurrence indicates a period of extensive phosphate deposition during Proterozoic times.

Asia

The detailed geology of large tracts of Asia remains incompletely known and it will be many years before phosphorite resources are fully evaluated in many areas. The bulk of the identified resources in the region are of Proterozoic to Cambrian age, by far the largest deposits occurring in the USSR, China, Mongolia and Vietnam. In addition to the Cambrian phosphogenic province referred to earlier, further exploration aimed at delineating Proterozoic sedimentary basins is likely to lead to additional discoveries.

1. USSR

The largest accumulations of Proterozoic age in the USSR are the Ukha–Gol deposits in Buryat (Fig. 2.1) (Volkov *et al.*, 1972; Semeikin *et al.*, 1976) where there are also discontinuous lenses ranging from a few centimetres to as much as 11 m in thickness in the Obrub Formation. The phosphorites closely resemble and probably represent a continuation of the Slyudyanka occurrences to the east in Irkutsk at the southern end of the Lake Baikal: the development of phosphorites in both these areas, within a sequence of dolomitic and calcitic marbles, quartz–diopside gneisses and diopside quartzites is strongly reminiscent of those found in Finland. The phosphorites usually contain only up to about 4% P_2O_5, with occasional pockets containing 15–20% P_2O_5 and, although some beds are traceable for considerable distances, they are too thin to be of commercial interest. In aggregate, however, resources must be substantial and the Buryat–Irkutsk area clearly contains a major, somewhat unusual, phosphogenic province of Proterozoic age.

Other Proterozoic occurrences in the Siberian USSR include thin phosphorite beds in the Uluntui Suite (Middle to Upper Riphean) along the western shore of Lake Baikal to beyond the upper reaches of the Lena River, and in the Baikal Highlands in north-eastern Irkutsk. Interbeds of siliceous phosphorite contain 11–16% P_2O_5. Phosphorites of Late Proterozoic to Early Cambrian age, containing up to 27% P_2O_5 are also reported to occur in the Soviet Far East (Zimin, 1980), including the Shantar Islands in the Sea of Okhotsk. These, as well as the minor occurrences in the Yenisei Ridge, Krasnoyarsk, and the Sangilen Mountains, Tuva, point to a very extensive distribution of Proterozoic phosphorites in eastern Siberia, although the phosphate potential appears to be largely unquantified.

Major deposits, in the Upper Riphean Yenisei Suite, are situated in the Gornaya Shoriya district at Kemerovsk, near Tashtagol, with probable reserves estimated at 274 million tonnes containing 6–11% P_2O_5. An additional 20 million tonnes averaging 21% P_2O_5 appear to be present in the associated residual karst deposits developed during Cenozoic times. The phosphorites lie between black, massive stromatolitic dolomites and partly dolomitised limestones. They are lenticular, rarely extending more than 1 km in length, although some lenses are up to about 150 m thick. Generally the beds contain 5–8% and, locally, 15% P_2O_5.

Undoubtedly the best-known Lower Cambrian phosphorites in Asia are those of the Karatau Basin, north-west of Dzambul in southern Kazakhstan (Gimmelfarb & Tushina, 1966; Kholodov, 1968; Tushina, 1968; Bushinsky, 1969; Eganov & Sovetov 1979; Eganov. Sovetov & Yanshin, this volume). The deposits, which were discovered in 1937, lie within a phosphatic interval of the Chulaktau Formation, the main ore horizon ranging from 5 m to 20 m in thickness and containing 23–32% P_2O_5.

The Karatau phosphate field is much complicated by folding and overthrusting, with phosphorites generally dipping at 30–40° to the north-west over an area of 250 km long and 60 km wide. Consequently, the area is divided into a series of narrow monoclinal 'horst' structures and as many as 45 individual deposits have been delineated. The character of the phosphorites varies considerably, reflecting facies changes within the basin of deposition and also the effects of metamorphism and various secondary changes. Reserves to a depth of about 250 m have been reported to exceed 1000 million tonnes containing 26–30% P_2O_5. Total resources are probably considerably greater, perhaps as much as 10 000 million tonnes. Early to Middle Cambrian phosphorites occur elsewhere in Kazakhstan and in Kirgiz (Bushinsky, 1969; Notholt, 1972), those near Karaganda and Baykonur perhaps representing the most northerly extension of the Cambrian phosphogenic province. It seems not inconceivable that this province continues south-westwards into northern Iran, where phosphorites of the Chopoghlu Shale Member (Soltanieh Formation) previously regarded as 'Infracambrian' are believed to have an Early Cambrian age (Ashkan & Salehi, 1982).

2. Mongolia

In northern Mongolia, the phosphorites of the Khubsugul Basin (Gimmelfarb & Egorova, 1969; Ilyin & Bjamba, 1980) compare closely in geological age, lithological and structural characteristics with those of Karatau. The Khubsugul Basin is 400–500 km in length from north to south and up to about 80 km wide. The phosphate series, which is of Late Vendian or Early Cambrian (Tommotian) age (Ilyin & Ratnikova, 1981), occurs in the lower part of a thick carbonate sequence, the Khubsugul Series.

As at Karatau, there are two distinct phosphorite units, the lower of which includes two beds, 7–10 m and 1–15 m thick respectively and containing up to 30% P_2O_5. These constitute the bulk of the phosphorite resources of the Khubsugul Basin. Eight major deposits have been delineated, each containing hundreds of millions of tonnes of rock averaging 20–22% P_2O_5. The largest deposit, situated near the south-western shore of Lake Khubsugul, is estimated to contain over 1000 million tonnes. 'Structureless' phosphorite (mudstone phosphorite or microsphorite) is a relatively abundant variety of phosphate rock in these deposits (Ilyin & Ratnikova, 1981).

3. Korea (D.R.)

Phosphorite in metamorphosed limestones and dolomites has been found at a number of localities in the northern part of the Democratic Republic of Korea since the first discovery in 1938. The deposits are among the oldest geologically to have been recorded, occurring principally in the thick and intensely deformed Mach'onnyang Series, to which an Early Proterozoic age has been assigned. The series consists mainly of crystalline limestone, dolomite, mica- and graphite-schist, amphibolite and other paragneisses, with a thick carbonate sequence in the middle part of the series containing extensive magnesite beds and a persistent horizon of limestone with *Collenia*. These features are characteristic also of the Liaoho System of southern Manchuria, with which the Mach'onnyang Series has been correlated.

Individual phosphorite-bearing lenses or zones ranging from 30 m to 40 m in thickness and averaging 4–5% P_2O_5 have been reported (Bushinsky, 1969), the largest being the Singpung deposits near Tanchon. One of these is about 1.7 km long and some 80–160 m wide, with several rich zones up to about 15 m wide containing 30–50% apatite (12–20% P_2O_5). Resources of phosphorite averaging 12.5% P_2O_5 have been estimated at 88 million tonnes (Anon, 1963). Similar deposits occur near Songjin where the principal zone is nearly 5 km long and up to 10 m thick. Resources have been estimated at several tens of millions of tonnes containing 10% apatite (4.2% P_2O_5). The Kacha-do deposit in Korea Bay comprises diopside marble which contains more than 20% apatite (8% P_2O_5). The deposit was estimated in 1940 to contain 3 million tonnes. These, and the Singpung occurrences, are generally regarded as representing phosphatic sediments laid down in a regionally extensive Early Proterozoic transgressive sea, although their close association with intrusive igneous bodies suggests that an origin by contact metasomatism cannot be entirely excluded.

4. *China*

Phosphorites of Proterozoic and Cambrian age are widely distributed in China, the numerous remnants of Lower Cambrian strata indicating the approximate extent of a major marine transgression during this period. They constitute the bulk of China's phosphate resources which were reported in late 1982 to have been officially placed at some 11 800 million tonnes. Since the search for phosphate rock has been intensified only relatively recently, however, it is very likely that further discoveries will be made. Earlier accounts of individual deposits (e.g. Bushinsky, 1969) have been superseded by more detailed descriptions and regional assessments (Yeh & Sun, 1980; Liang & Chang, 1981; China National Committee for IGCP, 1982; Li, this volume; Yeh *et al.*, this volume.)

The most important phosphorite deposits in China are located in the southern part of the country around the Yangzi Platform, including those in the Guizhou, Yunnan, Sichuan, Hubei, Hunan and Jiangxi provinces. They are particularly well developed in the Doushantuo (Upper Sinian) and Meishucun (Lower Cambrian) Formations. The Doushantuo Formation comprises dolomites, cherts and phosphorites the main are beds being usually several metres in thickness with grades exceeding 25% P_2O_5. In central Guizhou, for example, there are two ore zones in the Weng'an and Kaiyang districts, north of Guiyang, which are 8–25 m and 0–30 m thick respectively, each containing 25–35% P_2O_5. Proterozoic phosphorites also occur in northern China, notably at the base of the Lower Proterozoic Dongjiao Group. Phosphorites associated with schists, quartzites and marbles include those found near Haizhou, Jiangsu Province, and Hefei in southern Anhui. Similar deposits occur in southern Liaoning Province, north-west of Antung, all these sequences representing a mode of occurrence closely resembling that of the phosphorites of northern Korea. In the Hara Navin Ula and Yin Shan ranges of Inner Mongolia, individual beds contain up to 13% P_2O_5 but are less than 1 m thick. Because of its remoteness, this region has been only partly explored.

Phosphorites in the Lower Cambrian Meishucun Formation have the same general geographical distribution, those in Yunnan near Kunming being particularly well known and among the largest in China. Reserves are possibly of the order of 4000 million tonnes, the main orebodies occurring in the Zhongyicun Member. The major Kunyang and Haikou deposits are located on the southern and northern portions of an east–west trending anticlinal structure. At Kunyang the Zhongyicun Member is up to 10 m thick, containing two phosphorite ore beds averaging 6 m and 3 m in thickness separated by 2 m of shale.

The Lower Cambrian phosphogenic province so well developed and exposed in Yunnan extends into northern China, where Lower Cambrian phosphorites were discovered in the Kelpin Tagh and Kuruk Tagh ranges in 1958. The most characteristic deposits of this age in the area, however, are those located east of Fengtai, in Anhui Province. There lenticular beds are 0–2.4 m thick with a reported average of 20% P_2O_5. Phosphorites of Middle Cambrian age are also relatively widespread, being found mainly south of the Talimu–Sino-Korean Platform and in the Qinling–Qilanshan–Kulunshan areas. The best developed are at the base of the Xinji Formation. Although total resources are probably substantial, most of the beds are too thin to be of economic value and no quantitative estimates of their resource potential appears to have been made.

5. *Vietnam*

The Lower Cambrian phosphogenic province so well developed in the Yunnan Province of south-western China probably has its counterpart in northern Vietnam where metamorphosed, complexly folded and faulted deposits were discovered in 1924 near Lao Cai, about 260 km north-west of Hanoi (Fromaget, 1941; An & Khoa, this voulme). The deposits extend for a strike length of 100 km along the south-west side of the Red River in a zone 1–4 km wide, and consist predominantly of apatite–quartz–feldspar–calcareous schists and quartzites ranging from 100 m to 150 m in thickness. These make up part of the Cam Duong Formation (Coc San Series) to which a probable Vendian to Early Cambrian age has been assigned (An & Khoa, this volume). The deposits contain 10–36% P_2O_5, the highest grades being associated with secondary deposits produced by weathering.

6. *India*

Several significant phosphorite discoveries of Late Proterozoic age have also been made in northern India, as the result of a systematic exploration programme begun by the Geological Survey of India and other agencies in 1966. Important deposits have been delineated within the Aravalli phosphogenic province, including those at Jhamarkotra and

Matoon near Udaipur in north-western Rajasthan and near Jhabua in neighbouring Madhya Pradesh. These are distributed over a distance of more than 250 km and have rapidly become the principal commercial source of indigenous phosphate rock.

The phosphorites of the Aravalli phosphogenic province are well documented (Banerjee, 1971; Sant, 1979, 1980; Choudhuri & Balasubramanian, 1980; Pant, 1980; Geological Survey of India, 1981a). Essentially the deposits are developed in dolomitic and siliceous limestones or brecciated, jaspery quartzites and sandy phyllites of the Matoon Formation (Aravalli Series) or its stratigraphical equivalents, a striking characteristic being their association with stromatolitic structures. Individual phosphorite beds range from a few centimetres to 40 m in thickness and extend over strike lengths of up to 14 km (Sant, 1979). The Matoon Formation is best developed near Udaipur, where the original basin topography is perhaps reflected in the broad synformal structure of the phosphorites and associated sediments at Jhamarkotra, 26 km SSE of Udaipur, where phosphorite was discovered by the Rajasthan State Directorate of Mines and Geology in 1968. At Jhamarkotra the deposits are exposed over a distance of about 16 km in a roughly SE–NW direction, the main phosphate-bearing zone being lenticular, ranging up to more than 11 m in thickness, with outcrop widths of 10–70 m. Reserves are estimated at about 64 million tonnes with more than 12% P_2O_5, of which 16.8 million tonnes are of high-grade (+30% P_2O_5) ore. Variations in grade and overall quality are believed to be due to variations in the development of phosphatic stromatolites, repeated structural deformation and weathering. The age of the Aravalli phosphorites has been a matter of controversy, the morphology of the stromatolites indicating a close affinity with Late Proterozoic (Late Riphean or Vendian) forms (see Choudhuri & Roy, and Banerjee, this volume).

Proterozoic phosphorites have been recorded from other parts of India, many of the occurrences having been reported only recently in the literature (Geological Survey of India, 1981b). Although generally too small to be of commercial interest as a source of phosphate rock, their presence points, nevertheless, to new phosphogenic sequences in India whose resource potential has yet to be fully assessed. Belts of stromatolitic carbonate are developed in the isolated tectonic blocks of the Kumaun Himalaya, including those in the Deoban/Garhwal Basin which extends from Uttar Pradesh eastwards into western Nepal and westwards into Himachal Pradesh. The phosphorites occur in the Gangolihat Dolomite Formation of the so-called 'Calc Zone' in the Pithoragarh district, now referred to as the Garhwal Group. In neighbouring Nepal, similar occurrences have been reported from the Bajang and Baitadi districts in cherty massive dolomites (Bashyal, 1981, Bashyal, Banerjee & Khan, 1981). In Andhra Pradesh, southern India, phosphorites, all associated with stromatolites, occur in the Cuddapah Basin, in the Kurnool and Cuddapah districts, in beds up to 1 m thick and containing about 20% P_2O_5 in the Cumbum Shales (Nallamali Formation). Other discoveries include those recorded from the Champaner Group of the Aravalli Series near Gandhra, Panchmahals district, Gujarat, and from the Delhi Supergroup at Barrod, Alwar district in northern Rajasthan, in the Serska and Kushalgarh Formations. In Madhya Pradesh, and southern Uttar Pradesh, Lower Proterozoic phosphorites are associated with the Bijawar Group near Lalitpur and Hirapur. At Lalitpur siliceous phosphorites occur in lensoid bodies from 5 m to nearly 35 m in thickness and extending from 50 m to 250 m in length. Based on preliminary work, reserves are placed at 5 million tonnes containing 15–25% P_2O_5 (Pant, 1980).

Phosphorites of Late Cambrian–Early Ordovician age occur in the Lower Tal sequence near Dehradun in the Kumaun Himalaya of Uttar Pradesh. The Lower Tal has until recently been regarded as of Late Mesozoic (Jurassic to Cretaceous) age (Geological Survey of India, 1981c). An Early Palaeozoic age would, however, place a Late Proterozoic or Cambrian age on the underlying Krol Formation in which thin phosphorites have also been recorded. The Lower Tal phosphorites may represent an important link, therefore, with occurrences of comparable age in south-western China (Yunnan) and Vietnam and in the Hazara district of northern Pakistan.

A thin bed of phosphate nodules was first reported in the Mussoorie area north of Dehradun in 1884 (King, 1884) and led to the later discovery of phosphorite in 1966 (Sheldon, 1966) in the Tal Formation. The Krol–Tal Formations are part of the NW–SE trending Mussoorie Syncline, a doubly plunging structure, with a strike length of the Krol–Tal contact amounting to some 120 km. Individual phosphorite beds range from a few centimetres to about 10 m in thickness, and in places contain more than 30% P_2O_5. Eleven blocks have been delineated within the Mussoorie Syncline which together are estimated to contain about 45 million tonnes with 16–18% P_2O_5.

7. *Pakistan*

The discovery of Palaeozoic phosphorites in the Hazara district in the North West Frontier Province is the most significant and commercially the most promising to have been made to date in Pakistan. The phosphatic nature of the rocks of the Abbottabad Formation was first reported in 1969 (Latif, 1972) and investigations by the Geological Survey of Pakistan began in 1971 (Tayyab Ali, 1977; Bhatti, 1977). A number of deposits are now known to occur over a distance of 35 km in the Kakul–Dalola areas to the north-north-east of Abbottabad, occurring as thin lenses and stringers in the uppermost cherty dolomite of the Abbottabad Formation and the lower part of the overlying Hazira Formation (Asrarullah, 1980). Individual beds of grainstone phosphorite are generally less than 1 m thick, with at least 10–15% P_2O_5 and a maximum of 30% P_2O_5. Proved and probable reserves total nearly 23 million tonnes. Beyond the Hazara district, the phosphorite horizon is traceable along the syntaxial bend of the Himalaya into Azad Kashmir near Muzaffarabad and probably extends southwards through Kotli in Kashmir (Tayyab Ali, 1977) so that additional reserves undoubtedly exist, although individual occurrences are likely to be thin and scattered, and interbedded with thick dolomitic sequences.

Australia

Small deposits of Cambrian phosphate rock are known and have been worked for many years in South Australia, but resources are negligible. By far the most important are those discovered in northwestern Queensland in 1966 near Duchess, some 180km south-south-east of Mount Isa. Subsequent investigations led to the delineation of several large deposits, notably in the Middle Cambrian Beetle Creek Formation in other parts of Queensland and in neighbouring parts of the Northern Territory (Russell & Trueman, 1971; de Keyser & Cook, 1972; Howard, 1972; Cook, 1972, 1976; Howard & Perrino, 1976; Rogers & Keevers 1976; Cook & Shergold, 1979, 1980; Rogers & Crase, 1980). Their discovery is a striking example of the success that can attend systematic exploration, particularly in remote areas in which there are also very limited rock exposures. All the deposits are situated within the Georgina Basin, a Palaeozoic shallow-water, marine depositional structure, 320 000 km^2 in extent, which contains Cambrian and Ordovician marine sediments, as well as Devonian continental sediments. The so-called Burke River Outlier, south-east of Duchess, forms a subsidiary unit to the main body of the Georgina Basin. The Beetle Creek Formation is a richly fossiliferous sequence of siliceous claystone, chert, thin limestone and phosphorite which reaches a maximum thickness of 122 m. Beds of phosphorite may be up to 30 m, but usually they range from 3 m to 9 m in thickness, with an average of about 20% P_2O_5.

Total identified resources of phosphate rock in the Georgina Basin are placed at 2800 million tonnes averaging 17% P_2O_5, 10 separate orebodies having been recognised. The most northerly of the deposits, Lady Annie and Lady Jane, which are located approximately 225km south of Burketown on the Gulf of Carpentaria, together contain at least 400 million tonnes averaging 17–18% P_2O_5. The Duchess property, opened up in order to utilise its resources of 'direct shipping grade' rock, is estimated at some 45 million tonnes averaging 31% P_2O_5. Total resources in the Duchess area may amount to 1000 million tonnes of ore averaging 17–18% P_2O_5. Additional resources of phosphate rock exceeding 1500 million tonnes containing 17% P_2O_5 have been identified in the Yelvertoft Station area of Queensland and the Alexandria and Wonarah districts of the Northern Territory.

In the Northern Territory, 65 km south of Darwin, phosphorites were discovered near Rum Jungle in 1961 during a search for uranium (Pritchard & Cook, 1965). Although this relatively remote area has been incompletely explored for phosphate, drilling has outlined at least 2.3 million tonnes containing 5–38% P_2O_5, averaging 10–12% P_2O_5, distributed over 18 deposits. The deposits are poorly exposed, and they are associated with folded hematitic siltstone and quartz–breccia, hematitic sandstone, minor conglomerate and dolomite, together with lenses of chlorite–schist, all of which form part of a Lower Proterozoic sequence.

Economic considerations

Phosphorite deposits of Proterozoic and Cambrian age have been worked on a commercial scale for many years and currently account for an estimated 19 million tonnes of marketable phosphate rock (Table 2.2), equivalent to about 14% of the total world production, of both sedimentary and igneous phosphate rock of some 134 million tonnes. Virtually all the Proterozoic–Cambrian output was derived from deposits situated in China and the USSR and is believed to have been used mainly as fertilizer for direct application to the soil. In China, India, Korea (DR) and Vietnam, Proterozoic–Cambrian phosphorite deposits represent the prinicipal or sole source of supply of raw material for these countries' phosphate fertilizer needs. With the notable exception of the USSR and Brazil, the phosphate rock marketed averages 30% P_2O_5 or more.

Table 2.2. *Proterozoic–Cambrian phosphate rock production, 1983*

Country	Geological age	Production (thousand tonnes)	Average Grade P_2O_5 (%)
Brazil	Proterozoic	200	24
China (e)	Cambrian (a)	12500	30
India	Proterozoic (b)	600	32
Korea (D.R.)	Proterozoic	500	32
USSR	Cambrian (c)	5000	23
Vietnam	Cambrian	220	35

(e) Estimate
(a) Includes Proterozoic production
(b) Includes Cambrian production
(c) Approximate figure for Karatau Basin derived from estimates for other domestic sources of production

Proterozoic production

Commercial production of Proterozoic phosphorite is derived almost entirely from India and Korea (DR) and, to a lesser extent, Brazil. In India, the most important are the Jhamarkotra deposits south-east of Udaipur, in Rajasthan, which have been worked since 1969. At present, the deposits account for at least 80% of Indian phosphate rock production. Mining, at the rate of 400 000 to 500 000 tonnes per annum, is conducted along a series of benches using power shovels to remove both overburden and ore. Most of the marketable output to date has been of high-grade ore containing 30% P_2O_5 or more and 7–11% SiO_2, and this has been merely crushed and ground prior to shipment to fertilizer plants. An expansion programme is under way to raise production to over one million tonnes per annum of concentrate, in order to utilise resources that are of lower grade (14–20% P_2O_5) and also have a high carbonate content owing to the presence of dolomite. A semi-commercial beneficiation plant capable of treating 200 tonnes per day to yield 80 tonnes per day averaging at least 32% P_2O_5 by flotation has been commissioned. Deposits similar to those of Jhamarkotra are also worked at Matoon, also near Udaipur, where production is at the rate of 180 000 tonnes per annum of flotation concentrate. Since the beneficiation plant became fully operational in 1976, ore containing 20–25% P_2O_5 and up to 30% SiO_2 has been deslimed, conditioned and passed through a flotation unit to give a product with 32% P_2O_5 8–10% SiO_2. Only limited quantities of Proterozoic phosphorite are produced at Jhabua in Madhya Pradesh.

In Korea (DR) the low grade and inaccessible location have prevented successful commercial development of the Proterozoic deposits and only a few have been worked on a limited scale, notably where these have become enriched by karstic weathering. Surface mining appears to have been begun around 1960, current production being around 500 000 tonnes per annum. The Patos de Minas deposits in western Minas Gerais, Brazil, have been worked since 1976 and production is up to 200 000 tonnes per annum of phosphate concentrate averaging 24% P_2O_5 which is sold for direct application to the soil as fertilizer. There are plans to expand mining operations and to construct a beneficiation plant for the production of over one million tonnes per annum of marketable phosphate rock averaging 30% P_2O_5.

Cambrian production

Although there was some production of Cambrian phosphorite from south-eastern Poland between 1938 and 1970, by far the largest output of phosphate rock of this age has come from the well-known Karatau Phosphate Basin in southern Kazakhstan, USSR. Five mines are reported to be operating, in the area, of which four are known to represent open-pit operations using power shovels and draglines. However the fifth mine, at Chulak Tau, the oldest in the Karatau Basin, is an underground operation, two 400 m shafts having been sunk to access the ore. To date only the higher grade reserves with about 20–28% P_2O_5 have been exploited and the quantity of this ore that is available is much more limited. As in the case of the Proterozoic phosphorites of Jhamarkotra in India, for example, much of the ore is highly siliceous and dolomitic. Ore from the mines is either crushed and screened to 23% P_2O_5 for feed to the nearby elemental phosphorus plants at Chimkent and Dzhambul or carried by rail to a beneficiation plant at Chulak Tau. At the plant up to 1.6 million tonnes per annum of phosphate concentrate containing 28% P_2O_5 can be obtained by flotation and used in fertilizer manufacture.

The emergence of China as a major world producer is a significant development in recent years. Most of its marketable output of 12.5 million tonnes in 1983 is believed to have been derived from deposits of Cambrian age with some latest Proterozoic production, principally for direct application to the soil as fertilizer. Although production is reported to be based on a relatively large number of small-scale mines, there are also several major mines, with an output of up to 1.5 million tonnes per annum. For example, the Kunyang mine south of Kunming, Yunnan Province, is believed to be producing 1.5 million tonnes of phosphate rock containing 22–30% P_2O_5. Also near Kunming is the Haikou mine which is reportedly being developed with a capacity of 2 million tonnes per annum. To the south of Yunnan, the Lao Cai deposits in northern Vietnam were first worked by the Japanese in 1940 until the mining installations were destroyed at the end of 1943. Total output during that period amounted to about 200 000 tonnes, most of which was carried by rail to Haiphong and shipped to Japan. Commercial-scale production again started in 1955 and during the 1960s rose from 555 000 to nearly 1 million tonnes per annum. The ore, which contains a minimum of 16% P_2O_5, yields a marketable flotation concentrate averaging around 35% P_2O_5. The entire output, currently only 220 000 tonnes per annum, is used domestically for fertilizer manufacture, but in the past, small quantities have been exported to China and eastern Europe, notably Romania and Poland.

Probable Cambrian phosphorite is also produced, but on only a very limited scale, in northern India, near Mussoorie, north of Dehradun, in Uttar Pradesh. There, the Maldeota and Durmala mines, which were brought into production in 1970 and 1976 respectively on the northern and southern limbs of the Mussoorie Syncline, together supply about 60 000 tonnes per annum of phosphate rock averaging between 20% and 24% P_2O_5. The mines are characterised by being extremely labour-intensive, extracting ore by pick and shovel methods from steeply dipping mine sections. The output from the mines is blended and sold finely ground as 'Mussoorie Phos' for direct application on rice, tea and rubber plantations. In the neighbouring part of northern Pakistan, crushing and grinding equipment is to be installed near Kakul to treat ore from the Langarban deposit averaging 27% P_2O_5. The associated mine is to produce 650 tonnes per day of phosphate rock for use in superphosphate fertilizer manufacture.

A major potential source of phosphate are the Cambrian deposits near Duchess in north-western Queensland, Australia, which were briefly worked before the decision to close down operations at the end of June 1978. Since crushing facilities were first installed in March 1975, a total of 1.2 million tonnes averaging 32% P_2O_5 has been produced from the Phosphate Hill Zone. Plant capacity has been rated at 1 million tonnes per annum. From Duchess the phosphate rock was carried by rail for a distance of over 1000 km to Townsville on the east Queensland coast. The Duchess mine was re-opened in 1981 but closed for the second time at the end of 1982. In 1982 the Duchess mine produced 229 176 tonnes averaging 31.3% P_2O_5, of which 35 199 tonnes was exported, mainly to Japan and South Korea. Small amounts of phosphate rock (about 6000 tonnes averaging 18% P_2O_5) continue to be produced from Cambrian deposits in South Australia. Because of its high iron and aluminium content, the product is not suitable for superphosphate manufacture, being used instead in organic fertilizers for horticultural applications.

Concluding remarks

A large number of individual or groups of occurrences of either Proterozoic or Cambrian phosphorite is known, and it is evident that Late Proterozoic and Cambrian times represent a major global phosphogenic event that took place in all continents except, as yet, Antarctica. Many of the discoveries have been made in relatively recent times and, in terms of exploration potential, there seems to be no geological reason to suppose that further discoveries of Proterozoic or Cambrian age will not occur. Many areas remain to be systematically explored and, in this context, the shield areas of Africa, South America and Siberia offer considerable promise by virtue of their large size. Unfortunately, large tracts of the African and South American Shields occur in tropical regions with deeply weathered rocks, thick soils, and, in places, dense tropical forests. Prospecting is extremely difficult in all such regions. However, the development of fertilizer raw material resources in Africa and South

America is urgent in the light of their large and growing populations and poor soil fertility.

Estimates indicate that resources of Proterozoic–Cambrian phosphorite probably exceed the total of 14000 million tonnes of highly variable phosphate content, or approximately 10% of total identified world resources of phosphate rock. The bulk of the resources is located in sediments of Cambrian age, chiefly in the USSR, China, Mongolia, Vietnam and Australia. Significant resources of Proterozoic phosphorite again occur mainly in the USSR and, to a lesser extent, in Upper Volta and Brazil. The substantial resources of phosphorite in the USSR and China in rocks of Proterozoic and Cambrian age is a feature that contrasts sharply with the much younger geological age of known resources in most other countries. In Australia, the Georgina Basin contains the largest resource of Cambrian phosphorite to have been identified outside the Asian landmass. Because of their similar geological characteristics, the Cambrian Asian–Australasian phosphogenic province may be extended to include deposits of comparable age in the Himalayan areas of northern India and Pakistan.

A major portion of the known resources of Proterozoic–Cambrian phosphorite is, unfortunately, characterised by a high carbonate (dolomite) content, the presence of which in other than moderate amounts has deleterious effects on the performance of phosphate rock during processing. Because of this, several significant phosphate fields, notably those in the USSR and India, have not yet been fully developed on a commercial scale. Appropriate beneficiation techniques are being tested or developed which could conceivably result in increased production of either Proterozoic or Cambrian phosphorite in the relatively near future. However, commercial production is as yet relatively limited compared with the resources of Proterozoic and Cambrian that are known. Total production amounts to about 14% of total world production, based almost entirely on mines developed in China and the USSR.

Many Proterozoic phosphorite deposits occur in association with strata-bound deposits of uranium, lead, silver, zinc, manganese, iron and gold. Often there is a close association between the two groups of deposits, and it appears that the sedimentological controls on phosphogenesis are similar to those of strata-bound metallogenesis. Thus, further studies of these controls can be expected to improve exploration techniques in respect to these deposits. The discovery of one type of these strata-bound occurrences may also define areas favourable to the exploration of others.

Similarly, but on a much more speculative level, there appears to be an empirical relationship between geologically younger igneous apatite deposits, rift systems and orogenic belts, and older phosphorite deposits. Although an origin of igneous apatitic carbonatites and apatitic alkaline intrusives by differentiation of mantle-derived magmas is generally favoured, the possibility should not be overlooked that deeply buried phosphorite, carbonate rock, iron-formations and evaporites may individually or collectively contribute through anatexis, phosphorus and other major elements such as fluorine, carbon dioxide, calcium and sodium to these unusual magmas. If so, phosphorites in general may represent important 'pathfinder' rock for other kinds of strata-bound deposits and, possibly, igneous apatite deposits.

References

Anon (1963). *Mineral Resources of Korea. Vol. 6A. Non-metallics and Miscellaneous Metals*, 121pp. Mining Branch, Industry and Mining Division, USOM, Korea.

Äikäs, O. (1980). Uraniferous phosphorite and apatite-bearing gneisses in the Proterozoic of Finland. In *Uranium in the Pine Creek Geosyncline*, ed. J. Ferguson & A.B. Goleby, pp. 675–81. International Atomic Energy Agency, Vienna.

Äikäs, O. (1981). Proterozoic phosphorites in Finland. *International Geological Correlation Programme, Project 156, Newsletter*, **9**, 21–7.

Ashkan, F. & Salehi, H. (1982). New age assignment of the phosphatic Chapoghlu Shale Member (Soltanieh Formation). *International Geological Correlation Programme, Project 156, Newsletter*, **9**, 23–5.

Asrarullah (1980). Phosphate, potash and natural gas occurrences in Pakistan. In *Fertilizer Mineral Occurrences in the Asia–Pacific Region*, ed. A.I.N. Lee, pp. 66–7. East–West Resource Systems Institute, Honolulu.

Banerjee, D.M. (1971). Precambrian stromatolitic phosphorites of Udaipur, Rajasthan, India. *Bulletin, Geological Society of America*, **82(8)**, 2319–29.

Bashyal, R. (1981). Stromatolitic phosphorite occurrences in the Lesser Himalaya of far western Nepal. *Abstract, 4th International Field Workshop and Seminar on Phosphorite, International Geological Correlation Programme, Project 156, Udaipur, India*, p. 5.

Bashyal, R.P., Banerjee, D.M. & Khan, M.W.Y. (1981). Certain characteristics of the stromatolitic phosphorites from western Nepal. *International Geological Correlation Programme, Project 156, Udaipur, India*, p. 2.

Bhatti, N.A. (1977). Phosphorite deposits of the Kakul-Mirpur area, Hazara district, North West Frontier Province, Pakistan. *Phosphate Rock in the CENTO Region*, ed. A.J.G. Notholt, pp. 69–98. Central Treaty Organization, Ankara.

Bushinsky, G.I. (1969). Old phosphorites of Asia and their genesis. Israel Program for Scientific Translations, Jerusalem, 266pp. (Translation of *Akademiya Nauk SSSR, Trudy Geologicheskogo Instituta*, **149**, 192pp).

Cannon, W.F. & Klasner, J.S. (1976). Phosphorite and other apatite-bearing sedimentary rocks in the Precambrian of northern Michigan. *Circular US Geological Survey*, **766**, 6pp.

Cathcart, J.B. (1977). Phosphorite deposits near Patos de Minas, Minas Gerais, Brazil. *American Institute of Mining Engineers, Annual Meeting, Atlanta Georgia, Preprint*, **77-H-76**, 32pp.

China National Committee for IGCP (1982). *Guidebook to Field Excursions*, 50pp. (Fifth International Field Workshop and Seminar on Phosphorite, Kunming, China, November 17–24, 1982.)

Choudhuri, R. & Balasubramanian, N. (1980). Grade distribution pattern and its bearing on the formation of high and low grade in Jhamarkotra phosphorite deposit, Udaipur, Rajasthan. *Indian Journal of Earth Sciences*, **7(1)**, 89–93.

Cook, P.J. (1972). Petrology and geochemistry of the phosphate deposits of northwest Queensland. *Economic Geology*, **67(8)**, 1193–213.

Cook, P.J. (1976). Georgina Basin phosphatic province, Queensland and Northern Territory – Regional Geology. *Economic geology of Australia and Papua New Guinea. Vol 4. Industrial Minerals and Rocks*, ed. C.L. Knight, pp. 245–50. Monograph 8. Australasian

Institute of Mining and Metallurgy, Melbourne.

Cook, P.J. & Shergold, J.H. (Eds) (1979). *Proterozoic–Cambrian Phosphorites* 106pp. ANU Press, Canberra.

Cook, P.J. & Shergold, J.H. (1980). Proterozoic and Cambrian phosphorites of Asia and Australia – a progress report. In *Fertilizer Mineral Potential in Asia and the Pacific*, ed. R.P. Sheldon & W.C. Burnett, pp. 207–23. East–West Resource Systems Institute, Honolulu.

de Keyser, F. & Cook, P.J. (1972). Geology of the Middle Cambrian phosphorites and associated sediments of northwestern Queensland. *Bulletin of the Bureau of Mineral Resources, Geology and Geophysics of Australia*, **138**, 79pp.

Eganov, E.A. & Sovetov, Yu. K. (1979). Karatau-a model for phosphorite deposition. *Trudy Instituta Geologii i Geofiziki, Akademiya Nauk SSSR, Sibirskoe Otdelenie*, **427**, 192pp.

Fromaget, J. (1941). Etudes géologiques des gisements de phosphates (apatite) de Cam Duong (Tonkin) et des formations qui les contiennent. *Bulletin de la Service géologique de l'Indochine*, **26(1)**, 30pp.

Geological Survey of India (1981a). *Excursion Guide Book on Dehradun–Mussoorie Area.* 53pp. Fourth International Field Workshop and Seminar on Phosphorite, Calcutta.

Geological Survey of India (1981b) *Guidebook for Excursion: Aravalli Phosphorites around Udaipur, Rajasthan, India.* 57pp. Fourth International Field Workshop and Seminar, Calcutta.

Geological Survey of India (1981c). Fourth International Field Workshop and Seminar on Phosphorite, Abstracts, 54pp., Calcutta.

Gimmelfarb, B.M. & Egorova, O.P. (1969). Geology of the Khubsugul phosphorite deposit, Mongolian People's Republic. *Lithology and Mineral Resources*, **2**, 139–144.

Gimmelfarb, B.M. & Tushina, A.M. (1966). Principal phosphorite ore deposits of the Kara–Tau. *Lithology and Mineral Resources*, **4**, 483–95.

Howard, P.F. (1972). Exploration for phosphorite in Australia – a case history. *Economic Geology*, **67(8)**, 1180–92.

Howard, P.F. & Perrino, F.A. (1976). Wonarah phosphate deposits, Georgina Basin, Northern Territory. *Economic Geology of Australia and Papua New Guinea, Vol. 4, Industrial Minerals and Rocks*, ed. C.L. Knight, pp. 273–7. Monograph 8. Australasian Institute of Mining and Metallurgy, Melbourne.

Ilyin, A.V. (1978). Late Precambrian – Early Cambrian phosphorites in Africa. *International Geological Correlation Programme, Project 156, Newsletter*, **2**, 8–9.

Ilyin, A.V. & Bjamba, Z. (1980). *Guidebook for Excursion, phosphorites of the Khubsugul Basin, Mongolian People's Republic.* Field Conference of IGCP Project 156, Geologicheskiy Institut, Akademiya Nauk SSSR, Moscow, 135pp.

Ilyin, A.V. & Ratnikova, G. I. (1981). Primary, bedded, structureless phosphorite of the Khubsugul Basin, Mongolia. *Journal of Sedimentary Petrology*, **51(4)**, 1215–22.

Kholodov, V.N. (1968). The palaeogeography of the Lesser Karatau (Kazakhstan) during the deposition of the 'Lower' Dolomite phosphorite-bearing strata. *Byulleten Moskovskoye Obshchestvo Ispytately Prirody, Otdelenie Geologicheskiy*, **43**, 70–84.

King, W. (1884). Phosphatic beds, Musuri, *Records Geological Survey of India*, **17(4)**, 198–9.

Latif, M.A. (1972). An occurrence of Palaeozoic phosphate rock in Hazara district, West Pakistan. *Transactions of the Institute of Mining and Metallurgy, Section B*, **81**, 50–3.

Lee, A.I.N. (Ed) (1980). *Fertilizer Mineral Occurrences in the Asia–Pacific Region*, 156pp. East-West Resource Systems Institute, Honolulu.

Liang, T.Y. & Chang, A.C. (1981). *On the characteristics and genesis of Late Pre-Cambrian phosphorites associated with Gondwana*, 81pp. Geological Institute for Mining and Chemical Industries, Zhuo Zian.

Lucas, J., Prévôt, L. & Trompette, R. (1980). Petrology, mineralogy and geochemistry of late Precambrian phosphate deposits of Upper Volta (W. Africa). *Journal of the Geological Society*, **137(6)**, 787–92.

Notholt, A.J.G. (1972). Palaeozoic phosphate rock in southwest Asia and the Near East. *Transactions of the Institute of Mining and Metallurgy, Section B*, **81**, 158–63.

Notholt, A.J.G. (1979). Resources of Precambrian and Cambrian sedimentary phosphate rock. In *Proterozoic–Cambrian Phosphorites*, ed. P.J. Cook & J.H. Shergold, pp. 71–86 ANU Press, Canberra.

Pant, A. (1980). Resource status of rock phosphate deposits in India and areas of future potential. In *Fertilizer Mineral Potential in Asia and the Pacific*, ed. R.P. Sheldon & W.C. Burnett, pp. 331–57. East–West Resource Systems Institute, Honolulu.

Pritchard, P.W. & Cook, P.J. (1965). Phosphate deposits of the Northern Territory. *Geology of Australian Ore Deposits*, ed. J.McAndrew, pp. 219–20. 8th Commonwealth Mining and Metallurgical Congress, Melbourne.

Rehtijävi, P., Äikäs, O. & Makelä, M. (1979). A Middle Precambrian uranium- and apatite-bearing horizon associated with the Vihanti zinc ore complex, western Finland. *Economic Geology*, **74(5)**, 1102–17.

Rogers, J.K. & Crase, N.J. (1980). The Phosphate Hill rock phosphate deposit, north-west Queensland, Australia – an outline of geological development. In *Fertilizer Mineral Potential in Asia and the Pacific*, ed. R.P. Sheldon & W.C. Burnett, pp. 307–28. East–West Resource Systems Institute, Honolulu.

Rogers, J.K. & Keevers, R.E. (1976). Lady Annie–Lady Jane Phosphate deposits, Georgina Basin, Queensland. In *Economic geology of Australia and Papua New Guinea. Vol 4. Industrial Minerals and Rocks*, ed. C.L. Knight, pp. 251–65. Monograph 8. Australasian Institute of Mining and Metallurgy, Melbourne.

Russell, R.T. & Trueman, N.A. (1971). The geology of the Duchess phosphate deposits, northwestern Queensland, Australia. *Economic Geology*, **66(8)**, 1186–214.

Sant, V.N. (1979). Precambrian phosphorite of Rajasthan, India: a case history. In *Proterozoic-Cambrian Phosphorites*, ed. P.J. Cook & J.H. Shergold, pp. 39–40. ANU Press, Canberra.

Sant, V.N. (1980). Geology of the Indian Platform and its phosphate occurrences. In *Fertilizer Mineral Potential in Asia and the Pacific*, ed. R.P. Sheldon & W.C. Burnett, pp. 20–33. East–West Resource Systems Institute, Honolulu.

Semeikin, I.N., Kolesnikov, V.L., Belogoldov, V.G. & Khrapal, Yu.I. (1976). Geologic structure and ore types of the northern Ukha-Gol phosphorite deposits (E. Sayan). *Soviet Geology and Geophysics*, **17(9)**, 61–8.

Sheldon, R.P. (1966). Exploration for phosphate in the Eastern Hemisphere. *Professional Paper* **550-A**, *US Geological Survey*, pp. 108–9.

Slansky. M. (1980). West African Infracambrian phosphorites. In *Proterozoic–Cambrian Phosphorites*, ed. P.J. Cook & J.H. Shergold, p. 51. ANU Press, Canberra.

Sougy, J. (1964). Les formations paléozoïques du Zemmour noir (Mauritanie septentrionale). Étude stratigraphique, pétrographique et paléontologique. *Annales de la Faculté des Sciences, Université de Dakar*, **15(XII)**, 695pp.

Tayyab Ali, S. (1977). Phosphate deposits of Pakistan. In *Phosphate rock in the CENTO region*, ed. A.J.G. Notholt, pp. 42–62. Central Treaty Organization, Ankara.

Trompette, R. (1978). Occurrences of phosphorites in the Volta Basin. *International Geological Correlation Programme, Project 156, Newsletter*, **2**, 7–8.

Tushina, A.M. (1968). Lithofacies characteristics of the formation of the phosphorite stratum of Karatau. *Lithology and Mineral Resources*, **4**, 456–65.

Ullmer, E. (1981). A mid-Proterozoic phosphate occurrence in east-central Minnesota. *International Geological Correlation Programme Project 156, Newsletter*, **8**, 22–5.

Viland, J.-C. (1977). Présence d'horizons phosphatés à la base de l'Infracambrien supérieur du Haut Atlas de Marrakech (Maroc). *Notes et Mémoirs de la Service géologique du Maroc*, **38(268)**, 13–22.

Volkov, R.I., Zaitsev, N.S., Ilyin A.V. & Osokin, N.V. (1972). The Ukha-Gol phosphorite deposit in East Sayan Mountains. *Soviet Geology*, **2**, 94–107.

Yeh Lientsun & Sun Shu (1980). Salient features of Chinese phosphorite deposits. In *Fertilizer Mineral Potential in Asia and the Pacific*, ed. R.P. Sheldon & W.C. Burnett, pp. 81–2. East–West Resource Systems Institute, Honolulu.

Zimin, S. (Ed.) (1980). *Phosphates of the Far East*, 147pp. Akademiya Nauk SSSR, Dalnevostochnyi Nauchnyi Tsentr, Vladivostock.

3

Proterozoic and Cambrian phosphorites – regional review: Australia

P.F. HOWARD

Introduction

Phosphorite deposits are widespread in the Proterozoic and Cambrian sediments of Australia, although with the exception of those in the Georgina Basin they are generally small and uneconomic. Total production of rock phosphate, since the discovery of occurrences in South Australia in 1903, amounts to 1 455 000 tonnes, of which approximately 1 080 000 tonnes have been derived from Middle Cambrian sediments of the Georgina Basin, 350 000 tonnes from the Early Cambrian sediments of the Adelaide Geosyncline, and a minor tonnage from the Late Proterozoic sediments of the Adelaide Geosyncline (Fig. 3.1). During the past decade South Australian production has averaged 5500 tonnes per annum for direct application as fertilizer to garden soil. The Duchess deposit in the Georgina Basin produced 830000 tonnes of direct shipping grade rock phosphate in the period 1975–1978 and a further 250 000 tonnes in 1981–1982, prior to the cessation of operations.

The number of known phosphorite occurrences of Proterozoic and Cambrian age is approximately 110. They occur in four Proterozoic–Cambrian basins, namely the Adelaide Geosyncline, the Amadeus Basin, the Pine Creek Geosyncline, and the Georgina Basin (Fig. 3.1). Other sequences containing phosphatic sediments, but no presently known phosphorite, include the Early Proterozoic banded iron formations of the Hamersley Basin and the Willyama Inlier, and the Cambrian sediments of the Arafura, Bonaparte Gulf, Ord, and the Daly River Basins. To date no phosphatic Cambrian sediments are known in the Wiso or the Ngalia Basins.

Information on Proterozoic and Cambrian phosphate occurrences in Australia is summarised in Table 3.1. It is apparent from Table 3.1 and Fig. 3.1 that there is a concentration of occurrences in the Late Proterozoic sediments extending through the Adelaide Geosyncline and the Amadeus Basin as well as a similar distribution in the overlying Cambrian sediments of these basins. However, the only known Australian phosphorite deposits which have potential for economic development occur in Middle Cambrian sediments of the Georgina Basin. Cook & McElhinny (1979) suggest that there are four discrete peaks of phosphogenesis in world Precambrian systems corresponding to the Early, Middle, Late I and Late II Proterozoic, with the Late II Proterozoic (*c.* 640 m.y.) being the most important. In Australia Late II probably corresponds to the Rum Jungle, Amadeus Basin and the upper phosphorite bed of the Adelaide Geosyncline (725 m.y. and less), while the Late I Proterozoic probably corresponds to the lower phosphorite bed of the Adelaide Geosyncline (*c.* 725–800 m.y.), allowing for some uncertainty in the dating of the sedimentary succession. The Early Proterozoic occurrences in Australia, like those in Finland and Michigan, are contained in banded iron formations, but based on present knowledge can be regarded only as phosphate occurrences rather than deposits. No Australian occurrences are known in Middle Proterozoic successions.

This chapter reviews the geology of the basins containing phosphogenic sequences, placing major emphasis on the Late Proterozoic and Cambrian sediments of the Adelaide Geosyncline and the Middle Cambrian of the Georgina Basin where most of the deposits occur.

Proterozoic evolution

Cratonic nuclei of Archaean rocks are exposed over wide areas of Australia and include those of Pilbara, Yilgarn and Gawler, together with the Rum Jungle, Nanumbu and Litchfield Complexes which form the basement to the Pine Creek Inlier (Fig. 3.1). These cratons were basement to Early Proterozoic platform cover which included the quartz-rich turbidites, shallow-marine evaporite–carbonate, fluviatile sediments and volcanics of the Pine Creek Geosyncline, banded iron formations, shales, dolomites and volcanics of the Hamersley Basin and the Gawler region, and the mature meta-sedimentary sequence consisting of shales, sandstones, volcanics, banded iron formation and the lode horizon of the Willyama Complex, Broken Hill. Early Proterozoic sediments were deformed, metamorphosed, intruded and largely cratonised before the end of the Early Proterozoic, with the main period of cratonisation occurring as a well-defined period of tectonism between 1900 and 1700 m.y.

Following the cratonisation of the Early Proterozoic mobile belt sequences, Early to mid-Proterozoic platform covers of northern Australia were deposited in basins characterised by

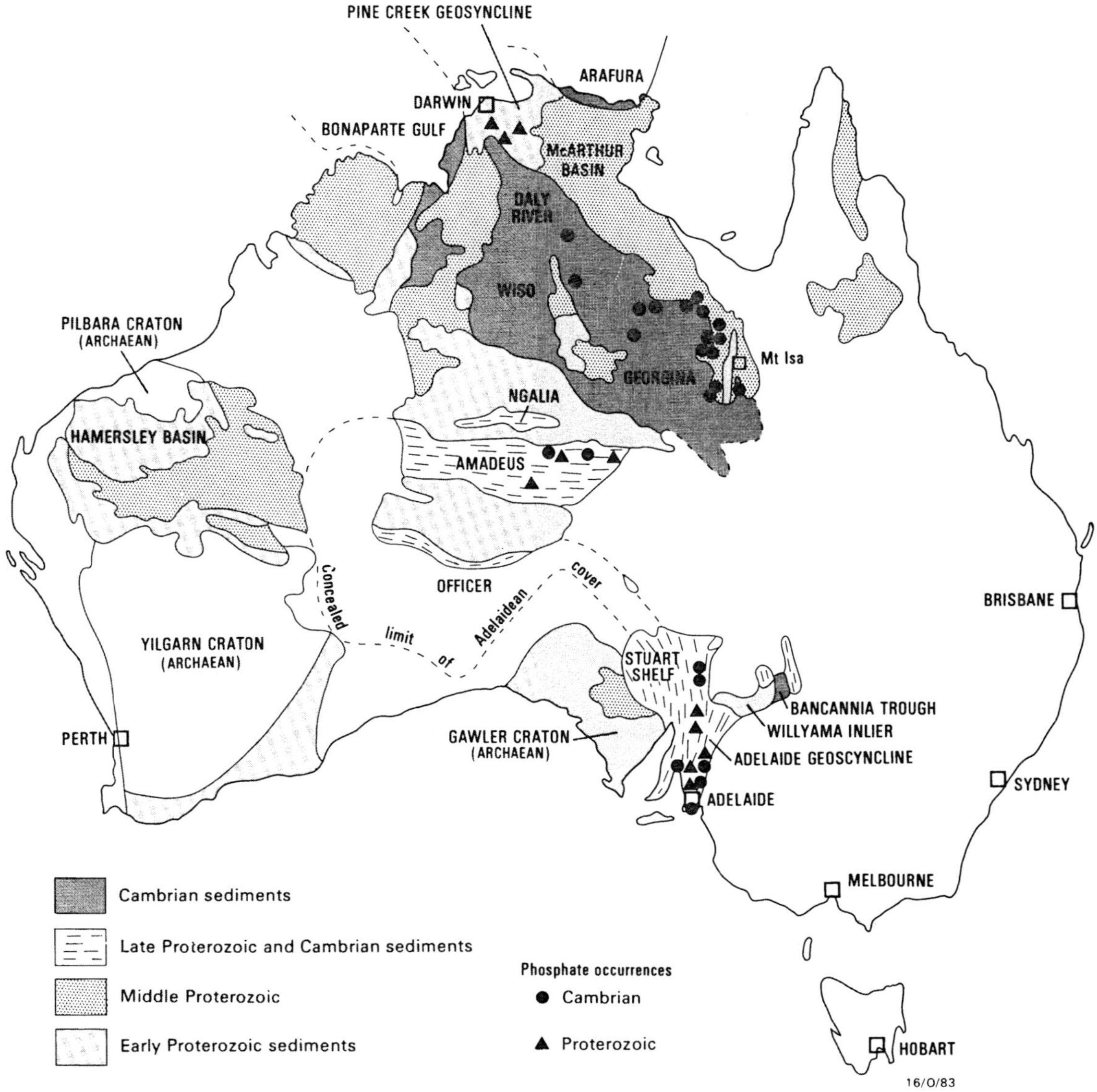

Fig. 3.1. Distribution of Proterozoic and Cambrian sediments and their contained phosphate deposits.

broad stable shelves which have been interpreted as intracontinental, and continent margin marine embayments separated by narrow intervening aulacogen-like mobile zones (Plumb *et al.*, 1981). The sequences in both the shelves and troughs include very shallow-marine, intertidal, supratidal and fluvial facies characterised by evaporite–carbonate complexes. Commencing in the mid-Proterozoic (1400 m.y.) and extending through to the close of the Middle Proterozoic (1000 m.y.), orogenesis and taphrogenic reworking of the pre-existing craton controlled the depositional sites of platform cover. Intense tectonism was restricted to narrow belts of metamorphism, migmatisation and granite emplacement about the major cratons. The stratotype for the Middle Proterozoic platform cover is the Carpentarian succession of the McArthur Basin (Fig. 3.1) (Plumb *et al.*, 1981). The period was marked by mafic magmatism, and the completion of cratonisation of the Precambrian rocks except those affected by the terminal Proterozoic Petermann Range Orogeny which was localised to a belt in central Australia.

Following final cratonisation (*c.* 1000 m.y.) Late Proterozoic platform cover was deposited, the stratotype of which is the Adelaidean succession in South Australia. These sequences pass up with only minor breaks into the Cambrian and characterise a period lacking intense tectonism (Plumb, 1979). The first Cambrian sediments were shallow-marine fluviatile sandstones, siltstones and carbonates localised over the most recently cratonised mobile belts in the Amadeus, Ngalia and southern Georgina Basins. It is likely that these basins are the structurally preserved remnants of more extensive and probably interconnected sedimentary covers which extend to the Officer Basin and the Adelaide Geosyncline to the south (Fig. 3.1, and Rutland, 1981). Later Adelaidean sedimentation spread through most of northern Australia, identified by the occurrence of Sturtian glaciogene sequences (*c.* 725 m.y.) (Rutland *et al.*, 1981).

Hamersley Basin

The Hamersley Basin sediments comprise thick clastic, chemical and volcanogenic sediments cropping out over an area of approximately 100 000 km^2 (Fig. 3.1). The basin is most noted

Table 3.1. *Proterozoic and Cambrian phosphate occurrences and their associated formations*

		Basin	Formation	Number of Occurrences	Deposits/localities
Late	Cambrian	Amadeus	Pacoota Sandstone	—	Scattered minor occurrences in northern part of Basin
Middle		Georgina	Beetle Creek, Burton Beds, Border Waterhole Formation, Wonarah Beds	18	Duchess, Quita Creek, Ardmore, Sherrin Creek, Engine Creek, Lily Creek, D Tree, Lady Annie, Lady Jane, Riversleigh, Phantom Hills, Mount Jennifer, Babbling Brooke Hill, Mt O'Connor, Highland Plains, Alexandria, Alroy, Wonarah
		Amadeus	Tempe	—	Scattered minor occurrences in northern part of basin
Early		Adelaide Geosyncline	Heatherdale Shale	—	
			Parara, Wilkawillina Limestone	4	Clinton, Ardrossan
			Kulpara, Wangkonda	—	—
			Koonunga Phosphorite Member	23	Moculta, St Johns, St Kitts, McCarthys, Koonunga,
			Angaston Marble, Mount Terrible		Toms, Myponga
			Parachilna	12	Southern Flinders Ranges (3), northern Flinders Ranges (9)
			Andamooka Limestone	—	—
		Amadeus	Todd River Dolomite	1	Ooraminna Anticline
Late	Proterozoic	Adelaide Geosyncline	Brighton Limestone	20	Orroroo: Napa Napa, Duns, Bendleby, Pekina, Lyons, Dodd, Stacey, Noarlunga: Oliver, Waits, Reynella: Marino. Kapunda: Mt Pleasant, Dermodys
			Skillogalee Dolomite	6	Martins, Burra, Baldina, Fairview, Hermitage
		Amadeus	Areyonga	1	Ringwood
		Pine Creek Geosyncline	Depot Creek	19	Rum Jungle, Sleisbeck, Coronation Hill, Saddle Ridge South
Middle		—	—	—	—
Early		Hamersley	Brockman Iron Formation	—	—
		Willyama inlier	Mine Sequence	—	Broken Hill

for major deposits of high-grade iron ore. Current estimates of reserves of high-grade iron ore ($>55\%$ Fe) are 18 000 million tonnes of low-phosphorus ($<0.08\%$ P_2O_5) ore and 17 000 million tonnes of high-phosphorus ($>0.10\%$ P_2O_5) ore (Goode, 1981). In addition there are immense submarginal resources of mineralisation adjacent to proven reserves and as banded iron formation. Thin bands of apatite account for the overall low phosphorus content of the ores. The geology of the Hamersley Iron Province has been described by Trendall & Blockley (1970), Goode (1981) and Ewers & Morris (1981).

Following the formation of the Pilbara Craton by about 3000 m.y. platform cover onlapped it approximately 2200–2400 m.y., ago. In the Hamersley Basin, three groups of the Mount Bruce Supergroup, namely the Fortescue, Hamersley and Wyloo, comprise about 16 000 m of sediment (Trendall & Blockley, 1970).

The initial sedimentation (Fortescue Group), which commenced with major flood basalt flows and minor fluvial sedimentation on an undulating surface, was followed by the development of a regional depression in the central portion of the basin and an initial shallow-water marine transgression. Some 4 000 m of lava, tuff, sandstone and shale were deposited in the centre of the basin. The conformably overlying Hamersley Group, about 2500 m thick, is characterised by abundant banded iron formations and interstratified acid lava flows. The major part of the sedimentation was probably deposited in water depths greater than 100–200 m, which is below wave base and the photic zone (Goode, 1981). Stromatolites and shallow-water sedimentary structures are absent. The complete lack of coarse terrigenous sedimentation and the great lateral continuity of individual mesobands attest to either great distance from land, low relief of the surrounding land and/or aridity (Goode, 1981). The Wyloo Group is up to 9000 m thick and comprises clastics and locally important dolomite and basalt. The sequence was deposited during a general regression, and chert breccias indicate extensive subaerial exposure. It is likely that lowermost clastics were derived from the Hamersley Group, although carbonate deposition prevailed later during marine conditions (Goode, 1981). The sedimentary history of the Hamersley Basin came to a close with an orogeny, accompanied by regional metamorphism and granite intrusion approximately 1700–1800 m.y. ago.

The Hamersley iron ore deposits and accompanying apatite

The Hamersley Group contains eight constituent formations, the most important of which in terms of iron ore reserves is the Dales Gorge Member of the Brockman Iron Formation. Trendall & Blockley (1970) divide the 150 m thick Dales Gorge Member into 33 macrobands varying from 1 m to 25 m thick. The macrobands are composed of alternating BIF and S (shale) bands, of which the BIF macrobands are essentially oxide facies banded iron formation, and the S macrobands dark green to black, iron-rich shales containing stilpnomelane, minnesotaite, siderite, ankerite, chert, magnetite and apatite. Trendall & Blockley (1970) further divide the BIF macrobands into mesobands, with an average thickness of less than 2 cm, and into microbands with a usual interval of about 1 mm. The macrobands and mesobands have basin-wide continuity.

Trendall & Blockley (1970) and Ewers & Morris (1981) have carried out detailed studies of the chemical and mineralogical distribution of phosphorus in the macrobands and mesobands within the Dales Gorge Member. Analyses of BIF macrobands yield P_2O_5 contents varying in the range 0.18–0.40% over thickness of 2.35–13.95 m, with a mean P_2O_5 of 0.21%. The range of P_2O_5 in the S macrobands is 0.05–0.25%, with a mean of 0.17% over widths of 0.25–3.51 m (Ewers & Morris, 1981). The highest phosphate content recorded is 0.86% P_2O_5 over 4 cm in a BIF macroband (Trendall & Blockley, 1970). Petrographic and X-ray studies suggest that the phosphorus is present entirely as apatite, occurring in a variety of textural forms, the most abundant of which is as amoeboid anhedral aggregates in chert of 'chert-matrix'. The apatite content of some mesobands may reach about 30% over widths of 5 mm and is preferentially streaked along bedding suggesting a primary depositional significance (Trendall & Blockley, 1970).

Ewers & Morris (1981) recorded systematic changes in the proportion of iron-bearing oxides, silicates and carbonate minerals contained in the macrobands across the basin (135 km), but discerned no changes in the percentage of associated apatite. They favour a landlocked depositional environment free from any terrigenous input, and with a negative water balance requiring iron and phosphorus to be introduced to the basin in surface water, and silica to be precipitated at least in part by evaporation. The change from BIF to S macroband style of precipitation is interpreted as due to a rise in pH and/or p_{CO_2}.

Willyama Inlier

The iron-bearing sediments of the Willyama Inlier and the Gawler Craton have long been correlated with those of the Hamersley Basin (Fig. 3.1) (Plumb, 1979). Available isotopic data place the sequence between 2400 m.y. and 1800 m.y. with the main metamorphic event at about 1700 m.y. Unlike the Hamersley Group, however, it has undergone a complex history of strong deformation and high-grade metamorphism (Rutland, 1981).

The deposition of the Willyama sediments and volcanics took place within a mobile belt which extended northwards from Broken Hill through the Mount Isa area along the eastern margin of a Proterozoic supercontinent (Stevens, 1980). The sequence is composed of psammitic, psammopelitic, pelitic and quartzo-feldspathic sediments and a small proportion of coeval rhyolitic, dacitic and tholeiitic volcanics. The dacitic metavolcanics (Potosi Gneiss) in the lower part of the sequence are well developed at Broken Hill and appear to be genetically associated with the major stratiform base-metal mineralisation of the region (Rutland *et al.*, 1981). Banded iron formation is likewise closely associated and is noted for its apatite content.

The depositional setting was tectonically unstable and the region was subsequently subjected to strong deformation, regional scale recumbent folding, and high-grade metamorphism followed by the development of retrograde schist zones, and intrusion of basic and ultrabasic bodies. Stevens (1980) interprets these various features as consistent with a rift zone developed in an area of thin continental crust or in an arc–trench complex in an advanced stage of development. The interpreted stratigraphic progression from immature sediments plus volcanics to mature sediments without volcanics indicates a progression from unstable to stable environments.

The BIFs and accompanying apatite

In the Broken Hill mine area, the principal stratigraphic units are as follows:

Unit 5 (top). Pelitic and psammopelitic sillimanite gneiss without volcanics

Unit 4. Psammopelitic sillimanite gneiss, about 1000 m thick, containing units of the 'Mine Sequence', including banded iron formation, amphibolite, Potosi Gneiss and the lode horizon (lead–zinc)

Unit 4 (bottom). Metavolcanic granite gneiss

The 'lode horizon' is associated with the Potosi Gneiss, a metavolcanic dacite, as is the banded iron formation. The BIFs in the Broken Hill district comprise only four principal mineral components in magnetite, garnet, quartz and apatite, and these occur in various proportions and demonstrate a fine-scale continuity. The fine-scale layering in the range 0.5–2 mm is a feature which marks the rock type as texturally distinct from the stratiform base metal 'lode horizon'. Stanton (1976) described the BIF as a phosphatic, manganiferous low-grade ironstone containing about 30% Fe, 3–6% MnO, 2.5–4.0% P_2O_5, 7–10% Al_2O_3 and 5% CaO. In all occurrences the BIF shows a consistent chemical stratigraphy such that Fe, Mn and P progressively increase from the margins of a bed to its centre. The trace-element content of apatite shows a corresponding change, with lead increasing from 1 to 8 weight per cent and copper and zinc decreasing from 6 ± to 4 ± weight per cent.

Apatite-rich bands, which are bimineralic, consist of approximately 70% apatite and 30% quartz and occur as laminae 0.5 mm to 1 cm thick (Stanton, personal communication).

Stanton (1976) likens the BIF to the Lahn–Dill type ironstones which occur within volcanic sediments in the Mesozoic iron-ore fields of Europe. The BIF is interpreted as an original chemical sediment consisting primarily of chamosite and hematite formed from submarine exhalation of hydrothermal solutions as was the spatially related stratiform lead–zinc mineralisation.

Pine Creek Geosyncline

In the early 1960s, 16 phosphate deposits were found in the Rum Jungle District in the north-west part of the Pine Creek

Geosyncline and three were found in the South Alligator Valley–Sleisbeck area some 200 km to the east-south-east.

The phosphate deposits in the Rum Jungle District are in an extensive breccia mapped as part of the Depot Creek Sandstone Member at the base of the Adelaidean platform cover which overlies Archaean complexes and Early Proterozoic sediments of the Pine Creek Geosyncline. The deposits in the South Alligator–Sleisbeck area are in similar breccias in a unit mapped as part of the Carpentarian Edith River Volcanics and in sandstone at the base of the Carpentarian Kombolgie Formation, both of which lie at the base of the Carpentarian platform cover overlying the Early Proterozoic rocks of the Pine Creek Geosyncline and its Archaean Basement.

It is generally accepted that the deposits were formed of phosphorus derived from Early Proterozoic sediments during a weathering period and incorporated into later cover rocks. However recent field work in the Rum Jungle District supports the conclusion that the breccias containing the phosphate deposits are part of the Early Proterozoic sequence and as a result there is uncertainty about the generally accepted explanation for the origin of the deposits (P.W. Pritchard, personal communication).

Stratigraphy The stratigraphy of the sediments related to the occurrences and possible source of phosphate in the Pine Creek Geosyncline deposits is illustrated in Figure 3.2. Approximately 14 km of Early Proterozoic continental, supratidal and shallow-marine sediments followed by turbidites were deposited in an intra-continental ensialic geosyncline (Plumb *et al.*, 1981). These sediments were intruded and cratonised by about 1800 m.y. Carpentarian cover rocks comprising 1600 m of fluviatile sediments and volcanics overlie them on the eastern side of the inlier where they are exposed, and Adelaidean cover rocks comprising 1000 m of continental sandstones and shallow-marine sandstones, siltstones and carbonates overlie them in the west.

Tectonic setting The Pine Creek Geosyncline is a north-trending intracontinental ensialic structure consisting of the central South Alligator Trough bounded by the Batchelor Shelf

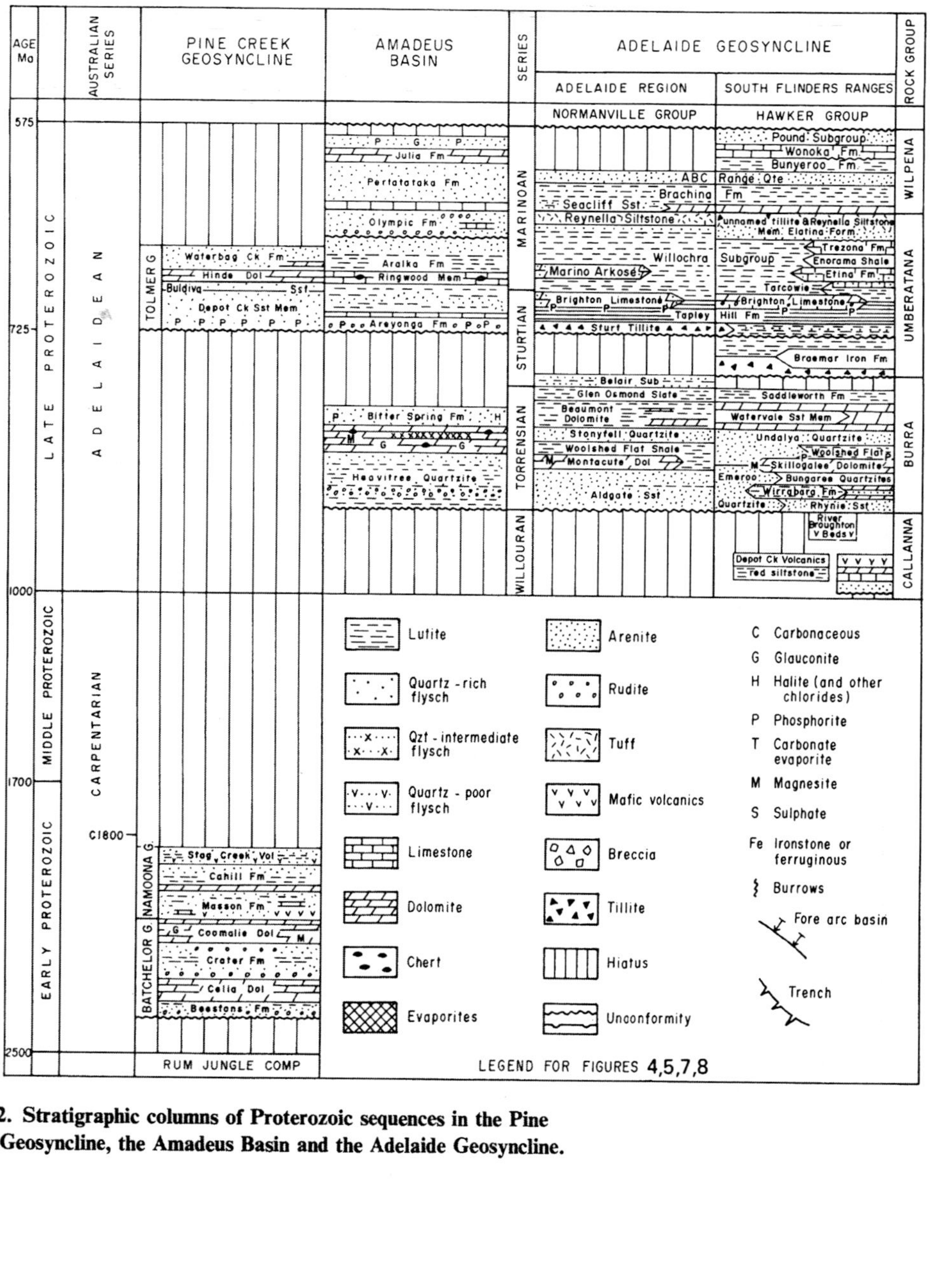

Fig. 3.2. Stratigraphic columns of Proterozoic sequences in the Pine Creek Geosyncline, the Amadeus Basin and the Adelaide Geosyncline.

to the west and the Kakadu Shelf to the east. The geosyncline was initiated by rifting and marginal subsidence of the Archaean crust, the evidence for which is based on the concentration of volcanism along the axis of the trough, the absence of deep-water deposits, marginal uplifted areas and inferred presence of a major bounding fault.

Sedimentation The basal sequence in the Pine Creek Geosyncline on the Batchelor Shelf, the Batchelor Group, surrounds Archaean complexes and contains arkosic psammite, conglomerate and minor shale deposits of the basal Beestons and the higher Crater Formations. The Group, however, is dominated by interbedded carbonates of the Celia and Coomalie Dolomites, interpeted by Crick & Muir (1980) as having been massive evaporite deposits later replaced by carbonate. These beds are succeeded by the Mount Partridge Group consisting of carbonaceous mudstones and shales, dirty carbonate rocks, turbidites and a basalt, which in turn is succeeded by the shales, ironstones, chert and volcaniclastic sediments of the South Alligator Group and finally by the turbidites of the Finniss River Group.

A similar sequence is present on the Kakadu Shelf, where the basal sequence which appears to be equivalent to the Batchelor Group on the Batchelor Shelf is referred to as the Kakadu Group and the overlying Namoona Group.

The principal uranium and base metal deposits in the Rum Jungle District on the Batchelor Shelf are in the lower part of the Mount Partridge Group. Those on the Kakadu Shelf are in the Namoona Group and the South Alligator Group.

Phosphate deposits

The 16 phosphate deposits in the Rum Jungle District have estimated reserves of 4 million tonnes, averaging 10% P_2O_5, based on limited drilling. The phosphate bodies have rock grading up to 30% P_2O_5, 4–8% Fe_2O_3 and 1–33% Al_2O_3. They are clustered in the outer parts of breccia bodies mapped as outliers of the Depot Creek Sandstone Member at the base of the Adelaidean Tolmer Group (Pritchard *et al.*, 1963).

The breccia bodies are estimated to be 300–600 m thick. They consist of fragments and blocks of massive cross-bedded quartz sandstone with pebble bands, hematitic mudstone, sandstone and quartz breccia, and chloritic mudstone and quartz breccia set in a matrix of hematitic mudstone or siltstone. In places these breccias pass down into breccias of Coomalie Dolomite fragments set in a matrix of hematitic or chloritic mudstone or siltstone.

These breccias have been regarded as remnants of an ancient and widespread regolith (Pritchard *et al.*, 1963) which was developed over Coomalie Dolomite in the Rum Jungle District and over the Edith River Volcanics in the South Alligator Valley. A similar palaeo-weathering profile exists below the Carpentarian Kombolgie Formation further east and is important in the genesis of the Alligator Rivers Uranium Field deposits.

In all the Rum Jungle District deposits the phosphate rock is associated with the hematitic breccias of clasts of mudstone, siltstone, sandstone and cherty quartz. The phosphate rock ranges in colour from lilac to red and consists of apatite containing finely divided hematite. Minor wavellite, millisite and variscite–strengite series minerals occur in the recent weathering profile. Drilling has outlined the Geolsec deposit as a lens-shaped body which thins and becomes poorer in grade at its peripheries. Its boundaries are defined by grade within the gradual transition from barren rock at its edges to higher grade in the centre (Pritchard *et al.* 1963).

Origin The Rum Jungle District deposits have been interpreted as (a) secondary enrichments of phosphorus leached from underlying Batchelor Group metasediments; and (b) a sedimentary lens in the upper part of the Coomalie Dolomite (Pritchard & Cook, 1965). The latter proposal is not supported by the drill hole data which demonstrate that unweathered dolomite contains only a trace of P_2O_5. Most dolomite breccia contains less than 1% P_2O_5 except where there is an appreciable content of ferruginous siltstone and mudstone matrix similar to that associated with overlying hematitic quartz breccia into which it grades, and here values as high as 9% P_2O_5 are recorded. It appears that the matrix to the dolomite breccia is a transported detrital infill.

The origin of the deposits is still unknown. Recent field work (P.W. Pritchard, personal communication) supports the conclusion that the breccias containing the phosphate deposits in the Rum Jungle District (west) and the South Alligator Valley (east) lie in the Early Proterozoic sequence either in the South Alligator Group or (at Rum Jungle) at the top of the Coomalie Dolomite. Pritchard speculates (personal communication) that the phosphate deposits may be hydrothermal deposits formed by the volcanic activity in progress during the deposition of the South Alligator Group.

Proterozoic of the Adelaide Geosyncline–Stuart Shelf

The distribution of the Late Proterozoic (Adelaidean) sediments of the Adelaide Geosyncline is shown in Fig. 3.3 together with the general location of some 26 phosphorite occurrences. The deposits centred near Noarlunga, Kapunda, Burra and Orroroo (Fig. 3.2) have an aggregate production tonnage of only a few thousand tonnes.

The term Adelaide Geosyncline in its original usage (Sprigg, 1952) was regarded as a miogeosyncline bounded on the west by shallow-water stable shelf sediments and on the eastern side by continental terrace sedimentation, thus implying the existence of a continental margin to the southeast. The recognition of the Torrens Hinge Zone (Thomson, 1969) led to the differentiation of three structural and tectono-stratigraphic zones, the Stuart Shelf, a transitional depositional Torrens Hinge Zone, and the Adelaide Geosyncline. Most recently, Rutland (1981) distinguished between a period of intermittent tensional tectonism (*c.* 1000–700 m.y.) up to the time of the deposition of the Tapley Hill Formation (Fig. 3.4), and a relatively quiescent period for the remainder of Sturtian and the succeeding Marinoan sedimentation. Rutland (*op. cit*) interpreted the protracted phase of tensional tectonics as corresponding to the development of a multiple-rifted arch system bounded on the western margin by the Torrens Hinge Zone and extending beyond the Broken Hill rift system perhaps 1000 km to the east. It is suggested that the late Sturtian and Marinoan sediments correspond to broad

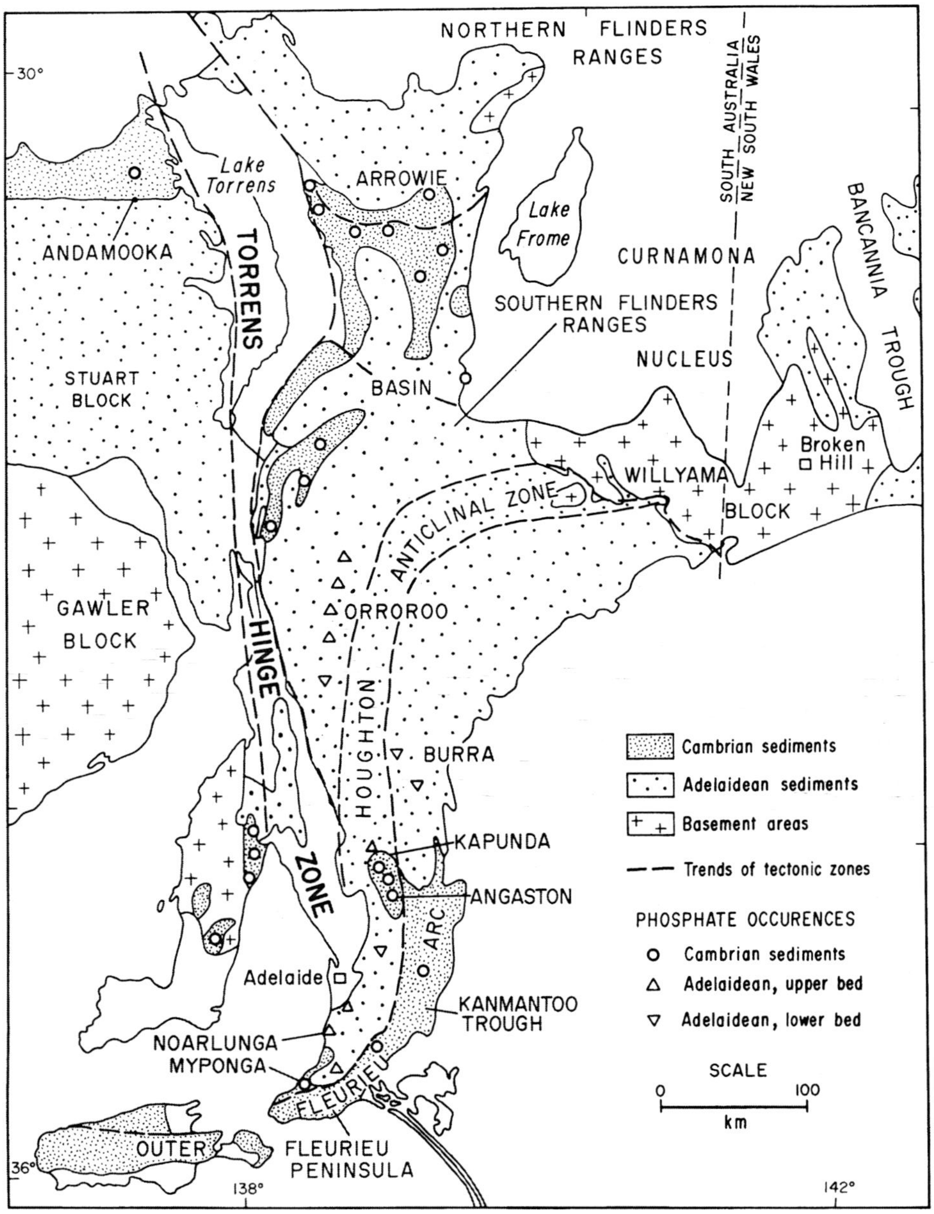

Fig. 3.3. The distribution of Late Proterozoic and Cambrian sediments, structural elements and phosphate deposits in the Adelaide Geosyncline.

inter-arch and subsequent rim basin deposition as demonstrated by the environment in the region of the Curnamona Nucleus (Fig. 3.3). The subsequent Cambrian carbonates are suggested as a likely development of a prograding carbonate wedge.

The Late Precambrian Adelaidean sediments are divided into four series which are used in a chronostratigraphic sense, namely, the Willouran, Torrensian, Sturtian and Marinoan Series (Fig. 3.2). Although their actual boundaries can be defined only at their type locations they have provided a useful framework for discussion of geological history.

Lithostratigraphic subdivision and correlation of the Adelaidean has been attained by Thomson *et al.* (1964) by tracing marker beds such as tillites, and making lithostratigraphic comparisons. The rock groups (Fig. 3.2) contain characteristic lithologies, for example, Callana Beds of Willouran age contain volcanics, carbonates and replaced evaporites, the Burra Group sedimentary magnesite and minor phosphorite, and the Umberatana Group, which unconformably overlies older groups, contains two phases of glacials at its base and a third at its top, in addition to an iron formation and a phosphorite horizon.

The palaeogeographic evolution of the Adelaide Geosyncline is summarised in Figure 3.4. The Willouran Series (Fig. 3.4a) consists of transgressive basal quartzites and carbonates, or red sandstone on the Stuart Shelf, followed by basic volcanics which are an expression of tensional rifting, and lastly evaporitic dolomitic carbonate, black shale, and quartz–arenites. The latter sequence is interpreted by von der Borch (1980) as having formed in a series of alkaline playa lakes in a continental-rift setting, as evidenced by the presence of chert-microcline pseudomorphs after sulfate and sodium carbonate evaporites, and the occurrence of magadiite-type chert.

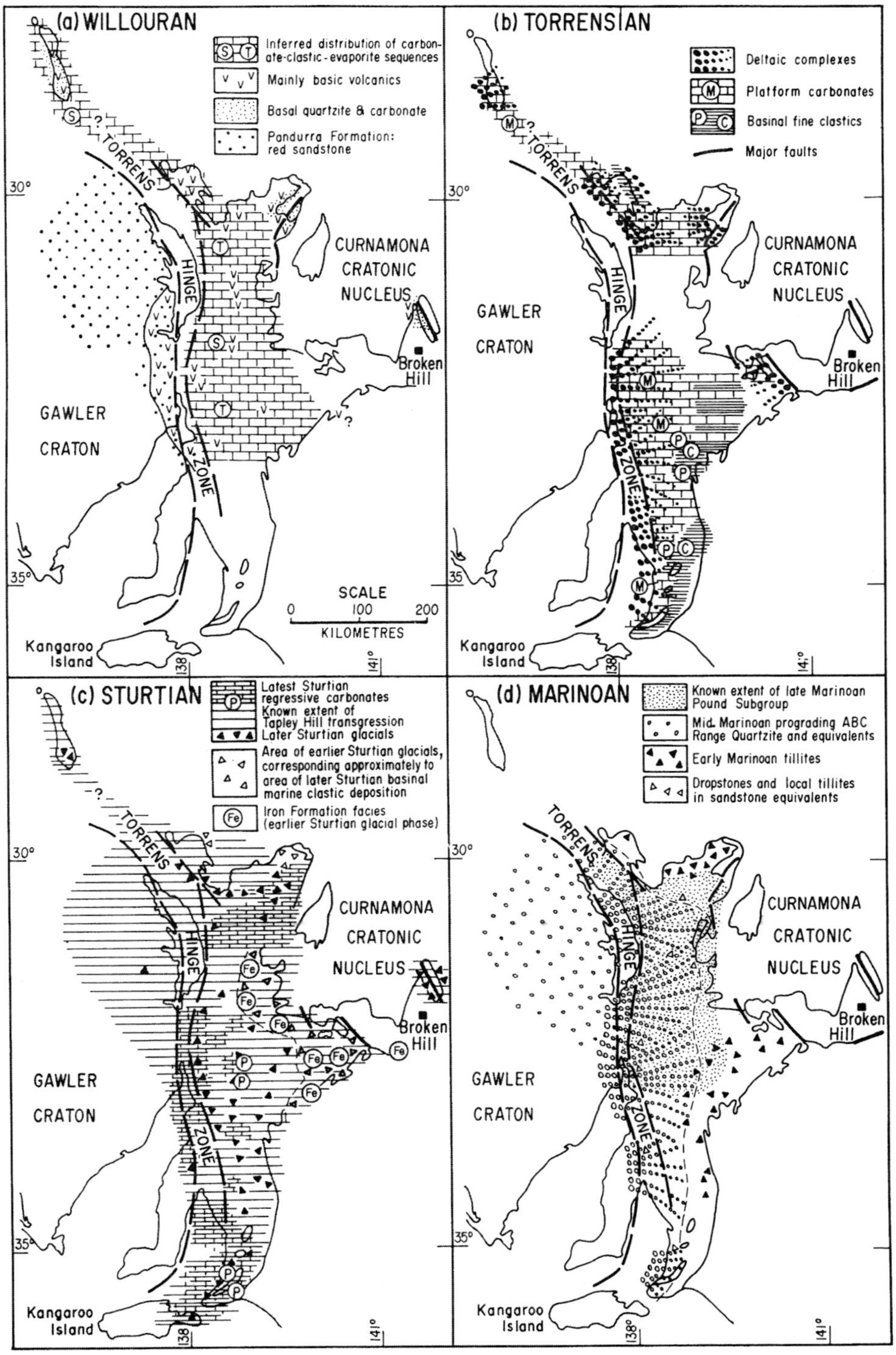

Fig. 3.4. Distribution of sedimentary facies during Adelaidean deposition (after Rutland *et al.*, 1981). Symbol legend, Fig. 3.2.

During Torrensian time (Fig. 3.4b), episodic clastic sedimentation was superimposed on widespread carbonate shelf sedimentation. The clastic sedimentation is interpreted as a series of four eastward and westward prograding deltaic complexes. The Skillogalee Dolomite which occurs throughout the basin is characterised by primary dolomite, commonly stromatolitic, and beds of sedimentary magnesite and diagenetic black chert, and shows a lateral basinal black-shale facies towards the east. The dolomite in some sections passes upwards into a phosphorite-bearing carbonaceous shale, breccia and black chert sequence, which is in turn overlain by siltstone–chert breccias and beds of 'quartzitic' chert.

In Sturtian times (Fig. 3.4c) a widespread marine transgression followed a glacial period and deposited up to 2000 m of Tapley Hill Formation – a finely laminated, carbonaceous and pyritic basinal siltstone which becomes increasingly calcareous in its upper section – followed by an overlying regressive ooid shoal–supratidal facies of the Brighton Limestone (Preiss, 1973; Preiss & Kinsman, 1978; and Rutland *et al.*, 1981). Phosphorite lenses occur within the sandstone facies of the Tapley Hill Formation at the base of, and as occasional lenses within the Brighton Limestone over a strike length of approximately 80 km in the vicinity of Orroroo.

The pronounced facies changes which developed across the basin in the late Sturtian persisted into the Marinoan during which time red beds formed on the western shelf and passed basinward into laminated grey-green silts of the Willochra Subgroup. Syn-depositional carbonates such as the Etina Formation were deposited in high to low energy conditions possibly on off-shore banks and in tidal channels. Shelf and basinal environments persisted through the Marinoan glaciation but were followed by basin-wide uniformity and lateral facies persistence suggesting deposition within an enlarged and tectonically less active rift basin (Rutland *et al.*, 1981). Adelaidean sedimentation concluded with sand sheets associated with major pulses of deltaic progradation during a final regressive phase (Fig. 3.4d).

Phosphorite deposits

Proterozoic phosphorite deposits crop out intermittently in two stratigraphic positions over a strike length of approximately 350 km, from Noarlunga in the south to areas north of Orroroo. The country is gently undulating, outcrop conditions are poor, and the relationship between phosphate and its enclosing sediments is often masked. Additionally, intense weathering during the Tertiary has oxidised surface exposures, resulting in leaching, solution and reprecipitation of apatite and, in some instances, its precipitation in sink holes on karst surfaces, and the replacement of limestones or dolomites. Outcrop of phosphorite is characterised by ferruginous and siliceous cappings which are high in iron, manganese and aluminium oxides and secondary wavellite, crandallite and vivianite. Drilling and pit exposures have demonstrated that pronounced zones of secondary enrichment underlie the cappings and that these in turn are underlain by low-grade primary phosphorites. The enriched zones consist of friable clays and sandstones and collapse breccias, and typically have grades in the range of 20% P_2O_5 to 30% P_2O_5 while the primary grade may be about 5% P_2O_5.

The Lower Phosphorite Unit at the top of the Skillogalee Dolomite contains six known widely separated prospects (Table 3.1). Two of these, Fairview and Hermitage, have a known production record of approximately 100 tonnes and less, grading 25% P_2O_5. Figure 3.5c shows schematically the vertical sequence at the Fairview Deposit and this may be related to the basin stratigraphy (Fig. 3.3). The phosphorite unit crops out over a strike length of 0.8 km, and has a maximum thickness of 22 m. It is underlain by fine-grained laminated dolomite and chert and overlain by siltstone–grey chert breccia and massive quartzitic chert. Below the siliceous and ferruginous capping, the phosphorite unit consists of oxidised phosphatic siltstone, breccia, thin black chert beds, dolomite intercalations and dolomitic siltstone with black chert bands. Phosphate occurs as secondary nodules and irregular bodies within breccia. Drilling demonstrates that the primary bed has a grade of only 2–3% P_2O_5, and consists of a cherty siltstone breccia (Johns, 1969).

In a basinal context the sequence is interpreted by Uppill (1979) as having been deposited in a marine lagoonal, shallow subtidal to intertidal environment based on lithological asso-

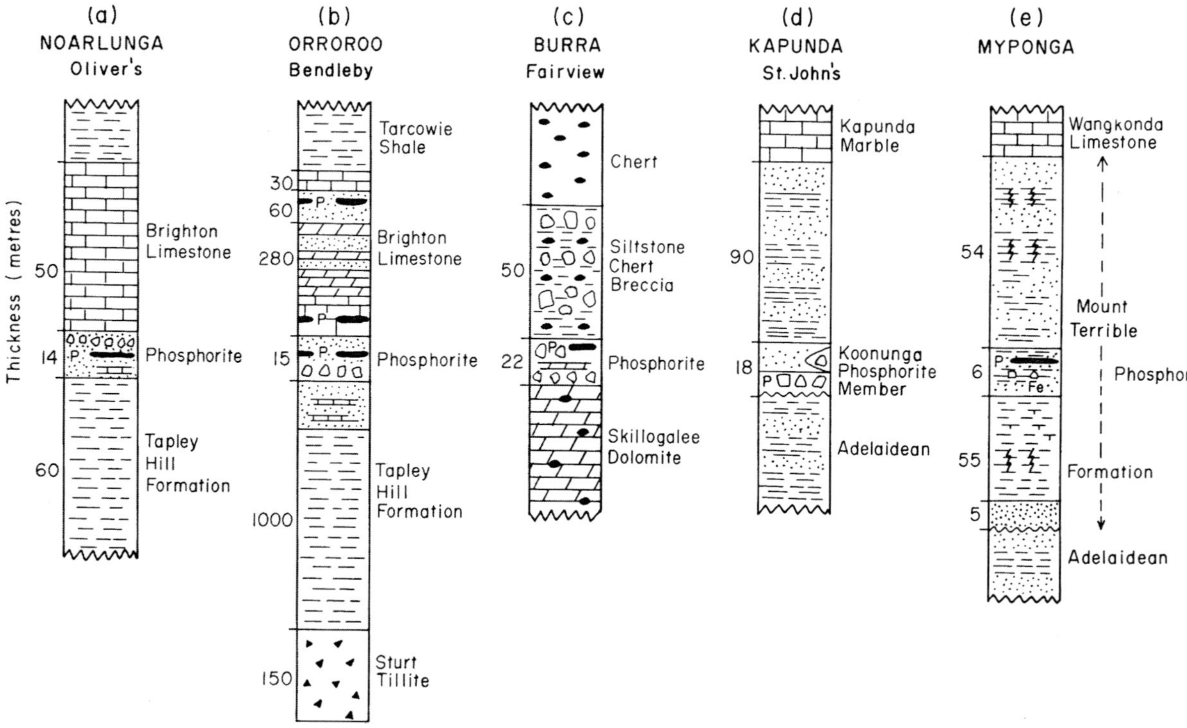

Fig. 3.5. Stratigraphy of selected Late Proterozoic and Early Cambrian phosphate deposits in the Adelaide Geosyncline. (a) and (b) Upper phosphate bed at Noarlunga and Orroroo (Adelaidean). (c) Lower phosphate bed at Burra (Adelaidean), (d) and (e) Cambrian phosphorites at Kapunda and Myponga. Localities, Fig. 3.3; Legend, Fig. 3.2.

ciation and the presence of microfossils in silicified stromatolites (Schopf & Fairchild, 1973).

The Upper Phosphorite Unit at the base of the Brighton Limestone contains some 20 known prospects, most of which are clustered along an 80 km long belt centred at Orroroo, two near Kapunda and three at Noarlunga (Table 3.1 and Fig. 3.4c). Fig. 3.5 (a, b) illustrates the vertical sequences found near Noarlunga and Orroroo.

At all localities there are a number of phosphorite beds, totalling as many as five, 0.7 m to 2.5 m thick. They occur within a unit 1.15 m or so thick, comprising phosphatic siltstone, sandstone, dolomitic limestone bands, intraformational breccia and chert. The phosphatic zones extend over strike lengths of up to two or three kilometres. In the Orroroo region, phosphorite also occurs in silty intercalations in the lower portions of the Brighton Limestone and above and below an upper oolitic limestone bed (Brook, 1962). The phosphatic unit and the overlying Brigthon Limestone is exposed over a strike length of 7 km in the vicinity of Olivers and Waits deposits near Noarlunga, which are described by Callen (1970a, 1971c). The underlying Tapley Hill Formation consists of a nearshore finely laminated silty limestone with occasional calcareous shale and slate beds exhibiting coarse cross-bedding and scour structures.

The Brighton Limestone which overlies the phosphorite consists of a lenticular sandy and silty limestone passing upwards into a thin limestone containing shale and dolomite fragments and finally into dolomitic oolitic limestone. Callen (1970a) believes the carbonates to have formed in an offshore shallow-water environment, while Preiss & Kinsman (1978) interpret them as having formed as offshore ooid banks. In a basinal context the sequence lenses out into deeper water environments to the east, while to the west it passes into intertidal and supratidal facies consisting of non-phosphatic interbedded oolitic limestones, and laminated silty micaceous and carbonaceous siltstones.

The phosphorite unit is finely cross-bedded, micaceous and feldspathic. It contains minor sandstone and dolomitic limestone bands, and is overlain by an intraformational breccia. Graphite is present, and goethite and hematite both replace pyrite and are pervasive through the phosphorite. Manganese, occurring as pyrolusite and cryptomelane, pervades the bed replacing the matrix (Callen, 1971c). Ooids are common to the phosphate, the breccias and the limestones; some have nuclei of rhombic apatite pseudomorphs after carbonate. Both apatite and silica replace carbonate ooids and some of the matrix of oolitic limestone. Because of severe weathering the original distribution of the apatite is indeterminate although Callen (1971c) favours a pre-diagenetic replacement of carbonate ooids.

Amadeus Basin

Following the final cratonisation of the Early Proterozoic sediments of Central Australia, up to 10 km of Adelaidean and Palaeozoic sediments were deposited in intracontinental basins. This belt was the site for much thicker deposition than the cover extensions into the Ngalia and southern Georgina Basins, or sequences which extend through the Officer Basin to the Adelaide Geosyncline in the south (Plumb *et al.*, 1981). Much of the cover to the north, beneath the Wiso, Georgina, Daly River Basins and overlying earlier Proterozoic terrains, has been eroded.

The Late Proterozoic stratigraphy in the Amadeus Basin, and its correlations with the stratotype of the Adelaide Geosyncline, is shown in Fig. 3.2. The correlation of the Heavitree Quartzite–Bitter Springs Formations with the Burra Group, the Areyonga and Aralka Formations with the Umberatana Group and the Olympic Formation–Arumbera Sandstone I with the Wilpena Group are proposed by Preiss *et al.* (1978). These three groups of formations are separated by unconformities and have characteristics which allow their recognition over a wide area, including the Amadeus, Ngalia and the southern portion of the Georgina Basins. Plumb *et al.* (1981) named the groups tectosomes, the lowermost of which consists of a widespread basal arenite passing up into a carbonate-rich succession. It is followed by the middle tectosome which comprises a basal diamictite and phosphorite bed which is succeeded by a lenticular cap dolomite, arenite, lutite and carbonate. The upper tectosome also commences with a basal diamictite followed by an extensive cap rock, lutite, arenite and carbonate.

The Adelaidean succession accumulated generally as a southward-thickening wedge, and commenced with a continuous sheet of fluvial and shallow-marine sand which passed up into marginal marine sequences including dolomite, magnesite and evaporites showing cyclical repetition, and finally into marine lagoonal stromatolitic dolomites and interbedded cherts and fine clastics (Wells *et al.*, 1970). Following the development of a westerly-trending hinge zone through the centre of the basin, the clastic sediments of the middle and upper tectosome continued to thicken as a widespread sheet to the south. However, to the north of the hinge zone the sediments are generally thinner and the middle tectosome exhibits discontinuous deposition.

Phosphate occurrences

Minor phosphate occurrences are known in the Bitter Springs, Areyonga and Pertatataka Formations and their equivalents.

Wells *et al.* (1970) recorded phosphate in the central-northern portion of the basin in a glauconitic and pyritic carbonaceous shale or silt facies of the Bitter Springs Formation associated with a cyclical repetition of beds representing stages of restriction which varied from shallow-marine to saline silled-basin environments. Thin phosphorites occur in the basal part of the Areyonga Formation in the central-eastern portion of the basin where it directly overlies the Bitter Springs Formation. It consists of a bed about 1.3 m thick comprising lenticular phosphatic bands containing microcrystalline apatite and laminae of chert, 8–30 cm thick and up to 3 m long, set in a conglomerate, the matrix of which is phosphatic. The phosphorite bands contain up to 30% P_2O_5 and the conglomerate up to 7% P_2O_5. To the south of the hinge zone, correlatives in the central-south part of the basin contain rare glauconite and phosphate grains in sandstone interbedded with carbonaceous shales near the base of the formation (Wells *et al.*, 1970). North of the hinge zone in the third tectosome, the Pertatataka Formation, Wells *et al.* (1970) recorded glauconite and phosphate in sandstones interpreted as having formed in a neritic environment on a stable shelf. To the south of the hinge zone in a subsiding depression they similarly noted glauconite and phosphate in a silt–sand shallow-marine facies.

Cambrian phosphorites of Australia

Phosphogenic sediments are widely distributed in Lower and Middle Cambrian sediments within the intracratonic basins of Australia (Fig. 3.1). In the Adelaide Geosyncline 39 small deposits and prospects occur in the Lower Cambrian sediments, while in the Amadeus Basin one prospect occurs in Lower Cambrian sediments and two other formations contain phosphatic sediments of Middle and Late Cambrian age. Table 3.1 and Figure 3.2 illustrate that in these basins phosphogenesis also extends back into the Proterozoic. The time and location of major phosphogenesis in Australia, however, was in the Middle Cambrian of the Georgina Basin where the 18 deposits found to date constitute a major world resource.

A fundamental basement fracture pattern was established throughout northern Australia by the end of the Early Proterozoic, and Plumb (1979) suggested that all subsequent tectonic development was controlled by periodic reactivation of major elements of this pattern. Figure 3.6 illustrates the fracture pattern and the configuration of Early Palaeozoic sedimentary basins in relationship to it.

Many of the Early Palaeozoic basins, including the Bonaparte Gulf, Ord, Arafura, Wiso, Daly River and Georgina Basins, developed following the close of the Proterozoic, and their tectonic style can be directly related to their basements (Plumb *et al.*, 1981). Those which lie over or within the Northern Australian Craton, such as the Georgina Basin, have much thinner fill and are relatively undeformed compared to those in mobile belts such as the Amadeus Basin. In the case of the Adelaide Geosyncline, sediments demonstrate a continuum of thin sequences within an enlarged and tectonically less active rift basin, through to thicker sequences deposited to the east on a continental margin. The last have considerable vertical relief (von der Borch, 1980), for example the Kanmantoo Trough trending through Fleurieu Peninsula (Fig. 3.3).

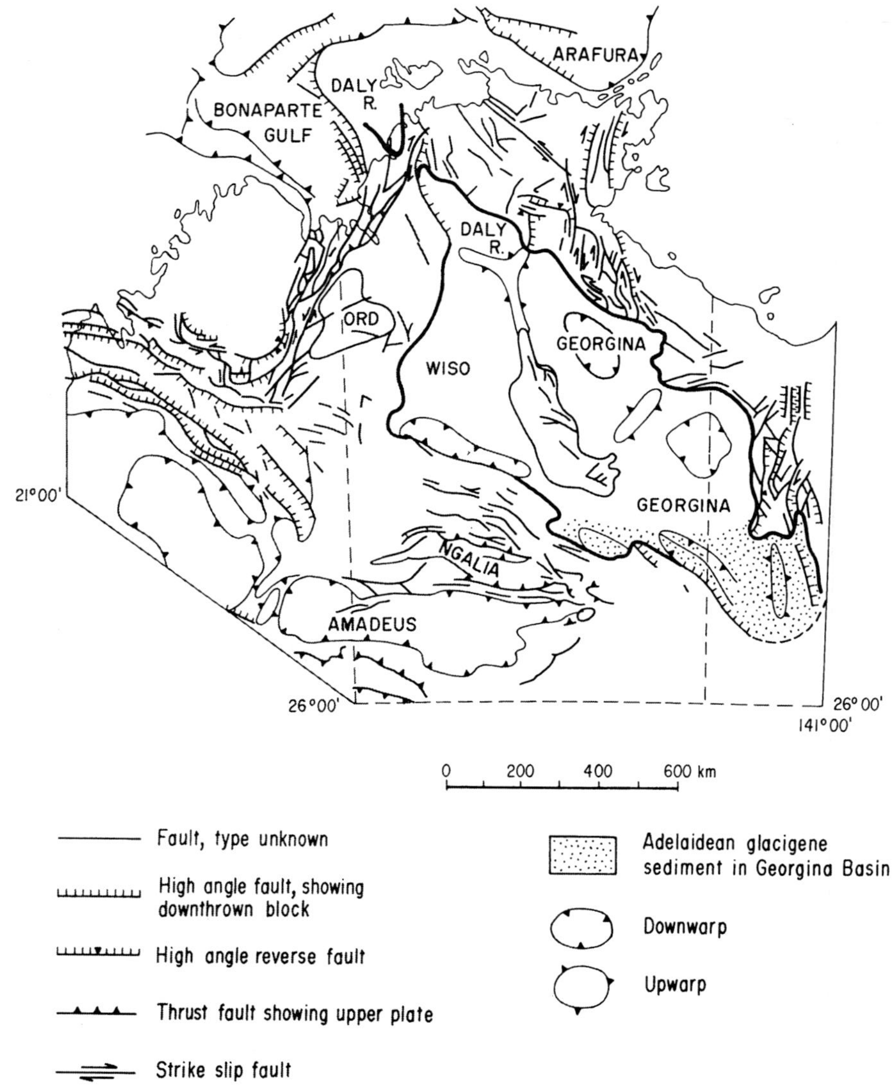

Fig. 3.6. The configuration of the Cambrian sedimentary basins, northern Australia, in relation to Early Proterozoic structures and later warping (modified after Plumb *et al.*, 1981), and also the extent of late Adelaidean cover in the Georgina Basin.

Sedimentation, stratigraphy and palaeogeography

Many of the platform cover sequences show no, or only slight, unconformity between the Cambrian and the Adelaidean. The sediments were dominated by shallow-water quartz-rich arenite and lutite and shelf carbonates, with a more abundant development of lutites occurring in the thicker sequences of the Amadeus Basin and in the Kanmantoo Trough of the Adelaide Geosyncline. To the north-east a thick sequence, including volcanics, occurs within the Bancannia Trough, while further east in the Tasman Geosyncline the sequence thickens further and includes a basal succession, more than 1500 m thick, of pyroclastics and cherts, followed upwards by 5000 m of flysch extending through the Middle and Late Cambrian.

Using data from a recent study, Cook (1982) constructed stratigraphic columns summarising the Cambrian stratigraphy of Australia, and these illustrate comparative time breaks, major lithological changes and inter- and intrabasinal correlations. A number of these columns characterise the stratigraphy of basins containing phosphogenic beds together with the stratigraphy in adjacent basins (Fig. 3.7). Cook (1982) identified six sedimentologically significant time intervals (TI 1 to TI 6) within the Cambrian varying from 3–4 to 30 million years and constructed palaeogeographic maps accordingly. Those pertinent to times of phosphogenesis are reproduced in this paper as Fig. 3.8. It is evident from these reconstructions that a marine transgression across the Adelaidean Basins and the Northern Australian Craton commenced in the Early Cambrian (TI 1, Tommotian–Botomian) and reaching its maximum extent by the Early Middle Cambrian (TI 3, Templetonian). It was in these time intervals that significant phosphogenesis occurred. By Middle Cambrian times (TI 4, Floran–Undillan) the seas regressed and phosphogenesis ceased.

Adelaide Geosyncline–Stuart Shelf

Thirty-nine phosphate deposits, prospects or occurrences are recorded in the Lower Cambrian sediments of the Adelaide Geosyncline. Twelve of these occur at the northern end of the Basin in the Flinders Ranges, 20 are concentrated in the Angaston–Kapunda area immediately north of Adelaide, while the remainder are scattered between the southern Flinders Ranges and the southern end of the basin (Fig. 3.3).

At the onset of the Cambrian, seas transgressed a depression which, in the north, comprised the subsiding Arrowie Basin and the western stable shelf areas of the Stuart Shelf and Yorke Peninsula. To the south, Cambrian sediments are preserved in widely separated occurrences extending from the

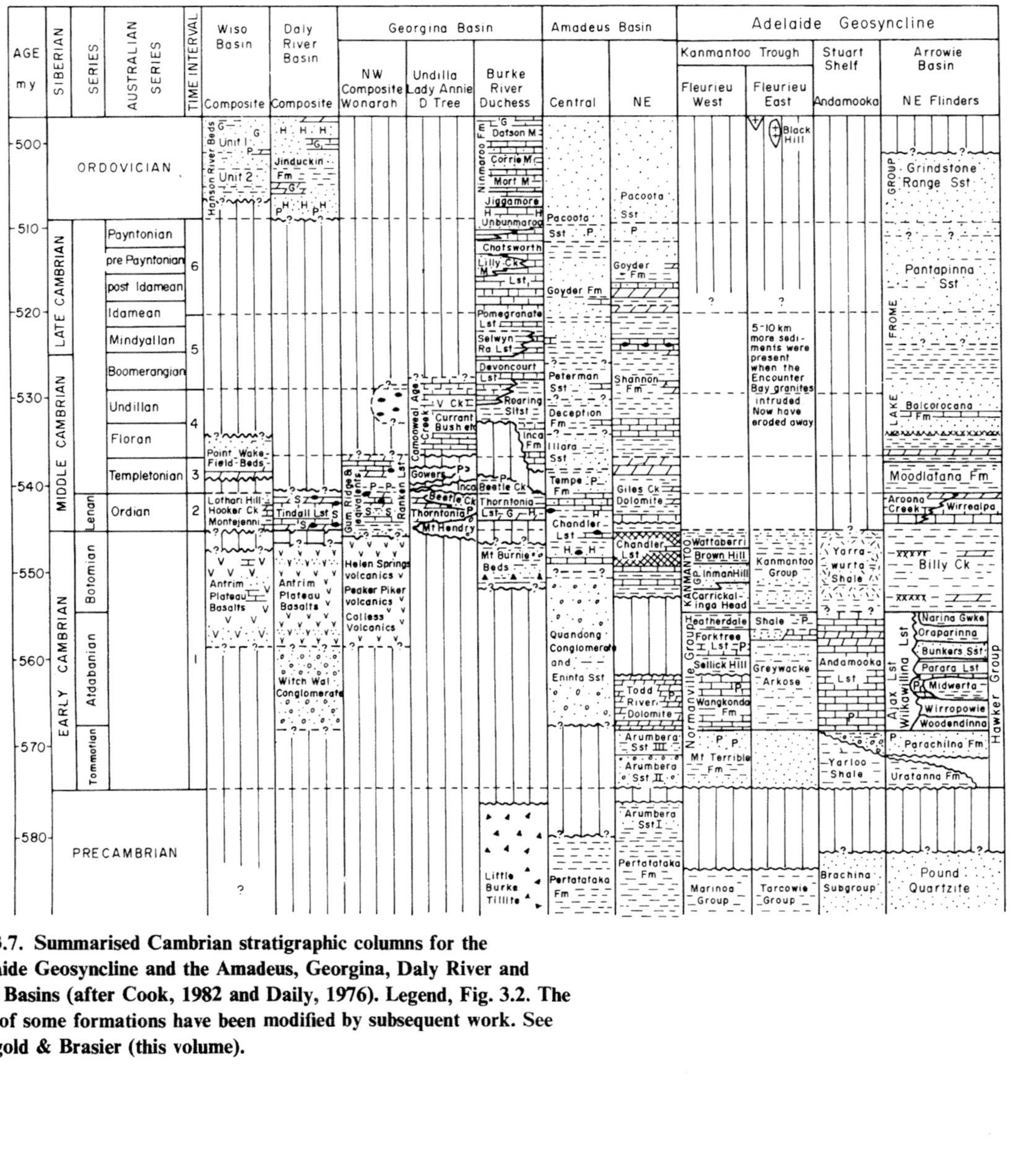

Fig. 3.7. Summarised Cambrian stratigraphic columns for the Adelaide Geosyncline and the Amadeus, Georgina, Daly River and Wiso Basins (after Cook, 1982 and Daily, 1976). Legend, Fig. 3.2. The ages of some formations have been modified by subsequent work. See Shergold & Brasier (this volume).

Kapunda–Angaston district to Kangaroo Island and are interpreted (Daily, 1956) to have been deposited over and on the western slope of a rising basement ridge, the Houghton Anticline Zone, and in the rapidly subsiding Kanmantoo Trough to the east (Fig. 3.3). Stabilisation of the basin occurred towards the end of the Early Cambrian and final deformation took place during the Cambro-Ordovician Delamerian Orogeny.

The Cambrian stratigraphy of the Adelaide Geosyncline shown in Fig. 3.7, comprises some 600 m of sediments on the western shelf areas, approximately 4000 m in the Arrowie Basin, 300–1000 m over the Houghton Anticlinal Zone, and in excess of 10 km to the east in the Kanmantoo Trough. The fossiliferous sequences of the Hawker Group of the northern regions and the Normanville Group in the south are described by Daily (1956, 1963), Abele & McGowran (1959), Horwitz *et al.* (1959), Dalgarno & Johnson (1964), and Parkin (1969). The large unfossiliferous turbidite sequence of the Kanmantoo Group sediments is documented by Flint (1978) and Parkin (1969).

The depositional environment varied widely in response to the tectonic framework, and the Lower Cambrian is characterised by intertonguing of facies and variation in formation thickness. To the north deposition commenced with the silt-rich Uratanna Formation which filled basins and channels in the underlying Adelaidean Wilpena Group. The succeeding Parachilna Formation represents a shallow-water transgressive facies consisting of richly fossiliferous sandstone, interbedded siltstone and shale, and correlates with the upper member of the Mount Terrible Formation in the south. The two formations are phosphogenic and host most of the Lower Cambrian phosphate deposits of the Adelaide Geosyncline. In the Myponga prospect area the Mount Terrible Formation consists of three members (Fig. 3.5e), the lowest of which comprises massive arkosic sandstone 5 m thick, a middle member which includes 55 m of burrowed grey-green finely laminated siltstone and calcareous beds equivalent to the Winulta Formation of the Flinders Ranges, and an upper member comprising 60 m of bioturbated fine-grained sub-arkose and sandy siltstone, and intraformational breccia. A phosphorite horizon occurs in the upper member at the base of a coarse-grained feldspathic siltstone–shale unit. Close to the rising basement, the Mount Terrible Formation is represented by a thin pebbly arkose bed. Elsewhere it is present as thin phosphatic transition beds ('Hyolithes sandstone') at the base of Wangkonda Formation and at the base of the Kulpara Formation on Yorke Peninsula. In the Kapunda–Angaston district, the equivalent succession (Fig. 3.5d) is unnamed and consists of several hundred metres of sandstone, siltstone and shale which include the lenticular Koonunga Phosphorite Member (Johns, 1967).

The phosphatic basal Cambrian is succeeded by a prolonged phase of mainly carbonate sedimentation, the most complete section of which is preserved in the northern Flinders Range and the Stuart Shelf region. The Ajax and Wilkawillina Limestones are thick biostromal limestones, characterised by

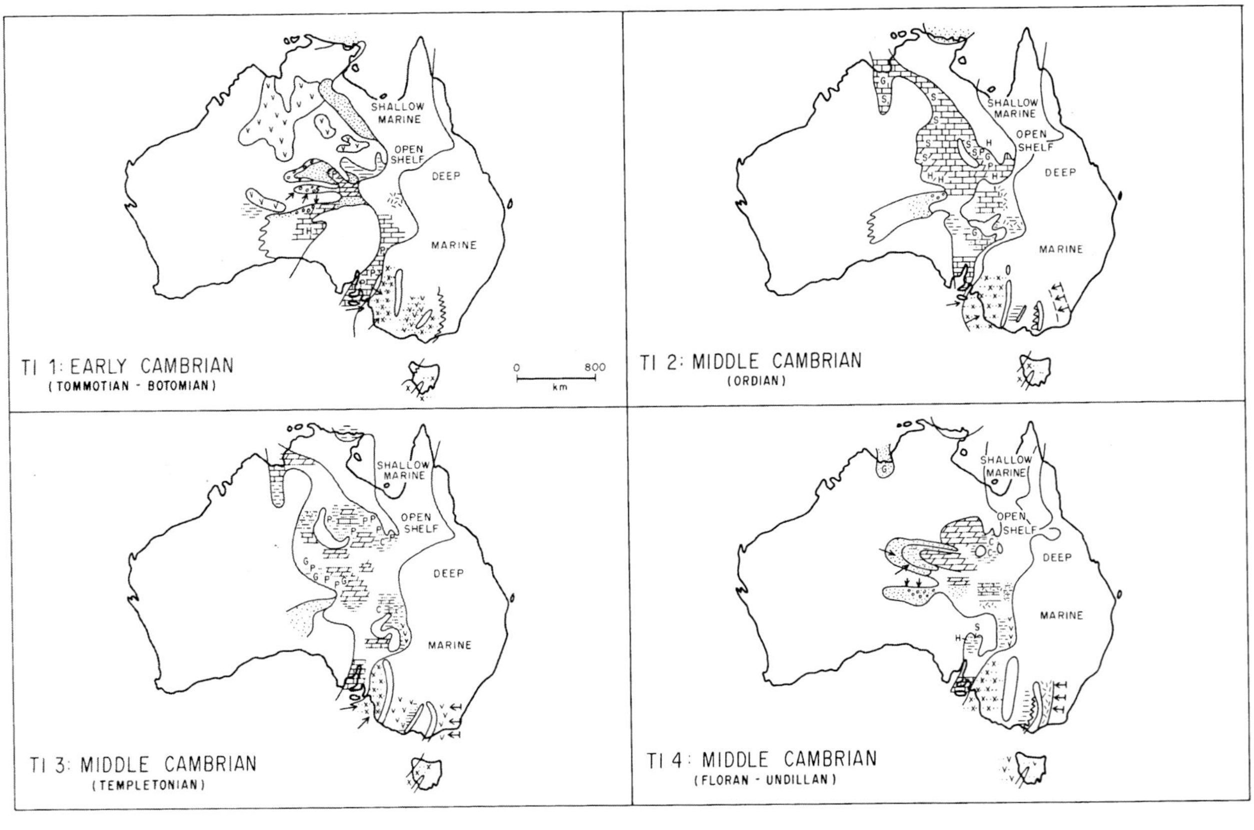

Fig. 3.8. Sedimentation, depositional environments and palaeogeography of the Early and Middle Cambrian: TI 1 Tommotian–Botomian, TI 2 Ordian, TI 3 Templetonian and TI 4 Floran–Undillan (after Cook, 1982). Legend, Fig. 3.2.

Archaeocyatha, hyolithids and phosphatic brachiopods, and were deposited on stable shelves marked by numerous islands and shoals. In the central basinal areas and hinge zones between the stable shelves, where there was more active subsidence and greater water-depth, thicker, dark and impure limestones (Parara) and interbedded shales were laid down (Daily, 1976). Lateral and vertical facies changes involving contrasting deeper, shallower and intertidal carbonate types and clastic sediments are illustrated in Fig. 3.9. Many of these carbonate units are sporadically phosphatic due to their fossil content or to pelletal apatite zones in glauconitic limestones, and the Parara Limestone, in particular, contains phosphorite beds 5–10 cm thick in oolitic and carbonaceous facies (Callen, 1971a).

On the Stuart Shelf, on the opposite side of the basin, a similar intertidal to supratidal carbonate succession is represented by the Andamooka Limestone. An offshore archaeocyathid biostromal facies containing phosphatic (7–12% P_2O_5, Broken Hill Proprietary Ltd, personal communication) shelly beds passes shoreward into phosphatic laminated cherty dolomite showing desiccation cracks, intraformational conglomerate oolitic and pelletal dolomites (Fig. 3.9).

The general regression of the Early Cambrian sea is indicated by widespread occurrence of dolomite and algal colonies at the top of the Andamooka, Ajax and Wilkawillina Limestones and the Moorowie Formation prior to the influx of red bed clastics of the Yarra-wurta Shale and the Billy Creek Formation (Fig. 3.7,; Thomson *et al.*, 1976).

To the south, on Yorke and Fleurieu Peninsulas, facies similar to the Hawker Group are recognised in the Normanville Group succession. The Kulpara and Wangkonda Formation and the Fork Tree Limestone were deposited as stable shelf facies while the Sellick Hill and Parara Formations represent carbonaceous and deeper water facies. Each of the formations is phosphatic, at least in part, and secondary phosphate deposits are developed on a karst surface of the Kulpara Limestone on Yorke Peninsula (Fig. 3.3). The uppermost formation is the

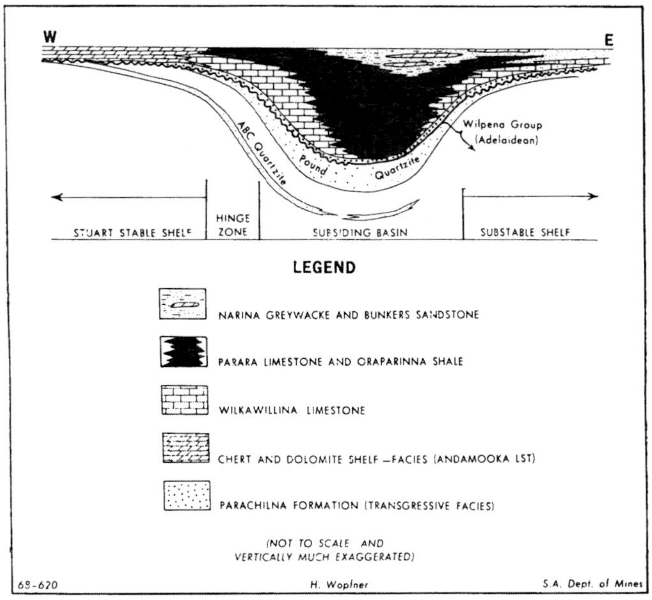

Fig. 3.9. Schematic cross-section and facies distribution – Lower Cambrian basin, Flinders Ranges (after Parkin, 1969).

Heatherdale Shale, a basinal facies which attains a thickness of 320 m. It consists of grey or black shales which are occasionally carbonaceous and pyritic, and contains small nodules and elongate segregations of apatite.

The Normanville Group exhibits pronounced changes of facies and thickness across the Houghton Anticlinal Zone, decreasing rapidly from a maximum thickness of some 940 m in its type area to 300 m close to the basement high. All formations of the Group decrease in thickness and the carbonate facies change mainly to siltstones and calcareous siltstones (Fleurieu West and East, Fig. 3.7). On the flanks of the Kanmantoo Trough to the east of the basement high the sequence thickens to 1000 m. While a thin basal arenite and upper pyritic phosphatic and graphitic shale persists, the carbonate facies is presumed to grade laterally into phyllite and greywacke (Horwitz, Thompson & Webb 1959). The Kanmantoo Group, a turbidite sequence, then prograded across the Normanville Group.

In the early Middle Cambrian a widespread shallow-water transgression led to the deposition of the Wirrealpa and Aroona Creek Limestones, and this sequence contrasts with the early Middle Cambrian successions of the Georgina Basin and Amadeus Basin in being barren of phosphate. The Wirrealpa Limestone is bioclastic and contains minor chert, but exhibits current-oriented stromatolite mounds attesting to its shallow-water depositional environment. The laterally equivalent Aroona Creek Limestone is largely dolomite, commonly stromatolitic and devoid of fauna and was deposited in a restrictive and regressive environment (Thomson *et al.*, 1976). The succeeding Lake Frome Group, which concludes the Cambrian sedimentation, consists largely of red beds deposited during deltaic, fluvial and terrestrial conditions.

Phosphate deposits and their origin

The depositional environment of the phosphatic Lower to Middle Cambrian sediments has been described above. Of the 39 prospects and occurrences listed in Table 3.1, production was derived from 20 quarries in the Kapunda–Angaston district and from quarries at Myponga on Fleurieu Peninsula and Clinton and Ardrossan on Yorke Peninsula. No production has been derived from occurrences in the Flinders Ranges.

Descriptions of the rock phosphate within quarries at Myponga are contained in Jack (1919), Johns (1962, 1976), Blisset & Callen (1969) and Callen (1970b, 1971b), while Callen (1971a) describes the phosphate occurrences of the Flinders Ranges.

All the deposits are small and each produced less than 50 000 tonnes. The largest are Moculta, St Johns, Toms and St Kitts. Apart from emergency production during World War II, the deposits were mostly worked early in the 1900 s and descriptions of deposits are necessarily poor.

Like the Proterozoic Adelaidean deposits previously described, all appear to have been derived as weathering products or by the replacement of limestone during intense Tertiary and Quaternary weathering cycles. Of the 25 deposits which have been quarried, all except those at Myponga and at St Johns and St Kitts in the Kapunda–Angaston district, are limestone replacement deposits in which the phosphate has been derived from an associated low-grade primary phosphorite or, in the case of Clinton, from the dissolution of the phosphatic Kulpara Limestone. At Clinton high-grade rock phosphate is related to a karst surface where limestone is replaced along prominent joints to depths of 45 m and is associated with depressions filled with phosphatic clays containing up to 8% P_2O_5 (Johns, 1976). Elsewhere secondary rock phosphate is described as nodules associated with limonitic earthy clays (Jack, 1919). In all cases the deposits are capped by siliceous ironstone or limonitic laterite.

The development of secondary deposits from primary phosphorites has occurred at Moculta, St Johns, Toms and St Kitts to depths of 33 m, with production grades in the order of 20–30% P_2O_5. Later studies by the South Australian Mines Department (Johns, 1976) showed the grades at St Johns and Toms over widths of 20 m to average 15.2% P_2O_5, 5.1% Fe_2O_3 and 8.0% Al_2O_3 and 20% P_2O_5, 15% Fe_2O_3 and 7% Al_2O_3 respectively. Myponga has an average grade of 27% P_2O_5 and low iron and aluminium content.

Lithological columns at Myponga and St Johns are illustrated in Figs. 3.5(d) and 3.5(e). In the Kapunda–Angaston district exploration has shown that the underlying primary phosphorite has grades less than 10% P_2O_5 and generally in the order of 1–5% P_2O_5 associated with variable but high iron and aluminium (Johns, 1976). Primary beds at Myponga are 3–5 m thick and 200–300 m long within a sandy siltstone. In the Kapunda area the primary bed has a strike length of up to 3 km, and at St Johns it consists of phosphatic shale, siltstone and intraformational breccia deposited on an undulating Adelaidean surface. At Myponga the phosphate is also contained in an intraformational breccia associated with coarse-grained siltstone, fine-grained arkose, feldspathic sandstone and iron-rich bands composed of detrital geothite. The environment is interpreted by Callen (1970b) as having been a shallow-water high-energy one in which low-grade unconsolidated beds were brecciated by wave and current activity and phosphate was concentrated during reworking.

Amadeus Basin

Thin bands of Cambrian phosphorite occur in the Todd River Dolomite, the Tempe Formation and Pacoota Sandstone (Fig. 3.7), all of which were deposited during phases of marine transgressions of the Amadeus Basin.

The Petermann Ranges Orogeny at the close of the Proterozoic caused widespread regression of the seas from the southern and western portions of the Amadeus Basin. Cambrian sediments were deposited in an embayment into an elongated intracratonic basin. They exhibit transgressive–regressive cycles first in the earliest Cambrian and late Middle Cambrian and secondly in the Late Cambrian.

The basal Cambrian deltaic sediments (Arumbera Sandstone II and III) were deposited on marine Adelaidean successions without a major break (Wells *et al.*, 1970). The detritus making up these rudites and arenites was derived from the south and west and grades into marine facies to the east, a pattern which persisted throughout the Cambrian as the seas transgressed from the east. (Fig. 3.8).

Following the deposition of detrital sediments in the Early Cambrian, carbonate and shale deposition persisted until the Late Cambrian, but with migration of the facies to the east during a regressive phase in the late Middle Cambrian. As the seas transgressed along the basin from the east in the Early and Middle Cambrian, well-defined east–west facies were developed which included continental and transitional red beds in the west passing eastward into shallow-marine carbonates and shales. During Late Cambrian times the seas become broad and shallow, and the well-defined east–west facies gave way to an extensive cover of thick sandstone which formed above wave base (Wells *et al.*, 1970).

Phosphate deposition

In the Early Cambrian, marine deposition was confined to the eastern portion of the Amadeus Basin. The Arumbera Sandstone passes up into the marine limestone, dolomite, sandstone and siltstone of the Todd River Dolomite. Unpublished exploration reports of Broken Hill Proprietary Ltd (personal communication) record that the dolomites and interbedded sandstones are phosphatic and in a drill hole yielded a 6 m intersection of calcareous silty sandstones associated with minor limestones, chert and ferruginous siltstone averaging 13.2% P_2O_5. Sporadic values up to 5% P_2O_5 also occur in the overlying red siltstones. The succeeding Chandler Limestone is far more widespread to the south and west and consists of sparsely fossiliferous evaporitic sediments including limestone, dolomite, interlaminated chert, evaporite and shale deposited in a restricted environment.

The basal Middle Cambrian Tempe Formation marks the first occurrence of richly fossiliferous marine sediment, and comprises siltstone, dolomite glauconitic sandy limestone with associated black shale, and sandstone. Thin grainstone phosphorites in glauconitic sandy limestone consist of phosphate replacement of rounded echinoderm plates and brachiopod fragments (Wells *et al.*, 1970). This is also common in the Beetle Creek Formation, its probable time equivalent in the Georgina Basin. Phosphate also occurs as concentric structures around quartz or calcite nuclei, but Wells *et al.* (1970) report that all bands are low grade and of limited lateral extent. No phosphate has been recorded from the deeper-water Giles Creek Dolomite facies to the east.

The Upper Cambrian Goyder Formation, which exhibits a wide range of facies from continental red beds through to deep-water marine shales and carbonates, is not known to contain phosphate. However, Wells *et al.* (1970) report that the succeeding Pacoota Sandstone contains phosphatic grains and clasts and rare granular and nodular bands containing up to 16% P_2O_5. These occurrences are compared to the lag deposits of the succeeding Ordovician Stairway Sandstone. Wells *et al.* (1970) conclude that sandstones were deposited in a broad shallow sea in which the sediments were reworked above wave base.

Georgina Basin

The early Middle Cambrian phosphorite deposits of the Georgina Basin, discovered in 1966–68, comprise 18 named deposits occurring over a distance of approximately 1000 km along the palaeo-periphery and insular portions of the basin (Fig. 3.10). These deposits contain reserves of 3.4×10^9 tonnes

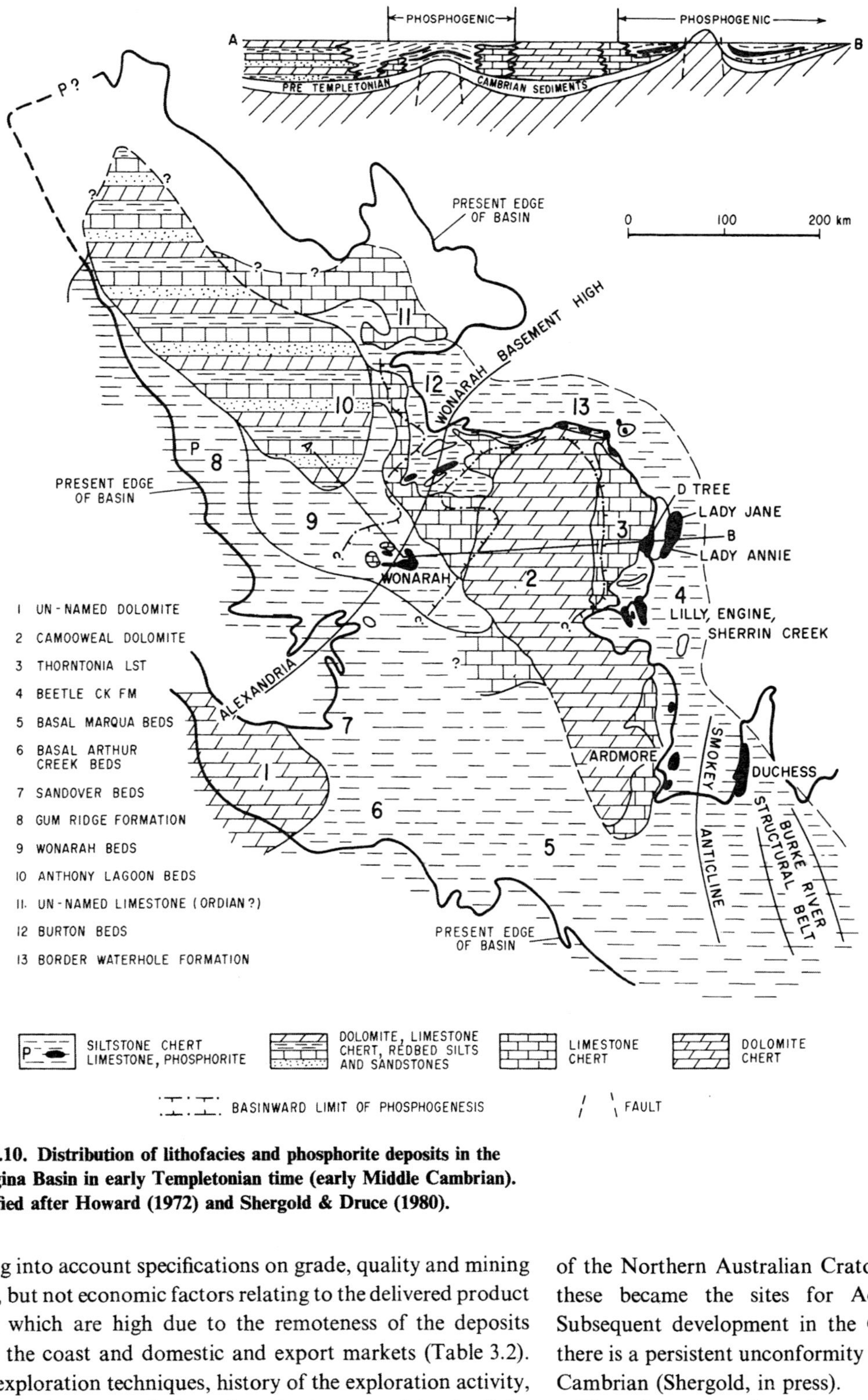

Fig. 3.10. Distribution of lithofacies and phosphorite deposits in the Georgina Basin in early Templetonian time (early Middle Cambrian). Modified after Howard (1972) and Shergold & Druce (1980).

taking into account specifications on grade, quality and mining ratio, but not economic factors relating to the delivered product costs which are high due to the remoteness of the deposits from the coast and domestic and export markets (Table 3.2). The exploration techniques, history of the exploration activity, and the discovery of these deposits is covered in detail by Howard (1971) and Thomson & Russell (1971).

The location of the Georgina Basin and the contiguous Wiso and Daly River Basins and their relationship to a terminal Early Proterozoic fracture pattern is illustrated in Fig. 3.6, together with the distribution of Adelaidean sediments in the Georgina Basin (Plumb, 1979). Diastrophism which occurred in the mid-Adelaidean in the Amadeus Basin coincided with the development of narrow, deep north-westerly-trending grabens or aulacogens which were incised into the southern edge of the Northern Australian Craton (Plumb *et al.*, 1981), and these became the sites for Adelaidean cover (Fig. 3.7). Subsequent development in the Georgina Basin shows that there is a persistent unconformity between the Adelaidean and Cambrian (Shergold, in press).

The palaeogeography and nature of the contained sediments indicate that the Georgina, Daly River and Wiso Basins are mild Cambrian structural downwarps which followed on from the Adelaidean depressions along the south-eastern margin (Fig. 3.6). Second-order pre-Templetonian (Middle Cambrian) upwarping produced a structural ridge which forms the divisions between the Georgina, Daly River and Wiso Basins (Fig. 3.6). The Alexandria–Wonarah Basement High (Figs. 3.6, 3.10), presently reflected by lineament patterns in Cambrian sediments and by magnetic data, may have originated at the

Table 3.2. *Phosphate reserves, Georgina Basin*

Deposit	Tonnes $\times 10^6$	Grade $\% P_2O_5$(a)
Duchess	1345	17.3
Quita Creek	30	17.0(b)
Ardmore	47	15.6
Lily Creek	192	13.4
Engine Creek	NA	—
Sherrin Creek	212	16.0
D Tree	220	18.6
Lady Annie	293	16.6
Lady Jane	193	17.6
Riversleigh	11	14.4
Phantom Hills	46	16.0
Mount Jennifer	20	15.3
Babbling Brooke Hill	38	16.8
Mount O'Connor	42	17.4
Highland Plains	84	13.4
Alexandria	NA	—
Alroy	NA	—
Wonarah	606	15.7(c)
Total	3379	16.6

(a) Source of information, Dreissen & Cook (1981) except for (c), Perrino (1970). (b), estimated. NA, not available

same time. However it was still rising during the Templetonian transgression as judged by the occurrence of synsedimentary drape structures (Howard, 1972). This second-order upwarping together with that of the Smokey Anticline isolated two broad depressions along the north-west trend of the Georgina Basin (Fig. 3.6).

Reactivation of Early Proterozoic faults (Fig. 3.6) peripheral to and within the basin during the Cambrian have been described by Smith (1972), de Keyser & Cook (1972), Rogers & Keevers (1976) and Howard & Hough (1979). Movement can be demonstrated to have continued during Middle Cambrian sedimentation, resulting in examples of pronounced facies changes. It also took place much later as indicated by the displacement of overlying Cretaceous beds (Howard & Cooney, 1976).

The stratigraphic succession of the Georgina Basin is characterised by complex facies variations which have resulted in the naming of 70 Upper Proterozoic and Cambro-Ordovician stratigraphic units. Shergold & Druce (1980) summarised their inter-relationships, thicknesses and temporal–spatial relationships and discussed how the multitude of names resulted from definition of units on both lithological and faunal criteria, and from evidence derived from rubble of uncertain provenance. An historical account of the classifications and a review of the problems is given by de Keyser (1973). In a step towards clarifying the inter-relationship of the defined units, de Keyser & Cook (1972) defined six often diachronous lithosomes, interdigitating beds of essentially uniform lithological character. Shergold & Druce (1980) have more recently defined three tectotopes or tectono-stratigraphic units which contain a dominant group of sediment types separated by disconformities or unconformities. The initial tectotope includes late Adelaidean to Early Cambrian glaciogene and siliciclastic sediments passing upwards into carbonates. The distribution of these sediments is confined to the deep southern and south-eastern portion of the basin (Fig. 3.6). The middle tectotope which includes the phosphogenic successions is widespread, covering most of the basin. Essentially it comprises a carbonate succession but with terrigenous clastic intervals in its lower part, and extends from Ordian (early Middle Cambrian) time to Early Ordovician. The upper tectotope comprises a largely siliciclastic sequence with minor occurrences of carbonates of Early Ordovician age, and like tectotope 1 is confined to the southern portion of the basin.

Cambrian and Ordovician sediments are thicker in depressions along the southern margin of the basin and in the Burke River Structural Belt where they reach approximately 3000 m and 2000 m respectively. Elsewhere in the basin the sediments are mostly Middle Cambrian and are less than 300 m thick.

In summarising the stratigraphic relationship for the areas of the principal phosphorite deposits, Howard (1972) used a classification based on data of Öpik & Pritchard (1960) and de Keyser (1968, 1969) and this is presented in a modified form by Cook (1982) (Fig. 3.7).

Generalised palaeogeographic maps of the Georgina Basin were first prepared by Smith (1972). Subsequently, Howard (1972), Howard & Hough (1979) and Cook (1979) produced palaeogeographic maps which included facies relationship of the sediments in that portion of the basin containing known phosphate deposits or phosphogenic beds. The most comprehensive maps are those of Shergold & Druce (1980) which are arranged by time series and faunal zones from Vendian to Early Ordovician, and illustrate the distribution of formations by their dominant facies. Most recently, Cook (1982) has constructed Cambrian palaeogeographic maps of the Australian continent on the basis of six sedimentologically significant time intervals which portray the sedimentation and depositional environments.

The first widespread event following the development of grabens along the southern margin of the developing basin was the deposition of Adelaidean glacial sediments comprising morainic material and lacustrine sediments which are capped by dolomite (Fig. 3.7). Preiss *et al.* (1978) have suggested a correlation with Adelaidean Sturt Tillite (*c.* 725 m.y.) Other Adelaidean sediments occur further north but their distribution transgresses the present basin boundaries and their deposition preceded its formation.

The basal Cambrian sequences are Lower Cambrian flood basalts such as the Colless, Peaker Piker, Helens Springs and Nutwood Downs Volcanics in the north and north-west portions of the basin (Fig. 3.8 TI 1). These extrusions occur at localities about the periphery of the basin, and are also known from drilling operations to be extensively distributed in the subsurface. At some localities they were deeply weathered prior to the initial Middle Cambrian transgression (Howard & Perrino, 1976).

The earliest Cambrian sediments consist of predominantly marine arenaceous sequences developed as high-energy scree slides and fluvial systems through to migrating intertidal or shallow subtidal systems. Their distribution and lithology reflect uplift and erosion of a hinterland featuring reactivated late Early Proterozoic fault systems. In the south-west, dolomite banks extended into the basin from the Amadeus Basin, and the wide extent of these intertidal and supratidal dolomite shoals indicate the onset of tectonic stability of the region (Shergold & Druce, 1980).

By the close of Ordian time marine conditions covered all of the Georgina Basin and the contiguous Wiso and Daly River Basins through to the present north-west coast of Australia (Fig. 3.8 TI 2). Local early Ordian clastics graded upwards into dolomite–limestone sequences (Camooweal Dolomite, Thorntonia Limestone, Ranken Limestone, Tindall Limestone) and siltstone–shale–chert sequences of the Gum Ridge Formation (Fig. 3.7). For the first time well-defined lateral facies trends were established between arenites, lutites, clastic and chemical carbonates and chert, and the deposition of phosphorite commenced. The Ordian closed with a minor regression which exposed carbonate banks and resulted in the replacement of the Thorntonia Limestone by phosphate in shoreline supratidal ponds and on karst surfaces. In adjacent areas hypersaline pools developed thin sheets of evaporites (Henderson & Southgate, 1978).

Renewed transgression in the Templetonian resulted in deposition of major deposits of phosphorite in the silt–shale–chert–carbonate facies of the Beetle Creek and Border Waterhole Formations, and of the Wonarah and Burton Beds (Fig. 3.8 TI 3). Fig. 3.10 shows the detailed facies distribution and the interpreted maximum extent of sedimentation in respect of the present margins of the basin. On the eastern margin of the basin, facies rapidly change basinward from siltstone–chert–phosphorite of the Beetle Creek Formation to intertonguing phosphatic limestone–dolomite limestone–chert of the Thorntonia Limestone into dolomite banks (Camooweal Dolomite) which were contained to the west by the Alexandria–Wonarah Basement High and to the south by the Smokey Anticline structural rise (Fig. 3.10 Inset). The unnamed dolomite bank to the southwest was separated from the Camooweal bank by subtidal channels.

The Alexandria–Wonarah Basement High was actively rising during sedimentation and the structure was the site of phosphogenesis. Phosphorite beds can be demonstrated to drape over positive elements which were related to reactivated Early Proterozoic faults (Fig. 3.10 Inset). To the north-west of the basement high, evaporite conditions prevailed (Anthony Lagoon Beds) and a red bed sequence of red and purple dolomite, shale, sand and gypsiferous sediments were deposited, and interfingered with marine limestone, siltstone, chert and phosphorite deposited in nearshore subtidal channels.

In the south-east quadrant of the basin, the Smokey Anticline flexed upwards during the Templetonian (Shergold & Druce, 1980). On the shelf to the south-east of it, in the Burke River area, transgression continued from late Templetonian time well into Middle Cambrian (Floran and Undillan) time, but only into Floran time along the eastern margin of the basin where a laminated silt–shale facies of the Inca Formation indicates deepening water conditions. Upward-shallowing sequences in the Marqua and Arthur Creek Beds to the south-west indicate upward movement of the basin. After this a gradual migration of all sedimentary environments to the south-east occurred as a major regression took place, and sandy facies developed (Shergold & Druce, 1980).

Phosphorite deposits

The Smokey Anticline and the Alexandria–Wonarah Basement High were structural elements which divided the Georgina Basin into three distinct depositional regimes exhibiting divergent sedimentary history and phosphorites of markedly different character. These are:

(1) A south-east open-marine shelf region, which includes the subsidiary Burke River Embayment. The Duchess phosphorites were deposited across the seaward flanks of the Smokey Anticline structural rise and in the depression of the Burke River Structural Belt (Figs. 3.6, 3.10).
(2) A central shallow-marine epicontinental basin (Undilla) containing large hypersaline carbonate banks surrounded by subtidal channels which were fringed by embayments, lagoons and estuaries. Related depositional environments include coastal subtidal, intertidal and supratidal. Phosphorites (Lady Annie, D Tree etc.) were deposited in these coastal environments (Figs. 3.6, 3.10). Ardmore in the south-east has affinities more akin to the Duchess phosphorite.
(3) A northern epicontinental basin (Brunette) which was dominated by an extensive central area of transitional red beds of both continental and marine character and partly fringed by marine subtidal channels. Phosphorites (Wonarah, Alroy) were deposited on the crest and slope of the submarine ridge forming the south-eastern margin of the basin. Within this basin, phosphatic sediments have been found at only two localities to date (Wells *et al.*, 1970) (see herein Fig. 3.10).

Duchess Duchess is the largest deposit in the Georgina Basin but ceased production in December 1982. The phosphorite has an average thickness of 11 m, attains a maximum of 38 m and extends over a strike length of 23 km. The deposit exhibits important differences compared to deposits which occur within the epicontinental portions of the basin. The geology of the deposit is covered in detail de Keyser (1968), Russell & Trueman (1971), de Keyser & Cook (1972), Cook & Shergold (1979), and Rogers & Crase (1980).

The Duchess phosphorites, and their related sediments which comprise the Beetle Creek Formation, are located within a shallow-water shelf embayment. The embayment occurs at the western margin of a graben located in the Burke River Structural Belt (Fig. 3.11) separated from the main basin by a series of basement ridges. The lithologies of the Beetle Creek Formation are dominated by chemical and biochemical sediments including phosphorite, chert and carbonate, together with siltstone.

The embayment sequence includes in upward order, the Mount Birnie Beds, consisting of ferruginous clastics and dolomite, unconformably overlain by the Thorntonia Limestone, which is in turn overlain by a chert horizon deposited in a shallow hypersaline environment (Southgate,

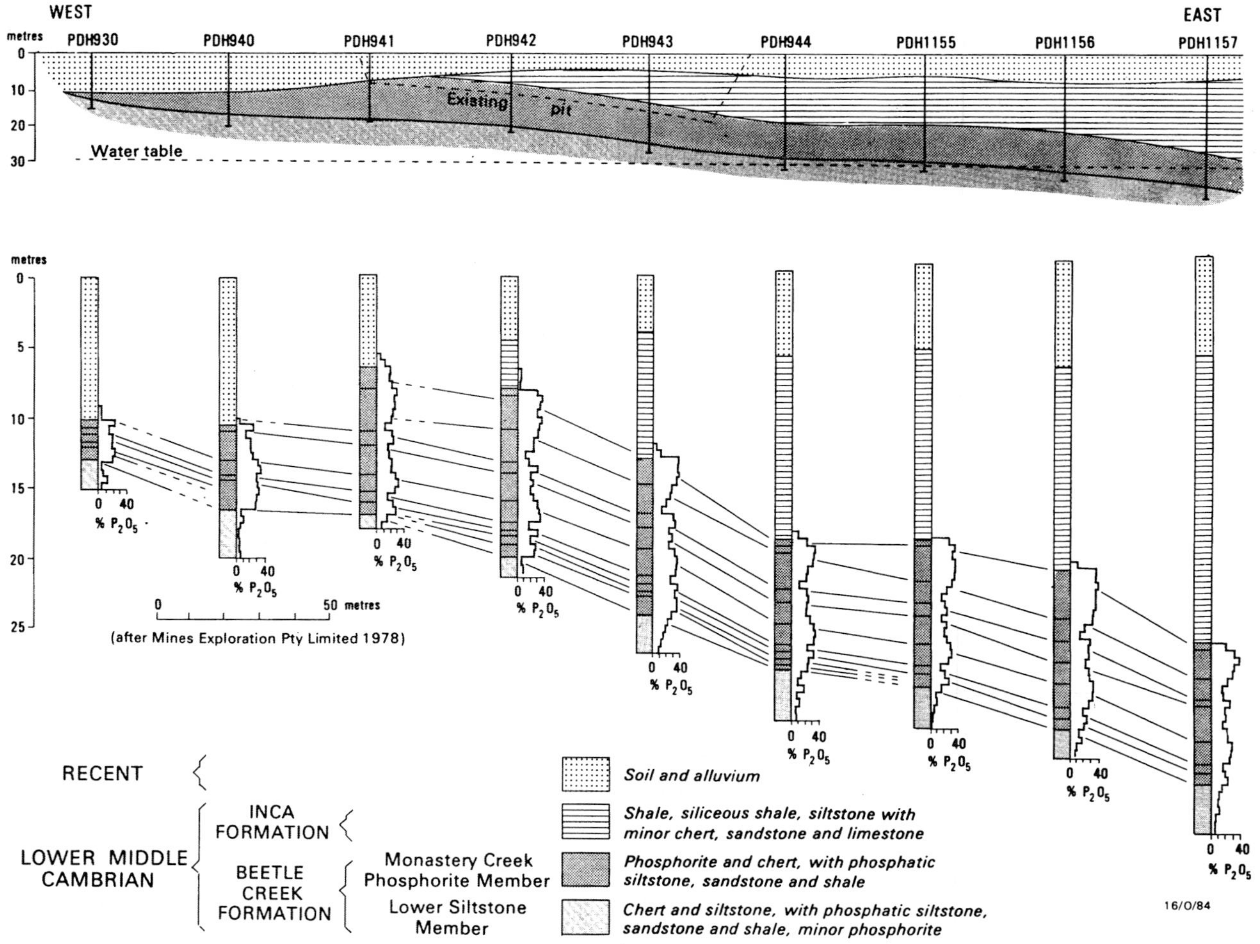

Fig. 3.11. Geological cross-section TI 8 zone, Phosphate Hill, Duchess (after Cook & Shergold, 1979).

1979). The overlying Beetle Creek Formation exhibits a number of cycles of phosphorite deposition, with maximum development in the Monastery Creek Phosphorite Member at the top of the formation. Deposition of phosphogenic sediments concluded with the Inca Formation which consists of a finely banded silt–shale–chert facies grading laterally into a carbonate chert facies.

The features which characterise the Duchess phosphorites include the cyclicity and distinctiveness of the contained lithological units, their lateral continuity and the nature of the phosphorite itself. In the Phosphate Hill mine area there are ten distinctive lithological units in the Monastery Creek Phosphorite Member and these are laterally continuous in fine detail, and are used for the purpose of correlation in the course of pit design and in grade control during mining (Fig. 3.11). The ore is dominantly a dense grainstone phosphorite consisting of phosphate grains in a cherty matrix, or apatite grains in calcareous phosphorite facies. The size and degree of sorting of the grains indicate that they were distributed by current transport, but their texture provides no evidence as to whether they were formed by replacement of carbonate pellets, or the comminution of a precursor phosphorite bed. There are only minor occurrences of thin beds of laminated mudstone phosphorite or microsphorite.

Lady Annie–Lady Jane The Lady Annie/Lady Jane deposit is an outlier about 20 km east of the D Tree deposit which crops out along the eastern margin of the Georgina Basin. The Lady Annie deposit is 25 km long, 5 km wide and the phosphorite has an average thickness of 5.5 m with a maximum of 20 m. The Lady Jane deposit is about 10 km long and has an average thickness of 4 m and a maximum of 12 m. The Lady Annie deposit has been described by Rogers & Keevers (1976) and, most recently, Elgueta (1981) has completed a detailed sedimentological study of the pelletal type of phosphorite occurrence. The deposit is described in this volume by Cook & Elgueta, and here only major differences between it, Duchess and D Tree are discussed.

The stratigraphy at Lady Annie is generally similar to that described at Duchess. However, the basal unit, the Mount Hendry Formation, is distinctly different from the Mount Birnie Beds. It consists of a cliff-shored submarine talus breccia, conglomerate and quartz sandstone fans and sheets, formed in response to reactivation of Early Proterozoic faults. The overlying Thorntonia Limestone generally passes up gradationally into the Beetle Creek Formation. Locally, however, a karst-like surface is filled with white chalky high-grade phosphate which can be seen to replace the underlying limestone and its contained chert nodules down planes or zones. The nature of this carbonate replacement is distinctly different from that at the D Tree deposit described later. There are important differences in the nature of the phosphorites at Lady Annie compared to those at Duchess, and these include a reduction in the number of phosphorite depositional cycles, the short-distance continuity of individual phosphorite beds, the nature of the phosphorite itself, and their depositional environment.

Rogers & Keevers (1976) recognised five significant beds of phosphorite in the deeper western portion of the Lady Annie basin. These aggregate 76.5 m of phosphorite but then interdigitate with phosphatic siltstones, and wedge down to a single 5 m thick bed on the eastern extremity of the deposit. Lateral continuity of a given lens may be less than 600 m. Elgueta (1981) has shown that phosphorites consist of relatively monotonous successions of well-bedded to laminated grainstone and packstone phosphorite interbedded with chert and minor wackestone and mudstone phosphorites. The grains or pellets resemble clastic carbonate grains and consist of varying proportions of rounded and well-sorted fragments (calcisponges and echinoderms), ovulitic grains of uncertain skeletal origin, rounded intraclasts and rare nucleated pellets and ooids (Elgueta, 1981). Previously Rogers & Keevers (1976) and Fleming (1977) estimated that 50% or more of the phosphate grains were phosphatised skeletal fragments and the remainder were structureless micritic apatite. Elgueta (1981) proposed that phosphate allochems were trapped in a semi-protected microtidal tectonic depression and that reworking, transport and winnowing played an initial role in determining grade. The close spatial relationship between phosphorite, carbonate and oncolite facies indicates that the deposit formed in water only a few metres in depth in a subtidal to intertidal environment, as summarised by Cook & Elgueta (Chapter II this volume).

D Tree The D Tree deposit lies on the eastern margin of the Georgina Basin, to the west of the Lady Annie–Lady Jane outlier. The deposit is 17 km long and a maximum of 4 km wide and the phosphorite beds have a maximum aggregate thickness of 15 m, tapering out on the periphery of the basin as well as basinward. The deposit has been described by Howard & Cooney (1976) and Howard & Hough (1979). Its setting in the broader aspects of the regional geology is presented by de Keyser & Cook (1972). Southgate (this volume) describes the phosphate hardgrounds in the Gowers Formation on the western limits of the deposit.

There are important differences in the geology of the D Tree deposits compared with those of Lady Annie even though they are only 20 km apart. The basal talus scree and conglomerate is distributed as fans, one of which can be demonstrated to originate from a reactivated Early Proterozoic fault. In places talus is shed across the palaeoslope of the basin without regard to topography. These faults can be shown to have been active during the deposition of the Thorntonia Limestone and the Beetle Creek Formation, causing slumping and intraformational breccias. Near the end of the deposition of the Beetle Creek Formation, sandstone fans with geographical locations and shapes identical with the earlier talus breccia and conglomerate fans also formed.

Thorntonia Limestone both underlies the Beetle Creek Formation and intertongues as a lateral equivalent (Fig. 3.10 Inset). It onlaps both the conglomerate–talus breccia and Proterozoic basement sediments as does the Beetle Creek Formation. As at Lady Annie, phosphate replaces the Thorntonia Limestone but not in the style of a karst surface. Grainstone and packstone carbonates a metre or more thick, were wholly replaced by phosphate prior to the deposition of cover sediments. Intermittent fault movement caused the replaced limestone to slide and form isoclinal folds against a submarine hillock also prior to the deposition of cover sediments (Howard & Cooney, 1976; Howard & Hough, 1979).

The Beetle Creek Formation and its phosphorite beds were deposited in an environment more closely confined in terms of proximity of shore-line to carbonate banks than is evident at either Duchess or Lady Annie. The phosphorite was deposited in a narrow lagoon and a shelving embayment or estuary which, basinward, was barred by carbonate banks represented by the Thorntonia Limestone. The distribution of underlying and overlying fans of unsorted conglomerates and sands, a similar distribution of detrital base metal minerals in the phosphorites, attest to the proximity of the shore line and its precise location. Phosphate replacement of coarse coquina is evident. Also apparent is subaerial exposure of the phosphorite and its erosion by steep-sided braided stream channels prior to the deposition of cover phosphorite, including replacement coquina and fine laminae of mudstone phosphorite.

Three phosphorite beds occur at D Tree, compared with five at Lady Annie and ten at Duchess. The principal bed has a maximum thickness of 12 m and an average of 6 m. All show sharp lateral and vertical variations in siltstone content. The phosphorite beds are finer-grained than at Lady Annie: grain size is in the range of wackestone to mudstone with well-sorted grainstone phosphorite present as intercalations. Beds show grading and fine upwards into a micritic mudstone phosphorite through a grainstone, boundstone or wackestone in which the grains are commonly angular and poorly sorted. Rounded phosphate grains at the base of a fining sequence, or individual laminae, may show all stages of peripheral micritisation of phosphatised fossil hash through to structureless phosphate grains.

Phosphate replacement of carbonate debris is a major feature of the D Tree deposit. This is interpreted to have occurred in shallow subtidal to intertidal environments. Reworking, transportation and winnowing was a local feature and confined to a depositional basin not significantly larger than the deposit itself (Howard & Cooney, 1976). It is likely that a now-eroded south-east extension of the deposit connected to the subtidal coastal channelway which trended north through to the Lady Annie–Lady Jane area, and that together these channels constituted a small segment of a basin-wide circumferential channelway (Fig. 3.10).

Other deposits Seventy-five kilometres to the south (Fig. 3.10), the Sherrin Creek, Lilly Creek and Engine Creek deposits lie in a southern embayment of the Flora Basin (Howard & Cooney, 1976). Probably originally one deposit, they have been separated by block faulting and subsequent stripping. Sherrin Creek is the largest of these deposits and is approximately 20 km long and up to 7 km wide. Phosphorite is present as a single bed which is underlain by and interfingers basinward with the Thorntonia Limestone. It is overlain by nonphosphatic siltstones and a conglomerate reflecting fault activity in the immediate hinterland. The conglomerate is an equivalent of the sandstone fans which overlie the phosphorites at D Tree.

Between Sherrin Creek and D Tree, wedges of phosphorite are common along the margin of the basin. Inliers of Precambrian basement, considered to be remnants truncated by reactivated Early Proterozoic faults, are also common. The uplifted portions of the beds and their accompanying sequence have been eroded subsequently.

Some of the northern deposits, including the Alexandria deposit, illustrate a lateral relationship of phosphorite interfingering with offshore carbonate banks. The Alexandria deposit consists of a single phosphorite bed 1–6 m thick which is in contact with the Precambrian rocks of an inlier along a post-depositional fault. The deposit, which is incompletely drilled because of its small size and low grade (10–16% P_2O_5), is perhaps 25 km long but only 300–1000 m wide. It shows a lateral facies sequence of phosphorite–minor chert, phosphorite–chert, phosphatic siltstone–minor chert and phosphatic siltstone intertonguing with limestone, dolomitic limestone or dolomite.

In the region along the Alexandria–Wonarah Basement High carbonate banks developed in the centre of troughs between intervening highs or palaeo-islands. At Wonarah two phosphorite beds extend along the flanks of a shallow submarine arch for a distance of 25 km and over widths ranging from 2 km to 6 km. The upper bed is the principal phosphorite and attains a maximum thickness of 19 m in a trough between a volcanic ridge and a carbonate bank (Fig. 3.10 Inset).

References

Abele, C. & McGowran, B. (1959). The goeology of the Cambrian south of Adelaide (Sellick Hill to Yankalilla). *Transactions of the Royal Society of South Australia*, **82**, 301–20.

Blissett, A.H. & Callen, R.A. (1969). Myponga phosphate deposits. *Mineral Resources Review, South Australia*, **127**, 93–102.

Brook, W.A. (1962). The geology of the phosphate deposits of the Orroroo district, Report to C.R.A. Ltd, Dept. Mines Env. 265 (unpublished).

Callen, R.A. (1970a). Oliver's phosphate deposits. *Mineral Resources Review, South Australia*, **128**, 127–50.

Callen, R.A. (1970b). Phosphate deposits in the Myponga Area. *Mineral Resources Review, South Australia*, **129**, 42–53.

Callen, R.A. (1971a). Sedimentary Phosphate Exploration Part 1 – The Cambrian of the Flinders Ranges. *Mineral Resources Review, South Australia*, **130**, 9–34.

Callen, R.A. (1971b). Sedimentary phosphate exploration, Part 2 – Mt Lofty Ranges and Yorke Peninsula. *Mineral Resources Review, South Australia*, **130**, 80–94.

Callen, R.A. (1971c). Wait's Noarlunga phosphate deposit workings, section 40 and 48, hundred of Noarlunga. *Mineral Resources Review, South Australia*, **130**, 110–20.

Cook, P.J. (1976). Georgina Basin phosphatic province, Queensland and Northern Territory – regional geology. In *Economic Geology of Australia and Papua New Guinea, 4. Industrial Minerals and Rocks*, ed. C.L. Knight, pp. 245–50. Monograph 8. Australasian Institute of Mining and Metallurgy, Melbourne.

Cook, P.J. (1982). The Cambrian palaeogeography of Australia and opportunities for petroleum exploration. *APEA Journal*, **22**, 42–64.

Cook, P.J. (1984). Spatial and temporal controls on the formation of phosphate deposits – a review. In *Phosphate Minerals*, ed. J.O. Nriagu & P.B. Moore, pp. 242–74. Springer-Verlag, Berlin.

Cook, P. & McElhinny, M.W. (1979). A re-evaluation of the spatial and temporal distribution of sedimentary phosphate deposits in the light of plate tectonics. *Economic Geology*, **74**, 315–30.

Cook, P.J. & Shergold, J.H. (1979). The Field Workshop. In *Proterozoic–Cambrian Phosphorites*. ed. P.J. Cook & J.H. Shergold, pp. 1–17. ANU Press, Canberra.

Crick, I.H. & Muir, M.D. (1980). Evaporites and uranium mineralisation in the Pine Creek Geosyncline. In *Uranium in the Pine Creek Geosyncline*. ed. J. Ferguson & A.B. Goleby, pp. 531–42. International Atomic Energy Agency, Vienna.

Daily, B. (1956). The Cambrian in South Australia. In *El Sistema Cámbrico, su Paleogeografía y el Problema de su Base; 20th Session International Geological Congress, Mexico*, vol. 2, pp. 91–147.

Daily, B. (1963). The fossiliferous Cambrian succession on Fleurieu Peninsula, South Australia. *Records of the South Australian Museum*, **14**, 579–601.

Daily, B. (1976). The Cambrian of the Flinders Ranges. In *Late Precambrian and Cambrian Geology of the Adelaide 'Geosyncline' and Stuart Shelf, South Australia*, ed. B.P. Thompson, B. Daily, R.P. Coats & B.G. Forbes, pp. 15–9. *25th International Geological Congress, Sydney, Excursion Guide 33A*.

Dalgarno, C.R. & Johnson, J.E. (1964). Wilpena Group (new name). *Quarterly Geology Notes, South Australian Geological Survey*, **9**, 12–5.

de Keyser, F. (1968). The Cambrian of the Burke River Outlier. *Record. Bureau of Mineral Resources, Geology and Geophysics*, **1968/67.**

de Keyser, F. (1969). On the genesis of the Georgina Basin phosphorites, northwest Queensland. *Record, Bureau of Mineral Resources, Geology and Geophysics*, **1969/79.**

de Keyser, F. (1973). A review of the Middle Cambrian stratigraphy in the Queensland portion of the Georgina Basin. *Bulletin, Bureau of Mineral Resources, Geology and Geophysics*, **139**, 13–28.

de Keyser, F. & Cook, P.J. (1972). The geology of the Middle Cambrian phosphorites and associated sediments of north-west Queensland. *Bulletin, Bureau of Mineral Resources, Geology and Geophysics*, **138**, 79pp.

Driessen, A. & Cook, P.J. (1981). The current status and long term outlook for Australia's phosphate resources. Presented at the Annual General Meeting of ISMA Ltd, Singapore, May 1981, 15pp.

Elgueta, S. (1981). Sedimentological study of the western zone of the Lady Annie phosphate deposit, Queensland, Australia, MSc. Thesis, Australian National University (unpublished).

Ewers, W.E. & Morris, R.C. (1981). Studies of the Dales Gorge Member of the Brockman Iron Formation, Western Australia. *Economic Geology*, **76**, 1929–53.

Fleming, P.J.G. (1977). Faunas, lithologies, and the origin of phosphorites in parts of the Middle Cambrian Beetle Creek Formation of northwest Queensland. *Queensland Geological Survey Publication*, **364**, 21pp.

Flint, D.J. (1978). Deep sea fan sedimentation of the Kanmantoo Group, Kangaroo Island. *Transactions of the Royal Society of South Australia*, **102**, 203–22.

Goode, A.D.T. (1981). Proterozoic geology of Western Australia. In *Precambrian of the Southern Hemisphere*, ed. D.R. Hunter, pp. 105–203. Elsevier, Amsterdam.

Henderson, R.A. & Southgate, P.N. (1978). Middle Cambrian stratigraphy and sedimentology of the eastern Georgina Basin. *3rd Australian Geological Convention Townsville, Excursions Handbook*, 85–110.

Horwitz, R.C., Thompson, B.P. & Webb, B.P. (1959). The Cambrian–Precambrian boundary in the eastern Mt Lofty Ranges Region: South Australia. *Transactions of the Royal Society of South Australia*, **82**, 205–18.

Howard, P.F. (1971). Discovery of phosphorite in northern Australia. *Transactions of the American Institute of Mining Engineers*, **250**, 269–73.

Howard, P.F. (1972). Exploration for phosphorite in Australia – a case history. *Economic Geology*, **67**, 1180–92.

Howard, P.F. & Cooney, A. (1976) D Tree phosphate deposit, Georgina Basin, Queensland, In *Economic Geology of Australia and Papua New Guinea, 4. Industrial Minerals and Rocks*, ed. C.L. Knight, pp. 273–8. Monograph 8, Australasian Institute of Mining and Metallurgy, Melbourne.

Howard, P.F. & Hough, M.J. (1979). On the geochemistry and origin of the D Tree, Wonarah, and Sherrin Creek phosphorite deposits of the Georgina Basin, northern Australia. *Economic Geology*, **74**, 260–84.

Howard, P.F. & Perrino, F.A., (1976). Wonarah phosphate deposits, Georgina Basin, Northern Territory. In *Economic Geology of Australia and Papua New Guinea, 4. Industrial Minerals and Rocks*, ed. C.L. Knight, pp. 273–7. Monograph 8, Australasian Institute of Mining and Metallurgy, Melbourne.

Jack, R.L. (1919). The phosphate deposits of South Australia. *Bulletin of the Geological Survey of South Australia*, **7**, 133pp.

Johns, R.K. (1962). South Australian rock phosphate deposits. *Mineral Resources Review, South Australia*, **114**, 22–30.

Johns, R.K. (1967). Cambrian phosphorite, St John's. *Quarterly Geology Notes, Geological Survey of South Australia*, **21**, 3–5.

Johns, R.K. (1969). Diamond drilling in Fairview phosphate deposit. *Quarterly Geology Notes, Geological Survey of South Australia*, **31**, 8–11.

Johns, R.K. (1976). Phosphate – South Australia. In *Economic Geology of Australia and Papua New Guinea, 4. Economic Minerals and Rocks* ed. C.L. Knight, pp. 282–85. Monograph 8, Australasian Institute of Mining and Metallurgy, Melbourne.

Öpik, A.A. & Pritchard, P.W. (1960). Cambrian and Ordovician. In *The Geology of Queensland*, ed. D. Hill & A.K. Denmead.*Journal of the Geological Society of Australia*, **7**, 89–114.

Parkin, L.W. (Ed.) (1969). *Handbook of South Australian Geology*. Geological Survey of South Australia, Government Printer, Adelaide.

Perrino, F.A. (1970). Progress Report on Prospecting Authorities 1766, 1897, 2159, 2160, 2161 and 2199 in the Alexandria Region. International Metals and Chemicals Development Corporation (unpublished report).

Plumb, K.A. (1979). The tectonic evolution of Australia. *Earth Science Reviews*, **14**, 205–49.

Plumb, K.A., Derrick, G.M., Needham, R.S. & Shaw, R.D. (1981). The Proterozoic of Northern Australia. In (Ed), *Precambrian of the Southern Hemisphere*, ed. D.R. Hunter, pp. 205–307. Elsevier, Amsterdam.

Preiss, W.V. (1973). Palaeoecological interpretations of South Australian Precambrian stromatolites. *Journal of the Geological Society of Australia*, **19**, 501–32.

Preiss, W.V. & Kinsman, J.E. (1978). Stratigraphy and palaeoenvironmental interpretation of the Brighton Limestone south of Adelaide and its equivalents in the Orroroo region. *Report of Investigations, Geological Survey of South Australia*, **49**, 34pp.

Preiss, W.V., Walter, M.R., Coates, R.P. & Wells, A.T. (1978). Lithological correlations of Adelaidean glaciogenic rocks in parts of the Amadeus, Ngalia and Georgina Basins. *BMR Journal of Australian Geology & Geophysics*, **3**, 43–53.

Pritchard, P.W., Barrie, J., Jauncey, W. & Fricker, A.G. (1963). Rum Jungle Phosphate Survey 1962/63. *Record, Bureau of Mineral Resources, Geology and Geophysics* **1963/73**.

Pritchard, P.W. & Cook, P.J. (1965). Phosphate deposits in the Northern Territory. In *Geology of Australian Ore Deposits*, ed. J. McAndrew, pp. 219–20. Eighth Commonwealth Mining and Metallurgical Congress, Melbourne.

Rogers, J.K. & Crase, N.J. (1980). The Phosphate Hill rock phosphate deposit, North-west Queensland, Australia–an outline of geological development. In *Fertilizer Mineral Potential in Asia and the Pacific*, ed. R.P. Sheldon & W.C. Burnett, pp. 307–30. East–West Resources Systems Institute, Honolulu.

Rogers, J.K. & Keevers, R.E. (1976). Lady Annie–Lady Jane phosphate deposits, Georgina Basin, Queensland. In *Economic Geology of Australia and Papua New Guinea, 4. Industrial Minerals and Rocks*, ed. C.L. Knight, pp. 251–65. Monograph 8. Australasian Institute of Mining and Metallurgy, Melbourne.

Russell, R.T. & Trueman, N.A. (1971). The geology of the Duchess phosphate deposits, northwestern Queensland, Australia. *Economic Geology*, **66**, 1186–214.

Rutland, R.W.R. (1981). Structural framework of the Australian Precambrian. In *Precambrian of the Southern Hemisphere*. ed. D.R. Hunter, pp. 1–32. Elsevier, Amsterdam.

Rutland, R.W.R., Parker, A.J., Pitt, G.M., Preiss, W.V. & Murrell, B. (1981). The Precambrian of South Australia. In *Developments in Precambrian Geology 2. Precambrian of the South Hemisphere*, ed. D.R. Hunter, pp. 309–60. Elsevier, Amsterdam.

Schopf, J.W. & Fairchild, T.R. (1973). Late Precambrian microfossils: a new stromatolitic biota from Boorthanna, South Australia. *Nature*, **242**, 537–538.

Shergold, J.H. & Druce, E.C. (1980). Upper Proterozoic and Lower Palaeozoic rocks of the Georgina Basin. In *The geology and geophysics of northeastern Australia*, ed. R.A. Henderson & P.J. Stephenson, pp. 149–74. Geological Society of Australia, Queensland Division, Brisbane.

Smith, K.G. (1972). Stratigraphy of the Georgina Basin. *Bulletin, Bureau of Mineral Resources, Geology and Geophysics*, **111**, 156pp.

Southgate, P.N. (1979). Middle Cambrian hypersaline environments along the eastern margin of the Georgina Basin. In *Proterozoic–Cambrian Phosphorites*, ed. P.J. Cook & J.H. Shergold, p. 28. ANU Press, Canberra.

Sprigg, R.C. (1952). Sedimentation in the Adelaide Geosyncline and the formation of the continental terrace. In *Sir Douglas Mawson Anniversary Volume*, ed M.F. Glaessner & R.C. Sprigg, pp. 153–9. University of Adelaide.

Stanton, R.L. (1976). Petrochemical studies of the ore environment at Broken Hill, NSW 1 – Constitution of 'banded iron formation'. *Transactions of the Institution of Mining and Metallurgy, Section B, Applied Earth Sciences*, **85**, B33–46.

Stevens, B.P.J. (Ed) (1980). A guide to the stratigraphy and mineralisation of the Broken Hill Block, NSW. *Records of the Geological Survey of New South Wales*, **20(1)**, 153pp.

Thomson, B.P. (1969). Precambrian basement cover – the Adelaide System. In *Handbook of South Australian Geology*, ed. L.W. Parkin, pp. 49–83. South Australian Geological Survey, Government Printer, Adelaide.

Thomson, B.P., Coats, R.P., Mirams, R.C., Forbes, B.G., Dalgarno, C.R. & Johnson, J.E. (1964). Precambrian rock groups in the Adelaide Geosyncline: a new subdivision. *Quarterly Geological Notes, Geological Survey of South Australia*, **9**, 1–19.

Thomson, B.P., Daily, B., Coats, R.P. & Forbes, B.G. (1976). Late Precambrian and Cambrian geology of the Adelaide 'Geosyncline' and Stuart Shelf, South Australia. *25th Session International Geological Congress, Sydney, Excursion Guide, 33a*, 65pp.

Thomson, L.D. & Russell, R.T. (1971). Discovery, exploration and investigations of phosphate deposits in Queensland. *Proceedings of the Australasian Institute of Mining & Metallurgy*, **240**, 1–14.

Trendall, A.F. & Blockley, J.G. (1970). The iron formations of the Precambrian Hamersley Group, Western Australia. *Geological Survey of Western Australia, Bulletin*, **119**, 366pp.

Uppill, R.K. (1979). Stratigraphy and depositional environments of the Mundallio Subgroup (new name) in the late Precambrian Burra Group of the Mt Lofty and Flinders Ranges. *Transactions of the Royal Society of South Australia*, **103(1–2)**, 25–43.

Von der Borch, C.C. (1980). Evolution of Late Proterozoic to Early Paleozoic Adelaide Foldbelt, Australia: comparisons with post-Permian rifts and passive margins. *Tectonophysics*, **70**, 115–34.

Wells, A.T., Forman, D.J., Ranford, L.C. & Cook, P.J. (1970). Geology of the Amadeus basin, central Australia. *Bulletin, Bureau of Mineral Resources, Geology and Geophysics of Australia*, **100**, 222pp.

4

Proterozoic and Cambrian phosphorites – regional review: China

LI YUEYAN

Introduction

Marine sedimentary phosphorite deposits occupy a pre-eminent position among the hundreds of known economic phosphate deposits in the world. This is particularly true in China where, throughout geological history, there have been several important global phosphogenic periods. Phosphorite deposits formed in each period have a specific temporal and spatial distribution. In China, Late Proterozoic–Early Cambrian sedimentary phosphorites are mainly distributed in two areas: in south China on the Yangzi Paraplatform and in the north and northwest of China within the Tarim and Sino-Korean Paraplatforms.

In south China (i.e. the Yangzi area), economic phosphorite deposits mainly occur in Guizhou, Yunnan, Sichuan, Hubei, Hunan and Jiangxi Provinces. This area is an important part of the global Late Proterozoic–Cambrian phosphogenic belt on the northern margin of Gondwanaland. It is also a principal area for the present-day and future production of phosphate. In the Yangzi phosphogenic area, the sedimentary sequence from the Doushantuo and Dengying Formations of the Upper Sinian Series to the Lower Cambrian Series is continuous. Although phosphorites are found throughout the sequence from the Doushantuo Formation to the Middle Cambrian Damao Formation, horizons of economic value are confined to the Doushantuo (700–650 m.y.) and Meishucun (615–612 m.y.) Formations.

While phosphorites were being deposited on a large scale in the Yangzi area, northern China was a broad arch. It was not until the middle part of the Early Cambrian (corresponding to the Canglangpu Stage in the Yangzi area, with the fossils *Pteroredlichia chinensis, Redlichia mansuyi, R. major, R. yunnanensis, Conoredlichia conica*, and *Megapalaeolenus deprati*) that phosphatic sediments were deposited on the margins of the Sino-Korean and Tarim Paraplatforms. Although quite extensive areas of phosphatic sedimentation have been noted on these paraplatforms, no phosphorite deposits of economic value have so far been found.

Late Sinian phosphorites of the Doushantuo Formation

The Upper Sinian Series occurs in the drainage basin of the Yangzi River. The well-known Kaiyang, Weng'an, Jingxiang, Yichang, Dongshanfeng and Chaoyang phosphorite deposits are concentrated in Guizhou, Hubei, Hunan and Jiangxi. However, in Sichuan, Shaanxi and other Provinces phosphorites of this stage are of only secondary importance. The division of the Sinian Series in the Yangzi area is shown in Table 4.1.

Phosphorites occur in the Doushantuo Formation, which unconformably overlies the pre-Sinian Shennongjia Group (in Hubei) or disconformably overlies the Early Sinian Nantuo tillites (in the broad area outside Hubei), and grades upward into the Dengying Formation. The probable reserves of phosphorite of the Doushantuo Formation amount to about one-third of the total phosphate reserves of China, and constitute

Table 4.1. *Subdivision of the Sinian System on the Yangzi Paraplatform*

	Age	Yangzi Paraplatform
	Early Cambrian	Meishucun Formation*
		615 ± 20 m.y.
	Late	Dengying Formation
		650 ± 20 m.y.
		Doushantuo Formation*
Sinian		700 ± 30 m.y.
	Early	Nantuo Formation (T)
		740 ± 30 m.y.
		Liantuo Formation (T)
		800 ± 30 m.y.
Pre-Sinian		

(T) Tillite.
*Major phosphorite.

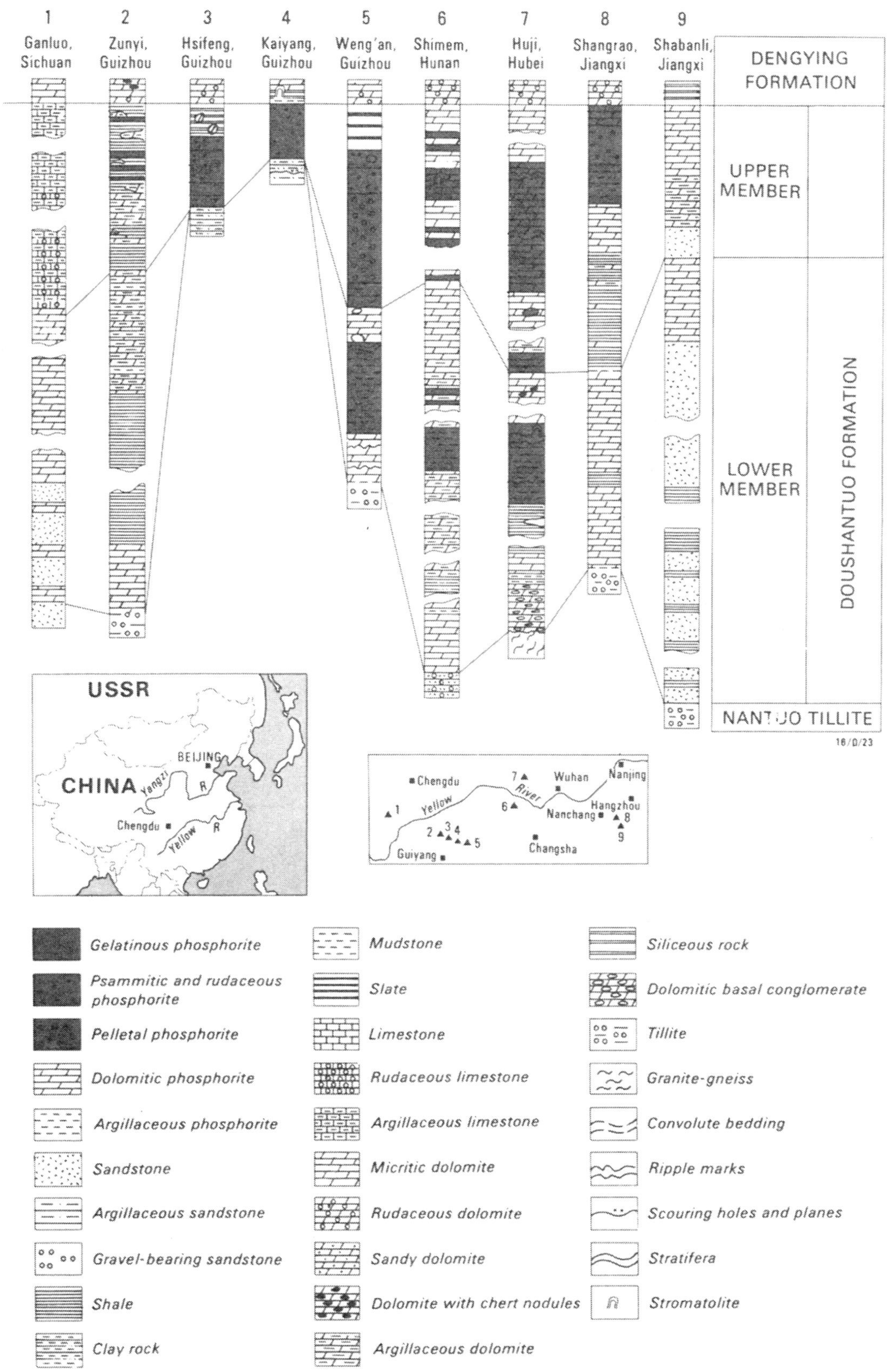

Fig. 4.1 Stratigraphic comparison of the Doushantuo Formation in the Yangzi area (scale not available).

more than two-thirds of the total high-grade phosphorite reserves. The Doushantuo Formation is, therefore, the single most important phosphatic horizon in the country. However, that part of the formation containing economic phosphorite beds is largely restricted to the pinch-out zone of the Lower Sinian Series.

The Doushantuo Formation is the oldest phosphatic sedimentary sequence in southern China. It is composed of phos-

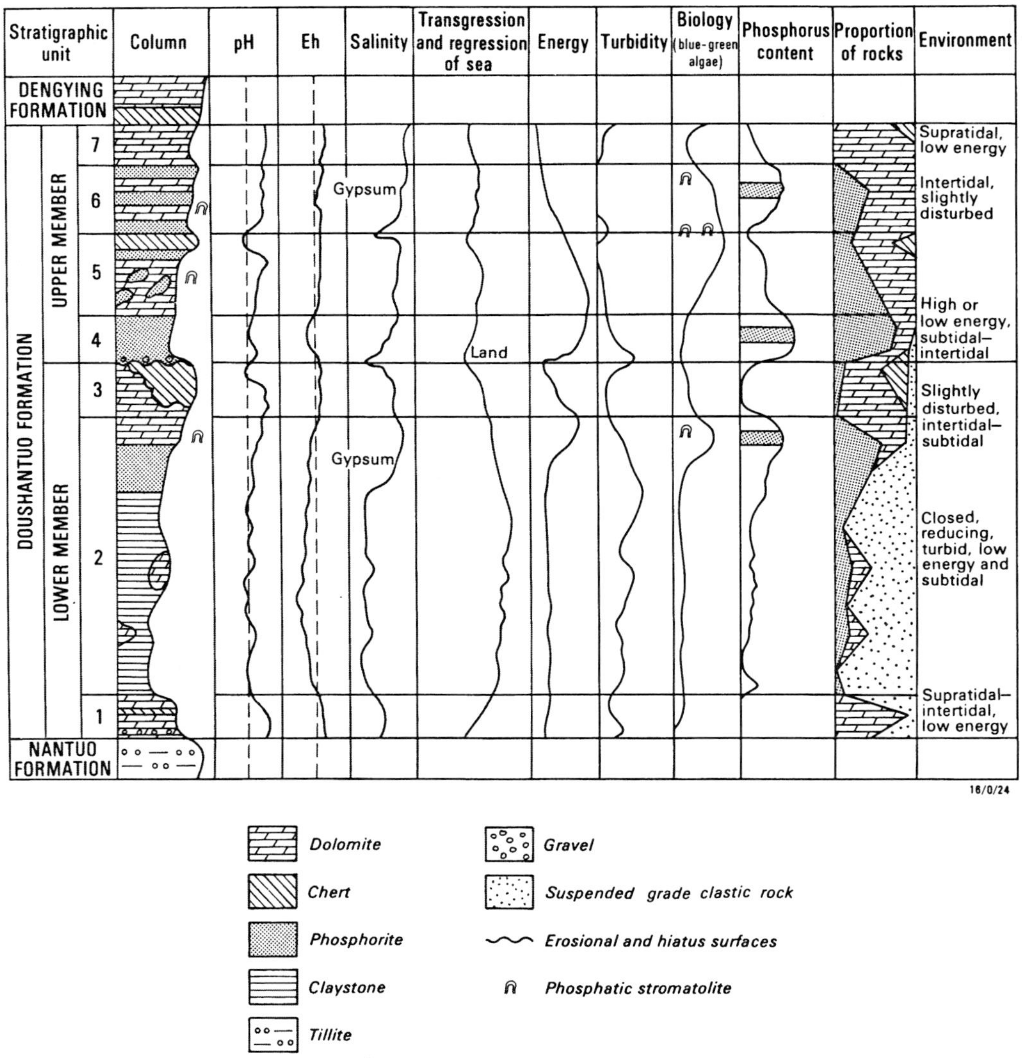

Fig. 4.2 Summary of the characteristics of the Doushantuo Formation.

phorites and fine-grained clastic and carbonate rocks with total thickness varying between 0 m and 250 m. The formation shows cyclic sedimentation, deduced from variation in composition, sedimentary structures and textures, phosphatic content and biota, related to changes in relative sea level and fluctuating geochemical and geophysical conditions subsequent to the Nantuo glacial event. The Doushantuo Formation contains an abundance of microfossils; those found in the stratotype section in the Sanmenxia area indicate a Late Proterozoic age. The formation varies greatly in thickness and relative ratio of phosphatic to other sediments (Fig. 4.1). It is divided into Lower and Upper Members comprising seven beds. These members may represent two discontinuous sub-cycles. The lithological sequence is summarised in Fig. 4.2. The stratigraphic sequence described below is in ascending order. Phosphorite characteristics are shown in Fig. 4.3.

Fig. 4.3 (A) Rudaceous phosphorite, crossed nicols, × 10. (B) Calcareous psammitic-silty phosphorite, crossed-nicols, × 40. (C) Sand-size fragment composed of microcrystalline carbonate fluorapatite, grain size 0.2–2 μm, electron microscope, × 6000. (D) Calcareous oolitic phosphorite, plane polarised light, × 16. (E) Oolitic crust composed of carbonate-phosphate (from lower left to upper right), scanning electron microscope, × 3000. (F) Radiating ooids composed of fibrous-foliated carbonate fluorapatite, grain size 4 μm, electron microscope, × 6000. (G) Irregular clasts, the radiating grains composed of concentrically arranged fibrous-foliated carbonate fluorapatite, grain size 7 μm, electron microscope, × 5000. (H) Calcareous and argillaceous grainstone phosphorite, plane polarised light, × 40. (I) Bioclastic texture (micritic limestone with phosphatic hyolithes), plane polarised light, × 10. (J) Bioclastic phosphorite, hand specimen. (K) Gelatinous and cracked phosphorite, plane polarised light, × 20. (L) Dolomite-matrix spherulitic phosphorite; phosphatic spherulites in micritic dolomite-matrix, crossed nicols, × 20. (M) Compound phosphatic grains, interstitially filled by micritic dolomite and other impurities, crossed nicols, × 20. (N) Crystals of phosphate cement between phosphatic grains, electron microscope, × 2750. (O) Tidal bedding in a psammitic phosphorite with a dolomite-matrix about 10 times reduced. (P) Herring bone cross-bedding of dolomite-matrix psammitic phosphorite, about 10 times reduced. (Q) Interference ripple marks of mud-matrix psammitic phosphorite with a mud-matrix. (R) Convolute bedding in a banded dolomitic phosphorite. (A–J) collected from the Lower Cambrian Series, and (K–R) collected from the Sinian Doushantuo Formation.

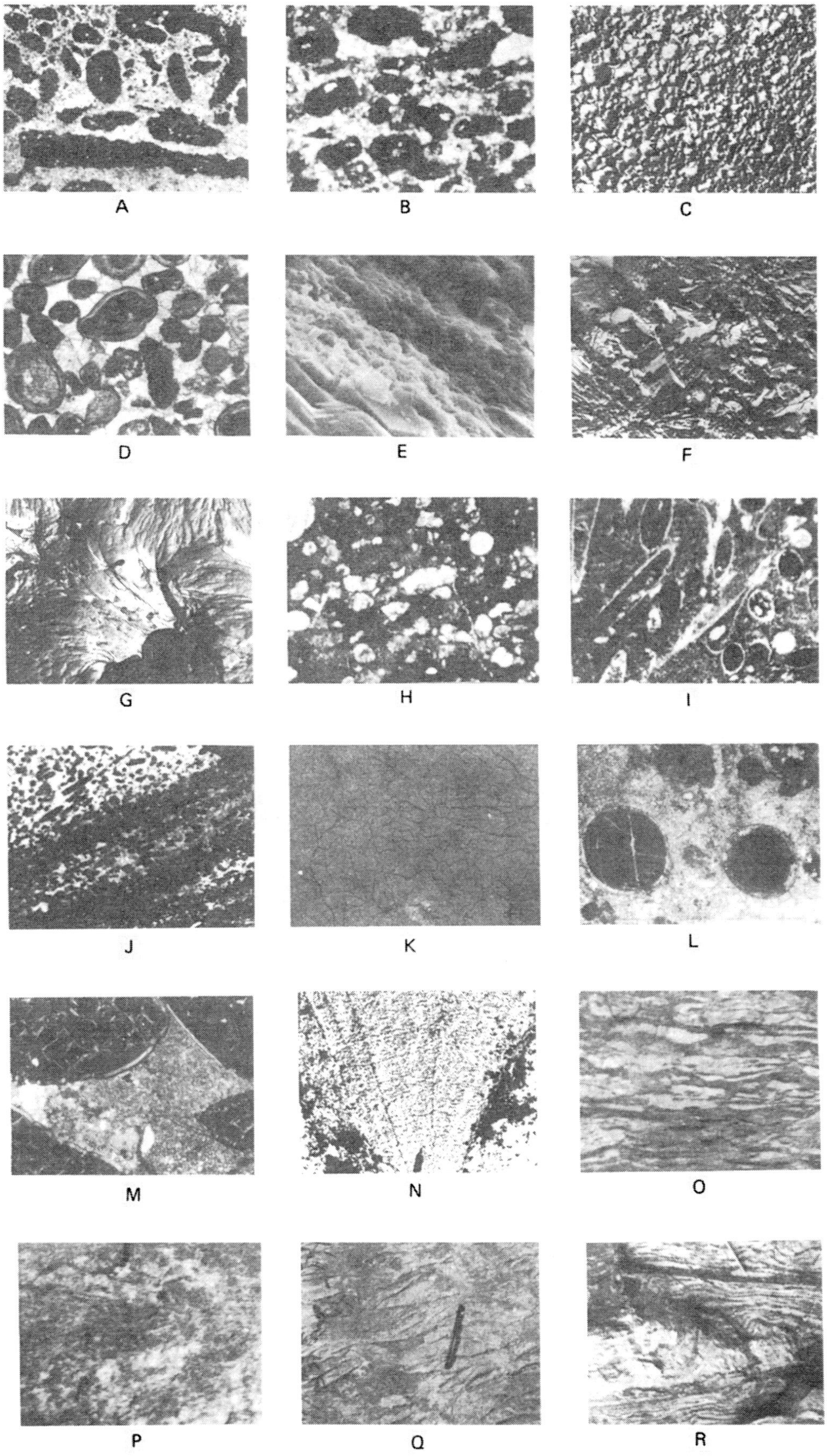
A
B
C
D
E
F
G
H
I
J
K
L
M
N
O
P
Q
R

Lower Member of the Doushantuo Formation

Bed 1. This is 0–20 m thick, averaging 3.5 m. It is a reddish-grey, medium to thick and massive-bedded manganiferous micritic dolomite. Horizontal bedding is developed; mud cracks and ripple marks are seen occasionally. The upper part of the bed is more argillaceous or clayey, while the lower part contains gravel-size clasts from the underlying rocks, forming a dolomitic basal conglomerate or sandy dolomite and dolomitic sandstone. Bed 1 is persistent in lithology and is considered a marker bed for the lower limit of the Doushantuo Formation. It is represented by lag deposits of intertidal–supratidal origin formed during a period which was unfavourable for phosphogenesis. There is a well-defined disconformity or unconformity at the base of Bed 1 where the Doushantuo Formation overlies the Lower Sinian tillites of the Nantuo Formation or metamorphosed pre-Sinian rocks.

Bed 2. This bed is 0–310 m thick, and is one of the major phosphorite horizons of Guizhou and Hubei. It consists of black claystone or black grainstone phosphorite with an argillaceous and dolomitic matrix, mudstone phosphorite, phosphatic claystone and phosphatic argillaceous dolomite. Claystone is interbedded with argillaceous-carbonate phosphorite in varying proportions. Pyrite is common. Generally, the lower part of the bed is dominated by argillite with a relatively high potassium content (3–10% K_2O), often intercalated with thin beds or lenses of argillaceous dolomite and marl, locally with more terrigenous debris. Generally bedding is horizontal and laminations of about 1 mm are common. The upper part of the bed is slightly more calcareous and phosphatic and locally may be intercalated with thin beds of gypsum. Wavy bedding, lenticular bedding, ripple marks, minor tabular cross-bedding, scour-and-fill structures and mud cracks are developed. In some sections, phosphatic columnar and stratiform stromatolites are found, forming reefs in Hubei and Guizhou. The sedimentary environment for Bed 2 varies greatly from place to place. In general, clastic sedimentation has occurred under restricted, reducing, low-energy and turbid conditions grading upward into a tidal flat facies of argillaceous carbonate and phosphorite.

Bed 3. 1–45 m thick, this bed is a greyish-white, thick-bedded and massive micritic dolomite. Many phosphatic veinlets (derived from the overlying phosphorite) and chert nodules occur at the top of this sequence. Horizontal or wavy bedding, and rare ripple marks and mud cracks are found. At the base of the unit there are phosphatic sandy fragments, and fine breccias derived from the underlying rocks (including the phosphorite). Convolute bedding is often observed. The upper boundary is wavy or karstic. Bed 3 is persistent both in thickness and lithology and is the product of an intermittent period of phosphate sedimentation and can be regarded as a marker bed for dividing the Doushantuo Formation. Bed 3 is mostly represented by clean low-energy intertidal to supratidal carbonates. The shallow water zone in which these sediments were deposited was subject to subsequent strong submarine scouring or local uplift and was sub-aerially exposed and subjected to karstification.

Upper Member of the Doushantuo Formation

This member consists mainly of chemical sediments such as carbonates, chert and phosphorites, together with some clastic sediments. From two to four phosphorite beds are recognised in this member as follows.

Bed 4. This is a sequence, 1–12 m thick, of dark, thin-to medium-bedded grainstone phosphorite, locally with phosphatic sandy shale. The grains are composed of phosphatic sand- and gravel-sized intraclasts. At the base of the unit, argillaceous materials and glauconite are rather abundant, and fragments and clasts from the underlying rocks are common. Pyrite is present locally. The content of phosphatic and dolomitic materials increases at the top at the unit. Sedimentary structures of the intertidal zone such as horizontal bedding, wavy bedding, intertidal ripple marks and scour-and-fill structures are common.Bed 4 consists mainly of residual clastic and intraclastic sediments of a low-energy intertidal zone, together with some sediments derived from shallow banks and sand bars of the high-energy subtidal zone. Reworking of phosphate grains within the basin is of primary importance in the formation of phosphorites. The thickness and grade of phosphorite beds varies greatly depending on the environment. Unit 4 is a very widespread phosphorite and is one of the major economic horizons of central Guizhou. At the base of Unit 4 there is a major erosional break, with karsting in places. This break is underlain by the Lower Member of the Doushantuo Formation which is dominated by clastic rocks, though it does contain a single phosphorite bed. It ranges from 0 m to 320 m in thickness and can be subdivided into three characteristic sequences (Beds 1 to 3, above).

Bed 5. This is a 14–40 m thick, grey or greyish-black, medium- to thick-bedded phosphatic micritic dolomite with various kinds of phosphatic stromatolite. Phosphatic materials show a gel-like texture or an intra-clastic texture of various grain sizes. Littoral sedimentary structures such as cross-bedding, ripple marks, and scour-and-fill and flutes are well developed. Erosional breaks, marked by chert and gravels, are noted locally at the top of the sequence. The content of P_2O_5 varies with the amount of dolomite present. In Hunan and Guizhou this interval comprises relatively thick but low-grade economic phosphorites but elsewhere the sequence is phosphatic but not economic. Bed 5 was deposited under high-energy, subtidal to intertidal conditions.

Bed 6. This is a 2–20 m thick, grey–dark grey, thin- to medium-bedded dolomite, and silty mudstone with interbedded mudstone phosphorite, intraclastic grainstone phosphorite and dolomite. Each boundary between the dolomite and phosphatic bands is marked by a microscopic plane of separation. Under the microscope, scattered clasts of dolomite are found on the undulatory to irregular erosion surfaces. Locally, there is a decrease in the dolomite content and Bed 6 is represented by greyish-white massive mudstone, grainstone and psammitic phosphorite, but there are still coarse- to fine-grained cycles associated with the erosion planes. These compositional and grading cycles are often fixed in number in given areas, the cycles ranging in frequency from 10 to more than 100 per centimetre. Flaser bedding, wavy bedding, lenticular bedding, herringbone cross-bedding, flat-top ripple marks and scouring planes of various sizes are well developed. In some places there are concentrations of phosphatic stromatolites forming conspicuous reefs in areas exposed to waves, as exemplified by those in Longdonggou–Yingping. Guizhou, and Zoumaping-Fengxiangpo, Hunan Provinces. The intertidal nature of Bed 6 is indicated by its composition, frequent variation of geophysical and geochemical conditions and high algal productivity. Bed 6 is a persistent phosphorite horizon in which economic phosphorite 'beds are found in virtually the whole Yangzi area except in a few parts of Hubei Province.

Bed 7. This bed ranges from 0 m to 50 m in thickness and is represented by dark, thin to medium-thick micritic dolomite, argillaceous dolomite and argillite. Striped or nodular networks of siliceous rocks are common, as are mud cracks and drapes and horizontal–wavy laminations. This bed represents inter-supratidal mud-flat sedimentation.

Sedimentary relationships within the Doushantuo Formation

The relationships between phosphorites and other sediments of the Doushantuo Formation, and the sedimentary cycles are described below.

Phosphorites or phosphatic sediments always occur at the beginning and end of every sub-cycle. Sediments formed at the beginning of the sub-cycle (e.g. Bed 2 of the Lower Member and Bed 4 of the Upper Member) are mostly associated with surfaces of erosion or a hiatus, and mainly show a granular texture. They are phosphatic intraclastic sediments formed under relatively unstable conditions. The accumulation of phosphorite is mainly controlled by geological conditions and the physical characteristics of the water column, provided sufficient phosphatic clasts and the conditions necessary for sediment accumulation exist. Deposition and concentration of phosphate to form phosphorites is not controlled by the geochemical environment. For example, in low-energy, turbid and acidic environments, phosphatic clasts and suspended nonphosphatic clasts were deposited simultaneously to form medium-grade, argillaceous, sandy and rudaceous phosphorites or phosphatic claystone. In a low-energy, clean and alkaline environment, phosphates were precipitated together with carbonates to form medium- to low-grade, sandy and rudaceous dolomitic phosphorites or were combined with gel-like phosphate to produce high-grade, sandy rudaceous phosphorites with a phosphatic matrix. In high-energy environments winnowing produces thick, high-grade phosphorites with grain-supported textures or dolomites with phosphate cements. Sediments formed at the end of the sub-cycle (e.g. the top portion of Bed 2 of the Lower Member and Bed 5 and Bed 6 of the Upper Member) are mostly conformable and have gradational contacts with the surrounding rocks and mainly exhibit a gel-like texture. They represent quiet-water chemical sediments, the formation of which is controlled by both geochemical and geophysical conditions. High-grade basal phosphorites may form in zones of algal mats or on back-reef shallow banks. Some are deposited as alternating bands of phosphate and dolomite in slightly disturbed platform lows and on tidal flats as banded, striped, interbedded and concretionary lower-grade phosphorites. Others are deposited together with suspended non-phosphatic clasts as low-grade argillaceous and gel-like phosphorites in low-energy turbid restricted zones. Both laterally and vertically, phosphorite is the transitional facies between clastic and chemical sediments. Phosphorites are associated with siliceous rocks and with dolomite in particular.

Lithofacies distributions

Sediments of the Doushantuo Formation occurring on the Yangzi Paraplatform can be clearly divided into two lithofacies: (1) a basin margin clastic rock facies; and (2) a basinal chemical lithofacies (Fig. 4.4).

Inner basin margin clastic lithofacies Restricted to the shoreward basin margin, this facies consists of terrigenous clastic and pyroclastic rocks deposited on the slopes and in the depressions of the back margin of a palaeo-island chain. Minor amounts of carbonate are occasionally present. Although no phosphorites are found in this facies, weak and sporadic phosphatisation occurs locally.

As shown in Fig. 4.4, the Region (I_1) represents an inner basin-margin clastic and carbonate subfacies, consisting of quartz sandstone and feldspar–quartz sandstone, intercalated with pyroclastic and carbonate rocks. Although this subfacies is generally thin, it varies greatly in thickness. The subfacies represents littoral sedimentation on the margins of a palaeo-island or old land area.

The Region (I_2), as shown in Fig. 4.4, represents a back-basin depression-molasse subfacies. This subfacies consists of low-grade metamorphosed gravelly feldspar–quartz sandstone, quartz sandstone and intercalated pyroclastic rocks, with a thickness of up to 1000 m.

Intrabasin chemical sediment lithofacies This facies is distributed within the area of sedimentation of the Doushantuo Formation and consists of carbonate rock, phosphorite, siliceous rock and claystone. The distribution of the shallow platform subfacies (II_1 in Fig. 4.4) is controlled by the location of submarine highs. The subfacies consists mainly of carbonate rock, phosphorite and siliceous rock. Varying amounts of claystone are found locally in the lower part of the subfacies which is light coloured, contains abundant phosphate intraclasts, phosphatic stromatolites and glauconite and is characterised by sedimentary structures consistent with a littoral environment of deposition. High contents of Br, Sr and I, also occur. This subfacies is in sharp contact with the adjacent shallow basin facies. It represents shallow-water platform-type carbonate-phosphate sedimentation on submarine high lands which have steep edges and flat tops. The wave-agitated zone is the most important area for the deposition of phosphorites in this instance.

The relatively depressed zone between the submarine highs, subfacies II_2 in Fig. 4.4, is largely composed of black claystone with only minor amounts of black carbonaceous clayey carbonate in the middle and upper parts. Stromatolites and sedimentary structures indicative of a shoreline are absent in this subfacies which is relatively persistent in both lithology and thickness. Notable amounts of pyrite and organic matter as well as terrigenous clasts and heavy minerals are present.

The subfacies represents suspended clastic sediments deposited in a relatively restricted and turbid environment below

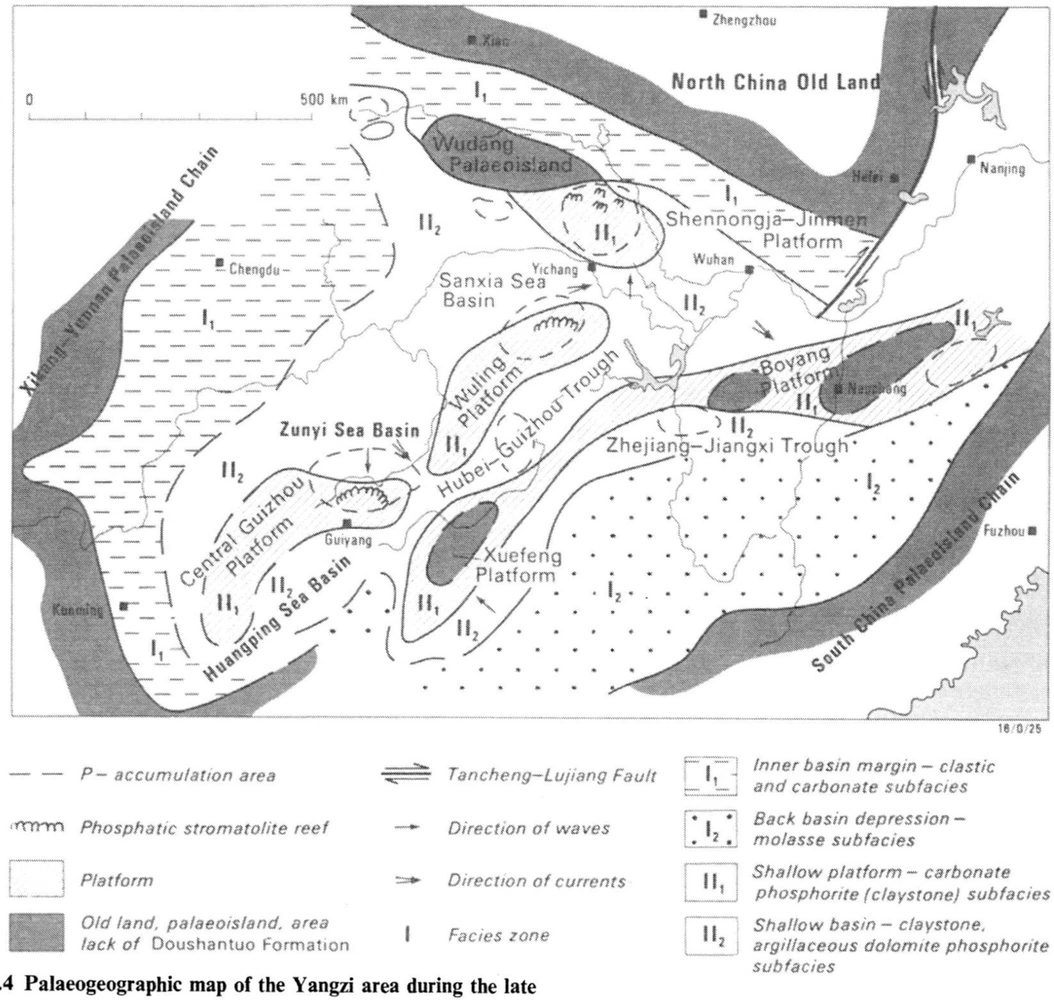

Fig. 4.4 Palaeogeographic map of the Yangzi area during the late Sinian.

Table 4.2 *Sedimentary characteristics of the major facies of the Doushantuo Formation, Yangzi Paraplatform*

Facies zone	Petrographic characteristics
II$_1^1$ Platform slope phosphatic dolomite and claystone facies	1. Thickness of this facies varies greatly. It is markedly thinner than the flanking facies. 2. Mainly represented by phosphatic intraclastic dolomite and claystone. Talus deposits are common. 3. Convolute bedding, high-angle oblique slump bedding, as well as other slumping and scouring sedimentary structures are well developed. 4. Mainly shows a micritic texture, but irregular lenses of intrabasin clastic sediments ranging in grain size from sand to gravel are fairly common. 5. Rich in glauconite, but mostly oxidised, showing evidence of repeated winnowing and reworking within the basin.
II$_1^2$ Platform reefs stromatolitic phosphorite and dolomite facies	1. Biological (blue-green algae) sedimentary structures are developed. Stromatolitic phosphorite (with phosphatic stromatolite as a 'skeleton') is predominant. 2. Phosphorite mainly shows sandy and pebbly rudaceous textures. Gel-like texture indicated by oncolite concretions, and other algal debris is of secondary importance. 3. The phosphorite beds are thick, high grade, light coloured and high in I, Sr, Ba, and Fe_2O_3. 4. Within the 'reef blocks' there are few scour features but littoral sedimentary structures are well developed in the phosphorites and associated sediments at the top and bottom.

Table 4.2 *(Contd.)*

Facies zone	Petrographic characteristics
II_1^3 Platform bank-grainstone phosphorite, dolomite and chert facies	1. Ratio of phosphorite to dolomite is quite high. 2. Various sedimentary structures characteristic of the intertidal zone are well developed, and hiatuses and erosional surfaces are abundant. 3. Grainstone textures, some grain-supported and others cemented, are predominant. The grain size distribution mainly is an asymmetrical probability distribution curve with several populations evident. In most cases, two or more populations are recognisable. 4. Phosphorites are rich in glauconite.
II_1^4 Platform basin-claystone, phosphorite and dolomite facies	1. The thickest of the shallow platform facies, with a well defined and continuous sedimentary sequence. 2. The ratio of claystone to argillaceous dolomite is high. 3. Phosphorites are very thick, multi-layered, low-grade, with essentially no iodine, and mostly intercalated with non-phosphatic rocks to form rhythmic bands. 4. Phosphorites show, in the main, gel-like textures, or plastic intraclastic textures with a clayey carbonaceous matrix and abundant pyrite. Granular textures showing abnormal distributions in keeping with autochthonous sediments. 5. Littoral sedimentary structures and algal-sedimentary structures are rare, but rhythmic bedding is pronounced. 6. Phosphorite and other sediments of this facies are dark coloured, rich in carbonaceous matter and pyrite, and locally containing barite, gypsum and celestine.
II_1^5 Platform flat-clayey dolomite, claystone and phosphorite facies	1. Complex sedimentary association, dominated by claystone and clayey dolomite. 2. A thin and impersistent unit. 3. Terrigenous and residual footwall clasts are abundant. 4. Mud cracks and mud partings, and sedimentary structures created by storm waves are well developed. 5. Ferric iron and manganese contents are fairly high. Phosphorites and associated sediments are red, purple and brown in colour. 6. Phosphorites are rare. Only poor-quality sandy and pebbly phosphorite occurs in the near shore sand flat sediments. The phosphorites have a variable phosphate content depending on the relative abundance of phosphate, clastic grains and clayey matrix.
II_2^1 Basin margin black shale, clayey dolomite and gel-like phosphorite facies	1. Thickness is relatively uniform (50–150 m). 2. Black shale is predominant, with minor amounts of clayey dolomite. 3. Contains abundant pyrite and organic carbon (more than 2%). 4. Thin and multi-layered phosphorites occur in the middle and upper parts of the facies where dolomite is present. 5. The phosphorites which contain many impurities have a gel-like texture.
II_1^2 Basin-black shale and argillaceous dolomite facies	With the same characteristics as above, but phosphorites completely absent.

II_1 Shallow platform-carbonate rock and phosphorite (claystone) sub-facies
II_2 Shallow basin-claystone, clayey dolomite and phosphorite sub-facies

wave base and the redox interface. Conditions favourable for phosphate sedimentation in this facies only existed in the middle and upper parts of the Doushantuo Formation at the margins of the basin. But thin-bedded and low-grade phosphorite beds were formed locally.

The main characteristics of the various zones of the intrabasin–chemical sedimentary rock facies are summarised in Table 4.2. The wave-agitated zone of the platform reef facies, the shallow bank facies and the platform basin facies constitute the major zones for accumulating high-grade and thickly bedded phosphorites. The well-known Kaiyang. Weng'an Dongshanfeng, Yichang, Jingxiang and Chaoyang phosphorite deposits are all found in these facies. The amount of phosphate accumulated in them accounts for more than 90% of the total amount of phosphate deposited in the shallow Yangzi sea. The shallow basin marginal facies adjacent to the platform (or near an estuary or the old land mass) has yielded only medium to small phosphorite deposits in northern and eastern Guizhou, northern Sichuan, southern Shaanxi and central Hunan.

Most of the facies zones on the marginal slope and tidal flats of the shallow basins did not experience conditions favourable for phosphate deposition. Although they may have occurred, phosphorites were not preserved owing to subsequent erosion.

Palaeogeography of the Doushantuo Formation

During Doushantuo deposition the Yangzi area was covered by a shallow epicontinental sea, between the southern margin of the palaeo-Asian sea and the northern margin of Gondwanaland. Preliminary palaeomagnetic data indicate that the area was situated at a palaeo-latitude of 20° S to 30° S. A topographic high divided this shallow sea into an east–west series of troughs, basins and platforms (Fig. 4.4).

(a) In the southern section of the western series, the central Guizhou Platform formed a nose-shaped north-east-trending submarine high that was steep to the north and gentle to the south; high in the west and low in the east. The location of this high was controlled by the central Guizhou–Wuling fault blocks. Away from the margins the high was relatively flat, except at the eastern end of the northern side where there was a series of NNE- and NNW-trending swales. Among the obvious features were the depressions constituting the platform-basin at the eastern end, and topographic highs on the south-western margin of the platform-basin, as manifested by the local absence of the Doushantuo Formation and by a steep bank built up by the sand flat facies.

During early Doushantuo time the sea only flooded the eastern end of the depression. The central part was relatively restricted and characterised by anoxic water. Grainstone phosphorite with a black mud matrix of the Lower Member was deposited in a low-energy reducing environment. The north-western part of the depression was affected by relatively strong waves. There, algal mats form mounds and the Lower Member consists mainly of dolomite with phosphatic stromatolitic patch reefs and lenses of nodular and pebbly phosphorite.

During deposition of the middle and upper parts of the Doushantuo Formation the sea flooded the whole platform except for some elevated areas. Algal flats were developed on the northern and eastern margins. Near the open sea and in estuaries both areas were affected by strong waves.

Shallow bank facies and platform reef facies with variable energy conditions, as well as thick and high-grade grainstone phosphorite (pebbly in places), stromatolitic phosphorite and dolomite were deposited. In the southern part of the central Guizhou Platform in the intertidal-supratidal zones, variegated claystone and clayey dolomite were mainly deposited, associated with lag deposits. The north-eastern margin of the central Guizhou Platform controlled the location of the well-known phosphorite accumulation area north of Guiyang in central Guizhou (Fig. 4.4).

(b) In the middle section of the western series occurs the Wuling Platform (Fig. 4.4). The location of this platform was governed by the fold- and fault-induced central Guizhou–Wuling High which separated the Sanxia and Zunyi Basins from the Hubei–Guizhou Trough. This trough linked up with two deeply incised channels, lying between the Shennongjia–Jingmen and central Guizhou Platforms. The Wuling Platform as a whole was comparatively low, with a subdued topography. Its southern margin was bounded by a fault. To the north, in the Dongshanfeng area, there was a relatively low north-north-east platform. There the platform-basin facies consists of thick black shale, black clayey carbonate and a minor amount of low-grade phosphorite which was deposited early in Doushantuo time. During late Doushantuo time, Dongshanfeng gradually became a shallow bank with algal mats. There, the Upper Member of the Doushantuo Formation consists of rhythmically interbedded stromatolitic phosphorite and organic-rich clayey dolomite. This sequence constitutes the north-western Hunan phosphorite accumulation area which contains relatively thick medium-grade phosphorites.

The remaining areas formed either tidal flats or marginal slopes away from the platform. Throughout the whole of Doushantuo time in the Dayong–Taoyuan area of western Hunan there appears to have been a marginal slope away from a shallow platform which was controlled by a fault. Here a sequence of sandy and pebbly dolomite, which includes talus deposits, large convolute bedding, and high-angle oblique bedding developed. This sequence represents a typical marginal slope facies.

(c) The Shennongjia–Jingmen Platform of the northern section of the Yangzi area represents a submarine extension of the Wudang Palaeo-Island chains. The island chain was high in the west and low in the east and formed a NWW-trending submarine slope. The northern part of the platform was relatively gentle and flat, while the southern part was a series of escarpments resulting from a set of earlier faults.

During the early part of Doushantuo sedimentation the platform margin was represented by tidal flats where blue-green algal mounds were developed. The southern side of the margin represented a fault-bounded tidal-flat or platform slope. During subsequent Doushantuo time, the sedimentary environments on both the northern and southern sides of the platform became almost identical. Taking an axis through Shennongjia and Jingmen, tidal flat, shallow bank and platform slope environments were arranged successively from south to north. It is in this area that the important western Hubei phosphorite deposits are located.

(d) The Xuefeng–Boyang Platform on the eastern side of the Yangzi region was represented by a series of long shallow platforms inherited from the Xuefeng–Boyang zone of folding and faulting which lies between the Hubei–Guizhou Trough and the South China Depression. Internally, the platform was undulating, with topographic culminations in south-eastern Guizhou and north-eastern Jiangxi, and two depressions in western Hunan and western Jiangxi. Carbonaceous mudstone similar to that of the basin facies was deposited in these depressions. The Doushantuo Formation is absent either because it was never deposited or because it was removed by erosion. Thin-bedded, but persistent, claystone, clayey dolomite and other rocks of the low-energy tidal facies were deposited in most of the remaining areas. Frequently, only the lower part of the Doushantuo Formation was deposited. Low-grade phosphorite was deposited locally in channel locations, e.g. in the central Hunan phosphorite accumulation area on the southern side of the eastern extremity of the Xuefeng Platform, and in the north-eastern Jiangxi–western Zhejiang phosphorite accumulation area on the eastern side of the southern part of the Boyang Platform. During the Caledonian orogenic period all these platforms were uplifted and formed land, referred to as the 'Jiangnan Old Land'.

The marine basins and troughs between the platform mostly represented relatively low-lying, quiet low-energy offshore environments where suspended clastic sediments were predominant. Thin-bedded and low-grade phosphorites were deposited only locally on the margins of shallow platform basins under suitable conditions. Deposits of this type include the small and medium phosphorite deposits of western Hunan and eastern Guizhou.

Thus it can be seen that during Doushantuo time, the palaeogeographic environment and ancient geological setting were the major factors determining the sedimentary facies. Phosphorites mostly accumulated in the wave-agitated zone of the margins of the shallow platforms flanking broad seas and channels. The outer margins of the platforms, e.g. on the north-western margin of the central Guizhou–Wuling Platform and on the south-eastern margin of the Xuefeng–Wuling Platform were particularly favourable locations. Depending on their

Table 4.3. *Summary of depositional characteristics of phosphorites in the Doushantuo Formation*

Characteristics	Type of deposit: Shallow marine platform	Shallow marine basin
Known phosphorite accumulation region	Central Guizhou, north-west Hunan, west Hubei and north-east Jiangxi–west Zhejiang	North and central Guizhou, west and central Hunan, north Sichuan and south Shaanxi
Size of deposit	Large to giant economic deposit	Mostly small deposits
Thickness of the Doushantuo Formation	Extremely variable in thickness (0–520 m)	Relatively persistent (50–150 m)
Rock–phosphorite association of the Doushantuo Formation	Phosphorite, dolomite, chert and claystone	Black shale, argillaceous dolomite and phosphorite
Economic phosphorite horizon	Upper Member (major) and Lower Member (local)	Top (major) and middle part (secondary) of Doushantuo Formation
Thickness of economic phosphorite bed	Persistent, thick-bedded, several to tens of metres	Thin lenses of one to several metres thick
Grade of economic phosphorite bed	More than 25% P_2O_5 in general	Mostly less than 25% P_2O_5
Composition of phosphorite	Monophosphate; dolomitic and argillaceous in part	Primarily carbonaceous and argillaceous, secondarily dolomitic in part
Texture of phosphorite	Primarily grainstone, secondarily gelatinous with cracks	Gelatinous texture with no cracks but many impurities
Structure of phosphorite	Laminated, striped, concretionary and massive	Laminated, thin-bedded, and concretionary
Sedimentary structures	Shoreline sedimentary structures developed	
Biological relics	Reef and patch reef built up by phosphatic stromatolites	
Facies associated authigenous mineral	Glauconite, barite and gypsum (celestine)	Abundant pyrite
Trace elements	High Sr/Ba, Mn, Cu and B contents and rich in I	Total organic carbon more than 2%

palaeogeographic settings, the deposition of phosphorite was in general more intense on the western side of the platform than on the eastern side.

Characteristics of Doushantuo phosphorite deposits

According to analytical results, the Doushantuo phosphorites are primarily carbonate–fluorapatite. They are composed predominantly of cryptocrystalline collophane with fibrous, bedded-fibrous and columnar-granular apatites (depending on the degree of crystallisation), forming up to 5% of the total phosphate. All of the crystalline apatites formed by syn-sedimentary or diagenetic recrystallisation of collophane. A summary of depositional characteristics of the Doushantuo phosphorites is given in Table 4.3.

Early Cambrian phosphorites of the Meishucun Formation

The Early Cambrian Meishucun Stage is another important phosphogenic period following that of the Doushantuo Formation. The resources of phosphorite formed in the Meishucun Stage account for about two-thirds of the total

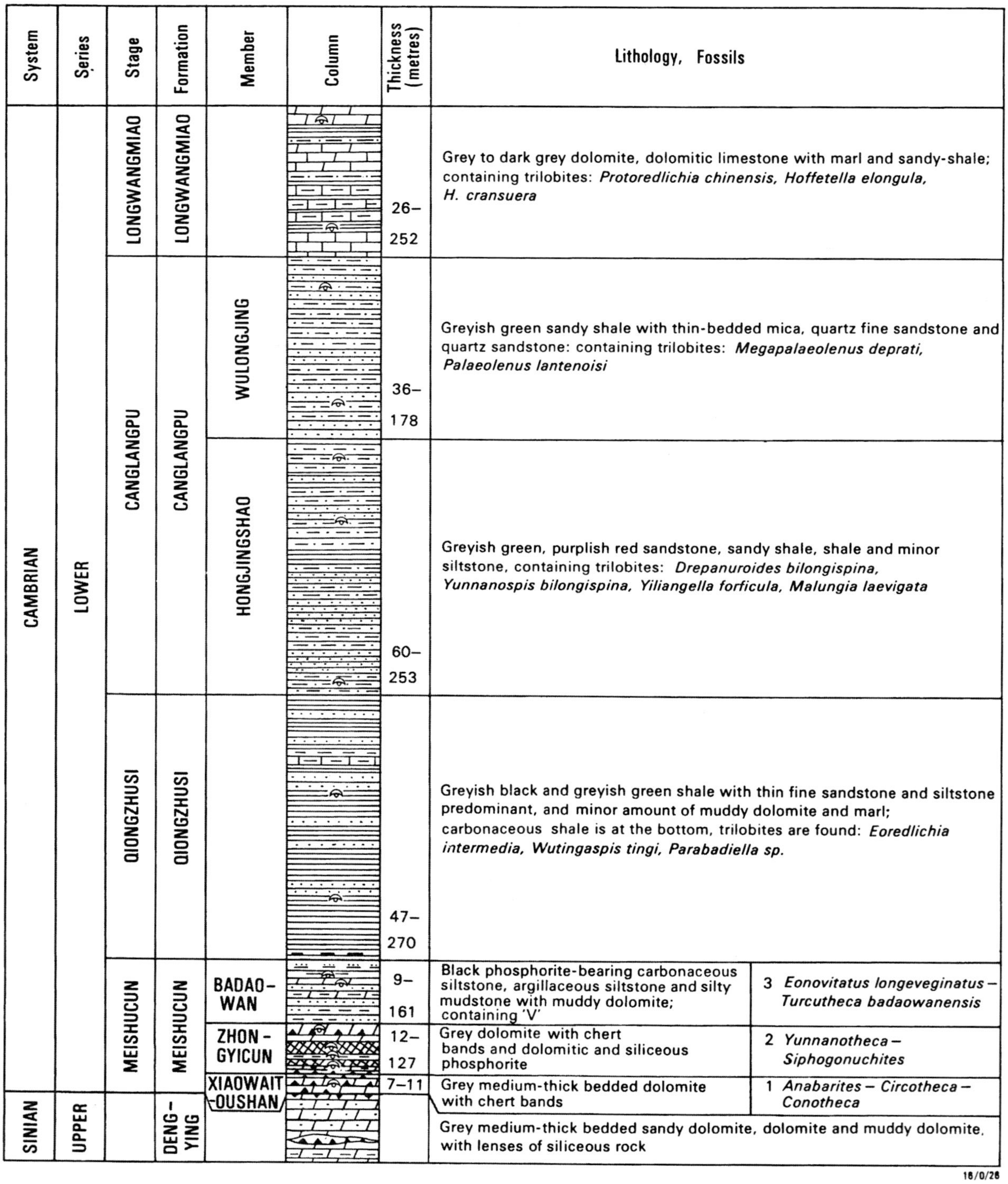

Fig. 4.5 Stratigraphy of the Lower Cambrian in eastern Yunnan Province.

phosphate resource of China. Deposition of the Meishucun Formation occurred in essentially the same area as that of the Doushantuo Formation. Economic phosphorite deposits mainly occur in eastern Yunnan, western Guizhou, southern Shaanxi and at the western margin of the Sichuan Basin of the western Yangzi area. Small phosphorite deposits also occur sporadically in the eastern Yangzi area at a stratigraphically higher level than those in the western Yangzi area.

The Lower Cambrian is mainly developed in the south-western Yangzi area, and is divided from bottom to top into the Meishucun, Qiongzhusi, Canglangpu and Longwangmiao Formations. A typical stratigraphic column for eastern Yunnan is shown in Fig. 4.5.

The Meishucun Formation is subdivided into three members. In ascending order these members are the Xiaowaitoushan, Zhongyicun and Badaowan Members. The Zhongyicun

Table 4.4 *Correlation of the Early Cambrian biostratigraphy of eastern Yunnan with the Siberian Platform*

Eastern Yunnan (China)					Siberia (USSR)		
Qiongzhusi Stage	Qiongzhusi Formation	Lower Member	*Eoredlichia–Wutingaspis* Zone *Parabadiella* Zone		Atdabanian Stage	*Profallotaspis* Zone	
Meishucun Stage	Meishucun Formation	Badaowan Member	3. *Eonovitatus lonevaginatus–Turcutheca badaowannensis* Assemblage		Tommotian Stage	3. *Dokidocyathus lenaicus–Majatheca tumefacta* Zone	
		Zhongyicun Member	2. *Yunnanotheca–Siphogonuchites* Assemblage	2. *Carinochites–Siphogonuchites* Sub-assemblage		2. *Dokidocyathus regularis* Zone	1. *Lapworthella bella* Sub-zone
				1. *Yunnanotheca–Quadrotheca* Sub-assemblage			2. *Lapworthella tortuosa* Sub-zone
				4. *Circotheca–Anabarites* Sub-assemblage		1. *Aldanocyathus sunnaginicus–Tiksitheca licis* Zone	
		Xiaowaitoushan Member	1. *Anabarites–Circotheca–Conotheca* Assemblage	3. *Pupoella–Barbitositheca* Sub-assemblage	Yudomian Formation	Sabelliditids *Anabarites trisulcatus* *Protohertzina* sp.	
				2. *Circotheca–Conotheca* Sub-assemblage			
				1. *Anabarites primitivus* Sub-assemblage			
Upper Sinian Series	Dengying Formation	Baiyanshao Member	Fossil algae–microplaflora Assemblage				

Member is the major phosphorite horizon, but the Badaowan Member is also important. The isotopic age of the Formation is 615 m.y. The upper limit of the Meishucun Formation is marked by the lower surface of the Qiongzhusi Formation which contains trilobites (e.g. *Eoredlichia* and *Parabadiella*) and Ostracoda (*Meishucungolla*). The base of the Meishucun Formation is marked by the lowest horizon of the Xiaowaitoushan Member which contains the oldest primitive hyoliths (*Anabarites primitivus*). Taking the typical section of eastern Yunnan as an example, the Meishucun Formation is made up of the following sequence (from top to bottom).

Xiaowaitoushan Member: This member is 7–11 m thick. It consists of grey medium- to thick-bedded sandy dolomite with black chert bands and lenses. The fossils *Anabarites, Turcutheca, Circotheca, Hyolithellus, Cassidina and Artimycta* occur.

Zhongyicun Member: This member is 13–127 m thick. Its upper part is characterised by grey thick-bedded phosphatic and manganiferous sandy dolomite with cherty bands or beds. The fossils *Palaeosulcachites, Solenotia, Sachites, Parasachites, Chancelloria* and *Mobergella* occur in this interval. Phosphorites occur in the middle part of the member and are intercalated with a 1.6 m thick, grey, shaly, phosphatic, sandy claystone containing the fossils *Pupoella, Circotheca, Barbitositheca, Hyolithellus, Cassidina, Kunyangotheca, Turcutheca, Oralitheca,* and *Anabarites*. The lower part of the member is composed of rhythmic units of dark grey thin-bedded chert argillaceous dolomite and siliceous silstone, intercalated with black chert and phosphorite.

Badaowan Member: This member is 9–161 m thick. The upper part consists of grey to dark grey, thin- to medium-bedded, argillaceous siltstone and dolomitic siltstone, with intercalated phosphatic silty dolomite. The lower part is composed of black thin- to medium-bedded, phosphatic, argillaceous quartz siltstone. This sequence is underlain by a 0.4-m thick interval of glauconite-bearing concretionary phosphorite and shale. The fossils *Circotheca, Allatheca, Chancelloria, Allonnia* and *Camena* are present.

Detailed biostratigraphy and correlation between the Meishucun Stage in China and its equivalents in the USSR is shown in Table 4.4. At present, small shelly fossils have been reportedly found in quite a number of Early Cambrian 'non-trilobite zones' throughout the world. It is apparent from comparison between the assemblages of small shelly fossils that the Tommotian Stage of Siberia in the USSR corresponds with the Meishucun Stage of China. The sequence below the trilobite zone of the Welsh Cambrian system is equivalent to the middle and upper parts of the Meishucun Formation.

Phosphatic formations and associated sedimentary sequences

The Early Cambrian phosphogenic cycle on the Yangzi Paraplatform coincides with the appearance of small shelly invertebrates associated with a large-scale transgression in the upper part of the Dengying Formation. Siliceous, carbonate phosphorite formation occurred during the early Meishucun Stage, in the Zhongyicun and Xiaowaitoushan Members.

The terrigenous clastic rock phosphorite assemblage was formed during the deposition of the Badaowan Member, late in the Meishucun Stage.

The Early Meishucun siliceous carbonate phosphorite facies consists of dolomite, dolomitic limestone, black chert and phosphorite. The facies can be divided into three successive subunits. The lower subunit contains abundant silica or chert with low carbonate content. It is dark-coloured with an organic carbon content of 0.27%. Pyrite is present. Horizontal lamination occurs and wavy bedding is developed in places. Phosphorite is common, but generally low-grade due to silty intraclasts and rhythmic intercalations. Fossils are rare. This subunit represents a low-energy subtidal environment. The middle phosphorite subunit contains abundant phosphate and carbonate grains in the lower part. Shallow-water sedimentary structures such as cross-bedding, ripple marks, wavy bedding, ripple laminations, graded bedding, flat pebble conglomerates and syndepositional erosional planes are common. The middle subunit is light-coloured due to a low organic carbon content, but contains much shelly debris. Its upper part is dominated by light-coloured micritic dolomite intercalated with a few thin beds or lenses of chert. Algal laminations, fenestrae and horizontal micro-bedding are present. Although the upper subunit is similar to the middle one, phosphorites commonly occur as extremely irregular lenses, cavities and fenestrae. The content of organic carbon is very low (0.08%), and micritic algal sandy limestones occur locally.

Laterally the siliceous carbonate phosphorite can be divided into five subunits:

(1) The Yunnan–Sichuan subunit accumulated on the tidal flats of a bay on the eastern side of the western palaeo-island chain. The base of the subunit is topographically quite varied and cut by major north–south faults and east-west structural rises, resulting in a number of restricted basins, and three different vertical rock associations of dolomite-phosphorite, chert, limestone-dolomite, and terrigenous clastic-carbonate rock. The principal features of the subunit include the dominance of carbonate rock, rock composition and texture, clearly developed sedimentary cycles, moderate thickness of accumulation (100–200 m) and slow sedimentation rate. The Yunnan–Sichuan subunit is the most persistent and economically important.

(2) The central Guizhou subunit accumulated away from sources of terrigenous sediment in areas controlled by ancient submarine topographic highs. It is composed of chert, dolomite, minor terrigenous quartz grains, and phosphorite. The sediments are characterised by thin accumulations (less than 20 m), poorly developed cycles and the presence of intraclasts. This subunit is of secondary importance as a phosphatic formation.

(3) The southern Shaanxi–northern Sichuan facies accumulated adjacent to the Qinling geosyncline. Its distribution is controlled by the Longmenshan Fault and Ningqiang Fault in south Qinling. The base of the facies has relatively marked topography and the sedimentary environment is variable, consisting of black phosphatic limestone, sandy limestone, dolomitic limestone, limy dolomite, siliceous limestone, sandstone, siltstone, siliceous claystone and carbonaceous shale. The cycles are complex and highly variable. This subunit is more

than 2000 m thick, low in P_2O_5 content and is persistent.

(4) The central Sichuan–western Hubei subunit is restricted in distribution by ancient submarine highs. This thin subunit has a relatively flat base: it is composed of limestone and dolomite, and contains no phosphorite.

(5) The eastern Yangzi subunit extends through eastern Guizhou, central Hunan, north-eastern Jiangxi and northern Zhejiang. Its south-eastern area adjoins the strongly active zone of the South China geosyncline and its base is poorly defined. The subunit represents a structural basin with a long history of slow subsidence. It is composed of black chert, minor limestone and shale and is characterised by poorly defined sedimentary cycles up to 10 m thick and a low P_2O_5 content.

The late Meishucun terrigenous clastic rock phosphorite facies was developed when the western palaeo-island chain was emergent and the old land provided a major source of clastic materials, producing clayey and silty clastic rocks rich in organic matter, associated with 'stone coal' and polymetallic sulfides. The facies normally varies between 50 m and 150 m thick, but may reach a maximum of 480 m in the Longmenshan Depression. The P_2O_5 content is generally low where phosphorites were formed locally.

Palaeogeography of Meishucun sedimentary facies

Palaeogeography of Zhongyicun Member The Zhongyicun Member is the most important phosphatic horizon of the Meishucun Formation. Its palaeogeography reflects to a large extent that of the Doushantuo Formation, with a shallow marine basin which deepened to the east (Fig. 4.6). The

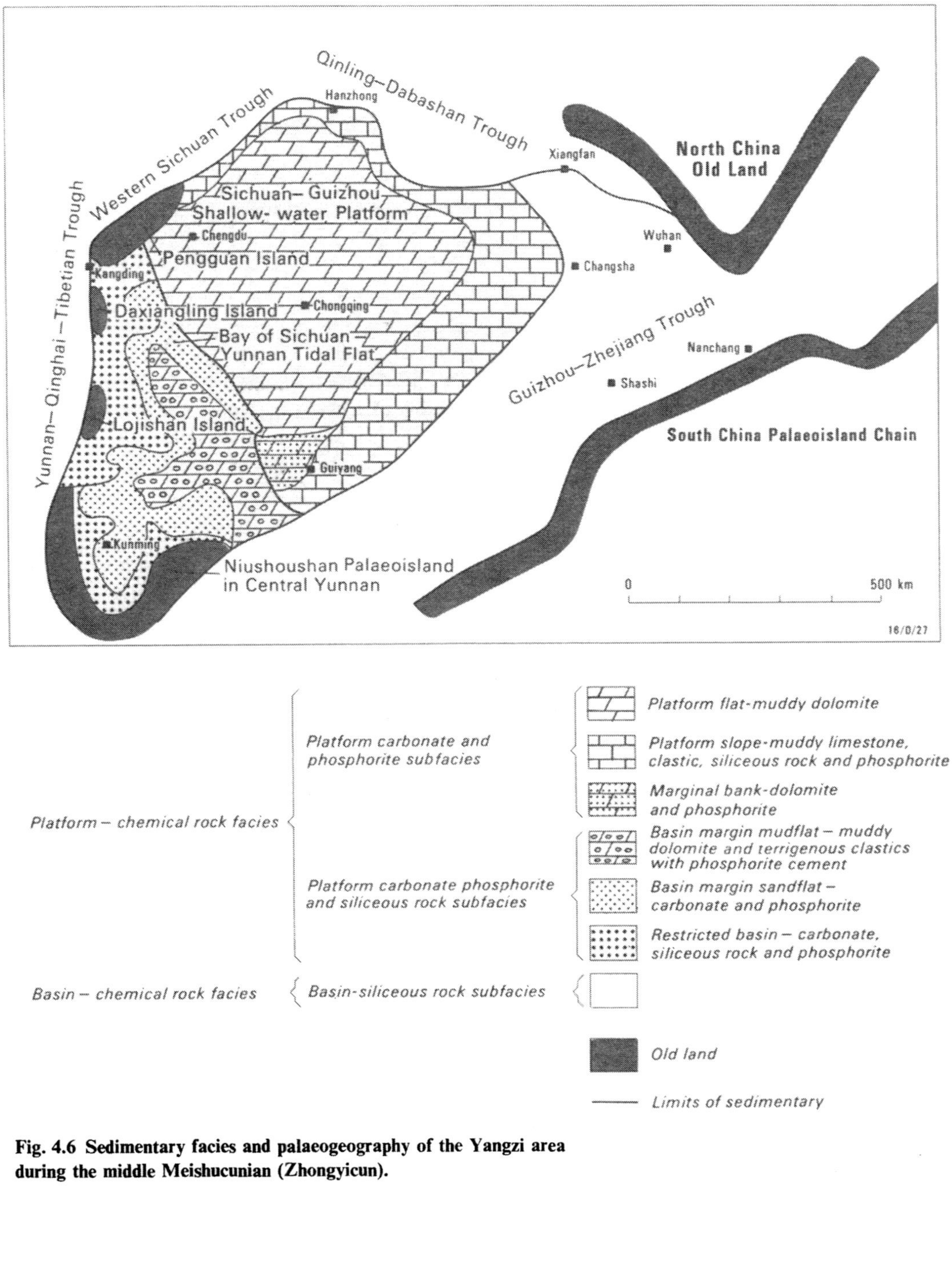

Fig. 4.6 Sedimentary facies and palaeogeography of the Yangzi area during the middle Meishucunian (Zhongyicun).

Table 4.5. *Classification and relationship of facies zones of the Zhongyicun Member, Meishucun Formation*

I. Platform chemical sediment	I_1 Platform carbonate phosphorite facies	I_1^1 Platform flat argillaceous dolomite
		I_1^2 Platform slope argillaceous limestone clastics, siliceous sediments and phosphorite
		I_1^3 Marginal bank dolomite and phosphorite
	I_2 Platform basin carbonate phosphorite and siliceous facies	I_2^1 Basin margin mud flat–argillaceous dolomite and phosphatic clastic sediments
		I_2^2 Basin margin sand flat–carbonate siliceous sediments and phosphorite
II. Basin chemical sediment	Basin siliceous facies	

western part of the basin is represented by shallow platform chemical sediments, whereas the eastern part is represented by shallow basin chemical sediments. The classification and relationship of the various facies zones are shown in Table 4.5.

The platform chemical sediments occur in the western Yangzi area. This region of shallow water was controlled by the Sichuan–Guizhou shallow platform and the tidal flats of the Sichuan–Yunnan embayment.

The platform flat argillaceous dolomite subfacies (I_1^1) (Fig. 4.6) constitutes the main body of the platform-carbonate rock and phosphorite facies, extending over the whole of the Sichuan 'basin' and the northern part of Guizhou. The rocks are light-coloured and generally thin, but locally they may vary in thickness. The dominant lithology is micritic dolomite; laminations, cavities and fenestrae are common. Small-scale wavy cross-bedding is found occasionally. This facies represents sediments deposited under supratidal conditions.

The platform slope subfacies (I_1^2) is located on the northern and eastern margins of the Sichuan–Guizhou Platform. The northern slope was relatively steep and adjoined the Qinling Geosyncline. As a consequence, the sediments of that region are mainly composed of argillaceous limestone, dolomitic and argillaceous limestone with chert bands, lenticular limestone and sandy shale. These lithologies vary greatly in thickness and relative proportions. They are generally dark-coloured because of their high content of organic matter and contain pyrite concretions and sporadic phosphate nodules. The gentle eastern slope consists mainly of black siliceous rock in gradational contact with the Guizhou–Zhejiang Trough. Phosphorites mostly occur as lenses. Small economic phosphorite deposits formed locally in the slope facies.

The platform marginal bank subfacies (I_1^3) occurs in western Guizhou and consists of intraclastic dolomite, phosphorite and phosphatic dolomitic and silicified conglomerate. Bioclasts account for more than 80% of the sediments in some places, where they form shell banks. This subfacies shows tidal bedding and ripple laminations and obviously represents a very shallow high-energy environment. The facies is not only rich in P_2O_5, but also high in rare earth elements.

The platform-basin carbonate, phosphorite and siliceous rock facies constituted the most important zone of phosphorite accumulation during the early Meishucun Stage.

Depending on the local palaeogeography the northern section of the basin margin mud flat–argillaceous dolomite subfacies (I_2^1) consists mainly of micritic dolomite, showing many supratidal structures such as angular and irregular fenestrae, ripple marks and mud cracks. The southern section consists of clastic deposits such as phosphatic sandstone, fine sandy and pebbly phosphorite and phosphatic dolomite. Terrigenous material was derived from the central Yunnan–Niushoushan palaeo-island.

The basin margin sand flat–carbonate and phosphorite subfacies (I_2^2) which contains abundant carbonate-type phosphorite, constitutes the most important phosphatic zone. Phosphorite and carbonate grains are abundant. Littoral sedimentary structures such as oblique bedding, herringbone bedding, horizontal bedding, wavy and lenticular tidal bedding, ripple marks and scoured surfaces are common.

The carbonate chert phosphorite subfacies (I_2^3) is up to 200 m thick. Phosphorite is also very thick but is low grade and mostly siliceous. Because of the high content of organic carbon the lower part of the unit is dark-coloured. It also contains abundant pyrite. The upper part of the unit shows a high content of silica and limestone in the dominant carbonate rock. The sedimentary environment was probably a sand flat.

The basin-chemical sediments (II) occur in the eastern Yangzi area in a depression within the shallow Yangzi sea. Rocks of this category consist mainly of siliceous rocks intercalated with siliceous phosphorite and siliceous carbonaceous shale. They are thinner than the platform-chemical sediments, dark in colour, high in organic carbon content, contain pyrite and are manganiferous. Fossils are rare. Horizontal micro-bedding is present. The thickness and lithology are laterally persistent. These characteristics are taken

Table 4.6. *Depositional characteristics of the Early Cambrian Meishucun Formation in the Yangzi area*

Time		Sedimentary environment	Thickness of phosphatic sequence (m)	Thickness of phosphorite bed (m)	Form of phosphorite bed	P_2O_5 (%)	Dominant texture	Potentially useful associated constituents	Examples of deposits
Late stage	Badaowan Member	Relatively marine basin	more than 10	0.1–0.9	concretionary	8–20	argillaceous phosphatic nodule	Mo, Ni, V, U	Hunan, east Guizhou
Late stage	Badaowan Member	Nearshore marginal bank	50–180	2–10	bedded	11–23	oolitic sandy and bioclastic	K_2O	western part of Sichuan Basin
Late stage	Badaowan Member	Platform sand flat	40–20	6–16	bedded	25	oolitic pelletal sandy bioclastic and gel-like	—	east Yunnan
Early stage	Zhongyicun Member	Platform restricted basin	105–200	28–40	bedded	15–18	oolitic pelletal sandy bioclastic and gel-like	heavy rare earths	east Yunnan, south Sichuan Basin
Early stage	Zhongyicun Member	Platform marginal slope	2–190	0.05–19	lenticular	10–17	silty clastic and pelletal	heavy rare earths	south Shaanxi
Early stage	Zhongyicun Member	Platform marginal bank	4–30	1.2–27.9	bedded concretionary	8–30	Sandy, pebbly and bioclastic	rare earth (Y)	central Guizhou

to indicate that these sediments were deposited in relatively deep, stagnant water and reducing aqueous environments. Phosphorites found with this association of lithologies are non-economic.

2. Palaeogeography of Badaowan Member The Badaowan Member is a subordinate phosphatic horizon in the Meishucun Formation. After the deposition of the Zhongyicun Member, major changes took place in the topography of the Yangzi shallow sea. The major feature was the further development of the palaeo-island chain into a narrow north–south land mass due to the uplift of the western part which had provided abundant terrigenous clastic materials for basin sedimentation and resulted in the gentle elevation of the whole Yangzi basin in the west and a low depression in the east.

The Badaowan Member is generally represented by sandy clastic sediments in the west and by carbonaceous, siliceous and argillaceous sediments in the east. To the west it is light-coloured with a low content of organic matter and is intercalated with stratiform sandy and K-feldspar-bearing phosphorite. In the east where chemical sediments predominate it is dark-coloured with a high content of organic matter, with 'stone coal' and phosphate concretions in places. Whereas the western area represents a littoral bank environment, the eastern area represents an outer shelf environment.

Characteristics of Meishucun phosphorite deposits

Phosphorites of the Meishucun stage are mainly composed of collophane with a small amount of francolite. The associated minerals include: dolomite, calcite, quartz, chalcedony, kaolinite, illite, sericite, glauconite, organic matter, barite and pyrite. X-ray diffraction and infrared spectral analyses indicate that the collophane is carbonate–fluorapatite. Under the electron microscope, the apatite is in the form of discrete crystal grains 0.1–1 μm in diameter and radial spherulites 4–10 μm in diameter. The phosphorite is characterised by five types of grains: intraclasts, ooids, ovules, gel and bioclasts. These make up the major components of the grainstone phosphorites. Compact massive, striped, brecciated and concretional structures are common. The main characteristics of the deposits are summarised in Table 4.6.

Middle Early Cambrian phosphorites

Geotectonically, phosphorite deposits occur in the Qinling–Qilianshan–Kunlunshan folded geosynclinal region, the Tarim and Sino-Korean Paraplatform regions, the

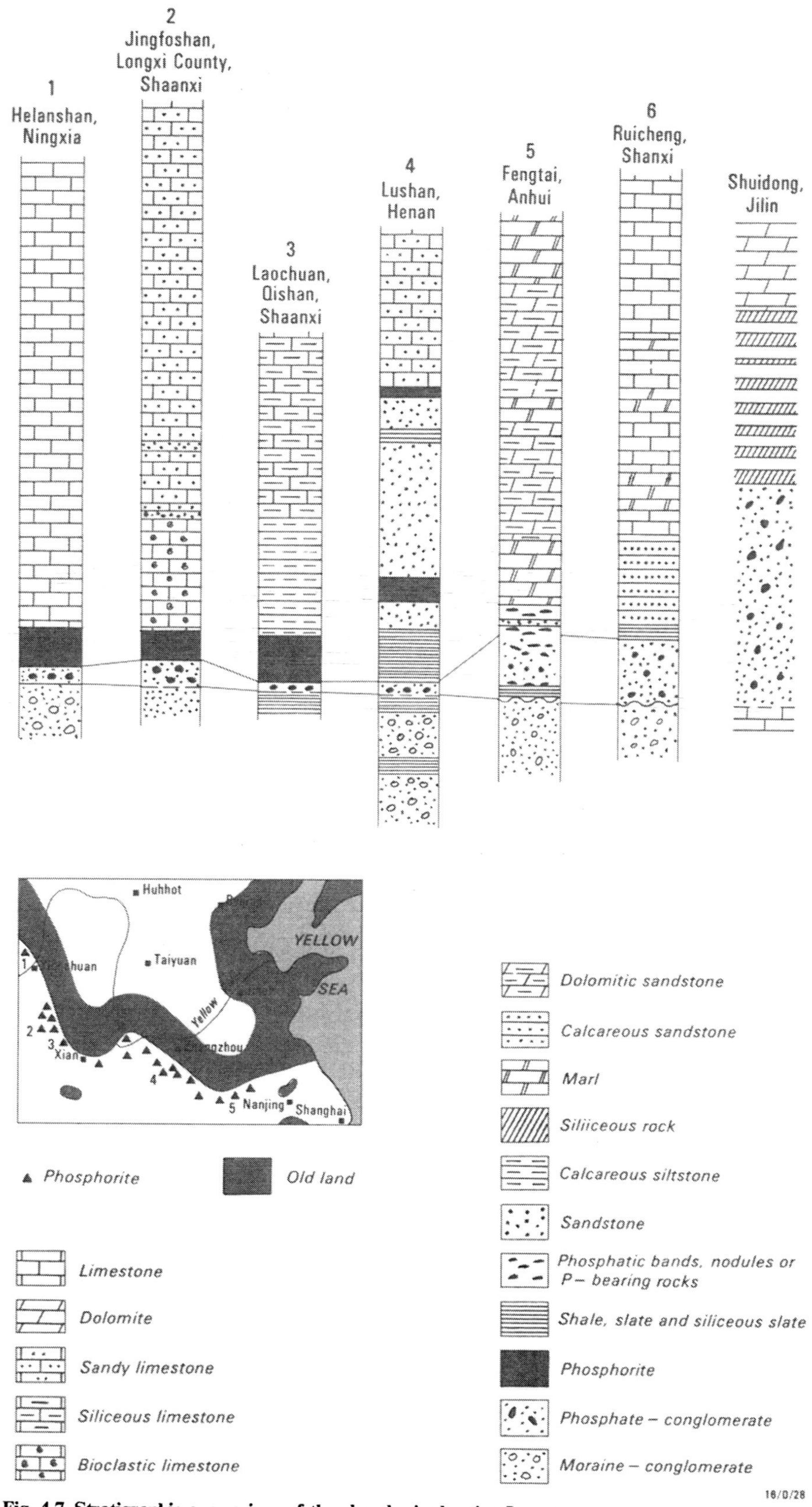

Fig. 4.7 Stratigraphic comparison of the phosphorite-bearing Lower Cambrian sequences of the Sino-Korean Paraplatform. (Scale not available).

Table 4.7. *Correlation of the Lower Cambrian formations of the Yangzi, Sino-Korean and Tarim Paraplatform.* indicates phosphatic formations. No correlation is implied for the underlying strata*

Horizon		Kuruktag (Xinjiang Region)	Beishan (Gansu Province)	Longshoushan (Gansu Province)	Sino-Korean Paraplatform	Yangzi Paraplatform
Lower Cambrian	Upper	Xidashan Formation	Shuangyishan Group*	Hanmushan Group	Mantou Formation*	Longwangmiao Formation
	Middle	Xishanburake* Formation			Xinji Formation	Canglangpu Formation
						Qiongzhusi Formation
	Lower					Meishucun Formation*
Underlying Stratum		Hangeerquioke Formation	Tillite Bed	Yanhuotonggou Formation	Loquan Formation	Dengying Formation

Urutau–Tianshan–Hinganling folded geosynclinal region, and the Altai geosyncline region in northern China. Phosphorites occur on the marginal slopes of the geosyncline or on the central blocks of the folded regions, but they mostly occur in depressions or on the margins of the platform regions. There are many phosphorite occurrences in northern China but they are generally low-grade and small in size.

The Early Cambrian Meishucun phosphorite deposits mainly occur south of the Tarim and Sino-Korean Paraplatforms. The northern limit of the Yangzi transgression approximately reached the area south of the Qinling Mountains (i.e. 34° N) during the beginning of the Early Cambrian. During the middle Early Cambrian, the transgression extended northward, passing through the Qinling Trough to reach the vicinity of Tarim in the west and the Longxian (Shaanxi)–Helanshan (Gansu)–Lingbao (Henan)–Fengtai (Anhui) areas on the margin of the Sino-Korean Paraplatform in the north. The Xinji Formation is a representative phosphatic formation that contains *Bergeroniellus* sp. and *Hsuaspis*, which indicate equivalence with the Canglangpu Formation of the middle Early Cambrian of the Yangzi area. The Xinji Formation represents the major phosphorite horizon of the Late Proterozoic–Early Cambrian of northern China. The distribution of the phosphorite deposits of this age in northern China is shown in Fig. 4.7. Correlations of the Lower Cambrian of the Tarim and Sino-Korean Paraplatforms and the folded geosynclinal region of Qinling–Qilianshan–Kunlunshan with the Lower Cambrian of the Yangzi area are given in Table 4.7.

Phosphatic sequences

In the Tarim and Qinling–Qilianshan–Kunlunshan areas, the Lower Cambrian phosphatic sequence consists of argillaceous limestone with quartz sand, thin-bedded chert and limestone, ten to hundreds of metres thick, and disconformably overlying Sinian tillites. Phosphorites occur above chert or between the carbonaceous and siliceous mudstones and limestones which occur in bedded, lenticular or concretionary forms. This mode of occurrence represents a carbonaceous siliceous and argillaceous carbonate unit which formed in a restricted basin on the marginal slope of the old land mass.

On the southern edge of the Sino-Korean Paraplatform, the phosphatic sequence of the Xinji Formation represents a terrigenous-carbonate (mainly limestone) unit, underlain by tillite. Phosphorite is located at the base of the formation. Comparison of the phosphatic sequences of the various areas is shown in Fig. 4.8. The Fengtai phosphorite deposit in Anhui is a typical example of a Lower Cambrian phosphorite deposit located on the margin of the Sino-Korean Paraplatform. The sequence there, from top to base, is as follows:

4. Yellowish pink dolomitic breccia phosphatised at the top. Gravels are angular, composed of siliceous limestone and cherty quartzite, cemented mainly by calcite and dolomite.
3. Bluish-grey micaceous (muscovite) silty shale with phosphatic bands, 0–1 m thick.
2. Phosphorite with thin-bedded silty shale. Cross-bedding is well developed. The phosphorite is thin-bedded or lenticular and 0–2.4 m thick.
1. Phosphatic dolomitic sandstone, marl, sandy and dolomitic marl intercalated with thin beds of purple shale, and black phosphorite at the base. Ripple marks and cross-bedding are present. This unit is 2–20 m thick.

As a generalisation, the middle Early Cambrian phosphatic formations occurring on the margin of the Sino-Korean Paraplatform have the following characteristics: (1) the phosphatic sequences are thin and commonly red, and exhibit littoral sedimentary structures; (2) clastic sediments are more abundant than chemical sediments and carbonates are dominated by limestone with dolomitic and chert nodules but pure limestone or pure dolomite is rare; (3) dolomitic material is common in the clastic rock.

Characteristics of Xinji phosphorite deposits

The middle Early Cambrian phosphorite deposits of

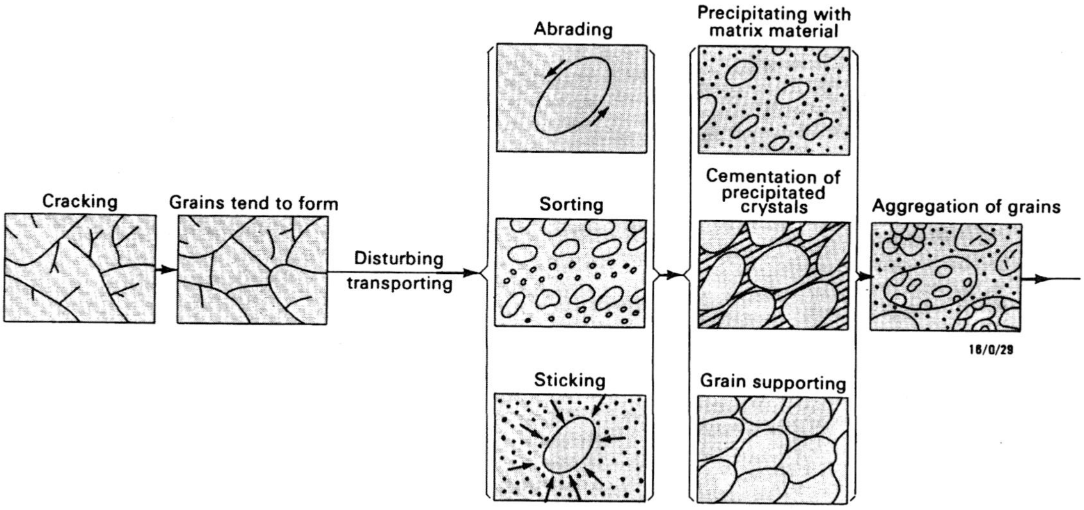

Fig. 4.8 Model for the origin of phosphate grains.

northern China are mostly small, oolitic, sandy and pebbly, and have a low P_2O_5 content. They are generally thin and are either planar-bedded or lenticular. The phosphate largely occurs as cement, but some occurs as clasts or ooids with a nucleus of sand-size quartz grains. The Xinji-type deposits are believed to represent a nearshore sedimentary facies.

Phosphogenesis

The Late Proterozoic–Cambrian phosphorites of China were formed firstly through the aggregation of phosphatic colloids of marine origin followed by the redeposition of phosphatic grains within the basin. Phosphorus, whether as in gel-like phosphorites, cements for the granular phosphorites (from pore solutions precipitation), or as interstitial gelatinous phosphorus, was initially derived from phosphorus-rich sea water. The phosphatic grains of the redeposited granular phosphorites were derived from previously or penecontemporaneously deposited intrabasin phosphatic clasts. Although the source of the phosphorus is marine, the phosphorus in the sea water may have many sources and may have undergone long-term storage. Most of the Late Proterozoic–Early Cambrian phosphorites of China have undergone one of the phosphogenic processes shown in Fig. 4.8.

(a) *Colloidal aggregation*: This is the primary form and an essential stage for the formation of phosphorites by providing the materials for the redeposition of the grainstone phosphorites of the shallow platform facies. The phosphorus 'segregation' from solution is the primary stage for the accumulation and sedimentation of phosphatic material and is in agreement with the theory of oceanic upwelling. Because accumulation and sedimentation of colloidal material is a slow process, various impurities become incorporated, depending on the phosphogenic environment. As a consequence these phosphorites vary greatly in thickness and grade.

(b) *The deposition of phosphatic grains*: Granulation took place under shallow high-energy conditions, in which the phosphatic colloids or consolidated phosphorites were broken-up, winnowed and redeposited *in situ* or in other environments with different energy levels. The redeposition of phosphatic grains within the basin thus occurred by mechanical reworking. This process is an important stage for the formation of large high-grade phosphorite deposits, because it is during this phase of clastic sedimentation that the rate of deposition is fast; repeated winnowing takes place, thus minimising the input of non-phosphatic grains. Most of the large high-grade phosphorite deposits of the shallow platform facies, i.e. the various granular phosphorite deposits, are formed by the repetition of this process. Apart from the intraclasts, phosphatic clasts composed of weathered residual grains of earlier phosphorites can be found on supratidal flats and sedimentary hiatus surfaces.

(c) *Biological processes*: During the formation of phosphorites photosynthesis of organisms consumed large amounts of CO_2 and increased the oxygen content of the water column. This resulted in a series of changes in the geochemical environment producing oversaturation, segregation and precipitation of phosphate. Binding and capturing of phosphatic particles by organisms were also important phosphogenic processes. Such processes not only speed up the precipitation of phosphate, but also can prevent grains from being lost due to winnowing and erosion. Whilst biological processes are important, as are other phosphogenic conditions, they are by no means essential. The bedded reef facies phosphorite deposits formed as algal concentrations on the marginal banks of shallow platforms in the Yangzi area during Doushantuo time, and produced high-grade phosphorite deposits. However, in other places, there are also plenty of thick high-grade phosphorite deposits which show few signs of algal activity.

(d) *Diagenesis*: During diagenesis, cementation and then differentiation and aggregation of interstitial phosphate may also play a major part in the enrichment of phosphate, and in the concentration of trace elements in phosphorites, such as iodine and rare earths.

The relative importance of these various phosphogenic processes differs from one area to the next. However, high-grade phosphorite deposits are mostly formed by repetition of these

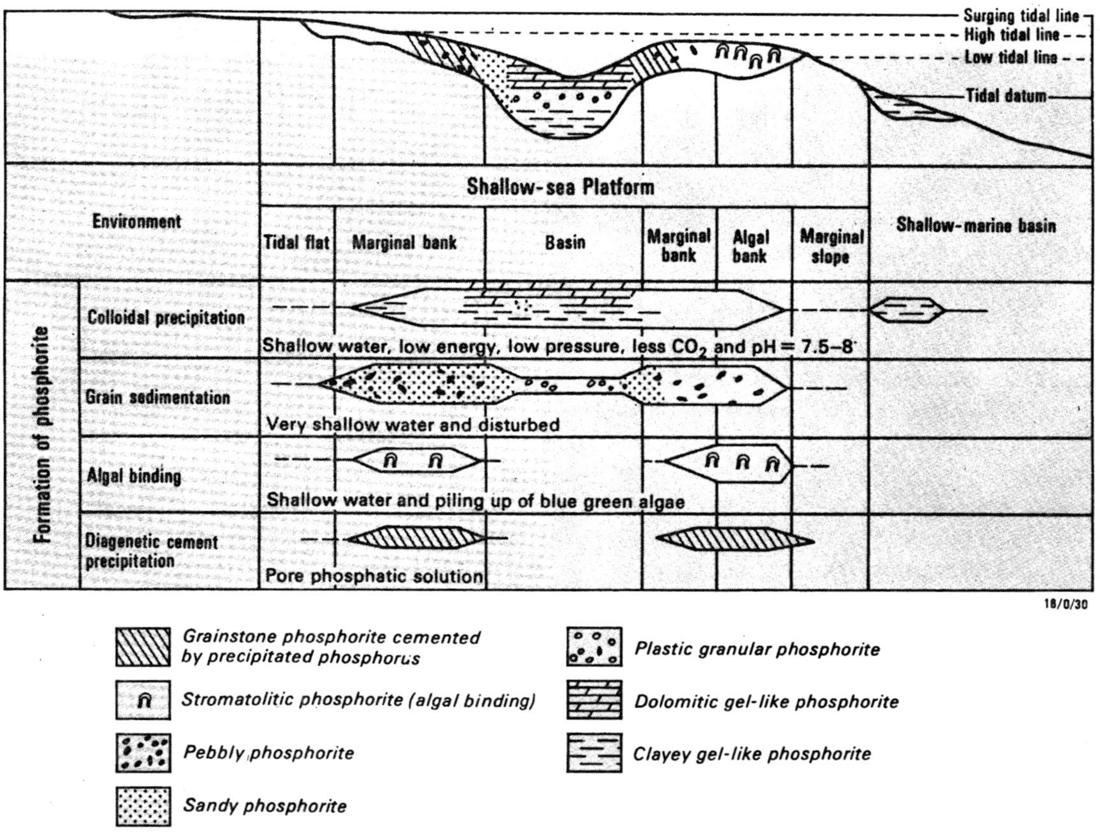

Fig. 4.9 Model for sedimentation of phosphorites in the Doushantuo Formation.

processes. A sedimentary model for Late Proterozoic–Early Cambrian phosphorite deposits in China is summarised in Fig. 4.9. This model is also considered to be applicable to marine sedimentary phosphorite deposits of other ages.

Temporal and spatial distribution of Late Proterozoic–Cambrian phosphorite deposits in China

(1) Late Proterozoic–Cambrian time is an important global phosphogenic period. Many major economic phosphorite deposits were formed during this period in the various Provinces of southern China on the northern margin of Gondwanaland. A strict temporal control of phosphogenesis occurred in south China during a relatively quiescent period in the Late Proterozoic–Early Cambrian following strong crustal movements, extensive magmatic activity, and abrupt changes which took place in palaeoclimate and biological evolution after a global glacial event represented by the Nantuo Tillite. In southern China, the most important economic phosphorite deposits of Late Proterozoic–Early Cambrian age occur in the Late Sinian Doushantuo Formation and the early Early Cambrian lower Meishucun Formation. Within Guizhou Province, phosphatisation is very extensive and economic ore-bodies are also locally present in the Late Sinian Dengying Formation which separates the Doushantuo and Meishucun Formations. This distribution indicates that although phosphate accumulations have their own major preferred horizons, phosphogenesis was fairly continuous. Similarly, phosphates deposited in the upper Meishucun Formation in the central and eastern Yangzi area, and in the Xinji Formation on the southern margin of the Sino-Korean Paraplatform, may also represent the temporal continuation of phosphogenesis over a broad area.

(2) According to the preliminary palaeomagnetic data available, the Doushantuo, Dengying and Meishucun Formations near Kunming in Yunnan Province, were situated at the northern palaeolatitudes of 20°, 26° and 43° respectively; the Doushantuo Formation near Yichang, in Hubei province, was located at a palaeolatitude of about 24° S. This indicates that phosphorite is mainly concentrated in low to mid-latitude areas, which is in accordance with the hypothesis that phosphorites generally occur in low latitude areas, in response to oceanic upwelling.

(3) The geotectonic framework has a major influence on the sedimentation and distribution of phosphorites. The Yangzi Paraplatform, controlling the deposition of Late Sinian–Early Cambrian phosphorites, is a relatively young platform formed prior to the phosphogenic event. The subsequent Wuling (1300–1450 m.y.), Jinning (800 m.y.) and Chengjiang (740 m.y.) Movements that occurred over a period of 600 m.y. produced a series of rises and depressions by early Doushantuo time. These rises and depressions resulted in a complicated topography in the old land mass, the palaeo-islands and the palaeo-sea floor in the Yangzi shallow sea. This topography provided favourable sites for the continuous supply of phosphorus-rich sea water and for the precipitation of phosphate. Following deposition of the Dengying Formation, these favourable sites in the central and

eastern Yangzi sea were mostly subdued and even levelled off. The only depressed zone favourable for the deposition of Early Cambrian phosphorites was on the western margin of the Yangzi epicontinental sea. Sites on the ocean-facing marginal zone of the shallow platform or within marine channelways are considered to have been favourable areas for the concentration of phosphorite during the Late Proterozoic–Early Cambrian of southern China. Such environments not only provided sufficient phosphorus of marine origin (with minimal terrigenous input) but also provided suitable locales for the precipitation of phosphorus from the sea water, for the granulation of phosphate, and for the subsequent winnowing and accumulation of the phosphate grains. Phosphate may be deposited in other shallow areas, especially in the vicinity of estuaries or straits, but the deposits are generally small, thin and low-grade due to the abundant terrigenous input near land areas and the lack of conditions suitable for intra-clastic reworking of phosphate grains. Although thin poor-quality phosphorites were deposited as a type of colloidal aggregation in the Yangzi Basin at depths of tens of metres to more than 100 m below the wave base, large high-grade phosphorite deposits were mainly deposited in the intertidal zone on the marginal bank of the shallow platform and its surrounding shallow waters. It is in these very shallow waters that conditions were favourable not only for the initial segregation of phosphorus but also for the formation and subsequent accumulation of phosphate grains. In sharp contrast to the Yangzi Paraplatform, the Sino-Korean Paraplatform in northern China remained a typical old platform until the Mesozoic Yanshanian Movements. Prior to the middle stage of the Early Cambrian, it was an old land area and its surface had by and large been peneplaned. During the middle of the Early Cambrian, a broad sea transgressed on to the margins of this paraplatform. However, barrier topography was poorly developed and there were no major topographic features on the sea floor, so that mechanical deposition of terrigenous grains extended over a large part of the sea floor. Therefore although phosphatisation was extensively developed over the flat palaeogeographic environment, no large economic deposits are found along the margin of the old land mass in northern China.

(4) The economically important phosphorite deposits formed in the late Sinian Doushantuo Formation occur vertically above glacial sediments and are distributed laterally around the margin of the area over which the glacials are distributed.

(5) The claystone–chert–phosphorite–carbonate assemblage is the most important phosphatic sequence: the claystone–phosphorite–argillaceous dolomite assemblage is of secondary importance. Generally the clastic rock–carbonate rock assemblage is only slightly phosphatic.

5

Proterozoic and Cambrian phosphorites – regional review: Asian part of USSR and Mongolian People's Republic

A.L. YANSHIN

Introduction

There are two spatially isolated and tectonically heterogenous phosphorite-bearing regions in central Asia: the Mongolian–Siberian region in the east; and the Kazakhstan region in the west. The former lies within an east–west belt extending from the Altai–Sayan Mountains in the west to the Shantar Islands in the Okhotsk Sea in the east. The latter occupies the central and southern parts of the Kazakhstan SSR and adjacent areas (Fig. 5.1). The Mongolian–Siberian region represents the peripheral part of a mobile Palaeozoic belt fringing the southern margins of the Precambrian Siberian Platform. By contrast, the Kazakhstan region is tectonically a crustal block, variously called the Central Kazakhstan Massif or the pre-Riphean Block (Tectonics of Kazakhstan, 1982). The exact palaeotectonic position of Karatau, which is the principal locus of phosphogenesis during the late Proterozoic–Early Cambrian time span, basically remains unknown. Because of geological and tectonic heterogeneity and geographic isolation, the two areas are treated separately here, although phosphogenesis appears to have proceeded in a similar manner in both areas. Since specific chapters of this book consider in detail the Karatau deposits of southern Kazakhstan and the Khubsugul deposit of northern Mongolia, these will only be considered in a general fashion here.

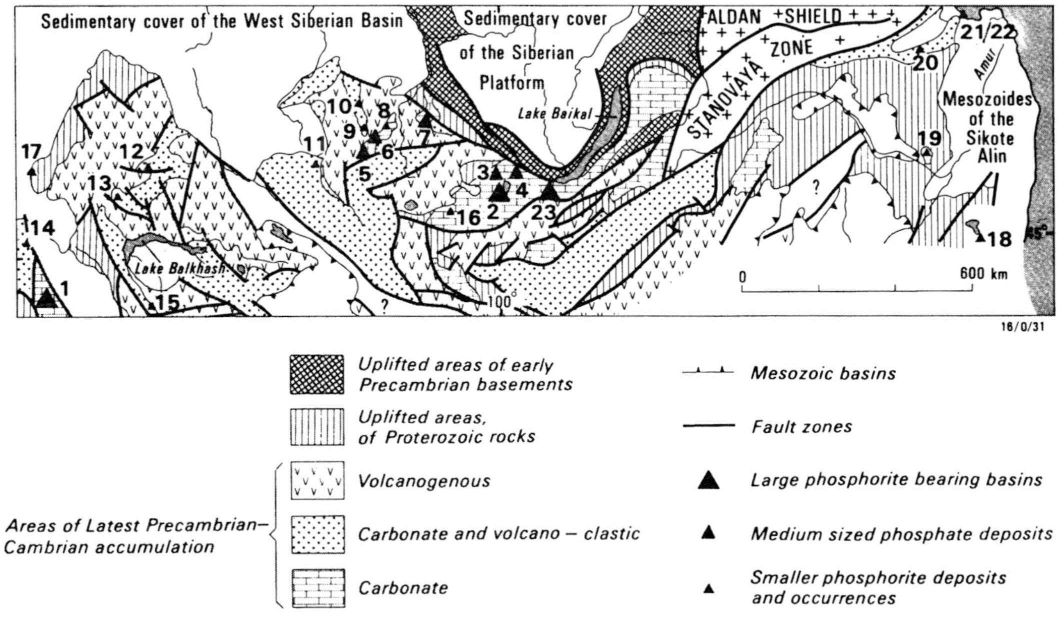

Fig. 5.1. Latest Precambrian–Cambrian tectonic pattern and phosphorite deposits of Soviet Asia and Mongolia (after Volkov 1974). 1, Karatau; 2, Khubsugul; 3, Ukha-gol; 4, Kharanur; 5, Belka; 6, Tamalyk; 7, Seiba; 8, Obldzhan; 9, Pyzas; 10, Ust-Kundat; 11, Gorny Altai; 12, Shiderty; 13, Ata-su; 14, Greater Karatau Ridge: 15, Kendyktas; 16, Sangilen; 17, Ulutau (Baikonur); 18, Khanka; 19, Lesser Khingan Mountains, Khabarovsk; 20, Lesser Khingan Ridge: 21–22, Udo–Shantar Group; 23, Sludyanka metamorphics.

Mongolian–Siberian region

Phosphorites were originally discovered in the Altai–Sayan Mountains during the early 1950s (Krasilnikova, 1962). Although these early discoveries were small in scale, they served to provide the impetus to examine the whole miogeosynclinal belt bordering the southern Siberian Platform for ancient phosphorite deposits. In so doing, the general idea of a phosphogenic episode which occurred during 'latest times of Precambrian and earliest Cambrian' proposed by Yanshin (1964) was tested.

During the subsequent decade new discoveries were made, including the Khubsugul phosphorite basin in the northern part of the Mongolian People's Republic in 1965. This has proved to be by far the largest in the region (Donov, Edemsky & Ilyin, 1967). The phosphorites of the Mongolian–Siberian region occur mostly in the miogeosynclinal zone, and are underlain by pre-Riphean crystalline basement which crops out at the surface in several blocks in northern Mongolia and Eastern Sayan. The basement and the generally carbonate-hosted phosphorite sequences are separated by thick clastic/arkosic sequences. The clastic sequence rests on deeply eroded basement and includes extrusive rocks, bimodal in composition, generally rhyolites and basalts, with intermediate types absent. Alkaline subvolcanic rocks in this sequence have given a radiometric age of 752 m.y. (Ilyin & Volkov, 1977), and granite pebbles from the basal conglomerate of the clastic series are dated at 823 m.y. Thus the clastic sequence underlying the phosphorites most probably has a late Riphean to Vendian age. In the overlying carbonate sequences, with which the phosphorites are associated, oncolites and stromatolites are present. Above the phosphorites, skeletal remains of archaeocyathids, hyolithids and trilobites of Early Cambrian age are present. Thus the stratigraphic sequence probably extends from Late Riphean to Middle Cambrian. The largest phosphorite occurrences are, however, clearly confined to the late Vendian–initial Cambrian interval, and although older strata do contain phosphatic sediments, the occurrences are on a much smaller scale (Fig. 5.2).

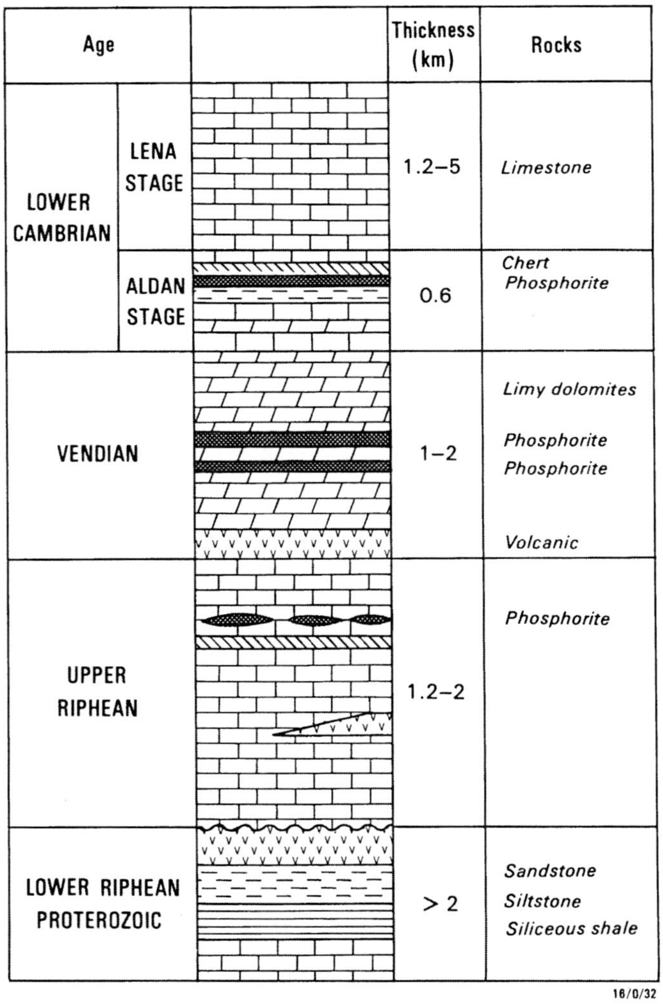

Fig. 5.2. Generalised Riphean to Cambrian stratigraphy, showing phosphogenic intervals, Altai–Sayan Mountains.

Phosphorite deposits

The largest deposits of the Mongolian–Siberian region are those of the *Khubsugul Basin* in the northern part of the Mongolian People's Republic (Ilyin *et al.*, this volume). The productive phosphorite unit of this Basin occurs at the base of the Khubsugul Group and is associated with limestone, dolomite and chert. Late Aldanian (Atdabanian)–Early Cambrian trilobites have been found some 500 m above the phosphatic interval in a continuous sequence (Korobov, 1980). Thus the Khubsugul phosphorites may have either a Vendian or initial Cambrian (pre-trilobite, Tommotian) age.

Phosphate-bearing strata continue northwards from Khubsugul and cross the international border into the Soviet Union. The *Ukha-gol deposit* occupies the axial portion of the Eastern Sayan Ridge straddling the USSR/MPR border. In this area, arkose, volcanogenic sediments and volcanics of the Upper Riphean Sarkhoi Series are overlain by the predominantly carbonate Boxon Series of Vendo-Cambrian age. Phosphorites, as at Khubsugul, are associated with dolomites in the lower part of the Boxon Series, although carbonates with Early Cambrian fossils occur higher in the sequence than at Khubsugul. A Vendian age has been proposed for the lower part of the Boxon Series by Semikhatov & Serebryakov (1967), and this has been subsequently substantiated. Medusoid remnants have been found by Zhabin (1974) close to the phosphatic interval which B.S. Sokolov (personal communication) considers to be similar to those of the Australian Ediacarian System. Immediately below the phosphorites Ilyin (1973) reported the stromatolite *Jurusania cylindrica* which I.N. Krylov considers to indicate a late Riphean age, and Vologdin studied other stromatolites, *Luella* and *Johella*, collected from within the phosphatic interval which indicate a correlation with the Dsin-lin Suite of the Sinian System of China, once more suggesting a Vendian age for the Ukha-gol phosphorite. The productive phosphatic unit of the Ukha-gol deposit contains three sub-units: lower and upper phosphorite horizons separated by dolomites with chert nodules. In the lowest phosphatic subunit there are three phosphorite beds. As at Khubsugul, the basal one is thickest (8–10 m), most persistent (being traceable for more than 4 km), and is relatively rich (20–28% P_2O_5, average 22%). The phosphorite beds are intercalated with black, siliceous, phosphatic shale which resembles that in the middle portion of the Chulaktau phosphatic unit of the Karatau Basin. The phosphatic unit at Ukha-gol is quite conspicuously black compared to the pale underlying and overlying dolomites. Phosphatic dolomite and sandstone also occur stratigraphically higher than the productive unit, but such horizons have a rather low P_2O_5 content. Structurally, the deposit is a narrow overturned syncline, with both limbs inclined westwards. Phosphorite has been found on both limbs, and can be traced for 17 km along strike. Phosphorite reserves amount to 483 million tonnes with a cut-off P_2O_5 grade of 7–14.7% on average (Kazarinov & Krasilnikova, 1980). However, the area is highly mountainous and access is rather difficult.

Some 60 km east of Ukha-gol, the *Kharanur deposit* is situated in a similar geological position, but is associated with a large fault zone. The geology in this area is obscured by a considerable thickness of glacial deposits. The phosphate deposit is apparently smaller and of lower grade than that of Ukha-gol.

The *Boxon deposit* is located some 30 km north of Ukha-gol, and again has a similar stratigraphy. Here, the phosphatic interval is dominated by phosphatic dolomite containing little more than 7–8% P_2O_5. About 200 m of non-phosphatic dolomite overlies the phosphatic sequence, and this is in turn overlain by high-grade bauxite beds. All the phosphorite deposits in the south-eastern portion of the Eastern Sayan Basin occur with dolomitic sequences, but younger strata may also consist of limestone.

Several deposits are located in the western part of the Altai–Sayan Mountains, but are all much smaller than the Khubsugul deposits. They are concentrated in two areas: the Gornaya Schoriya–Batenev Range; and the north-western part of the Eastern Sayan Ridge. The Belka deposit of Gornaya Schoriya and Tamalyk deposit of the Kuznetsk Alatau region are typical of the first area; and the Telek deposit of the second.

The phosphorites of the *Belka deposit* are underlain by late Riphean dolomites of the western Siberian series, and overlain by carbonates of the Early Cambrian Usa Suite. Accordingly, they are probably of Vendian age. The phosphatic interval is 100–400 m thick, and is composed of up to 30 beds, individually up to 50 m thick, with a P_2O_5 content of 9–12%. The beds are of phosphatic dolomite which change very gradually to weakly phosphatic dolomite both along strike and across the section. Brecciated phosphorites frequently occur, as well as stromatolitic varieties resembling those of the Jhamarkotra deposit of Rajasthan, India. Inclusions of secondary calcite and fluorite are also characteristic. Smaller deposits are also known in the Gornaya Schoriya area.

Unlike the Belka deposit, which is carbonate-hosted, the phosphorites of the *Tamalyk deposit* in the Kuznetsk Alatau, are overlain and underlain by siliceous sediments. At Tamalyk, the phosphorites occur in the Usa Suite which contains Early Cambrian fossils. The phosphatic interval is 85–220 m thick, and consists of phosphatic limestone and dolomite. The number of phosphatic horizons varies from one to eight, each ranging in thickness from 2 m to 12 m, with a P_2O_5 content of 10–20%. Manganiferous limestones, containing up to 10% Mn, are common.

Nearly all the deposits to the west of the Altai-Sayan area are associated with secondary phosphorite in a Mesozoic weathering profile. These are referred to here as karst phosphorites. The *Telek deposit*, in the north-western part of the Eastern Sayan Basin, has the largest concentration of secondary phosphorite with proven reserves amounting to 180 million tonnes with an average P_2O_5 grade of 14.2% (Kazarinov & Krasilnikova, 1980). The secondary phosphorites develop over the slightly phosphatic dolomites of Vendian age. The main phosphorite body fills a topographic depression; it has a width of 6–700 m, is up to 200 m thick, and has been traced for about 6 km. Uncemented varieties comprise the main part of the ore body, but consolidated rocks also occur, and these may contain as much as 38% P_2O_5 (compared to the uncemented forms which have 14–15%).

In other parts of the Mongolian–Siberian region, small deposits occur to the north-west and south-east of Lake Baikal. In both areas they appear to be associated with Upper Riphean or Vendian carbonate sequences.

The easternmost occurrences are located between the Uda and Selemzha Rivers, and on the neighbouring Shantar Islands in the Okhotsk Sea. These deposits are associated with cherts and volcanic rocks, thought to be of Middle Cambrian age. Exploration has revealed that the phosphorites are laterally non-continuous.

General observations

(1) In the Mongolian–Siberian region phosphogenesis appears to have occurred on three occasions: during the late Riphean, Vendian and Early to Middle Cambrian. In general, phosphorites have not been found in rocks older than about 1000 m.y. or younger than early Middle Cambrian. However, the age of the Sludyanka apatite-bearing high-grade metamorphics is a subject of considerable controversy and could range from Archaean to Riphean. There are also Ordovician shelly phosphorites on the Yenisei Ridge.

(2) The Tamalyk deposit appears to be younger than that at Khubsugul, these deposits seemingly being indicative of Vendian and Early Cambrian phosphogenic events in the region.

(3) In most cases, phosphorites are associated with carbonate sediments, especially where only small deposits are found. The total amount of phosphate contained in carbonate sediments is probably more than that in the calculated phosphorite reserves (Yanshin, 1982).

(4) Early Cambrian phosphogenesis coincided with the radiation event of the organic world. At this time a number of organisms used phosphate to build skeletal structures. More organisms used carbonate fluorapatite at this time than at any other during the Phanerozoic. The ability of calcium phosphate minerals to change their physical and chemical properties depending on P–T conditions seems to have been readily exploited by some organisms enabling them to expand into new ecological niches, thus facilitating the evolution of life. To provide for organic consumption, as well as mineral deposition, it is necessary to envisage a considerable reserve of dissolved phosphorus in the latest Proterozoic and Early Cambrian seas.

(5) To solve the problems of provenance of this phosphorus it is pertinent to analyse the palaeogeography of the Mongolian–Siberian region at this time. There are three principal lithological associations during the late Proterozoic

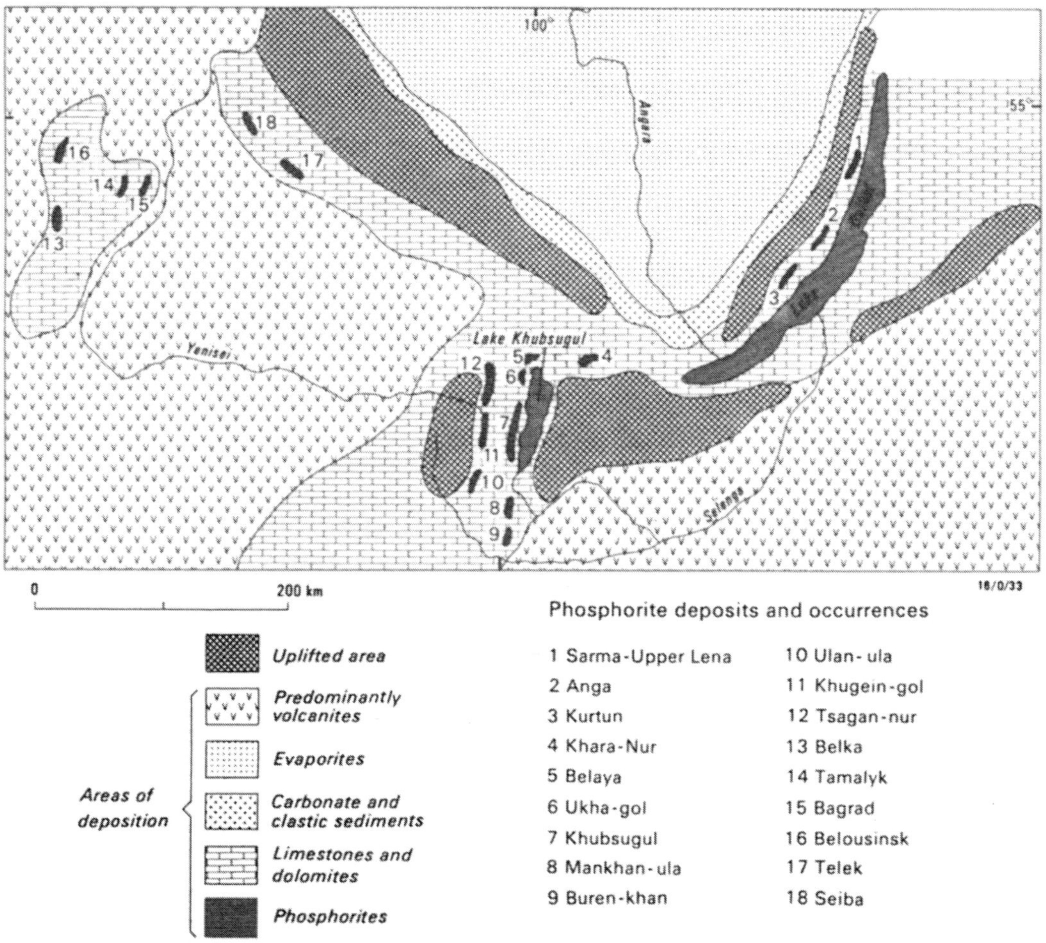

Fig. 5.3. General palaeogeography and facies of the Latest Precambrian–Cambrian of southern Siberia and Mongolia showing location of phosphorite deposits and occurrences.

and Early Cambrian (Fig. 5.3). During the Riphean, volcanoclastic formations accumulated in the so-called Irkutsk Amphitheatre, and carbonates, generally dolomite, began to form during the late Riphean. By the Cambrian the dolomite deposition had given way largely to limestone. During Vendian time both terrestrial and shallow marine clastic sediments were deposited. In the middle part of this sequence glauconites give radiometric ages of 615–609 m.y. Towards the end of the Vendian, the Baikalian tectonic event occurred, affecting the peripheral part of the Siberian Platform and adjacent miogeosynclinal belt. The Baikalian folding and uplift caused the isolation of the Siberian Platform from the more southerly eugeosynclinal areas, and led to the formation of a broad evaporitic basin which existed until Middle Cambrian time. Immediately south of the area of Baikalian uplift, the miogeosynclinal belt subsided continuously. Volcanogenic rocks, including ophiolite, andesite–greywacke–chert, basalt, and other essentially magmatic formations are associated with the Baikalian tectonic event.

(6) Phosphorite deposits are confined mostly to the shallow-water carbonate facies of the miogeosynclinal belt, except for Belka and Tamalyk which were located on a broad uplift where carbonate sedimentation was isolated by areas of volcanogenic sediment accumulation. Thus Vendo-Cambrian palaeogeography indicates a phosphorus source related to deep water rather than continents. It is possible that Baikalian magmatism may account for some phosphorus supply to the seawater. The Vendian–Early Cambrian basaltoides of northern Mongolia are known to include basanites and limburgites which are abnormally enriched in phosphorus (up to 2% P_2O_5). Halmyrolysis of basaltoides on the sea floor could well contribute to phosphorus in sea water.

(7) Palaeoclimatic and palaeomagnetic data indicate that the southern part of the Siberian Platform was located close to the equator during the latest Precambrian (Brovkov, 1974). Such a position was favourable for the initiation of phosphogenesis through the upwelling of deep phosphorus-rich water and its spread over the continental shelf, where, after warming and degassing, phosphate sedimentation took place. It is likely that these waters were also rich in magnesia making possible the contemporaneous deposition of dolomite.

Kazakhstan region

The Karatau Basin, located in the Lesser Karatau Mountain Range, is one of the world's largest sites of phosphorite deposition. Several smaller occurrences in central and southern Kazakhstan also belong to the same depositional event. Since the Karatau deposit is described in detail in this

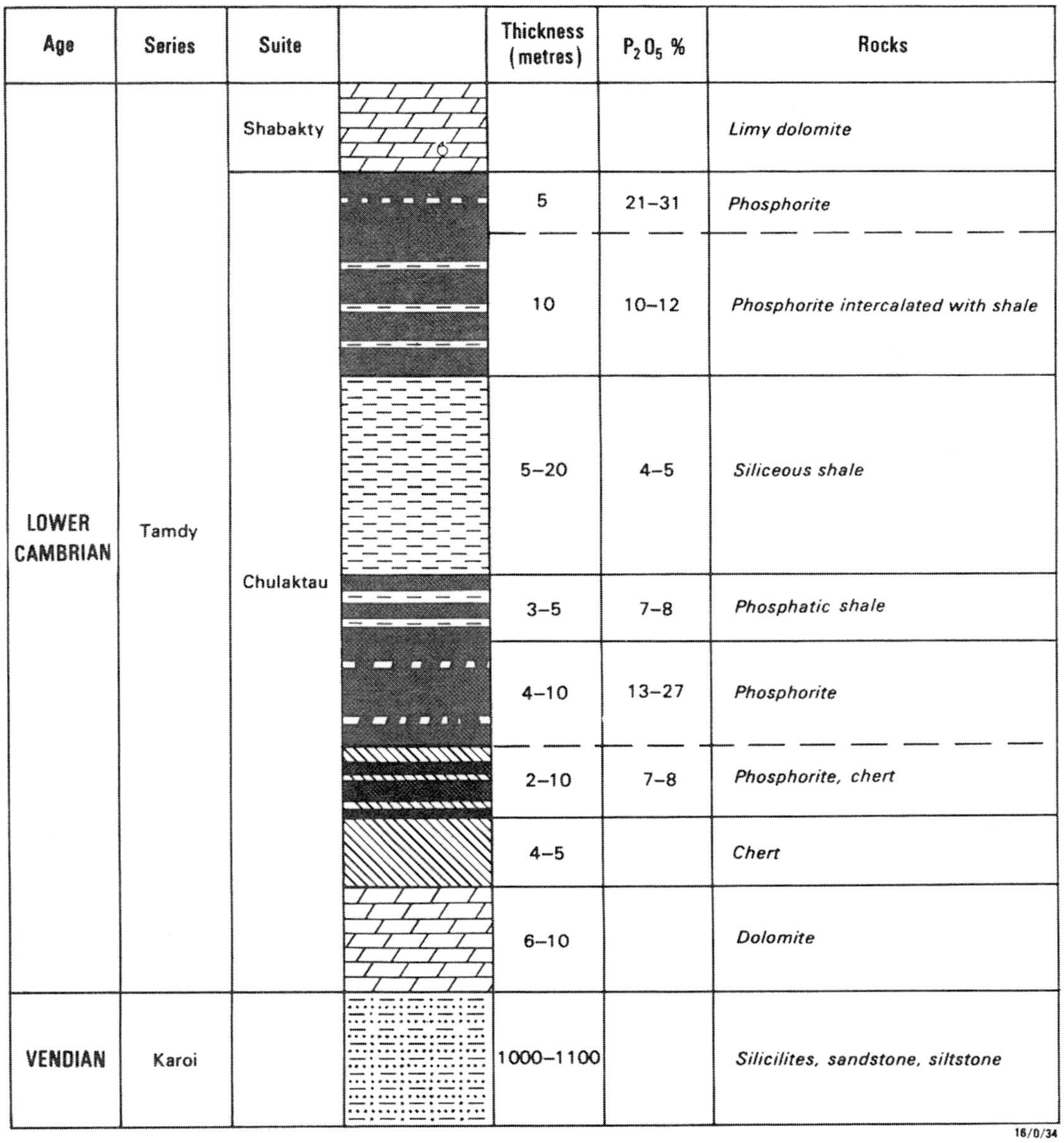

Fig. 5.4. Stratigraphy of the phosphatic unit, Koksu Deposit, Karatau Basin.

volume (Eganov *et al.*), only general points will be considered here.

The productive phosphorite-bearing series, the Chulaktau Suite, forms the lower part of the Tamdy Group, a thick carbonate sequence with an age span from terminal Vendian to Early Silurian. Numerous small phosphatised and phosphatic shelly fossils occur just below and within the phosphatic interval, and indicate an initial Cambrian (Tommotian) age. The Chulaktau Suite commonly includes three phosphatic members, with lower and upper phosphorite units separated by phosphatic shale (Fig. 5.4) with a combined thickness of 50–70 m. As a whole the phosphatic interval is continuous and persistent, but its individual beds vary in thickness, lithology, chemistry and petrology. This variation has a negative effect on beneficiation and further treatment of mined phosphorites. Structurally, the phosphatic and associated strata are steeply dipping, vertical or even overturned in places. In the north-eastern corner of the Karatau Basin, the Chulaktau Suite has been severely metamorphosed by granite intrusions.

Other late Proterozoic–Cambrian phosphate occurrences in Kazakhstan are of negligible importance when compared with Karatau. At several localities, however, the Cambrian Baikonur Group is notable for its alumino-phosphatic mineralisation: wavellite and variscite have been described by Popova & Sarbasov (1963).

In attempting to analyse the broad geological setting, the distribution of essentially Precambrian massifs should be noted. The most important of these are the Kokchetau-Ulutau, Chu-Kendyktas and Buruntau massifs, which together with several lesser ones, occur in central, southern and western Kazakhstan. During Riphean time, the whole of the western half of Kazakhstan was occupied by a single continental block (Tectonics of Kazakhstan, 1982). This block suffered strong faulting and rifting during Vendian time when several individual

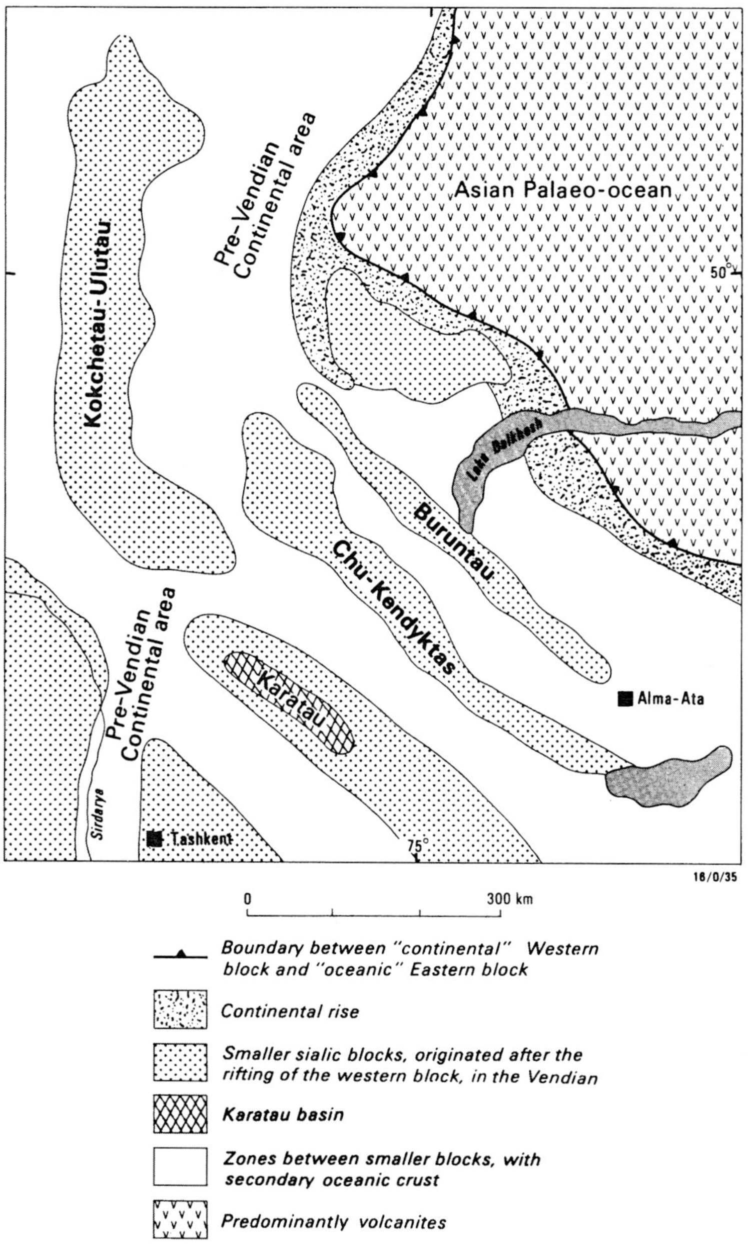

Fig. 5.5. Vendian–Cambrian tectonics of western, central and southern Kazakhstan (after Tectonics of Kazakhstan, 1982).

blocks or massifs developed and secondary oceanic crust was generated in intervening zones of subsidence. Ophiolite and basalt–andesite–greywacke complexes were widely developed in such areas as Dzalair-Naiman, northern Kisilkum and Tecturmas. Simultaneously, Vendian and early Cambrian carbonate sediments accumulated on the continental blocks. The Tamdy Series of Karatau is considered to be a good example of this kind of shelf sedimentary sequence.

The Lesser Karatau Range represents an island of older rocks surrounded by thick Mesozoic and Cenozoic sedimentary cover which makes it difficult to establish the palaeogeographical environment of the Karatau Basin. It may be that the portion of the Basin exposed today is in fact part of a much larger one deeply buried under younger sediments. In attempting to understand the palaeogeographic position of the Karatau Basin, it is necessary to note that sialic crust was not developed on the eastern block of Kazakhstan during Vendo-Cambrian time (Fig. 5.5).

This block is considered by some to have been a part of a late Proterozoic–Early Palaeozoic palaeo-Asian ocean. Kazakhstan was close to the equator during the Early Cambrian (Ziegler, 1982). The spatial relationship of the palaeo-Asian ocean to the Karatau Basin has still to be decided, but it is believed that this ocean was the most probable source of deep water which gave rise to the Karatau phosphogenesis.

References

Brovkov, G.N. (1974). Climatic environment of the southern part of Mid-Siberia in the later Precambrian–Palaeozoic. *Trudy Sibirskogo Nauchno-issledovatelyskogo Instituta Geologii, Geofiziki*

Mineralynogo Syriya, 182pp. Novosibirsk.

Donov, N.A., Edemsky, H.B. & Ilyin, A.V. (1967). Cambrian phosphorites in Mongolia. *Soviet Geology*, **3**, 55–60.

Ilyin, A.V. (1973). Khubsugul phosphorite-bearing basin. *Trudy Sovmestnaya Sovetsko-Mongolskaya Nauchno-Issledovatelskaya Geologicheskaya Ekspeditsiya*, **6**, 168pp. Geologicheskiy Institut, Akademiya Nauk SSSR; Nilzarubezhgeologiya Ministerstovo Geologii SSSR; Geologicheskiy Institut, Akademiya Nauk MNR; Izdatelstvo 'Nauka', Moscow.

Ilyin, A.V. & Volkov, R.I. (1977). A comparative description of the Khubsugul and other phosphorite-bearing basins. Pp. 167–179, In *Principal problems of the geology of Mongolia*, ed. N.S. Zaitsev & A.L. Yanshin, pp. 167–79. *Trudy Sovmestnaya Sovetsko-Mongolskaya Nauchno-Issledovatelskaya Geologicheskaya Ekspeditsiya*, **22**, 236pp. Geologicheskiy Institut, Akademia Nauk SSSR, Geologicheskiy Institut, Akademiya Nauk MNR.

Kazarinov, V.P. & Krasilnikova, N.A. (Eds) (1980). *Phosphorite and apatite in Siberia*. Institut Geologii i Geofiziki, Akademiya Nauk SSSR, Sibirskoe Otdelenie, 230pp. Ministerstovo Geologii SSSR & Ministerstvo Khimicheskoy Promyshlennosti SSSR, Izdatelstvo Nauka Sibirskoe Otdelenie, Novosibirsk.

Korobov, M.N. (1980). Lower Cambrian trilobites and the age of the phosphorites of the western Khubsugul (Mongolia). *Izvestiya Akademii Nauk SSSR, Seriya Geologicheskaya*, **1980(9)**, 97–107.

Krasilnikova, N.A. (1962). Phosphorite of the Krasnoyarsk district. *Minerals of the Krasnoyarsk district*. ed. L.V. Gromov, pp. 190–201. Akademiya Nauk SSSR, Moscow.

Popova, Z.D. & Sarbasov, Sh. Ch. (1963). Baikonur type deposits of alumophosphate ore and phosphorite in the Dzhezkazgan area. *Vestnik Akademii Nauk Kazakhskoy SSR*, **6**, 62–5.

Semikhatov, M.A. & Serebryakov, S.N. (1967). Vendian and Lower Cambrian of the southeastern part of Eastern Sayan. *Izvestiya Akademii Nauk, Seriya Geologicheskaya*, **1967(4)**, 87–102.

Tectonics of Kazakhstan (1982). Explanatory Notes to the Tectonic Map of Eastern Kazakhstan, Scale I:25 000 000. *Objyasnitelnaya zapiska k tektonicheskoi karte Vostochnogo Kazakstana*, 140pp. 'Nauka', Moscow.

Volkov R.I. (1974). *Phosphorite of the western part of the Khubsugul basin*. Publication of the Geological Prospecting Institute, Moscow.

Yanshin, A.L. (1964). Perspectives and scientific problems in the search for potassium salts and phosphorites in Siberia. *Byulleten Moskovskoye Obshchestvo Ispytateley Prirody, Otdelenie Geologicheskiy*, **39(5)**, 1–17.

Yanshin, A.L. (1982). Evolution of geological processes and phosphorites at the Precambrian-Cambrian boundary. *Geology of phosphorite deposits and problems of phosphoritogenesis*, ed. A.V. Ilyin, A.L. Yanshin & Yu.N. Zanin, Sbornik Nauchnykh Trudov, Institut Geologii i Geofiziki, Akademiya Nauk SSSR, Sibirskoe Otdelenie, Novosibirsk.

Zhabin, V.V. (1974). New data on Bokson bauxite deposit. In *Seminar po genezisu boksitov, Tezisy Dokladov*, ed. G.I. Bushinsky Akademiya Nauk SSSR, Ministerstvo Geologii SSSR. Moskow, 259pp.

Ziegler, A.M. (1982). Cambrian world paleogeography and phosphogenesis. In *Geology of phosphorite deposits and problems of phosphoritogenesis*, ed. A.V. Ilyin, A.L. Yanshin & Yu.N. Zanin, pp. 36–7. Sbornik Nauchnykh Trudov, Institut Geologii i Geofiziki, Akademiya Nauk SSSR, Sibirskoe Otdelenie, Novosibirsk.

6

Proterozoic and Cambrian phosphorites – regional review: Indian subcontinent

D.M. BANERJEE

Introduction

Phosphate deposits of the Indian subcontinent are small: nevertheless, some of them are unique because of their interesting lithofacies associations. Serious studies on phosphorite started on the Indian subcontinent only after the discovery of (Lower ?) Palaeozoic deposits near Birmania in the Jaisalmer district of western Rajasthan (Sheldon, 1966), and another discovery in the Tal Formation of Mussoorie in the Garhwal Himalaya. These discoveries marked the beginning of concerted exploration programmes in India under the aegis of the Geological Survey of India. With the finding of new deposits in the Aravalli sequence of Udaipur in south-central Rajasthan (Mukti Nath *et al.*, 1969), a further breakthrough was made in this field which ultimately triggered an extensive follow-up search programme in many Precambrian terrains. As a result new deposits were discovered in the Aravalli dolomites of Rajasthan (Udaipur, Banswara) and Madhya Pradesh (Jhabua), and in the ferruginous formations of the Bijawar Group in Uttar Pradesh (Lalitpur or Sonrai) and Madhya Pradesh (Hirapur–Bassia and Barwah). Smaller occurrences were recorded from Vindhyan dolomite and porcellanite, Gangolihat Dolomite (Calc Zone) of Kumaun Himalaya; in the Delhi Group of rocks of Rajasthan and Haryana; the Raipur Formation of the Chattisgarh area in Madhya Pradesh; and in the carbonate rocks of the Cuddapah Group of southern India (see Banerjee & Khan, 1981 for references). Meticulous reviews by Mukti Nath & Chakravarty (1972), and Mukti Nath (1974) laid a framework for further papers distinguishing potential areas for future exploration. Most comprehensive amongst these were the summaries by Sant & Pant (1980), Sant (1980) and Pant (1980). These authors demonstrated the possibilities of finding phosphorites in all the identified sedimentary basins of India. An exploration programme for phosphorite in Pakistan began around 1964, where initially Stanin & Hasan (1966) presented a not too optimistic picture of phosphate prospects. However, in 1970, with the discovery of phosphorites in the Kakul area, near Hazara, in the Northwestern Frontier Province, a major advance was made in this field. These phosphorites are confined to the lower parts of the Cambrian Hazara/Galdanian Formation and upper parts of the Abbottabad Formation. Important occurrences have been recorded from Sirbun, Kakul–Mirpur–Lagarban and Dalola areas (see Hasan, this volume; Bhatti, 1979; Asrarullah, 1980). Phosphorites in Upper Proterozoic dolomitic rocks have also been recorded recently from the Bajang and Baitadi regions of western Nepal (Bashyal, 1981). To date, however, no sedimentary phosphorite deposits have been recorded from the remaining countries of the Indian subcontinent: Bhutan, Bangla Desh and Sri Lanka.

The distribution pattern of phosphorites in the subcontinent is discussed under two regional headings; the Himalayan and the Peninsular regions, a scheme of description which involves both geographic as well as geological bias (Fig. 6.1). The Himalayan phosphogenic province includes the Proterozoic Krol Formation and Pithoragarh Formation in Kumaun Himalaya; Bajang and Baitadi Formations in western Nepal; the Lower Palaeozoic Tal Formation in Garhwal Himalaya; and the Hazara and Abbottabad Formations in Pakistan. In Peninsular India, phosphorite deposits belong to two major phosphogenic periods or cycles in the Lower Proterozoic and Upper Proterozoic. While phosphorites have been recorded from several Proterozoic horizons significant mineable deposits have been found only in the Aravalli and Bijawar sedimentary cycles of Peninsular India.

Phosphorites in all these instances occur in one or more lithofacies associations and forms. The principal ones are: (1) as primary-bedded cryptocrystalline phosphorite associated with dolomitic stromatolites; (2) as intraclastic and 'pelletal' microcrystalline phosphorite associated with carbonates, black shales and dark cherts; and/or (3) as leached, residual weathering products associated with ferruginous sediments, leached carbonates, manganese-rich argillites and as localised lateritic cappings.

Proterozoic phosphorites

Outer Himalayan belt

The outer belt of Proterozoic phosphorite in the lesser Himalayan region incorporates minor occurrences in the dolomites of the Krol Belt which is an important allochthonous

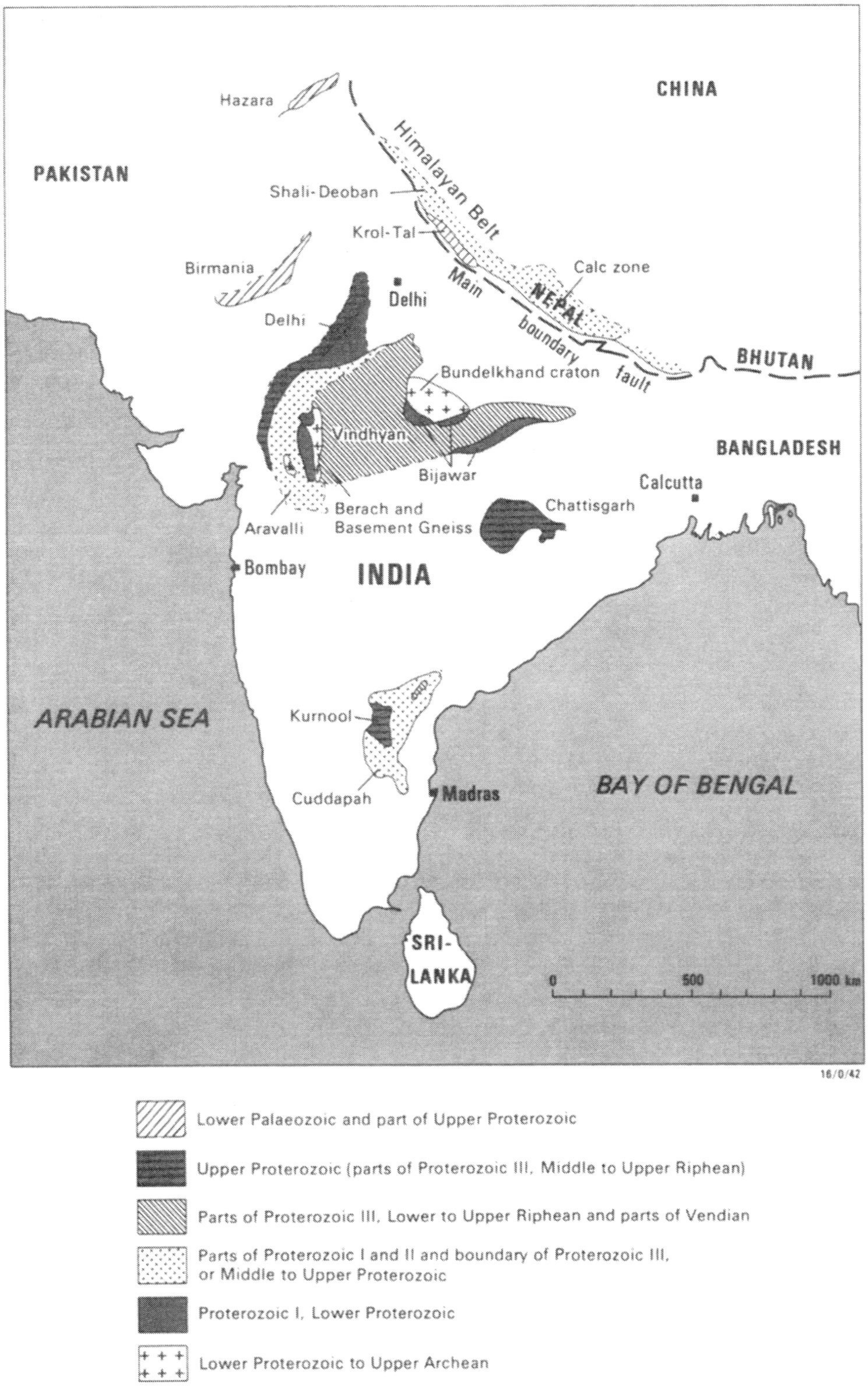

Fig. 6.1. Geographical distribution of Upper Proterozoic and Lower Palaeozoic phosphorites of the Indian subcontinent.

sequence of rocks in the Simla, Garhwal and Kumaun Hills (Gansser, 1964), extending over a total length of 350 km. The Krol Formation, which contains phosphatic beds at several levels, is a thick succession of carbonate with minor shales and rare sandy intercalations. It was divided into five units by Auden (1934) as Krol A, B, C, D and E. Bhargava (1972), however, preferred to group these as Lower Krol (A), Middle Krol (B) and Upper Krol (C, D, E). These are dominantly tidal deposits (Awasthi, 1970; Kharakwal & Bagati, 1976; Singh & Rai, 1977 a, b). Due to its position above the Infra-Krol and Blaini Formations, and the incorrect correlation of boulders of Blaini Formation with the glacial Talchir boulders of the Peninsular India, and also due to the dubious identification of fossils in the horizons of controversial stratigraphic disposition, the Krol

Formation has been assigned several ages: Triassic (Ghosh & Srivastava, 1962), Jurassic (Tewari, 1969), Late Jurassic to Late Cretaceous (Sinha, 1975), Permo-Triassic (Kumar, 1978), Permian (Valdiya, 1980; Kumar, 1980) and Precambrian (Gundu Rao, 1970; Singh & Rai, 1977 a, b; and Singh, 1979 a, b, c). A Precambrian age had been suggested on the basis of the occurrence of the stromatolite form species *Conophyton*; the possibility of correlating conglomerates in the Blaini Formation with Proterozoic glacial sediments; and the lack of recognition at that time of identifiable megafossils which should be abundant in such Phanerozoic tidal flat deposits. Younger ages assigned by most workers have inadequate geological, palaeontological or sedimentological support (Singh, 1981). On the basis of the above reasoning, and the recent discovery of a fauna close to the Cambrian–Ordovician boundary in the overlying Tal Formation (Azmi, Joshi & Juyal, 1981), and also an Early Cambrian fauna in thin phosphatic layers in the uppermost Krol Formation (Azmi & Pancholi, 1983), the suggestion of a Late Proterozoic (Vendian) age for the major part of the massive Krol Formation appears closer to the truth. Dr M.E. Raaben, USSR Academy of Sciences, Moscow, has recently studied the stromatolites of the Upper Krol Formation which are in the collections of the Universities of Delhi and Lucknow, and has concluded (*in litt.*) that

> Evidence now available indicates that the possibility of a Precambrian (Uppermost Precambrian ?) age for the Krol stromatolite-bearing beds cannot be excluded. Suggestive in this aspect is the presence of *Conophyton* Maslov in these beds which now is known in Proterozoic only. The form-species in Krol is not *Conophyton garganicus* but seems to be a new one. A very low rate of thickening of the lamina in the axial zone as well as peculiarities in the texture of the lamina have to be noticed. Columnar branching structures in Krol stromatolite-bearing beds do not belong to *Baicalia* or to any other Riphean form-genus and affinities to Phanerozoic (Cambrian and Ordovician mostly) stromatolites are seen in the shape and especially in the inner structure of the columns and also in lamina texture. The outlook of some of these branching structures is slightly reminiscent of the *Tungussida* Supergroup (*sensu lato*), but the mode of branching has to be studied more in detail. As to the non-columnar stromatolites of Krol, the morphology of the structures is not suggestive of any definite age, and the microstructures have not yet been studied. The Krol assemblage as a whole cannot be compared to any other stromatolitic assemblage known in Lesser Himalaya, nor in the Vindhyan or the Riphean. It will be of interest to try a comparison with some assemblages of Latest Precambrian–Lower Cambrian age.

Nainital–Solan areas. Phosphorite in the Krol Formation of the Nainital area was first recorded by Choudhuri & Banerji (1966) and around Solan in the Simla hills by Joshi (1970). The exposures at Nainital, along the Tiffin Top Section, are perhaps the most instructive. Lithostratigraphy along this section is given in Fig. 6.2. Phosphorite occurs as small discontinuous beds within a black shale–dolomite lithofacies. Phosphorite beds are also seen on the Lands-End road, where they were mapped as greyish black thinly bedded shaly rocks, which on polished surfaces show an uneven distribution of darker phosphate grains against a groundmass of dark, shaly material. Phosphorite-rich beds are 3–4 cm thick and are interbedded with laminated shaly carbonates 1–2 cm in thickness. These beds pinch out within a short distance and grade into shaly carbonate layers. There are two basic types of phos-

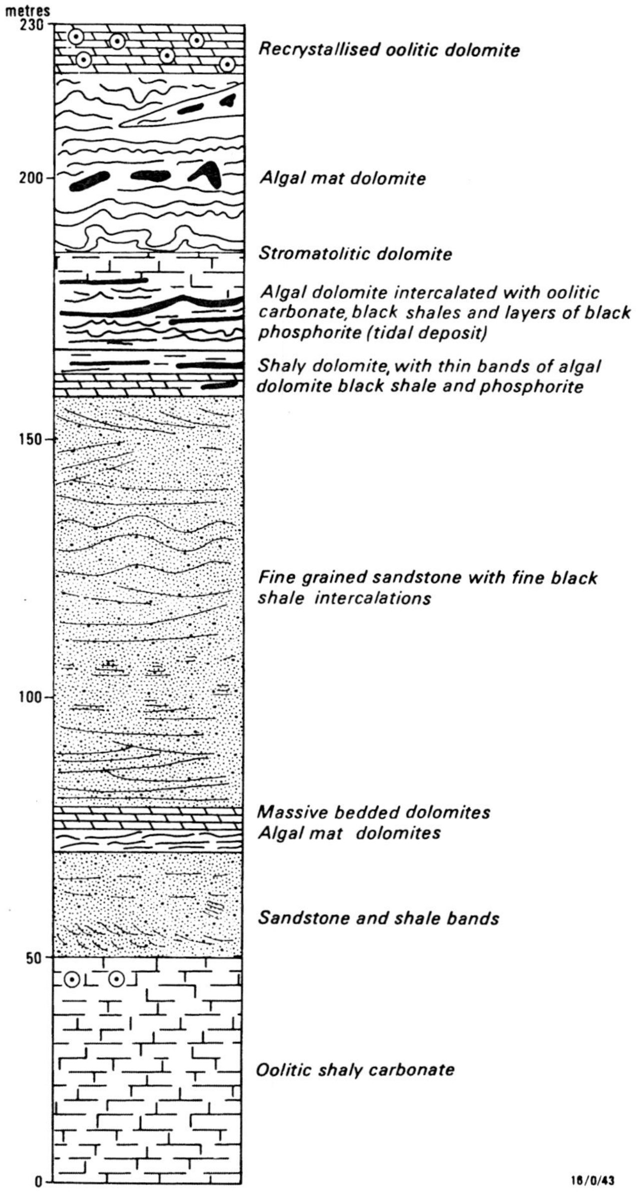

Fig. 6.2. Lithological section along the Tallital–Tiffin top road.

phorite: Type-A contains 23% P_2O_5 and contains carbonate as the major gangue, while Type-B is argillo-calcareous, occurs with black cherts and thin dolomites, and contains about 10–15% P_2O_5. Minor occurrences have also been recorded in dolomites of this belt from different areas, but no significant geological information is available about them.

Inner Himalayan belt

The inner belt of sedimentary rocks in the Lesser Himalaya is located south of the Central Crystalline Belt and north of the crystalline and metamorphic belt of the Kumaun and Garhwal Himalaya. It includes the Calc Zones of Tezam, Pithoragarh, Badolisera, Bageshwar and Chamoli in Kumaun Himalaya. No megafossils have been found in these rock formations, but they do contain well-developed columnar stromatolites and are correlated with Middle to Upper Riphean cycles of the Siberian Platform (Valdiya, 1964; Banerjee, 1967; Jain, Banerjee & Mithal, 1971). This group of rocks is now designated as the Garhwal Group by the Geological Survey of India.

Pithoragarh–Bageshwar areas. Phosphorite in these areas of Kumaun Himalaya is associated with stromatolitic dolomites (Fig. 6.3A) and occurs as thin beds, streaks and veinlets along the rims of stromatolitic columns and as surface encrustations. The stromatolite columns are invariably enveloped by dark phosphate layers. Short and stumpy stromatolites are more phosphatic than the taller ones (Valdiya, 1972). A common association of crystalline magnesite with these phosphorites, in the form of inter-laminated early diagenetic layers, could be significant for the interpretation of their genesis. Occasionally, these phosphorites are granular (0.07–0.7 mm) and occur in association with agglutinated micritic pellets in a micritic groundmass. Filamentous micro-organisms occur in abundance in these phosphorites which are similar to algal and cyanobacterial assemblages found in Precambrian cherts. In spite of their close association with the stromatolites, these phosphorites are low in C_{org} content (*c.* 0.2%) and the $\delta^{18}O_{carb}$ value of $-17‰$ and $\delta^{13}C_{org}$ value of $-4.7‰$ are somewhat anomalous and difficult to interpret.

The phosphorites of the Pithoragarh region have been studied by Valdiya (1972), Khan (1978) and Israili & Khan (1980).

Bajang–Baitadi areas. The Calc Zone of Pithoragarh and Tezam of the Kumaun Himalaya extends eastward into western Nepal (Remy, 1975) where similar stromatolitic phosphorites (Fig. 6.3B) have been recorded in the Bajang–Baitadi region (Bashyal, 1981; Bashyal, Banerjee & Khan, 1981). Two stromatolitic-bearing horizons have so far been identified in the carbonates of the Baitadi area, of which the lower one, the Dhungad stromatolitic dolomite, is non-phosphatic, and the upper part is mildly phosphatic. In the Bajang Carbonate Formation, only the upper part of the stromatolitic carbonate is phosphatic. The Baitadi Formation is the physical continuation of the Gangolihat Dolomite Formation of the Pithoragarh area (Valdiya, 1964) and the Bajahang Formation is the eastward continuation of the Kapkote Dolomite (Banerjee, 1967) of the Calc Zone of Tezam in the Kumaun Himalaya. In the Baitadi area, columnar stromatolites are phosphatised along their peripheral parts and contain phosphate intraclasts, chert and quartz fragments. They analyse at 10–15% P_2O_5. In the Bajang area, phosphatic beds are 32–60 m thick, but the stromatolitic phosphorites are only 6 m thick, with sandy matrix, and analyse 8–15% P_2O_5. A few massive 1-m thick beds of phosphorite with 24% P_2O_5 have also been located in the Juil Gad section near Khet village.

Peninsular Indian Platform

Aravalli Group of Rajasthan. The type area of the Aravalli Group is around the township of Udaipur, and phosphorite deposits in this area are located near Kanpur, Matoon, Jhamarkotra, Neemuch Mata, Badgaon and Dakankotra. These phosphorites are found associated with phosphatic and dolomitic stromatolites. The Aravalli stratigraphy *vis-à-vis* phosphorite beds has been interpreted differently by present and past workers. While Banerjee (1971a, b, 1975, 1980) and Chauhan (1973, 1979, 1980) interpreted two distinct stromatolitic horizons within the Matoon Formation, Roy & Paliwal (1981) believe that only a single stromatolitic phosphorite bed exists in the area which is repeated by regional folding.

Serious differences of opinion exist about the ages of these rock formations. The classical viewpoint tends to propagate an Early Precambrian to Late Archaean age for the erstwhile 'Aravalli System', which may have influenced Crawford (1970) to conclude that the Aravalli metasediments must be younger than 2500 m.y., but older than 2250 m.y., in spite of his own statement and admission that the material on which the Rb–Sr ratios were calculated and used for interpreting the age of the Aravalli Group were not suitable materials for age determination (Crawford, 1970, pp. 100–101). K–Ar analysis of minerals, however, indicates that the gneisses from the Sarara Inlier, south of Udaipur – correlatable with part of the Banded Gneissic Complex of Heron (1953) and Gupta (1934) and the Bundelkhand Granite of the Chittorgarh area (= Berach Granite of Pascoe, 1950) underlying Aravalli Group metasediments, are only 1500 m.y. old (Vinogradov *et al.*, 1966). Likewise, Sarkar *et al.* (1964) assumed a 953–1020 m.y. age for the micas in the 'Aravalli System'. Except for these controversial and confusing age data, no rock has been dated in or around the phosphorite-bearing Aravalli metasediments. Consequently, Banerjee (1971b, 1980) and subsequently Chauhan (1973, 1979) used stromatolites as a tool for biostratigraphic zonation and concluded that the Aravallian phosphorites are Middle to Late Riphean in age on the basis of characteristic stromatolite form species. Identification of the form species *Baicalia* was however questioned by Raaben (1981) and more recently by M.R. Walter (personal communication), who during a field inspection recognised forms closely resembling the late Proterozoic stromatolites *Acaciella* and *Linella avis* (Fig. 6.3C) in these Aravalli phosphorites, and accepted the earlier identification of *Kussiella* and *Minjaria*-like forms in the same assemblage. These form identifications, if correct, and properly correlated with other evidence, would move the Aravalli Group much higher in the Proterozoic scale, much younger than Banerjee (1971b, 1973) had earlier interpreted. In contrast to these viewpoints, Raaben (1980, 1981) described some new forms of stromatolites from the Kanpur area of the Udaipur Basin and ascribed the Aravalli phosphorites an Early Proterozoic age.

The Udaisagar Granite, in the Ahar river section near Udaipur, has been dated by Crawford (1970) at 2250 m.y. Previously, it was believed (Heron, 1953) to be a post-Delhi Group granite of the Erinpura intrusive event, and accordingly instrusive into the Aravalli sediments. The basic presumption of its disposition has been controversial. While Heron (1953, p. 328) favoured a post-Aravalli age of these granites he was not fully convinced (Heron, 1953, p. 325). Sharma (1953) also believed that these granites belonged to the Aravalli cycle. Recently, rounded rolled-up pebbles and boulders of the same granite have been recorded in the carbonate–phosphorite beds of the Middle Matoon Formation at the Matoon phosphorite mines (Dutta, 1980). This evidence clearly suggests a pre-Aravalli or early Aravalli age for the granites. The view that

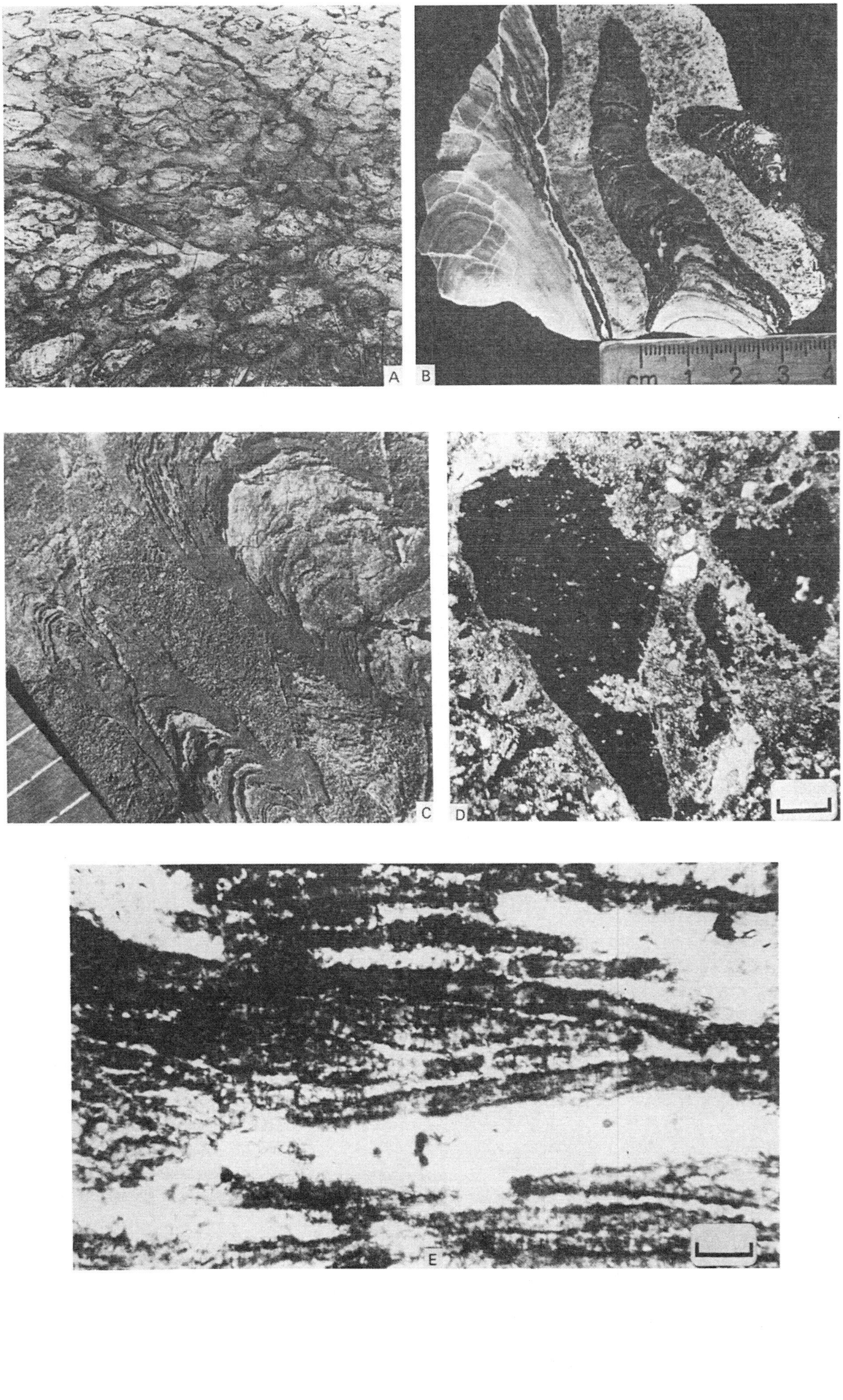
A
B
cm 1 2 3 4
C
D
E

most of the Udaisagar granitic mass is part of the basement-complex and shows localised intricate contact effects along certain carbonate–granite margins, giving a false impression of an intrusive body, has been expressed by Banerjee (1980), Roy *et al.* (1980) and Saigal (1983). Hence, the Aravalli sediments are certainly younger than the Udaisagar Granite (2250 m.y.), and not older as concluded by Crawford (1970). Stable Pb isotope data for the pre-Aravalli or Lower Aravalli or Banded Gneissic Complex synsedimentary stratiform lead–zinc ores of Rajasthan suggest a maximum age of 1500 m.y. (Chernyshev *et al.*, (1980). Using the data of Chernyshev *et al.* (1980), Deb (1982) has calculated the age of the pre-Aravalli (or possibly Lower Aravalli) lead from the Rajapura–Dariba deposits, east of Udaipur, at 1700 m.y. The *sensu-stricto* Aravalli sediments with phosphorite occur hundreds of metres above these Pb–Zn-bearing horizons, hence must be considerably younger.

Phosphorite in these rocks is chiefly confined to algal stromatolitic bodies, in upright *in-situ* columns (Fig. 6.3C), as fragmental stromatolitic debris (Fig. 6.3D), as discontinuous microsphorite layers, as leached products in the form of veins and cavity fillings, or as surface cappings on weathering profiles. The petrology of stromatolitic phosphorites has been published by Banerjee (1971a, b), Chauhan (1979), Banerjee, Basu & Srivastava, (1980) and Saigal (1983), and a number of recent workers. The weathered apatitic cappings which occur as dense porcellaneous phosphorites (Fig. 6.4A) have been described by Choudhuri (1981). Except for this specific variety, all other types of phosphorite show a certain consistency in petrological, mineralogical and geochemical characteristics. The dominant phosphatic fraction of the Aravallian phosphorite is microcrystalline and is mineralogically carbonate fluorapatite, tending to become carbonate-poor fluorapatite in some deposits, particularly at Jhamarkotra and Hirapur. Microcrystalline apatite occurs as needle-shaped bundles showing low birefringence. Hexagonal and pseudohexagonal forms are common along voids and fracture fillings, having formed due to solution, redeposition and recrystallisation. Minor calcite and abundant dolomite are the most common minerals of the cement and the matrix. Quartz and chalcedonic silica are the most frequently occurring contaminant. Micaceous minerals are rather uncommon. Replacement of apatite by secondary calcite is frequently observed. Opaque Mn-minerals, iron oxides and iron-carbonates occur more commonly in the associated dolomites rather than in the phosphorites, although their close proximity to each other makes it difficult to identify the true lithofacies associations. Filamentous micro-organisms of cyanobacterial affinity (Fig. 6.3E), commonly described from many Proterozoic cherts of North America and Africa, have been recorded in the stromatolitic phosphorites of Neemuch Mata and Dakankotra (Banerjee, 1983).

Stromatolitic phosphorites are also found in the Banswara district of southern Rajasthan, around Sallopat and Ram-Ka-Munna (Ghosh & Biswas, 1973), and in the southern extension of the Aravalli Group in Gujarat State, known as the Champaner Formation in the Panchmahal district. The phosphorites of Banswara and Panchmahal are characteristically pink-brown in colour, contrasting to the grey-white hue of the Udaipur phosphorites. In all other mineralogical and petrological characteristics they are similar.

Aravalli Group of Madhya Pradesh. The Aravalli phosphogenic basin extends southward into Madhya Pradesh State where outcrops of phosphatic and non-phosphatic stromatolitic dolomite occur within massive thicknesses of metamorphosed argillite and cross-stratified lithic-arenite bordering an old cratonic gneissic mass. The dolomite–chert bands form the eastern branch of a regionally asymmetrical syncline. The best developed sections are those around Khatama (Khatamba), Kelkua and Amlamal and these have been studied by Munshi, Khan & Ghosh (1974), Basu & Banerjee (1975), Basu (1976), Banerjee & Basu (1979, 1980), Khan (1979) and Banerjee *et al.* (1980). In a recently compiled geological map of the Aravalli belt (Gupta *et al.* 1980), the Aravalli Group of the Jhabua area has been shown to be younger than the Aravalli Group of the Udaipur area, although there is no sound geological argument or basis to propose such a correlation. There are no major differences between the two types of phosphorite or the stromatolites, and hence they could be of similar ages. Complex folding and faulting has rendered the Jhabua area quite difficult for proper stratigraphic interpretation. Placement of litho-units on a stratigraphic scale without considering the complex tectonics of the area has led to erroneous lithological groupings by several workers. Furthermore, weathering has resulted in the creation of extensive phosphatic soil cappings over the fragmental debris of stromatolitic phosphorites. *In situ* growth position stromatolites, with phosphorite confined to their columnar body, occur towards the lower part of the sequence and constitute a prominent litho-unit. Fragmental stromatolitic debris is, however, richer in P_2O_5 content following the removal of associated carbonate by leaching. A vertical sequence from a typical section near Khatama is given in Fig. 6.5.

These phosphorites are characterised by layers of microcrystalline apatite, a large amount of silt-sized quartz detritus, unresolved pelitic matter and some iron-oxides. Stromatolitic columns are made up of banded boundstone phosphorite layers frequently bordered by chalcedonic silica bands. Alternating

Fig. 6.3. (A) Dolomitic stromatolites with marginal and interlaminar dark phosphorite, Chandak, near Pithoragarh. (B) A single complete phosphatic stromatolite column in a non-phosphatic dolomitic groundmass, Bajang area, Nepal. (C) *Linella*-type stromatolitic phosphorite showing marginal concentration of phosphate, Neemuch Mata, Udaipur. Scale, 1 cm. (D) Fragmental stromatolitic phosphorite, Matoon Mines, Udaipur. Scale, 0.25 mm. (E) Group of filamentous microfossils (dark) filled with dark phosphatic matter in a light-coloured carbonate groundmass. Scale, 0.25 mm.

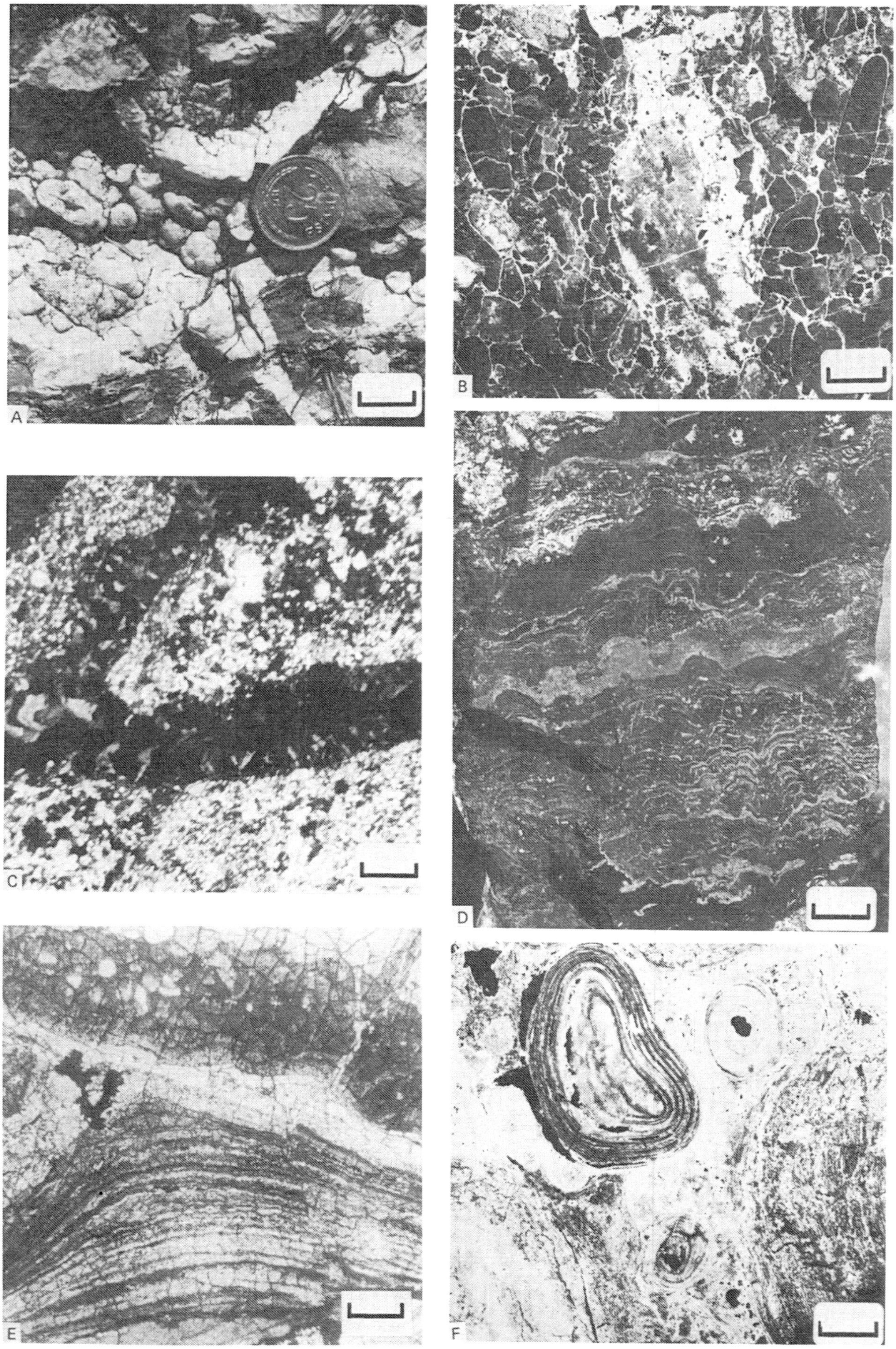
A
B
C
D
E
F

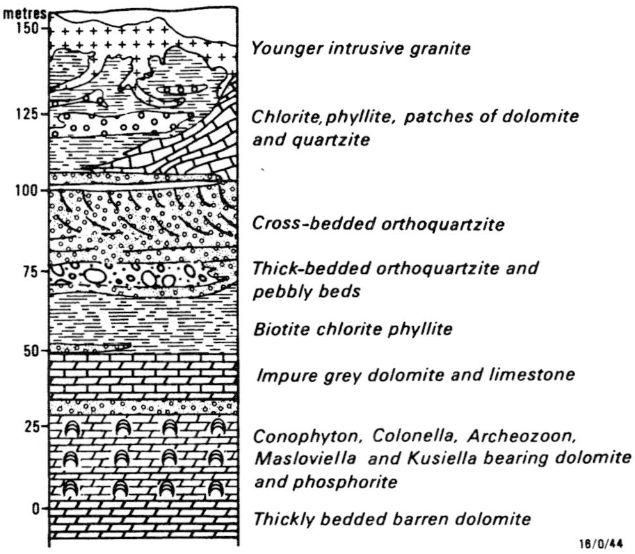

Fig. 6.5. Lithological section of the Khatama block, Jhabua phosphorite deposit.

apatite- and carbonate-rich layers show rhythmic banding in some samples (Banerjee, 1978). Brecciated, fragmental phosphorites contain angular to subangular fragments of quartz and early-formed microsphorite fragments. While most of the brecciation appears to be a result of solution-collapse and recementation after the consolidation of the rocks, some could well be early diagenetic. Cavity fillings by apatite grains, through solution and recrystallisation, are common features. Dense, laminated, white phosphorites containing a high P_2O_5 content are somewhat restricted to some parts of the Khatama deposit and are distinctly different from the stromatolitic and fragmental phosphorites.

Aravallian phosphorites of both Udaipur and Jhabua districts contain an average of 0.22% organic carbon. This value goes up to 0.59% in some stromatolitic samples. $\delta^{13}C_{carb}$ values of about −0.7‰ to −1.1‰ and $^{18}O_{carb}$ values of −10‰ to −11‰ suggest normal marine conditions for their formation. The $\delta^{13}C_{org}$ values of −11‰ to −15‰ are also of interest and are being interpreted in relation to their genesis (Banerjee, in preparation).

Bijawar Group of Madhya Pradesh. The Bijawar Group of rocks, considered older than the Aravalli Group or lateral equivalents (Crawford & Compston, 1970; Crawford, 1970, p. 107–8), named after Bijawar town in Madhya Pradesh, comprise ferruginous breccia, quartzite, dolomite, limestone (locally stromatolitic) and basic effusive lavas. This metasedimentary cover overlies the southern margins of the ancient Bundelkhand Granitic Basement Complex. A large deposit of low- to medium-grade phosphorite was discovered within the ferruginous breccia and red shales around the Bassia–Hirapur–Mardeora areas, in the Sagar and Chattarpur districts, during exploration for radioactive minerals in the area. The phosphorite horizon overlies a thick sequence of Lower Bijawar Dolomite and is in turn overlain unconformably by the Vindhyan sandstones of Semri age. This deposit represents the western part of the northern Bijawar Basin. The phosphorites are of platform type and contain 15–20% P_2O_5, but being rich in ferruginous matter are not suitable for super-phosphate manufacture. They are therefore used for direct fertilizer application to acidic soils. There is no direct radiometric age control for these phosphorites, hence the true ages of these rocks are quite speculative. Brief descriptions of the country rocks (phosphorite was not identified at that time) by Mathur (1960), Mathur & Mani (1978) and Rajarajan (1978) are the only sources of published information about these rocks. It is interesting to note that Rajarajan (1978) correlated the Bijawar Group with the Cuddapah Group, and not with the older Dharwar Group (as per convention) because of its position in the local stratigraphy and lower metamorphic grade. If such a correlation receives more support from other lines of evidence, the suggested correlation of Bijawar with the Lower Aravalli Group (parts of the pre-Aravalli of some authors) would seem valid (Banerjee, 1982). The position of phosphorite beds within the stratigraphic framework of the area has been given by Pant (1980), and the geology of the deposit, mineralogy and petrology of the ore has been discussed by Banerjee (1982) and Banerjee *et al.* (1982). An idealised sequence across the Bassia–Hirapur deposit is shown in Fig. 6.6.

The bulk of phosphorite in the area is of residual type and shows interesting mineralogical transformations, first approximating the composition of fluorapatite from a basic carbonate fluorapatite precursor and then to crandallite and wavellite. All these phosphorites show evidence of extensive leaching of carbonate and phosphate minerals from their matrix during episodes of weathering. Four types of phosphorite (designated Types I–IV) have been distinguished in this area (Banerjee *et al.*, 1982). Type-I is ferruginous, shaly and laminated; Type-II is massively bedded and ferruginous; Type-III is brecciated and conglomeratic; and Type-IV is a vein and pore-filling type with recrystallised crystalline apatite (Fig. 6.4C). Opinions differ as regards the nature and origin of the brecciated/conglomeratic phosphorites. While these are the most common type in the Hirapur area they are not so prevalent in the Sonrai area of Uttar Pradesh. The present author believes that the brecciation is basically a primary synsedimentary phenomenon, formed

Fig. 6.4. (A) Dense porcellaneous surface encrustation type of phosphorite, Jhamarkotra-E mining block, Udaipur. Scale, 2 cm. (B) Pelletal phosphorite of Chelima. Scale, 0.25 mm. (C) Pore-filling type of recrystallised apatite crystallites in a predominantly siliceous groundmass, Hirapur. Scale, 0.25 mm. (D) Chert-phosphorite-pyrite association in the laminated stromatolitic rock at Durmala, Mussoorie Syncline. Scale, 1 cm. (E) Phosphatic stratiform stromatolite at Chelima, Cuddapah. Scale, 0.25 mm. (F) Pisoliths in phosphorites from the Hazara district, Pakistan. Scale, 0.25 mm.

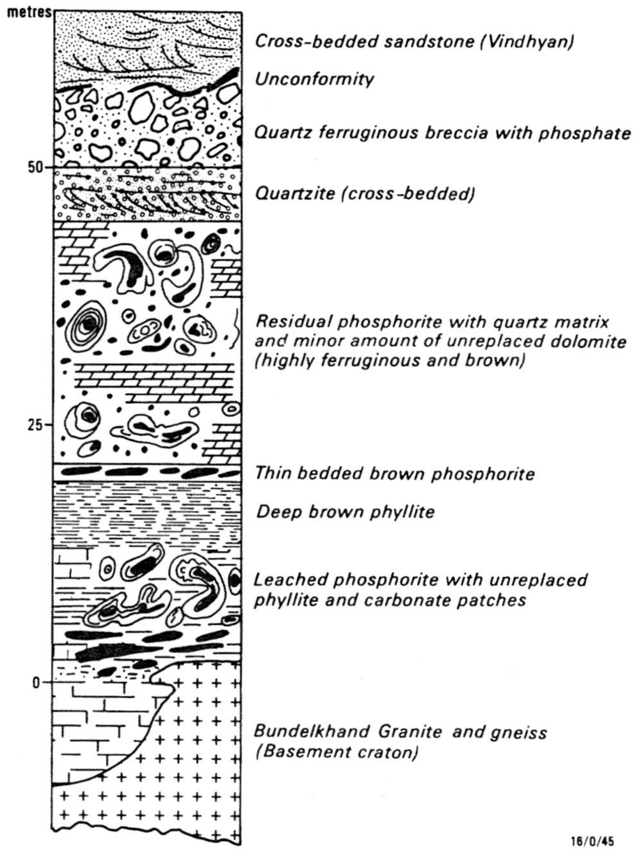

Fig. 6.6. Lithological section of the middle part of the Bassia phosphorite deposit, Hirapur.

through the processes of slumping, collapse and fragmentation of the clastic units and their recementation by circulating ferruginous solutions, and later by phosphatic solutions. The bedded character of this unit can be shown by its near-conformable boundaries with the overlying and underlying shales and carbonates. An alternative hypothesis, invoking secondary collapse of chert fragments into voids left by solution of possible evaporites, has been suggested by Sheldon (personal communication), and it would be worthwhile examining the deposits again with this possibility in mind. It has however been shown that the weathering was a Proterozoic phenomenon and was enhanced during the sub-Recent and Recent weathering cycles (Banerjee *et al.*, 1982).

South-east of Indore township in Madhya Pradesh, a new phosphate deposit has recently been discovered by the Directorate of Geology and Mining M.P., in the vicinity of Mondri village, close to the Barwah and Chandgarh areas to the east. Detailed exploration work in these areas is continuing, but the preliminary geology indicates deposition of the Bijawar Group over an Archaean gneissic basement. The Barwah deposit is quite similar to that of Hirapur–Bassia, except that the secondary apatites in the Barwah area were fewer compared to the Hirapur area (Sheldon, personal communication). The association of bedded manganese deposits in the Barwah area makes it more similar to the Aravallian phosphorites of Jhabua, which have a more or less similar lithological association.

Bijawar Group of Uttar Pradesh. The eastern extension of the Bijawar Basin is exposed near Sonrai in the Lalitpur district. During the early 1970s this area remained a centre of vigorous geological activity due to involvement of the United Nations in an exploration programme for base-metals in these rocks. Although voluminous data on the base-metals were generated, phosphorite in these rocks remained unrecorded until similar phosphorites were discovered in the Hirapur area of Madhya Pradesh. The exploration area in Sonrai is divided into several mining blocks. In one of these blocks, near Tori, it was possible to prepare a more or less complete section (Fig.6.7). As in the case of the Hirapur–Bassia areas, the phosphorite is banded, leached and residual (Jain & Pant, 1981). The lithological successions in the two areas are not identical, and their correlation is open to question.

It has been proposed that the Upper Bijawar Group of the type-area, designated as the Gangau Ferruginous Formation in the east, seems to have been overlapped by a regressive cycle of sediments represented by clastic sequences of the western part of the Bijawar Basin. If such was the case, then the Sonrai Formation and Gangau Ferruginous Formation could be considered facies variants, reflecting changes from an argillaceous, phosphatic, conglomeratic facies in the Hirapur–Bassia area to a dolomitic, sandy, phosphate facies in the Sonrai area. Unconformable overlapping by Lower Vindhyan sandstone in the Hirapur area and by the Upper

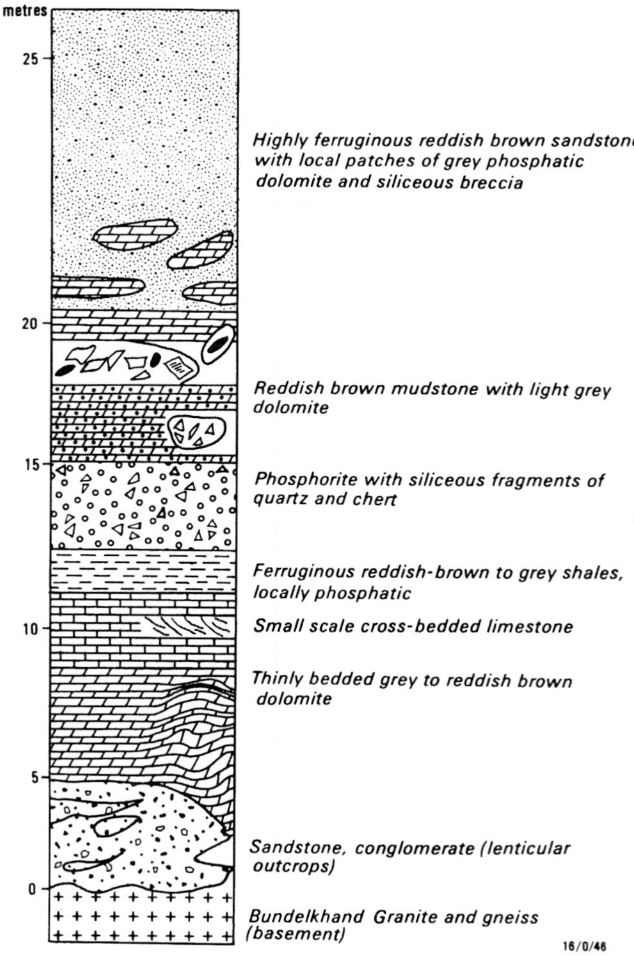

Fig. 6.7. Lithological section across the mining area at Sonrai.

Vindhyan rocks in the Sonrai area, suggests an accretionary type of prograding deposition, gradually shifting its locus outward and westward from the cratonic borderland of the Bundelkhand Granitic Massif (Banerjee, 1982, pp. 52–54).

Petrographically, the phosphorites are identical to those of Hirapur–Bassia area, except that the paragenetic weathering sequence is more extended, so that millisite and wavellite are found in abundance. These deposits are being studied by the Directorate of Geology and Mining and its associates in Uttar Pradesh.

Delhi Group of Western India. Recent reports of 2.5-m thick beds of phosphorite with 25% P_2O_5 in the Delhi Group (Dayal, Chaurasia & Hore 1980; Dayal *et al.*, 1981) around Achraul, in the Siriska Forest area of Rajasthan and minor occurrences in Haryana, has expanded our outlook beyond the boundaries of the Aravalli and Bijawar Groups. The Delhi Group of rocks is considered Upper Proterozoic, even in the classical scheme, and is often correlated with the Gwalior Group (confidently) and the Lower Cuddapah Group (less confidently). If the suggested dating of the Aravalli Group as Middle to Late Proterozoic is correct, the phosphorite bearing Delhi rocks could then be placed around the Proterozoic–Cambrian boundary.

Vindhyan Group of Central India. The Vindhyan Group is also generally correlated with the Delhi and Gwalior Groups. Correlation of the Lower Vindhyan Group with the upper parts of the Aravalli Group was suggested by Banerjee (1971b) on the basis of similarities of some stromatolitic forms and a overall similarity of the geological depositional pattern. However, it now appears that the major part of the Vindhyan Group (Middle and Upper) is directly correlatable with the Delhi Group (Banerjee, 1982), hence is likely to have a latest Proterozoic–Early Palaeozoic age.

Stromatolitic carbonates, containing chert lenses and analysing 20% P_2O_5, have been reported from the Kajrahat Limestone Formation. Phosphatic beds have also been recorded from the Dala, Susnai, Argarh, Chandi and Basuhari areas. A new occurrence has been found near Bakhsawa (Sheldon, personal communication) where porcellanite beds in the Lower Vindhyan Group were found to be phosphatic.

Chattisgarh Basin of Madhya Pradesh. Highly phosphatic stromatolitic limestone has been recorded from the Nandini-Drug district, occurring in the Raipur Formation (Series) of the Chattisgarh Basin (Vishwanathiah & Sastri, 1973). Further investigation by the Geological Survey of India has indicated new occurrences in the Charmuria Limestone which occurs at the base of the Raipur Group (Series) overlying the Chandarpur Sandstone. Phosphorite occurs with stromatolites and as small intraclasts within the laminae of weathered limestones.

Upper Cuddapah Group of Southern India. The Upper Proterozoic Cuddapah Group in Andhra Pradesh has yielded promising results as regards phosphorite exploration, after only a brief survey by the Geological Survey. Though small in extent, the phosphorites discovered near Chelima, Pachcherla and Peddasettipalle (Basu, Sood & Venkatesh, 1981), associated with stromatolitic dolomite, quartzite and shale, are significant from the point of view of understanding the episodes of phosphate formation during Proterozoic times. Most are of the algal-mat variety (Fig. 6.4E), but some are found associated with columnar stromatolites. Some pelletal varieties occur with quartzose rocks and khaki-green shales, and sometimes with lead–zinc-bearing light-grey dolomite. Stromatolitic phosphorites resemble the Neemach Mata phosphorites and show discontinuous apatitic laminae and an outer rim of grey microcrystalline phosphate. Grainstone phosphorite (Fig. 6.4B) is made up of ovules of variable sizes and shapes and is embedded in a siliceous, ferruginous and calcareous matrix. Recrystallisation of carbonate into sparry dolomite is common. Diverse assemblages of unicellular and multicellular filamentous micro-organisms can be observed in these phosphorites.

Lower Palaeozoic phosphorites

Several lines of evidence, including those based on the global distribution of phosphorite, indicate that the Early Palaeozoic (in particular Early Cambrian, close to the Proterozoic–Cambrian boundary) was a period of world-wide phosphogenesis (Cook & Shergold, 1980). Identification of Early Cambrian fauna in the phosphorites of the Abbottabad and Hazira Formations of Pakistan, indicates that the western Himalayan belt was an area of phosphate deposition during the Early Palaeozoic. Phosphorites in the Lesser Himalayan Tal belt of Kumaun and Garhwal in India, have been known for a long time, but it was only recently that these have been considered related to the Early Palaeozoic global event. So far, definite marine Lower Palaeozoic rocks have not been recorded from the Peninsular part of the Indian subcontinent, and it is generally believed that after the Precambrian metallogeny, Peninsular India remained a land area until localised marine transgressions took place in the central parts of India during Permo-Carboniferous times. There are however, several sequences which have not been precisely dated palaeontologically. One of these possibly Palaeozoic (?) sequences, around Birmania in western Rajasthan, contains low-grade phosphorite, and it would not be surprising if the Birmania phosphorite is also proved to have an Early Palaeozoic age, synchronous perhaps with the Tal phosphorites of the Lesser Himalaya.

Hazara phosphorites of Pakistan. In the Hazara division of the North West Frontier Province of Pakistan, phosphorite occurs in the Abbottabad and Hazira Formations of Early Cambrian age, over an area of approximately 160 km^2. Two levels of phosphatisation have been identified. While the Abbottabad Formation contains phosphorite in its upper cherty dolomitic horizon, the Hazira Formation contains phosphorite only in the lower silty layers. The Kakul–Mirpur, Lagarban, Kaludi–Banda, Sirbun and Dalola areas are the prominent mining districts where the exploration and exploitation programme is in operation. The phosphorite contains about 35% P_2O_5 grade in the Kakul–Mirpur areas, while in most other places the P_2O_5 values are less than 25% (Asrarullah, 1980).

The Hazara phosphorites are markedly dense and dark with poorly to well-developed grains. Some of these grains resemble ooids (fig. 6.4F); some show the characters of carbonate peloids. Grain size varies from locality to locality and also within a single sample. Quartz, minor chert, dolomite, minor calcite, pyrite and clay minerals are associated gangue, which occurs intermingled with microcrystalline apatite. Phosphatised and calcareous hyolithids of Early Cambrian age are characteristic. While total carbon content in these phosphorites is fairly high (up to 6%), the C_{org} content is only 0.16% and the C_{carb}/C_{org} ratio is rather high (36.5). Isotopic values of $\delta^{13}C_{carb}$ and $\delta^{18}O_{carb}$ are about $-3.7‰$ and $-6.5‰$ respectively. $\delta^{13}C_{org}$ is about $-20.4‰$. These values are quite significant and are comparable with the values obtained for the Lower Cambrian phosphorites of Urandush Mountain in Mongolia. They are significantly different from the values available from other deposits in the Indian subcontinent.

Tal phosphorites of Mussoorie (India). Phosphatic nodules from the Lower Tal Formation have been known for more than a hundred years, yet mineable phosphorite deposits were discovered only in the late 1960s. The phosphorite of the Mussoorie Hills occurs in the basal parts of a 2000-m thick sequence, stratigraphically denoted as Tal Formation. The Tal Formation rests conformably on the dolomites of the Upper Krol Formation which has yielded *Conophyton* of possible Vendian or Early Palaeozoic affinities, and Early Cambrian (Tommotian) conodonts with some shelly fauna (Azmi & Pancholi, 1983). The Krol Formation, minus the uppermost algal-mat dolomite, can therefore be classified as Upper Proterozoic, as suggested by Singh (1976, 1979a, b). Precise determination of the age of the Tal Formation, which contains phosphorite, chert, and black shale at its base in the Mussoorie Synform of the Garhwal Himalaya, is a controversial topic. Bhatia (1980) tried to summarise the 'Tal Tangle', and this was followed by a critical review of fossil records from Tal by Singh (1981). The basal Tal Formation is devoid of easily recognisable fossils, and apart from some sporadic layers within the Tal Formation which have yielded fossils of doubtful affinity, the Shell Limestone which occupies a supra-Tal position, is the only horizon in this belt which contains clearcut, unambiguous fossil assemblages. As described by Singh (1979a, b), except for the basal member of the Tal Formation, the complete succession provides suitable conditions for the development and preservation of a benthonic faunal community, especially because the major part of the Tal siltstone facies is a mixed tidal flat deposit which normally should support benthonic fauna in abundance; yet there are no records of any unquestionable megafossil assemblages. This absence has been interpreted as being due to the fact that the Tal Formation has a Proterozoic age (Singh, 1979a, b). The discoveries of Cambro-Ordovician conodonts from the phosphorites of the lowermost Tal Formation (Azmi *et al.*, 1981), possible Vendian–Early Palaeozoic stromatolites in the Upper Krol Dolomite, and Tommotian fauna from similar Upper Krol horizons, seem to indicate that the Lower Tal–Upper Krol phosphatisation was broadly related to the Proterozoic and Cambrian phosphogenic episode.

An alternative view that the Tal Formation is Jurassic–Cretaceous in age, is based on the presence of a definitely Mesozoic fauna in the Shell Limestone unit, which occurs at the top of the Tal succession. This unit is now considered separate from the Tal Formation for a number of tectonic, sedimentological and faunal reasons (Singh, 1976, 1979a, Acharya & Ray, 1979) and is believed to be an independent transgressive unit which overlies different rock units in the Krol belt and the Garhwal Nappe. However, it has been claimed that the Lower Tal is Early Cretaceous (Bhatia, 1980), Late Jurassic (Srivastava, 1972), or Permian (Patwardhan, 1978) and that the Upper Tal is Late Jurassic (Bhargava, 1976), Permian (Kalia, 1974), Late Cretaceous (Singh, 1980; Bhatia, 1980), or even Late Cretaceous to Palaeocene (Tewari & Singh, 1976, Tewari & Gupta, 1978). Singh (1981), in a critical analysis, discussed these uncertain fossil records pointing out that (a) fossils should be fully described before being granted any biochronological credence; (b) the presence of a specific group of species should be consistent with sedimentological and ecological niches indicated by the overall sedimentary package; (c) every record of the fossils should be precisely located in the sequence; and (d) the precise structural relationships of the fossil-bearing unit should be known. Except for the well-documented descriptions of Cambro-Ordovician protoconodonts, none of the fossils described so far satisfies these basic criteria to qualify as age indicators. The tectonostratigraphic position of various lithofacies units in which fossils of Mesozoic and Late Palaeozoic ages have been found, have not been fully taken into consideration when assigning an age to the 2000-m thick sequence of the Tal Formation. With numerous regional and local thrust sheets in the Lesser Himalaya, it would not be surprising if fossils of variable ages occurred in an apparently continuous sedimentological sequence. However, this palaeontological variety can also be explained by invoking a series of marine transgressions over a basement comprising Upper Proterozoic platform of the Nagthat–Blaini and Lower Krol carbonates (Singh, 1979b, 1981), perhaps synchronous with the well-documented transgressions in the central part of Peninsular India.

The lower part of the Tal Formation in the Mussoorie Hills is primarily a chert–phosphorite unit (Fig. 6.8) which vertically grades into greyish shale, siltstone, impure carbonate and minor sandstone (Ghosh, 1968; Shanker, 1971, 1976). Greyish to black shales overlie the chert–phosphate unit and are invariably finely laminated. Thin- and thick-bedded sand layers are intercalated within these black shales. This succession was essentially deposited in a moderately to strongly reducing milieu, under low-energy conditions. Sedimentological and geochemical studies indicate that the energy increased upward from the basal chert–shale–phosphorite unit, and that sedimentation must have occurred in protected lagoons or embayments with poor circulation, only a limited supply of clastic material being available. Shanker (1971, 1975) and Singh (1979a) made similar environmental interpretations for the Lower Tal sediments. The Lower Tal phosphorites are rich in total carbon content (4–9%), but the total C_{org} values are not very high (0.3–0.9%): however, a few samples from the Durmala area analyse up to 2.2% C_{org}.

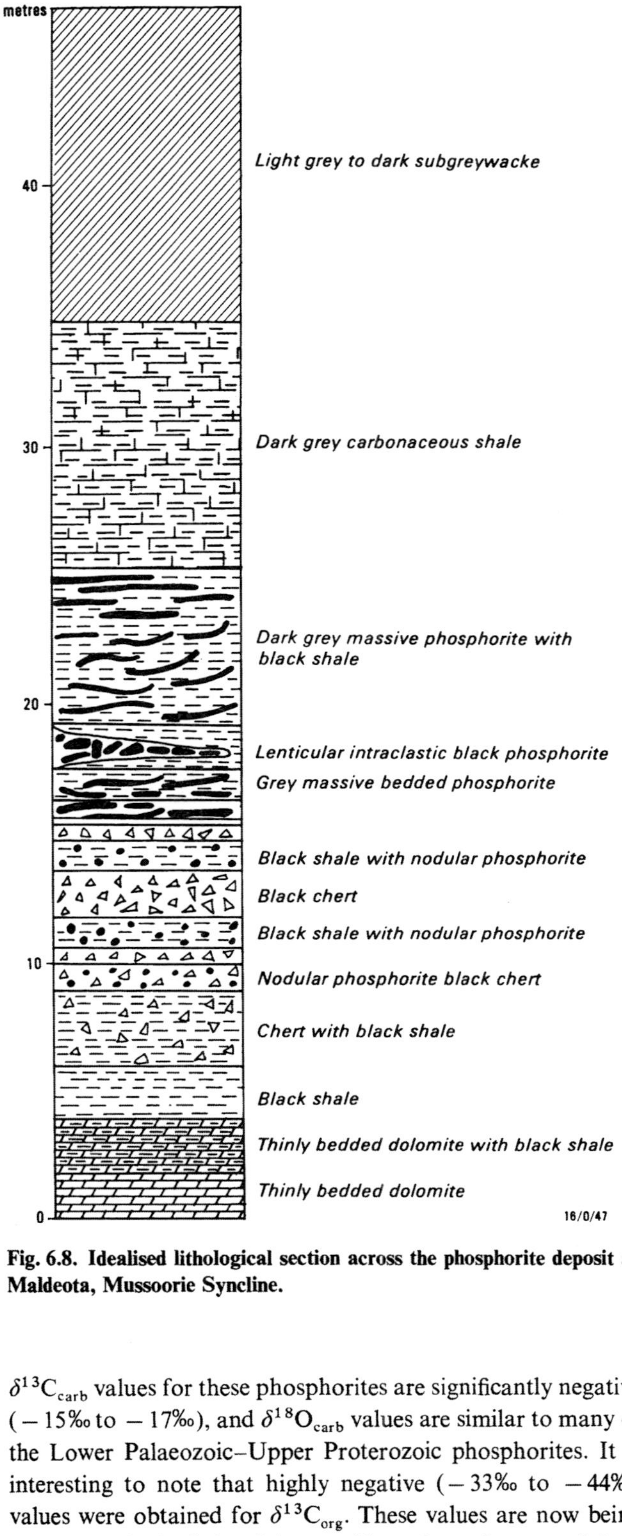

Fig. 6.8. Idealised lithological section across the phosphorite deposit at Maldeota, Mussoorie Syncline.

$\delta^{13}C_{carb}$ values for these phosphorites are significantly negative ($-15‰$ to $-17‰$), and $\delta^{18}O_{carb}$ values are similar to many of the Lower Palaeozoic–Upper Proterozoic phosphorites. It is interesting to note that highly negative ($-33‰$ to $-44‰$) values were obtained for $\delta^{13}C_{org}$. These values are now being interpreted in the light of the new ideas about the ages of these rocks.

At some localities, e.g. Durmala, black pyritic phosphorites are intimately associated with columnar stromatolites (Fig. 6.4D). Some of these stromatolites have sharp apical zones, hence resemble some form-species of *Conophyton*. No specific identification of these stromatolites has so far been made. At other places, like Maldeota, the phosphorites are nodular and locally pulverised due to local tectonics, weathering and deep-seated leaching by circulating groundwater.

Phosphate intraclasts vary in size and shape, and these have been classified under different textural groups (Rao & Rao, 1971). These intraclasts are mostly oval, medium-grained, made up of microcrystalline apatite and occur in a chert–shale matrix. Martix chert is frequently recrystallised into flamboyant growth of specular and orbicular, radiating crystals. Apatites also occur as dispersed interstitial to rounded and oval intraclasts, some of which are polynucleated.

Composition of phosphorites

Carbonate fluorapatite is the principal phosphate mineral in all the phosphorites described in this paper, although the amount of structural carbonate is low in some phosphorites, especially those from the Aravalli and Bijawar Groups. Stromatolitic phosphorites show slightly larger crystallite size ($a = 0.64$ nm, $c = 0.80$ nm) compared to non-stromatolitic fragmental phosphorites ($a = 0.64$ nm, $c = 0.67$ nm). The average crystallite size of all Proterozoic phosphorites from India show near constancy in their values (Srivastava & Banerjee, 1982). The average cell dimensions of these phosphorites vary in the range of $a = 0.935$ to 0.937 nm and $c = 0.685$ to 0.689 nm. In the leached phosphorites of Sonrai and Hirapur, the crystallite size shows an increase in the c dimension (0.90–1.0 nm). Crandallite and wavellite have also been recorded in these phosphorites (Banerjee, 1982, Banerjee *et al.*, 1982). Most of them contain free but mechanically inseparable and chemically leachable calcium and magnesium carbonates. In addition, these phosphorites contain clays, organic matter and iron oxides; minor kaolinite, goethite, siderite, calcite and quartz are other accessories.

Mineralogical studies on Indian phosphorites have been carried out by Choudhuri *et al.* (1968), Rao & Rao (1971), Mukherjee & Chatterjee (1972), Dubey & Parthasarathy (1975) and Banerjee *et al.* (1980). No detailed mineralogical data are available for the Hazara phosphorites of Pakistan. The Bajang phosphate of Nepal has similar mineralogical characteristics to those of the Calc Zone of Kumaun Himalaya (Bashyal *et al.*, 1981).

The mean elemental concentrations compared to standard deviation of several hundred sets of samples, representing specific geological and geographical provinces and petrographic types, are given in Tables 6.1 to 6.3. Trace element data for a limited number of phosphorite types have also been summarised in these tables. Data presented have been obtained by the present author and his research group in the University of Delhi. Wherever the values have been obtained from other sources, this has been acknowledged. Statistical correlation charts and parameters have not been included. The geochemistry of phosphorites of the Indian subcontinent and their inter-element correlations have been discussed by Saraswat *et al.* (1970), Raha (1972), Banerjee *et al.* (1980), and Patwardhan (1980), Banerjee (1982), Banerjee, Srivastava & Saigal (1982), Banerjee *et al.* (1982), Srivastava, Saigal & Banerjee (1982), Saigal (1983), Srivastava (1983).

Depositional environment

All known phosphorite occurrences on the Indian subcontinent are of probable marine origin, occurring typically as

Table 6.1. *Mean and standard deviation (S.D.) of major and trace elements in Precambrian–Cambrian phosphorites of the Indian subcontinent*

Element	Average Aravalli Group (Jhamarkotra, Kanpur, Matoon, Neemuch Mata, Dakan Kotra, Sallopat, Khatama, Kelkua, Amliamal) (Middle to Upper Proterozoic) $n = 82$		Average Bijawar Group (Hirapur, Bassia, Mardeora, Sonria, Tori). (Lower to Middle Proterozoic) $n = 32$		Average Calc Zone of Lesser Himalaya (Pithoragarh, Bageshwar, Kanda Baitadi-Bajang Nepal) (Middle to Upper Proterozoic) $n = 11$		Average Cuddapah and Vindhyan Group (Chelima, Pachcherla, Mirzapur) (Proterozoic) $n = 11$		Average Tal Formation (Maldeota, Durmala, Chamasari, Kolti, Masrana) (Proterozoic to Cambrian–Ordovician) $n = 65$	
	Mean	SD	Mean	SD	Mean	SD	Mean	SD	Mean	SD
SiO_2	4.46	3.58	18.75	10.9	9.45	7.03	7.17	4.85	20.55	25.00
TiO_2	0.02	0.03	0.04	0.01			0.14	0.15		
Al_2O_3	1.68	1.25	2.56	2.45	0.83	0.41	5.32	3.32	1.20	1.85
Fe_2O_3	1.11	1.19	4.89	7.02	1.28	1.66	4.46	4.50	2.49	1.66
MgO	2.99	4.19	0.62	0.34	6.77	4.95	3.09	4.57	5.38	7.32
CaO	45.55	6.98	40.63	8.34	42.5	5.24	38.2	9.3	34.11	13.03
Na_2O	0.9	0.9	1.48	0.93	0.26	0.27	0.57	0.53	0.18	0.19
K_2O	0.17	0.15	1.15	0.95	0.28	0.22	2.49	1.44	0.19	1.18
P_2O_5	25.49	7.75	25.32	6.91	19.31	7.11	22.3	8.14	19.27	7.28
CO_2	1.64	3.73	0.74	0.47			1.48	0.41	15.00	7.58
F	2.63	1.00	2.10	1.21	2.6	0.59	3.20	0.25	1.56	0.53
H_2O	0.17	0.08	0.003*	—	0.28	0.11	0.20	0.16	01.67	0.85
Ag	31.9	16.54	49	40.5	49	28	154	236.3	2	
Cd							381	286	450	312
Ce	115.0	54.0	227	112	43	29	111	36.5		
Co	63.0	—					—	—		
Cr	84	38.8	159	44.9	656	36	101	49.7	5	
Cu	149.0	36.2	213	186	121	135	160	94	60	36.0
Ga	107.5	56.14	221	106	193	44.37	91	25	148	139.2
La	1.5*						6.0*			
Mn	45*						46.0*			
Mo	599.5	371.35	619	1420	512	403	491	290		
Ni	8.0*						850*		60	85.2
Pb	116.4	64.78	372	132	192	86	165	403	115	69.5
Sc	18.5*				15.20*		110	72.6	60	78
Sr	3.0				50*	3.5	68	22		
Th	444.0	215.0	49	27.5	160*	521	432	392	500	423
U	22*		15	13			8*			
V	320*		120	27.5			25*			
Y	15*		15.8*				58*		300	216.5
Yb	14.5*		14.0*		11.88*		305*			
Zn	0.64*									
Zr	147.14	35.63	570	497	159	61	380	234	110	78.5
	$*n = 3$		$*n = 5$		$*n = 2$		$*n = 2$			

stratified deposits with individual phosphate beds up to 20 metres in thickness and containing 10–35% P_2O_5. These sediments, therefore, can be grouped as a set of anomalous sediments deposited under very specialised environmental conditions, resulting in a concentration which is often 100 times that of average crustal abundance of 0.23% P_2O_5. All the phosphatic sequences are characterised by a marked deficiency in terrigenous components and consequently the phosphate-bearing strata are either made up of carbonaceous shale, slate, chert or dolomite in condensed sequences. In most areas, the trapping mechanisms, which created favourable restricted depositional sites necessary for the accumulation of large quantities of phosphate, included the basement topography, which in turn determined the regional topography of the depositional basin (Banerjee & Deb, 1981). The sedimentation of the Aravalli phosphorite was largely controlled by shallow carbonate embankments along a Precambrian basement ridge. The Himalayan Mussoorie phosphorites, with Cambrian–

Table 6.2. *Major and some minor elements in average stromatolitic phosphorite and associated country rocks*

	Phosphorites				Country rocks					
	Stromatolitic columns		Inter-columnar spaces		Dolomite		Quartzite		Phyllite	
	Mean	SD*	Mean	SD	Mean	SD	Mean	SD	Mean	SD
P_2O_5	30.66	2.70	4.66	1.81	1.68	1.46	1.21	0.91	0.74	0.67
CaO	58.03	3.21	44.64	11.99	35.0	3.69	8.4	1.98	12.14	1.62
MgO	4.22	4.84	21.58	3.83	16.15	5.08	1.86	1.91	1.73	0.125
Na_2O	0.66	0.43	—	—	1.34	0.76	—	—	—	—
K_2O	0.63	0.37	—	—	0.89	0.66	—	—	—	—
Fe_2O_3	0.58	0.37	1.25	1.18	1.02	0.63	1.85	0.59	6.0	0.7
Al_2O_3	1.90	0.28	2.18	0.36	3.2	0.76	2.33	1.25	—	—
SiO_2	5.84	1.12	3.82	2.3	8.84	6.4	34.2	4.93	—	—
Mn	286	171	1060	—	454	232	116	32	550	241
Cu	26	8.43	35	12.71	150	139	218	114	175	34.96
Zn	60	21.25	170	15.70	64	13.2	54	11.6	121	36.36
Ni	310	167	280	149	260	73	198	25.3	204	29.26
Cr	180	127	120	68	227	170	—	—	—	—
Cd	40	7.10	30	7.03	170	29.7	133	21	192	19.69
Co	200	18.75	170	18.4	180	40.2	158	28.05	213	17.61
Ag	18	7.88	20	12.01	32	9.34	20	14	19	6.02

*Standard deviation, n = number of samples (112).

Table 6.3. *Comparison of minor trace element concentrations in phosphorites*

Element	Crustal abundance (Taylor, 1964)	Worldwide average (Tooms *et al.*, 1969)	Phosphoria Formation (Gulbrandsen, 1960)	Georgina Basin (Howard & Hough, 1979)	Aravalli Group ($n = 10$)	Bijawar Group ($n = 10$)	Calc Zone of Pithoragarh & Nepal ($n = 6$, *$n = 2$)	Cuddapah Super Group ($n = 5$, *$n = 2$)	Tal Formation (Mussoorie) Abbottabad Formation (Hazara) ($n = 10$)
V	135	167	300	50–180	16	15	—	55	300
Cr	100	285	1000	2–260	350	250	100	150	60
Zn	70	90	300	100–2000	150	550	165	300	125
Sr	375	1900	1000	50–300	500	50	150*	400	500
Y	33	0–50	300 (1000)	50–500	15	10	10*	300*	—
U	1.8	190	90	5–60	30	100	—	25	?
Cd		—	50–300	-	120	225	40	110	—
Cu		22	100	9–260	105	220	110	90	150
Mn		428	30	100–1800	550	600	490		
Ni		13	100	50–300	102	375	190	165	110
Co		3	-	84	150	65	100	—	

Ordovician fauna, were similarly deposited over the palaeotopography of the Upper Proterozoic Lower Krol Formation which has similar characteristics to the Upper Proterozoic Vindhyan dolomites of Peninsular India. The phosphatic dolomite of the Bajang–Baitadi region in Nepal, and the stromatolitic phosphorite of the Pithoragarh region were clearly formed along shallow stromatolite carbonate banks prior to major influx of terrigenous debris and immediately following the depositional episode of a quiet-water argillaceous facies. The Cambrian phosphorite of Pakistan reflects the same story as that of the Mussoorie area. However, subsequent tectonism, involving major uplift, thrusting, erosion and marine transgressions during the late Palaeozoic and Cretaceous have caused confusion in the stratigraphic relationships of the strata. The Bijawar phosphorites are clearly products of differential weathering and leaching.

Divergent views have been expressed on the precise mechanism of phosphate precipitation. The association of most Precambrian deposits with stromatolites suggests some role for micro-organisms in the concentration of phosphate from sea water, and in bringing about the formation of phosphates within the anoxic, organic-rich bottom sediments (as in the Mussoorie area). Further mobilisation of trapped phosphate in the sediments, involving early diagenetic mobilisation and reactions, leading to inter-columnar precipitation of apatite, has also been seen as a major factor responsible for phosphate concentration (as in some parts of the Aravalli Group of Rajasthan and Madhya Pradesh, and the Calc Zone of Kumaun and Nepal). Weathering, leaching, replacement and precipitation in voids and fissures (as in the Bijawar Group of central India) were also important mechanisms of phosphate generation in selective areas. Existing data on the palaeogeography and palaeo-oceanography of these old phosphogenic basins are insufficient to establish definite links with the zones of possible marine upwelling and deep-water sinks. Most of the presently known phosphogenic areas appear to be related to shallow Late Proterozoic basins whose biota, as far as is known, is restricted to micro-organisms, with algae predominating. There is little evidence for the kind of high organic productivity associated with oceanic upwelling, nor would the known palaeogeography favour the classic upwelling situation. No deep-water sedimentary basins can be linked directly to the shallow-water phosphate sedimentation basins. Nevertheless, it would be worthwhile considering an alternative hypothesis for the phosphogenesis, linking most of the primitive phosphogenic episodes in the subcontinent to a continuous progradational sedimentation event. This hypothesis is based on the idea of continuation of the Peninsular Indian rocks into the Himalayan basins beneath the alluvial cover of the Indo-Gangetic Plain. The idea of Peninsular continuation into the Himalayan belt is not new (Holland, 1908), but has received recent support from the identification of Proterozoic Vindhyan rocks in deep drill cores for petroleum exploration on the margins of the Himalaya. It therefore now appears that the northern extent of the Indian Peninsular mass extended into the areas of the present Himalaya and possibly much further. Complicated tectonics and sedimentation in this northern Himalayan region completely modified the Peninsular geological and structural elements. Only a few recognisable tectonic elements and lithologies retained their parent characters and also retained identifiable sedimentary signatures. These signatures are now gradually being understood and interpreted. Identification of similarities in the stromatolitic assemblages, petrological characters and synsedimentary mineralisation, have attracted the attention of geologists interested in regional correlation. Similarly, with the generation of more quantified data on the comparative geological ages of the rocks of the Peninsula, it is becoming easier to correlate Proterozoic sediments and phosphogenesis over the entire subcontinent.

Basing our hypothesis on evolution of Proterozoic sedimentation and continental evolution, it is possible to visualise a broad expanse of undulating Archaean basement topography composed of rocks which are represented today by the Bundelkhand granites, granites and gneisses of the Banded Gneissic Complex (= BGC-II) and Berach granites in the central and west-central parts of India. As postulated by Banerjee (1982, p. 53, fig. 3), sedimentation started during Bijawar times along shallow, starved, land-locked epicontinental marine basins, bordering the ancient Bundelkhand granite massif. Contemporaneously, sedimentation also commenced further to the west in Rajasthan, on the margins of similar (or possibly the same) Berach and Sarara granitic and gneissic massifs, with more open-sea type tidal-flat conditions (= lowermost Aravalli Group, or BGC-I, or pre-Aravalli Group of Geological Survey of India publications). Although there was no phosphate formation in the open tidal-flat type sediments in the lower Aravalli Group of Rajasthan, phosphorites formed in the Bijawar Basin because of its land-locked nature, and also possibly because of evaporation and residual concentration (Sheldon, personal communication). With the passage of time, the Bijawar and lower Aravalli (or pre-Aravalli) sedimentary pile provided basement for the deposition of the Lower Vindhyan Group (ample subsurface evidence is available for this idea) around the Bundelkhand massif in the east and north, and Lower Aravalli in the west. Hence, it can be envisaged that the locus of sedimentation prograded from east to west: the Aravalli Group graded west into the Delhi Group, and the Lower Vindhyan Group graded into the Upper Vindhyan Group in central India and the Gwalior Group in west-central India. It is worth recording that the Delhi Group is predominantly an arenaceous facies, while the Upper Vindhyan Group is mostly a carbonate-bank facies. The trans-Aravalli/Vindhyan (?) sediments of the Jodhpur area appear to be contemporaneous with the Upper Delhi Group sedimentation event, but with its own basinal characteristics. Phosphorite formed only in the Bijawar Group or uppermost levels of the Aravalli Group, possibly because of the restricted nature of these sub-basins. Sporadic occurrences of phosphatic rocks in the Vindhyan and Delhi Groups, etc., could be related to local, abnormal sedimentation conditions generated by the palaeotopography and other factors.

This east–west model for Peninsular India is much less complex than that required for the area to the north. However, there are reasons to believe that sedimentation took place along the northern borders of Bundelkhand massif, but the evidence for this is concealed below the great thicknesses of Indo-Gangetic alluvium. The sedimentary basin to the north was possibly a more open-oceanic type, with shallow, protected banks and a continental slope inclined towards the north and north-west, approximating the linear extent of the present Lesser Himalaya (allowance having been made for crustal shortening and tectonic irregularities caused during the late Mesozoic Himalayan Orogeny). If such a hypothesis could be supported by other geological and geophysical data, then the Lesser Himalayan Cambrian–Ordovician phosphorites of Mussoorie and Hazara should be palaeogeographically related. Possibly they were deposited on the same ocean-facing continental slope over which upwelling waters were spreading.

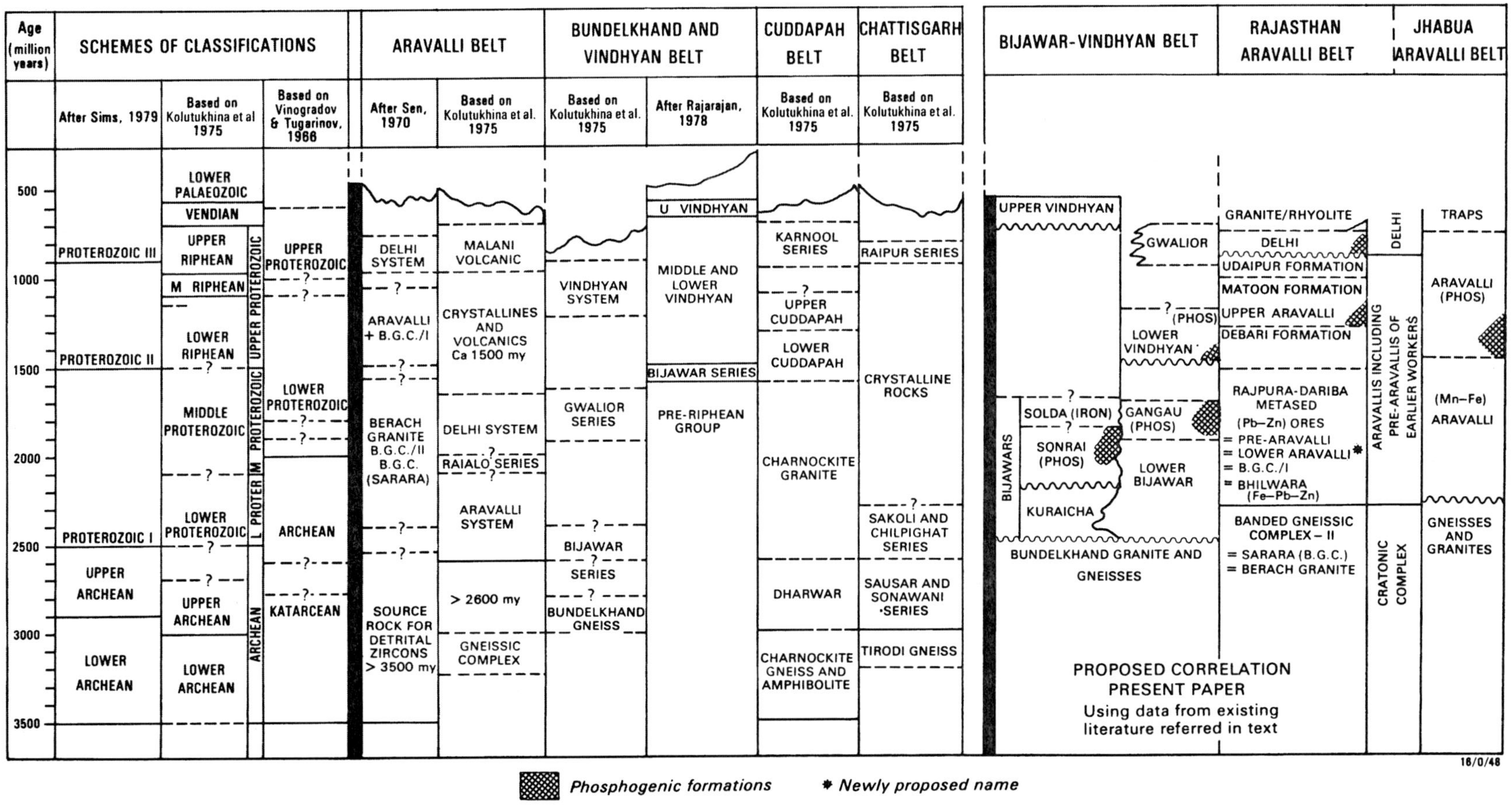

Fig. 6.9. Comparative stratigraphic scheme for the Proterozoic sedimentary formations of the Indian Peninsula (west-central part only) in relation to phosphorite distribution patterns.

Phosphogenic episodes

It is highly desirable to discuss the various problems related to defining the ages of the phosphorite-bearing strata, *vis-à-vis* rocks above and below these units, and this is presented in summary form in Figs. 6.9 and 6.10. These figures, in combination with the literature referred therein, make it quite apparent that the correlation of phosphorite-bearing sequences over the entire Indian subcontinent is in a state of flux. The geological, palaeontological, sedimentological, environmental and radiometric data are often at odds and preconceived ideas have helped to perpetuate certain dogmas or play down new ideas. It is extremely necessary to generate new, reliable and systematic data on all the phosphorite-bearing strata. Due to the style of cyclic tectonic evolution and igneous activity, it would be worth attempting to date these sediments, especially the carbonates, both radiometrically and palaeontologically.

The existing information and an intelligent summary of the available data may prompt us to regroup the phosphogenic cycles of the Indian subcontinent and surrounding areas and relate them to the world-wide phosphogenic episodes suggested in Cook & Shergold (1980). The oldest record of iron-rich sedimentary phosphorite comes from the Bijawar rocks of central India, around the cratonic complex of Bundelkhand. This was followed by phosphate sedimentation within the carbonate-stromatolite complex of the Upper Aravalli Group (*sensu stricto*). The Lower Aravalli, or the pre-Aravalli or the BGC-I and the BGC-II (Banded Gneissic Complex) were the source rocks for the phosphorite and other sediments to the main Aravalli sedimentation basin. Phosphorites in the Lower Vindhyan carbonates and porcellanites, and in the Upper Cuddapah Group of the southern Indian platform can be considered as belonging to the same phosphogenic episode. The stromatolitic phosphorites of the Pithoragarh and Bajang–Baitadi regions of the inner Himalayan Proterozoic belt also belonged to the Aravalli–Vindhyan cycle. Minor phosphogenesis in the Lower Delhi Group, Kurnool Group, Chattisgarh (Raipur) Group and Lower Krol Formation in the Solan, Nainital and Mussoorie areas were therefore strictly Late Proterozoic events. The uppermost Krol Formation, transitional to the Lower Tal Formation, and the Lower Tal

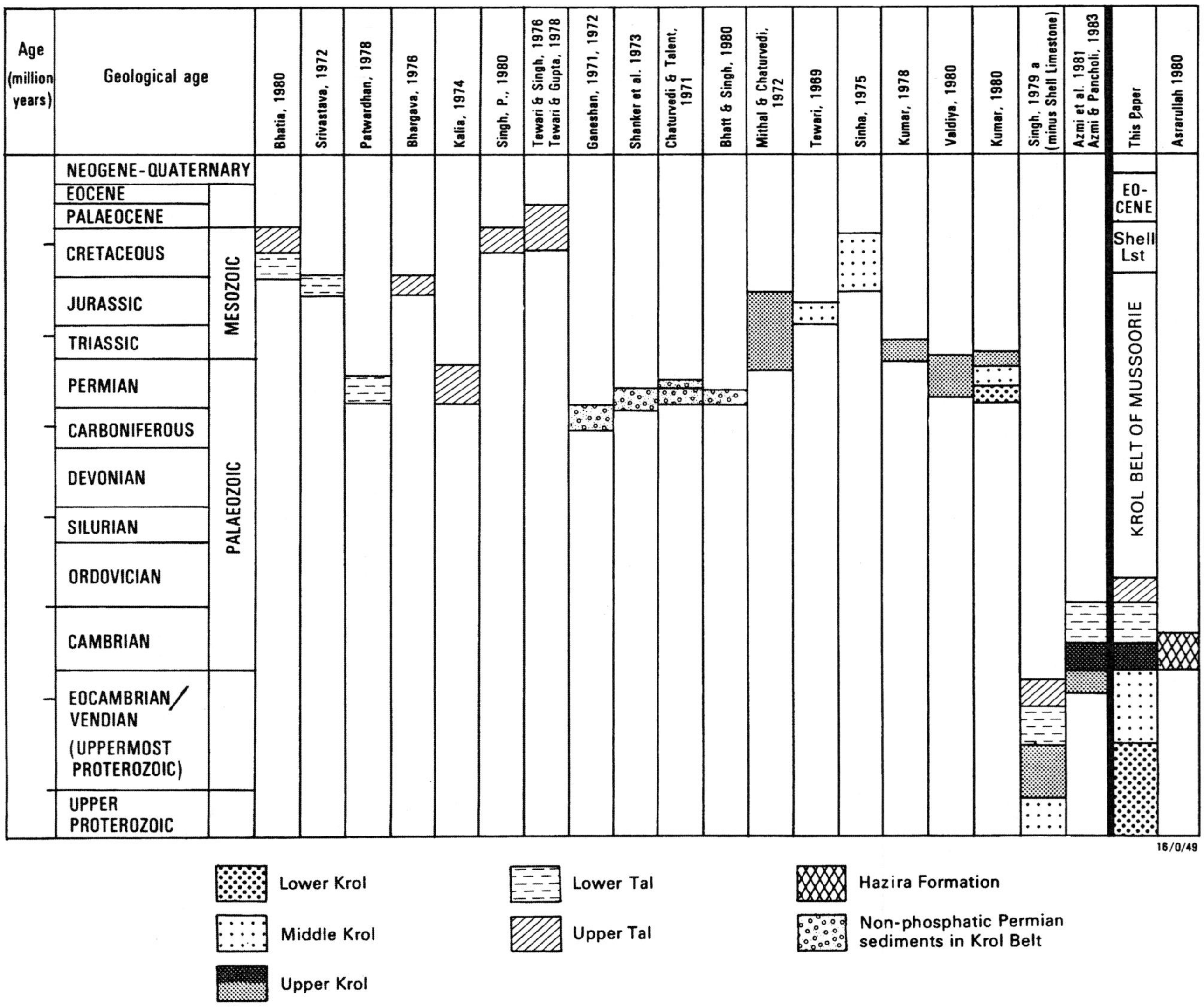

Fig. 6.10. Comparative stratigraphic schemes for the Krol–Tal sequences of Lesser Himalaya and Hazara.

phosphorite of Kumaun–Garhwal Himalaya and the Abbottabad–Hazara phosphorites of Pakistan belong to successively younger phosphogenic episodes which culminated approximately at the beginning of the Ordovician.

References

Acharya, S.K. & Ray, K.K. (1979). Age of the 'Tal Shell limestone' Garhwal Himalaya. *Seminar on Correlation of the Lesser Himalayan Formations*. Nainital, April, 1979.

Asrarullah, (1980). Phosphate, potash and natural gas occurrences in Pakistan. In *Fertilizer Mineral Occurrences in the Asia-Pacific Region*. ed. A.I.N. Lee, pp. 66–7. East–West Resource Systems Institute, Honolulu.

Auden, J.B. (1934). The geology of the Krol Belt. *Records of the Geological Survey of India*, **67**, 357–433.

Awasthi, N. (1970). Some aspects of the Krol Formation of the Himalaya. *Contributions to Mineralogy and Petrology*, **28**, 192–220.

Azmi, R.J., Joshi, H.N. & Juyal, K.P. (1981). Discovery of Cambro–Ordovician conodonts from the Mussoorie Tal Phosphorite: its significance in the correlation of the Lesser Himalayas. *Contemporary Scientific Researches in the Himalaya*, ed. A.K. Sinha, pp. 245–50. Hindustan Publishing Corporation, Delhi.

Azmi, R.J. & Pancholi, V.P. (1983). Lower Cambrian (Tommotian) conodonts and shelly fauna from the Upper Krol of the Mussoorie Syncline, Lesser Garhwal Himalaya, with remarks on the Precambrian–Cambrian boundary. *Himalayan Geology*, **11**, 360–72.

Banerjee, D.M. (1967). Stratigraphy, petrology and tectonies of Bageshwar and adjacent areas – Districts of Almora and Pithoragarh U.P., Ph.D. Thesis, Lucknow University, 246pp.

Banerjee, D.M. (1971a). Aravallian stromatolites from Udaipur, Rajasthan, India. *Journal of the Geological Society of India*, **12(4)**, 349–55.

Banerjee, D.M. (1971b). Precambrian stromatolitic phosphorites of Udaipur, Rajasthan, India. *Geological Society of America Bulletin*, **82**, 2319–30.

Banerjee, D.M. (1973). Microfossils from Late Precambrian phosphatic Aravalli stromatolites of Udaipur, Rajasthan, India. *Recent Researches in Geology*, **1**, 263–8.

Banerjee, D.M. (1975). Environment of formation of some older and younger phosphorites of India. *Proceedings of the IXth International Sedimentological Congress*, Nice (France), Theme-9, 1–7.

Banerjee, D.M. (1978). Chemical rhythmicity in the Precambrian laminated phosphatic stromatolites and its bearing on the origin of algal phosphorite. *Indian Journal of Earth Sciences*, **5(1)**, 102–10.

Banerjee, D.M. (1980). Surface characteristics of Precambrian stromatolitic phosphorites of a part of the Indian shield. *Remote sensing and Mineral Exploration*. ed. W.D. Carter, L.C. Rowan & J.F. Huntington, pp. 109–19. COSPAR Special Publication, Pergamon Press, London.

Banerjee, D.M. (1982). Lithotectonics, phosphate mineralization and regional correlation of Bijawar Group of rocks in the Central India. In *Geology of Vindhyachal*, ed. K.S. Valdiya *et al.*, pp. 47–54. Hindustan Publishing Corporation, Delhi.

Banerjee, D.M. (1983). Filamentous microorganisms in the late Precambrian phosphatic stromatolites from India. *7th ISSOL Congress*, Mainz, 10–15 July, [B], 1–10.

Banerjee, D.M. & Basu, P.C. (1979). Geology and structure of Precambrian Jhabua phosphorite deposit, Madhya Pradesh. *Indian Mineralogist*, **20**, 32–42.

Banerjee, D.M. & Basu, P.C. (1980). Stromatolites in the Jhabua Phosphorite–Lithostratigraphy, age and palaeoenvironment. Proceedings, Workshop on Stromatolites 1978, *Geological Survey of India, Miscellaneous Publication*, **44**, 240–54.

Banerjee, D.M., Basu, P.C. & Srivastava, N. (1980). Petrology, mineralogy, geochemistry and origin of Precambrian Aravallian phosphorite deposits of Udaipur and Jhabua, India. *Economic Geology*, **75**, 1181–99.

Banerjee, D.M. & Deb, M. (1981). Some thoughts on phosphorite-basemetal deposition in relation to basinal evolution in South Central Rajasthan. *Abstract, 4th International Field Workshop and Seminar on Phosphorite, International Geological Correlation Programme, Project 156, Udaipur, India*, p. 38.

Banerjee, D.M. & Khan, N.W.Y. (1981). Annotated bibliography of phosphate geology in India. *Department of Geology, University of Delhi, Publication*, **3**, 43pp.

Banerjee, D.M., Khan, M.W.Y., Srivastava, N. & Saigal, G.C. (1982). Precambrian phosphorites in the Bijawar Rocks of Hirapur-Bassia-area, Sagar district, Madhya Pradesh, India. *Mineralium Deposita*, **17(4)**, 349–62.

Banerjee, D.M., Srivastava, N. & Saigal, G.C. (1984) Element variation patterns in the phosphatic stromatolites and intercolumnar carbonates and the countryrocks, around Udaipur, India. *Geological Survey of India, Special Publication*, **17**, 63–77.

Bashyal, R.P. (1981). Stromatolitic phosphorite occurrences in the Lesser Himalaya of Far Western Nepal. *Abstract, 4th International Field Workshop and Seminar on Phosphorite International Geological Correlation Programme – Project 156, Udaipur, India*, p. 5.

Bashyal, R.P., Banerjee, D.M. & Khan, M.W.Y. (1981). Certain characteristics of the stromatolitic phosphorites from Western Nepal. *Abstract, 4th International Workshop and Seminar on phosphorite. International Geological Correlation Programme–Project 156, Udaipur, India*, p. 2.

Basu, P.C. (1976). Sedimentary environment of phosphorite formation of Jhabua dist. M.P. Unpublished Ph.D. Thesis, University of India, 210pp.

Basu, P.C. & Banerjee, D.M. (1975). Petrographic studies of Precambrian phosphorites of Jhabua dist., M.P. In *Proceedings, Symposium on Sedimentology, Sedimentation and Sedimentary Environment*, pp. 268–75. University of Delhi.

Basu, P.C., Sood, N.K. & Venkatesh, V. (1981). Precambrian phosphorites of Cuddapah Basin – a phosphogenic model. *Abstract, 4th International Field Workshop and Seminar on Phosphorite, International Geological Correlation Programme–Project 156. Udaipur, India*, p. 39.

Bhargava, O.N. (1972). A reinterpretation of the Krol Belt. *Himalayan Geology*, **2**, 47–81.

Bhargava, O.N. (1976). Geology of the Krol Belt and associated formations – a reappraisal. *Memoirs, Geological Survey of India*, **106(I)**, 168–234.

Bhatia, S.B. (1980). The Tal Tangle. In *Stratigraphy and Correlations of Lesser Himalayan Formations*, ed. K.S. Valdiya, pp. 79–96. Hindustan Publishing Corporation, Delhi.

Bhatt, D.K. & Singh, M.P. (1980). A note on the fauna of Artinski (Lr Permian) affinity from Garhwal Synform Lesser Himalaya. *Journal of the Palaeontological Society of India*, **26**, 1–5.

Bhatti, N.A. (1979). Hazara Phosphate Deposits – Pakistan. In *Proterozoic–Cambrian Phosphorites*, ed. P.J. Cook & J.H. Shergold pp. 44–5. ANU Press, Canberra.

Chaturvedi, R.S. & Talent, J.A. (1971). A preliminary note on the discovery of richly fossiliferous Permian rocks in Garhwal Himalayas, in Landsdown area, Second Seminar on Himalayan Geology. *Wadia Institute of Himalayan Geology, Special Report*.

Chauhan, D.S. (1973). Stromatolites from the Precambrian phosphorite bearing stratum of the Aravalli region of Udaipur, Rajasthan (India). *Doklady Akademii Nauk, SSSR.*, **208**, 1429–31.

Chauhan, D.S. (1979). Phosphate-bearing stromatolites of the Precambrian Aravalli phosphorite deposits of the Udaipur region, their environmental significance and genesis of phosphorite. *Precambrian Research*, **8**, 95–126.

Chauhan, D.S. (1980). Stromatolites of Nimach Mata and Bargaon phosphorite deposits and genesis of rock phosphate. Proceedings,

Workshop on Stromatolites 1978. *Geological Survey of India, Miscellaneous Publication*, **44**, 314–29.

Chernyshev, I.V., Safonov, Yu. G., Radhakrishna, B.P., Krishna Rao, B., Vasudev, V.N. & Deb, M. (1980). Study of lead-isotopes for investigating the origin of endogenic deposits with special reference to some ore deposits. *Journal of the Geological Society of India*, **21**, 107–16.

Choudhuri, R. (1981). Botryoidal apatite (staffelite) in the Jhamarkotra area of Udaipur district, Rajasthan, *Abstract, Symposium on Three Decades of Development in Petrology, Mineralogy and Geochemistry in India.* Geological Survey of India, Jaipur, May 1981, p. 28.

Choudhuri, R. & Banerji, K.C. (1966). Phosphorite occurrences in Nainital Area, U.P. *Technology*, **3(4)**, 219.

Choudhuri, R., Chakraborty, A.K., Ghosh, S.K. & Banerjee, K.C. (1968). Studies in Indian Phosphates Part I. Technology, **5(4)**, 285–93.

Cook, P.J. & Shergold, J.H. (1980). Proterozoic and Cambrian phosphorites of Asia and Australia – a progress report. In *Fertilizer Mineral Potential of Asia and the Pacific*, ed. R.P. Sheldon & W.C. Burnett, pp. 207–33. East–West Resource Systems Institute, Honolulu.

Crawford, A.R. (1970). The Precambrian geochronology of Rajasthan and Bundelkhand, Northern India. *Canadian Journal of Earth Sciences*, **7**, 91–110.

Crawford, A.R. & Compston, W. (1970). The age of the Vindhyan System. *Quarterly Journal of the Geological Society of London*, **125**(3), 351–71.

Dayal, B., Chaurasia, P.K. & Hore, M.K. (1980). Phosphorite in Rajasthan–Haryana. *Geological Survey of India, Newsletter*, **11(10)**, 5.

Dayal, B., Chaurasia, P.K., Jamwal, R.S., Negi, R.S., Sinha, A.P. & Hore, M.K., (1981). Phosphorite occurrences in Delhi Supergroup in Northeastern Rajasthan. *Abstract, 4th International Field Workshop and Seminar on Phosphorite, International Geological Correlation Programme – Project 156, Udaipur, India*, pp. 10–11.

Deb, M. (1982). Crustal evolution and metallogenesis in Precambrians of Western India. *Revista Brasileira de Geociências São Paulo*, **12(1–3)**, 94–104.

Dubey, R.R. & Parthasarathy, A. (1975). Mineralogical studies of phosphorites from Durmela, Mussoorie, U.P. *Recent Researches in Geology*, **2**, 289–95.

Dutta, M.K. (1980). Petrology of the rocks around Jhamarkotra and its bearing on the environment of phosphorite formation. Unpublished M.Sc. Thesis, University of Delhi, 51pp.

Ganeshan, T.M. (1971). Clay mineral constituents of the Krol-Tal sequence of the Mussoorie Syncline, U.P. Seminar on Recent Studies in the Himalaya, Abstract, Geological Survey of India.

Ganeshan, T.M. (1972). Fenestellid bryozoa from the Boulder Slate Sequence of Garhwal. *Himalayan Geology*, **2**, 431–51.

Gansser, A. (1964). *Geology of the Himalaya*, 289pp. Wiley Interscience, New York.

Ghosh, A.K. & Srivastava, S.K. (1962). Microfloristic evidence of the age of Krol Beds and associated formations. *Proceedings of the National Institute of Science, India*, **28A**, 710–17.

Ghosh, R.N. (1968). A note on the phosphorites of Chaunpa–Kumali–Areas of Tehri–Garhwal District, U.P. *Journal of Mines, Metals and Fuels*, **6(2)**, 358–62.

Ghosh, R.N. & Biswas, C. (1973). Precambrian stromatolitic phosphorites of Sallopat, Banswara district, Rajasthan. *Abstract, Seminar on Recent Advances in the Geology of Rajasthan and Gujarat, Jaipur*, p. 52.

Gulbrandsen, R.A. (1966). Chemical composition of phosphorites of the Phosphoria Formation. *Geochimica et Cosmochimica Acta*, **30**, 769–78.

Gundu Rao (1970). A note on the oncolites in the Krol Series and their age significance. *Publication of the Centre for Advanced Studies in Geology, Chandigarh*, **7**, 127–9.

Gupta, B.C. (1934). The geology of Central Mewar. *Memoirs, Geological Survey of India*, **65(2)**, 1–116.

Gupta, S.N. Arora, Y.K., Mathur, R.K., Iqballuddin, Prasad, B., Sahai, T.N. & Sharma, S.B. (1980). *Lithostratigraphic map of Aravalli region.* Geological Survey of India, Hyderabad.

Heron, A.M. (1953). Geology of Central Rajputana. *Memoirs, Geological Survey of India*, **79**, 1–389.

Holland, T.H. (1908). On the occurrence of striated boulders in the Blaini Formation of Simla, with discussion on the geological age of the rocks. *Records of the Geological Survey of India*, **37**, 129–35.

Howard, P.F. & Hough, M.J. (1979). On the geochemistry and origin of the D-Tree, Wonarah and Sherrin Creek phosphorite deposits of the Georgina Basin, North Australia. *Economic Geology*, **14**, 260–84.

Israili, S.H. & Khan, M.W.Y. (1980). Trace-element studies of phosphate-bearing Gangolihat Dolomites, Pithoragarh, U.P., India. *Chemical Geology*, **31**, 133–43.

Jain, A.K. Banerjee, D.M. & Mithal, R.S. (1971). Correlation of unfossiliferous Himalayan formations of Garhwal. *Himalayan Geology*, **1**, 92–110.

Jain, S.C. & Pant, A. (1981). Stratigraphy and sedimentation of the phosphate deposits of Lalitpur basin, U.P. *Abstract, 4th International Field Workshop and Seminar on Phosphorite, International Geological Correlation Programme, Project 156, Udaipur, India*, p. 14.

Joshi, V.K. (1970). Phosphate deposits in Krol sandstone. *Indian Minerals*, **24(4)**, 359–63.

Kalia, P. (1974). Upper Permian fusulinids from Garhwal Himalaya. *Proceedings of the 2nd Indian Colloquium on Micropalaeontology and Stratigraphy, Lucknow University*, 107–10.

Khan, H.H. (1979). An appraisal of Precambrian phosphorites of Jhabua, M.P., India. *Proterozoic–Cambrian Phosphorites*. ed. P.J. Cook & J.H. Shergold, pp. 42–3. ANU Press, Canberra.

Khan, M.W.Y. (1978). Geology and Geochemistry of phosphorites of Pithoragarh, U.P. Unpublished Ph.D. Thesis. Muslim University, Aligarh, 235pp.

Kharakwal, A.D. & Bagati, T.N. (1976). Diagenetic and epidiagenetic silicification in carbonate rocks of the Krol Formation. *Bulletin of the Indian Geologists Association*, **7**, 35–41.

Kolotukhina, S.E., Moralev, V.N., Topunova, G.A., Klapovskaya, L.I., Burkov, V.V. & Rozanitz, A.V. (1975). *Geology and economics of the rare element deposits of India and Sri Lanka.* Nauka, Moscow, 152pp. (in Russian).

Kumar, S. (1978). Calcareous algae from the Krol Formation of Himalaya. *Indian Journal of Earth Sciences*, **6**, 19–23.

Kumar, A. (1980). A note on the new forms of stromatolites and trace fossils from Nainital area; Kumaun Himalaya, India. *Geophytology*, **10**, 265–6.

Mathur, S.M. (1960). A note on the Bijawar Series in the eastern part of the type area, Chattarpur area, M.P. *Records of the Geological Survey of India*, **86**, 539–44.

Mathur, S.M. & Mani, G. (1978). Geology of the Bijawar Group in the type area, M.P. *Proceedings of a Symposium on the Purana Formations of Peninsular India, Sagar, M.P.*, 313–20.

Mithal, R.S. & Chaturvedi, R.S. (1972). Possible algal structures in the Upper Krol limestones of Mussoorie area, Uttar Pradesh. *Geological Survey of India, Miscellaneous Publication*, **15**, 265–8.

Mukherjee, B. & Chatterjee, P.K. (1972). Mineralogy of phosphorite from Udaipur, Rajasthan. *Indian Minerals*, **44(2)**, 101–3.

Munshi, R.L., Khan, H.H. & Ghosh, D.B. (1974). Algal structure and phosphorite in the Aravalli rocks of Jhabua, M.P. *Current Science*, **10**, 446–7.

Nath, Mukti, Sant, V.N., Das, N., Chakravarty, D.K., Saxena, S.P., Chakravarty, B. & Singh, G.B. (1969). Precambrian Phosphorite near Udaipur, Rajasthan. *Journal of Mines, Metals and Fuels*, **6(2)**, 335–9.

Nath, Mukti (1974). Phosphorites – its present status and guides for further search in India. *61st Session Indian Science Congress, Presidential Address*, 91–114.

Nath, Mukti & Chakravarty, S.C. (1972). Rock phosphates. *Records of the Geological Survey of India*, **102(II)**, 117–30.

Pant, A. (1980). Resource status of rock phosphate deposits in India and areas of future potentials. In *Fertilizer Mineral Potential in Asia and the Pacific*, ed. R.P. Sheldon & W.C. Burnett, pp. 331–58. East–West Resource Systems Institute, Honolulu.

Pascoe, E.H. (1950). *A manual of the geology of India and Burma*. vol. 1 (3rd edn), 485pp. Government of India Press, Calcutta.

Patwardhan, A.M. (1978). First moravamminids from the Himalaya. *Indian National Academy of Science Letters*, **1**, 7–8.

Patwardhan, A.M. (1980). Phosphate–pyrite association and the genesis of stromatolitic and pelletal phosphorites. *Proceedings, Third Indian Geological Congress, Poona*, 347–65.

Raaben, M.E. (1980). Microstromatolites – characteristic element of Lower Proterozoic stromatolitic complex. *Doklady Akademii Nauk, SSSR*, **250(3)**, 734–7.

Raaben, M.E. (1981). Riphean stromatolite Vs. Lower Proterozoic. *Izvestiya Akademii Nauk, SSSR, Seriya Geologicheskaya*, **6**, 51–64.

Raha, P.K. (1972). Mineralogical and crystal chemistry of the phosphorite of the Chamasari-Partibba and Chipaldi-areas, Mussoorie Syncline, Districts, Dehradun and Tehri Garhwal, U.P. *Indian Minerals*, **24(2)**, 192–3.

Rajarajan, K. (1978). Geology of Sagar district and western parts of Damoh district, M.P. *Memoirs, Geological Survey of India*, **109**, 99pp.

Rao, N.K. & Rao, G.V.U. (1971). Petrographic study of some Indian phosphorites. *Journal of the Geological Society of India*, **12(4)**, 356–67.

Remy, J.M. (1975). *Geology of Nepal in West Nepal Himalaya*, 72pp. CNRS, Paris.

Roy, A.B., Nagori, D.K., Golani, P., Dhakar, S.P. & Choudhuri, R. (1980). Structural geometry of the rock phosphate bearing Aravalli rocks around Jhamarkotra mines area, Udaipur district, Rajasthan. *Indian Journal of Earth Sciences*, **7(2)**, 191–202.

Roy, A.B. & Paliwal, B.S. (1981). Evolution of Lower Proterozoic epicontinental deposits–stromatolite bearing Aravalli rocks of Udaipur, Rajasthan, India. *Precambrian Research*, **14**, 49–74.

Saigal, G.C. (1983). Geochemistry, petrology and mineralogy of Precambrian phosphorite around the base-metal bearing carbonates of Udaipur, Rajasthan state, India. Unpublished Ph.D. Thesis, Delhi University, 115pp.

Sant, V.N. (1980). Geology of the Indian Platform and its phosphate occurrences. In *Fertilizer Mineral Potential in Asia and the Pacific*, ed. R.P. Sheldon & W.C. Burnett, pp. 29–46. East–West Resource Systems Institute, Honolulu.

Sant, V.N. & Pant, A. (1980). A note on the known resources of phosphate and potash in India. In *Fertilizer Mineral Occurrences in the Asia–Pacific Region*. ed. A.I.N. Lee, pp. 20–33. East–West Resource Systems Institute, Honolulu.

Saraswat, A.C., Varada Raju, H.N., Teneja, P.C., Bargaja, V.B. & Sankaran, A. (1970). Geochemical data on the uraniferous phosphorites of Mussoorie, Dehradun district, Uttar Pradesh, India. Proceedings 2nd Seminar on Geochemical Prospecting Methods & Techniques. *UNESCO Mineral Development Series*, **38**, 179–87.

Sarkar, S.N., Polkhanov, A.A., Gerling E.K. & Chakrov, F.V. (1964). Geochronology of the Precambrians of Peninsular India – a synopsis. *Science and Culture*, **30**, 527–37.

Sen, S. (1970). Some problems of Precambrian geology of the Central and Southern Aravalli range, Rajasthan. *Journal of the Geological Society of India*, **11(3)**, 217–31.

Shanker, R. (1971). Stratigraphy and sedimentation of Tal Formation, Mussoorie Syncline, U.P. *Journal of the Palaeontological Society of India*, **16**, 1–15.

Shanker, R. (1975). Stratigraphy and analysis of the Chert Member, Lower Tal Formation, Dehradun and Tehri district, U.P. *Records, Geological Survey of India*, **106(II)**, 54–74.

Shanker, R. (1976). Maldeota phosphorite prospects, Dehra Dun district, U.P. *Geological Survey of India, Miscellaneous Publications*, **24**, 465–74.

Shanker, R., Dhaundiyal, J.N. & Kapoor, H.M. (1973). The fossiliferous bed (Boulder Slate Member) of Garhwal Syncline. *Journal of the Palaeontological Society of India*, **17**, 50–4.

Sharma, N.L. (1953). Problems in the Correlation of the pre-Vindhyan igneous rocks of Rajasthan. *Presidential Address, 40th Indian Science Congress (Geology and Geography Section)*, 1–27.

Sheldon, R.P. (1966). *Phosphate deposits of India – their discovery, geology, potential and proposed development*. US Geological Survey, Technical Letter–Phosphate-I, 8pp.

Sims, P.K. (1979). Precambrian Subdivided. *Geotimes*, **24**(12), 15.

Singh, I.B. (1976). Evolution of Himalaya in the light of marine transgressions in Peninsular and extra-Peninsular India. *Proceedings, 125th Anniversary Celebration Symposium*, Geological Survey of India, Lucknow (Preprint).

Singh, I.B. & Rai, V. (1977a). Some observations on the sedimentology of the Krol succession of Mussoorie area, U.P. *Journal of the Geological Society of India*, **21**, 232–8.

Singh, I.B. & Rai, V. (1977b). Some observations on the stromatolites in the Krol Formation of Nainital area and its implications on the age of the Krol Formation. *Current Science*, **46**, 736–8.

Singh, I.B. (1979a). Environment and age of the Tal Formation of Mussoorie and Nilkanth areas of Garhwal Himalaya. *Journal of the Geological Society of India*, **20**, 214–25.

Singh, I.B. (1979b). Some thoughts on the evolution of Himalaya and the northern limits of the Indian Shield. *Geologische Rundschau*, **68**, 342–50.

Singh, I.B. (1979c). Recognition of sedimentological break between Quartzite and Limestone members of the Tal Formation, Lesser Himalaya, India. *Current Science*, **48**, 206–8.

Singh, I.B. (1981). A critical review of the fossil records in the Krol Belt succession and its implications on the biostratigraphy and palaeogeography of Lesser Himalaya. *Journal of the Palaeontological Society of India*, **25**, 148–69.

Singh, P. (1980). Microfauna, age, palaeoenvironment and palaeobiogeography of the Tal Formation, Lesser Hamalaya, Garhwal, U.P., India. *Current Science*, **49**, 255–61.

Sinha, A.K. (1975). Calcareous nannofossils from the Simla Hills. *Journal of the Geological Society of India*, **16**, 69–77.

Srivastava, N. (1983). Distribution and behaviour of trace-elements in the Precambrian phosphorites of India, Unpublished Ph.D. Thesis, University of Delhi.

Srivastava, N. & Banerjee, D.M. (1982). Apatite crystallite size of Precambrian phosphorites of India. *Economic Geology*, **77(1)**, 193–4.

Srivastava, N., Saigal, G.C. & Banerjee, D.M. (1982). Element variation patterns in the phosphatic stromatolitic columns and intercolumnar carbonates and bulk phosphorites from the rocks of Udaipur Rajasthan, India. In *Geology of Phosphorite Deposits and Problems of Phosphoritogenesis*, ed. A.V. Ilyin, A.L. Yanshin & Yu.N. Zanin, pp. 137–45. Institut Geologii Geofiziki, Sibirskoe Otdelenie, Akademiya Nauk SSR, Novosibirsk.

Srivastava, R.N. (1972). Fossil lamellibranch from the Lower Tal Shales of Mussoorie area. *Geological Survey of India, Miscellaneous Publication*, **15**, 269–72.

Stanin, X. & Hasan, M.S. (1966). *Phosphorite occurrences in Pakistan*. Geological Survey of Pakistan, Information Bulletin.

Taylor, S.R. (1964). Chondritic earth model. *Nature*, **202(4929)**, 281–2.

Tewari, B.S. (1969). Nannofossils from the Krols. *Bulletin of the Indian Geologists Association*, **2**, 22.

Tewari, B.S. & Gupta, V.J. (1978). The stratigraphic position of the Tal Formation of the Lesser Himalaya. *Recent Research in Geology*, **4**, 128–39.

Tewari, B.S. & Singh, R.Y. (1976). The significance and occurrence of Late Palaeozoic plant remains in the Infra-Krol sequence of Nainital, U.P. *Bulletin of the Indian Geologists Association*, **12**, 263–6.

Tooms, J.S., Summerhayes, C.P. & Cronan, D.S.L. (1969). Geochemistry of marine phosphate and manganese deposits. *Oceanographic Marine Biology Annual Review*, **7**, 49–100.

Valdiya, K.S. (1964). The unfossiliferous formations of the Lesser

Himalaya and their correlation. In *22nd Session International Geological Congress, Proceedings of Section 11, Himalayan & Alpine Orogeny, New Delhi*, 15–36.

Valdiya, K.S. (1972). Origin of phosphorite of the Late Precambrian Gangolihat dolomites of Pithoragarh, Kumaun Himalaya, India. *Sedimentology*, **17**, 115–28.

Valdiya, K.S. (1980). Discovery of late Palaeozoic brachiopod in the Upper Krol of the Nainital Hills, Kumaun Himalaya. *Journal of the Geological Society of India*, **21**, 97–101.

Vinogradov, A.P., Tugarinev, A.I., Zykov, C.I., Stapnikova, N.I., Bibikova, E.V., Khorre, K.G. & Melnikova, G.L. (1966). Geochronology of the Precambrian of India. *Akademiya Nauk SSR, Kammissiya po Opredeleniyu Absolyutnogo Vozrasta geologicheskikh Formatsiy, Trudy*, **13**, 394–408.

Vishwanathiah, M.N. & Sastri, J.C.V. (1973). High phosphorus stromatolitic limestone from Nandini-Drug district, M.P. India. *Current Science*, **42(1)**, 738–41.

7

Proterozoic and Cambrian phosphorites–regional review: Europe

A.J.G. NOTHOLT AND M.D. BRASIER

Introduction

Phosphorites associated with Precambrian or Cambrian sediments have been known in Europe since the latter part of the 19th century, the earliest discoveries, of Cambrian age, being described from the United Kingdom and Sweden (Hicks, 1875; Andersson, 1896). Similarly, the phosphatic nodules in shales of Late Riphean age (Peach *et al.*, 1907) are the first to have been recorded from Precambrian sediments. Since then, discoveries of either Precambrian or Cambrian age have been made in most other European countries, with the notable exceptions of Italy and the Netherlands (Fig. 7.1). Although there are no phosphorites of economic importance known from the classic Cambrian successions of Europe at the present time, there are numerous phosphorite occurrences which yield valuable information on the nature and timing of Cambrian sedimentary cycles and phosphogenic episodes. Also the recent discovery of Late Precambrian phosphorites in Spain (Perconig *et al.*, this volume), which may be of economic significance, indicates that opportunities exist for phosphate exploration.

A review of the Lower Cambrian of western Europe (Brasier, 1980) indicated that the present continent is made up from the remnants of four separate platforms of Cambrian age, each associated with intracratonic and marginal basins (Fig. 7.1). These platforms were divided by seas or oceans of uncertain width, of which the best studied is the Iapetus Ocean ('proto-Atlantic Ocean') which lay between the North American Platform and the other platforms. According to Bjørlykke, Elvsborg & Høy (1976), the Iapetus Ocean was formed in late Precambrian times as a result of the development of a rift zone to the west of the Baltic Shield.

The purpose of the present paper is to review the Precambrian and Cambrian geology of each of the four platforms and their associated phosphate occurrences, with particular reference to major palaeogeographic features. Full use is made also of the biostratigraphic correlations of Shergold & Brasier (this volume).

North American Platform (part) and basins

Facies studies show that Scotland was formerly part of the Precambrian and Cambrian North American plate, prior to the opening of the present Atlantic, and that Scottish sequences may be correlated with similar successions in eastern Labrador, western Newfoundland, East Greenland and Spitsbergen (e.g. Cowie, 1974; Brasier, 1980). A Durness platform along the north-west seaboard may be distinguished from the Dalradian basinal sediments of the Scottish Highlands and north-west Eire. In the former, a transgressive Cambrian succession of typical North American Cordilleran type (quartzites–shales–limestones and dolomites with an olenellid fauna of the Pacific faunal province) unconformably overlies Riphean (Torridonian) sediments and basement gneisses. The Dalradian succession is a thick Riphean to Ordovician succession with Late Precambrian tillites, rare fossils and complex folding and metamorphism.

During Cambrian time this region and the rest of the present North American continent has three major concentric facies belts:

(1) an inner detrital belt containing some glauconitic and rare phosphatic sediments, endemic trilobites of Pacific type (Cowie, 1971) and a variety of other fossils;

(2) a middle carbonate belt with ooid shoals, archaeocyathid bioherms, calcarenites and peritidal rocks with stromatolitic or fenestral fabric;

(3) an outer detrital belt on the outer shelf, slope and basin, bearing more siliceous sediments, a more cosmopolitan 'open ocean' trilobite assemblage (including some Acado-Baltic taxa), hexactinellid sponges and inarticulate brachiopods (Palmer, 1973; Brasier, 1980). The carbonate belt may have effectively separated warmer water, endemic trilobites from cooler, deeper water, open ocean assemblages, necessitating separate zonal schemes in nearby areas (Cook & Taylor, 1975).

Phosphatic sediments

Phosphorites are scarce in this region. Two factors may have contributed to this: a high rate of clastic sedimentation in the basins (e.g. Torridonian and Dalradian), and the exclusion of open marine influences by the development of extensive middle carbonate barriers on the platforms (e.g. Durness Platform: Brasier, 1980). However, brown to black phosphatic

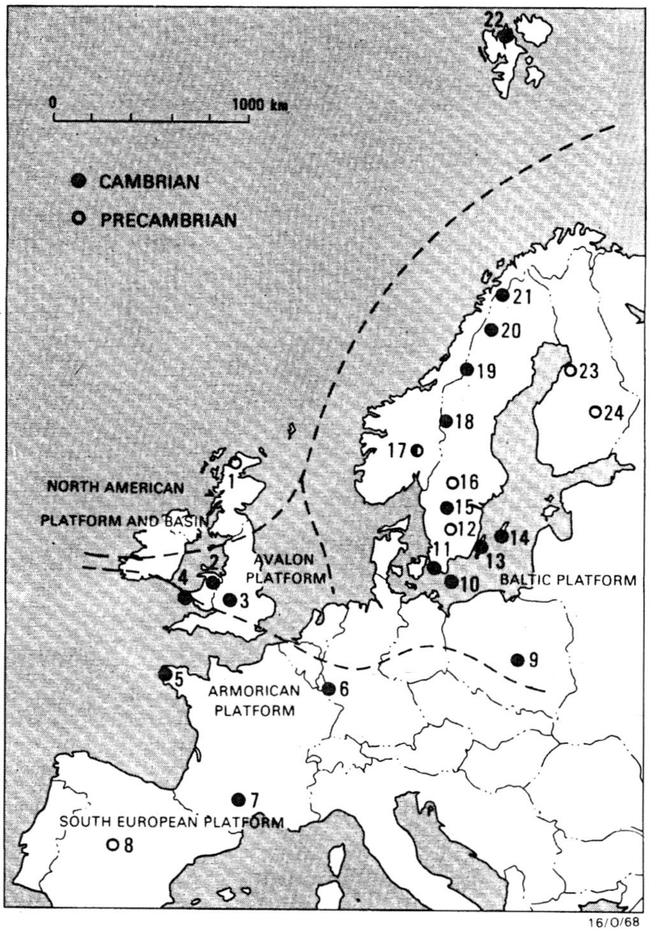

Fig. 7.1 Distribution of some Precambrian and Cambrian phosphatic sediments in Europe.
***United Kingdom*: 1, Cailleach Head, Loch Broom, Scotland; 2, Criccieth–Harlech Dome, Wales; 3, Shropshire, and Warwickshire, England; 4, St David's Wales.**
***France*: 5, Contentin Peninsula, Manche Department; 7, Monts de Lacaune, Tarn Department.**
***Spain*: 8, Cuidad Real.**
***Germany* (FR): 6, Hohe Venn, Aachen.**
***Poland*: 9, Sandomierz, Lublin.**
***Denmark*: 10, Bornholm Island.**
***Sweden*: 11, Scania; 12, Västmanland and Värmland (Eryhthytte Slate); 13, Öland; 14, Gotland; 15, Västergotland; 16, Lake Vättern (Visingsö Series); 18, Vassbo; 19, Tasjö; 20, Laisrall; 21, Luopakte.**
***Norway*: 17, Mjøsa (Cambrian and Biskôpasen Conglomerate); 22, Ny Friesland, Spitzbergen (Svalbard).**
***Finland*: 23, Lampinsaari, Oulu; 24, Temo, Kuopio.**

nodules and lenses of Late Riphean age occur in shales of the Cailleach Head Formation, Torridon Group, in north-west Scotland (Peach *et al.*, 1907; Downie, 1975).

One distinctly laminated variety, comprising minute grains of quartz, feldspar, epidote, muscovite and biotite, contains thin black lenses of similar composition, but with a phosphate cement. The phosphatic nodules and lenses have a matrix which contains spherical bodies about 0.01 mm in diameter interpreted either as 'spores' or as a type of marine phytoplankton (Downie, 1975).

The Cambrian rocks which unconformably overlie the Torridonian are generally devoid of phosphatic material, although mudstones near the top of the Lower Cambrian Fucoid Beds (*Olenellus* Zone) are occasionally phosphatic, as are the fossiliferous black limestones of comparable age near Callander. Phosphatic sediments also occur in the thicker and more complete sequences of East Greenland. There, the Lower and Upper Bastion Formations, comprising respectively sandstones and limestones, contain phosphatic nodules and conglomerates (Cowie & Adams, 1957; Cowie & Spencer, 1970; Haller, 1971; Henriksen & Higgins, 1976). These compare lithologically with occurrences in western Europe and may have formed prior to the development of an effective carbonate barrier during *Bonnia–Olenellus* Zone times.

On the margin of the North American Platform bordering the Iapetus Ocean a shallow-water, tidal environment is suggested (Swett & Crowder, 1982) for the Lower Cambrian Topiggane Shale (Tokammane Formation) on Ny Friesland in northern Spitsbergen (Svalbard). The Cambro-Ordovician strata of Spitsbergen are regarded as coeval and possibly contiguous with equivalent sequences in central East Greenland, north-west Scotland and western Newfoundland. The shales are interbedded with thin glauconitic, phosphatic and dolomitic sandstones, nearly all the detrital grains of which are coated with oolitic phosphate rims. Phosphate coatings constitute 18% of the total rock composition (Swett & Crowder, 1982). The Topiggane Shale is of interest also because of its anomalously high potash content (up to 12.5% K_2O) in the form of finely crystalline adularia, a characteristic feature shared with the 'Fucoid' Beds of the same age in north-west Scotland.

Avalon Platform and basins

During Cambrian, and probably also late Precambrian times, the North American and Avalon Platforms are thought to have been separated by an Iapetus Ocean, subsequently destroyed by continental margin subduction. The Avalon Platform formerly stretched from present-day Massachusetts, through New England, Nova Scotia, Cape Breton and south-east Newfoundland to Wales and central England. Mesozoic–Cenozoic continental drift has now separated these formerly contiguous regions.

Vendian calc-alkaline volcanics, marine tuffs, intrusives and molasse of the Avalon Platform were folded and faulted into basins and ridges before or during the course of the Cambrian transgression. The Lower Cambrian typically comprises red basal conglomerate, quartzite and sandstone followed by red shale and pink, fossiliferous, glauconitic limestone; shale and limestone bands are of darker colour in the Middle and Upper Cambrian successions. Marginal basins such as the Bray Basin in Eire, and the Welsh Basin, subsided very rapidly and received copious turbidites at intervals during the Cambrian. Oolitic and peritidal carbonates such as those of north-west Scotland, which were probably formed in tropical seas, were never developed in the Cambrian of the Avalon Platform and it seems likely that the region was swept by cool currents from the Iapetus Ocean (e.g. Theokritoff, 1979; Brasier, 1980). The Lower Cambrian successions of Shropshire, on the south-east flank of the Longmynd–Wrekin horsts, and the contemporaneous succession in Warwickshire (central England) on the south-west flank of the Charnian Block (Fig. 7.2) exhibit four

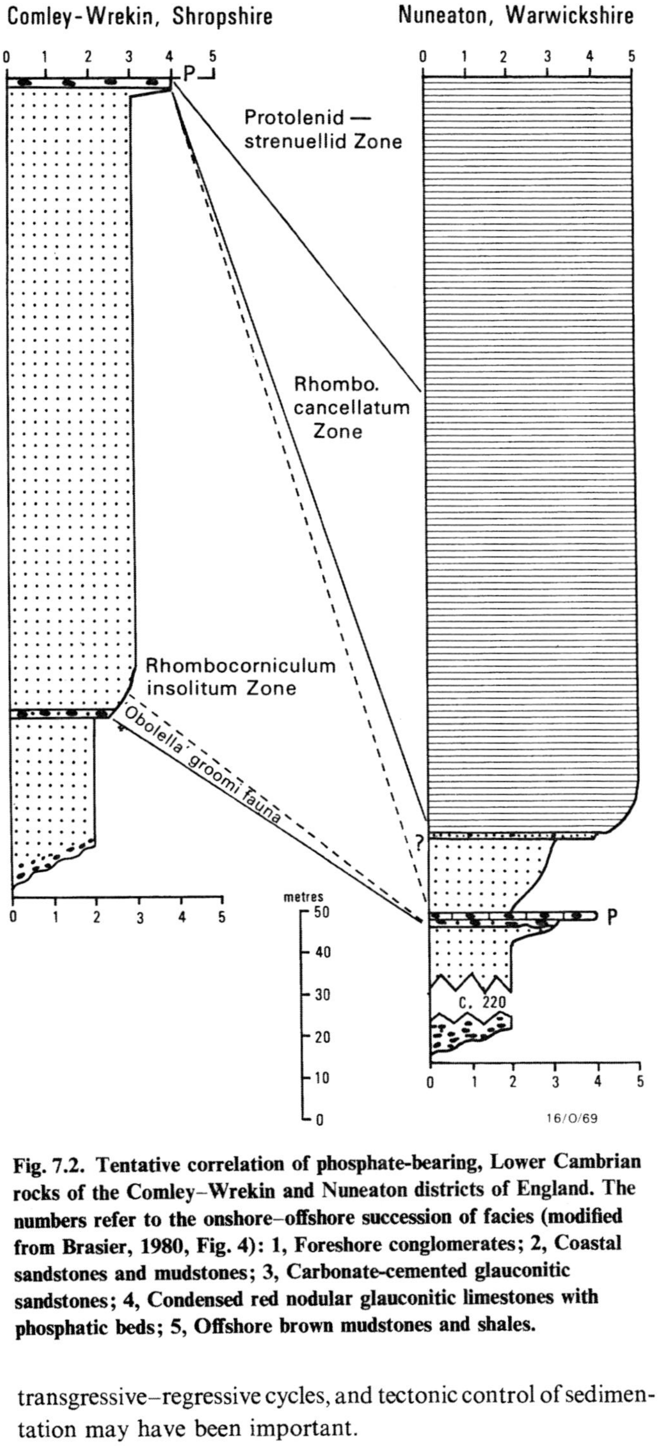

Fig. 7.2. Tentative correlation of phosphate-bearing, Lower Cambrian rocks of the Comley–Wrekin and Nuneaton districts of England. The numbers refer to the onshore–offshore succession of facies (modified from Brasier, 1980, Fig. 4): 1, Foreshore conglomerates; 2, Coastal sandstones and mudstones; 3, Carbonate-cemented glauconitic sandstones; 4, Condensed red nodular glauconitic limestones with phosphatic beds; 5, Offshore brown mudstones and shales.

transgressive–regressive cycles, and tectonic control of sedimentation may have been important.

Phosphatic sediments

Phosphatic sediments are unknown from the Precambrian (mainly Vendian) of the Avalon Platform probably because of the mobile tectonic setting of the region at this time. The Late Precambrian Cadomian Orogeny ultimately introduced a slightly more stable tectonic situation for the Cambrian locally, with the more rigid blocks recognisable by their thin, condensed or incomplete sequences. It is around the margins of these various positive blocks that phosphatic sediments and conglomerates are best preserved, apparently related to fluctuations in the rate of deposition and of relative sea level. In the Shropshire and Warwickshire areas referred to above, phosphatic sediments occur in 2 m thick nodular limestones with condensed faunal sequences, a fining upward of clastics and a culmination in faunally improverished but glauconitic manganiferous and phosphatic limestones with algal structures (Rushton, 1974; Brasier & Hewitt, 1979). Both phosphatic limestones are therefore interpreted as having formed in shallow water after an extensive transgression over a static or rising block, ending in final emergence. The phosphorites in these condensed limestones are clearly a secondary replacement of carbonate shells, carbonate cement or carbonate mud, previously lithified to hardgrounds (Brasier & Hewitt, 1979). In Warwickshire, phosphatisation is most pronounced in beds with scouring, winnowing, exotic and intraformational clasts. Robust apatite-shelled microfossils predominate and show evidence of transport and abrasion. A sample of the phosphatic limestone in Shropshire contained as much as 5.3% P_2O_5 (Oakley, 1942).

Phosphatic sediments were more widespread during the late Early Cambrian, occurring not only in the Comley Limestone Shropshire (Rushton, 1974), but also in shales at St Tudwals, North Wales (Bassett & Walton, 1960; Bassett, Owens & Rushton, 1976), south-east Newfoundland (Howell, 1925) and New Brunswick (Walcott, 1900). It is worth noting the association of these Lower Cambrian phosphorites with pandemic faunas.

The setting of Middle Cambrian phosphorites is similar, being associated with structural highs and sea level fluctuations. In *Oelandicus* times phosphatic intervals were limited but may include the phosphatic nodules at the base of the *Paradoxides bennetti* beds in south-east Newfoundland and the layer at the base of the Middle Cambrian disconformity in Shropshire (Howell, 1925; Cobbold & Pocock, 1934). Manganiferous beds were more widely deposited at this time, occurring in England, Wales and Newfoundland (e.g. Bassett *et al.*, 1976). The *Paradoxissimus* transgression deposited phosphatic sandstones in Shropshire (Cobbold, 1910, 1911; Rushton, 1974) and shales with phosphate enriched nodules in Warwickshire (Illing, 1916), South Wales (Hicks, 1875) and south-east Newfoundland (Howell, 1925; Poulsen & Anderson, 1975). This was a period of widespread deposition of black shales with phosphorite nodules, particularly during *Hypagnostus parvifrons* and *Ptychagnostus punctuosus* times.

A regression in early *Forchhammeri* times is thought to have resulted in non-deposition and erosion of underlying beds, particularly in areas on the margins or tops of more stable blocks. Conglomerates at the base of the *Lejopyge laevigata* Zone may have formed in earlier times and include phosphatic clasts perhaps derived from the *Paradoxissimus* Siltstone below (e.g. south-east Newfoundland: Poulsen & Anderson, 1975; and possibly Warwickshire: Taylor & Rushton, 1972). Nearer the centre of the basins in North Wales and south-west Newfoundland shelly and algal limestones with conglomerates were deposited (Nicholas, 1915; Poulsen & Anderson, 1975) including some phosphatic clasts. These facies resemble Lower Cambrian limestones discussed previously and also the broadly contemporaneous *Hyolithes* and Andrarum Limestones and *Exporrecta* conglomerates of Sweden (see below) but their age is not precisely known.

In the Upper Cambrian, phosphatic nodules occur scattered through dark shales, especially near the base and top of the Dolgellau Member of the Harlech Dome area of North Wales (Allen, Jackson & Rushton 1981), near the base and top of Upper Cambrian shales in Shropshire (Cobbold, 1927; Stubblefield, 1930) and throughout the Upper Cambrian shales of Warwickshire (Taylor & Rushton, 1972). Nodules of phosphorite contain up to 14% P_2O_5 (Oakley, 1942).

Baltic Platform

The Baltic Platform is probably the best known of all Precambrian areas, occupying most of Sweden, Finland, southern Norway and the region of Karelia in north-western USSR, an area of at least 500 000 km^2. Tectonically it is divisible into three major provinces. Most of the Platform is occupied by the Svecofennide mobile belts which stabilised in early Proterozoic times. This is bordered to the north and east by older (Archaean) rocks; to the west is part of a later Proterozoic (including the Riphean) mobile belt, much of which has since been incorporated in the Caledonian belt. The basement comprises a broad and stable dome of Proterozoic gneiss and granite, flanked by the Finnmark and Moscow Basins where nearly continuous marine Riphean, glacial-marine Vendian and marine Cambrian sediments accumulated. Rift-like sparagmite basins containing marine and fluvial Riphean, glacial-fluvial Vendian and marine Cambrian also cut the Baltic Platform in south and central Norway. In the Visingsö Basin of south-east Sweden Upper Riphean–Vendian clastics and algal carbonates accumulated.

The Riphean–Vendian history of the Baltic Platform compares more closely with that of the Scottish Dalradian and East Greenland sequences than with the Avalon Platform and there is some speculation that the Dalradian Basin and the Baltic Platform were more closely located in Vendian times (Anderton, 1980). There is no evidence of a significant Cadomian Orogeny in the Baltic region and the Vendian passes conformably into Cambrian rocks in the major basins. However, the lithofacies and biofacies of the Avalon and Baltic regions become very similar during the Cambrian and it therefore seems likely that they were adjacent and isolated from the North American continent at this time. An Acado-Baltic fauna apart, other similarities include glauconitic–phosphatic facies and absence of tropical-type oolitic and peritidal carbonates. In contrast, the Baltic Platform sequences are thinner, with extensive hiati, and they generally lack red sediments and Fe–Mn deposits, probably because of the lack of weathered volcanic source rocks (e.g. Ziegler & McKerrow, 1975). Typically, the Baltic Lower Cambrian comprises quartzites, glauconitic sandstones, siltstones and thin condensed limestones with a sparse fauna that contains relatively few pandemic elements. The Middle Cambrian consists predominantly of dark alum shales with some conglomerates and condensed shelly limestones (often phosphatic). Pandemic agnostids are present and also occur in Upper Cambrian shales and anthraconite limestones.

Mens (1981) and Mens & Pirrus (1975) regard the Lower Cambrian transgression in the Baltic region as comprising three transgressive–regressive cycles, each of which was glauconitic–phosphogenic and notably so in those thinner and discontinuous deposits on the crystalline basement in Estonia and Scandinavia. The first, or Baltian, cycle deposited the blue clays of the Lontova Formation, in the Moscow Basin, Poland and Estonia, and is followed by the Liivian cycle in Estonia. The third, or Aisciaian, cycle follows a widespread regression.

Non-sequence or erosion occurs at the top of the Lower Cambrian or base of the Middle Cambrian in most of Scandinavia, even in the largely continuous succession of Mjøsa (Martinsson, 1974). This event may be correlated with the Hawke Bay regression across northern and eastern North America (Bergström, 1981; Palmer, 1981) though this hiatus is less obvious in Avalonian sequences. It may be significant that the closest parallel is in the Shropshire succession, which has many other similarities with the Lower Cambrian of Scania. Regressive facies would therefore appear to have been caused by a eustatic drop in sea level.

The faunas of the early *Paradoxissimus* Stage have a relatively cosmopolitan distribution which seems to be connected with a eustatic rise in sea level (Robison, 1976, Robison *et al.*, 1977), and the succeeding late *Paradoxissimus* assemblages also occur around the world (*ibid*; Taylor & Rushton, 1972) and may be taken to imply relatively high sea levels.

Alum shale deposition was halted by another Middle Cambrian regression in late *Paradoxissimus* or early *Forchhammeri* times, which correlates partly with a similar event on the Avalon Platform. Shoaling led to some alum shale erosion followed by deposition of the *Exporrecta* Conglomerates, with phosphatic pebbles, or *Hyolithes* and Andrarum Limestones with autochthonous phosphate (Fig. 7.3), e.g. Scania and Bornholm (Hadding, 1958; Poulsen, 1966), Östergotland (Westergard, 1944) and Västergötland (Hadding, 1958). The *Exporrecta* Conglomerate facies was relatively widespread, occurring also in Närke, Jämtland, Ångermanland and southern Lappland (Martinsson, 1974) while the Andrarum Limestone fauna is found in deposits transgressive over Precambrian basement (Martinsson, 1968).

During the Late Cambrian, euxinic and probably deeper water conditions encouraged the deposition of alum shales with anthraconite limestone concretions, some of which contain phosphatised microfossils and faecal pellets (Müller, 1979). However, no further actual phosphorites were formed until Tremadocian (Ordovician) time.

Phosphatic sediments

Phosphogenic episodes are evidently more widespread and synchronous on the Baltic Platform than in the other regions discussed here. Although they are not common in the Precambrian, significant discoveries of Late Proterozoic age have, nevertheless, been made recently in the eastern part of the platform. Phosphatic sediments of Cambrian, notably Early Cambrian, age are particularly numerous. The conclusion by Smith, Hurley & Briden (1981) and Ziegler *et al.* (1979) that in Early Cambrian times most of the Baltic Platform lay at low latitudes in the southern hemisphere (to about 15°S) and, in addition, had a north–south orientation similar to the present, is of special significance in that it implies a western margin of the Baltic Platform facing the Iapetus Ocean. A depositional environment favourable to the formation of phosphorite com-

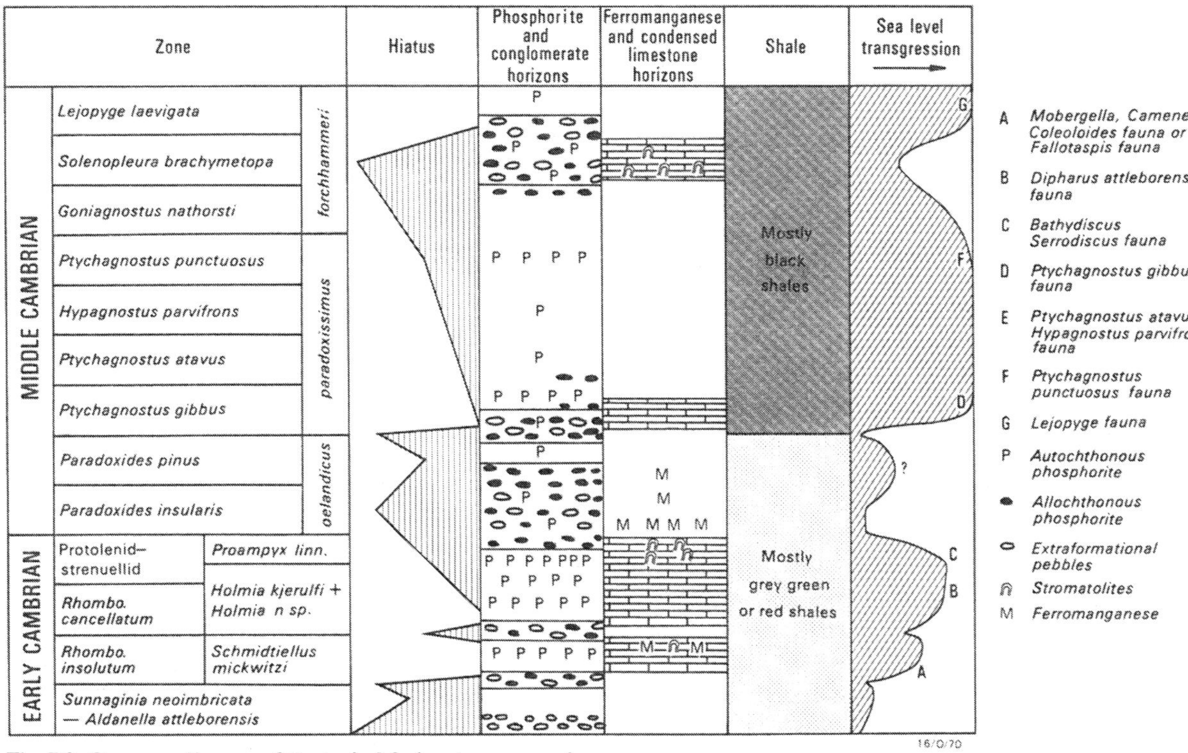

Fig. 7.3. Summary diagram of the typical facies changes seen in Lower and Middle Cambrian rocks of the Avalon and Baltic Platforms. A–G refer to pandemic faunas, perhaps reflecting the higher sea level of a transgressive–regressive cycle.

parable with that associated with the upwelling of deep, phosphorus-rich waters on westward-facing coasts is therefore conceivable.

1. Precambrian. In the eastern part of the Baltic Platform, uraniferous, phosphatic beds associated with dolomite and calcitic marbles, skarn rocks, quartz–feldspar gneisses and graphitic schists occur in the Lower Proterozoic (Karelian–Svecofennian) schist belts of central and northern Finland (Äikäs, 1980, 1981). The most important occurrences identified to date lie in the marine Upper Jatulian (Karelian) sequence of metasediments. An age of around 1900–2080 m.y. has been suggested (Vaasjoki *et al.*, 1980) for these apatite-bearing rocks. At Lampinsaari, central Finland (Fig. 7.1, locality 23), a phosphatic zone forming part of the Vihanti ore complex is up to 50 m thick. It consists of lenticular beds, up to 0.5 m thick, of apatite-bearing dolomite, skarn and fine-grained apatite–quartz–plagioclase gneiss interpreted as a metamorphic derivative of phosphatic acid tuff (Rehtijärvi, Äikäs & Makela, 1979). Samples of the richer layers (phosphorite) contain up to 25.5% P_2O_5 over a thickness of 0.3 m. However, overall, the phosphate zones average only 3–5% P_2O_5 and 0.3% U. The phosphorite consists of apatite, pyrite, pyrrhotite, graphite, quartz, phlogopite and accessory uraninite and rutile. Uraninite occurs interstitially or as inclusions in apatite. According to Rauhamaki, Makela & Isomaki (1978) the Vihanti sulfide ores and uraniferous apatite-bearing rocks are of syngenetic volcano-sedimentary origin, deposited in an environment where carbonate sediments were also forming. A similar body, 3.5 km long and up to 90 m wide, occurs at Temo, to the south-east (Fig. 7.1, locality 24). Fine grained apatite–quartz–feldspathic schist predominates, usually in layers less than 20 cm thick but locally up to 1 m thick. Up to 40% apatite may be present with apatite-bearing diopside skarn containing up to 13% P_2O_5 (Äikäs, 1981). Minor phosphorite interbeds occur in Proterozoic banded iron formations (Laajoki & Saikkonen, 1977) and appear to be of comparable age.

Phosphatic pebbles occurring in the Grythytte Slate and the Upper Visingsö Series in Sweden are the two world occurrences of Precambrian phosphate recorded by Geijer (1962). The Grythytte Slate in Västmanland and Värmland (Fig 7.1, locality 12), southern Sweden (Sundius, 1923), lies within the upper sedimentary section of the well-known Leptite Formation of Svecofennian (Early Proterozoic) age. Leptites are essentially quartzo-feldspathic, pale and usually fine-grained rocks, generally regarded as metamorphosed acid volcanics but frequently containing also metasedimentary material in the form of psammitic sediments, limestones or dolomites and banded iron formations common in Precambrian sequences elsewhere in the world. The nodules are generally less than 3 cm in diameter, sometimes coalescing to form lenses, and containing in their central parts aggregates of fine-grained apatite and graphite, surrounded by apatite. According to Sundius (1923) the nodules were originally composed of phosphorite and amorphous silica (chert). Specimens of the lower part of the Grythytte Slate show on analysis up to 9% P_2O_5. Organic structures have not yet been identified.

Both the Upper Visingsö Series of south-east Sweden (Vidal, 1976) and the Biskopåsen Conglomerate of South Norway, appear to be of Early Vendian (pre-tillite) age (Vidal, 1981)–phosphatic sediments do not occur in the Vendian of the Moscow Basin (Mens & Pirrus, 1975) nor in the Finnmark Basin.

The relatively undisturbed Visingsö Formation, which is virtually restricted to the Lake Vättern Basin (Fig. 7.1, locality 16), consists of a basal sandstone, arkose, feldspathic sandstone and shale, with thin beds of limestone, the entire formation having a maximum thickness of 1040 m. Some of the undoubted fossil remains are interpreted as chitinous foraminifera, while algae and remnants of terrestrial plants have also been reported (Timofeev, 1960). Both the fossil assemblage and the phosphatic nodules are similar to those found in the Cambrian. The smooth, flattened and somewhat ellipsoidal nodules are up to 4.5 cm in length, and are composed chiefly of apatite, calcite, opal and chalcedony, chemical analysis of four samples showing a P_2O_5 content of 16.1%, 23.7%, 21.6% and 13.0% (Hedström, 1930). At Lake Mϕjsa in south-central Norway (Fig. 7.1, locality 17), phosphatic pebbles occur in the lowest part of the Biskopasen Conglomerate. This area has been well-known since the beginning of the present century for providing a complete section of the coarse feldspathic sandstone and conglomerates referred to as 'sparagmites'. The Biskopasen Conglomerate is interpreted as fluviatile-deltaic in origin, having developed during a regressive phase of deposition along the southern margin of the Sparagmite Basin (Skjeseth, 1963; Bjϕrlykke, 1966). The formation forms a tongue in the Biri Shale and Limestone, in places resting also on the Brϕttum Formation. Pebbles and boulders of the Brϕttum Formation, which consists of dark greywackes alternating with black, calcareous and pyritic shales, are common in the Biskopåsen Conglomerate. The formation is probably the source of the associated phosphatic pebbles.

According to Spjeldnaes (1967), the Biskopasen Conglomerate comprises well-rounded, often spherical pebbles and boulders up to 2 m in diameter of quartzite, pegmatite quartz and granite, and, less commonly, angular pebbles of very variable size which consist primarily of limestone or sandstone. The well-rounded phosphatic pebbles, which are up to about 10 cm diameter but mostly 1.5–3 m in diameter, range from limestone to sandstone in composition, with only minor amounts of phosphate cement, to phosphorite composed of apatite, quartz and traces of clay minerals. The more phosphatic varieties show micro-brecciation and some, because of their thick coal-like appearance, probably have a high carbon content. They may also contain remains of *Papillomembrana compta* and minute spherical structures of 'Sporomorphs' (Spjeldnaes, 1967).

2. Cambrian. Phosphatic sandstone of Early Cambrian age occurs at a relatively large number of localities along the eastern border of the Caledonides in Sweden (Fig. 7.1), being preserved below a cover of overthrusted nappes (Grip, written communication, 1981). The phosphatic bed is of variable thickness, generally between a few decimetres and 1 m, the phosphate occurring as black, elongated pebbles and as a diagenetic cement in sandstone. The phosphorite bed is always associated with, and stratigraphically overlain by, shales or siltstones and occurs in or at the beginning of a transgressive cycle (Wallin, 1982). However, phosphate deposition was not isochronous along the margin of the Baltic Platform but occurred at intervals during the later part of the Early Cambrian (Fig. 7.4). According to Thorslund (1960), the transgression which extended across the Baltic Platform reached its maximum during the middle part of the Lower Cambrian.

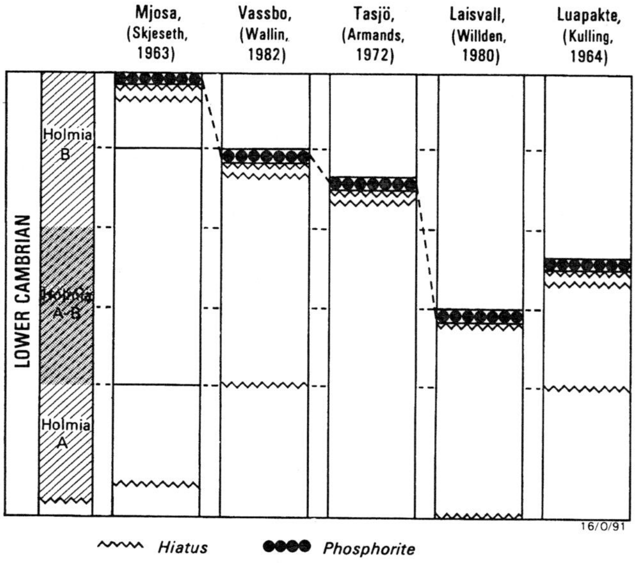

Fig. 7.4. Tentative correlation of Lower Cambrian phosphorites in Norway and Sweden. (Based on Wallin, 1982.)

A stillstand or regression at the end of the first or Baltian transgressive cycle may account for phosphatic conglomerates found at the base of the succeeding Liivian cycle in Estonia (Fig. 7.3) (Mens & Pirrus, 1975), although contemporaneous phosphorites of this zone are also widespread and could have been a source of the phosphatic clasts. For example, autochthonous phosphorites of the same age occur as both nodular and tabular beds in the glauconitic Norretorp Siltstone of Scania (Hadding, 1929, 1932) and the Green Shales of Bornholm (Poulsen, 1967). However, allochthonous phosphorites are more common, occurring in the Brantevik glauconitic sandstones and higher Norretorp siltstones of Scania (Hadding, 1927, 1929, 1932), the Green Shales of Bornholm (Vidal, 1981), the *Mobergella* beds of the West Baltic (Bengtson, 1968, 1970), in the neptunean dykes of Västergötland (Hadding, 1927), and in the basal and topmost laα beds of Mjϕsa, south Norway (Vogt, 1924; Bjϕrlykke *et al.*, 1976).

The third or Aiscianian cycle of Mens & Pirrus (1975) which follows a widespread regression (Fig. 7.3), is represented by phosphatic conglomerates at the base of the Kakumagi Beds in Estonia (Öpik, 1956), the Bråstad Shale of southern Norway (Vogt, 1924, Skjeseth, 1963) and perhaps the Asijatj Member below the Grammajuku Formation of southern Lappland (Willdén, 1980). Autochthonous phosphorite nodules also occur in the upper part of the Rispbjerg Sandstone of Scania

(Hadding, 1958; Poulsen, 1967; Bergström & Ahlberg, 1981) and the Upper Sandstone unit of northern Lappland and Troms (Hadding, 1958): these may all be referred to the supposed *Holmia* n. sp. Zone (e.g. Bergström, 1981). Intense bioturbation has destroyed most of the depositional structures of the Lower Cambrian Rispbjerg Sandstone on Bornholm Island, Denmark (de Marino, 1980b). The sandstone, which contains layers of mudstone phosphorite, is believed to have formed in a shallow-water tidal environment in a barred lagoon, more or less complete diagenetic replacement of calcite by phosphate having taken place penecontemporaneously. The uppermost 40 cm of the sandstone averages about 15% P_2O_5. The sandstone is about 4 m thick (Poulsen, 1960).

The more generally recognised *Holmia kjerulfi* part of the Aisciaian transgression (Fig. 7.3) locally resulted in thin limestones and shales which are often phosphatic: e.g. the Gislöv Formation, Scania (de Marino, 1980a, Bergström & Ahlberg, 1981), Lingulid Sandstone of Västergötland (Hadding, 1927), shales, sandstones and conglomerates of northwest Dalecarlia and southern Lappland (Hadding 1929, Willdén, 1980) and the Upper Shale of Troms and northern Lappland (Hadding, 1958).

The transgression in the final *Proampyx linnarssoni* Zone of the Lower Cambrian was reflected by phosphatic conglomerates lying upon basement gneiss, as at Ustaoset in south-west Norway (Størmer, 1925). A regressive conglomerate at the top may parallel the shoaling-upward sequence in Scania, where allochthonous phosphorites occur in glauconitic sandstones (Hadding, 1932, 1958; de Marino, 1980a; Bergström & Ahlberg, 1981). The *Lapworthella* Limestone at the top of the Lower Cambrian in Shropshire may record this same shoaling event. The basal Middle Cambrian shales of the *Oelandicus* Stage have basal phosphatic conglomerates at Mjøsa (Skjeseth, 1963), Östergotland and Närke (Martinsson, 1974). In Scania, Bornholm and Västergötland, however, the basal phosphatic deposits, the *Exsulans* Limestones or *Acrothele granulata* Conglomerates, are younger (early *Paradoxissimus* Stage). Allochthonous phosphorite clasts show many cycles of phosphatisation and erosion, although there is also some autochthonous phosphorite in these beds (Poulsen, 1963; Hadding, 1958).

Armorican Platform

Sediments of the Armorican (Variscan) Platform extend from northern France, through Belgium to northern-central Europe (Doré, 1974). Precambrian sedimentary rocks are best known from the Brioverian of Brittany, where there are tilloids, turbidites and volcanics, terminating in a Cadomian Orogeny dated at about 660–620 m.y. (Roach, 1977). The Cambrian rocks transgress the Brioverian with unconformity and may contain post-orogenic volcanic rocks. Gabbros and granites were also emplaced during the Cambrian (Roach, 1977; Le Gall *et al.*, 1975). Differential subsidence then led to the development of basins, gulfs and positive blocks (Doré, 1972). The Cambrian facies belts on this platform generally compare with those outlined for North America, so that the following can be distinguished: an inner detrital belt (Brittany, Vendée); a middle carbonate belt with rare archaeocyathids (Normandy, Maine, Fichtelgebirge, Lausitz, Görlitz): and an outer detrital belt (Ardennes, Thuringia, Frankenwald) continuing into the Holy Cross Mountains (Doré, 1974). This platform may have been independent of the South European Platform, though it shares with it an archaeocyathid and redlichiacean trilobite fauna in the Early Cambrian (Sdzuy, 1967).

Phosphatic sediments

In Normandy, a transgressive sequence with basal conglomerates and arkoses is succeeded by glauconitic sandstones with phosphorites, shales, and nodular limestones (Doré, 1972). A *Chancelloria*-hyolithid fauna occurs in these beds, suggesting an Early Cambrian *Rhombocorniculum insolutum* Zone age, a period of phosphogenesis in many other areas (Shergold & Brasier, this volume).Phosphatic sediments were not formed after the establishment of a middle carbonate belt, perhaps because this formed a barrier to oceanic influence. In the Frankenwald (north-eastern Bavaria), the flysch-like Tiefenbach Formation of Early Cambrian age is overlain by relatively regressive facies of the Galgenberg and Wildenstein Formations of *Oelandicus* age. Both are arenaceous with phosphatic nodules (Ludwig, 1969). Paproth & Zimmerle (1980) refer to phosphate nodules from the Upper Cambrian in the Hohes Venn area, in the north-western part of the Federal German Republic. In the Armorican Massif of north-western France, two phosphatic horizons have been found in the Lower Cambrian succession. In the eastern part of the Massif blocks of phosphate rock containing more than 25% P_2O_5 have been found associated with Cambrian limestones and dolomites (Slansky, written communication, 1983).

South European Platform

Cambrian sequences in north-west Spain, the Montagne Noire (France) and Sardinia are sufficiently similar to infer that these are the remnants of a formerly contiguous outcrop, although Cambrian faunas in the Sierra Morena of southern Spain have similarity with those in Morocco. The Precambrian is very variable over this region, ending with volcanics in the Sierra Morena and Montagne Noire but with clastics in other areas, where they may pass conformably upwards into Cambrian rocks. Unfortunately, the relationship between Precambrian and Cambrian rocks is still very poorly known in the Mediterranean area. During the Early Cambrian, the inner detrital redlichiacean trilobite fauna has Tethyan and Asian affinities (Cowie, 1971). Middle Cambrian rocks comprise mainly shales with some glauconitic sandstones and condensed nodular limestones succeeding tropical carbonate facies. *Paradoxides* and *Conocoryphe* suggest cooler and more open ocean conditions (Palmer, 1973) although much of the fauna is restricted to the region (Sdzuy, 1967).

Phosphatic sediments

Reported occurrences of phosphorites are relatively few, though glauconite occurs in facies attributable to both inner and outer detrital belts (Brasier, 1980). The Lower Cambrian of the south-eastern Montagne Noire, around the Hérault district, contains a sequence of basal conglomerates, sandstone and limestone with archaeocyathids, and towards the top, some phosphatic sandstones of late Early Cambrian age with phosphatic shells and steinkerns of the bivalved *Heraultipegma* (i.e.

Heraultia of Müller, 1975), a genus which is widely distributed in the Early Cambrian (e.g. Runnegar, 1981). Cambrian outcrops on the north flank of the Montagne Noire contain a variety of phosphatic facies brought together in complex fault slices. These occurrences range from phosphatic nodules in euxinic black shales or deep-water sponge-bearing limestones of Early Cambrian age, to phosphatised algal mats and oncoliths near the top of the Schistes à *Paradoxides* facies of Middle Cambrian age (Prian, 1979). Locally there are two beds totalling 6–10 m in thickness which contain 5–20% P_2O_5. The more phosphatic bed consists of a sandy, feldspathic and micaceous phosphatic limestone which is about 70 cm thick and contains more than 12% P_2O_5 (Notholt, Highley & Slansky, 1979). A Palaeozoic inlier in the region of Vigan (east of the Montagne Noire) contains a Lower Cambrian shale–dolomite sequence, associated with ancient karstic features (Orgeval & Capus, 1978). The phosphorites occur above a basal sandstone.

Summary

(1) The Precambrian and Cambrian sedimentary successions are known in considerable detail for most of Europe. The present Cambrian areas of the European continents may be regarded as consisting essentially of remnants of four separate and relatively stable platforms (the American, Avalon, Baltic and Armorican), each associated with relatively unstable intracratonic basins. These platform–basin complexes have histories closely linked with the development of the Iapetus Ocean ('proto-Atlantic Ocean') in late Precambrian times.

(2) Phosphorites of Precambrian and, particularly, Cambrian age are fairly widespread in the European continent. Most of the occurrences are scattered and, are too thin to be of commercial interest. Particularly well developed are those of Early Cambrian age that occur along the eastern border of the Caledonides in Sweden. Although phosphate precipitation was not isochronous along this Caledonian zone, they are almost identical in terms of their phosphate content and overall physical composition, the phosphorites usually being associated with the early stages of a trangressive cycle.

(3) Similarly, the numerous uraniferous phosphatic occurrences in the marine Upper Jatulian of Finland indicate a period of extensive phosphate deposition during Proterozoic times, probably in separate small basins. Other occurrences of comparable age in Sweden give support to the view (Äikäs, 1980) that a Lower Proterozoic phosphogenic province exists within the Baltic Platform. The recent discovery of phosphate of Late Proterozoic age in Spain (Perconig *et al.*, this volume) suggests that Late Precambrian phosphate may be more widespread in southern Europe than has hitherto been supposed.

(4) The Avalon and Baltic Platform sequences contain the majority of phosphorite occurrences in Europe. They are associated with glauconitic sediments or black shales, in successions that otherwise seem to lack tropical oolitic and peritidal limestones, dolomites or archaeocyathid facies. The phosphorites occur mostly on top of blocks of variable tectonic stability (e.g. England), prolonged stability (e.g. Scandinavia) and on the flanks of depositional basins (e.g. Wales, south-east Newfoundland). Phosphorites occur mainly as microsphorite replacement of $CaCO_3$ in limestones or of $CaCO_3$ cement in sandstones and conglomerates.

(5) The lithofacies and fauna of the phosphatic sediments in the Avalon and Baltic regions suggest relatively shallow, cool waters, with connections to open seaways or oceans, such as the Iapetus Ocean. Elsewhere in Europe, platforms had warmer-water carbonate developments and the phosphorites tend to either precede the establishment of such facies (e.g. Normandy, East Greenland) or they occur near the top of such facies, followed by more open-ocean trilobites and lithologies (e.g. Montagne Noire).

(6) Transgressive–regressive sedimentary cycles are typical of the Lower and Middle Cambrian on the Avalon and Baltic Platforms. Phosphatic sediments may occur throughout such cycles but they appear to be more allochthonous and conglomeratic in the shallower, basal and uppermost part of a cycle and occur as autochthonous nodules, tabular beds or hardgrounds in the deeper middle part of a cycle. Some of the clasts in conglomeratic phosphorites are derived from one or more of the previous cycles of sedimentation.

(7) Present evidence suggests that there were perhaps six major sedimentary and phosphogenic cycles in the Lower and Middle Cambrian (Fig. 7.3), each initiated by a transgression and followed by a stillstand or regression. Regressive episodes are recognised by hiati or shoaling facies changes that include exotic and phosphorite conglomerates, sandstones and condensed shelly limestones with algal structures or fenestal fabric. On the Avalon Platform Fe–Mn deposits occur. Deeper transgressive episodes contain limestones and shales with autochthonous phosphorites and elements of a cosmopolitan open ocean fauna. The synchroneity of these cycles is more difficult to demonstrate in the Early Cambrian, where biostratigraphic correlation is less certain, but the possibility remains that these cycles may reflect eustatic changes in sea level.

References

Äikäs, O. (1980). Uraniferous phosphorite and apatite-bearing gneisses in the Proterozoic of Finland. In *Uranium in the Pine Creek Geosyncline*, ed. J. Ferguson & A.B. Goleby, pp. 675–81. International Atomic Energy Agency, Vienna.

Äikäs, O. (1981). Proterozoic phosphorites in Finland. *International Geological Correlation Programme Project 156 Phosphorites, Newsletter*, **9**, 21–7.

Allen, P.M., Jackson, A.A. & Rushton, A.W.A. (1981). The stratigraphy of the Mawddach Group in the Cambrian succession of North Wales. *Proceedings of the Yorkshire Geological Society*, **43**, 295–330.

Andersson, J.G. (1896). Uber cambrische und silurische phosphoritfuhrende Gesteine aus Schweden. *Bulletin of the Geological Institutions of the University of Uppsala*, **2**, 133–238.

Anderton, R. (1980). Distinctive pebbles as indicators of Dalradian provenance. *Scottish Journal of Geology*, **16**, 143–52.

Armands, G. (1972). Caledonian geology and uranium-bearing strata in the Tasjö Lake area, Sweden. *Geologiska Föreningens i Stockholm Forhandlingar*, **94**, 321–45.

Bassett, D.A. & Walton, E.K. (1960). The Hell's Mouth Grits: Cambrian greywackes in St. Tudwal's Peninsula, North Wales. *Quarterly Journal of the Geological Society of London*, **116**, 85–110.

Bassett, M.G., Owens, R.M. & Rushton, A.W.A. (1976). Lower

Cambrian fossils from the Hell's Mouth Grits, St. Tudwal's Peninsula, North Wales. *Journal of the Geological Society*, **132**, 623–44.

Bengtson, S. (1968). The problematic genus *Mobergella* from the Lower Cambrian of the Baltic area. *Lethaia*, **1**, 325–51.

Bengtson, S. (1970). The Lower Cambrian fossil *Tommotia*. *Lethaia*, **3**, 363–92.

Bergström, J. (1981). Lower Cambrian shelly faunas and biostratigraphy in Scandinavia. *United States Department of the Interior, Geological Survey, Open-File Report*, **81–743**, 22–25.

Bergström, J. & Ahlberg, P. (1981). Uppermost Lower Cambrian biostratigraphy in Scania, Sweden. *Geologiska Föreningens i Stockholm Forhandlingar*, **103**, 193–214.

Bjørlykke, K. (1966). Studies on the latest Precambrian and Eocambrian of Norway, No. 1: Sedimentary petrology of the sparagmites of the Rena district S. Norway. *Norges Geologiske Undersøkelse*, **238**, 5–53.

Bjørlykke, K., Elvsborg, A. & Høy, T. (1976). Late Precambrian sedimentation in the central sparagmite basin of south Norway. *Norsk Geologisk Tiddskrift*, **56**, 233–90.

Brasier, M.D. (1980). The Lower Cambrian transgression and glauconite-phosphate facies in western Europe. *Journal of the Geological Society*, **137**, 695–703.

Brasier, M.D. & Hewitt, R.A. (1979). Environmental setting of fossiliferous rocks from the uppermost Proterozoic–Lower Cambrian of central England. *Palaeogeography, Palaeoclimatology, Palaeoecology*, **27**, 35–57.

Brasier, M.D. & Hewitt, R.A. (1981). Faunal sequence within the Lower Cambrian "Non-trilobite Zone" (*s.l.*) of central England and correlated regions. *United States Department of the Interior, Geological Survey, Open-File Report*, **81–743**, 26–28.

Cobbold, E.S. (1910). On some further excavations among the Cambrian rocks of Comley, Shropshire, 1908. *Report of the British Association for 1908*, 231–42.

Cobbold, E.S. (1911). Trilobites from the *Paradoxides* beds of Comley (Shropshire). *Quarterly Journal of the Geological Society of London*, **67**, 282–311.

Cobbold, E.S. (1927). The stratigraphy and geological structure of the Cambrian area of Comley (Shropshire). *Quarterly Journal of the Geological Society of London*, **83**, 551–73.

Cobbold, E.S. & Pocock, R.W. (1934). The Cambrian area of Rushton (Shropshire). *Philosophical Transactions of the Royal Society, Ser. B*, **223**, 305–409.

Cook, H.E. & Taylor, M.E. (1975). Early Palaeozoic continental margin sedimentation, trilobite biofacies and the thermocline, western United States. *Geology*, **3**, 559–62.

Cowie, J.W. (1971). The Cambrian of the North American Arctic Regions. *Cambrian of the New World.* ed. C.H. Holland, pp. 325–84. John Wiley, London.

Cowie, J.W. (1974). The Cambrian of Spitsbergen and Scotland. In *Cambrian of the British Isles, Norden and Spitsbergen.* ed. C.H. Holland, pp. 123–56. John Wiley, London.

Cowie, J.W. & Adams, P.J. (1957). The geology of the Cambro-Ordovician rocks of central East Greenland. Part I. Stratigraphy and structure. *Meddelelser øm Grønland*, **153**, 150pp.

Cowie, J.W. & Spencer, A.M. (1970). Trace fossils from the late Precambrian–Lower Cambrian of East Greenland. *Special Issue of the Geological Journal*, **3**, 91–100.

de Marino, A. (1980a). The upper Lower Cambrian strata south of Simrishamn, Scania, Sweden. *Sveriges Geologiska Undersökning, Ser. C*, **771**, 1–22.

de Marino, A. (1980b). Sandstones and phosphatized calcareous sediments of the Lower Cambrian Rispbjerg Sandstone, Bornholm, Denmark. *Danmarks Geologiske Undersøgelse. III Raekke* **113**, 39pp.

Doré, F. (1972). La transgression majeure du Paleozoique inférieur dans le Nord-Est du massif armoricain. *Bulletin de la Société Géologique de France, 7th series*, **14**, 79–93.

Doré, F. (1974). L'Europe moyenne cambrienne. Les modèles sédimentaires, leur zonalité, leur controle. *Colloque International du Centre National de la Recherche Scientifique*, **243**, 143–55.

Downie, C. (1975). Precambrian of the British Isles – Palaeontology. In *Precambrian*, ed. A.L. Harris, R.M. Shackleton, J. Watson, C. Downie, W.B. Harland & S. Moorbath, *Special Report of the Geological Society*, **6**, 113–5.

Geijer, P. (1962). Some aspects of phosphorus in Precambrian sedimentation. *Arkiv for Mineralogi och Geologi*, **3**, 165–86.

Hadding, A. (1927). The Pre-Quaternary sedimentary rocks of Sweden. I. A survey of the Pre-Quaternary sedimentary rocks of Sweden. II. The Paleozoic and Mesozoic conglomerates of Sweden. *Lunds Universitets Arsskrift. N.F. Avd. 2*, **Bd 23(5)**; *Kungliga Fysiografiska Sallskapets Handlingar. N.F.*, **Bd38(5)**, 171pp.

Hadding, A. (1929). The Pre-Quaternary sedimentary rocks of Sweden. III. The Paleozoic and Mesozoic sandstones of Sweden. *Lunds Universitets Arsskrift, N.F. Avd. 2*, **25(3)**; *Kungliga Fysiografiska Sällskapets Handlingar, N.F.*, **40(3)**, 287pp.

Hadding, A. (1932). The Pre-Quaternary sedimentary rocks of Sweden. IV. Glauconite and glauconitic rocks. *Lunds Universitets Arsskrift. N.F. Avd 2*, **Bd 28(2)**; *Kungliga Fysiografiska Sallskapets Handlingar. N.F.*, **Bd 43(2)**, 175pp.

Hadding, A. (1958). The Pre-Quaternary sedimentary rocks of Sweden. VII. Cambrian and Ordovician limestones. *Lunds Universitets Arsskrift. N.F. Avd 2*, **Bd 54(5)**; *Kungliga Fysiogarfiska Sallskapets Handlingar. N.F.*, **Bd 63(5)**, 26pp.

Haller, J. (1971). *Geology of the East Greenland Caledonides.* Interscience, New York. 381pp.

Hedström, H. (1930). Fosforitbollar fran Visingsöserien? *Sveriges Geologiska Undersökning*, **361**, 8pp.

Henriksen, N. & Higgins, A.K. (1976). East Greenland Caledonian Fold Belt. In *Geology of Greenland*, ed. A. Escher & W. Watt, pp. 182–246. Geological Survey of Greenland, Copenhagen.

Hicks, H. (1875). On the occurrence of phosphates in the Cambrian rocks. *Quarterly Journal of the Geological Society of London*, **31**, 368–83.

Howell, B.F. (1925). The faunas of the Cambrian Paradoxides beds at Manuels, Newfoundland. *Bulletin of American Palaeontology*, **11(43)**, 9–140, 3 pls.

Illing, V.C. (1916). The Paradoxidian Fauna of a part of the Stockingford Shales. *Quarterly Journal of the Geological Society of London*, **71**, 386–450.

Kulling, O. (1964). Norra Norrbottensfjällens Kaledonberggrund *Sveriges Geologiska Undersökning, Ser. Ba*, **19**, 166pp.

Laajoki, K. & Saikkonen, R. (1977). On the geology and geochemistry of the Precambrian iron formation in Vayrylankylä, South Puolanko area, Finland. *Bulletin of the Geological Survey of Finland*, **292**, 137pp.

Le Gall, J., Doré, F., Giordano, R. & Pottier, Y. (1975). Position stratigraphique et cadre tectono-sédimentaire des manifestations volcaniques cambriennes dans le Nord-Est du massif armoricain. *Bulletin Société Géologique de France*, 7ᵉ Ser., **17(6)**, 1101–9.

Ludwig, V. (1969). Lithologische Untersuchung des Kambriums im Frankenwald (Bayern). *Geologisches Jahrbuch*, **87**, 89–156.

Martinsson, A. (1968). Cambrian palaeontology of Fennoscandian basement fissures. *Lethaia*, **1**, 137–55.

Martinsson, A. (1974). The Cambrian of Norden. In *Cambrian of the British Isles, Norden and Spitsbergen*, ed. C.H. Holland, pp. 185–283. John Wiley & Sons, London.

Mens, K.A. (1981). The Lower and Middle Cambrian boundary in the eastern Baltic area. *United States Department of the Interior, Geological Survey Open File Report*, **81–743**, 130–131.

Mens, K.A. & Pirrus, E.A. (1975). The basal conglomerate of the Lower Cambrian Lukati Beds in Estonia. In *Geologiya Kristallicheskogo fundamenta i osadochnogo chekhla pribaltiki*, 132–143. Zinatne (In Russian).

Müller, K.J. (1975). *"Heraultia" varensalensis* Cobbold (Crustacea) aus dem unteren Kambrium, der älteste Fall von Geschlechtsdimorphismus. *Paläontologische Zeitschrift*, **49**, 168–80.

Müller, K.J. (1979). Phosphatocopine ostracodes with preserved appendages from the Upper Cambrian of Sweden. *Lethaia*, **12**, 1–27.

Nicholas, T.C. (1915). The geology of the St. Tudwals Peninsula. *Quarterly Journal of the Geological Society of London*, **71**, 83–143.

Notholt, A.J. (1979). Resources of Precambrian and Cambrian sedimentary phosphate rock. In *Proterozoic–Cambrian Phosphorites*, ed. P.J. Cook & J.H. Shergold, pp. 71–86. ANU Press, Canberra.

Notholt, A.J.G., Highley, D.E. & Slansky, M. (1979). *Raw Materials Research and Development IV. Dossier on Phosphate.* 234pp. Commission of the European Communities, Brussels.

Oakley, K.P. (1942). Occurrences of phosphate in pre-Cretaceous rocks. *Wartime pamphlet, Geological Survey of England & Wales* **8(4)**, 29pp.

Öpik, A.A. (1956). Cambrian (Lower Cambrian) of Estonia. *20th Session International Geological Congress, Mexico*, **1**, 97–126.

Orgeval, J.J. & Capus, G. (1978). Existence d'un horizon phosphaté uranifère et thorifère à la base des formations carbonatées cambriennes de la région du Vigan (Gard, France): son remaniement karstique. *Compte Rendu Société Géologique de France*, **1978**, 115–117.

Palmer, A.R. (1973). Cambrian trilobites. *Atlas of Palaeobiogeography*, ed. A. Hallam, pp. 3–11. Elsevier, Amsterdam.

Palmer, A.R. (1981). Subdivision of the Sauk sequence. *United States Department of the Interior, Geological Survey, Open-File Report*, **81-743**, 160–2.

Paproth, E. & Zimmerle, W. (1980). Stratigraphic position, petrography, and depositional environments of phosphorites from the Federal Republic of Germany. *Mededelingen Rijks Geologische Dienst*, **32-11**, 81–95.

Peach, B.N., Horne, J., Gunn, W., Clough, C.T., Hinxman, L.W. & Teall, J.J.H. (1907). The geological structure of the North-West Highlands of Scotland. *Memoirs of the Geological Survey of Great Britain*. **10**, xviii, 668pp.

Poulsen, C. (1960). The Palaeozoic of Bornholm. *21st Session International Geological Congress, Norden, Guide to Excursions A46 and C41*, 15pp.

Poulsen, C. (1967). Fossils from the Lower Cambrian of Bornholm. *Det Kongelige Danske Videnskabernes Selskab, Matematisk-fysiske Meddelelser*, **36**, 1–48.

Poulsen, V. (1963). The Lower Middle Cambrian Kalby-Ler (Kalby Clay) on the Island of Bornholm. *Det Kongelige Danske Videnskabernes Selkab, Biologiske Meddelelser*, **23**, 1–13.

Poulsen, V. (1966). Cambo-Silurian Stratigraphy of Bornholm. *Meddelelser fra Dansk Geologisk Forening*, **16**, 117–37.

Poulsen, V. & Anderson, M.M (1975). The Middle–Upper Cambrian transition in southeastern Newfoundland, Canada. *Canadian Journal of Earth Science*, **12**, 2065–79.

Prian, J.-P. (1979). Caractéristiques des paléoenvironnements des phosphorites cambriennes du versant septentrional de la Montagne Noire (Sud du Massif Central, France). In *Géologie Comparée des Gisements de Phosphates et de Pétrole*. Colloque International, Orléans 6–7 Novembre, 1979. Documents du BRGM, **24**, 93–111.

Rauhamaki, E., Makela, T. & Isomaki, O.P. (1978). Geology of the Vihanti mine. In *Metallogeny of the Baltic Shield*, Helsinki Symposium.

Rehtijärvi, P., Äikäs, O. & Makela, M. (1979). A Middle Precambrian uranium-and apatite-bearing horizon associated with the Vihanti zinc ore deposit, western Finland. *Economic Geology*, **74**, 1102–17.

Roach, R.A. (1977). A review of the Precambrian rocks of the British Variscides and their relationships with the Precambrian of N.W. France. *Colloque International du Centre National de la Recherche Scientifique*, **243**, 61–79.

Robison, R.A. (1976). Middle Cambrian trilobite biostratigraphy of the Great Basin. *Brigham Young University, Geology Studies*, **23**, 93–109.

Robison, R.A., Rosova, A.V., Rowell, A.J. & Fletcher, T.P. (1977). Cambrian boundaries and divisions. *Lethaia*, **10**, 257–62.

Runnegar, B. (1981). Biostratigraphy of Cambrian mollusks. *United States Department of the Interior, Geological Survey, Open-File Report*, **81–743**, 198–202.

Rushton, A.W.A. (1974). The Cambrian of Wales and England. In *The Cambrian of the British Isles, Norden & Spitsbergen*, ed. C.H. Holland, pp. 43–121. John Wiley, London.

Sdzuy, K. (1967). The Tethys in Cambrian time. *Systematics Association Publication*, **7**, 5–9.

Skjeseth, S. (1963). Contributions to the geology of the Mjφsa districts and the classical sparagmite area in southern Norway. *Norges Geologiske Undersφkelse*, **220**, 126pp.

Smith, A.G., Hurley, A.M. & Briden, J.C. (1981). *Phanerozoic Paleocontinental World Maps*. Cambridge University Press, Cambridge, 102pp.

Spjeldnaes, N. (1967). Studies on the latest Precambrian and Eocambrian rocks in Norway, 6. Fossils from pebbles in the Biskopasen Formation in southern Norway. *Norges Geologiske Undersφkelse*, **251**, 53–82.

Stφrmer, L. (1925). On a Lower Cambrian fauna at Ustaoset in Norway. *Fennia*, **45**, 12–22.

Stubblefield, C.J. (1930). A new Upper Cambrian section in south Shropshire. *Summary of Progress, Geological Survey of Great Britain for 1929*, 54–62.

Sundius, N. (1923). Grythyttefältets geologi (English Summary). *Sveriges Geologiska Undersökning, Ser. C*, **312**, 354pp.

Swett, K. & Crowder, R.K. (1982). Primary phosphatic oolites from the Lower Cambrian of Spitzbergen. *Journal of Sedimentary Petrology*, **52**, 587–93.

Taylor, K. & Rushton, A.W.A. (1972). The pre-Westphalian geology of the Warwickshire Coalfield, with an account of three boreholes in the Merevale Area. *Bulletin of the Geological Survey of Great Britain*, **35**, vii + 150pp.

Theokritoff, G. (1979). Early Cambrian provincialism and biogeographic boundaries in the North Atlantic region. *Lethaia*, **12**, 281–95.

Thorslund, P. (1960). The Cambro-Silurian of Sweden. *Sveriges Geologiska Undersökning, Ser. Ba*, **16**, 69–110.

Timofeev, B.V. (1960). Sur la caractéristique micropaléontologique de la formation de Visingsö. *Geologiska Föreningens i Stockholm Forhandlingar*, **82**, 28–42.

Vaasjoki, M., Äikäs, O. & Rehtijärvi, P. (1980). The age of mid-Proterozoic phosphatic sediments in Finland as indicated by radiometric U-Pb dates. *Lithos*, **13**, 257–62.

Vidal, G. (1976). Late Precambrian microfossils from the Visingsö Beds in southern Sweden. *Fossils & Strata*, **9**, 57pp.

Vidal, G. (1981). Micropalaeontology and biostratigraphy of the Upper Proterozoic and Lower Cambrian sequence in East Finnmark, northern Norway. *Norges Geologiske Undersφkelse*, **362**, 1–53.

Vogt, T. (1924). Forholdet mellem sparagmitsystem og det marine underkambrium ved Mjφsen. *Norsk Geologisk Tidsskrift*, **7**, 281–385.

Walcott, C.D. (1900). Lower Cambrian terrane in the Atlantic Province. *Proceedings of the Washington Academy of Sciences*, **1**, 301–339.

Wallin, B. (1982). Sedimentology of the Lower Cambrian sequence at Vassbo, Sweden. *Stockholm University Contributions in Geology*, **39(1)**, 111pp.

Westergard, A.H. (1944). Borrningar genom alunskifferlagret pa Östergötland. *Sveriges Geologiska Undersökning, Ser. C*, **463**, 22pp.

Willdén, M.Y. (1980). Paleoenvironment of the autochthonous sedimentary sequence at Laisvall, Swedish Caledonides. *Stockholm University Contributions in Geology*, **33**, 100pp.

Ziegler, A.M. & McKerrow, W.S. (1975). Silurian marine red beds. *American Journal of Science*, **275**, 31–56.

Ziegler, A.M., Scotese, C.R., McKerrow, W.S., Johnson, M.E. & Bambach, R.K. (1979). Paleozoic paleogeography. *Annual Review Earth and Planetary Sciences*, **7**, 473–502.

8

Proterozoic and Cambrian phosphorites – regional review: North America

R.L. CHRISTIE AND R.P. SHELDON

Introduction

Sedimentary phosphate deposits are known in several Precambrian and Cambrian basins of North America but none so far has proved economic. North America can be described as comprising four geological 'components', each of which contains phosphogenic sedimentary basins of various ages (Fig. 8.1).

The Canadian Shield forms a central 'core' of North America. The Canadian Shield is a complex of Archaean and Proterozoic basins that includes several relatively undisturbed Proterozoic basins overlying metamorphosed basement (Baer, 1970). The Phanerozoic Interior Platforms, where the Precambrian basement is covered by relatively undeformed Phanerozoic sedimentary rock, surrounds the Canadian Shield. The Canadian Shield and the Phanerozoic Platforms combine to form the North American Craton. The Interior Platform includes the Arctic Platform in the north and the St Lawrence Platform in the east. Geosynclines of late Precambrian and Phanerozoic ages occur outward from the North American Craton. The Cordilleran Geosyncline occurs on the western side, the Appalachian Geosyncline on the east, and the Franklinian Geosyncline on the north. Several orogenic events of different ages have affected these marginal geosynclines to form the Cordilleran, Innuitian and Appalachian Orogens. Finally, the Avalon Accreted Terrain (Williams, 1979) forms part of the eastern margin of North America (Fig. 8.1). Other accreted terrain occurs on the western margin of the continent, but is not indicated on Fig. 8.1 because Cambrian and Proterozoic phosphatic rocks are unknown in them. The sedimentary basins of such terrain may have formed at palaeo-positions now considerably removed from those of adjacent, autochthonous North American basins so that contrasting sedimentary and tectonic features may be expected. The Avalon Accreted Terrain appears to have been part of north-western Africa in earlier times.

Many sedimentary basins of the Interior Platform are foreland basins of the adjacent geosynclines and it is convenient, therefore, to describe the various geosynclinal and adjacent platformal phosphorites under single headings. The descriptions that follow will be grouped under: Canadian Shield; Cordilleran Geosyncline; Franklinian Geosyncline and Arctic Platform; Appalachian Geosyncline and St Lawrence Platform; and Avalon Accreted Terrain.

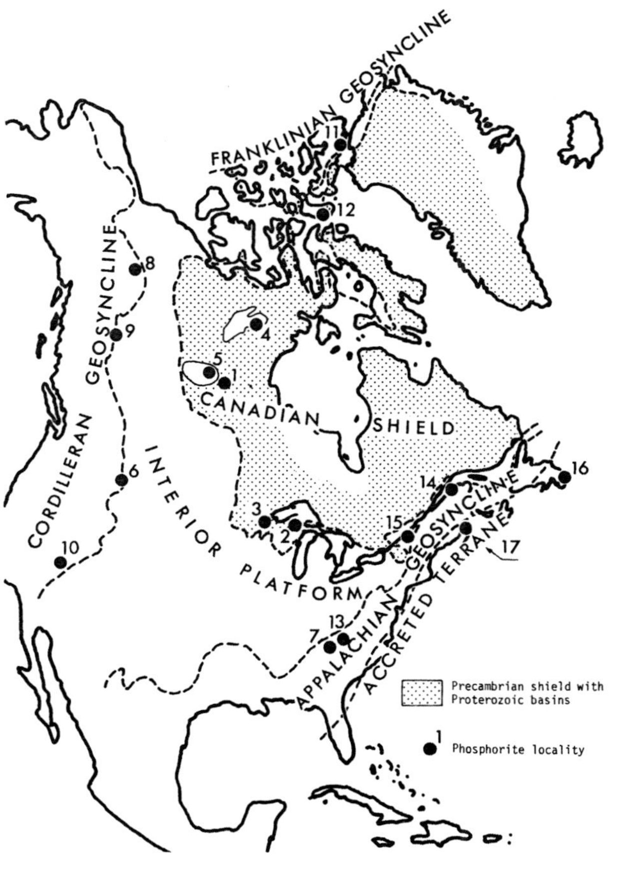

Fig. 8.1. Geological provinces of North America, with some Late Precambrian and Cambrian phosphorite localities.

Canadian Shield

The Canadian Shield comprises several geological subprovinces or orogenic belts that represent orogenies of Archaean and Proterozoic ages, and certain overlying, undisturbed basins of middle and late Proterozoic ages. The principal orogenies of the deformed belts took place about 2500, 1750, and 1000 million years ago according to isotopic dates

Table 8.1. *Late Precambrian and Cambrian phosphorite localities of North America*

Locality*	Tectonic province basin	Age: estimated ages of boundaries** Canada		USA
17	Avalon Zone			
16	Avalon Zone			
15	St Lawrence Platform			
14	St Lawrence Platform			
13	Appalachian Geosyncline	Cambrian		
12	Arctic Platform			
11	Franklinian Geosyncline			
10	Cordilleran Geosyncline			
9	Cordilleran Geosyncline			
8	Cordilleran Geosyncline			
			570–570	
7	Appalachian Geosyncline	Hadrynian		Precambrian Z
6	Cordilleran Geosyncline		-800	
			1000	
5	Athabasca Basin	Helikian		Precambrian Y
4	Thelon Basin		-1600	
			1750-	
3	Penokean Fold Belt			Precambrian X
2	Penokean Fold Belt	Aphebian		
1	Wollaston Lake Fold Belt			
			2500–2500	

*Localities in Fig. 8.1.
**Note the approximate nature of the correlation between Canadian and US divisions of Precambrian times.

(Stockwell, 1982). Apatite minerals at greater than accessory concentrations have been found in metasedimentary rocks of the Wollaston and the Penokean Fold Belts and in the post-orogenic Athabasca and Thelon basins. The apatite deposits were found in the course of prospecting for uranium and in no case have economic grades and thicknesses yet been found. Aphebian and Paleohelikian ages are represented.

Aphebian

Wollaston Lake Fold Belt Apatite has been noted (L.P. Tremblay, personal communication) in metasedimentary rocks of Aphebian age in the Wollaston Lake Fold Belt of northern Saskatchewan (Fig. 8.1, locality 1, Table 8.1). The rocks appear to have been shallow-water marine, impure quartz arenites (Money, Baer & Scott 1970). The apatite is associated with uranium mineralisation at the profound unconformity at the base of the overlying Athabasca Basin. Whether the apatite was synsedimentary or was due to a weathering cycle and formation of a regolith beneath the basal Athabasca sandstones is not clear.

Penokean Fold Belt Phosphorites of Aphebian age have been found in northern Michigan (Fig. 8.1, Locality 2) (Cannon & Klasner, 1976) and in east-central Minnesota (Locality 3) (Ullmer, 1981; McSwiggen, Morey & Weiblen 1981). The Michigan phosphorite occurs in the basal units of the Marquette Range Supergroup, which overlies the Archaean crystalline basement. A conglomeratic bed about 15 m thick averages about 15% P_2O_5 at one locality, where also many thinner beds are of comparable grade. The phosphorite is associated with black to light-grey chert, argillite and arkosic quartzite. Apatite in nearby areas is associated with carbonate iron-formation (Mancuso, Lougheed & Shaw, 1975), a unit of the Marquette Range Supergroup that lies lower in the section than the major economic iron-formation of this region. In east-central Minnesota, the phosphorite occurs in the Thomson Formation, which consists of interbedded greywacke, siltstone, shale and lesser amounts of quartzite, graphitic slate and phyllite, sulfide facies iron-formation, and volcanic rocks. The formation is a part of the Animikie Group, a series of dominantly clastic sediments and iron-formations.

Helikian

Thelon Basin (Paleohelikian) Phosphate minerals have been found at several localities in the Thelon Basin of the Canadian Shield, north-west of Hudson Bay (Fig. 8.1., Locality 4). The phosphates occur in the Thelon Formation, a wide-spread unit in the upper part of the Dubawnt Group, and also in regolith below the Thelon Formation.

The Dubawnt Group comprises nearly undisturbed, mainly

continental sedimentary and volcanic rocks, now preserved in a large structural basin and in fault-controlled outliers. Maximum thickness is uncertain, but may exceed 5200 m. The basement beneath the sedimentary rocks is a gneiss–granodiorite complex of Archaean and Aphebian ages.

The Thelon Formation is a very extensive, flat-lying sheet or blanket of orthoquartzite. The quartzites, about 300 m thick, disconformably overlie older volcanic and clastic rocks of the Dubawnt Group. A pre-Thelon regolith is developed beneath the Thelon Formation on granitic basement rocks in areas where older Dubawnt units are missing. Regolithic patches are also locally preserved along contacts between Thelon strata and underlying volcanic rocks of the group, demonstrating a post-volcanic age for the regolith (Fraser *et al.*, 1970, p. 233). Phosphate minerals occur in the regolith, in basal beds of the Thelon Formation, and in beds higher in the unit (A. Miller, personal communication).

The phosphate minerals, apatite and goyazite[1], occur in two forms: (a) as a matrix with hematite and potassic mica, and (b) as a matrix, but uncontaminated by hematite and with only small amounts of mica (A. Miller, personal communication). The phosphates form a cement in fine- to medium-grained quartz arenite; the quartz grains are rounded to subrounded. Other interstitial minerals include quartz, hematite, white mica, and an unidentified chlorine-bearing mineral. One sample contained 10% P_2O_5 and 53.8 ppm uranium (Le Cheminant *et al.*, 1981).

The textural maturity of the sand grains and the mineral components of the rock are taken (Le Cheminant *et al.*, 1981) to indicate a shallow marine depositional environment for this part of the Thelon Formation.

Exposures are limited in the Thelon Subprovince, which is a heavily glaciated upland of low relief, and the areal and stratigraphic distribution of the phosphate-bearing rocks is as yet unknown.

Athabasca Basin (Paleohelikian) Apatite occurs as grains and as matrix in fine-grained sandstone, siltstone, and conglomerate in the Wolverine Point Formation in the upper part of the Athabasca Group (Fig. 8.1, locality 5); grades up to 20% P_2O_5 are reported. Data reported here for the Athabasca Basin were obtained by P. Ramaekers of the Saskatchewan Geological Survey and by J.E. Mathison of Eldorado Nuclear Limited. Ramaekers (1979, 1980) has carried out a sedimentological study of the entire Athabasca Basin and Mathison (personal communication) has completed surface and diamond drill investigations of a locality in the north central part of the basin.

The almost entirely clastic Athabasca Group has been divided into subgroups and formations by Ramaekers (1979, 1980). The group is considered to be composed of detritus from the slow erosion of a tectonically stable source area (the Canadian Shield), the sediments having been deposited in a fluvial, partly marine environment. The Wolverine Point Formation is made up of interbedded sandstone, siltstone and mudstone with granules in some beds; the unit is quartzose, but contains arkosic bands and shard-like forms taken to be altered volcanic fragments. The environment of deposition for the Wolverine Point Formation is interpreted to be that of a relatively quiet marine shelf.

The phosphates of the Wolverine Point Formation were reported by Ramaekers (1980, p. 101) as follows:

> The upper clay and silt rich unit of the Wolverine Point Formation contains scattered, probably thin phosphorites throughout the entire Athabasca Basin. Phosphatic zones with values of several per cent P_2O_5 over intervals up to 6 m thick contain thin (30 cm) zones with P_2O_5 values of up to 20 per cent. The phosphates consist largely of fluorapatite and carbonate fluorapatite... the apatite occurs as collophane deposited in concentric layers around sand grains or as distinct grains, as a subhedral cement in crystals radiating from quartz grains, or as a euhedral cement randomly oriented in pores, or as vein fillings in fracture zones. The apatite crystals are often zoned, with hematite coatings separating growth stages. The oldest generation of apatite (collophane and radiating euhedral crystals) are generally fluorapatite and seem to be the most radioactive. Vugs are often lined with carbonate fluorapatite. The other zones of the euhedral apatite crystals generally have a lower uranium content, as witnessed by a lower natural fission track density...'

Ramaekers (1980) adds that in parts, but not all of the Athabasca Basin, the apatitic zones may easily be found by detection of radioactivity. The bedded phosphates locally are known to reach potentially commercial grades (*op. cit.*, p. 105); it is yet to be established, however, whether economic thicknesses, extent, or mining conditions exist.

The Wolverine Point phosphatic beds were tested by drill in a limited area by Eldorado Nuclear Limited (J.E. Mathison, personal communication). The phosphate occurs in middle grades over thin intervals. Certain intraformational conglomerates contain apatite-rich siltstone clasts and are particularly phosphatic. Some fragments in the conglomerate consist almost entirely of apatite and hematite. In addition, isolated grains and acicular spherules of apatite were observed in thicker sections of slightly phosphatic, fine-grained sandstone. Matrix in the sandstone may be fibrous quartz with hematite or apatite with hematite, according to Mathison (personal communication), these matrix types perhaps replacing an earlier silica matrix of the sandstone. Remobilised phosphate is represented by apatite-rich veins in sandstones of the underlying Manitou Falls Formation.

Cordilleran Geosyncline

The Canadian Shield is overlain on the south-west and west by mainly Phanerozoic rocks of the stable interior of the continent. The Phanerozoic Platform is a region of plains and plateaux, from 300 km to over 1200 km wide, underlain by nearly horizontal beds. In the western United States, however, the cover rocks have been affected by basement-controlled tectonic activity. The cratonic cover increases in thickness away from the Shield to merge westward with the thick sequence of the Cordilleran Geosyncline (Fig. 8.1), with the boundary commonly marked by an overthrust belt. Helikian and Hadrynian (Middle and Upper Proterozoic) sediments are common in the eastern parts of the geosyncline and some slightly deformed cratonic cover of these ages occurs in the northern Interior Platform (northern mainland). Lower

[1] Aluminium-strontium phosphate; first reported by Cecile (1973).

Cambrian sediments are restricted to the geosyncline; Middle and Upper Cambrian deposits onlap the eroded Shield but are truncated eastward in the subsurface by younger erosion surfaces. Much of the exposed Precambrian and Cambrian sequences are of shelf and miogeosynclinal facies; these facies pass westward into basinal facies, with the 'hinge-line' or boundary between the sedimentary belts forming an irregular line with numerous 'embayments'. Phosphorites of Hadrynian (Precambrian Z) age occur in western Montana in the Belt Series, and also appear to be related to hinge-line, transgressive facies. Phosphorites of Cambrian and Ordovician age occur in east-central Yukon, north-eastern British Columbia, and eastern Nevada, the favoured sites of deposition apparently relating to the hinge-line and to certain arches (e.g. the Mackenzie Arch) (Cecile & Norford, 1979; Christie, 1979).

Hadrynian

Thin beds of phosphorite in the Belt Series in Montana were found and described by Gulbrandsen (1966) (Fig. 8.1, Locality 6). These beds occur as a 6-mm to 25-mm thick layer within a 20-cm unit of light brownish-grey to light-green coarse siltstone. Discontinuous laminae and individual grains of phosphate occur in the siltstone both above and below the phosphorite layer. The phosphorite occurs in the central part of the Spokane Formation, a predominantly red-bed unit of the Middle Belt Series.

The phosphatic rock is a medium-grained grainstone phosphorite, dark grey to black in the richer phosphatic parts and speckled light and dark in the leaner ones. The principal minerals are carbonate fluorapatite, quartz, muscovite and plagioclase, and minor amounts of chalcocite, malachite, and heavy detrital minerals are present.

Harrison (1972, p. 1230) observed that the Belt Basin was an epicratonic re-entrant of a sea to the west. The thick sediments had source areas around the re-entrant. The Spokane Formation is shown by Harrison as possibly a facies of the St Regis Formation, which underlies the transgressive Middle Belt carbonate sequence. Alternatively, the Spokane Formation may be the lowest unit of the transgressive sequence, and the phosphorite would then be part of the transgressive sedimentation, a common stratigraphic position for phosphorites. Lange & Eby (1981) have shown that the phosphorite is associated with strata-bound copper and silver.

Cambrian

Phosphorites have been found in Cambrian and Ordovician rocks at a few localities in the Cordilleran Geosyncline. These localities are widely scattered and phosphate contents are not well known (Rogers *et al.*, 1970; Christie, 1979, 1980).

Phosphate is present in more than accessory amounts in Cambrian shale near the base of the Cambro-Ordovician Road River Formation in the District of Mackenzie near the Yukon border (Fig. 8.1, Locality 8). Average phosphate values in intervals as large as several tens of metres range as high as 1.5%, and one specimen of 'intrasparite' contained 8.2% P_2O_5. The phosphatic beds are underlain by bedded barite and overlain by a unit of thin-bedded argillaceous limestone. The occurrence lies in the Misty Creek embayment, a north-eastern incursion of basinal rocks in a north-west-trending carbonate belt that flanks the Mackenzie Arch. Eastward, toward the arch, the unit with phosphatic beds is truncated by an unconformity at the base of an Upper Cambrian carbonate unit. The phosphatic beds themselves are enclosed by beds with shelly fauna that are dated as late Early Cambrian and late Middle to early Late Cambrian (Cecile, 1978).

Two new Cambrian (?) phosphorite occurrences are reported in condensed Lower Palaeozoic basinal sections associated with the Niddery High, a relatively positive palaeogeographic feature that separates the Misty Creek Embayment (mentioned above) and the Selwyn Basin of east-central Yukon (Cecile, 1982). The phosphorite occurrences (Fig. 8.1, also shown as Locality 8) are in basinal limestones of the Cambro-Ordovician Rabbitkettle Formation. At one locality, irregular black nodules up to 1 cm in diameter and similarly thin, discontinuous laminae of phosphatic rock lie scattered over a 10-m interval of thick-bedded limestone. An X-ray diffraction assay as high as 10% fluorapatite has been obtained from these phosphorites. At the second locality, at the base of the Rabbitkettle, a unit of limestone breccia about 20 cm thick contains a black, phosphatic cement. About 2% fluorapatite is present in this rock according to X-ray diffraction analysis.

Black phosphate nodules, 0.5 cm to 1 cm in diameter, occur in Upper Cambrian basinal shale near the head of Muskwa River, north-eastern British Columbia (Fig. 8.1, Locality 9; Cecile & Norford, 1979, p. 225). The nodules are scattered throughout 100 m of section. The top of the phosphatic section lies 120 m below an Ordovician carbonate unit, the Kechika Formation. No analyses are available.

Phosphate also occurs in Cambrian rocks in the southern part of the Cordilleran Geosyncline, in Nevada (Rogers *et al.*, 1970). The largest of these deposits occurs in the upper 3 m of the Bonanza King Formation (of Middle and Late Cambrian age) on the west flank of Banded Mountain in southern Nevada (Fig. 8.1, Locality 10). The phosphatic rock is a dark brownish-grey limestone with disseminated apatite, apatite as small, irregular grains, and apatite as a replacement of abundant fossil debris that appears to be made up largely of trilobite fragments. Several other occurrences are in Cambrian rocks in central Nevada, and Cambrian rocks of the House Range in Utah were found to be slightly phosphatic by geologists participating in IGCP Project 156, during a field conference in 1979.

Franklinian Geosyncline and Arctic Platform

Sedimentary rocks of Late Cambrian and younger ages unconformably overlie the northern margin of the crystalline Canadian Shield (Fig. 8.1). The Phanerozoic rocks are disposed as nearly flat-lying structural basins (the Arctic Platform: Christie, 1972) that pass northward into the very much thicker Franklinian Geosyncline. (Trettin, 1972). The basins of the Arctic Platform are separated by arches, uplifts, and topographic highs of Precambrian rocks, both crystalline and sedimentary.

Remnants of little-disturbed Hadrynian and Neohelikian sedimentary basins are exposed as large and small inliers in the

Platform region where the thin cover of Palaeozoic beds has been stripped away. The Precambrian basins are outlying parts of the huge Canadian Shield to the south. Unfossiliferous sedimentary units of east central Ellesmere Island, totalling up to 1700 m in thickness, are overlain disconformably by Lower Cambrian to Siluro-Devonian beds of the Franklinian Geosyncline and form part of the geosyncline. The unfossiliferous, possibly Precambrian, units thin out and are overstepped by Cambrian beds toward the Canadian Shield, and the base of the thicker, basinal part of the column is not known (Thorsteinsson & Tozer, 1970). Concentrations of sedimentary apatite have been noted in Cambrian clastic beds at a locality within the Franklinian Geosyncline and at another in the Arctic Platform as described below.

Cambrian

Franklinian Geosyncline Apatite occurs both as grains and as matrix in fine-grained sandstone of the Rawlings Bay Formation (Fig. 8.1, Locality 11). The grains are visible as black, rounded clasts in a hand specimen of dark sandstone. The phosphatic beds lie at the top of the unit, near the contact with the conformably overlying Kane Basin Formation (H.P. Trettin, personal communication). The upper boundary of the Rawlings Bay is in some places transitional but in other places distinct, and the phosphate may represent sedimentary 'condensation' and a para-conformity. A Cambrian age is supposed from the conformable relationship with the fossiliferous (Lower Cambrian) Kane Basin Formation and correlation with Cambrian beds of the Platform succession.

Arctic Platform Phosphatic matrix and ooids have been identified in sandstone of the Cambrian Rabbit Point Formation of Devon Island, in the eastern part of the Arctic Platform (U. Mayr, personal communication). The grades and extent of these phosphatic beds are not known (Fig. 8.1, Locality 12).

Appalachian Geosyncline and St Lawrence Platform

The Appalachian and St Lawrence regions lie south-east of the Canadian Shield: the St Lawrence Platform borders the Shield as a narrow, north-east-trending extension of the stable interior platform of North America, and the Appalachian Orogen lies beyond the platform, along the Atlantic seaboard. The Appalachian region extends from Alabama to Newfoundland, some 3500 km, and before the opening of the Atlantic Ocean was continuous with the Caledonian mountain chain of Ireland and Britain (Poole *et al.*, 1970; Schenk, 1978; Williams, 1979).

The Appalachian Orogen comprises a complex of tectonic zones that result from orogenies in Ordovician (Taconic) and Devonian (Acadian) times. The 'model' for pre-Taconic palaeogeography and tectonics used here follows Williams (1979): a western zone (adjacent to the St Lawrence Lowlands) is taken to be the deformed remnants of an Atlantic-type continental margin, and a more central zone to represent vestiges of a pre-Atlantic ocean, Iapetus.

Hadrynian (Z)

Appalachian Geosyncline Phosphorite occurs in the Ocoee Supergroup of Hadrynian age in Sevier County, Tennessee (Fig. 8.1, Locality 7), in a unit called the Yellow Breeches Member. Phosphorite in the Ocoee has been traced by Helmuth Wedow (Wedow & Stansfield, 1968, p. 283) over a large part of the Blue Ridge Province of eastern Tennessee and into North Carolina. The phosphorite in the Ocoee Supergroup occurs in a unit of the Walden Creek Group, which forms the upper part of the supergroup. The Walden Creek is about 3400 m thick and the supergroup about 9000 m thick in the Sevier County area. The phosphatic beds lie in the Wilhite Formation (1100 m), about the middle of the group. The phosphorite occurs as nodules and thin lenses in several black slate units and as pebbles in associated clastic beds in the lower part of the Yellow Breeches Member. The phosphorite is closely associated with manganiferous beds (rhodochrosite where unweathered); this association is evident in the East Fork manganese mine, which is less than a kilometre from the principal phosphorite locality (Wedow & Stansfield, 1968, p. 283; H. Wedow, Jr, personal communication). The phosphorite has been metamorphosed to chlorite grade (Hamilton, 1961, p. A36).

The lower part of the Walden Creek Group is characterised by greywacke and shale in graded beds, which suggests turbidity currents and deep water. The upper part of the group contains the phosphatic Yellow Breeches Member, a predominantly carbonate unit. Hamilton (1961, p. A-28) proposed that deposition of the upper part of the group took place near sea level and that the Yellow Breeches Member may have originated as lime sand, perhaps formed on shallow carbonate banks similar to the Bahama Banks.

The Ocoee Supergroup is regarded as Hadrynian (Precambrian Z, 570–800 m.y.: US Geological Survey, 1981, P. 419). This age is based on stratigraphic position as well as on the presence of acritarchs. The supergroup is overlain by the Chilhowee Group of Cambrian and Cambrian? age and is underlain by basement rocks of Precambrian Y age (800–1600 m.y.). The phosphate-bearing Walden Creek Group is the youngest unit of the Ocoee Supergroup, but the exact age of the phosphorite is not yet known. The Ocoee Supergroup is a part of an eastward thickening wedge of Precambrian clastic sedimentary rocks of the classic Appalachian geosyncline, and marks the hinge of major downwarping of the continental margin in late Precambrian time. The sediments were laid down in deep orogenic basins or troughs on an older Precambrian basement complex (Hadley, 1970, p. 257–258). After deposition, the rocks were subjected to metamorphism and structural deformation.

The phosphorite deposits of the Ocoee Supergroup are not known to be large, and they are structurally disturbed. The deposits cannot yet be considered to be a resource.

Cambrian

Appalachian Geosyncline Phosphate has been noted in the Cambrian Helenmode Formation of the Chilhowee Group in north-eastern Tennessee and south-western Virginia. The phosphate occurs as greenish phosphatic nodules and grains, green phosphatic calcareous quartzite, and phosphatic arkosic sandstone (Wedow & Stansfield, 1968, p. 282–284).

Phosphate also occurs in the Lower Cambrian Shady and Rome Formations as pockets and irregular bodies concentrated as residual (weathered) material from phosphate-bearing dolomite or dolomitic limestone. Black oolitic, probably marine, phosphorite also is reported at a locality in the Rome Formation. These deposits were prospected and worked early in the century, but production was very small (Wedow & Stansfield, 1968, p. 282–284).

These occurrences are in close proximity and are shown as locality 13 on Fig. 8.1.

St Lawrence Platform The St Lawrence Platform is separated along its length into three parts by south-east-trending arches of crystalline rocks of the Canadian Shield. A south-western part is marginal to the Michigan Basin and Allegheny Trough of the north-eastern United States; a central part lies between the arches, flanking the upper St Lawrence River; and an eastern part underlies the Gulf of St Lawrence and western Newfoundland. The maximum thicknesses of the platformal sequences are as follows: nearly 1500 m (Cambrian to Devonian) in the south-western part; 3100 m (Upper Cambrian to Upper Ordovician) in the central part; and 4300 m (Cambrian to Lower Devonian) in the eastern part. The platformal sequences of Cambrian to Lower Devonian rocks lap on to the eroded Shield, but locally the lower Palaeozoic rocks lie with fault contact adjacent to the Precambrian terrain, which rises as a scarp above the platformal lowlands. The platform in Quebec is bounded on the south-east by thrusts of 'Logan's Line'. In western Newfoundland, large westward-transported klippen overlie the eastern edge of the St Lawrence Platform (Poole *et al.*, 1970).

Low-grade phosphates occur as phosphatic sandstone in the Cambrian 'Potsdam' beds of the platform, e.g. at Rivière Ouèlle, lower St Lawrence River (Fig. 8.1, Locality 14) (Spence, 1920, p. 106), and as phosphatic dolomite of the Cambrian Middleton Formation of the Lake Champlain region (Fig. 8.1, Locality 15) (L.P. Tremblay, personal communication). In each case, the phosphate occurs as black grains and nodules, in the past often described as 'coprolites' or taken to be black cherty fragments. At no locality are grades known that could approach economic levels.

Avalon accreted terrain

The eastern zones of the Appalachian Orogen represent a deformed continental margin and accreted terrain–the Avalon Zone–that formerly was part of an African craton (Williams, 1979; Pique, 1981). The Avalon Zone consists of relatively unmetamorphosed and undeformed upper Precambrian volcanic and sedimentary rocks that in places pass upward conformably into Cambrian and Lower Ordovician shales and sandstones. The Cambrian beds, which evidently represent stable platform conditions of sedimentation, contain phosphatic rocks in eastern Newfoundland, Nova Scotia, and southern New Brunswick.

Phosphate nodules and bands occur in Cambrian beds at widely scattered localities in eastern Newfoundland (grouped as Locality 16 in Fig. 8.1), and grades of samples up to 4.6% P_2O_5 are reported. The phosphatic nodules or bands form only up to 5% of the rock, however, so that none of the mineral deposits can be described as more than a minor occurrence.

Phosphate nodules in sandy shale beds in the Cambrian St John Group were noted by W.D. Matthew (1893) and G.F. Matthew (1909). Nodules at Hanford Brook, New Brunswick, were described (Matthew, 1893) as about 12 mm in average diameter, round or oval, black, and both associated with and containing trilobite fragments (Fig. 8.1, Locality 17). Nodules at a locality in the city of St John were described (Matthew, 1893) as 'irregularly rounded or flattened... of small size... composed of comminuted Lingulae... Matthew described from thin- section the appearance of the nodules and adjacent 'green, coarse-grained sandstone'; the nodules, it was noted, contain abundant tests of Protozoa and the sand grains near the nodules in one bed are 'coated with a dark layer of phosphate, and the cement is largely or entirely... phosphate'. Matthew discussed the possible origins of the phosphate and concluded that the phosphate probably formed either 'as organic bodies or by some form of concretionary action'.

Summary

The phosphatic deposits of Proterozoic and Cambrian age in North America are uneconomic to mine due to small size, low grade or complex structures, although limited mining has occurred in the past in the eastern United States. The distribution of the phosphatic rocks appears to be controlled by the accretionary growth of the North American Craton. The Proterozoic rocks older than 800 m.y. are found in the Canadian Shield in ancient basins and continental shelves. The Proterozoic phosphatic rocks younger than 800 m.y. and the Cambrian phosphatic rocks are found at the edges of the North American Craton, in the geosynclinal sequences or on adjacent platforms. The ages and palaeogeographic distribution of Proterozoic and Cambrian phosphogenesis of North America craton match those of other continents, but no large phosphorite deposit of these ages has yet been found.

References

Baer, A.J. (Ed.) (1970). Basins and geosynclines of the Canadian Shield (9 papers). *Geological Survey of Canada, Paper*, **70–40**, 265pp.

Cannon, W.F. & Klasner, J.S. (1976). Phosphorite and other apatite bearing sedimentary rocks in the Precambrian of northern Michigan. *US Geological Survey, Circular*, **746**, 6pp.

Cecile, M.P. (1973). Lithofacies analysis of the Proterozoic Thelon Formation, Northwest Territories (including computer analysis of field data). MSc. Thesis, Carleton University, Ottawa.

Cecile, M.P. (1978). Report on Road River stratigraphy and the Mistry Creek Embayment, Bonnet Plume and surrounding map-areas. In *Current Research, Part A, Northwest Territories. Geological Survey of Canada, Paper*, **78–1A**, 371–7.

Cecile, M.P. (1982). The lower Paleozoic Mistry Creek Embayment, Selwyn Basin, Yukon and N.W.T. *Geological Survey of Canada, Bulletin*, **335**.

Cecile, M.P. & Norford, B.S. (1979). Basin to platform transition, Lower Paleozoic strata of Ware and Trutch map areas, northeastern British Columbia. In *Current Research, Part A, Geological Survey of Canada, Paper*, **79–1A**, 219–26.

Christie, R.L. (1972). Central Stable Region. In *The Canadian Arctic*

Islands and the Mackenzie Region. Guidebook for Field Excursion A-66, 24th Session International Geological Congress, Montreal, 40–87.

Christie, R.L. (1979). Phosphorite in sedimentary basins of western Canada. *Geological Survey of Canada, Current Research, Paper*, **79–1B**, 253–8.

Christie, R.L. (1980). Paleolatitudes and potential for phosphorite deposition in Canada. *Geological Survey of Canada, Current Research, Paper*, **80–1B**, 241–8.

Fraser, J.A., Donaldson, J.A., Fahrig, W.F. & Tremblay, L.P. (1970). Helikian basins and geosynclines of the northerwestern Canadian Shield. In *Basins and Geosynclines of the Canadian Shield*. ed. A.J. Baer, *Geological Survey of Canada, Paper*, **70–40**, 213–38.

Gulbrandsen, R.A. (1966). Precambrian phosphorite in the Belt Series in Montana. *US Geological Survey, Professional Paper*, **550-D**, 199–202.

Hadley, J.B. (1970). The Ocoee Series and its possible correlatives. In *Studies of Appalachian Geology, Central and Southern*, ed. G.W. Fisher *et al.*, pp. 247–59. John Wiley & Sons, New York.

Hamilton, W. (1961). Geology of the Richardson Cove and Jones Cove Quadrangles, Tennessee. *US Geological Survey, Professional Paper*, **349-A**, 55pp.

Harrison, J.E. (1972). Precambrian Belt Basin of northwestern United States: its geometry, sedimentation, and copper occurrences. *Geological Society of America Bulletin*, **83**, 1215–40.

Lange, I.A. & Eby D.E. (1981). Stratabound copper-silver-bearing oolitic deposits in the Belt Supergroup of Montana. *Economic Geology*, **76**, 933–7.

Le Cheminant, A.N., Ianelli, T.R., Zaitlin, B. & Miller, A.R. (1981). Geology of Tebesjuak Lake Map Area, District of Keewatin: a progress report. *Current Research, Part B; Geological Survey of Canada, Paper*, **81–1B**, 113–28.

Mancuso, J.J., Lougheed, M.S. & Shaw, R. (1975). Carbonate-apatite in Precambrian cherty iron-formation, Baraga County, Michigan. *Economic Geology*, **70**, 583–6.

Matthew, G.F. (1909). Phosphate deposits of South Carolina and New Brunswick. *New Brunswick Natural History Society, Bulletin*, **27**, **VI(2)**, 121–6.

Matthew, W.D. (1893). On phosphate nodules from the Cambrian of southern New Brunswick. *New York Academy of Sciences, Transactions*, **XII**, 108–20.

McSwiggen, P.L., Morey, G.B. & Weiblen, P.W. (1981). Uranium in lower Proterozoic, apatite-bearing rock of east-central Minnesota. *Geological Society of America, Abstracts*, **13**, 508.

Money, P.L., Bear, A.J. & Scott, B.P. (1970). The Wollaston Lake Belt, Saskatchewan, Manitoba, Northwest Territories. In *Basins and Geosynclines of the Canadian Shield*, ed. A.J. Bear. *Geological Survey of Canada, Paper*, **70–40**, 170–200.

Pique, A. (1981). Northwestern Africa and the Avalonian plate: relations during late Precambrian and late Paleozoic time. *Geology*, **9**, 319–22.

Poole, W.H., Sanford, B.V., Williams, H. & Kelley, D.G. (1970) Geology of southeastern Canada. Geology and Economic Minerals of Canada, ed. R.J.W. Douglas, pp. 227–304. Economic Geology Report No. 1, (5th edn.). Geological Survey of Canada.

Ramaekers, P. (1979). Stratigraphy of the Athabasca Basin. In *Summary of Investigations, 1979*, pp. 154–60, Saskatchewan Geological Survey.

Ramaekers, P. (1980). Stratigraphy and tectonic history of the Athabasca Group (Helikian) of northern Saskatchewan. *Summary of Investigations, 1980; Saskatchewan Geological Survey, Miscellaneous Report*, **80–4**, 99–106.

Rogers, C.L., Kleinhample, F.J., Ziony, J.I. & Danilchik, W. (1970). Phosphate occurrences in Nye County and adjacent areas, Nevada. *US Geological Survey, Professional Paper*, **700-C**, 49–60.

Schenk, P.E. (1978). Synthesis of the Canadian Appalachians. In *IGCP Project 27, Caledonian-Appalachian Orogen of the North Atlantic region. Geological Survey of Canada, Paper*, **78–13**, 111–36.

Spence, H.S. (1920). Phosphate in Canada. *Mines Branch, Dept. of Mines, Canada, Publication*, **396**, 156pp, 31 pls.

Stockwell, C.H. (1982). Proposals for time classification and correlation of Precambrian rocks and events in Canada and adjacent areas of the Canadian Shield; Part 1: a time classification of Precambrian rocks and events. *Geological Survey of Canada, Paper*, **80–19**, 135pp.

Thorsteinsson, R. & Tozer E.T. (1970). Geology of the Arctic Archipelago, pp. 458–500, in *Economic Minerals of Canada*, ed. R.J.W. Douglas, pp. 458–590. Economic Geology Report 1, (5th edn). Geological Survey of Canada.

Trettin, H.P. (1972). Franklinian Geosyncline. In *The Canadian Arctic Islands and the Mackenzie region. 24th Session International Geological Congress, Montreal, Guidebook, Excursion A-66*, 87–109.

Ullmer, E. (1981). *A mid-Proterozoic phosphate occurrence in east-central Minnesota*. Rocky Mountain Energy Company, 4pp.

US Geological Survey (1981). Geologic names of the United States through 1975. *U.S. Geological Survey, Bulletin*, **1535**, 643pp.

Wedow, H. Jr & Stansfield, R.G. (1968). Fertilizer raw materials. In *Mineral Resources of the Appalachian Region. US Geological Survey Professional Paper*, **580**, 281–3.

Williams, H. (1979). Appalachian Orogen in Canada. *Canadian Journal of Earth Sciences*, **16**, 792–807.

9

Proterozoic and Cambrian phosphorites–regional review: West Africa

M. SLANSKY

Introduction

It is only during the last twenty years that Proterozoic and Cambrian phosphorites have been discovered in Africa, and it was only in the early 1970s that their economic feasibility began to be considered seriously. Present discoveries, which are certainly in no way exhaustive, indicate that the largest of these phosphate deposits are in West Africa. The geology of this part of Africa at the end of the Precambrian and beginning of the Palaeozoic is becoming increasingly better known, thanks to the detailed regional studies of the last few decades, recent syntheses and the development of isotopic dating methods. In the light of this knowledge it has become possible to place the phosphorites of this period within their general geological setting. The relationships existing between the geological setting of West Africa and the ancient phosphorites have already been considered regionally by Ilyin *et al.* (1979), and Trompette *et al.* (1980) who have dealt with the Volta Basin in particular. The present paper takes up the discussion, making use of the most recent data and unpublished results obtained by geologists of the Bureau de Recherches Géologiques et Minieres (BRGM): F. Barthelemy, A. Boujo, P. Boutin, M. Donnot, D. Giot, G. Maurin and M. Pascal, who have made a significant contribution to the knowledge of these phosphorites.

Major structural features of West Africa during the Proterozoic and Palaeozoic

The geological history of West Africa during the Proterozoic and the Palaeozoic is directly connected with the existence of a stable craton from the end of the Eburnean Orogeny (about 1700 m.y. ago) onwards (Clauer *et al.*, 1982) and with that of the two mobile zones surrounding it (Fig. 9.1). The West African craton, which is exposed in the Reguibat Chain in the north and in the Man Chain (or Leo Chain) in the south, created the conditions for the development of three large sedimentary basins. These basins, are the Taoudeni (covering the central part of the craton), Volta (on the south-eastern border), and Tindouf (on the northern border).

The eastern mobile zone produced that part of the Pan-African Chain known as the Dahomeyides in the south and the Pharusides in the Adrar des Iforas region further north. According to Caby, Bertrand & Black (1981), the history of this mobile zone began with an oceanic opening 800 m.y. ago between the craton to the west and the Touareg and Nigerian Shields in the east, the rifting being accompanied by a phase of basic magmatism. Passive margin sedimentation developed along the eastern boundary of the craton, attaining great thicknesses in the Gourma Aulacogen. The oceanic closing which followed was reflected by an initial collision 700 m.y. ago in the eastern part of the Pharusides. This closure became most marked at about 600 m.y. The collision between the passive border of the West African craton and the active border of the Touareg and Nigerian Shields caused deformation running parallel to the suture overthrusting, metamorphism and granitisation. In the Dahomeyides, for example, the collision folded the eastern border of the Volta Basin sedimentary deposits and immediately beyond emplaced the following units, from west to east (Trompette, 1980, pp. 23–24): (1) the Tiele or Buem Unit which is thrust over the earlier folded formations and has undergone low-grade metamorphism; (2) the Atacora Unit, which is thrust over the preceding unit and has undergone medium-grade metamorphism; and (3) the Benin Plain Unit, which is also thrust over the Atacora Unit, with mainly high-grade (gneissic) metamorphism. These last two units show several phases of folding and metamorphism.

The western mobile zone has a more complex tectonic history, having been affected successively by the Pan-African, Caledonian and Hercynian Orogenies. The folded and only slightly metamorphosed chain of the Rokelides, on the edge of the Man Shield, was synchronous or slightly younger than the Pan-African event. Further north, the Mauritanides are the product of three superimposed successive tectonic phases, the last of which – the Hercynican – was responsible for overthrusting towards the craton. Dia, Lecorche & Le Page (1979) suggested the following approximate divisions, from east to west: (1) a foreland area in which the Proterozoic and Lower Palaeozoic sediments of the western edge of the Taoudeni Basin were gradually folded towards the west; (2) an external zone, where rocks of volcanic and sedimentary origin, locally accompanied by mafic and ultramafic facies, were affected by a low- to medium-grade metamorphism; three types of deformation may

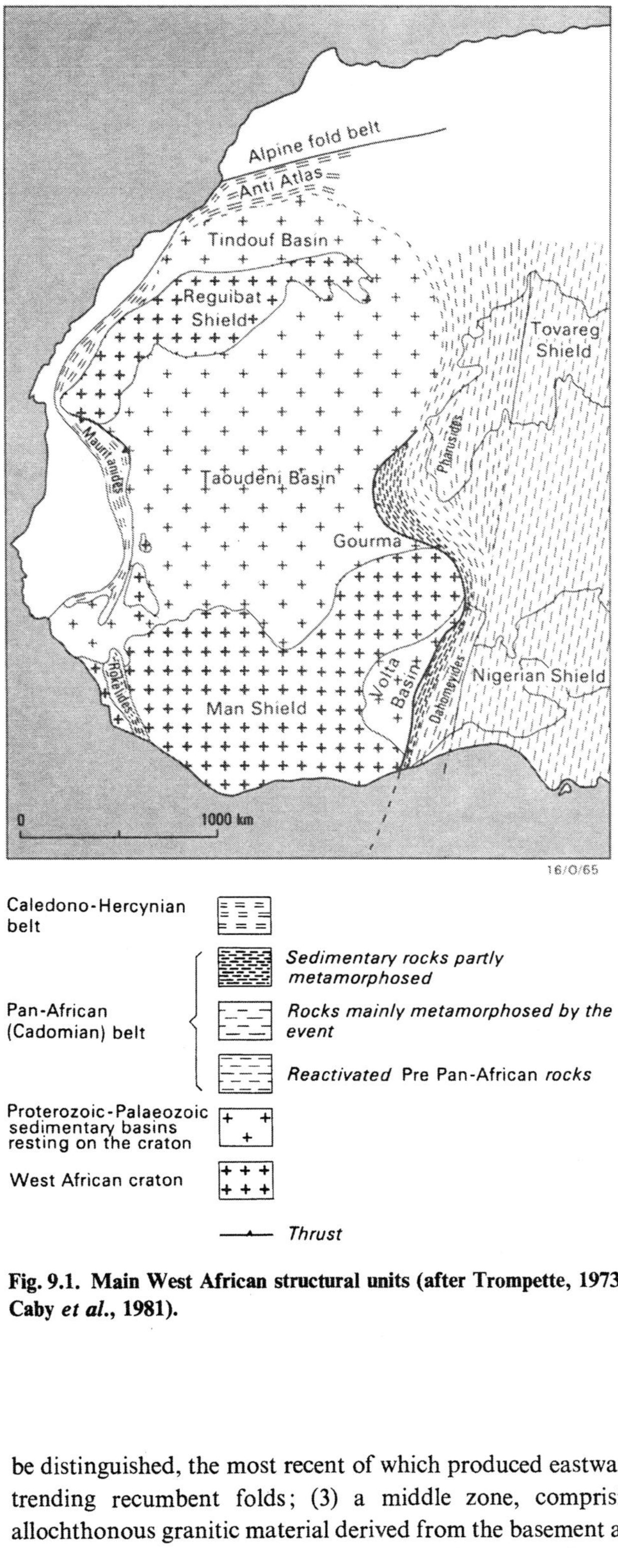

Fig. 9.1. Main West African structural units (after Trompette, 1973; Caby *et al.*, 1981).

be distinguished, the most recent of which produced eastward-trending recumbent folds; (3) a middle zone, comprising allochthonous granitic material derived from the basement and a quartzitic facies probably derived from Ordovician sediments; and (4) an internal zone of mixed composition, with an autochthonous basement and metamorphic cover in the south and a succession of metavolcanites and granitic gneiss probably derived from an earlier Precambrian basement in the north.

Further north, the effects of the Hercynian Orogeny can be discerned right up to the Anti-Atlas, albeit to a lesser extent than in the Mauritanides, and without any significant metamorphism. The Anti-Atlas region was also affected by the 'Pan-African' event (Clauer *et al.*, 1982).

General characteristics of the Proterozoic and Cambrian sedimentary formations

Palaeontological criteria are inadequate to date the pre-Silurian formations south of the Reguibat Chain. However, long-distance correlations were considerably improved when Zimmerman (1960) demonstrated that a very extensive tillite layer could be used as a marker bed. The map in Fig. 9.2 shows the main outcrops of this tillite layer according to Deynoux & Trompette (1976). These authors also assign the mixtites of the eastern Mauritanides and the western Dahomeyides to the same glacial episode.

The clay formations immediately overlying the tillite marker bed have been radiometrically dated by Clauer *et al.* (1982). In the Adrar region of Mauritania they are assigned an age of 595 ± 45 m.y., while in the Volta Basin they date from 660 ± 9 m.y., with a correspondingly slightly earlier age for the tillite. Deynoux *et al.* (1978) explain this possible age increase (which becomes more marked towards the south-east in the Congo Chain) by the displacement of the southern pole during the Proterozoic. The glacial episode marked by the tillite has made it possible to divide the pre-Ordovician sediments of the Taoudeni Basin into two Super-Groups, which are roughly applicable in the Volta Basin as well.

The Taoudeni Basin

Super-Group 1 is separated from the earlier Precambrian basement by a major angular unconformity. In the Adrar region of Mauritania (Atar zone) this Super-Group is about 1500 m thick according to Trompette (1980) and has the following succession:

Top. 3. Assabet el Hassian Group (470 m): Separated from the underlying Atar Group by an erosional unconformity; argillaceous or brecciated facies at the base, covered by alternating stromatolitic dolomite and chert followed by fine-grained micaceous sandstones often rich in glauconite

2. Atar Group (700 m): Overlying the Char Group with erosional unconformity; sandstone and conglomerate at the base giving way to alternations of stromatolitic carbonate, fine-grained sandstone and frequently glauconitic shale

Base. 1. Char Group (260 m): Sandstone and conglomerate at the base, giving way to sandstone and dolomite

These three groups have been identified trending north-eastwards along the southern border of the Reguibat Chain. Locally, the Atar Group may be directly transgressive over the basement and the upper group may be absent in the Hank area.

In the Affolé area, to the south of Adrar, Super-Group 1 is predominantly sandstone. In the Kedougou region of eastern Senegal it is over 2000 m thick (Bassot, 1966), and comprises a basal conglomerate, followed successively by 50–100 m of oolitic limestone, 600 m of fine-grained sandstone which becomes glauconitic towards the top, 500 m of sometimes calcareous pelites, 300–400 m of fine-grained quartzitic sandstone, and 600–800 m of pelites which are calcareous to varying degrees. Stromatolitic limestones are locally present. Isotopic measurements suggest that this sedimentation began some 1000 m.y. ago and finished about 650 m.y. ago.

In the Bamako region (Fig. 9.2), and further east, Super-Group 1 comprises predominantly sandstones with pelitic deposits – often glauconitic and with limestone intercalations – in the middle zone. Correlations in the Gourma area 700 km to the east of Bamako are open to doubt. It is likely that the equivalent of Super-Group 1 is the Ydouban Group, a sequence first defined by Reichelt (1972), which comprises a thick series of sandstone, shale and dolomite, folded in the northern area and metamorphosed in the east.

Super-Group 2 unconformably overlies Super-Group 1, and basement in places. Trompette (1980) further subdivides it into two series in the Adrar region of Mauritania where it comprises the following succession:

Super-Group 2

Top. Upper Series

6. Sandstone (600 m +)

Lower Series

5. Clayey sandstone (200 m)
4. Dolomitic limestone and stromatolitic dolomite (10–15 m)
3. Chert, siliceous shale, siltstone (250 m)
2. Dolomitic limestone with barite (2–5 m)

Base. 1. Tillite (1–50 m)

Super-Group 1

In the Hank area, to the north-east of Adrar, the tillite is thin and rather discontinuous, the overlying mudstones, shales, and siliceous shales and cherts often being directly transgressive over the units of Super-Group 1. South of Adrar, towards Tagant and Affolé, the succession of units is comparable with that of the Adrar zone proper, with a lower series ending in terrigenous deposits that may attain a considerable thickness. Towards the Mauritanides, Super-Group 2 beds become gently folded with horizontal fold axes. The folding then becomes isoclinal with overturning towards the east, and associated metamorphism results in the development of well-defined schistosity in the tillite.

In the southern part of the Kedougou region, Super-Group 2 is known as the Mali Series (Bassot, 1966) and consists of the following succession:

Top. 6. Mudstone with arkosic sandstone and greywacke (500–600 m)

5. Chert with a jasperlitic facies (5–15 m)
4. Calcareous dolomite (5–10 m)
3. Coarse quartzose sandstone (10–50 m)
2. Mudstone (10–50 m)

Base. 1. Tillite (1.5–30 m)

As in the north, the Mali Series becomes progressively affected in a westerly direction by the tectonics and then the metamorphism of the Mauritanides Chain.

The Volta Basin

The Volta Basin corresponds largely to the country of Ghana, although to the north it overlaps the territories of Togo, Benin, Upper Volta and Niger. Its stratigraphic series has been subdivided into three groups (Trompette, 1980):

Top. 3. The upper or Obosum Group is assigned to the Upper Proterozoic–Lower Cambrian. It includes coarse detrital facies which are generally interpreted as molasse derived from the Pan-African chain of the Dahomeyides

2. The middle group, also known as the Oti or Pendjari Group, with an age of between 675 and 660 m.y., unconformably overlies the lower group. Affaton (1975) distinguishes two component formations:
   (2b) the upper formation, known as the Oti or Pendjari Formation, is a soft unit, up to 2000 m thick, made up of shale, mudstone, siltstone and fine-grained sandstone, with thin beds of limestone or dolomite which are sometimes stromatolitic, and coarse-grained sandstone towards the top
   (2a) the lower or Kodjari Formation, starting with a thin lens of breccia, then a tillite, followed by dolomitic limestone, bedded cherts with intercalations of shale and tuffaceous beds, overlain by clayey and micaceous siltstones which are locally phosphatic. (In Ghana, as in the Adrar region of Mauritania, the dolomitic limestone locally hosts barite concretions.)

Base. 1. The lower group, dating from 1100 m.y. to 900 or 800 m.y. is predominantly sandstone with a clay–sandstone formation through the middle. This is known as the Dapango–Bombouaka Group and has a thickness varying between 250 m and 1500 m

The lower and middle groups are involved in the folded and overthrust Buem Series and in the metamorphic Atacora Series.

A comparison between the two basins reveals correlations between the lower group of the Volta Basin and Super-Group 1 of the Taoudeni Basin, and between the middle and upper groups of the Volta Basin and Super-Group 2 of the Taoudeni Basin, although the facies are in all probability diachronous.

The Tindouf and Anti-Atlas Basin

The stratigraphic series of the Tindouf Basin generally starts at the Ordovician or even the Silurian in the central part. Stromatolitic dolomites of the Upper Proterozoic, older than the Adrar tillite, have only been reported in the extreme south-west of the basin by Sougy (1964, 1975).

The sedimentary formations belonging to the Proterozoic and Cambrian are, however, extensively exposed in the Anti-Atlas zone, where Choubert (1963) distinguishes Precambrian III (or terminal), and Adoudounian Series beneath Lower Cambrian strata, dated by the presence of trilobites.

The Precambrian III, unconformable over all the earlier formations, represents a continental and volcanic facies. The Adoudounian, which is slightly unconformable, comprises four members:

Top. 4. Upper limestone member, attaining a maximum thickness of 500 m

3. Purple member, with a thickness of some 600 m, comprising mudstone, sandstone and conglomerate, with limestone or dolomite intercalations

2. Lower dolomite and limestone member, attaining a thickness of some 2000 m

Base. 1. Basal member, 100 m to 400 m thick, first calcareous and then schistose

Correlations between the Anti-Atlas and the Adrar region of Mauritania have yet to be clearly established. The upper limestones of the Adoudounian were first considered as Precambrian, but new palaeontological discoveries (Boudda & Choubert, 1972) have assigned them to the Lower Cambrian. Furthermore, the isotopic studies of Clauer (1980) have indicated that the Adoudounian is not more than 620 m.y. old and therefore the base of Super-Group 2 equivalent may be situated–at least provisionally–towards the base of this unit.

The Lower Cambrian, dated by its trilobites, attains a thickness of several hundred metres. It reflects, first a transgressive environment with limestone and shale, which later becomes regressive in the upper part of the unit with sandstones and volcanic tuffs.

The Middle Cambrian reflects a new transgression and attains a thickness of about 1000 metres. It comprises basal shales with *Paradoxides*, and then sandstone with conocoryphids and lingulids. A regressive phase after the deposition of these sediments was responsible for the absence of the top of the Middle Cambrian, the Upper Cambrian and the Lower Tremadocian.

Phosphatic beds of the Proterozoic and Cambrian

It is often very difficult to recognise ancient phosphorites in the field and their presence in western Africa has only been known for the last 20 years. J. Sougy, in his thesis presented in 1961 and published in 1964, appears to have been the first to report the presence of phosphatic beds in the Lower Palaeozoic sediments of Mauritania. Other occurrences were reported later by Deynoux (1968) in the Akjoujt region and by Pascal (1967) a little further south, while Trompette discovered occurrences in the Adrar region. In 1970, Pascal discovered three important occurrences in the Arli region, in Upper Volta and along the Mekrou river in Niger (Pascal, 1972). Affaton described several metres of phosphorite at Kodjari (Upper Volta) during work for his thesis presented in 1973 and published in 1975. In 1973, exploration for uranium mineralisation carried out by Dion for the Essex Iron Company revealed an extensive occurrence of aluminous phosphates in Niger, between the Tapoa River and the bend of the Mekrou River. Shortly afterwards, in 1974, Donnot (1975a, b) drew attention to the possibility of considerable thicknesses of phosphatic beds of this region. The first of the reconnaissance drillholes he sited there revealed the presence of about 70 m of phosphate rock. This discovery contributed to the development of prospecting in Niger, as well as in Benin and Upper Volta. In 1975, Pascal (unpublished BRGM reports) discovered several metres of phosphorite in the Kedougou region of eastern Senegal. In Morocco, phosphatic beds were found at the base of the Adoudounian south of Marrakech by Viland in 1977.

The zones in which the main phosphate occurrences discovered are situated are shown by an asterisk on Fig. 9.2. The main features of these occurrences are set out below.

Anti-Atlas

The occurrence discovered by Viland (1977) is in fact in the High Atlas, but the stratigraphic series which hosts it is close to the Anti-Atlas which geographically is situated a few kilometres further to the south. The occurrence is located near Taguedit, about 80 km south of Marrakech, in the basal part of the Adoudounian. Radiometric dating by Clauer (1980) and Clauer *et al.* (1982), suggests an age of about 600 m.y.

In the Taguedit region the base of the Adoudounian is a thin shallow-water sequence. The volcanogenic Precambrian III is unconformably overlain by the following series:

Top. 4. 20 to 30 m of sandstone, then massive dolomite with silica-filled fractures

3. 5 to 8 m of cupriferous clayey sandstone with white mica and a dolomitic cement

2. 2 to 3 m of yellow-weathering dolomite, with abundant silica and associated copper sulfide and barite

Base. 1. 3 to 4 m of volcanic tuff with carbonate lenses

The phosphorite occurs as a lens 50 m long, with a maximum thickness of 40–50 cm. It is intercalated between the clayey sandstone and a yellow-weathering dolomite unit, which itself contains some irregular phosphatic lenses a few tens of centimetres thick. The analysis of a sample from these inclusions shows 32.7% P_2O_5, 9.90% SiO_2, a CaO/P_2O_5 ratio of 1.46 and a high F/P_2O_5 ratio of nearly 0.12. According to Viland the phosphatic lens overlying the dolomite unit is more siliceous and less phosphatic, and also contains malachite, chalcopyrite, pyrite and barite.

Mauritanian Adrar

The phosphate mineralisation of this region was first identified by Trompette (1973) using a molybdate field test. Subsequent petrological work could not detect the phosphate. This would suggest that it is a very diffuse form of phosphate which is present fairly systematically in the clayey cherts overlying the tillite at the base of Super-Group 2. Some positive reactions were obtained in the tillite itself and in two layers of quartzose sandstone in Super-Group 1. Trompette (1973) also reports a thin horizon of iron phosphate in the Silurian. The phosphate mineralisation identified so far in the Proterozoic and Cambrian of the Adrar region of Mauritania is thus very tenuous and for this reason has not been included in Fig. 9.2.

Akjoujt region

About 30 km north-east of Akjoujt, in the Inchiri zone, the sedimentary formations of the Mauritanian Adrar are again exposed between the ancient basement to the north and the overthrust formations of the Mauritanides to the south. According to Deynoux (1968) the succession is as follows:-

Top. 4. A conglomerate with glacial and aeolian features

3. Conglomerate with a feldspathic sandstone matrix and flattened or elongated fragments of clayey cherts, sandy microsphorite and quartz (3–7 m)

2. Clayey chert (1–50 m)

Base. 1. Basal conglomerate with a feldspathic sandstone matrix, containing quartz pebbles and calcium phosphate nodules and grains (0.5–1.0 m)

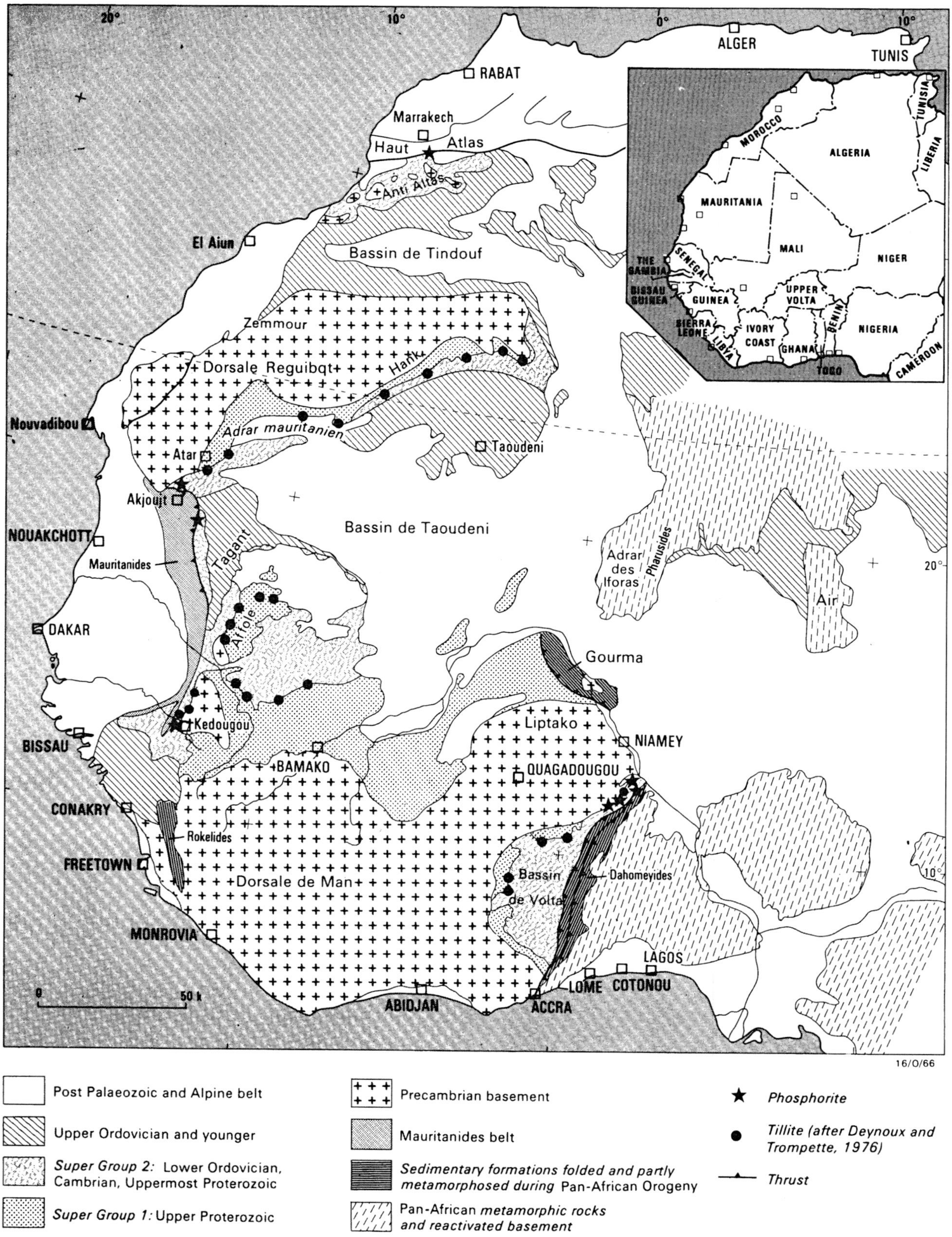

Fig. 9.2. Schematic geological map of West Africa (adapted from Choubert & Faure-Muret, 1975a, b).

Deynoux suggested that the upper conglomerate might be the equivalent of the Adrar tillite and thus the phosphatic beds would belong to the upper part of Super-Group 1. Trompette (1973), however, prefers to include these beds in the lower part of Super-Group 2.

About 150 km south-east of Akjoujt, in the Tamkarkart Zone, Pascal (1967) reports the presence of a phosphatic sandstone in the syncline of Koueriat el Khadra (eastern part of the Mauritanides Chain). This sandstone forms a lens with a maximum thickness of a few metres. The apatite forms small crystalline aggregates or scattered crystals and does not represent more than a few per cent of the total rock (a sample from the upper part gave 2.7% P_2O_5). A little further east, a small tectonic slice of this sandstone included within chloritic schists, does, however, contain almost 50% calcium phosphate. This sandstone appears to belong to the upper part of Super-Group 2.

The phosphatic facies discovered north-east and south of Akjoujt are relatively low grade, but the phosphatic inclusions are much more clearly identifiable than those of the Adrar region of Mauritania. It may also be noted that they are very closely associated with the mobile structural zone which produced the Mauritanides.

Kedougou region

The phosphatic formation of this region of eastern Senegal has been briefly described by Giot & Pascal (1978). The best outcrops are located in the vicinity of Namel, close to the western edge of the old Precambrian basement zone which is exposed around Kedougou.

In the best known cross-section, Super-Group 2, which unconformably overlies the basement, consists of the following sequence:

Top. 5. Cherty mudstone
4. Grainstone phosphorite assaying 22–32% P_2O_5 (several metres)
3. Siliceous and dolomitic phosphatic limestone (2–3 m)
2. Calcareous and dolomitic claystones with up to 8% P_2O_5 in the top third of the unit (25–30 m)
Base. 1. Tillite (3 m)

The petrographic studies by Giot (personal communication) show that the phosphate grains vary in size from 50 μm to several mm, most grains measuring between 1 mm and 2 mm. These grains are often oblong and angular, sometimes crescentic and rarely rounded. The included non-phosphatic material, which is not usually very abundant, may be either calcareous or siliceous in character. The phosphate is generally cryptocrystalline but sometimes microcrystalline. The matrix is siliceous, calcareous to varying degrees, and sometimes clayey.

Some beds display a finely laminated structure, sometimes wavy, in which beds of mudstone phosphorite with a thickness between 100 μm and 2 mm alternate with chalcedonic or sometimes calcareous beds. The presence of some stromatolitic beds suggests that a microbreccia structure evident in some layers was acquired by the reworking of pre-existent stromatolitic (boundstone) phosphorites. Chemical analyses indicate that samples with over 20% P_2O_5 all contain abundant SiO_2 varying between 12% and 44%; the CaO/P_2O_5 ratio varies from 1.2 in the upper zone, where alumino-phosphatic minerals are probably present, to 1.8 at depth due to the carbonate gangue. MgO varies from 0.75% to 5.2%, while iron is usually under 1% Fe_2O_3.

Volta Basin

It is in this basin that the phosphatic beds have been studied in the greatest detail. Only the general characteristics will be mentioned here for the sake of comparison. All the main phosphatic zones are located in the northern part of the basin in a strip 20–40 km wide between the Middle Precambrian outcrops of the Man Shield and the Dahomeyides Chain (Fig. 9.3). Three of these zones are situated in Upper Volta, in the vicinities of Arli, Kodjari and Aloub Djouana: the phosphorites of the last belong to the Buem structural unit. In Niger, the Tapoa deposit, near the Tapoa River, contains the thickest development of phosphorites in the Basin. Phosphatic layers are also found further south, near the Mekrou River, but the thicknesses are somewhat less. This latter zone is known as the Mekrou deposit (or the Mekrou Bend) in Benin. The Mekrou and Tapoa deposits are situated in the 'W Nature Reserve' in Niger which gets its name from the shape of the river which flows through this region.

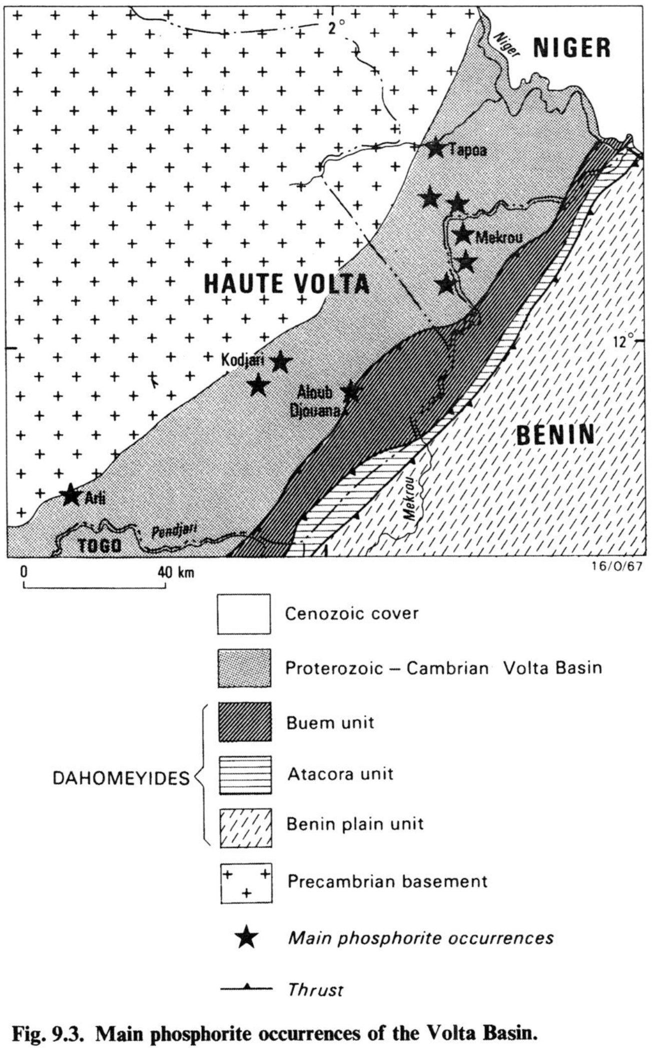

Fig. 9.3. Main phosphorite occurrences of the Volta Basin.

The Kodjari deposit provides the most characteristic cross-section and shows that, as in eastern Senegal, the phosphorite layer may be situated at the base of the lithostratigraphic equivalent to Super-Group 2. Trompette *et al.* (1980) describe the succession immediately overlying the lower group of Dapango–Bombouaka as follows:

Top. 5. Micaceous clayey siltstone; phosphatic in places (0–30 m) grading elsewhere into phosphorites ranging from 0 m to 15 m in thickness
4. Bedded chert with silt intercalations (25–30 m)
3. Dolomitic limestone (0.5–3 m)
2. Tillite (variable thickness up to tens of metres)
Base. 1. Lenticular breccia (0.1 m)

Unpublished BRGM reports suggest four main phosphatic facies within this deposit: laminate phosphorite at the base, fine-grained grainstone phosphorite (grains of 50–100 μm) forming the major part of the ore, quartz-arenites with phosphatic grains (including some ooids) of 100–250 μm at the top, and a variably calcareous and dolomitic conglomerate in which the clasts are reworked from the phosphatic facies and older formations. The presence of this conglomerate is consistent with the existence of tectonic movements also indicated by the underlying phosphatic quartz-arenites.

Analyses carried out on average samples have revealed three types of ore: (1) a surface ore assaying 29.3% P_2O_5, 17.6% SiO_2, with $CaO/P_2O_5 = 1.25$ and $F/P_2O_5 = 0.087$; (2) an intermediate ore assaying 25.4% P_2O_5, 25.2% SiO_2, with $CaO/P_2O_5 = 1.36$ and $F/P_2O_5 = 0.083$: and (3) a carbonate ore assaying 26.2% P_2O_5, 16.4% SiO_2, with $CaO/P_2O_5 = 1.59$ and $F/P_2O_5 = 0.09$. Studies of samples from this region by Lucas, Prévôt & Trompette (1980) and Lucas & Prévôt (1981) show F/P_2O_5 ratios of up to 0.101.

The Tapoa and Mekrou deposits are palaeogeographically situated on the external platform of the Proterozoic basin (Kühn, 1979), just beyond a tectonic high. They are thus further from the edge of this basin than the Kodjari deposit. The phosphatic formation, which in the Tapoa region is more than 70 m thick, overlies slightly phosphatic laminar mudstones. The phosphorites are generally coarser-grained in this area than at Kodjari. They also display an increasing grain size from bottom to top. Giot (1976) distinguishes two phosphatic zones of different grain size in the drill hole WC 92: an upper zone where the phosphate grains have an average size of 150–400 μm and a maximum size of 300 μm to 8 mm: and a lower phosphatic zone where the phosphate grains have an average size of 100–150 μm and a maximum size of 150–600 μm.

This suggests an increase in the energy of the environment from bottom to top and this is confirmed by the increase in the size of the quartz grains, which are generally abundant in these grainstone phosphorites, and by the conglomerate which disconformably overlies (and cuts into) the main phosphatic formation. Giot (personal communication) distinguishes several types of phosphatic grains: grains of homogeneous shape and size, simple or complex ooids with a size range of 100–150 μm, structureless ovular grains of between 200 μm and several mm, and phosphatic lithoclasts. The phosphorite layers of the Mekrou Deposit, which is further away from the craton, have been affected to a greater degree by the Pan-African Orogeny. This is also the case for the phosphorite beds of Aloub Djouana, which are often dipping at more than 60° and are occasionally almost vertical.

Summary

The end of the Proterozoic period is marked in West Africa by a probably diachronous glacial episode exemplified by the tillite deposits of the northern and western regions of the Taoudeni Basin and in the Volta Basin. These deposits were followed by fine-grained siliceous and calcareous sedimentation which, at the terminal Proterozoic or very early Cambrian, locally included phosphatic deposits, which may be very abundant in places.

The largest phosphorite deposits are located in the northern part of the Volta Basin where the ore reserves are evaluated at several hundred million tonnes. In eastern Senegal reserves total some ten million tonnes, but this region has not yet been prospected systematically. The phosphorite occurrences in the Akjoujt regions and those of the northern border of the Anti-Atlas are much smaller, while the phosphorites reported so far in the Adrar region of Mauritania are better termed 'showings' rather than actual occurrences.

The deposits and important occurrences are distributed around the West African craton, on the continental margins and highs, in the vicinity of mobile zones (probably ancient oceanic regions), which surround the craton.

In the Volta Basin, and probably in eastern Senegal, the phosphatic sedimentation is contemporaneous with the preliminary Cadomian (Pan-African) folding. This syn-sedimentary folding was probably directly responsible for the creation of the physiographic 'traps' which were essential for the accumulation of phosphate grains to form phosphorites.

References

Affaton, P. (1975). Etude géologique et structurale du Nord-Ouest Dahomey, du Nord Togo et du Sud-Est de la Haute Volta. *Travaux du Laboratoire des Sciences de la Terre, Saint-Jérôme, Marseilles, B*, **10**, 201pp.

Bassot, J.P. (1966). Etude Géologique du Sénégal oriental et de ses confins guinéo-maliens. *Memoire du Bureau de Recherches Géologiques et Minières*, **40**, 322pp.

Boudda, A. & Choubert, G. (1972). Sur la limite inférieure de Cambrien au Maroc. *Compte rendu de l'Academie des Sciences, Paris*, **275D**, 5–8.

Caby, R., Bertrand, J. & Black, R. (1981). Pan African ocean closure and continental collision in the Hoggar-Iforas segment, Central Sahara. *Precambrian Plate Tectonics*, ed. A. Kroner, pp. 407–34. Elsevier, Amsterdam.

Choubert, G. (1963). Histoire géologique du Précambrien de l'Anti-Atlas. *Notes et Memóires du Service géologique du Maroc*, **162**, 352pp.

Choubert, G. & Faure-Muret, A. (1975a). Grands bassins sedimentaires de l'Afrique occidentale. In *Tectonique de l'Afrique*, pp. 267–93. UNESCO, Paris.

Choubert, G. & Faure-Muret, A. (1975b). *Geological World Atlas, 1/10 000 000*, sheet 6. UNESCO, Paris.

Clauer, N. (1980). Nouvelles précisions sur la géochronologie isotopique et biostratigraphique du Précambrien et de l'Anti-Atlas au Maroc. *Compte Rendu Sommaire des Séances de la Société Géologique de France, lasc*, **3**, 84–87.

Clauer, N., Caby R., Jeannette, D. & Trompette, R. (1982). Geochronology of sedimentary and metasedimentary Precambrian rocks of the West African craton. *Precambrian Research*, **18(1–2)**, 53–71.

Deynoux, M. (1968). Etude du frond Nord des formations d'Akjoujt du Rag el Melgat à l'oued Jenné (Mauritanie occidentale). *Annales de la Faculté des Sciences, Université de Dakar*, **23**, 3.

Deynoux, M. & Trompette, R. (1976). Late Precambrian mixtites: glacial and/or non glacial? Dealing especially with the mixtites of West Africa. *American Journal of Science*, **276(10)**, 1302–14.

Deynoux M., Trompette R., Clauer N. & Sougy, J. (1978). Upper Precambrian and lowermost Paleozoic correlations in West Africa and in the western part of central Africa. Probable diachronism of the Late Precambrian tillite. *Geologische Rundschau*, **67(2)**, 615–30.

Dia, O., Lecorche, J.P. & Le Page, A. (1979). The principal relationship between northern central and southern regions of the main Mauritanides Belt. *Colloque de Géologie Africaine, Résumé*, **10**, 113–4.

Donnot, M. (1975a). Indices de phosphate dans le parc national du W du Niger (République du Niger). Point sur les résultats obtenus. Bureau de Recherches Géologiques et Minières, Orleans, unpublished report, 38pp.

Donnot, M. (1975b). Premières recherches de phosphate dans le bassin voltaien du Ghana. Bureau de Recherches Géologiques et Minièrès. Orleans, unpublished report, 21pp.

Giot, D. (1976). Etude sédimentologique du phosphate du parc du W du Niger. Bèreau de Recherches Géologiques et Minières. Orleans, unpublished report, 52pp.

Giot, D. & Pascal, M. (1978). Eocambrian phosphates of southeast Senegal. Faleme Series. *10th International Congress on Sedimentology, Jerusalem*, **1**, 249.

Ilyin, A.V., Vokov, R.I., Vydrin, V.N. & Zabrodin, V.E. (1979). West African late Precambrian–Cambrian phosphate province. *Lithology and Mineral Resources*, **14(3)**, 344–57.

Kuhn, A. (1979). Development of the phosphate bearing deposits of the Tapoa-Mekrou basin, southwestern Niger. International Geological Correlation Programme Project 156, Phosphorites, Newsletter, **5**, 11–3.

Lucas, J., Prévôt, L. & Trompette, R. (1980). Petrology, mineralogy and geochemistry of the late Precambrian phosphate deposits of Upper Volta (W. Africa). *Journal of the Geological Society*, **137**, 787–92.

Lucas, J. & Prévôt, L. (1981). Caractères pétrographiques et microchimiques de phosphorites précambriennes du Bassin des Volta (Afrique de l'Ouest). *Bulletin de la Société Géologique de France, Serie* 7, 23(5), 515–24.

Pascal, M. (1967). Géologie du NE du Tamkarkart. In Mission serie d'Akjoujt. Etude géologique et prospection de l'Arc des Mauritanides, by J. Marcelin M. Pascal. Bureau de Recherches Géologiques et Minières, Orleans, unpublished report.

Pascal, M. (1972). Réinterprétation de la géologie et examen des possibilités minières de l'ensemble voltaien–Buen–Atacorien au Togo, Dahomey, Haute-Volta et Niger. Bureau de Recherches Géologiques et Minières, Orleans, unpublished report, 77pp.

Reichelt, R. (1972). Géologie du Gourma (Afrique occidentale). Un 'seuil' et un bassin Précambrien supérieur. Stratigraphie, tectonique, métamorphisme. *Mémoire du Bureau de Recherches Géologiques et Minières*, **53**, 213pp.

Sougy, J. (1964). Les formations paléozoiques du Zemmour noir (Mauritanie septentrionale). *Annales de la Faculté des Sciences, Université de Dakar*, **15(XII)**, 695pp.

Sougy, J. (1975). Etude géologique du Zemmour noir (Mauritanie septentrionale) et correlation avec les régions voisines. In *Notice explicative de la carte géologique à 1/1 000 000e de la Mauritanie*, pp. 137–169. BRGM, Orleans.

Trompette, R. (1973). Le Précambrien supérieur et le Paléozoïque inférieur de l'Adrar de Mauritanie (bordure occidentale du bassin de Taoudeni, Afrique de l'Ouest). Un exemple de sédimentation de craton. Etude stratigraphique et sédimentologique. *Travaux du Laboratoire des Sciences de la Terre, Saint Jérôme, Marseilles, Serie B*, **7**, 702pp.

Trompette, R. (1980). La chaîne panafricaine des Dahomeyides et le bassin des Volta (bordure sud-est du craton ouest-africain). *Mémoire du Bureau de Recherches Géologiques et Minières*, **92**, 9–62.

Trompette, R., Affaton, P., Joulia, F. & Marchand, J. (1980). Stratigraphic and structural controls of Late Precambrian phosphate deposits of the northern Volta Basin in Upper Volta, Niger and Benin, West Africa. *Economic Geology*, **75**, 62–70.

Viland, J.C. (1977). Présence d'horizons phosphatés à la base de l'Infracambrien supérieur du Haut Atlas de Marrakech (Maroc). *Notes et Mémoires du Service Géologique du Maroc*, **268**, 13–22.

Zimmerman, M. (1960). Nouvelle subdivision des séries antegothlandiennes de l'Afrique occidentale (Mauritanie, Soudan, Sénégal). *21st Session International Geological Congress, Proceedings of Section 8, Late Pre-Cambrian and Cambrian Stratigraphy*, 26–36.

10

Proterozoic and Cambrian phosphorites – regional review: Brazil

M.A. DARDENNE, R. TROMPETTE
L.F. MAGALHÃES AND L.A. SOARES

Introduction

All phosphorite occurrences and deposits of Late Proterozoic age in Brazil are located in the sediments of the Bambuí Group of Central Brazil, in the States of Goiás and Minas Gerais (Figs. 10.1 and 10.2). Within this area, the Bambuí Group lies unconformably on various Precambrian sequences of the 1800–950 m.y. Espinhaço Supergroup, the 2500–1800 m.y. Transamazonian and the Archaean (Almeida *et al.*, 1973, 1981). The Bambuí Group of central Brazil is flat-lying and thin on the São Francisco craton, but becomes folded and slightly metamorphosed westwards (Fig. 10.2), within the external structural zone of the Brazilides Fold Belt ($\simeq$ 600 m.y.).

The age of the Bambuí Group is ill-defined as geochronological and palaeontological studies are often inconclusive. However an age of 950–600 m.y. is generally accepted (Bonhomme *et al.*, 1982). Sediments of the Bambuí Group dated by the K/Ar method give ages between 650 m.y. and 600 m.y. (Amaral & Kawashita, 1967; Amaral, 1966, 1968; Almeida & Hasui, 1969; Hasui & Almeida, 1970). These ages are interpreted as isotopic resetting ages associated with the tectono-metamorphic Braziliano events and provide only a minimum age for Bambuí sedimentation. Rb/Sr ages are bracketed between 950 m.y. and 600 m.y.; most of those between 700 m.y. and 600 m.y. are associated with successive late diagenesis synchronous with the Braziliano Orogeny. The older ones are possibly early diagenetic ages that can be roughly equated to ages of sedimentation. The Pb/Pb method gives ages between 780 $\pm$ 30 m.y. (Cassedanne, Duthou & Laserre, 1972) and 600 $\pm$ 50 m.y. (Parenti Couto *et al.*, 1981), this youngest age being associated with the Braziliano Orogeny.

A radiometric age for the Bambuí Group of 950–600 m.y. is roughly in agreement with chronostratigraphic information given by palaeontological studies on stromatolites. *Conophyton metula* Kirichenko (Cloud & Dardenne, 1973; Dardenne & Campos Neto, 1976; Dardenne, Faria & Andrade, 1976), *Conophyton cylindricus* Maslov (Moeri, 1972), *Gymnosolen* (Marchese, 1974), *Linella avis* (Bertrand Sarfati *in* Dardenne, 1979), *Baicalia* (Dardenne & Campos Neto, 1976; Dardenne, 1979) which are encountered in the Bambuí of central Brazil can be correlated with the mid-Riphean to Vendian (1050 $\pm$ 50 to 680–650 $\pm$ 20 m.y.) assemblages of the USSR.

The Bambuí Group

The Bambuí Group and correlative sequences are widespread in Minas Gerais, Goiás, Bahia, Mato Grosso and Mato Grosso do Sul States (Fig. 10.1). Presently, however, phosphorites have been found only in Minas Gerais and Goiás (Fig. 10.2) where two major structural units outcrop–the São Francisco craton to the east and the Brazilides Fold Belt to the west.

The flat-lying Bambuı cover on the São Francisco craton is progressively folded westwards to form the external units of the Brazilides Fold Belt. Folding is associated with low-grade anchimetamorphism to epimetamorphism. Further west, the internal structural units, made up of the Araí, Araxá and Canastra Groups (probably equivalents of the Espinhaço Supergroup) and the Ibiá Formation, equated by Campos Neto (1979) and Dardenne (1978a and 1979) with the Bambuí Group, are thrust over the external units.

From east to west, it is possible to subdivide the Bambuí Group of Central Brazil into three structural zones: (1) horizontal or sub-horizontal Bambuí cover of the São Francisco craton (Zone a); (2) folded anchimetamorphic or non-metamorphic Bambuí, west of the Abaeté reverse fault (Zone b); and (3) folded and epimetamorphic Bambuí, west of the Lagamar fault zone (Zone c). Phosphorite occurrences and deposits are all located in Zones b and c (Fig. 10.3).

Lithostratigraphy of the horizontal to sub-horizontal Bambuí Group.

The argillaceous carbonate-rich sediments cropping out in the São Francisco valley were first called the 'São Francisco Series' by Derby (1880) and later the 'Bambuí Series' by Rimann (1917), the latter having been formalised by usage in Brazilian literature.

The Bambuí Group lithostratigraphy was established by Branco & Costa (1961) and has been modified only slightly during the last 20 years by Oliveira (1967), Braun (1968),

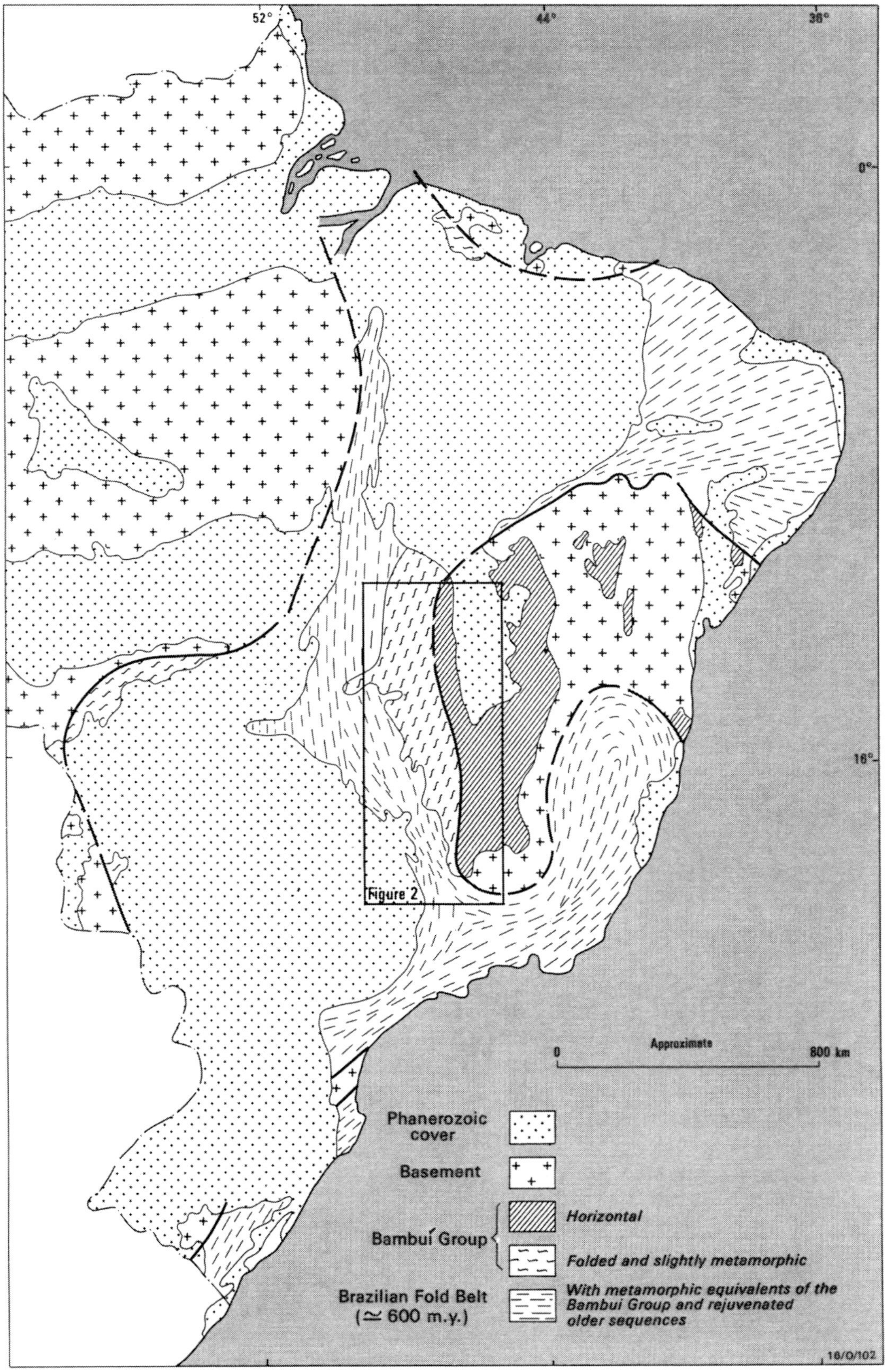

Fig. 10.1. Distribution of the Bambuí Group and equivalents (Upper Proterozoic) in Brazil. (After Almeida *et al.* 1981, simplified.)

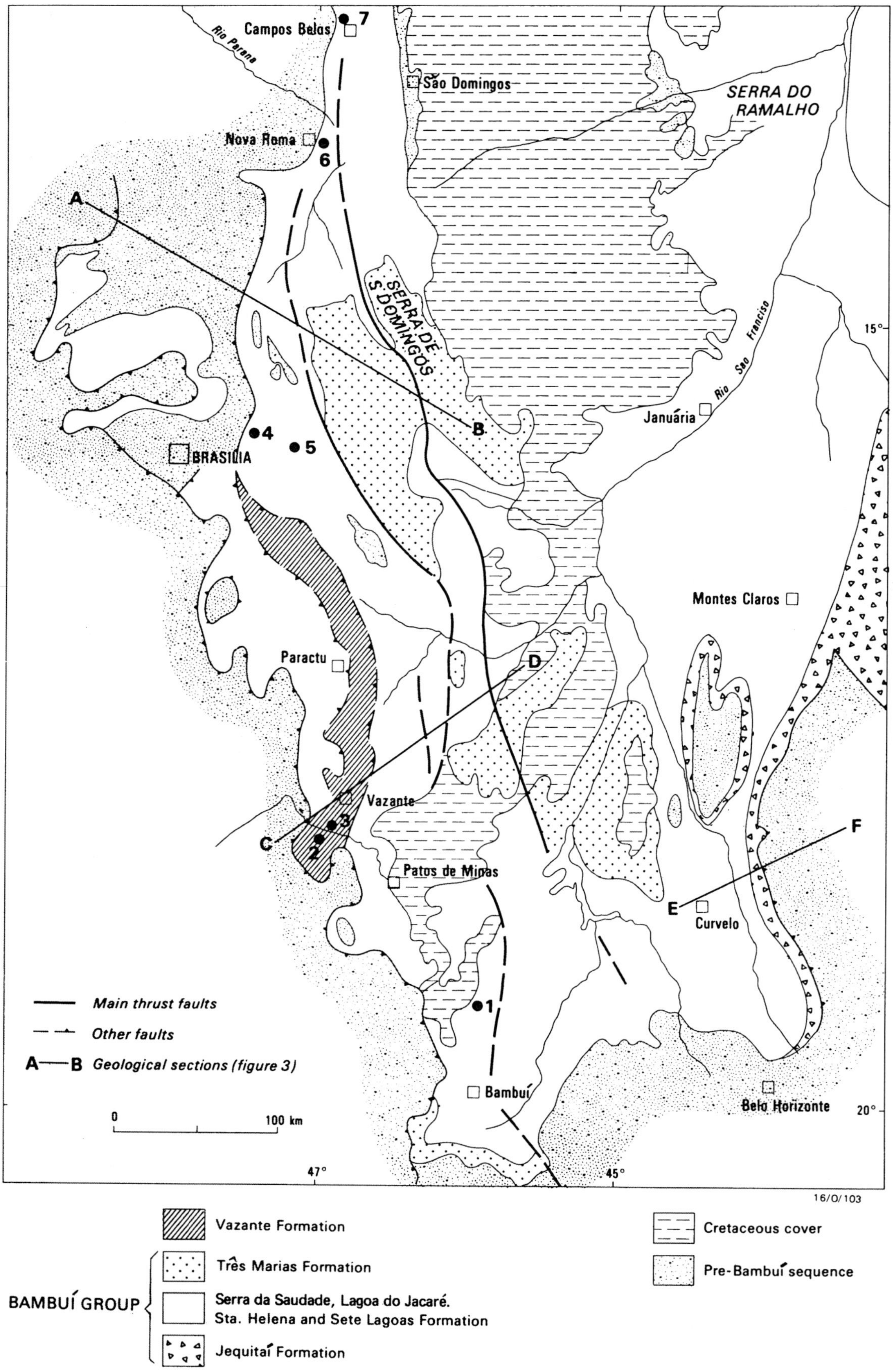

Fig. 10.2. Phosphorite occurrences of the Bambuí Group in central Brazil: (1) Cedro do Abaete; (2) Rocinha; (3) Lagamar; (4) Formosa; (5) Cabeceiras; (6) Nova Roma; (7) Campos Belos and Monte Alegre. (After Dardenne, 1979.)

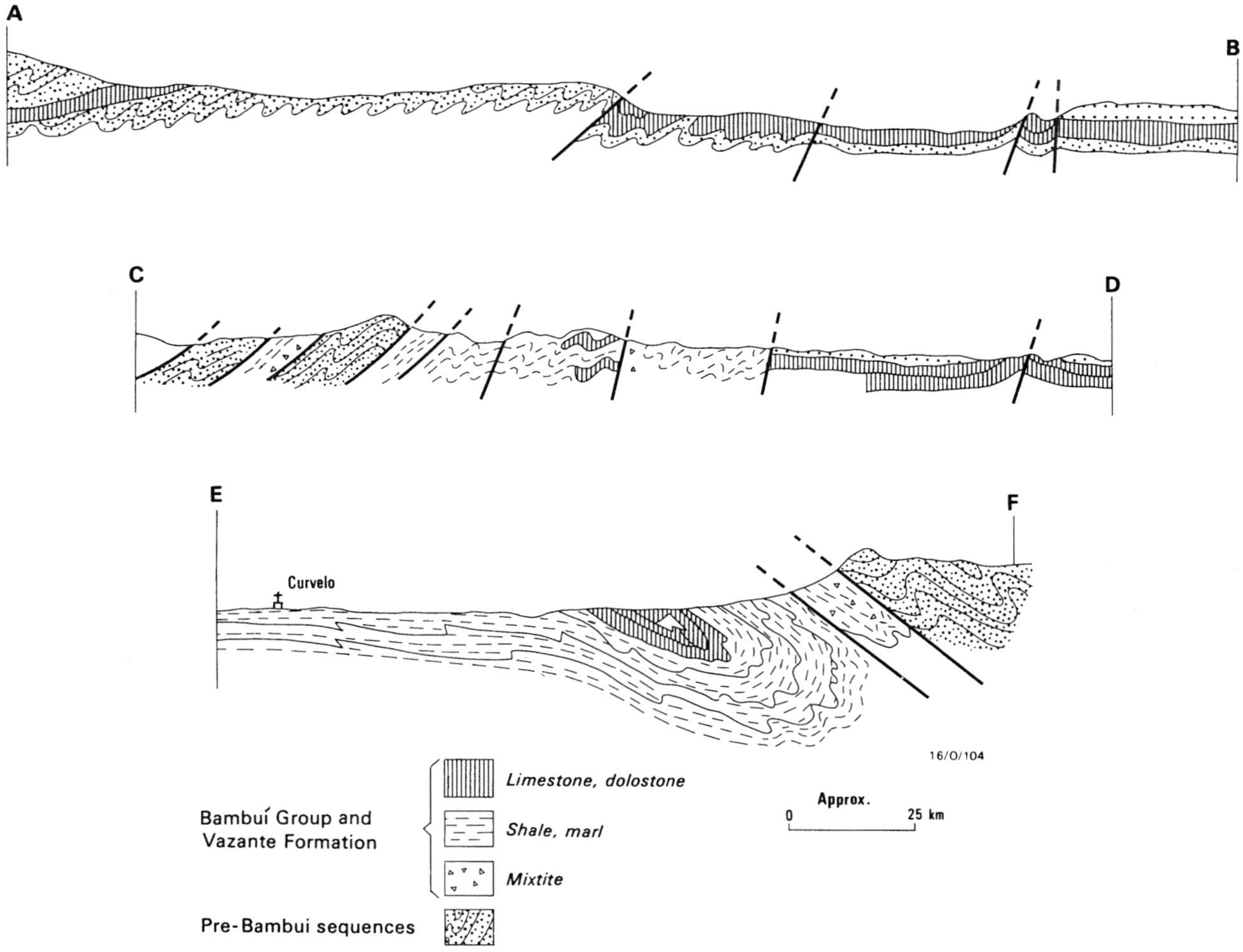

Fig. 10.3 Schematic sections of the borders of the São Francisco craton. (After Dardenne, 1978b.)

Barbosa *et al.* (1969, 1970), Costa *et al.* (1970), Costa & Angeiras (1971). Schöll (1972, 1973) and Dardenne (1978a, 1979). At present, six formations can be distinguished in the Bambuí of central Brazil (Fig. 10.4); these are, from bottom to top: Jequitaí, Sete Lagoas, Serra de Santa Helena, Lagoa do Jacaré, Serra da Saudade and Três Marias Formations.

(a) The Jequitaí Formation (Derby, 1880) lies unconformably on older Precambrian sequences. It is a glaciogenic sequence (Rocha-Campos & Hasui, 1981a, b), the age of which is estimated in Bahia State to be 950–900 m.y. (Macedo, 1982; Macedo & Bonhomme, 1981).

Generally the Jequitaí Formation is composed of mixtites or diamictites with a greenish argillaceous matrix and pebbles of quartzite, dolostone, limestone, chert, gneiss, granite, micashist and volcanics often faceted and/or striated. Glacial pavements with spectacular striations and grooves have been described on the quartzitic basement of the Espinhaço Supergroup in Minas Gerais (Isotta, Rocha-Campos & Yoshida 1969; Hettich, 1977; Walde, 1978). The overlying transgressive Sete Lagoas Formation lies with apparent conformity or with slight regional unconformity on the Jequitaí Formation. The contact is believed to represent a major hiatus (Dardenne & Walde, 1979) and there is presently a tendency to consider the Jequitaí as a distinct group below the Bambuí Group (Parenti Couto & Bez, 1981).

(b) The Sete Lagoas Formation (Branco & Costa, 1961) is composed of marls and siltstones with limestone and dolostone lenses of all sizes well exposed in the type area near Belo-Horizonte (Schöll, 1972, 1973). The transgression of the Sete Lagoas Formation is probably linked to the melting of Jequitaí continental ice. Phosphorites are present in this unit.

(c) The Serra de Santa Helena Formation (Branco & Costa, 1961) is formed of greyish or greyish-greenish marls, shales and siltstones with lenticular dolostone and sandstone intercalations.

(d) The Lagoa do Jacaré Formation (Branco & Costa, 1961) is characterised by greyish and greenish marls and siltstones with intercalations of black, organic-rich fetid, sparry limestone, a few intercalations being oolitic or pisolitic, with frequent cross-bedding.

(e) The Serra da Saudade Formation (Branco & Costa, 1961) is represented by greyish-greenish to green siltstones and shales with lenses of grey limestone. Economically, this is the most important phosphatic formation of the Bambuí Group.

(f) The Três Marias Formation (Branco & Costa, 1961), previously described by Eschwege (1932) under the name

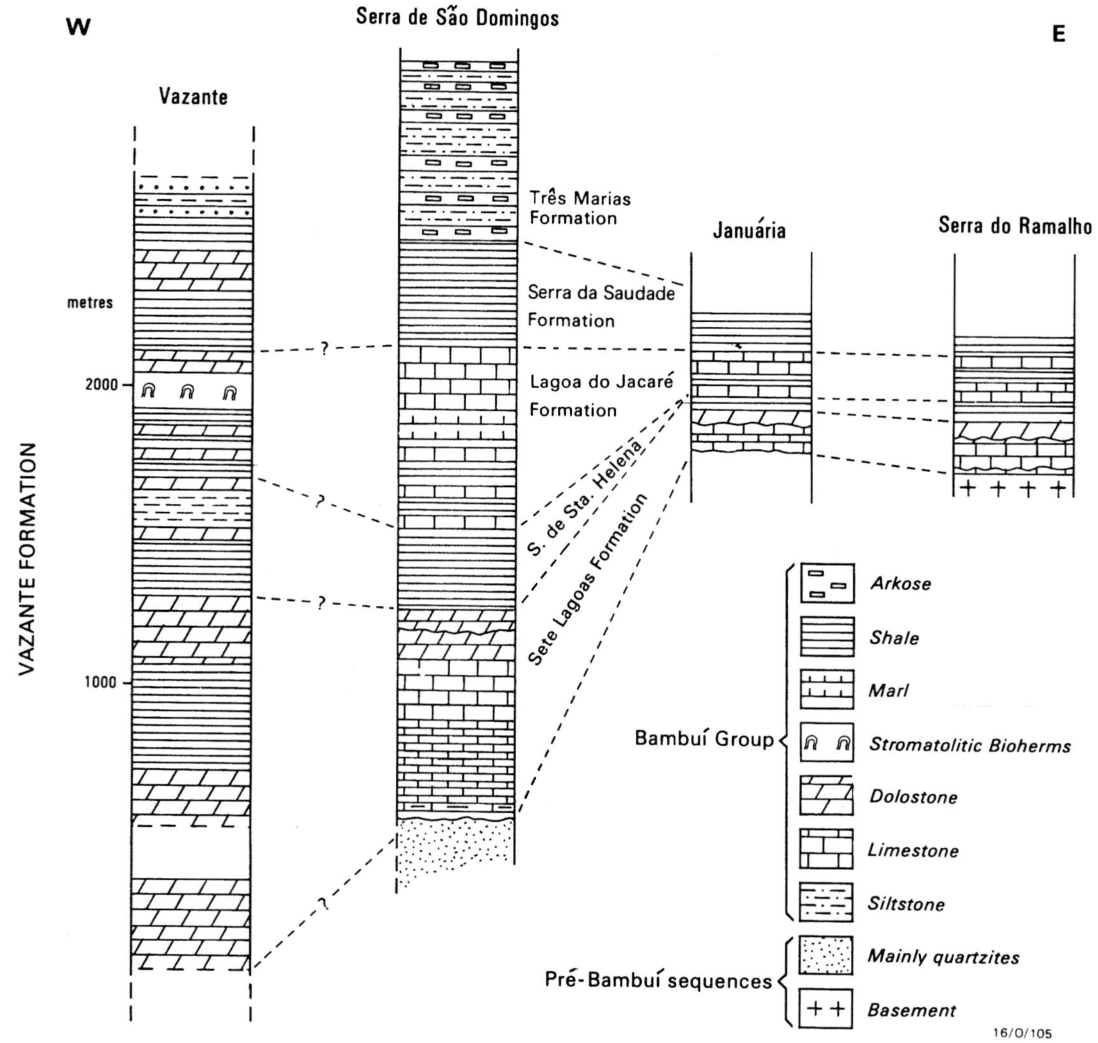

Fig. 10.4. Lithological variations of the Bambuí Group in central Brazil. (After Dardenne, 1978a.)

'Pirapora sandstones', is composed mainly of greenish siltstones and arkoses with a few reddish facies. Generally it is considered as the molasse deposit of the Brazilides Fold Belt (Costa *et al.*, 1970; Costa & Angeiras, 1971).

These six formations can be identified throughout central Brazil in spite of frequent variations in facies and thickness. All phosphorite occurrences and deposits are located in the lower parts of the Sete Lagoas and Serra da Saudade Formations (see below).

Lithostratigraphy and structure of the folded and slightly metamorphosed Bambuí Group.

In the external structural units of the Brazilides Fold Belt, the Bambuí Group is involved in complex structures such as tight folds, imbricated structures, thrust faults and nappes. As a consequence the lithostratigraphic succession is difficult to establish. In a few cases it is comparable with the lithostratigraphic sequences of the unfolded cover (see above). But more frequently the correlations are dubious, and it is safer to regroup the various sequences into a separate lithostratigraphic unit (Figs 10.2 and 10.4), the Vazante Formation (Dardenne, 1978a, 1979).

Palaeogeographic evolution of the Bambuí Group in central Brazil.

The Bambuí sedimentary basin developed upon a continental substrate which is separated by north–south trending faults into elongate, ribbon-like subdivisions, each with its own rate of subsidence. The borders of the basin, as well as the topographically higher parts of the sea floor, are marked by dolomitic facies. The distribution of these dolomitic facies suggests that the configuration of the sedimentary basin in Bambuí time was only slightly different from that of the present Bambuí Group outcrops. The transgression of the Bambuí sea was from west to east as indicated by the eastern wedging out of detrital units. The source of detrital materials was mainly from the south-west.

Various sedimentary environments can be distinguished with respect to the rate and mechanism of subsidence: (1) a stable, slowly subsiding platform characterised by thin sequences rich in carbonates, generally limestone, commonly stromatolitic, for example in the São Francisco Valley, especially in the vicinity of the city of Januaria; (2) a stable, moderately subsiding platform characterised by thicker and dominantly argillaceous sequences, for example, in the Serra de São Domingos region (Fig. 10.2)

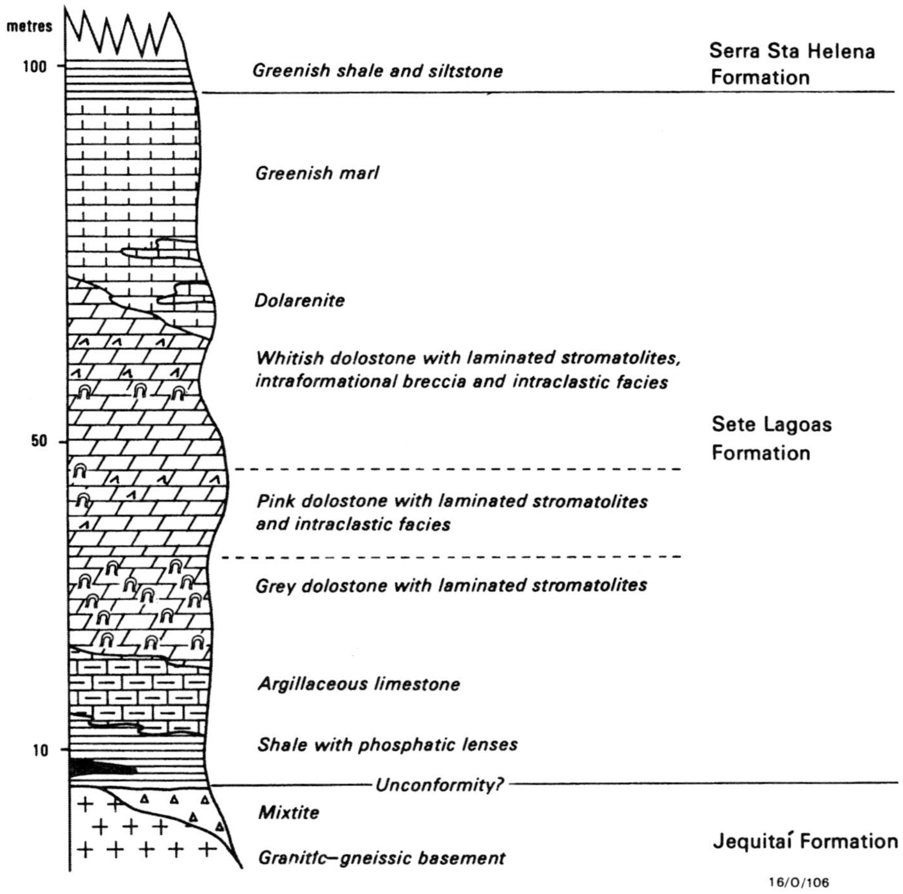

Fig. 10.5. Lithostratigraphic framework of the Campos Belos phosphorites (Goiás). (After Magalhães in Metago, 1977, 1979.)

which seems to be the principal zone of deposition in the Bambuí basin; and (3) basin margins and topographic highs within the basin marked by diverse topography and variable rates of subsidence such as in the Vazante area. This environment is characterised by algal dolostones grading laterally into argillaceous, more or less carbonate-rich, occasionally phosphatic, sequences.

Despite facies and thickness variations, the Bambuí sedimentary basin appears to have been a huge epicontinental basin characterised by shallow, quiet, warm (except during the initial stage of the transgression) waters. The sedimentation is finely detrital with a constant but subordinate carbonate contribution except on sea floor highs and shelfs where stromatolites built major dolomitic bioherms and biostromes.

Phosphorite occurrences and deposits

In the description of Bambuí Group phosphorites we shall use structural context as a guide to the geology of the deposits. We shall first describe occurrences in the folded but unmetamorphosed Bambuí Group and finish with those of the epimetamorphosed external units of the Brazilides Fold Belt.

Occurrences within unmetamorphosed sub-horizontal to folded Bambuí sequences.

None of these occurrences is of economic interest because the phosphorites are lenticular and/or low in P_2O_5 content.

1. State of Goiás Five occurrences, situated along the Parańa valley or immediately east of Brazilia (Fig. 10.2), have been studied by Metago (1977, 1979).

The occurrences at Nova Roma, Campos Belos and Monte Alegre are situated in the eastern part of Zone b (mentioned previously) which is characterised by open folds. The Formosa and Cabeceiras occurrences are located in the centre of Zone b and are affected by tight north–south folds showing sub-vertical axial-planes. All these occurrences, except at Cabeceiras which is found at the boundary between the Lagoa do Jacaré and Serra da Saudade Formations, are located in the lower part of the Sete Lagoas Formation (Fig. 10.5) overlying the glaciogenic deposits of the Jequitaí Formation or directly upon the basement. They are generally associated with shales and marls. At Cabeceiras, the phosphorites crop out within a greyish-greenish sequence made up of shales and siltstones with dolostone intercalations. Phosphorites form lenses reaching a few tens of metres in length and two metres in thickness within shales and siltstones as at Campos Belos (Fig. 10.6). The average grade of the phosphorite ore varies between 15% P_2O_5 and 17% P_2O_5 with a maximum value of 22% P_2O_5 at Nova Roma, Campos Relos and Monte Alegre. At Cabeceiras, the phosphorites contain grey, siliceous units rich in phosphate grains and intraclasts with a P_2O_5 content between 10% to 15% as compared to the adjacent argillaceous host rock which averages about 4% P_2O_5. All occurrences have been affected by lateritic weathering which has

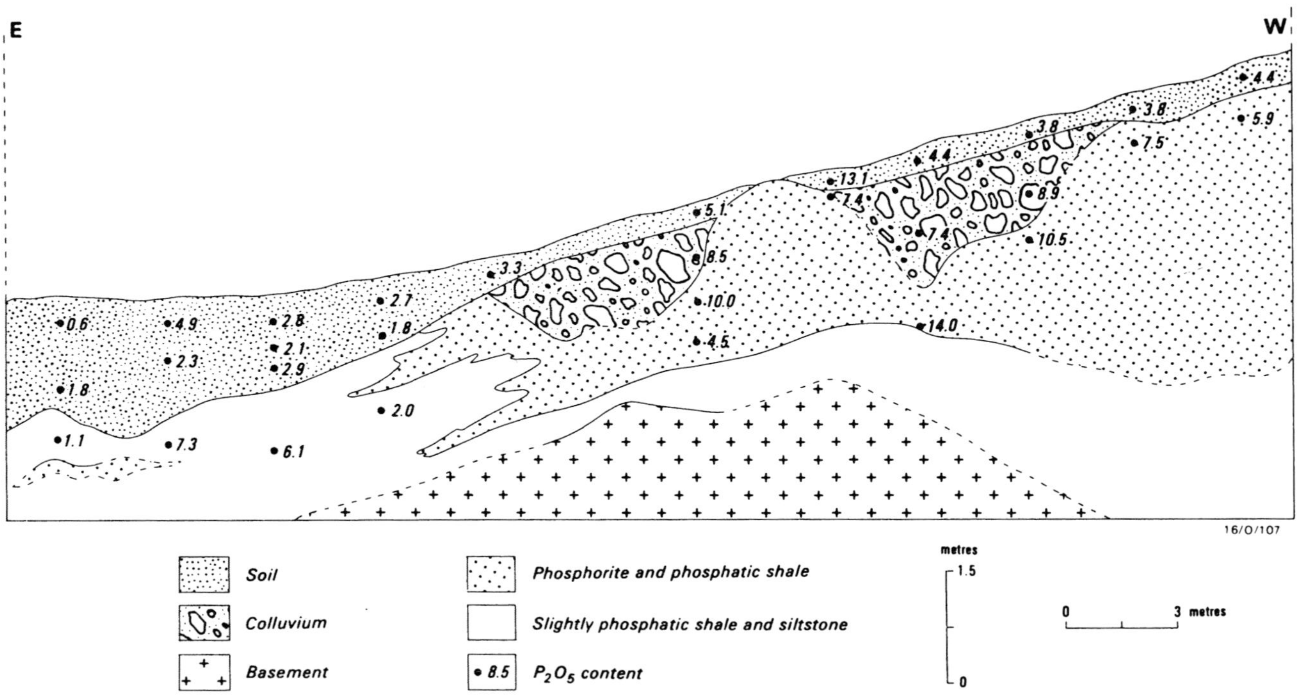

Fig. 10.6. Geometry of the phosphorite occurrence of Campos Belos, Goiás'. (After Magalhães in Metago, 1977, 1979.)

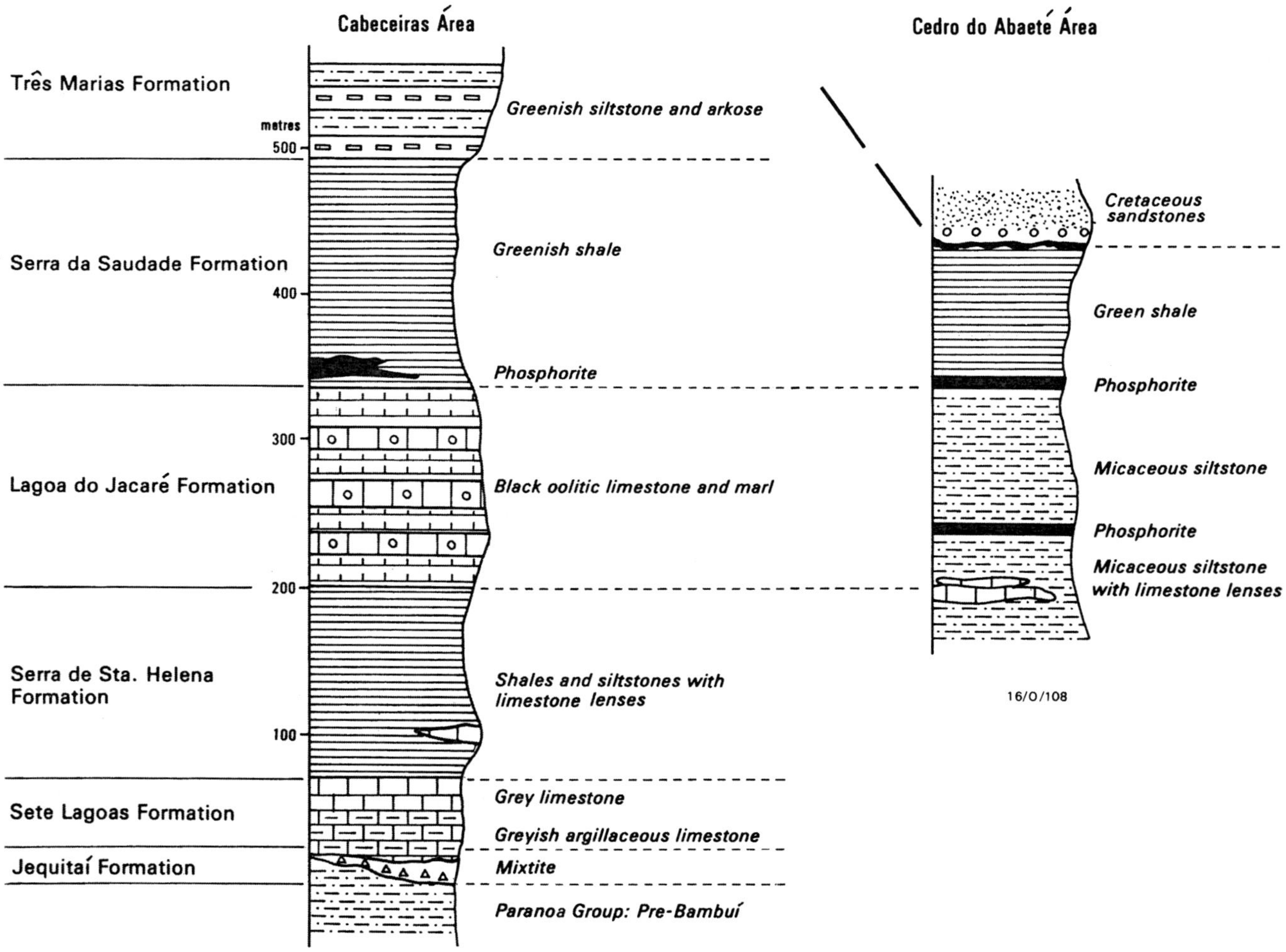

Fig. 10.7. Lithostratigraphic framework of the Cabeceiras (Goiás) and Cedro do Abaeté (Minas Gerais) phosphorites. (After Magalhães in Metago, 1977, 1979, and Dardenne, 1979.)

generated an earthy, dark-brown, friable ore, with local development of siliceous crust containing wavellite.

The Goiás phosphorites, located in the lower part of the Sete Lagoas Formation (i.e. all occurrences except that at Cabeceiras), were deposited in shallow basement depressions (the Jequitái Formation has a lenticular character in this area), in shallow, quiet, possibly cold waters. The phosphorites are capped by stromatolitic dolostones which may indicate a general warming of the climate following the Jequitaí glaciation. All these occurrences are located in an area of eastward thinning of the basal argillaceous sequence of the Sete Lagoas Formation. The Cabeceiras occurrence is associated with a lithostratigraphical and sedimentological context very similar to that of the Cedro do Abaeté phosphorites described below (Fig. 10.7).

2. Cedro do Abaeté (Minas Gerais) Discovered in 1961, this occurrence was first described by Guimarães (1967). It crops out in an elongate belt, trending N 20° E, 8 km wide and 30 km long. Reserves are about 1 million tonnes of phosphorite ore varying between 10% P_2O_5 and 34% P_2O_5 (Guimarães & Dutra, 1969). This occurrence has no economic value. The structural and lithostratigraphic contexts of the phosphorites have been described by Chaves *et al.* (1971). The Cedro do Abaeté occurrence is situated west of the N–S trending reverse fault which separates the subhorizontal or slightly folded Bambuí Group from folded Bambuí Group (Fig. 10.2) and which can be followed northwards over hundreds of kilometres. This reverse fault coincides with a westward deepening of the Bambuí basin. The occurrence is located near the transition between the Lagoa do Jacaré and Serra da Saudade Formations (Chaves, Heineck & de Taveres, 1971; Dardenne, 1979) like the Cabeceiras (Goiás) phosphorite. In the Cedro do Abaeté area, the Bambuí is unconformably overlain by flat-lying detrital sequences of Cretaceous age (Mata da Corda Formation) which themselves contain minor phosphorite occurrences with a very low P_2O_5 content. (These Cretaceous phosphorites, associated with volcanic tuffs, breccias and ashes, are very different from the Bambuí phosphorites).

The Cedro do Abaeté phosphorites occur as intercalations, a few centimetres to 3 metres thick, in dark-green shales and siltstones (the 'Verdetes' or 'green-rocks'). The phosphorites are distinguished from the adjacent 'Verdetes' by their grey-brown to ochre colour and their finely banded structure. In the 'Verdetes', the argillaceous fraction consists predominantly of an association of aluminium and/or iron–magnesium illites with minor smectites and chlorites (Bonhomme, 1976; Bonhomme & Cordani, 1975). The glauconite mentioned by Chaves *et al.* (1971) has not been identified by X-ray study. The green argillaceous mineral is probably an illite as suggested (Table 10.1) by the high (6.2–10.0%) K_2O content and the relatively low (6.4%) FeO content (Costa & Oliveira, 1970). Quartz of silt size and detrital micas are present in variable quantities. Phosphorites and adjacent 'Verdetes' exhibit decimetric to metric 'chevron' type folds ranging in size from tens of centimetres to metres, but are unmetamorphosed according to Guimarães (1967) or metamorphosed to anchimetamorphic grade according to Bonhomme (1976), who based his conclusions on a study of illite crystallinity.

Table 10.1. *Chemical analysis of phosphorite and green pseudo-glauconitic shales ('Verdete') from Cedro do Abaeté*

	Phosphorite	'Verdete'
SiO_2	11.41	63.39
TiO_2	0.05	0.07
Al_2O_3	2.75	16.57
FeO	1.70	6.40
MnO	0.34	0.46
CaO	44.84	trace
MgO	0.10	0.11
K_2O	0.96	6.20*
Na_2O	0.22	0.20
P_2O_5	33.73	0.04
F	0.22	—
SO_3	0.18	0.02
Cl	0.97	0.22
$H_2O + CO_2$	2.59	6.33
BaO	—	0.03

*4% to 8% according to Ladeira & Brito (1968).—, not determined. After Guimarães (1967).

The phosphate mineral is a brown, poorly crystallised apatite (Guimarães, 1967; Ladeira & Leal, 1972) associated with small euhedral, translucent, cavity-filling apatite crystals identified as dahllite, but probably a carbonate-apatite (Guimarães, 1967). However the chemical analysis (Table 10.1), though showing a low fluorine content (0.22%), indicates a CaO/P_2O_5 ratio of 1:33, a normal value for fluorapatite. Quartz of silt size and detrital micas are rare. Pyrite, generally weathered to a mixture of hematite and limonite, is uncommon. Wavellite has been identified in the more weathered facies.

The Cedro do Abaeté phosphorites were deposited on a stable platform covered by shallow waters in a partly reducing pyrite-rich environment which received an influx of fine detrital material with a low carbonate contribution. Ladeira & Leal (1972) have clearly demonstrated that these phosphorites have a sedimentary, probably early diagenetic, origin. Until this study, various origins had been proposed for the Cedro do Abaete phosphorites: (1) metasomatic enrichment contemporaneous with the emplacement of alkaline basic and ultrabasic volcanic pipes of Cretaceous age which are common in this part of Minas Gerais; (2) enrichment by weathering of the overlying Cretaceous volcano-sedimentary Mata da Corda Formation locally containing 2–3% P_2O_5 (Ladeira & Brito, 1968; Ladeira & Leal, 1972) and redeposition by precipitation in the Bambuí Group. The volcanic intercalations of the Mata da Corda Formation are considered as roughly synchronous with the alkaline pipes mentioned above. Thus, the Cedro do Abaeté area has suffered two distinct phosphogenic episodes: the first in Late Proterozoic time and the second in Cretaceous time. Presently the main source of phosphorus in Brazil is represented by the commonly apatite-rich alkaline pipes of Cretaceous age (Gonçalves de Lima, 1967: Albuquerque & Giannerini, 1981).

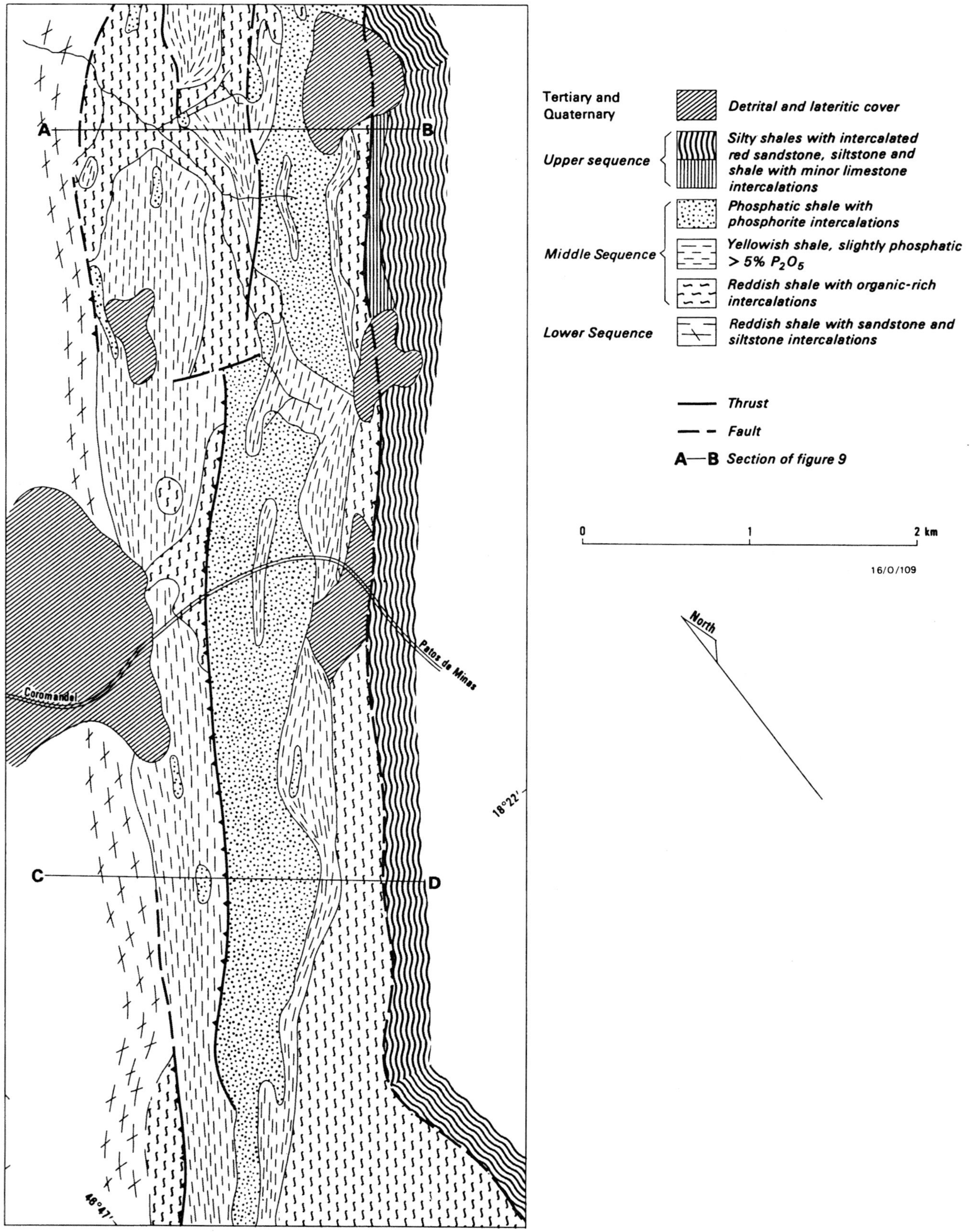

Fig. 10.8. Schematic map of the central part of the Rocinha phosphorite mine. (After CPRM, 1976.)

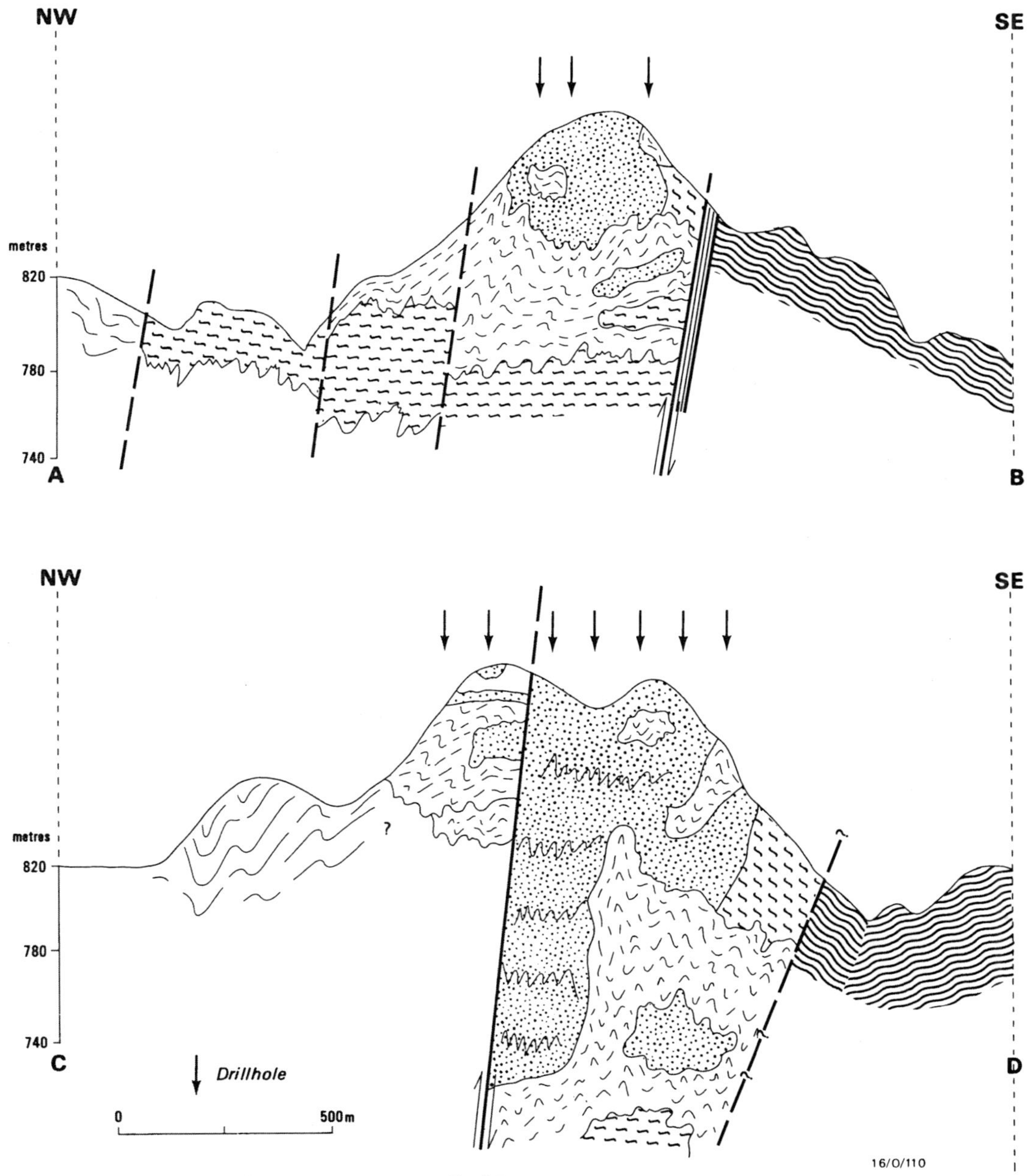

Fig. 10.9. Schematic sections showing the geometry of the Rocinha phosphorites (legend and localisation of sections in Fig. 10.8). (After CPRM, 1976.)

Occurrences within folded and metamorphic Bambuí sequences.

The phosphorite deposits of Rocinha and Lagamar, (CPRM, 1976; Cathcart, 1979), are situated 25 km apart, east of the city of Patos de Minas (Fig. 10.2). Here, the Bambuí is characterised by anchimetamorphism or, more generally, epimetamorphism and an important tangential tectonism with the development of imbricated structures and small nappes, both of which have been thoroughly studied a few kilometres north of this area near Vazante (Campos Neto, 1979).

The Rocinha and Lagamar occurrences are the only economic deposits presently known in the Bambuí Group of Brazil. The Rocinha reserves have been estimated at 200 million tonnes (Gonçalves de Lima, 1976) and 400 million tonnes (Albuquerque & Giannerini, 1981), and consist of an ore with 11–13% P_2O_5. The Rocinha deposit, the only one that is regularly mined, has been producing 200 000 to 250 000 tonnes per year of a concentrate with a P_2O_5 content of 34% since 1976. This concentrate, though considered as insoluble, is finely crushed or pulverised and used in farming for direct application to acidic soils. The Lagamar deposit, which can be considered as the extension of the Rocinha deposit, has reserves of about 5 million tonnes of ore with a P_2O_5 content of 35–38% (DNPM, 1975).

Lithostratigraphic and structural framework

In the Lagamar area, three lithostratigraphic units have been distinguished (Dardenne *et al.*, 1976). They are: (1) a lower fine detrital unit made up of black, organic-rich and pyritic slates locally with alternating, rhythmic laminations of siltstone and carbonate; (2) a middle dolostone unit made up of grey and pink dolostone with numerous oncolites and stromatolites such as *Baicalia, Conophyton metular, Kirichenko* and *Jacutophyton*. Laterally this unit grades to marl with lenses of black and grey-blue limestone; (3) an upper fine detrital unit made up of black to dark-grey slates with lenses of quartzite which is sometimes conglomeratic.

These three units form part of the Vazante Formation, a probable epimetamorphic equivalent of the Bambuí Group (Fig. 10.4).

The Rocinha and Lagamar deposits are sited in the folded and epimetamorphosed parts of the Bambuí Group (Zone c). Westwards, the Vazante Formation has been overridden by the internal structural units of the Brazilides Fold Belt which are from east to west: (1) the schists and quartzites of the Canastra Group; (2) the schists and mixtites of the Ibiá Formation; (3) the schists, micaschists, quartzites and amphibolites of the Araxa Group; and finally (4) the basement. The Ibiá Formation is correlated with the Bambuí Group, whereas the Araxá and Canastra Groups are considered as pre-Bambuí and correlated with the Espinhaço Supergroup.

The Rocinha and Lagamar deposits consist of two large lenses of phosphorite and phosphatic shale elongated in a north-north-east direction and $10 \times 0.3 \times 0.4$ km and $0.25 \times 0.3 \times 0.2$ km in size, respectively. Phosphorite and host rock are folded into synclinoria and anticlinoria a hundred metres wide. At the Rocinha mine, the phosphorites are located in a north-north-east syncline, slightly refolded and faulted by transverse east–west structures (Figs 10.8 and 10.9). The ore and host rock are affected by 'chevron type' folds up to tens of centimetres in size, with an axial-plane strain-slip cleavage locally associated with transposition. Metamorphic grade varies between anchi-epizonal to true epizonal greenschist facies.

Description of phosphorites

In the Rocinha mine, three main facies (CPRM, 1976) can be distinguished (Figs. 10.8 and 10.9): (1) reddish shales with intercalations of brown to greyish organic-rich shales; (2) yellowish, slightly phosphatic ($P_2O_5 \leqslant 5\%$), finely laminated, banded shales where apatite is concentrated in intercalations with a striped or banded appearance; (3) yellowish phosphatic shales with intercalations up to tens of centimetres thick of hard, dense, dark-grey, often banded phosphorite. These phosphorite intercalations are particularly abundant in the Lagamar mine and account for the high P_2O_5 content of the ore, the highest known from the Bambuí Group. They wedge out southwards and, at Rocinha, become subordinate to the argillaceous facies.

A few other less abundant facies are present in the Rocinha mine. Fault breccias are often phosphatic and silicified. Locally (Fig. 10.8) lateritic weathering has generated hard siliceous crusts or box-work structures with fissures and cavities partly filled by quartz crystals and small, fragile, white needles of wavellite.

Two main phosphorite facies can be distinguished. The first is a hard, massive, grey to ochre phosphorite (Figs. 10.10, A, B, C and C′; Figs. 10.11, A and A′), composed of thin, striped intercalations with phosphatic shales. Under the hand lens, rocks of this facies can be seen to be siltstone or very fine sandstone but they may be conglomeratic with clasts of quartz, schist, slate and phosphorite. The second facies is dominantly argillaceous and is made up of thin (millimetres to centimetres) phosphorite laminations within greenish shales (Figs. 10.11, B and B′). Phosphorite can be distinguished from the associated shales by its yellowish, ochre to greyish colour and its banded structure. Thus, in Rocinha, the excavated ore consists of a mixture of shales with very low P_2O_5 content, and intercalations of phosphatic shales (facies 2) and true phosphorite (facies 1). The average phosphate content of this primary ore is 11–13% P_2O_5.

Several thin sections of material from the Rocinha mine have been studied and revealed four apatite habits: (1) brown laminations; (2) intraclasts and grains; (3) brown amorphous masses, and (4) small, transparent crystals developed in cavities and fissures. According to an unpublished report consulted at the Rocinha mine, the apatite is apparently a fluorapatite.

Detrital quartz (of silt-size or very fine sand-size or rarely coarser) and micas are infrequent. Secondary silicification is responsible for the crystallisation of small euhedral quartz in geodes and, in a few cases, for the partial silicification of the phosphorite matrix (Figs 10.11, C and C′). Hematite–limonite pseudomorphs after pyrite are abundant. Carbonates have been identified only in the unweathered samples coming from the bottom of the mine and the deepest drill holes. The determination of two generations of carbonates suggests that the primary phosphorite was carbonate-bearing; and that the second generation of the carbonate results from the dissolution of primary carbonate in the surficial weathering zone, followed by accumulation at depth.

In the Lagamar deposits, drill holes show a gradation from phosphorite into dark argillaceous limestone intercalated in dark-greyish, carbonate-bearing slate containing 1–10% P_2O_5. The phosphorites exhibit the four facies described from Rocinha but with a predominance of granular and intraclastic facies. At Rocinha, secondary quartz fills up cavities and veins. The more weathered facies are characterised by the hematite–limonite–wavellite association. The shales of the host rock are made up predominantly of illites with kaolinite, detrital micas and calcite were identified in an unweathered sample.

As a general rule, weathering is intense in the whole area, reaching depths of up to 200 metres beneath the surface as observed in a few drill holes. This weathering is responsible for the secondary paragenesis (carbonate–silica–wavellite–limonite–hematite) observed in both the Lagamar and Rocinha phosphorites.

Sedimentary framework of phosphogenesis

Although geographically very close, the Rocinha and Lagamar deposits have developed in different sedimentological environments. At Lagamar, the phosphorites seem to outline the transition between a shallow- and deep-water facies. The former facies is associated with the shelf or with sea-floor highs, and is characterised by small stromatolitic bioherms developed in clear, oxidising, relatively warm and rough waters. The latter is deeper and quieter, with reducing zones where thicker argillaceous sequences were deposited. A few tens of kilometres

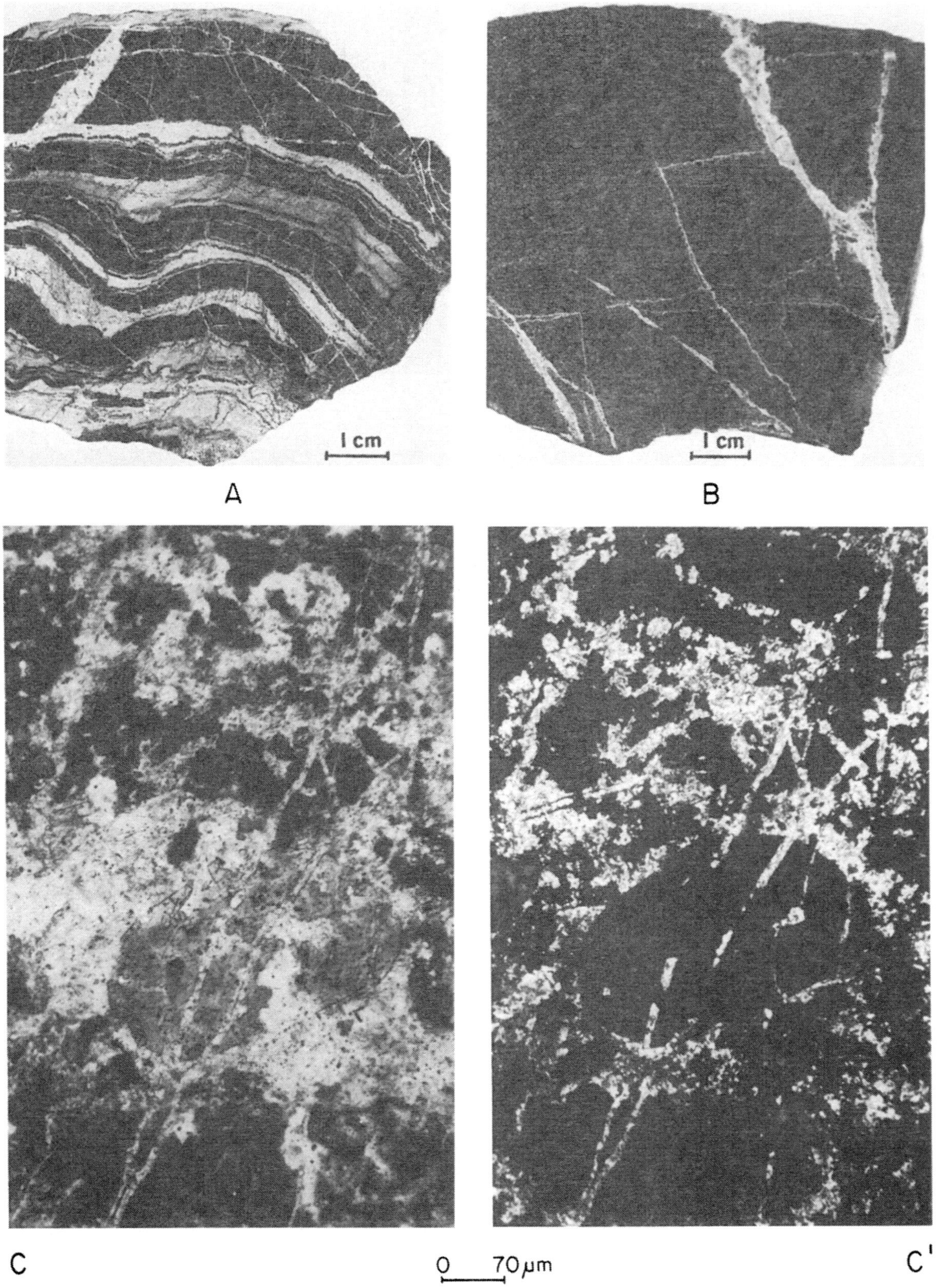

Fig. 10.10. Photographs and photomicrographs of various phosphorite facies. (A) Banded phosphorite, with light laminations representing argillaceous phosphorite and dark ones high grade phosphorite. Lagamar Mine (Minas Gerais). (B) Fine, homogeneous, slightly argillaceous phosphorite with quartz-filled veins. Lagamar Mine (Minas Gerais). (C and C′) Unweathered carbonate phosphorite with carbonates forming part of the matrix and filling in veins: (C) plane polarised light, (C′) crossed nicols – Rocinha Mine (Minas Gerais). (Photomicrographs courtesy of L. Prévôt, Strasbourg, France.)

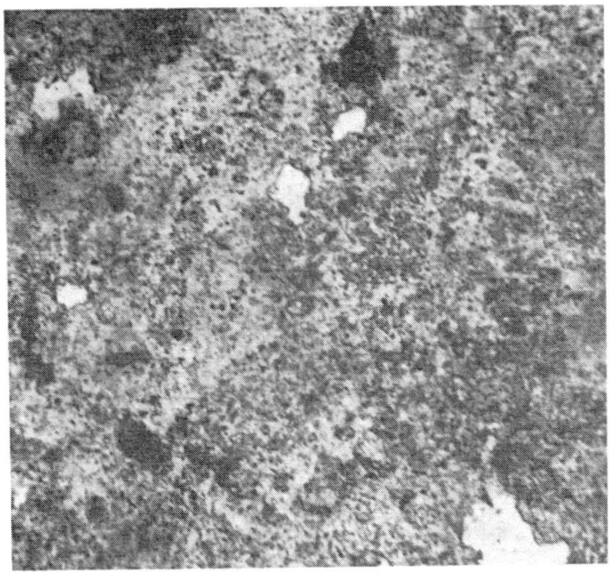

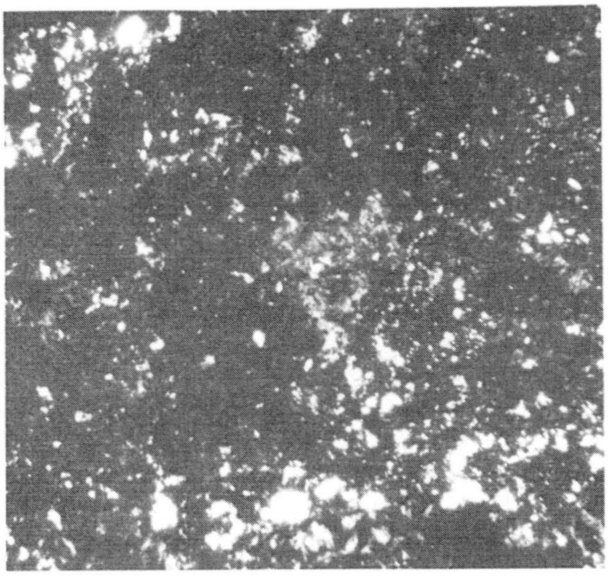

A

0 140 μm

A′

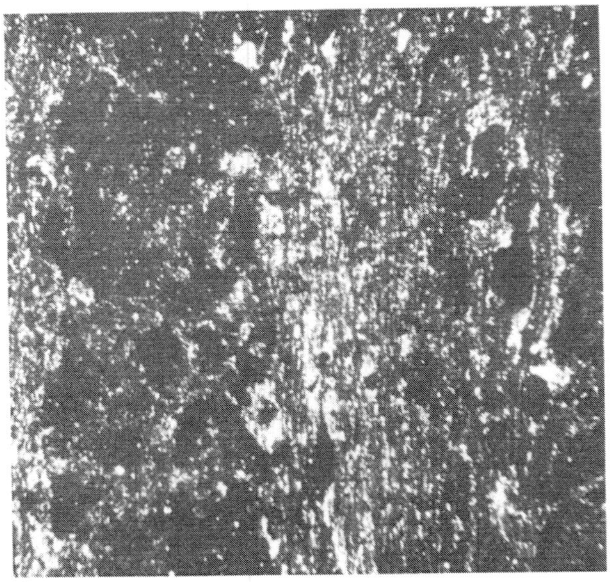

B

0 550 μm

B′

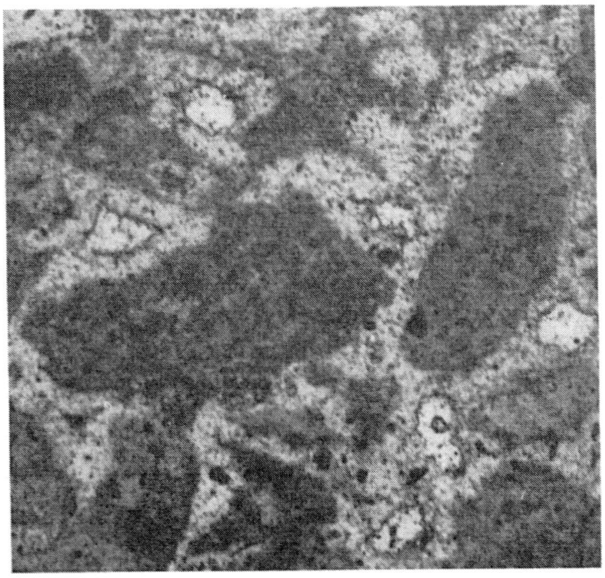

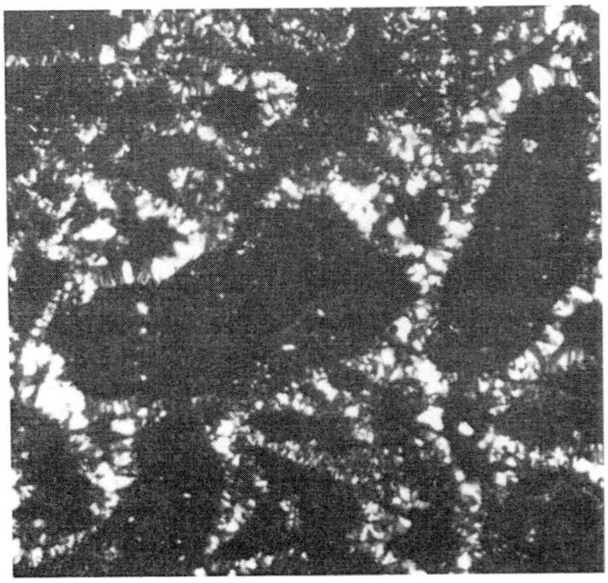

C

0 225 μm

C′

northwards, near Vazante, this shaly sequence is made up of 80%–100% illites, 0–20% chlorites and 0–15% smectites (Bonhomme, 1976). The smectites, like the carbonate-rich facies of Lagamar, suggest sea water with a relatively high Ca and Mg content. On the other hand, the Rocinha deposit is associated with a fine detrital argillaceous sequence quite similar to that of Lagamar, but phosphatic. At Rocinha, finely laminated phosphorites predominate, indicating that an *in situ* phosphogenesis occurred, probably by early diagenetic substitution of organic-rich, argillaceous, carbonate-bearing sediments. At Lagamar, reworked granular and intraclastic facies are more abundant, indicating a more energetic environment, which would account for the higher sedimentological maturity of phosphorite ore there.

Comparisons with Upper Proterozoic West African phosphorites

Phosphorite occurrences and deposits of the Bambuí Group in central Brazil are stratigraphically controlled. The Campos Belos, Monte Alegre, Formosa and Nova Roma occurrences are located in the lower part of the Sete Lagoas Formation which marks the onset of the Bambuí transgression. They are of no economic interest. In contrast, the Cabeceiras, Cedro do Abaete, Rocinha and Lagamar occurrences and deposits are situated higher in the sequence, in the lower part of the Serra da Saudade Formation close to the boundary with the Lagoa do Jacaré Formation. Of these, the Rocinha and Lagamar occurrences are presently mined.

Phosphorites are also structurally controlled. They are all on the western border of the São Francisco craton. The Nova Roma, Campos Belos and Monte Alegre occurrences are connected with the appearance of the first folds in the Bambuí Group. The Formosa, Cabeceiras and Cedro do Abaeté occurrences are in folded and anchimetamorphosed or unmetamorphosed Bambuí. The deposits at Rocinha and Lagamar, the only economically important ones, are situated in the highly tectonised and epimetamorphosed zone of the Bambuí Group. Thus, it seems that the best chance of finding phosphorite in the Bambuí Group of central Brazil is defined by the coincidence of a lithostratigraphic factor–the Serra da Saudade Formation–and a structural factor–a highly tectonised epimetamorphic setting.

The age of Bambuí Group phosphorites can be roughly estimated at 800 ± 100 m.y. Thus they are clearly older than the uppermost Proterozoic phosphorites of West Africa, specifically those of the Volta Basin (Benin, Upper Volta, Niger), which overlie a tillite (*sensu stricto*) and have been dated at close to 660 m.y. (Trompette *et al.*, 1980). As has been suggested by Parenti Couto & Bez (1981) and Trompette (1982), the Jequitaí Formation can be correlated with the 'Grand Conglomerat' or the Lower 'Tillite' of central Africa. However, in this part of Africa, phosphorites are unknown in sequences overlying the tillite/mixtite formations. One should note the close association of phosphorites with glaciogenic sequences in the lowermost Bambuí Group of Brazil and in the uppermost Proterozoic of western Africa, where, in addition, phosphorites are associated with volcanism. The diachronous character of the mixtite–phosphorite association shows the difficulties of Upper Proterozoic intercontinental correlations based on lithostratigraphy.

Bambuí Group phosphorites were deposited in shallow, slightly reducing waters together with fine detrital sediment and a minor carbonate contribution. In this environment phosphate deposition is primary and probably early diagenetic. Reworked facies, with intraclasts and grains, are subordinate which accounts for the generally low P_2O_5 content of these occurrences. Phosphorites are unknown on the eastern border of the São Francisco craton. This concentration of phosphorite on one side of the craton suggests, by comparison with present phosphogenesis, a possible relation with upwelling developed on a western margin. However the palaeogeographic position of the São Francisco craton in Upper Proterozoic time is unknown.

The nature of phosphate minerals is ill defined although at Rocinha it appears to be fluorapatite.

References

Albuquerque, G.S.C. & Giannerini, J. F. (1981). Phosphate: new research areas for Brazil. 2nd National Meeting on Phosphate Rock, Instituto Brasil Fosfato, Brasilia, 16pp. (unpublished).

Almeida, de F.F.M. & Hasui, Y. (1969). Geocronologia do Centro Oeste brasileiro. *Mineração, Metalurgia, Rio de Janeiro*, **50**, 43–66.

Almeida, de F.F.M., Amaral, G., Cordani, U.G. & Kawashita, K. (1973). The Precambrian evolution of the South American cratonic margin south of the Amazon river. In *The Ocean Basins and Margins. 1, The South Atlantic*, ed. A.E.M. Nairn & F.G. Stehli, pp. 441–6. Plenum Press, New York.

Almeida, de F.F.M., Hasui, Y., Brito Neves, B.B. & Fuck, R.A. (1981). Brazilian structural provinces: an introduction. *Earth Science Reviews*, **17**, 1–29.

Amaral, G. (1966). Isotopos de chumbo e gênese das jazidas de Vazante e Itacarambi. *Publicaçiones, XX congresso, Sociedade Brasileiro de Geologia, Rio de Janeiro*, **1**, 45–46.

Amaral, G. (1968). Resultados preliminares do estudo sobre a mineralização e composicão isotópica do chumbo em galenas de depósitos encaixados no Grupo Bambuí. *Publicaçiones, XXII Congresso, Sociedade Brasileiro de Geologia, Belo Horizonte*, pp. 51–3.

Amaral, G. & Kawashita, K. (1967). Determinação da idade do Grupo Bambui pelo método Rb/Sr. *Publicaçiones, XXI Congresso, Sociedade Brasileiro de Geologia, Curitiba*, pp. 214–7.

Barbosa, O., Baptista, M.B., Braun, O.P.G., Dyer, R.C. & Cotta, J.C. (1969). Geologia e inventário dos recursos minerais do Projeto Brasilia. Relatório Prospec–Departmento Nacional Produção Mineral Rio de Janiero, 225pp. (unpublished).

Fig. 10.11. Photomicrographs of phosphorites from the Rocinha Mine (Minas Gerais). (A and A′) Massive, structureless phosphorite: (A) plane polarised light, (A′) crossed nicols. (B and B′) Argillaceous phosphorite: (B) plane polarised light, (B′) crossed nicols. (C and C′) Siliceous intraclastic phosphorite: (C) plane polarised light, (Ć) crossed nicols. (All photomicrographs kindly provided by L. Prévôt).

Barbosa, O., Braun, O.P.G., Dyer, R.C. & Rodrigues da Cunha, C.A.B. (1970). Geologia da região do Triângulo Mineiro. *Boletim, Departmento Nacional Produção Mineral, Rio de Janiero*, **136**, 140pp.

Bonhomme, M. (1976). Minéralogie des fractions fines et datations rubidium-strontium dans le Groupe Bambuí, MG, Brésil. *Revista Brasileira de Geociencias, São Paulo*, **6(4)**, 211–22.

Bonhomme, M. & Cordani, U.G. (1975). Minéralogie des argiles et geochronologie des Séries précambriennes sédimentaires et légèrement métamorphiques, discordantes sur le Craton du São Francisco. Relat. Int., CPGeo/USP, Sao Paulo, 3pp. (unpublished).

Bonhomme, M.G., Cordani, U.G., Kawashita, K., Macedo, M.H.F. & Thomaz Filho, A. (1982). Radiochronological age and correlation of Proterozoic sediments in Brasil. *Precambrian Research*, **18**, 103–118.

Branco, J.J.R. & Costa, da M.T. (1961). Roteiro da excursão Belo Horizonte – Brasília. *Publicaçiones, XIV Congresso Sociedade Brasileiro de Geologia, Belo Horizonte*, **15**, 25pp.

Braun, O.P.G. (1968). Contribuição a estratigrafia do Grupo Bambuí. *Publicaçiones, XXII Congresso, Sociedade Brasileiro de Geologia, Belo Horizonte*, pp. 155–66.

Campos Neto. M.C. (1979). Contribution a l'étude des Brasilides. Lithostratigraphie et structure des groupes Canastra. Paranoc et Bambui dans l'Ouest-Nord-Ouest de l'état de Minas Gerais (Brésil). These Doctoral, 3e Cycle. Université de Paris VI, 155pp. (unpublished).

Carthcart, J.B. (1979). Phosphorites deposits, Bambuí Group, near Patos de Minas, Brazil. In *Proterozoic–Cambrian phosphorites*, ed. P.J. Cook & J.H. Shergold, pp. 51–2. ANU Press, Canberra.

Cassedanne, J.P., Duthou, J.L. & Lasserre, M. (1972). Analise isotópica pelo método chumbo de uma terceira série de galenas brasileiras. *Mineração, Metalurgia, Rio de Janeiro*, **335**, 12–9.

Chaves, A.G., Heineck, C.A. & Taveres, de W. P. (1971). Projeto fosfato de Cedro do Abaeté. *Convênio Departmento Nacional Produção Mineral/Companhia de Perquisa de Recursos Minerais S.A., Belo Horizonte*, **2**, 72pp.

Cloud, P.E. & Dardenne, M.A. (1973). Proterozoic age of Bambuí Group in Brazil. *Geological Society of America, Bulletin*, **84(5)**, 1673–6.

Costa, da L.A.M. & Angeiras, A.C. (1971). Geosynclinal evolution of the epi-Baycalian platform of Central Brazil. *Geologische Rundschau*, **60(3)**, 1024–50.

Costa, da L.A.M., Angeiras, A.G., Valenca, J.G. & Stevenazzi, A. (1970). Novos conceitos sobre o Grupo Bambuí e sua divisão em tectonogrupos. *Instituto de Geociencias, Rio de Janeiro, Boletim*, **5**, 3–34.

Costa, F.G. & Oliveira, de F.A. 1970. O 'Verdete' rocha potássi ca na região de Cedro do Abaeté. *Publicaçiones, XXIV Congresso, Sociedade Brasileiro de Geologica, Brasilia*, **1**, 115–16.

CPRM (Companhia de Pesquisa de Recursos Minerais) (1976). Projeto Patos de Minas. *Relatorio Final de Pesquisa, Departmento Nacional Produção Mineral, Belo Horizonte*, **2**, 78pp.

Dardenne, M.A. (1978a). Síntese sobre a estratigrafia do Grupo Bambuí no Brazil Central. *Publicaçiones, XXX Congresso, Sociedade Brasileiro de Geologia, Recife*, **2**, 597–610.

Dardenne, M.A. (1978b). Zonação tectônica na borda ocidental do Craton São Francisco. *Publicaçiones, XXX Congresso, Sociedade Brasileiro de Geologia, Recife*, **1**, 299–308.

Dardenne, M.A. (1979). Les minéralisations de plomb, zinc, fluor du Proterozoique supérieur dans le Brésil central. Avec une mise au point sur la chronologie du Précambrien brésilien. Thèse Doctoral, Université de Paris VI, 251pp. (unpublished).

Dardenne, M.A. & Campos Neto, M.C. (1976). Geologia da região de Lagamar, MG. *Publicaçiones, XXIX Congresso, Sociedade Brasileira de Geologia, Ouro Preto*, p. 17.

Dardenne, M.A., Faria, A. & Andrade, G.F. (1976). Occurrence de stromatolites columnaires dans le Groupe Bambui (Goias, Brésil). *Anais, Academia Brasileira de Ciençias, Rio de Janeiro*, **48(3)**, 556–66.

Dardenne, M.A. & Walde, D.H.G. (1979). A estratigrafia dos Grupos Bambuí e Macaubas no Brasil Central. *Atas I Simposium Geologico Minas Gerais, Geol. Espinhaço, Diamantina*, 1, 43–54.

Derby, O.A. (1880). Contribuçãoes para o estudo da geologia do vale do São Francisco. *Archivos do Museo Naçional, Rio de Janeiro*, **4**, 89–119.

DNPM (Departmento Naçional Produção Mineral) (1975). Fosfatos Lagamar. Relatório pesquisa no 807795, 807753, 815072 e 815073, Departmento Naçional Produção Mineral, Brasilia (unpublished).

Eschwege, von W.L. (1932). *Beitrag zur Gliederung des Praekambriums in Minas Gerais e Nordeste*. Departmento Naçional da Produção Mineral Divisão de Geologie e Mineralogia Relatório anual do director, Rio de Janeiro, pp. 79–89.

Gonçalves de Lima, J.M. (1976). Perfil analítico dos fertilizantes fosfatados. *Departmento Naçional Produção Mineral, Boletim*, **39**, 55pp.

Guimarães, D. (1967). Ocorrência de fosforita no município de Abaete, Minas Gerais. *Departmento Naçional da Produção Mineral, Ministerio da Minas e Energia, Notas preliminares e estudos*, **144**, 18pp.

Guimarães, D. & Dutra, C.V. (1969). Contribuição ao estudo da Série Bambuí. *Departmento Naçional Produção Mineral, Ministerio da Minas e Energia, Boletim*, **243**, 27pp.

Hasui, Y. & Almeida, de F.F.M. (1970). Geocronologia do Centro Oeste Brasileiro. *Boletim, Sociedade Brasileira de Geologica, São Paulo*, **19**, 5–26.

Hettich, M. (1977). A glaciação proterozóica no centro-norte de Minas Gerais. *Revista Brasileira de Geociencias, São Paulo*, **7(2)**, 87–101.

Isotta, C.A.L., Rocha-Campos, A.C. & Yoshida, R. (1969). Striated pavement of the Upper Precambrian glaciation in Brazil. *Nature*, **222(5192)**, 466–8.

Ladeira, E.A. & Brito, O.E.A. (1968). Contribuição à geologia do Planalto da Mata da Corda. *Publicaçiones, XXII Congresso, Sociedade Brasileiro de Geologia, Belo Horizonte*, pp. 181–99.

Ladeira, E.A. & Leal, E.D. (1972). Phosphate rock of the Cedro de Abaeté Region, Minas Gerais State, Brazil. *24th Session International Geological Congress, Montreal, Section 4, Mineral Deposits*, pp. 435–444.

Macedo, M.H.F. (1982). Les systèmes isotopiques rubidium-strontium et potassium-argon dans les argiles extraites de sédiments carbonatés. Application à la datation du Protérozoique sédimentaire du Brésil les états de Bahia et Santa Catarina. These Doctoral de l'ingénieur Université Louis Pasteur, Strasbourg, France, 119pp. (unpublished).

Macedo, M.H.F. & Bonhomme, M. (1981). Datações Rb/Sr e K/Ar das Formaçōes Bebedouro e Cabloco na Chapada Diamantina. In *Simp. S.B.G., Núcleo Bahia, Salvador: O Crato:: do São Francisco e suas faixas marginais*, pp. 98–9.

Marchese, H.G. (1974). Estromatólitos 'Gymnosolenidos' en el la-do oriental de Minas Gerais, Brazil. *Revista Brasileira de Geociencias, Sao Paulo* **4(4)**, 257–71.

Metago (1977). Região de Cabeceiras: Geologia e Propecção Geoquimíca de semi-detalhe. Unpublished Report, Goiânia, Brazil, 43pp.

Metago (1979). Região da Fazenda Convanca: Geologia e Prospecção geoquímica de semi-detalhe. Unpublished Report, Goiânia, Brazil, 71pp.

Moeri, E. (1972). On a columnar stromatolite in the Precambrian Bambuí Group of Central Brazil. *Eclogae Geologicae Helvetiae*, **65(1)**, 185–95.

Oliveira, de M.A. (1967). Contribuição à geologia da parte sul da bacia do São Francisco e areas adjacentes. *Petróleo Brasiliero, S.A., Rio de Janiero, Coletaneas Relatórios Exploração*, **1**, 71–05.

Parenti Couto, J.G. & Bez, L. (1981). A glaçiação Jequitai: um guia estratigrafico para o Pre-Cambriano superior no Brasil. *Revista Brasileira de Geociencias, Sao Paulo*, **11(1)**, 17–21.

Parenti Couto, J.G., Cordani, U.G., Kawashita, K., Iyer, S.S. & Moraes, N.M.P. (1981). Considerações sobre a idade do Grupo Bambuí com base em análises isotópicas de Sr e Pb. *Revista*

Brasileira de Geociencias, Sao Paulo, **11(1)**, 5–16.

Rimann, E. (1917). A Kimberlite in Brasil. *Anuário da Escola Minas, Ouro Preto*, **15**, 27–32.

Rocha-Campos, A.C. & Hasui, Y. (1981a). Proterozoic diamictites of Western Minas Gerais and Eastern Goiás, Central Brazil. In *Earth's Pre-Pleistocene Glacial Record*, ed. M.J. Hambrey & W.B. Harland, pp. 920–3. Cambridge University Press, Cambridge.

Rocha-Campos, A.C. & Hasui, Y. (1981b). Tillites of the Macaubas Group (Proterozoic) in Central Minas Gerais and Southern Bahia, Brazil. In *Earth's Pre-Pleistocene Glacial Record*, ed. M.J. Hambrey & W.B. Harland, pp. 933–8. Cambridge University Press, Cambridge.

Schöll, W.U. (1972). Der sudwestliche Randbereich der Espinhaço zone, Minas Gerais, Brasília. *Geologische Rundschau*, **61(1)**, 201–16.

Schöll, W.U. (1973). Sedimentologie der Bambuí Gruppe in Seteil des São Francisco Beckens (Minas Gerais, Brasília). *Münstersche Forschungen zur Geologie und Paläontologie*, **31/32**, 71–91.

Trompette, R. (1982). Brasil e a margem atlantica da África no Proterozóico superior–importância do fraturamento espinhaço (kibariano?). *Publicaçiones, XXXI Congresso, Sociedade Brasileiro de Geologia, Salvador*, **1**, 386–94.

Trompette R., Affaton, P., Joulia, F. & Marchand, J. (1980). Stratigraphic and structural controls of late Precambrian phosphate deposits of northern Volta Basin in Upper Volta Niger & Benin, West Africa. *Economic Geology*, **75**, 62–70.

Walde, D.H. (1978). Desenvolvimento faciológico do Precambriano entre a Serra Mineira e a Serra do Cabral (Regiâo Sudoeste da Serra do Espinhaço, MG.), *Publicaçiones, XXX Congresso, Sociedade Brasileiro de Geologia, Recife*, **2**, 711–24.

11

Proterozoic and Cambrian phosphorites – deposits: Lady Annie, Queensland, Australia

P.J. COOK AND S.A. ELGUETA

Introduction

The Lady Annie deposit, together with its northern extension Lady Jane, occurs in an elongate north-trending Cambrian outlier covering an area of approximately 400 km^2 within Proterozoic rocks. It is located 130 km north-west of Mount Isa, on the eastern margin of the Georgina Basin (Fig. 11.1). The deposit was discovered by Mines Exploration Proprietary Limited (MEPL) in 1967. Since that time it has been extensively trenched and drilled and 486 million tonnes of phosphate rock averaging 17% P_2O_5 have been delineated (Rogers & Keevers, 1976). The most detailed study undertaken to date on the Lady Annie deposit is by Elgueta (1981) who studied the western zone of the deposit and this chapter is based on that work. Other previous publications on the Lady Annie deposit include those of de Keyser (1969b) who carried out a regional study of Cambrian formations on the north-eastern margin of the Basin, and Trueman, Robinson & Hankel (1969) who studied the mineralogy and mineral chemistry of the main phosphorite types. Rogers & Keevers (1976) undertook the most comprehensive study of the deposit. They described the geology of the deposit as a whole, correlated formations in the outlier and provided the first sub-surface information. Papers relating to other deposits in the Georgina Basin include those on Duchess (de Keyser, 1968; Russell & Trueman, 1971; Fleming, 1974, 1977; Rogers & Crase 1980), D Tree (Howard & Cooney, 1976; Howard & Hough, 1979) and Wonarah (Howard & Perrino, 1976). Papers relating to the Georgina Basin phosphorite province as a whole include those by Russell (1967, 1968); de Keyser (1969a, 1973); Thomson & Russell (1971); Howard (1971, 1972); Cook & Armstrong (1972); de Keyser & Cook (1972); Cook (1972, 1976a, b); Thomson & Rogers (1974); and Cook & Shergold (1979, 1980).

In this paper we will briefly examine the stratigraphic setting of the western Lady Annie deposit, then consider the phosphorites in some detail. Finally we will discuss the depositional environments of the phosphorites and attempt to develop a depositional model.

Stratigraphy of the Lady Annie outlier

The Lady Annie phosphorites, like the other major phosphorite deposits of the Basin, occur in the Beetle Creek Formation, of probable Templetonian (Middle Cambrian) age, so that they are younger than most of the other Cambrian phosphorites considered in this volume.

The description of the Lady Annie–Lady Jane phosphate deposit by Rogers & Keevers (1976) provides the general stratigraphic framework. Figure 11.1 shows the interpretative Cambrian geology. In the south and east of the outlier the contact of Cambrian with Proterozoic rocks is unconformable. On the western side of the outlier the contact is faulted along the 'Western Fault'. West of the fault, an area of basement separates the Lady Annie and D Tree deposits. A northerly-trending line of basement highs known as 'Truemans Bank' divides the outlier into western and eastern zones (Fig. 11.2). The outlier contains up to 55 m of Cambrian sediments which are unconformably overlain by up to 71 m of Mesozoic sediments. The Middle Cambrian sedimentary sequence (Fig. 11.3), commences with a thin (3 m) discontinuous unfossiliferous basal conglomerate and sandstone unit, the 'Mount Hendry Formation', which was deposited over the eroded Precambrian basement. The Mount Hendry Formation is conformably and gradationally overlain by a carbonate facies, the Thorntonia Limestone, which like the underlying 'Mount Hendry Formation' is best exposed along the eastern margin, and in the southern part of the Lady Annie area (Fig. 11.1). It is variable both in distribution and thickness, having a maximum thickness of 16 m. The Thorntonia Limestone (perhaps more correctly referred to as a dolostone) is sandy and dolomitic at its base, but its dominant lithology is dolomite and dolomitic limestone with chert bands becoming increasingly common towards the top. In general, fossils are rare but occasional beds contain abundant fossils including species of the trilobite *Redlichia*, brachiopods, gastropods and hyoliths. The trilobites indicate an Ordian (early Middle Cambrian) age for this carbonate unit and by inference the 'Mount Hendry Formation' has a similar age. The increasing component of chert in the uppermost part of the carbonate facies has been used to define an upper 'Chert Member' of the Thorntonia Limestone in other parts of the Georgina Basin (de Keyser & Cook, 1972). More recently, Henderson & Southgate (1978) have described chert beds with

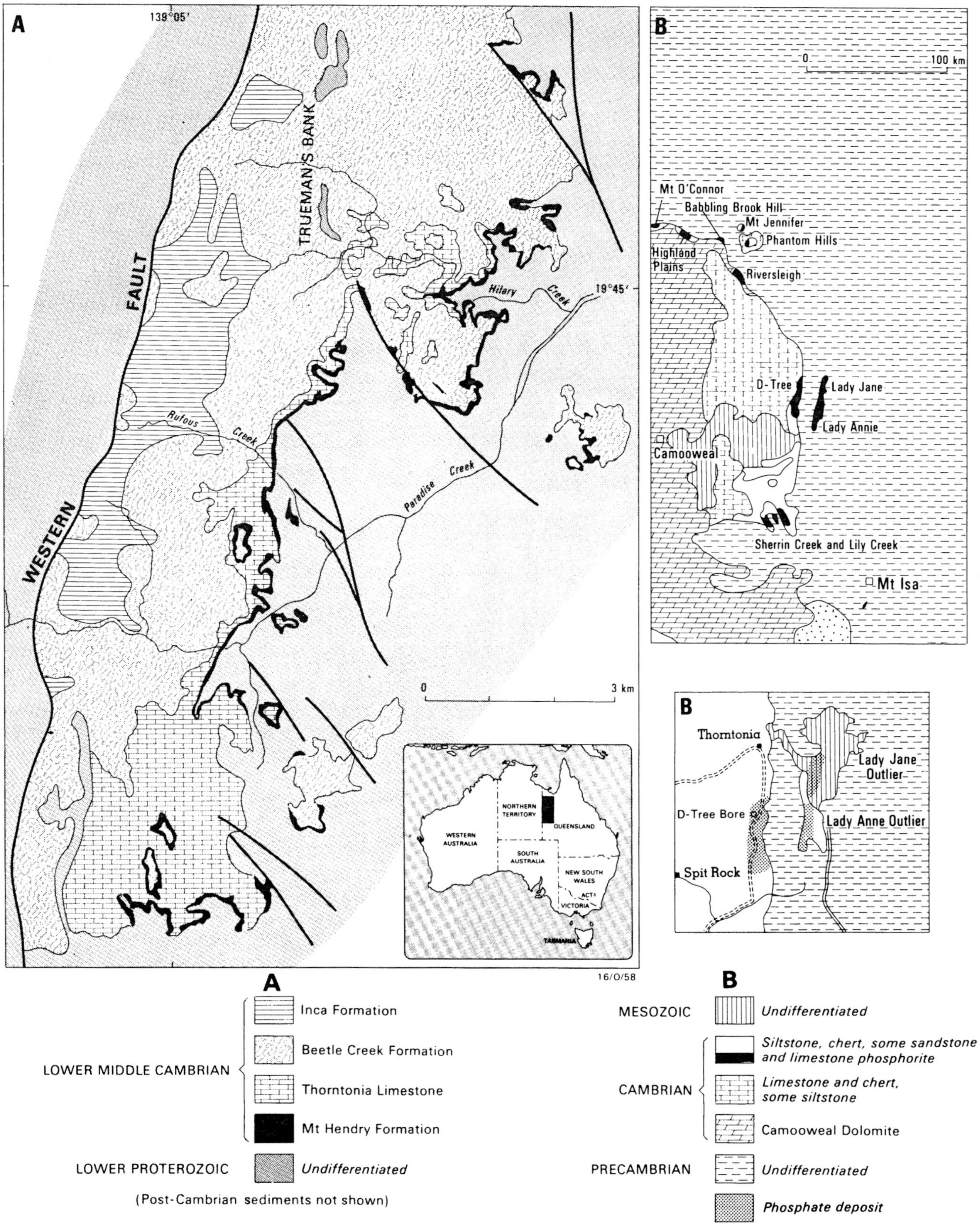

Fig. 11.1. Location and generalised geology of the Lady Annie phosphate deposit. (After Rogers & Keevers, 1976 and Elgueta, 1981.)

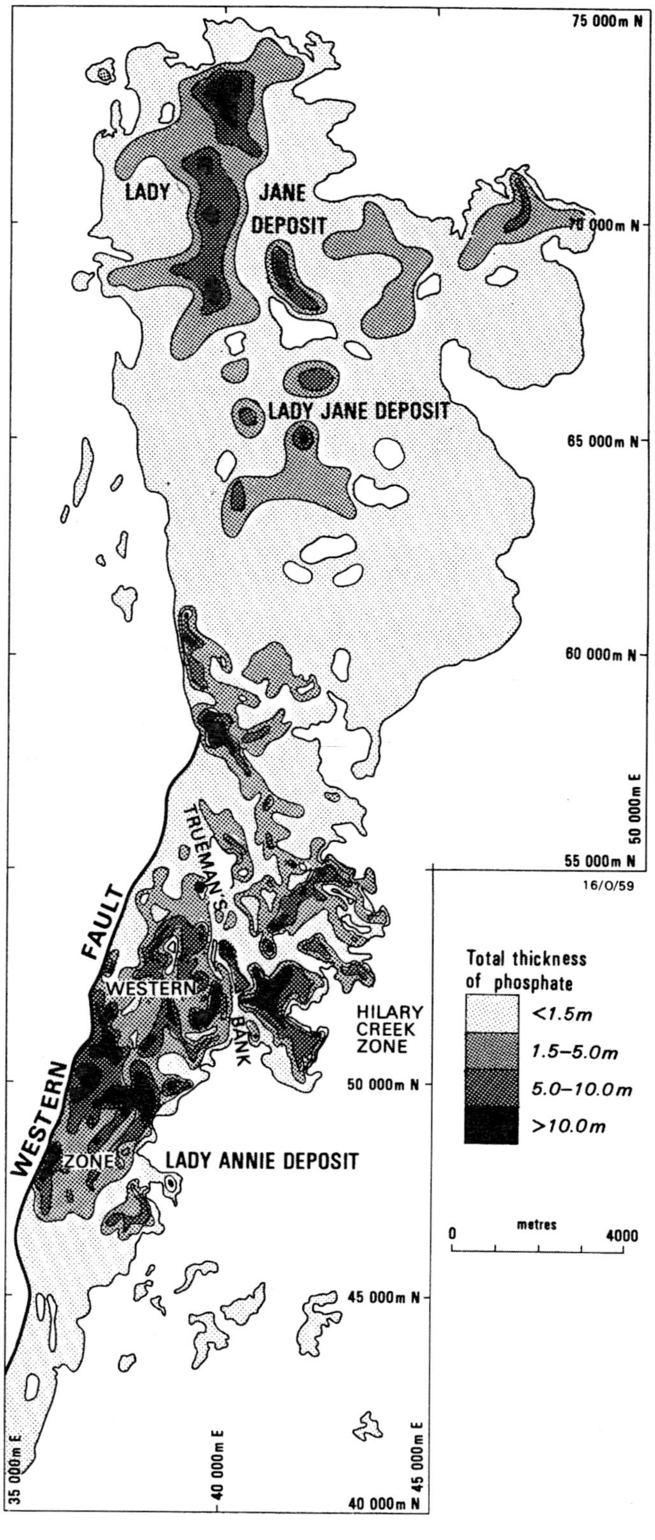

Fig. 11.2. Map showing the total composite thickness (in metres) of phosphorite in the Lady Annie and Lady Jane deposits. (After Rogers & Keevers, 1976.)

evidence of hypersaline conditions (stromatolitic textures and pseudomorphs after evaporites) overlying carbonates of the Thorntonia Limestone. However, they regard this chert as the basal member of the Beetle Creek Formation. The Beetle Creek Formation, in which the phosphate deposits occur, overlies the Thorntonia Limestone in most parts of the outlier. It consists of interbedded coarse- to fine-grained phosphorites, siltstone, claystone and chert. According to Rogers & Keevers (1976) a maximum of five phosphorite beds separated by phosphatic siltstone have been recognised from drill holes close to the Western Fault where the Beetle Creek Formation reaches its maximum thickness of 47 m with more than 10 m of phosphorites in this same area (Fig. 11.2). The Beetle Creek Formation intertongues with limestones and dolomites in places. Because of this relationship Southgate (1983) has proposed that the phosphorites and associated sediments of the Undilla Basin west of Lady Annie should all be included within a single unit. However in the Lady Annie area the two formations are retained, with some thin carbonates being recognised as lenses within the Beetle Creek Formation. The Beetle Creek Formation is conformably overlain by the Inca Formation which reaches a maximum thickness of 24 m (Rogers & Keevers, 1976). It is composed of laminated siltstones and shales (pale grey when weathered; black if unweathered) with frequent intercalations of fine-banded and spicular chert. It has a fauna of agnostid trilobites and siliceous tetraxon sponge spicules which indicate a Templetonian age for the formation. There are no younger Cambrian sediments overlying the Inca Formation in this area but in the main part of the Georgina Basin, to the west of Lady Annie, the Inca Formation is overlain by younger Middle Cambrian limestones (Southgate, this volume).

Facies distribution

Whilst the formalised stratigraphy provides a guide to the range of lithologies present at Lady Annie, the Middle Cambrian interval in the area comprises a complex interplay of facies and is best discussed from that viewpoint. The facies distribution is closely linked to the distribution of phosphate deposits. Consequently, each of the related facies will be discussed briefly, relating it where possible to the formalised stratigraphic names discussed in the previous section. In many cases the facies distribution is linked to the palaeo-shoreline or the location of a basement high. This is apparent from Figure 11.4 which demonstrates the facies distribution on the east side of the Lady Annie outlier. De Keyser & Cook (1972) recognised six diachronous lithosomes (intertonguing sedimentary units of essentially uniform lithological character) and used these as a basis to discuss the Cambrian stratigraphy of the Georgina Basin and clarify the interrelationships of the various rock-units proposed in earlier schemes. The descriptions of the facies at Lady Annie are based to some extent on the system used by de Keyser and Cook (1972) but modifications to their scheme were necessary because of the more detailed facies information available for Lady Annie through the work of Elgueta (1981). The basal *conglomerate–sandstone facies* which is given the informal name 'Mount Hendry Formation' (Fig. 11.3) represents the oldest Cambrian sediments in the study area. It unconformably overlies eroded basement rocks along the eastern margin of the outlier and around inliers of Precambrian basement. The facies has a highly irregular and patchy distribution and varies in thickness from zero to 3 metres. It is characterised by vertical and lateral gradational transitional relationships with the undifferentiated carbonate facies with

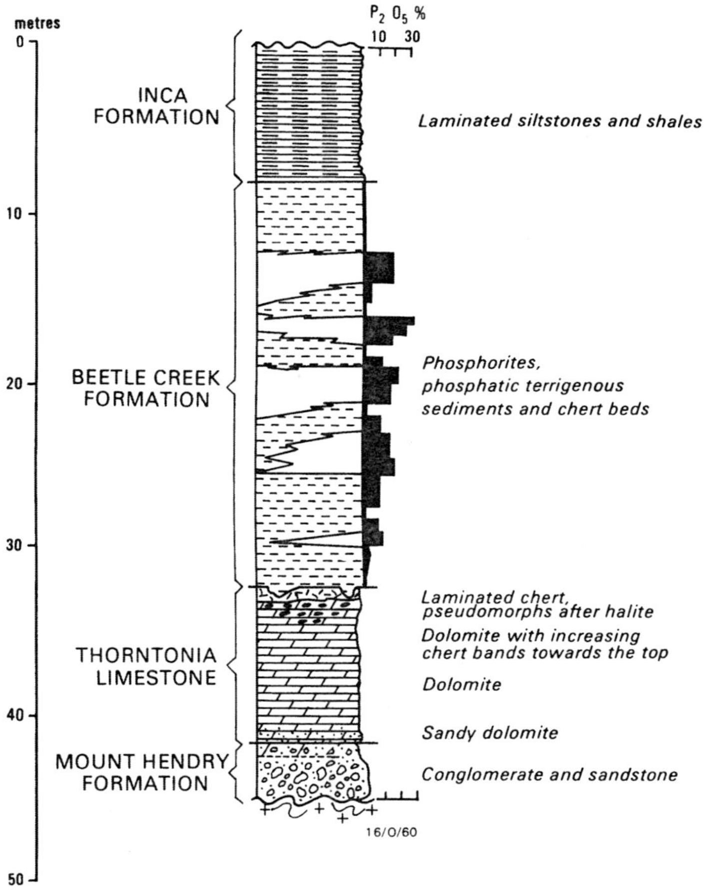

Fig. 11.3. Schematic Middle Cambrian sequence at Lady Annie using, for the most part, formalised stratigraphic names.

which it is associated. The facies typically consists of a poorly sorted polymict sedimentary breccia of angular to sub-angular sedimentary and metasedimentary clasts in a sandy dolomitic to dolomitic matrix. Some beds show imbrication structures. In places the sequence grades upwards into the carbonate facies with progressively diminishing clast size and increasing dolomite.

The *carbonate facies* which appears to be widespread, crops out adjacent to basin margins and basement highs, and has also been reported from many drill holes in the area underlying the phosphatic sequence. This facies mainly comprises the Thorntonia Limestone but in addition there are discrete carbonate lenses within the Beetle Creek Formation which do not constitute continuous mappable units and are therefore mapped as part of the Beetle Creek Formation. The base of the facies is either transitional with the underlying clastic facies or lies directly on basement. In many areas its top is taken as the first occurrence of a laminate vuggy chert although thin lenticular dolostone beds of similar composition to the carbonate facies occur interbedded within the laminate facies. The undifferentiated carbonate facies consists of a sequence of thickly bedded, massive, yellowish-grey dolomites and dolomitic limestones with only very minor phosphate (maximum P_2O_5 content of 0.5%) interbedded with thin beds of laminated dolomite. Irregular chert nodules occur in the upper part. A few beds contain fossil fragments including disarticulated *Redlichia*, hyoliths, gastropods, brachiopods and rare sponge spicules. Probable mudcracks indicate subaerial exposure at times. Medium- to large-scale planar cross-stratification and thin beds of abundant current-oriented trilobite and hyolith fragments suggest current action.

The *laminate facies* locally caps the main carbonate facies. It consists of a thin and discontinuous layer of distinctive unfossiliferous laminated chert which rarely exceeds 1 m in thickness and is restricted to basement margins and highs. This facies consists of several distinctive types of laminated vuggy chert. The most common type is a planar-bedded, poorly laminated vuggy chert in which vugs have a random distribution. A second type consists of highly contorted and wavy, very well laminated fenestral vuggy chert. Halite pseudomorphs are present in places. The distinctive contorted bedding present in the chert may be due to slumping and/or differential compaction of dolomite and chert, but the possibility of algal origin cannot be discounted.

The *oncolite facies* occurs marginal to basement and is in places up to 5 m thick. The oncolite facies is laterally equivalent to part of the phosphorite section (Fig. 11.4) and appears to be overlain by siliceous siltstones and chert beds. Oncolites have previously been described from the Thorntonia Limestone (Smith, 1972) and from the 'Chert Member' (de Keyser & Cook, 1972). The oncolite facies consists principally of poorly bedded brecciated oncolitic chert in which well-preserved spherical to

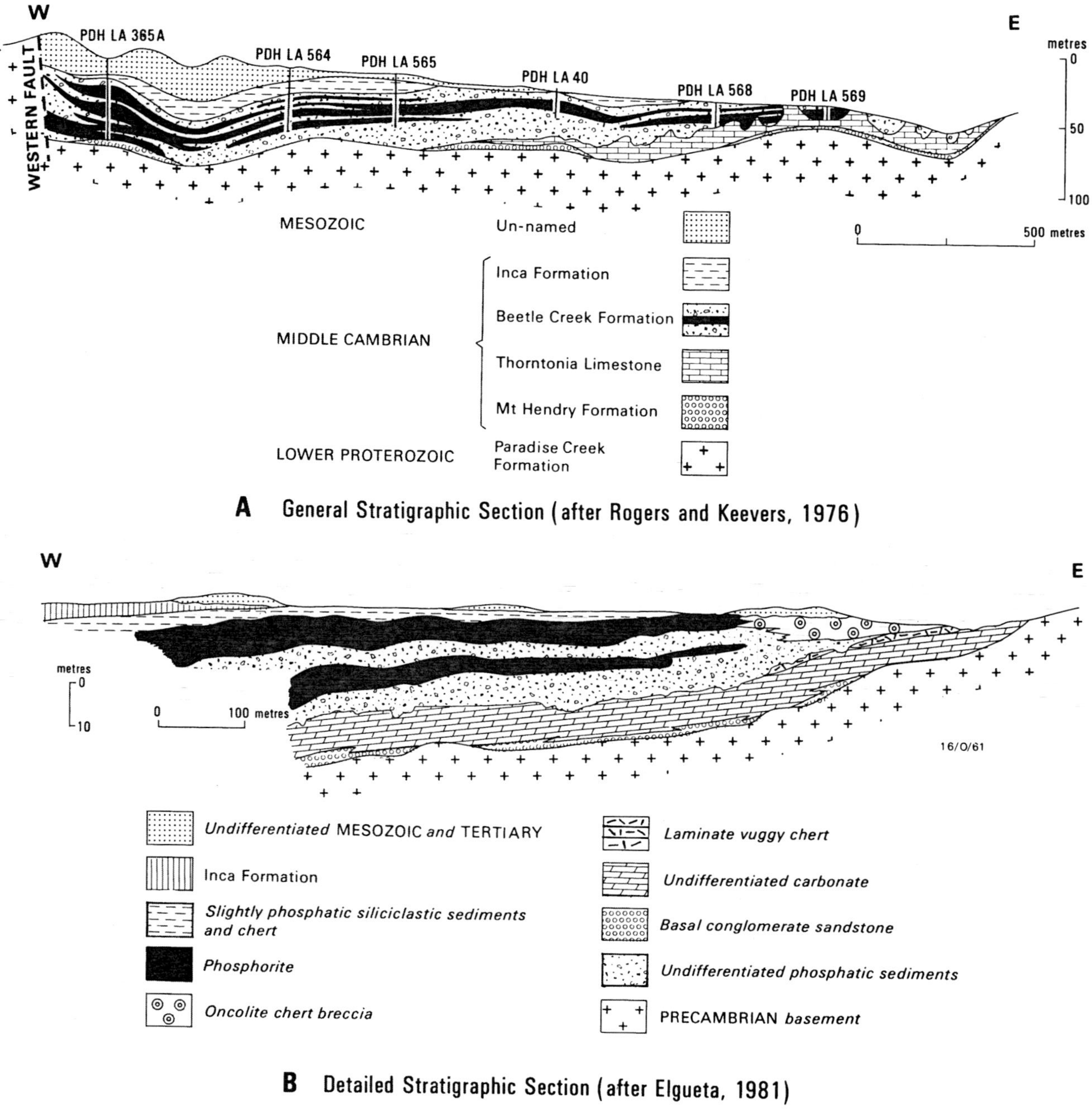

Fig. 11.4. Cross sections through parts of the western Lady Annie deposit. (A) Generalised E–W section across the deposit (after Rogers and Keevers, 1976). (B) Detailed E–W section across a part of the deposit (after Elgueta, 1981).

ellipsoidal oncolites, varying in size between 0.5 cm and 4 cm, generally form a self-supporting fabric. The chert is almost certainly replacing carbonate. Oncolites have a concentrically well-laminated internal structure which is frequently built around a central angular fragment of the gastropod *Helcionella* which also occurs rarely in the matrix and was the only fossil observed in this facies. Some of the concentric laminae are phosphatic and thin interbeds of coarse phosphatic allochems are locally present.

The laminate facies and the chert rubble surface of the carbonate facies are conformably overlain in places by the *skeletal facies* which consists mainly of silicified coquinas. This facies is distributed throughout the areas marginal to the basement, where it has a maximum observed thickness of about 4 m. To the west, this facies becomes finer-grained and interfingers with phosphorites and phosphatic sediments, although the contact between these units is difficult to establish since coquinitic chert and coquinas with phosphate grains are common in places within the phosphorites. It is also likely that the skeletal facies has a gradational relationship with the oncolite facies since both occur at the same stratigraphic level (Fig. 11.4) and the few fossils present in the oncolite facies are also found in the silicified coquina. The coquina is composed of a massive, very coarse-grained, poorly sorted, self-supporting framework of randomly-oriented angular to slightly abraded fossil fragments which have been extensively silicified. The gastropod *Helcionella*, the trilobite *Xystridura* (which is also found in the phosphorite facies), hyoliths, and brachiopods are commonly well preserved. However in most cases the original skeletal debris has been replaced by silica.

The *phosphorite facies*, which is included within the Beetle Creek Formation, crops out in a broad north–south-trending belt in the western zone of the Lady Annie outlier, extending to the south and west under chert and fine terrigenous Middle Cambrian sediments and a thick sequence of Mesozoic and Tertiary sediments (Figs 11.1 & 11.4). The phosphorite facies consists of several distinct lenticular high-grade phosphorite horizons interbedded with undifferentiated phosphatic sediment which has variable but low P_2O_5 content. The maximum thickness of this sequence observed in drill holes is 46.5 m in the area close to the Western Fault (Fig.11.2) where five phosphorite horizons were recognised by Rogers & Keevers (1976).

The lower part of the sequence, which is known only from drillhole information, overlies dolostones of the carbonate facies with a highly irregular contact. The sequence thins rapidly to the east and grades laterally into both the oncolite facies and the skeletal facies. Thin interbeds of grainstone phosphorite have been observed within the oncolite facies. The skeletal facies and the phosphorite facies show a completely transitional relationship in places, with the phosphorite facies becoming progressively more silicified and diluted with angular fossil fragments. Thin laminae of grainstone phosphorite are also fairly common in the skeletal facies for some distance away from the main area of the phosphorite facies.

The facies is characterised by a relatively monotonous succession of well-bedded to laminate grainstone and packstone phosphorite with minor mudstone phosphorite. The phosphorites are interbedded with chert and very minor lenticular beds of fine siliciclastic sediments. Because of deep post-Cambrian weathering, no unoxidised carbonaceous phosphorite crops out in the Lady Annie Outlier. The phosphorites are considered in detail later.

The *chert–mudstone facies* overlying the phosphorite facies is a widespread lithofacies which is also included within the Beetle Creek Formation. No interfingering or gradational relationships of this unit with the other facies is observed, but it does contain coquinitic beds similar to those of the skeletal facies, as well as thin beds of grainstone phosphorite. The chert–mudstone facies consists of a variable association of lithologies, including thin- to medium-bedded non-phosphatic quartzose siltstone interbedded with grey chert beds and nodules; siliceous, slightly phosphatic, very fine sandstones are also well represented. Rare, thin beds of silicified coquina containing a few well-preserved fossils occur in places but the facies is predominantly non-fossilliferous.

The *siltstone facies*, which comprises the Inca Formation, is poorly exposed, but reaches a maximum thickness of 24 m near the Western Fault (Fig. 11.1). The basal contact is indicated by the appearance of well-laminated siltstones with a faunal assemblage markedly different from the underlying units. The youngest Cambrian lithofacies in the area, it is unconformably overlain by Mesozoic and Tertiary sediments. The facies comprises well-bedded to laminate, fissile, white (when weathered; black if unweathered) siltstone and shale with interbedded thinly banded spicular chert and minor white friable siltstone. The chert beds are thin (generally less than 3 cm) but continuous, and are composed of alternating dark and light laminae. Although this facies is lithologically similar to the underlying facies, it can be recognised by its characteristically banded chert and faunal assemblage. The fauna characteristic of the underlying phosphatic sequence (*Xystridura*, brachiopods and hyoliths) is replaced by an assemblage consisting of very well preserved *Ptychagnostus gibbus* and large siliceous tetraxon sponge spicules. Small, disseminated (1–2 mm), cubic pseudomorphs after pyrite are common. Sedimentary structures are restricted to well-developed planar bedding and laminations.

Chert breccias (phosphatic in places) are widespread throughout the area but most are not considered to be a primary sedimentary facies. Phosphatic chert breccias consist of abundant angular chert and minor siltstone and phosphorite fragments set in a matrix of phosphate or phosphatic siltstone. They generally have a clast-supported matrix in which the clasts have a random arrangement. Chert breccias have a similar texture but a highly silicified matrix. Intraformational breccias, collapse breccias, tectonic breccias, and breccias associated with weathering have been recognised. Due to their similar appearance and the absence of clear field relationships, it is often difficult to distinguish between them.

Phosphorites

The phosphorites of the western zone at Lady Annie are characterised by a more or less monotonous succession of thin-bedded to laminate sediments. Major breaks in sedimentation are not apparent, but minor breaks may be represented by thin but persistent, non-phosphatic mudstone laminae. Beds pinch-out laterally and vary compositionally, making detailed correlations difficult. The most persistent units within the phosphorite section are the chert beds and non-phosphatic mudstone beds and laminae.

The phosphorite sequence is dominated by coarse- to medium-grained skeletal-ovulitic grainstone and packstone phosphorites. These tend to be coarser and better developed in the lower part of the sequence (Fig. 11.5). Finer-grained phosphorites (packstones, wackestones and mudstones) tend to occur in the upper portion of the sequence where they are associated with an increasing content of terrigenous material. Therefore there is an overall although irregular fining-upward character. Up to three subordinate fining-upward sequences were also observed within the phosphorite sequence itself (Fig. 11.5). The interbedded cherts are characteristically well bedded, conquinitic or granular depending on the dominant allochem. More or less persistent zones, up to 1 m thick, of intercalated chert bands and coarse phosphorites form useful marker units. Bedded siliciclastic sediments within the phosphorites make up a small proportion of the total sediments. They consist of lenticular beds of fine sandy siltstone and relatively continuous thin beds or laminae of non-phosphatic mudstone with sharp upper and lower contacts.

In the northern part of the western deposit the phosphorites appear to have more complex lateral variations than those in the south. Units tend to be laterally less continuous and more lenticular, and there are no marker beds. Fine-grained phos-

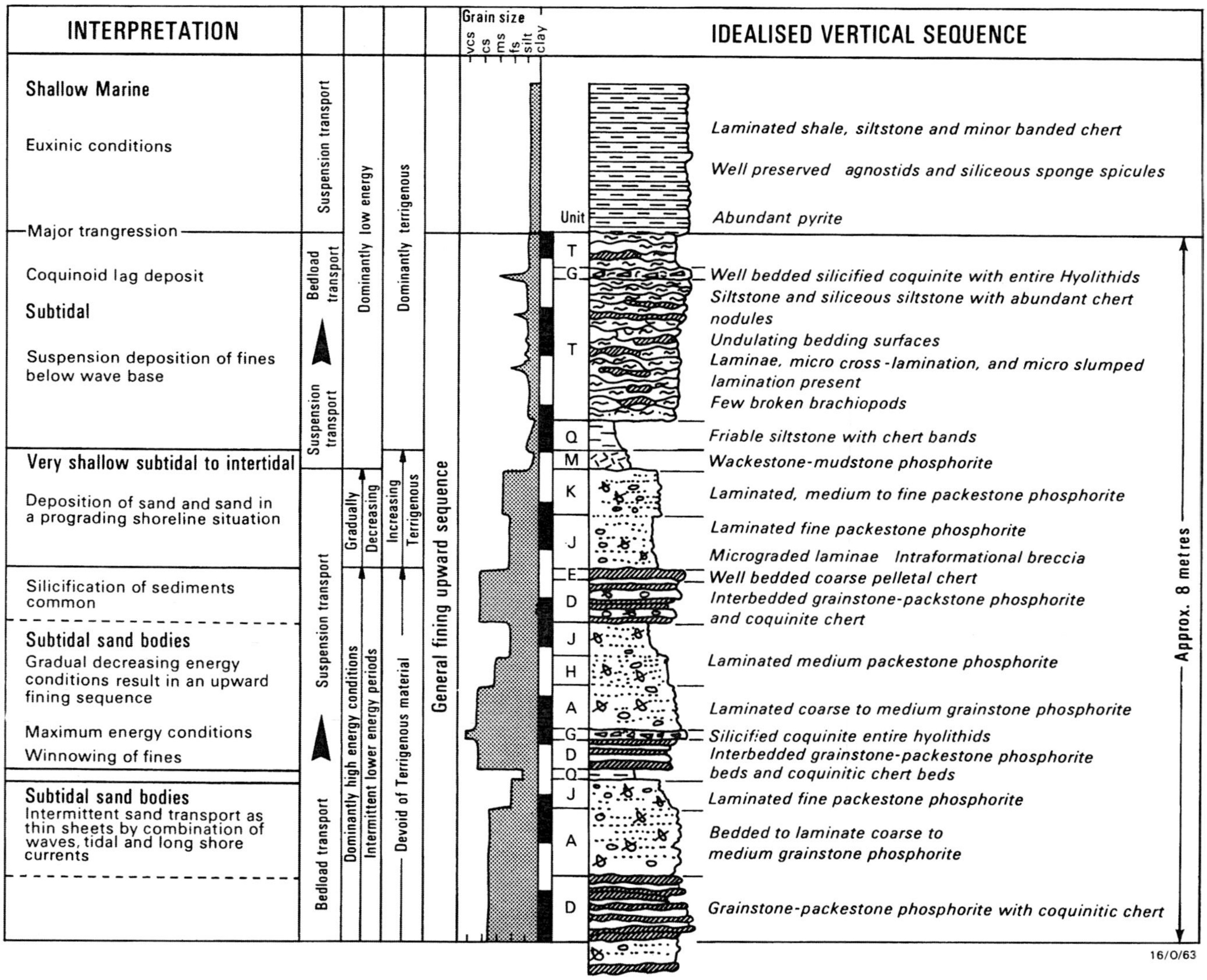

Fig. 11.5. Representative vertical section in the Rufous Creek trench in the southern part of the Lady Annie deposit. Reference for symbols is given in Fig. 11.8.

phorites (fine packstones, wackstones and mudstones) are more abundant in the north, though overall the grain types, sedimentary structures and thickness are relatively uniform throughout the deposit. There is also a much higher proportion of fine terrigenous material, and cherts are less common. Sedimentary structures observed within the phosphorites include planar wavy and lenticular bedding, well-defined laminations, medium-scale low-angle cross-bedding, soft sediment folding, intraformational brecciation, and trough structures. Bioturbation is characteristically absent from phosphorites and the associated cherts. Brecciation is a very common feature of many of the phosphorites. Some of the breccias are clearly synsedimentary flat pebble-type breccias. However, many appear to be collapse breccias, some are fault breccias, and many are probably a weathering phenomenon. Some of the trough-like features within the phosphorites are also somewhat problematical. Some of these can reasonably be interpreted as tidal or subaerial erosional channels cut into the underlying phosphorite and analogous to the erosion channels described from the D Tree deposit by Howard & Cooney (1976). However some of them may be a weathering phenomenon associated with the collapse of underlying units due to leaching.

A typical vertical section through one of the phosphatic sequences from the Rufous Creek section in the south-eastern part of the Lady Annie deposit is shown in Fig. 11.5. This section shows the high phosphate content of the sequence. The base of the phosphatic part of the sequence is not exposed but has been shown by drilling to extend for several metres below the base of the section in Fig. 11.5. A sharp cut-off in the phosphate content is apparent at the top. Fining-upward sequences are well defined and the relative abundance of bioclasts and phosphate grains clearly mirrors this.

Grain types

In the Lady Annie deposits, all varieties of phosphorite are present, ranging from coarse grainstone through coarse to medium packstone to very fine packstone, coarse to fine wackestone (depending on allochem size) and mudstone phosphorite. The grainstone and packstone phosphorites have a self-supporting framework of sand-sized phosphate grains which are characteristically moderately to well sorted and rounded (Fig. 11.5). Coarser varieties are frequently very well sorted and well rounded except where they contain angular and platy bioclast lenses. Grainstones are invariably coarse-or medium-grained and packstones tend to be fine grained. Wackestone and mudstone phosphorites contain minor coarse- to fine-grained allochems and rounding and sorting is less well developed than in grain-supported varieties; fine sand- to silt-sized detrital terrigenous grains occur in variable proportions. Five types of phosphate grains recognised by Mabie & Hess (1964) are present at Lady Annie–skeletal fragments (bioclasts), ovules, rounded intraclasts, nucleated and polynucleated pellets, and ooids. Irregular platy phosphate particles occur in thin laminae within the normal grainstone and packstone phosphorites. The proportion of the various grain types varies throughout the vertical sequence (Fig. 11.5), but in most grainstone and packstone phosphorites, rounded skeletal and ovular grains together make up over 90% of all grains.

Skeletal fragments (bioclasts) are typically round to subround, sand-sized phosphate grains which can be identified on the basis of their external form or internal structure as fossil fragments. Rounded bioclasts are the dominant grain type in most of the coarser phosphorites, often forming 60–80% of total grains. Angular phosphate bioclasts are rare in the coarser phosphorites but occur as scattered grains in wackestone and mudstone phosphorites or are concentrated in lenses.

It is impossible to assign most skeletal fragments to a particular fossil type because they are too abraded. Where it is possible, the dominant biological components are phosphatised (originally carbonate) rays of *Chancelloria* and related species, and echinoderm debris. Trilobite fragments do not appear to make up a significant proportion of grains overall, though locally they may be abundant. Hyoliths, on the other hand, are widespread and mostly are not fragmented. Other groups, including molluscs and brachiopods, are also present. This faunal assemblage is similar to that recognised by Fleming (1977) in the Duchess area. The assemblage suggests a carbonate precursor for a high proportion of the allochems. Ovules, which comprise the major component of most phosphorites, are collophane grains with ellipsoidal to spherical forms showing no identifiable internal structure (Fig. 11.6). They are generally the same size as the associated skeletal fragments, and appear to increase in proportion relative to bioclasts in finer-grained phosphorites but this may be merely a measure of increasing difficulty in identifying the nature of grains as they become smaller.

Rounded intraclastic grains of very fine grained phosphatised bioclasts and ovules, and siliceous sponge spicules set in a fine, structureless collophane matrix occur in places. They are most common in coarse grainstone and packstone phosphorites where they can make up to 10% of the total phosphatic material.

Nucleated and polynucleated pellets consist of a collophane rim, structureless or with weak concentric bands formed around a nucleus of one or more silt-sized quartz grains, skeletal fragments or, less commonly, an unidentified opaque mineral. This type of grain is extremely rare. Phosphate ooids, consisting of spherical grains with well-developed concentric bands are rare in many phosphate deposits and Lady Annie is no exception: Elgueta (1981) reports that only one ooid was observed in thin-sections of Lady Annie phosphorites, despite the fact that more than 100 such sections were examined.

Overall the faunal assemblage and the phosphatic allochems of the Lady Annie phosphorites are very similar to those described from the other Georgina Basin phosphate deposits (de Keyser & Cook, 1972; Cook, 1972), Duchess (Russell & Trueman, 1971; Fleming, 1977) and D Tree (Howard & Cooney, 1976; Howard & Hough, 1979). However, the proportion of bioclasts is higher at Lady Annie than in most other deposits. No grains of definite faecal origin (coprolites) such as those described by Fleming (1977) from the Duchess deposit, have been recognised at Lady Annie.

Apart from its occurrence as allochems, fine-grained phosphate is present as the principal component of mudstone and wackestone phosphorites, and also in the matrix of packstone phosphorites and phosphatic mudstones, siltstones and sandstones. Both mudstone and wackestone phosphorites have a matrix-supported fabric, mudstones having less than 10% of sand-sized grains and wackestones more than 10%. They form friable lenticular beds, usually with diffuse laminations and occasionally with bedding-parallel bands or irregular chert nodules containing included coquinitic material. In thin-section these phosphorites are seen to be composed of scattered medium to coarse phosphate grains set in a matrix of microcrystalline phosphate, with varying proportions of silt-sized detrital quartz, micas and clay minerals, which may be scattered or form discrete laminae. The phosphate grains are poorly sorted, with platy angular skeletal fragments dominant (Fig. 11.6). No other sedimentary micro-structures were observed, and no evidence was found for an algal origin as suggested by Fleming (1977), though Southgate (this volume) has clearly demonstrated the role of algae in places in the Georgina Basin. The quantity of detrital material varies considerably, both within single beds and between beds. Fine primary layering within these phosphorites is observed in thin-section to be the product of several different phenomena. Diffuse lamination is most commonly caused by the alternation of slightly coarser and finer sediments and by the bedding-parallel alignment of elongate grains, particularly bioclasts; less commonly, it is the result of micro-graded laminae in which fine-grained packstone grades upwards into phosphatic mudstone.

Chemical composition of the phosphorites

At the present time only limited geochemical data are available for the Lady Annie phosphorites apart from P_2O_5 content which ranges as high as 40% P_2O_5. Elgueta (1981, Fig. 12) was able to show a fairly well defined bimodal distribution in the phosphate content of the sediments of Lady Annie with modes at about 30% P_2O_5 and less than 2% P_2O_5. There are secondary modes at 36% P_2O_5, 18% P_2O_5 and 6% P_2O_5 with a well-defined minimum at around 15%. For this reason Elgueta (1981) proposes that 15% P_2O_5 be used as 'cut-off' for a phosphorite. The high-grade phosphorites are composed predominantly of francolite (McClellan & Saavedra, this volume) with a comparatively low lattice CO_2 content of 2% or less (determined from cell parameters). Silica is common, with clays (mainly kaolinite and lesser illite), micas (muscovite and

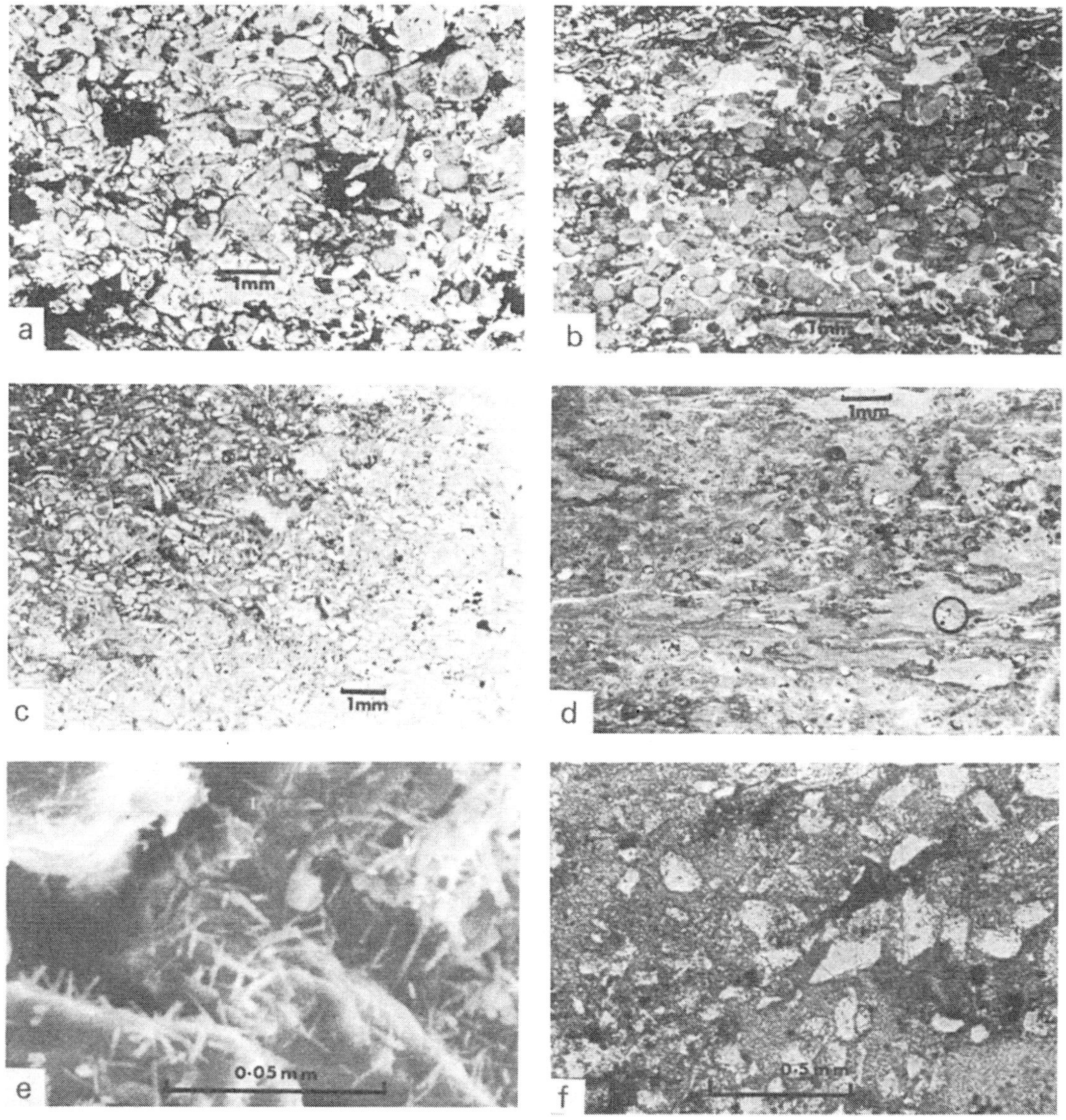

Fig. 11.6. Miscellaneous photomicrographs of Lady Annie phosphorites. (a) Typical coarse skeletal-ovulitic grainstone phosphorite composed largely of abraded skeletal debris and other structureless ovulitic grains. Iron-stained zones are commonly present. (Plain light). (b) Typical coarse skeletal-ovulitic well-sorted and well-rounded grainstone phosphorite from Rufous Creek. (Plain light). (c) Packstone phosphorite composed of ovulitic grains and rounded elongate skeletal grains. (Plain light). (d) Mudstone/packstone phosphorite composed of fine elongate angular and platy skeletal debris orientated parallel to bedding and other ovulitic grains settled in a fine-grained phosphorite matrix. (e) Mudstone phosphorite (scanning electron photomicrograph) with fine needles of apatite adhering to the surface of rounded detrital quartz grains. (f) Sandy-silty mudstone phosphorite containing abundant angular detrital grains set in a darker fine phosphatic matrix.

possibly sericite), dolomite and goethite. Analyses of three high-grade phosphorites are given in Table 11.1. The high grade grainstone/packstone phosphorite (Analysis 1) is similar to the 'average' Georgina Basin value (Analysis 4). Both the grainstone and mudstone phosphorites at Lady Annie (Analyses 1 and 2) are notable for their high P_2O_5/CO_2 ratios pointing to a fluorapatite rather than a carbonate fluorapatite. This is consistent with the cell parameters reported by Cook (1972). Elsewhere in the basin where extensive crystal chemistry data are available there do appear to be subtle compositional differences between apatite in the grainstone phosphorites and the mudstone phosphorites. Sufficient data are not presently available for Lady Annie; however, limited scanning electron microscopy of Lady Annie material indicates that the fluorapatite crystallites in grainstone phosphorites are broad hexagonal plates, whereas the crystallites in mudstone phosphorites are fine acicular hexagons. These differences point to possible variations in the mode of phosphate precipitation or replacement in the two types of phosphate at Lady Annie. The low organic carbon content is almost certainly a function of Cenozoic weathering rather than of a low initial organic carbon content, by analogy with locations elsewhere in the Basin where

Table 11.1. *Chemical composition of phosphorites and phosphatic mudstones from the Lady Annie deposit*

	Grainstone or packstone phosphorite (Rogers & Keevers, 1976) Analysis 1	Mudstone phosphorite (Rogers & Keevers, 1976) Analysis 2	Phosphatic mudstone (Cook, 1972) Analysis 3	Mean composition of Georgina Basin grainstone phosphorites (Cook, 1972) Analysis 4
P_2O_5	34.6	35.0	13.20	35.6
CaO	47.9	48.2	19.30	49.6
SiO_2	8.2	10.3	57.80	4.78
Al_2O_3	1.82	1.59	3.75	0.58
Fe_2O_3	1.07	0.14	1.16	1.37
FeO	n.d.	n.d.	0.01	0.16
MgO	0.29	0.15	0.14	0.31
Na_2O	0.18	0.04	0.11	0.52
K_2O	0.35	0.09	0.46	0.13
TiO_2	0.08	0.05	0.33	0.03
MnO	n.d.	n.d.	0.02	0.04
H_2O^+	0.84	0.7	1.48	1.02
H_2O^-	0.67	0.53	0.63	0.35
CO_2	1.1	1.2	0.55	1.67
SO_3	0.5	0.07	n.d.	n.d.
F	3.55	3.16	1.06	3.4
Organic matter	0.4	0.6	n.d.	n.d.
Less O equi F	1.49	1.33	0.45	1.43
Totals	100.06	100.49	99.71	98.16

n.d, not determined

samples of phosphorite have been obtained below the weathering profile. In such cases the phosphorites are black and organic-rich. Analysis 3 (Table 11.1) is of a low-grade phosphorite or phosphatic mudstone containing abundant detrital quartz and other minor detrital components. If Analysis 3 is recalculated on a quartz-free basis, then the composition is similar to Analyses 1 and 2.

Depositional environments

The western Lady Annie deposit is a part of a larger phosphate deposit which in turn belongs to a major phosphogenic province in which several deposits have similar stratigraphic, lithological, mineralogical and chemical characteristics (Howard, this volume). There is fairly general acceptance that the Georgina Basin phosphorites were deposited in very shallow nearshore marine environments either adjacent to or on topographic highs, in environments ranging through lagoonal, estuarine, littoral intertidal or even supratidal in places (de Keyser, 1969a; Thomson & Russell, 1971; de Keyser & Cook, 1972; Howard, 1972; Cook, 1976a; Cook & Shergold, 1979; Southgate, 1982). Howard & Cooney (1976) and Howard & Hough (1979) suggest that the D Tree deposit was formed partly in a lagoonal trough with its southern end opening on to a shallow open marine shelf. Howard & Perrino (1976) similarly suggested that the phosphorites at Wonarah formed in a low-energy nearshore environment. By analogy the phosphorites and associated sediments of the nearby western Lady Annie area were likely to have been deposited in shallow-marine conditions (de Keyser, 1969b; Rogers & Keevers, 1976). This is supported by the occurrence of a shallow-marine benthonic faunal assemblage and by the types of sedimentary structures present. The lateral relationships observed between phosphorites and the oncolite facies, and the typical carbonate–phosphorite–chert–siliciclastic association suggest that sedimentation took place in relatively shallow water in a nearshore situation close to basement areas. The association of the skeletal, oncolitic and carbonate facies and the disposition of these facies around the eastern basement areas (Fig 11.7) suggest that here the Middle Cambrian shoreline may have been fairly close to the present limit of outcrop in the Lady Annie area. However, the presence of small outliners of Middle Cambrian sediments within Proterozoic rocks to the east of the main outlier suggests that at times shallow seas extended further east. Since the phosphorite facies wedges out and interfingers with the skeletal and oncolite facies to the east, and since all these facies are conformably overlain by the chert-terrigenous facies, the phosphorite facies probably has a present distribution which is approximately coincident with its original extent. The series of north-trending inliers which make up 'Trueman's Bank' (Figs 11.1 & 11.2) and form the north-eastern

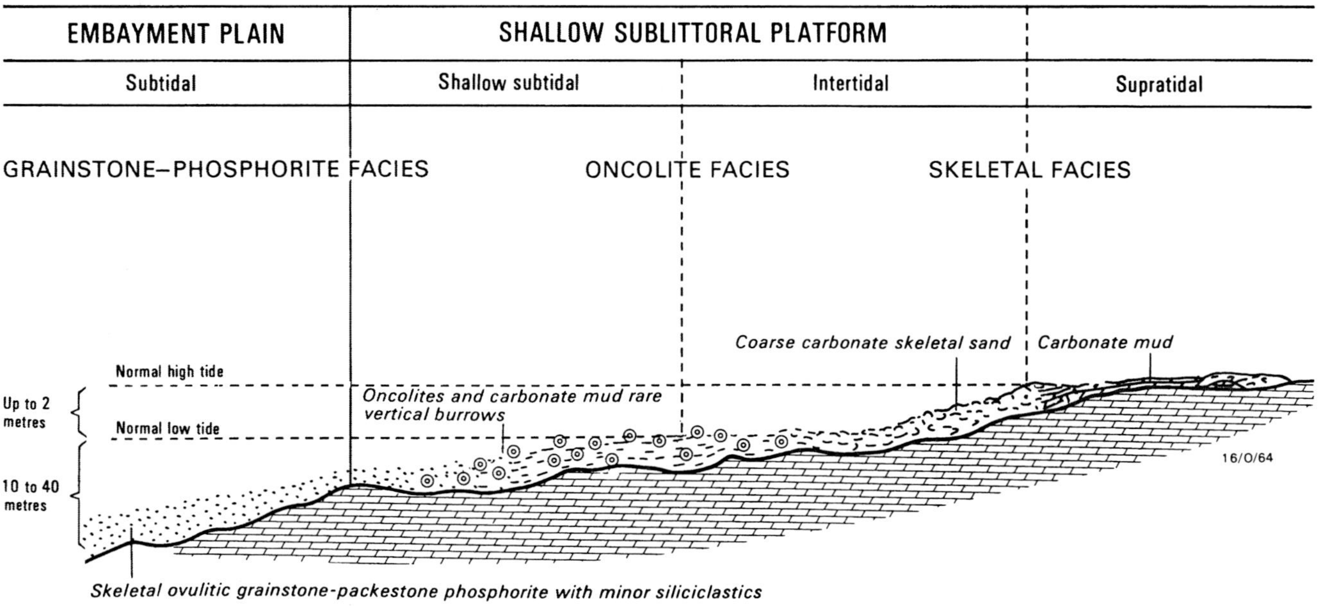

Fig. 11.7. Environmental interpretation of the depositional conditions prevailing during formation of a grainstone phosphorite deposit showing facies changes.

limit of the western Lady Annie deposit, are interpreted as positive topographic features which constituted shoal areas or possibly even islands during the Middle Cambrian.

To the west the basement rocks probably formed an area of partially emergent topographic highs during the Middle Cambrian. Rogers & Keevers (1976) consider that the Western Fault has had a long history of movement and was probably an important influence on sedimentation. They suggest that breccia along the Western Fault zone developed as a submarine talus breccia adjacent to a cliffed-shore environment along the fault escarpment. Cook & Armstrong (1972) suggest that the distribution of kaolinite and illite in the Beetle Creek Formation is consistent with a source of fine terrigenous sediments to the west. However, the presence of small, high-grade outliers of phosphorite between the Lady Annie and D Tree deposits, indicate that this area too was submerged at times.

Rogers & Keevers (1976) have suggested that there is a relationship between the distribution of phosphorite and palaeobathymetry within the Lady Annie–Lady Jane outlier. They suggest that the major phosphorite accumulations (Fig. 11.2) are associated with structural lows. In the western deposit the major phosphorites occur in a depression trending north-north-west parallel to the Western Fault and it is here that the maximum thickness of the phosphatic sequence (46.5 m) is recorded (Fig. 11.2). The sequence is dominated here by grainstone phosphorites. North and east of 'Trueman's Bank' (Fig. 11.2) grainstone phosphorites are subordinate to finer-grained varieties and the carbonate facies is best developed. Therefore the regional and tectonic setting of the western Lady Annie deposit is interpreted to have formed as a shallow, southerly plunging depression at the margin of the Georgina Basin originally flanked on the basinward side (the present west and south-west) by semi-emergent shoals and ridges. The shoals and ridges protected an asymmetrical embayment developed in the depression from the direct influence of open-marine swell and wave action.

Depositional model

The basal conglomerate–sandstone facies, the undifferentiated carbonate facies and the laminate chert facies, represent the oldest sediments of an early Middle Cambrian marine transgression–regression which rapidly invaded the entire Georgina Basin (Henderson & Southgate, 1978; Shergold & Druce, 1980). This basal facies has the characteristics of high-energy, probably braided, fluvial sedimentation. This is indicated by imbrication of angular polymict basement-derived clasts, medium-scale planar cross-bedding and a unipolar current pattern (Conybeare & Crook, 1968; Selley, 1970, 1976). However these features are also consistent with coastal alluvial fans. The close association and transitional relationships of the carbonate facies with the basal conglomerate–sandstone facies suggests a nearshore setting. The presence of a shallow, normal marine fauna, medium-scale planar cross-bedding and ?bioherms in the carbonate facies are considered by Rogers & Keevers (1976) to be characteristic of very shallow marine water deposition under moderate- to low-energy conditions. Emergent features are represented by possible mudcracks and pagoda halite (Southgate, 1982) in the immediately underlying laminate facies. The presence of a normal marine fauna (trilobites, siliceous sponges, gastropods and brachiopods) suggests that marine conditions during the formation of the carbonate facies were apparently normal, but not favourable for a great diversity or abundance of fauna. This may indicate semi-restricted, limited-circulation, nutrient-deficient conditions.

The thin laminate chert facies locally overlying the carbonate facies contains abundant pseudomorphs after halite. In places this facies is represented by laminate, vuggy cherts containing possible pseudomorphs after framboidal pyrite. On the basis of both local and regional correlations the laminate facies is taken to represent the culmination of a regressive phase of a transgressive–regressive cycle in which high intertidal to supratidal conditions prevailed. Subsequently there was a marked change in the environment, possibly in response to the influx of more normal marine conditions. These are represented by a sudden increase in faunal abundance and diversity and the commencement of phosphate accumulation. Carbonate muds, oncolites, skeletal sand, siliciclastics and phosphorites were deposited as a mosaic of complex facies. The sedimentary relationships of the likely geometry of the embayment and particularly the intertonguing of phosphorites and associated sediments with now-silicified oncolites (Fig. 11.7) suggest very shallow water conditions. Modern oncolites form by the lime-mud trapping by blue-green algae in areas in which currents are sufficiently strong to overturn the grains but not strong enough to winnow the mud (Wilson, 1975). They occur in warm waters of variable salinity and at very shallow depths (intertidal to shallow subtidal) in for example the Persian Gulf on subtidal offshore shoals (Hughes-Clarke & Keij, 1973), and in Shark Bay, Western Australia (Logan, Rezak & Ginsburg, 1964). In Florida, the Bahamas and Bermuda (Wilson, 1975; Gebelein, 1976) they occur on shelves, shoals, and around basement highs, in shallow lagoons and bays and in tidal channels, and are frequently associated with skeletal and pelletal sands. By analogy, the oncolites found at Lady Annie were deposited in shallow subtidal to intertidal conditions. Although oncolites are not common in phosphatic sequences, Bushinsky (1964) has described some occurrences from Cambrian phosphorite deposits in Asia, and Prian (1980) has documented their occurrence in Cambrian phosphorites in the Montagne Noire, France.

The skeletal facies has an extensive distribution along the eastern margin of the study area, and its faunal assemblage and variety suggest unrestricted marine conditions. Coarse angular, poorly sorted skeletal fragments, a grain-supported fabric and the apparent lack of mudstone suggest winnowing and reworking under relatively high-energy conditions. The coquinas are similar to those found today in skeletal sand beaches where constant wave or current action accumulates large quantities of skeletal debris and where mud is removed by winnowing (Wilson, 1975). The distribution of the skeletal facies fringing the basement areas is consistent with shallow nearshore waters. Therefore, both the oncolites and skeletal sands can reasonably be interpreted as having been deposited as thin beds on a semi-emergent shallow carbonate shelf, with energy conditions low to moderate, and intermittent storm-induced high-energy periods possibly indicated by the alternation of coarse silicified coquinas with dolomitic packstones and wackestones. There is no clear evidence of supratidal deposits, such as could be expected shorewards of this environment.

The thin shallow-water carbonate shelf facies interfinger and grade laterally basinwards into phosphorites (Fig. 11.7). No sedimentary structure found in the phosphorites is by itself diagnostic of a particular depositional environment. Evidence of tidal influence might be expected, given the apparent stratigraphic position of the phosphorites. However, macrotidal and mesotidal features normally associated with tidal-dominated subtidal to intertidal sand bodies, such as well-developed cross-bedding, current ripples, and tidal channels (Reineck, 1972; Wilson, 1975; Klein, 1977), are generally absent, through a bimodal current pattern is shown by oriented hyoliths. It seems likely that deposition in the western Lady Annie area took place in a dominantly micro-tidal setting (tidal range of less than 2 m), analogous perhaps to the carbonate-depositing environments of Shark Bay (Logan & Cebulski, 1970; Davies, 1970a, b) and the Trucial Coast (Purser, 1973). The characteristic bedding type observed in phosphorites of the study area – planar and wavy lenticular bedding with thin beds and laminae defined by slight variations in grainsize and, occasionally, by mudstone laminae – is compatible with bedding observed in subtidal and intertidal zones of modern and ancient siliciclastic tidal systems (Ginsburg, 1975; Klein, 1977) and carbonate tidal systems (Wilson, 1975; James, 1979), and in particular to areas of bedload and alternating bedload and suspension transport. Thin laminations of alternating coarse and fine sand-sized sediments and occasional micrograded bedding are evidence of deposition in a setting where current velocity varied, perhaps involving current bedload transport with suspension settlement during slack-water periods (Klein, 1977). Preservation of these laminae suggests the possibility of anoxic conditions within the sediments which lack bioturbation. Similar features are observed in mid-intertidal flats of lagoonal-back barrier siliciclastic environments (Reinson, 1979). Micrograded bedding in thin sand laminae has been described from siliciclastic tidal deposits by Reineck (1972) but micrograding may occur wherever sediments settle from suspension, and is generally not taken as evidence of any particular environment. Evidence of intraformational brecciation, though rare at Lady Annie, provides stronger evidence for an intertidal setting (Pettijohn, 1957; Conybeare & Crook, 1968). Soft sediment deformation is one of the most common sedimentary features observed in the phosphorite sequence. Its presence is commonly taken to indicate a palaeo-slope, though Klein (1977) suggests that soft sediment deformation is an early diagenetic feature developed in response to differential cohesiveness, and water content. However, taken in total, the sedimentary structures and the pattern of facies relationships at Lady Annie suggest that the phosphorites of the study area were deposited in shallow subtidal conditions in a microtidal setting. Possible intertidal conditions are suggested by local penecontemporaneous brecciation at the top of the phosphorite section in the southern trenches. Water depths appear to have increased westward from low intertidal to shallow subtidal conditions adjacent to a shallow carbonate shelf. An idealised vertical section (Fig. 11.8) summarises the sedimentary succession in trenches in the southern part of the deposit and provides an environmental interpretation of the phosphatic interval. An overall upward-fining character is an important feature of the sequences. Energy conditions decrease from a maximum associated with matrix-free coarse grainstones to a minimum associated with the mudstone and fine packstone phosphorite and increasing terrigenous material at the top. Suspension transport becomes dominant over bedload transport. This decrease in energy conditions may be attributable to progressively greater dissipation of current and wave action on

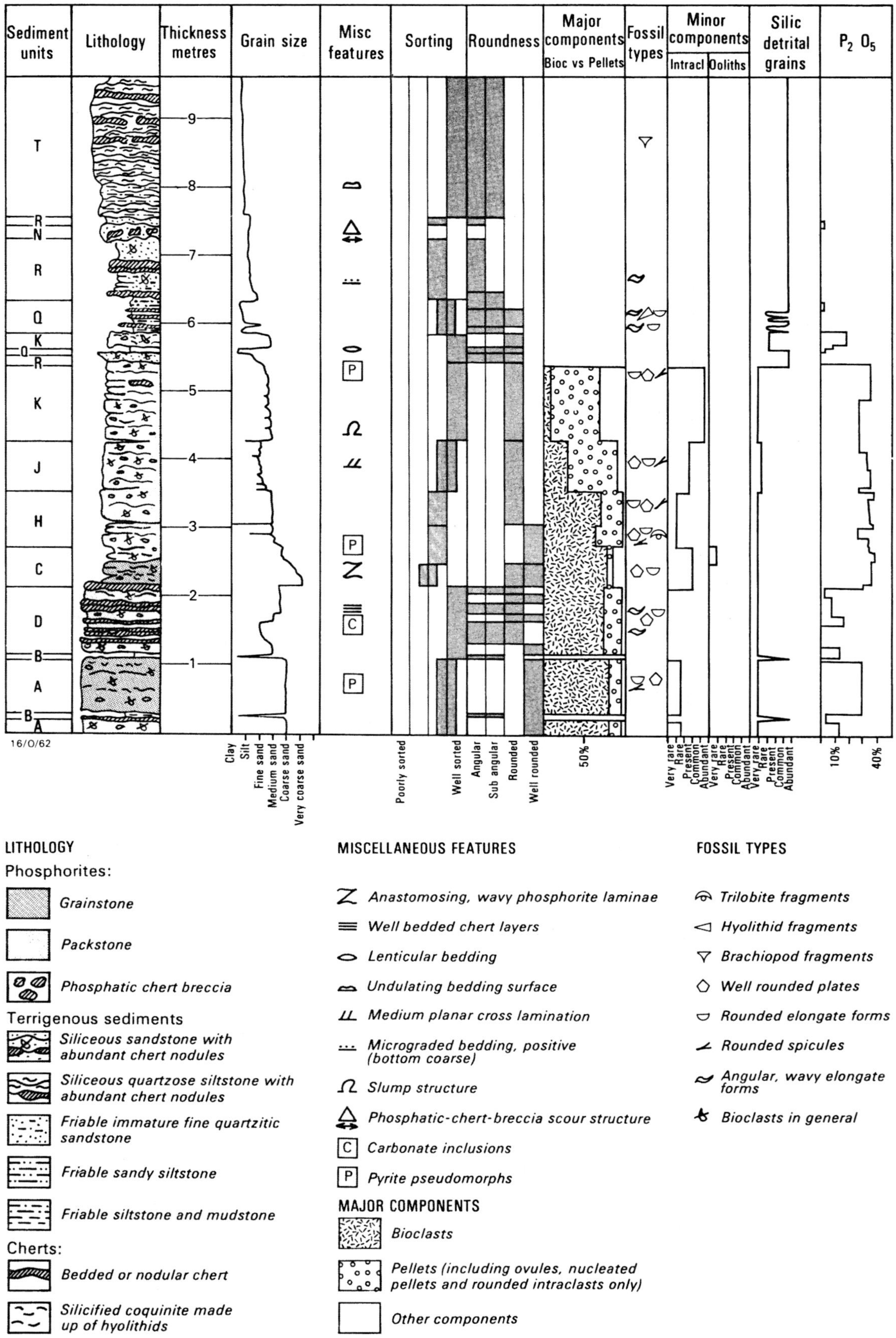

Fig. 11.8. Environmental interpretation of an idealised vertical sequence from the southern part of the deposit.

shoaling sediments as the embayment filled. The termination of phosphorite sedimentation may be related to the decrease in energy conditions. As energy decreased, the ability of currents to transport, rework and winnow coarse phosphate grains diminished.

Turning to the phosphate grains themselves, there are, as pointed out earlier, abundant coarse phosphatised fossil fragments in the deposit. The diversity and abundance of the fauna and especially the presence of stenohaline forms such as echinoderms and brachiopods (Heckel, 1972) indicates normal marine conditions. Ovules, the other major granular component of the phosphorites, have a disputed origin and give no firm indication of the depositional conditions from their petrology. A few may be of faecal origin. Others may have originated as highly reworked intraclasts of mudstone phosphorites (Riggs, 1979a).

A further possible origin suggested for the ovules is accretionary growth during constant depositional conditions in a similar way to ooids (Sheldon, 1981), though the lack of accretionary growth pattern in most ovules at Lady Annie does not seem to support this. The mature sorting and rounding of the phosphate grains, the lack of angular or entire fossils (with the exception of discrete beds of silicified coquinas) indicate that the sediments are essentially clastic accumulations of phosphate grains which accumulated to form phosphorite sand bodies. The importance of transport and reworking undergone by the sediments is indicated by the fact that there is a variety of types of phosphate grains yet all are well sorted and rounded. In addition, quartz detrital grains (when present) are also well sorted. Therefore the textures indicate that the environment was one in which there was an efficient method of accumulating sand-sized grains, first by their transport and accumulation in favourable areas to form phosphate sand shoals and bars, and second by the winnowing of the fine-grained fraction and its removal to lower-energy areas. It is inferred from sedimentary structures that tidal currents, and to a lesser extent wave processes, were the main agents influencing the phosphate sand bodies. In lower-energy conditions the finer grained packstones accumulated. Fines winnowed from the sand bodies were redeposited in relatively quiet waters where settling from suspension was possible. Such quiet conditions may be achieved in deeper waters below wave base and on protected tidal flats. The absence of bioturbation suggests that bottom conditions were not suitable for burrowing fauna, perhaps due to anoxia below the sediment–water interface. Bioturbation is an extremely common phenomenon in most nearshore carbonate environments, such as those of Shark Bay (Davies, 1970a; Logan & Cebulski, 1970) or the Trucial Coast (Purser & Evans, 1973). However, bioturbation is not commonly recorded in association with phosphorites in other deposits, perhaps because of strongly anoxic conditions in the phosphate sands below the sediment–water interface.

In summary then, the evidence for transport and reworking of the allochems of the packstone and grainstone phosphorites suggests that the allochems were derived from outside the area of study. This is compatible with the proposed palaeogeographic setting of the Lady Annie area of a semi-protected shallow depression off a major epeiric sea. The embayment would have been a suitable area for the entrapment of sediments analogous to a modern-day embayment such as Broad Sound (Cook & Mayo, 1977). Coarse detrital skeletal sand and finer fractions derived from offshore, possibly together with terrigenous sediments, would have been swept into the Lady Annie embayment by longshore and/or tidal currents and subsequently winnowed and deposited according to the prevailing energy conditions and morphology of the embayment. In the first stages of the evolution of sedimentation, sediments may have accreted around basement topographic highs in a similar manner to those found today in the central province of the Trucial Coast. The time and location of the earliest phosphatisation remains uncertain. It is clear, however, that if phosphatisation took place within the environment in which the grains accumulated, it was a very early diagenetic process because it occurred before the early diagenetic silicification of coarse grainstone or packstone phosphorite beds. It is considered more likely that grains were phosphatised prior to transport, i.e. as phosphatic allochems. Both accumulation of phosphate grains in favourable areas and winnowing were part of the same depositional system. This is similar in concept to the 'entrapment basin' proposed by Riggs (1979b) for the Florida deposits, with the major qualification that while the Florida deposits are largely composed of intraclasts derived from the reworking of semi-lithified mudstone phosphorites, the phosphorites of the Lady Annie deposit consist dominantly of phosphatised skeletal grains and ovulitic grains. By inference, the mudstone phosphorites may have formed by accumulation from suspension of very fine detrital phosphate particles derived from the biological and/or mechanical attrition of coarser grains. However, an argument against this option is that several authors (Russell & Trueman, 1971; Cook, 1972) have reported differences in the crystal forms and crystal chemistry of apatite in the grainstone phosphorites and in the mudstone phosphorites. Precipitation from the water column in very shallow supratidal areas Southgate, this volume) coupled with the effects of subaerial weathering (Cook, 1972) may be responsible for these differences. However, it is important to stress that at this stage the deposits which are of potential economic significance are grainstone phosphorites and these probably formed through the reworking transport and accumulation of previously phosphatised allochthonous grains. The environments of deposition of the mudstone and wackestone phosphorites at Lady Annie cannot be deduced with complete confidence. Very early diagenetic replacement of carbonate muds by phosphate has been suggested as a possible origin for some mudstone phosphorites (Cook, 1972; Fleming, 1977; Howard & Hough, 1979). An orthochemical origin, by the direct precipitation of phosphate in quiet energy conditions, has been proposed by Riggs & Freas (1965) and Riggs (1979a, b) for the microsphorite of the Florida deposits, although whether precipitation took place as a result of chemical or biochemical (bacterial) processes is uncertain. Some of the mudstone phosphorites at Lady Annie are possibly Cambrian phoscretes that formed under partly subaerial conditions (Southgate, this volume). It may be that superficially similar fine-grained phosphorites can form by several processes.

Palaeogeography

We are now able to build up an accurate picture of the environment in which phosphate grains accumulated to form the western part of the Lady Annie deposit (Fig. 11.9). The broad setting appears to be that Lady Annie–Lady Jane was a discrete deposit on the eastern margin of a broad stable epicontinental sea. Cambrian sedimentation took place in a complex system of offshore bar, lagoon-embayment and shoreline environments developed in a shallow asymmetric depression flanked by arid low-relief emergent and semi-emergent basement. Sedimentation was strongly controlled by basement topography and tectonics. The embayment was deepest along its faulted western flank and shallowed progressively eastwards towards an irregular basement shoreline. The outlier was probably open to the Georgina Basin epeiric sea along its long western flank but protected from direct open-marine influence by the emergent basement areas of the Western Fault block (Fig. 11.9). Maximum initial water depth was of the order of 40 m (Fig. 11.7) but was probably considerably less than this during the bulk of phosphorite deposition. The presence of oncolites provides good evidence of very shallow (shallow subtidal to intertidal) conditions in facies laterally equivalent to the phosphorites. Salinity in the deeper embayment areas was normal but may have varied from normal to hypersaline in intertidal to supratidal areas according to the restriction of the environment. Tidal range appears to have been microtidal (less than 2 m) and the current pattern is indicated to have been bipolar and parallel to the shore with a dominant southerly component. Coarse skeletal and ovulitic phosphatic sand was swept into the embayment by waves and tidal and longshore currents (Fig. 11.9) and formed the source of the allochthonous phosphatic material of the embayment plain. It is inferred from sedimentary structures that tidal currents and to a lesser extent wave processes within the embayment were the main agents influencing the formation of sand bodies in deeper subtidal areas. Currents strong enough to transport, rework and winnow coarse sediments produced the typical mature grainstone and packstone phosphorites of the subtidal embayment plain. Finer sediments (carbonate muds, phosphatic siltstone, mudstone phosphorites and fine siliciclastic material) settled from suspension in quieter areas of the eastern part of Lady Annie.

There does not appear to be a good modern analogue in any of the present day phosphogenic regions such as off Chile–Peru or off south-western Africa. However, the phosphate grains of the grainstone and packstone phosphorites (which comprise the main deposits of Lady Annie) have similar morphology and textures to carbonate grains in a number of Holocene analogues. Present-day carbonate environments which seem to be good potential analogues for Lady Annie include Shark Bay in Western Australia and the Trucial Coast of the Persian Gulf. It is envisaged therefore that the phosphate grains were formed, or more probably in this instance the pre-existing carbonate grains were phosphatised, outside the Lady Annie embayment. They were then transported by current action into the embayment where they were trapped in shallow nearshore locations. The vertical sequence of the phosphorites in the southern part of the deposit as exemplified in Figs. 11.5 and 11.8 shows a fining-upward sequence. This is interpreted as the consequence of a gradual decrease in energy conditions of the environment, probably due to the reduction of water depth as the embayment filled. Finer-grained sediments became progressively more important as currents became less able to transport and winnow coarse sediment. Finally, conditions suddenly changed and became such that the fine siliciclastic sediments dominated.

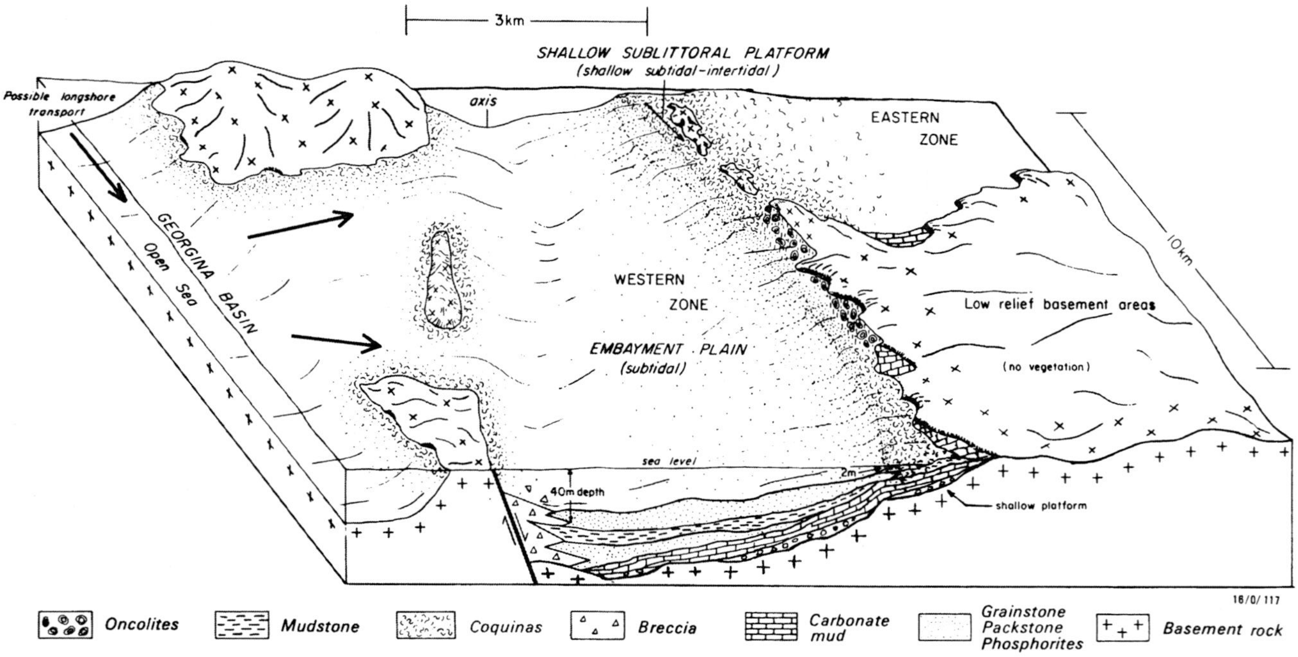

Fig. 11.9. Possible reconstruction of the palaeogeography, depositional conditions, and facies relationships prevailing in the Lady Annie region at the time of phosphate deposition.

There may also have been a rise of sea level. Bottom conditions became anoxic and there was a marked change from benthonic to pelagic faunas. This in turn brought phosphogenesis and the accumulation of phosphate grains at Lady Annie to a close.

References

Bushinsky, G.I. (1964). On the shallow-water origin of phosphorite sediments. In *Deltaic and Shallow Marine Deposits*, ed. L.M.U. van Straaten, pp. 62–70, Elsevier, Amsterdam.

Conybeare, C.E.B. & Crook, K.A.W. (1968). Manual of sedimentary structures. *Bureau Mineral Resources Geology & Geophysics Australia Bulletin*, **102**, 327pp.

Cook, P.J. (1972). Petrology and geochemistry of the phosphate deposits of NW Queensland, Australia. *Economic Geology*, **67**, 1193–213.

Cook, P.J. (1976a). Georgina Basin phosphatic province, Queensland and Northern Territory regional geology. In *Economic Geology of Australia and Papua New Guinea, 4, Industrial Minerals and Rocks*, ed. C.L. Knight, pp. 45–50. Australasian Institute of Mining and Metallurgy, Melbourne.

Cook, P.J. (1976b). Sedimentary phosphate deposits. In *Handbook of Strata-bound and Stratiform Ore Deposits*, **7**, pp. 503–35, Elsevier, Amsterdam.

Cook, P.J. & Armstrong, K.A. (1972). Clay mineralogy of the Middle Cambrian Beetle Creek Formation, Georgina Basin, NW Queensland. *Bureau Mineral Resources Geology & Geophysics Australia Bulletin*, **125**, 7–16.

Cook, P.J. & Mayo, W. (1977). Sedimentology and Holocene history of a tropical estuary (Broad Sound, Queensland). *Bureau Mineral Resources Geology & Geophysics Australia Bulletin*, **170**, 206pp.

Cook, P.J. Shergold, J.H. (Eds) (1979). *Proterozoic–Cambrian Phosphorites*, 106pp. ANU Press, Canberra.

Cook, P.J. & Shergold, J.H. (1980). Proterozoic and Cambrian phosphorites of Asia and Australia – a progress report. In *Fertilizer mineral potential in Asia and the Pacific*, ed. R.P. Sheldon & W.C. Burnett, pp. 207–24. Resource Systems Institute, Honolulu.

Davies, G.R. (1970a). Carbonate bank sedimentation, eastern Shark Bay, Western Australia. *American Association Petroleum Geologists, Memoir*, **13**, 85–168.

Davies, G.R. (1970b). Algal-laminated sediments, Gladstone embayment, Shark Bay, Western Australia. *American Association Petroleum Geologists, Memoir*, **13**, 169–205.

de Keyser, F. (1968). The Cambrian of the Burke River Outlier. *Bureau Mineral Resources Geology & Geophysics Australia Record*, **1968/67**

de Keyser, F. (1969a). On the genesis of Georgina Basin phosphorites, northwestern Queensland. *Bureau Mineral Resources Geology & Geophysics Australia Record*, **1969/79**.

de Keyser, F. (1969b). The phosphate-bearing Cambrian formations of the Lawn Hill and Lady Annie districts, north-western Queensland. *Bureau Mineral Resources Geology & Geophysics Australia Record*, **1969/147**.

de Keyser, F. (1973). A review of the Middle Cambrian stratigraphy in the Queensland portion of the Georgina Basin. *Bureau Mineral Resources Geology & Geophysics Australia Bulletin*, **139**, 13–27.

de Keyser, F. & Cook, P.J. (1972). Geology of the Middle Cambrian phosphorites and associated sediments of northwestern Queensland. *Bureau Mineral Resources Geology & Geophysics Australia Bulletin*, **138**, 79pp.

Elgueta, S. (1981). Sedimentological study of the western zone of the Lady Annie phosphate deposit Queensland, Australia. M.Sc. Thesis, Australian National University, 238pp. (unpublished).

Fleming, P.J.G. (1974). Origin of some Cambrian bedded cherts, and other aspects of silicification in the Georgina Basin. *Geological Survey of Queensland Publication*, **358**, 9pp.

Fleming, P.J.G. (1977). Faunas, lithologies, and the origin of phosphorites in parts of the Middle Cambrian Beetle Creek Formation of northwest Queensland. *Geological Survey of Queensland Publication*, **364**, 21pp.

Gebelein, C.D. (1976). Open marine subtidal and intertidal stromatolites (Florida, the Bahamas, and Bermuda). In *Stromatolites – Developments in Sedimentology*. ed. M.R. Walter, pp. 381–8. Elsevier, Amsterdam.

Ginsburg, R.N. (Ed.) (1975). *Tidal Deposits: A Casebook of Recent Examples and Fossil Counterparts*, 428pp. Springer-Verlag, New York.

Heckel, P.H. (1972). Recognition of ancient shallow marine environments. *Society Economic Paleontologists & Mineralogists, Special Publication*, **16**, 226–86.

Henderson, R.A. & Southgate, P.N. (1978). Middle Cambrian stratigraphy and sedimentology of the eastern Georgina Basin. *3rd Australian Geological Convention, Townsville, Excursions Handbook*, pp. 85–110.

Howard, P.F. (1971). Discovery of phosphorite in northern Australia. *Transactions of the American Institute of Mining Engineers*, **250**, 269–73.

Howard, P.F. (1972). Exploration for phosphorite in Australia – a case history. *Economic Geology*, **67(8)**, 1180–92.

Howard, P.F. & Cooney, A. (1976). D Tree phosphate deposit, Georgina Basin, Queensland. In *Economic Geology of Australia and Papua New Guinea. 4, Industrial Minerals and Rocks, ed.* C.L. Knight, pp. 265–73. Institute Mining & Metallurgy, Melbourne.

Howard, P.F. & Hough, M.J. (1979). On the geochemistry and origin of the D Tree, Wonarah, and Sherrin Creek phosphorite deposits of the Georgina Basin, Northern Australia. *Economic Geology*, **74(2)**, 260–84.

Howard, P.F. & Perrino, F.A. (1976). Wonarah phosphate deposit, Georgina Basin, N.T. In *Economic Geology of Australia and Papua New Guinea. 4, Industrial Minerals and Rocks*, ed. C.L. Knight, pp. 27–8. Australasian Institute Mining & Metallurgy, Melbourne.

Hughes Clarke, M.W. & Keij, A.J. (1973). Organisms as producers of carbonate sediments and indicators of environment in the southern Persian Gulf. *The Persian Gulf*, ed. B.H. Purser, pp. 33–56. Springer, New York.

James, N.P. (1979). Shallowing-upward sequences in carbonates. In *Facies Models*, ed. R.G. Walker, pp. 109–20. Reprint Series 1, Geoscience Canada, Toronto.

Klein, G. de V. (1977). *Clastic Tidal Facies*, 149pp. Continuing Education Publication Company, Boston.

Logan, B.W. & Cebulski, D.E. (1970). Sedimentary environments of Shark Bay, Western Australia. *American Association Petroleum Geologists, Memoir*, **13**, 1–37.

Logan, B.W., Rezak, R. & Ginsburg, R.N. (1964). Classification and environmental significance of algal stromatolites. *Journal of Geology*, **72**, 68–83.

Mabie, C.P. & Hess, H.D. (1964). Petrographic study and classification of Western phosphate ores. *U.S. Bureau of Mines, Report*, **6468**, 95pp.

Pettijohn, F.J. (1957). *Sedimentary Rocks*, 718pp. Harper & Row, New York.

Prian, J.P. (1980). Les phosphorites Cambriennes du versant Septentrional de la Montagne Noire au sud du Bassin Permien de Camares (Aveyron). Unpublished Thesis, Université Pierre et Marie Curie, Paris VI.

Purser, B.H. (1973). *The Persian Gulf: Holocene carbonate sedimentation and diagenesis in a shallow epicontinental sea*, 471pp. Springer, New York.

Purser, B.H. & Evans, G. (1973). Regional sedimentation along the Trucial Coast, south-east Persian Gulf. In *The Persian Gulf*, ed. B.H. Purser, pp. 211–31. Springer, New York.

Reineck, H.E. (1972). Tidal flats. In Recognition of Ancient Sedimentary Environments, ed. J.K. Rigby & W.K. Hamblin. *Society of Economic Paleontologists and Mineralogists Special Publication*, **16**, 146–59.

Reinson, G.E. (1979). Barrier island systems. *Facies Models*, ed. R.G.

Walker, pp. 57–74. Reprint Series 1, Geoscience Canada, Toronto.

Riggs, S.R. (1979a). Petrology of the Tertiary phosphate system of Florida. *Economic Geology*, **74(2)**, 195–220.

Riggs, S.R. (1979b). Phosphorite sedimentation in Florida – a model phosphogenic system. *Economic Geology*, **74(2)**, 285–314.

Riggs, S.R. & Freas, D.H. (1965). Stratigraphy and sedimentation of phosphorite in the Central Florida Phosphate District. *American Institute of Mining Engineers Preprint* **65-H-84**, 13pp.

Rogers, J.K. & Keevers, R.E. (1976). Lady Annie–Lady Jane phosphate deposits, Georgina Basin, Queensland. In *Economic Geology of Australia and Papua New Guinea, 4. Industrial Minerals and Rocks*, ed. C.L. Knight, pp. 251–61. Monograph Series 8, Australasian Institute Mining & Metallurgy, Melbourne.

Rogers, J.K. & Crase, N.J. (1980). The Phosphate Hill rock phosphate deposit, Northwest Queensland, Australia. *Fertilizer Mineral Potential in Asia and the Pacific*, ed R.P. Sheldon & W.C. Burnett, pp. 307–27. Resource Systems Institute, Honolulu.

Russell, R.T. (1967). Discovery of major phosphate deposits in northwest Queensland. *Queensland Government Mining Journal*, **68**, 153–7.

Russell, R.T. (1968). Discovery of major phosphate deposits in northwest Queensland, Australia. In *Sources of mineral material for the fertilizer industry in Asia and the Far East*. U.N. Mineral Resources Development Series, **32**, 241–8.

Russell, R.T. & Trueman, N.A. (1971). The geology of the Duchess phosphate deposits, northwest Queensland, Australia. *Economic Geology*, **66**, 1186–214.

Selley, R.C. (1970). *Ancient Sedimentary Environments*, 237pp. Reprinted in Science Paperback, (1976). Chapman and Hall, London.

Selley, R.C. (1976). *An Introduction to Sedimentology*, 408pp. Academic Press, London.

Sheldon, R.P. (1981). Ancient marine phosphorites. *Annual Review of Earth and Planetary Sciences*, **9**, 251–84.

Shergold, J.H. & Druce, E.C. (1980). The Upper Proterozoic and Lower Palaeozoic rocks of the Georgina Basin. In *The Geology of Northeast Australia*, ed. R.A. Henderson & P.J. Stephenson, pp. 149–74. Proceedings 3rd Australian Geological Convention, Townsville.

Smith, K.G. (1972). Stratigraphy of the Georgina Basin. *Bureau Mineral Resources Geology & Geophysics Australia Bulletin*, **111**, 156pp.

Southgate, P.N. (1982). Cambrian skeletal halite crystals and experimental analogues. *Sedimentology*, **29**, 391–407.

Southgate, P.N. (1983). Middle Cambrian phosphatic and calcareous environments, the Undilla region of the Georgina Basin, Ph.D. Thesis, Australian National University, 363pp. (unpublished).

Thomson, L.D. & Rogers, J.K. (1974). The discovery and development of Queensland phosphate by BH South Limited. Field Conference, pp. 25–8. Australasian Institute Mining & Metallurgy (NW Queensland Branch), Brisbane. (unpublished).

Thomson, L.D. & Russell, R.T. (1971). Discovery, exploration and investigations of phosphate deposits in Queensland. *Australasian Institute Mining & Metallurgy Monograph*, **240**, 1–14.

Trueman, N.A. Robinson, R.R., & Hankel, M.R. (1969). The mineralogy and mineral chemistry of the main phosphorite types, Lady Annie, Queensland. *Report* MP, pp. 626–70 Australian Mineral Development Laboratories, Adelaide (unpublished).

Wilson, J.L. (1975). *Carbonate Facies in Geologic History*, 471pp. Springer, New York.

12

Proterozoic and Cambrian phosphorites – deposits: Kunyang, Yunnan, China

YEH LIENTSUN, SUN SHU, CHEN QIYING AND GUO SHIZENG

Introduction

The Kunyang phosphorite deposit of Yunnan Province is one of the largest, best known, and best documented Early Cambrian deposits in China. The deposit is located on the south-west bank of Lake Dianchi, 64 km by road south-south-west from Kunming and 7 km north-north-west of Jinning (Fig. 12.1).

Originally discovered by Cheng Yuqi and Huang Hanqiu in early 1939 (Cheng, 1939), the deposit and associated rocks have been subsequently studied by Wang Yuelun, Lu Yanhao and Wang Hongzhen. Following the foundation of the People's Republic of China, reconnaissance and exploration have been undertaken by scientists of Academia Sinica and the Chinese Academy of Geological Sciences, and by workers in provincial geological bureaux.

The Kunyang deposit is mined as an open-pit operation 2.5 km along strike. Activity is concentrated in the eastern part of the deposit, near the village of Meishucun. The stratigraphic

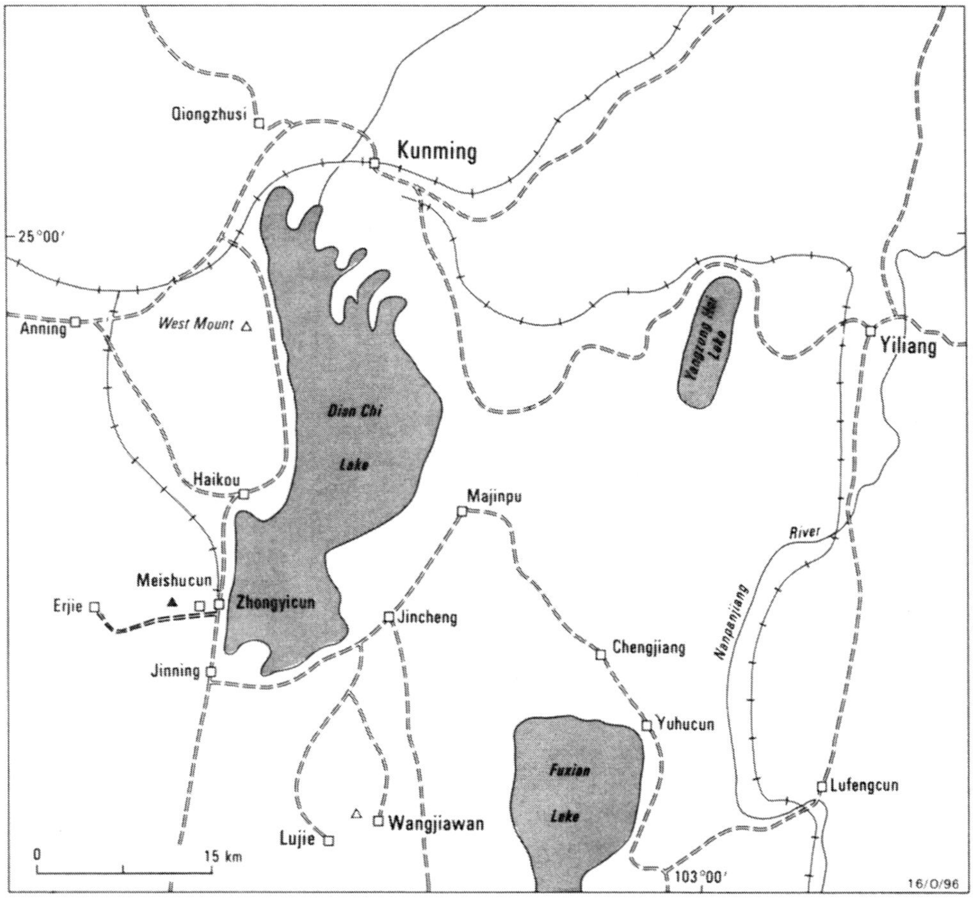

Fig. 12.1 Geographic location of the Kunyang phosphorite deposit.

section exposed by mining operations at Xiaowaitoushan, which extends from the Late Proterozoic to the middle of the Early Cambrian, is particularly important since it contains the Precambrian–Cambrian boundary. As such, it has become the focus of attention of the International Union of Geological Sciences Precambrian/Cambrian Boundary Working Group and International Geological Correlation Programme Project 29, the Precambrian/Cambrian Boundary. Accordingly, this section has been exhaustively studied by many scientists (*inter alia,* Lu, 1979; Zhang 1980; Luo *et al.* 1980, 1982; Luo, Jiang & He, 1982; Xing, Ding & Luo, 1982; Xing *et al.*, 1983), and has been proposed as a potential stratotype section for the Precambrian/Cambrian Boundary.

The Early Cambrian phosphorites of China are best developed in south-west China, along the western margin of the Yangzi Platform. Commerical grade deposits are also encountered on the flanks of syndepositional highs or platforms within the epicontinental seas (Li, this volume). They were mostly deposited and mechanically enriched in coastal to nearshore shallow-marine environments.

The phosphorite deposits of Kunyang are intertidal to shallow subtidal deposits. Deposits to the west of Lake Dianchi may be associated with carbonate platform sedimentation, while those to the east of the Lake, and the Wangjiawan deposit, may well represent more seaward sedimentation. Associated black shales represent periods of oxygen restriction and anoxia. The phosphorite-bearing series in Yunnan was deposited during the early phase of the initial Cambrian transgressive event. There are several erosion surfaces within the Lower Cambrian sequence which most probably reflect episodic sea level fluctuation associated with the tectonic development of the basin (Yeh & Sun, 1980; Yeh, Sun & Chen, 1980).

Stratigraphy

In eastern Yunnan Province, there are numerous occurrences of phosphatic sediments of various ages, including Sinian, Early Cambrian, Ordovician, Devonian, and Jurassic, but only those of the Early Cambrian are of commerical and industrial value. Early Cambrian deposits are clustered along the western margin of the Yangzi Platform. Their wider distribution and palaeogeographic setting has been described in detail by Li Yueyan (this volume). Essentially, the Yunnan

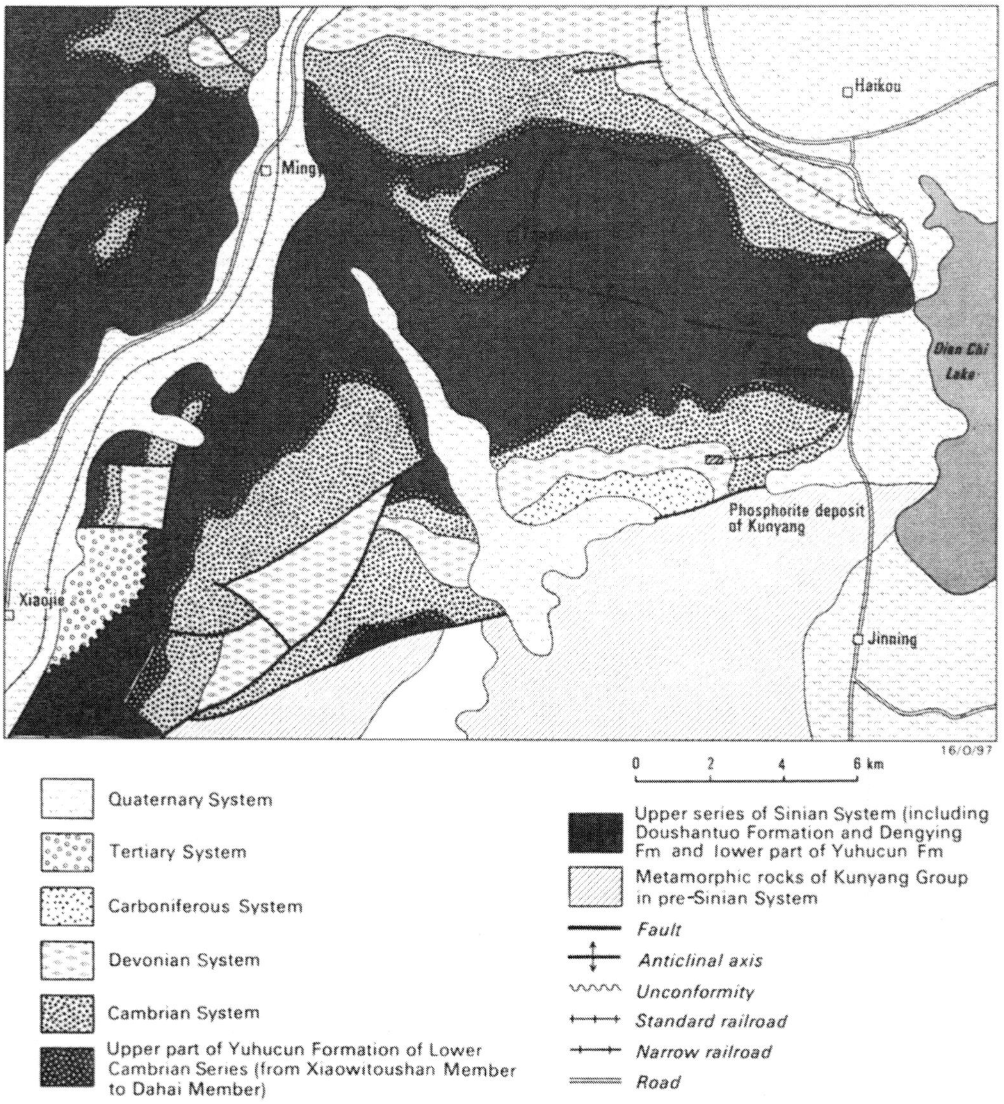

Fig. 12.2. Geological map of the Xiangtiaochong Anticline, Jinning, Yunnan Province.

sediments were deposited in an embayment bounded to the west by the Kang-Dian Oldland, and to the south-east by the Niushoushan Palaeo-island (Oldland) in central Yunnan. The sediments are in general of shallow water origin and apart from phosphorite, include dolomite (often sandy, silty or siliceous), mudstone, siltstone, shale and sandstone.

The Kunyang deposit is located on the eastern segment of the southern limb of the Xiangtiaochong Anticline (Fig. 12.2) which has an east–west orientation. At Kunyang Mine the Xiangtiaochong structure is monoclinal, and the sequence dips south at approximately 15–20°.

The detailed stratigraphy of the phosphatic sequence, and its relationship to overlying and underlying rocks, is demonstrated by two sections. At Xiaowaitoushan, the sequence extends from the late Sinian Jiucheng–Baiyanshao Member of the Yuhucun (formerly Dengying) Formation to the lower part of the Badaowan Member of the Qiongzhusi Formation (Fig. 12.3). Some 1000 m to the south-west, the Badaowan section exposes the Qiongzhusi Formation up to the level of the unconformity with the overlying Middle Devonian Haikou Formation. The Xiaowaitoushan section is the most important phosphate sequence.

The stratigraphic units used in this paper are those most recently proposed (Luo *et al.*, 1980) (Fig. 12.4). Essentially the Dengying Formation and the lower part of the Meishucun Formation of earlier writers has been redefined as the Yuhucun Formation which contains five members. From the base upwards, these are the Jiucheng, Baiyanshao, Xiaowaitoushan, Zhongyicun and Dahai Members. The upper part of the Meishucun Formation has been incorporated into the Qiongzhusi (Chiungchussu) Formation which is made up of a basal Badaowan Member and an upper Yuanshan Member.

Macrofossils first appear in the Xiaowaitoushan Member of the Yuhucun Formation and occur in all successive members, thus permitting precise biochronological dating. Phosphorites occur at the base of the Zhongyicun Member of the Yuhucun Formation and extend to the base of the Yuanshan Member of the Qiongzhusi Formation.

Xiaowaitoushan Section (Fig. 12.4) (from the base upwards)

(1) This section commences with muddy, sandy and siliceous dolomite referred to the Baiyanshao Member of the Yuhucun Formation. It is overlain, apparently conformably, by pale thick-bedded, sandy dolomite with chert layers and nodules, and thin phosphorite layers, referred to the

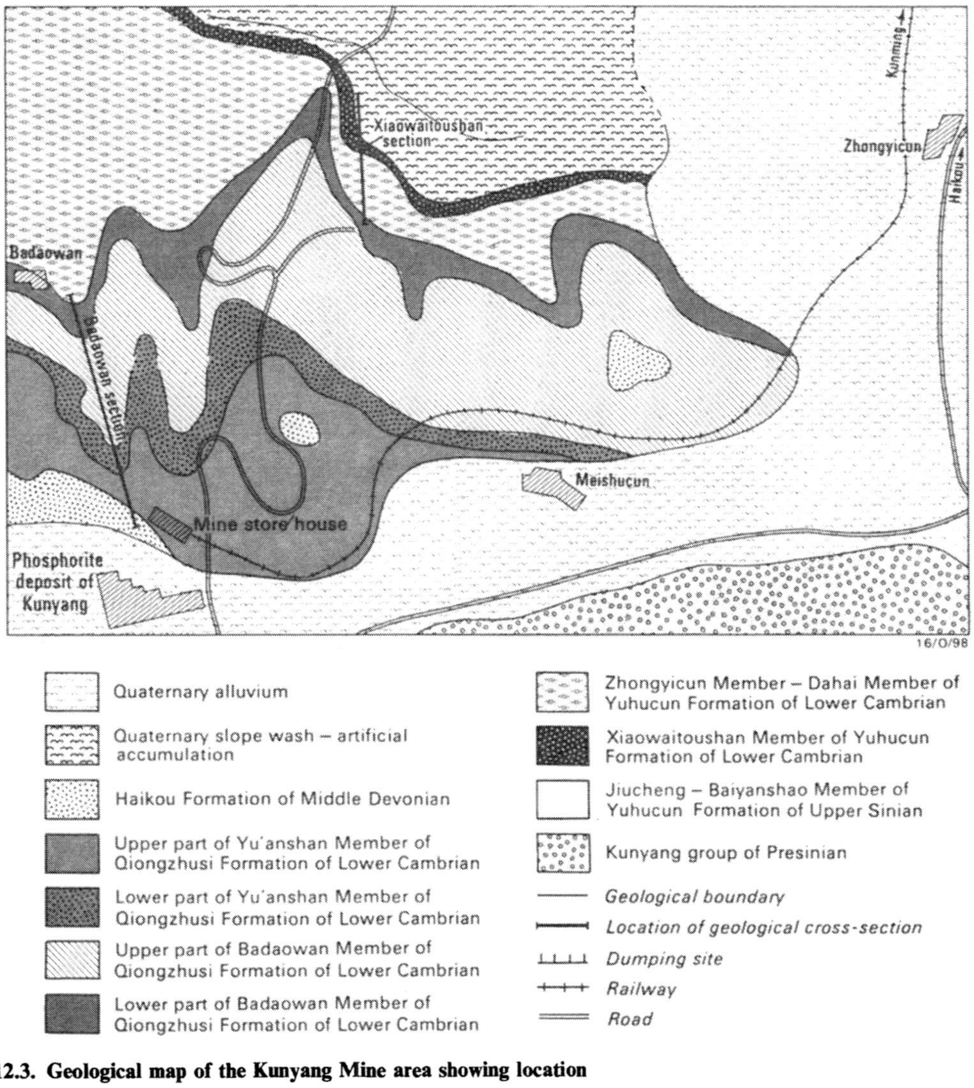

Fig. 12.3. Geological map of the Kunyang Mine area showing location of sections.

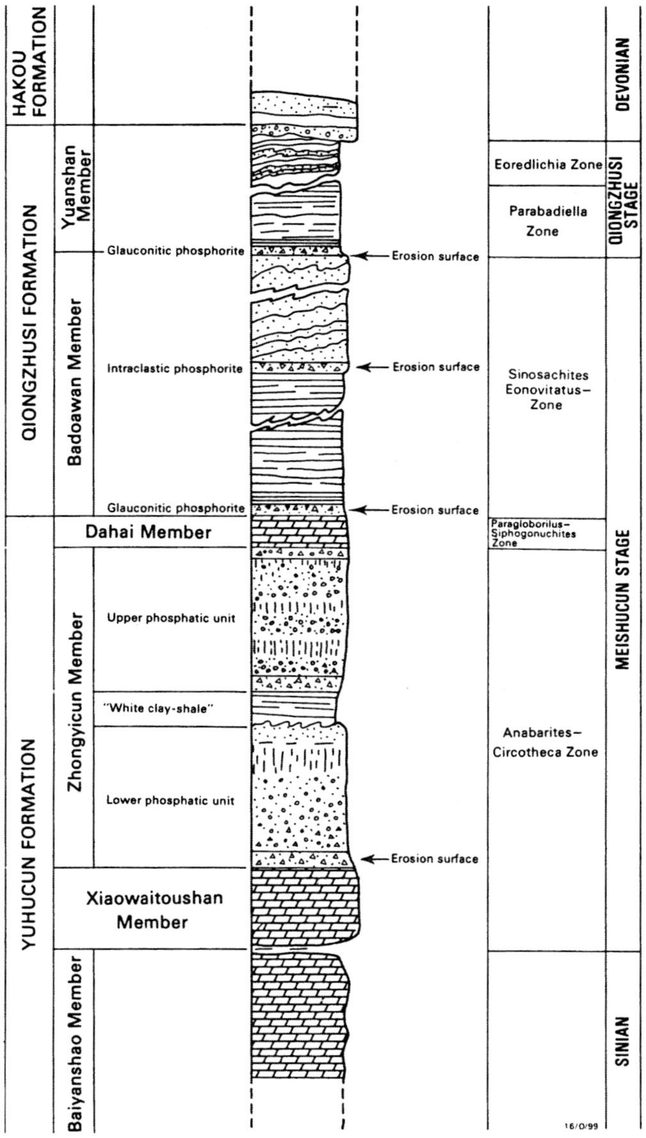

Fig. 12.4. Composite stratigraphic section based on the stratigraphy of the Xiaowaitoushan and Badaowan sections.

Xiaowaitoushan Member. This unit contains Hyolithida, Brachiopoda, chancelloriid sponge spicules and Hyolithelminthes referrable to the *Anabarites–Circotheca* Zone. It has a thickness of 8.3 m.

(2) The overlying Zhongyicun Member contains two phosphorite horizons separated by a clayey shaly layer, aggregating up to 28 m. An indurated erosion surface ('hardground') occurs at the base of the Zhongyicun Member.

This member contains small shelly fossils and trace fossils which are referred to the *Anabarites–Circotheca* Zone.

The lowermost phosphorite, ranging to 7.8 m in thickness, is one of the important ore horizons at Kunyang. It is rudaceous at the base, becomes dolomitic and laminated in the middle part, and slabby and siliceous at the top. Not infrequently the basal part is cross-laminated.

The shaly interval separating the two phosphorite layers is up to 7 m thick, but pinches out at the northern end of the Kunyang Mine. At Xiaowaitoushan it is 1.6 m thick. It may be clayey, sandy, calcareous or phosphatic and its bedding may be convoluted. The basal layers contain phosphatic fragments –representatives of Hyolithida, Monoplacophora and other small shelly fossil groups.

The upper phosphatic unit is the most important in the Kunyang area. It ranges from 2 m to 14 m in thickness, 5.3 m on the Xiaowaitoushan section, and is generally a bluish-grey, moderate- to thick-bedded dolomitic grainstone phosphorite. The basal part of this unit is consistently siliceous, at times rudaceous; the middle part is composed of massively-bedded fine-grained phosphorite; and the upper part, 0.3 m thick, consists of grainstone phosphorite with coated grains in a dolomitic matrix. The unit is highly fossiliferous, containing hyoliths, conodonts, monoplacophorans, gastropods, brachiopods, various problematical phosphatic and phosphatised shelly fossils, and trace fossils, again indicative of the *Anabarites–Circotheca* Zone.

To the north and south of the Kunyang Mine area, the upper phosphorite unit becomes increasingly dolomitic.

(3) An overlying dolomite layer, 1.1 m thick, formerly included in the Zhongyicun Member, is now separated out as the Dahai Member. It is represented by a greyish, thin- to medium-bedded phosphatic manganiferous silty dolomite with intercalated chert layers. It contains hyoliths, conodonts, monoplacophorans, brachiopods, chancelloriids and other shelly fossils which indicate the second zone of the Meishucun Stage – the *Paragloborilus–Siphogonuchites* Zone.

(4) The contact between the Dahai Member of the Yuhucun Formation and the overlying Badaowan Member of the Qiongzhusi Formation is an erosion surface marked by a thin layer, 0.4 m thick, of glauconitic siliceous phosphorite. Within this unit, there is an initial layer of glauconitic clay with a limonitic base, 0.05 m thick at Xiaowaitoushan. This is overlain by glauconitic nodular phosphorite and intraclastic phosphorite cemented by clay and glauconite, 0.25 m thick. This phosphorite is the third ore horizon at Kunyang.

Badaowan Section (Fig. 12.4)

(1) The Badaowan section commences at the base of the Qiongzhusi Formation as described above. The basal glauconitic clay unit is overlain by the 'lower black shale' which consists of black carbonaceous siltstone with chert nodules, 1.20 m thick, and black nodular phosphorite, 0.4 m thick. The remainder of the lower part of the Badaowan Member comprises grey-black thin-bedded phosphatic argillaceous dolomitic quartz siltstone, often containing large septarian nodules. It is up to 26 m thick.

(2) The upper part of the Badaowan Member consists of grey to dark grey, thin- to medium-bedded, argillaceous siltstone interbedded with dolomitic siltstone, dolomite, and in places thin-bedded quartz sandstone, some 26 m thick. An erosion surface marked by 'brecciated' phosphorite separates the lower and upper parts of the Badaowan Member.

The Badaowan Member contains a uniform fauna which includes hyoliths, monoplacophorans and acritarchs indicative of the late Meishucun Stage *Sinosachites–Eonovitatus* Zone.

(3) The content with the overlying Yuanshan Member of the upper Qiongzhusi Formation is marked by a coarse glauconitic bioclastic grainstone phosphorite layer, 0.20 m thick. This layer is in turn overlain by greyish-black thin-bedded carbonaceous

Table 12.1. *Chemical composition of two high grade phosphorites from the Kunyang deposit*

	1	2
P_2O_5	39.71	39.57
CaO	55.71	55.54
MgO	0.09	0.09
Na_2O	0.26	0.26
H_2O	0.61	0.90
CO_2	1.51	1.47
S	0.08	0.16
F	3.42	3.49
Chemical formula	$Ca_{10}P_{5.63}C_{0.35}O_{23.5}F_{1.81}(OH)_{0.71}$	$Ca_{10}P_{5.63}C_{0.34}O_{23.3}F_{1.86}(OH)_{0.87}$

Analyst: Chemical Laboratory, IGAS.

argillaceous quartz siltstone, 12–24 m thick. The first trilobites occur in the lower Yuanshan Member which is referred to the *Parabadiella* Zone.

(4) The upper part of the Yuanshan Member consists of yellowish-green shale with calcareous and quartzose sandston intercalations, containing trilobites and other fossils of the *Eoredlichia* Zone, 58 m thick.

(5) The Qiongzhusi Formation is truncated at the top by the basal conglomerate of the Haikou Formation – a quartz sandstone and shale sequence containing Middle Devonian fish fossils.

Petrography and geochemistry

Petrography

At Kunyang Mine, three basic types of phosphorite have been recognised in the lower and upper ore horizons of the Zhongyicun Member of the Yuhucun Formation. While all are either light grey, dark grey or bluish grey in colour, they differ in texture from intraclastic, through grainstone to mudstone phosphorite.

The predominant variety is an intraclastic phosphorite in which phosphate clasts may range in size from a fraction of a millimetre to some three centimetres in diameter. Intraclastic grainstone phosphorite constitutes the main part of the ore horizons. However, pebble-sized intraclastic phosphorite occurs as thin beds and lenses a few centimetres thick in the basal parts of both lower and upper horizons, and can be found occasionally in the topmost part of the upper horizon. In general, the phosphatic grains comprising the intraclastic phosphorite are subangular to moderately rounded.

Grainstone phosphorite is the next most important variety, and is recorded mainly in the upper parts of both ore horizons. The thickness of grainstone phosphorite is greatly variable both along strike and down dip. The most common grains are ovoid and semispherical, and show little or no evidence of transportation or abrasion. They can, therefore, be distinguished from the highly abraded and finer-grained intraclastic phosphorites.

Mudstone phosphorite is of minor occurrence in the upper parts of both ore horizons. The thickness of mudstone phosphorite layers is also quite variable.

The Yuhucun Formation is highly fossiliferous and contains originally phosphatic skeletal organisms, as well as those whose skeletal parts have been phosphatised. Grains derived from biogenic sources have been found in all of the phosphorite varieties.

Geochemistry

Examinations made by chemical analysis, X-ray diffraction, optical microscope, and infrared analysis reveal that the Kunyang apatite is a carbonate-fluorapatite with an approximate chemical formula of $Ca_{10}P_{5.63}C_{0.34}F_{1.84}(OH)_{0.79}$ (Table 12.1).

Whole-rock chemical analysis of the Kunyang phosphorites (Table 12.2) shows variations consistent with relative amounts of their mineral phases: apatite, dolomite, calcite, authigenic silica, and quartz sand and silt etc. Close to the surface, the P_2O_5 content of the phosphorites increases due to supergene weathering and leaching of carbonates.

Table 12.2 *Chemical composition of selected phosphorites from the Kunyang deposit*

	CS-9	CS-17	CS-19
SiO_2	13.58	8.62	3.44
TiO_2	n.d.	n.d.	n.d.
Al_2O_3	0.48	0.19	0.23
Fe_2O_3	0.73	0.32	0.22
MnO	n.d.	n.d.	n.d.
MgO	2.52	1.65	1.96
CaO	43.59	48.77	51.21
Na_2O	—	0.28	0.36
K_2O	0.24	0.12	0.2
P_2O_5	26.13	30.84	32.22
F	1.4	2.54	3.22
CO_2	8.55	6.08	6.62
SO_3	0.23	0.03	0.05
H_2O^+	2.05	1.2	0.78
H_2O^-	0.28	0.28	0.39
$-F=O$	0.59	1.07	1.36
Total	99.48	99.85	99.54

Analyst; Chemical Laboratory, IGAS.
n.d., not determined.
-, below detection limit.

References

Cheng Yuqi (1939). A brief report on geology of the phosphorite deposit at Zhongyicun–Waitaoshan district, Kunyang, Yunnan. *Geological Review (Dizhi Lunping)*, **4**, 185–92 (in Chinese).

Lu Yenhao (1979). *Cambrian mineral deposits in China and the bioenvironmental control theory*. Geological Press, Peking, 75pp., 11 pls (in Chinese).

Luo Huilin, Jiang Zhiwen, & He Tingui (1982). On the Sinian–Cambrian boundary in the Sichuan–Yunnan region. *Scientia Geologica Sinica*, **1982(2)**, 215–9 (in Chinese with English abstract).

Luo Huilin, Jiang Zhiwen, Wu Xiche, Song Xueliang, & Ouyang Liu (1982). *The Sinian–Cambrian boundary in Eastern Yunnan, China*. People's Publishing House, Yunnan, 265pp., 36 pls.

Luo Huilin, Jiang Zhiwen, Xu Zhongjiu, Song Xuiliang & Xue Xiaofeng (1980). On the Sinian–Cambrian boundary of Meishucun and Wangjiawan, Jinning Country, Yunnan. *Acta Geologica Sinica*, **2**, 95–111 (in Chinese with English abstract).

Xing Yusheng, Ding Qixiu, & Luo Huilin (1982). Biotic characteristics of the Sinian–Cambrian boundary bed in China and the boundary problems. *Precambrian Research*, **17(2)**, 77–85.

Xing Yusheng, Ding Qixiu, Luo Huilin, He Tinggui, & Wang Yangeng (1983). The Sinian–Cambrian boundary of China and its related problems. *Chinese Academy of Geological Sciences, Special Issue*, **10**, 260pp., 29 pls.

Yeh Lientsun & Sun Shu (1980). Salient features of Chinese phosphorite deposits. In *Fertilizer mineral potential in Asia and the Pacific*, ed. R.P. Sheldon & W.C. Burnett, · pp. 81–2. East-West Resource Systems Institute, Honolulu.

Yeh Lientsun, Sun Shu & Chen Qiying (1980). Origin and characteristics of lithologic associations of Late Sinian and Early Cambrian phosphorites in eastern China. In *Scientific Papers on Geology for International Exchange, Vol. 3, Metallogenesis and Mineral Ores*, pp. 113–20. Geological Publishing House, Beijing (in Chinese with English abstract).

Zhang Wentang (1980). The boundary between the Sinian and the Cambrian in China. *Precambrian Geology. Scientific papers on geology for international exchange*, **7**, Geological Publishing House, Beijing (in Chinese).

13

Proterozoic and Cambrian phosphorites – deposits: Lao Cai, Vietnam

TRAN QUOC AN AND NGUYEN DANG KHOA

Introduction

The Lao Cai phosphorite deposit is one of the largest phosphate deposits of south-east Asia. Phosphorites of marine sedimentary origin have been metamorphosed to greenschist facies to give crystalline phosphate rock averaging 10–36% P_2O_5. Lao Cai phosphate rock is classified on the basis of mineral and chemical composition into four grades known as First Grade Ore (34–36% P_2O_5), Second Grade Ore (23–24% P_2O_5), Third Grade Ore (15–17% P_2O_5), and Fourth Grade Ore (10–12% P_2O_5). Phosphate rock reserves have been estimated at about 260 million tonnes and hypothetical reserves about 1.5 billion tonnes.

The main ore bed, which is known as Coc San 5, is also the principal marker bed. The deposits are folded and faulted into complex inclined and inverted synclines, cut by thrusts, strike-slip faults, and lamprophyre dykes and veins. The phosphatic member is essentially apatite-bearing calcareous quartz and mica-quartz schist. Also in the weathering crust there is an

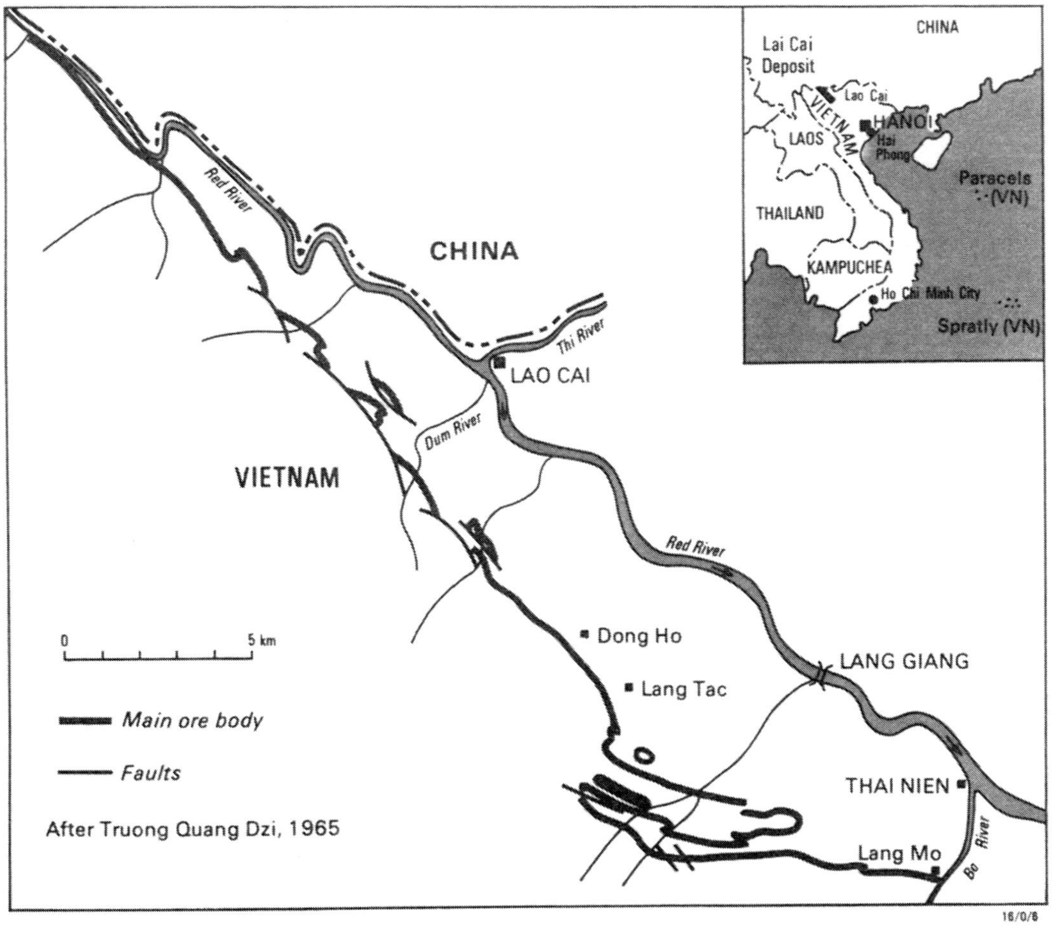

Fig. 13.1. Location of the Lao Cai phosphorite deposit.

apatite-bearing feldspar–quartz–carbonate rock whose carbonate minerals have been leached out during weathering leading to phosphate enrichment and the formation of secondary phosphatic ores. The Cam Duong phosphorite-bearing formation is considered by the authors to be Vendian to Early Cambrian in age. However, because of the lack of palaeontological evidence further study is required.

Location

The Lao Cai phosphate deposit is located in north-west Vietnam, about 300 km north-west of Hanoi, in Hoang Lien Son Province, near the border with China. It extends for more than 100 km strike length along the south-west side of the Red River in a 1–4 km wide zone. The average altitude of this area is about 300 m above sea level (Fig. 13.1).

The deposit was discovered in 1924 and explored by French geologists in 1931. Exploitation commenced in 1940 with open-pit and underground mining. The regional geology of the deposit was first described in detail by Fromaget (1941). However, the stratigraphy was not studied in detail and there were some misinterpretations of the sequence.

The Coc San Series was studied first by Jacob (1921) and then by Fromaget (1941) who distinguished an Algonkian lower series – the Sapa Series, and a Cambrian–Ordovician upper series – the Koksan Series.

During the period 1955–1962, the Lao Cai deposit was

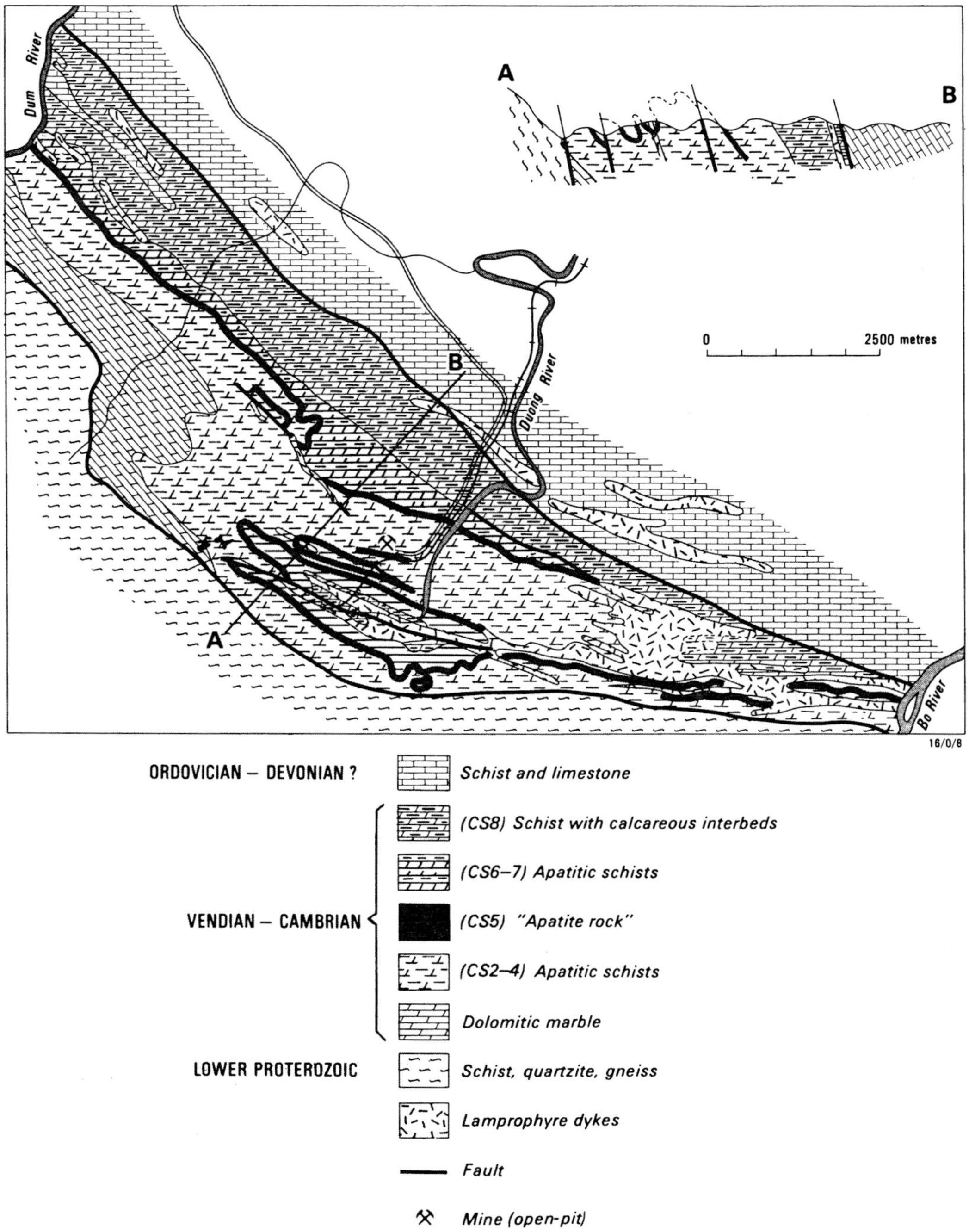

Fig. 13.2. Geological map of the central part of the Lao Cai phosphorite deposit.

studied by Vietnamese and Soviet geologists. Kalmykov *et al.* (1959, unpublished report) and Kalmykov, Sokolov & Tushina (1962) described in detail the stratigraphy and geological structure of the deposit. Although its name was later changed to the Cam Duong Formation, the Coc San Series set up by Kalmykov *et al.* (1959 unpublished report) retains its significance in exploration.

The stratigraphy and age of the Cam Duong Formation was studied by Glazek & Juskowiak (1964), Dovzhikov *et al.* (1965), Tran Van Tri (1967), Tran Van Tri *et al.* (1979), Bui Phu My *et al.* (1967, 1972 unpublished reports) Bui Phu My (1977), Hoang Hoa Cuong & Do Hoang Luong (1977), and Hoang Thai Son & Le Phung Le (1977). The structure and microtectonics of the Lao Cai deposit was described by Onokhin (1961), and Vo Nang Lac *et al.* (1975 unpublished report). The process of metamorphism of the phosphorite-bearing sediments was investigated by Hoang Thai Son (1973), and Do Hoang Luong (1974, 1976).

The Cam Duong phosphorite-bearing formation was dated as Cambro-Ordovician by Fromaget (1941) and Kalmykov *et al.* (1962), Middle Cambrian–Lower Ordovician by Tran Van Tri *et al.* (1979), Early Cambrian by Bui Phu My (1977), Late Sinian by Kalmykov *et al.* (1959, unpublished report), Riphean by Hoang Hoa Cuong & Do Hoang Luong (1977), and Vendian by Phan Truong Thi (1980) and Nguyen Dang Khoa (1981).

The area is structurally complex, with the limbs of inclined and inverted folds dipping to the north-east at angles of 50° or greater. Fold axes are often parallel, and strike SE–NW. They are also separated and displaced by numerous thrust faults and strike-slip faults. Intrusives (lamprophyre dykes) and veins following or cutting bedding planes are fairly common (Fig. 13.2).

Vo Nang Lac *et al.* (1975) have shown that there are five types of folds which range in length from some tens of metres to several kilometres. There are also four categories of faults, which range from a few metres to some tens of kilometres in length. The folds and faults are parallel to the major Red River Fault, which is located in the north-west part of this area. This is an old fault, which was initiated in the Palaeozoic and continued to be active until the Cenozoic.

Stratigraphy

There is overall agreement amongst the various authors on the regional stratigraphic sequence but there are differences of interpretation amongst them on the age, extent and the nature of the contacts between various formations. The stratigraphic sequence is summarised in Table 13.1 and outlined below (see also Figs 13.2, 13.3).

Basement in the vicinity of Lao Cai is a Lower Proterozoic rock sequence in excess of 3000 m thick consisting of strongly metamorphosed garnetiferous micaschists, micaceous quartzites, amphibolite-plagiogneisses with garnet, graphitic micaschists, hornblende–biotite plagiogneisses, and magnetite quartzites.

This sequence is unconformably overlain by the Cam Duong Group, of Late Proterozoic (Vendian)–Early Cambrian age. The Cam Duong Group consists of two formations: the Da Dinh Dolostone at the base and the Cam Duong Formation at the top.

Table 13.1 *Stratigraphy of the Lao Cai area*

Ordovician or Devonian		Dong Ho Formation (750 m)*	
		------------------ Unconformity ------------------	
	Vendian-Cambrian	Cam Duong Formation	
		Supra-ore Member	
		Coc San 8	(40 m)
		Ore-bearing Member	
		Coc San 7	(10 m)
		Coc San 6	(30 m)
		Coc San 5	(12 m)
		Coc San 4	(30 m)
		Infra-ore Member	
		Coc San 3	(15 m)
		Coc San 2	(100 m)
		Coc San 1	(15 m
		Da Dinh Dolostone Member	(300 m)
		------------------ Unconformity ------------------	
Lower Proterozoic		Garnetiferous mica schist (3000 m)	

*Numbers in brackets indicate maximum thickness of unit.

The Da Dinh Dolostone contains dolostone and dolomitic marble, with minor biotite–quartz schists and sericite–chlorite quartz schists of greenschist facies. These rocks have a thickness of 250–300 m. A radiometric age of 600 m.y. has been obtained by Zukov *et al.* (1979).

The relationship of the Da Dinh Dolostone to the overlying Cam Duong Formation is apparently conformable, although could be disconformable in places. This formation contains the phosphorite deposits of the Lao Cai region and is subdivided into three members.

(1) The Infra-ore member (Coc San Units 1–3) at the base consists of 3–400 m of mainly calcareous sericite–quartz schists with intraformational gravelly beds or lenses; calcareous sericite–quartz schists locally containing actinolite, quartzites, sericite–chlorite quartz schists with banded limestone intercalations. Microfossils found in this unit include microphytolites found in the topmost part, *Laminarites* sp., *Octaedrixium* sp., *Bavlinella* sp., *Phycomicetes* sp., and *Protosphaeridium* sp.

(2) The Ore-bearing member (Coc San Units 4–7) contains a total thickness of 100–150 m of sediments. Coc San Unit 4 is apatite-bearing mica–quartz calcareous schists with minor graphite and pyrite. Coc San 5 is the main ore bed – a crystalline phosphate rock 1–12 m thick. Coc San 6 contains apatite-bearing calcareous quartz schists locally bearing tremolite. Coc San Unit 7 at the top of the sequence comprises arkosic apatite-bearing feldspar–quartz calcareous schists and sandstone which contain conglomeratic lenses at the top. Microfossils found in this unit include oncolites, *Medularites limolatus*, microphytolites, *Protosphaeridium* sp., *Laminarites* sp., *Tetraedrixium* sp., *Scaphita* sp., and *Bavlinella* sp.

(3) The Supra-ore member (Coc San Unit 8) is a 40-m thick sequence of sericite–quartz and sericite–biotite–quartz cal-

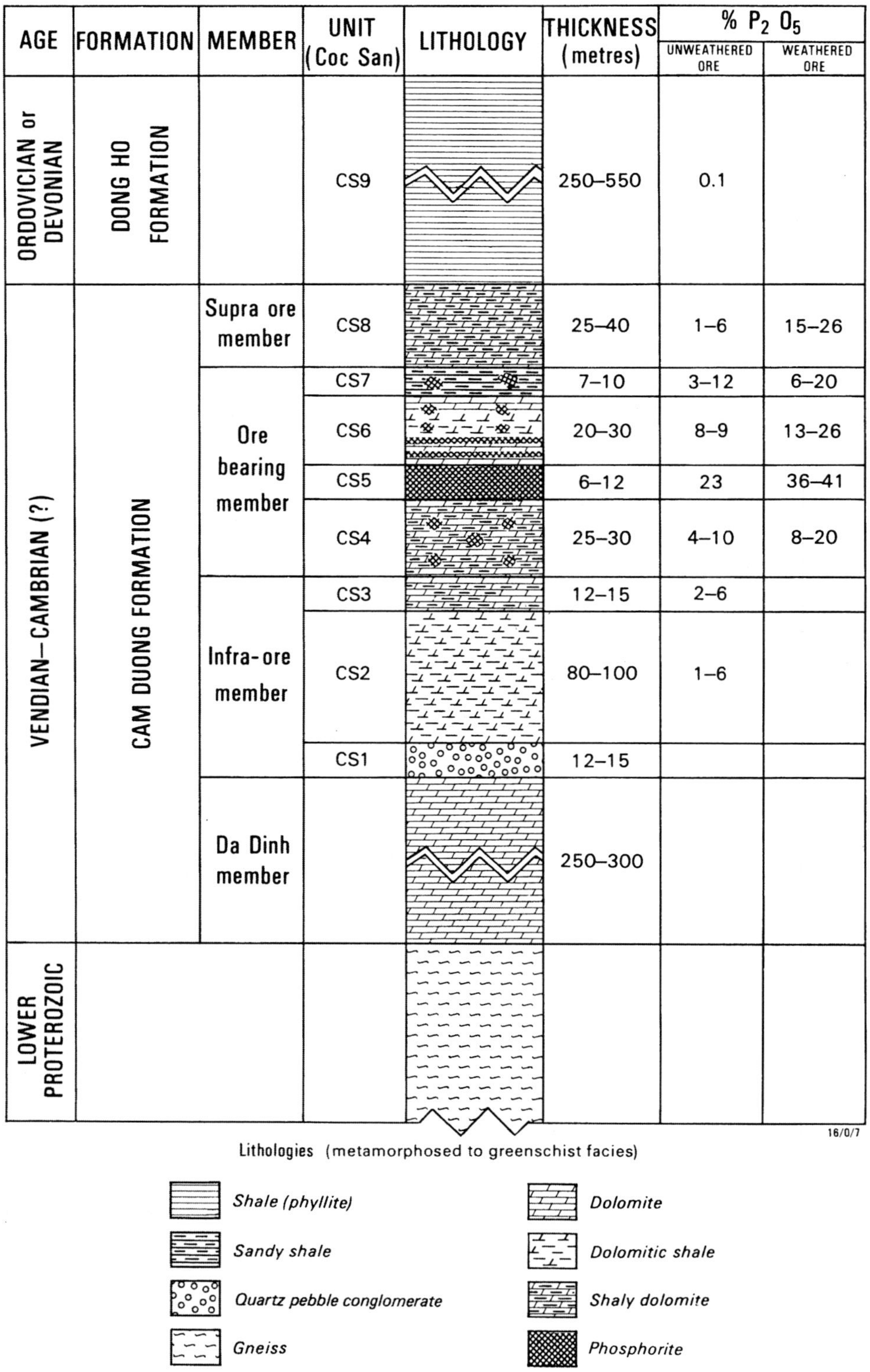

Fig. 13.3. Stratigraphy of the phosphorite-bearing section in the Lao Cai district.

careous schists with dolomitic limestone intercalations. Microfossils include *Ambigolamellatus* sp., and *Vermiculites* sp.

The Upper Proterozoic sequence at Lao Cai is overlain in turn by the Dong Ho Formation with tectonic unconformity. It consists of 550–750 m of actinolite schists (metamorphosed basic effusives), sericite–quartz schists, and slightly metamorphic limestones containing fossils, including *Calathium* (?), *Orbiculoidea* sp., *Lingula* (?) sp., *Desmodonta* gen. indet., Cryptostomata gen. indet., etc. An Ordovician or dubious Devonian age has been assigned to this formation.

Details of phosphatic members

A more detailed sequence of the Cam Duong ore-bearing member is summarised below (Fig. 13.4, Table 13.1).

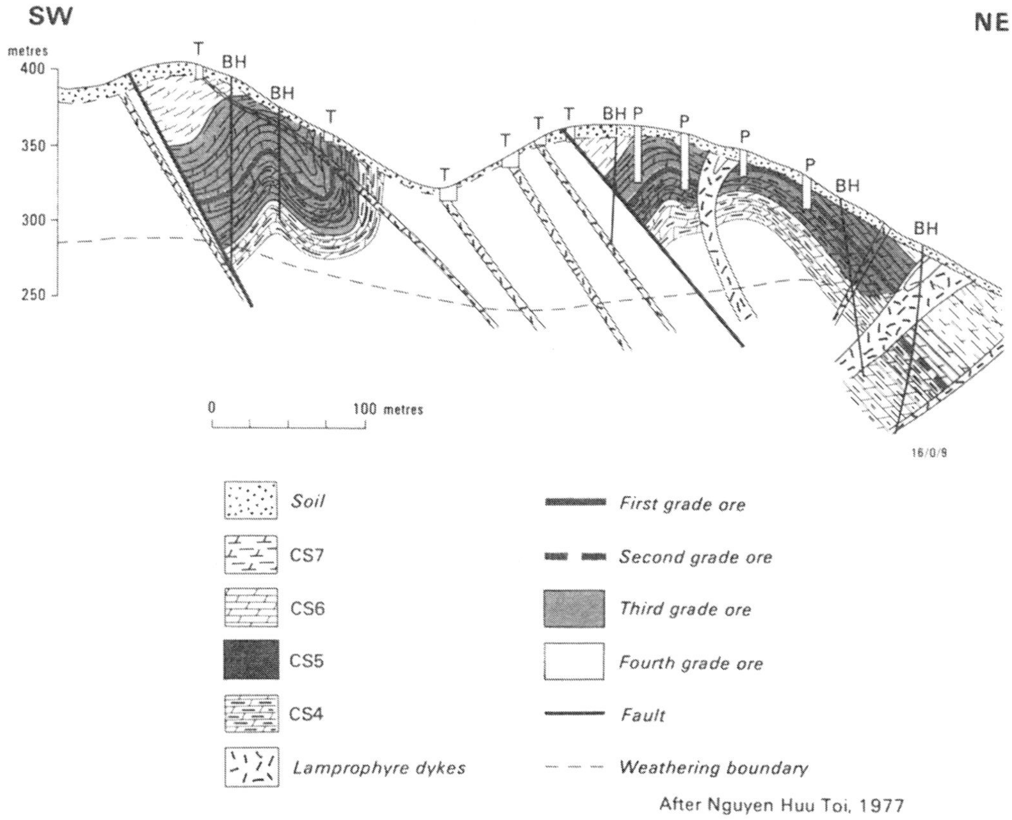

Fig. 13.4. Cross-section of the Lang Tac Block, Lao Cai.

Unit CS-4, the basal portion of the member, is composed of apatite-bearing pyritic mica–quartz calcareous schists 25–30 m thick. The rocks are finely banded and dark grey in colour due to their organic carbon (now graphite) content. The average mineral composition of the unit is 30–40% dolomite, 5–10% calcite, 10–30% quartz, 10–30% apatite, 2–10% organic carbon, 1.5–4% pyrite, and 1–2% manganite.

The phosphate grains are round and 0.03–0.15 mm in diameter. The overall rock texture is fine-grained, banded, and lepidogranoblastic. Some of the more phosphatic intervals consitute low-grade (Fourth Grade) ore. These comprise dolomitic–apatite–quartz schists and apatite–quartz dolomitic schists containing up to 35–40% apatite. In the weathering crust carbonates are leached out and the primary Fourth Grade ore is upgraded by weathering to a Third Grade ore.

Unit CS-5, the main phosphorite bed, is a light grey, fine-grained apatite–carbonate schist which discontinuously crops out, and is displaced by faulting in many places. Thickness varies from 1 m to 12 m. Bedding ranges from massive to poor. The phosphate grains are irregular and up to 0.05–0.07 mm in diameter. The average mineral composition is 65–70% apatite, 10–30% dolomite plus calcite, 5–10% quartz, scapolite, muscovite and pyrite.

Unit CS-6 is an apatite-bearing calcareous quartzose schist 20–30 m thick. It is difficult to distinguish from the underlying rock unit CS-4, the only difference being in the greater abundance of quartz, and a more coarsely banded texture. The phosphate grains are irregular to round with a diameter of 0.04–0.10 mm. The average mineral composition is 25–45% dolomite, 5–15% calcite, 20–30% quartz, 10–30% apatite, 2–15% muscovite, 1.5–2% pyrite, and 0.4–1.5% manganese oxides.

Unit CS-7, the upper unit, is an apatite-bearing calcareous sandstone. It differs from the underlying units in the presence of abundant microcline and orthoclase. It is light grey in colour and is characterised by more massive bedding, a coarse-grained texture and an arkose-like character. The average mineral composition is 15–70% dolomite and calcite, 10–30% microcline or orthoclase, 25–30% quartz, 3–30% apatite and 7–10% biotite and muscovite. At the topmost part of this unit a feldspathic sandstone intercalated with intraformational conglomerates is developed in places. These pass conformably into the non-phosphatic calcareous sericite–quartz schist of overlying Unit 8 which probably formed in a rather different environment of deposition.

Orebody morphology and ore classification

Apart from its metamorphism the Lao Cai phosphate deposit has the usual characteristics of a sedimentary phosphorite. The main phosphatic unit (CS-5) has typical sedimentary characteristics with parallel bedding, a well defined boundary between the overlying and underlying units and gradual changes of bed thickness (from 1 m to 12 m) along strike and dip. The low-grade ore bodies within units CS-4, 5 and 6 are more complex due to the irregular distribution patterns of the grades. They are commonly lens-shaped with fewer phosphatic intercalations. They are generally thicker than Unit CS-5, varying from 2 m to 30 m or more. Because of the greenschist facies

Table 13.2. *Partial chemical composition of average phosphate ores, Lao Cai deposit*

	Primary ores (non-weathered)		Secondary ores (weathered)	
	Second grade %	Fourth grade %	First grade %	Third grade %
P_2O_5	24–26	10–12	34–36	15–17
MgO	2–6	5–12	0.3–1.0	1.3
Fe_2O_3	1.3–1.5	2–5 (R_2O_3)	1–3	3.8
Al_2O_3	0.8–1.2		1–3	5.1
CO_2	5–11	21	—	—
SiO_2	6–7	—	—	—
Acid insolubles	5–9	26	4–9	48.9

— results not available.

regional metamorphism, phosphorites, grade into calcareous apatitic rocks (or crystalline phosphate). In the weathering zone the carbonate minerals are leached out, leading to the formation of weathered secondary phosphate rock (?phoscretes).

On the basis of their mineral and chemical composition the Lao Cai phosphorites are classified into four grades of phosphate ore.

Primary ores (unweathered) (Table 13.2)

(1) *Second Grade ore.* This is a dolomite–apatite ore from Unit CS-5. It is characterised by a high carbonate content and a cut-off grade of 15% P_2O_5. It is typically a grey, massive unit composed of apatite with admixed dolomite, quartz, calcite, pyrite, muscovite and scapolite. This ore is used primarily in the manufacture of thermophosphate as it requires extensive beneficiation (by flotation) to be usable for the manufacture of superphosphate and phosphoric acid.

(2) *Fourth Grade ore.* This is a low-grade calcareous quartzose apatite 'ore'. This ore type from Unit CS-4 is typically a micaceous calcareous apatite 'ore' whereas for Unit CS-6 it is typically calcareous quartzose apatite 'ore'. The characteristic of this ore type is a high quartz and mica content and a cut-off grade of 8% P_2O_5. This grade of phosphate rock is not currently being exploited because of the lack of a satisfactory beneficiation technique. It does, however, constitute a potential resource.

Secondary ores (weathered) (Table 13.2)

(3) *First Grade.* This is a high grade ore from Unit CS-5 with a cut-off grade of 25% P_2O_5. It is typically black and grey, and porous. It is used for the manufacture of superphosphate and phosphoric acid; a limited amount is exported.

(4) *Third Grade.* This is a medium-grade dark, very porous, partly friable quartzose apatite ore characterised by a high quartz content and a cut-off grade of 8% P_2O_5. It takes various forms – micaceous quartz apatite from Unit CS-4, quartzose apatite from Unit CS-6, and feldspathic and quartzose apatite from CS-7. The ore is beneficiated by flotation to 32% P_2O_5 and less than 3% MgO. It is used for the manufacture of various phosphorus compounds and phosphatic fertilizers, and for export.

Biostratigraphy and age of the phosphatic sequence

The Coc San sequence was first defined by Jacob (1921). Subsequently, Fromaget (1941) divided the sequence which overlies the orthogneisses of the Fan-Si-Pan complex into a lower part consisting of Algonkian amphibolite schists, chlorite schists, micaschists and calcareous shales, and an upper part consisting of metamorphosed shales and sandstones of Caledonian age. This upper unit consists of micaceous–sericitic–siliceous calcschists in part phosphatic; ferruginous, tourmaline and graphitic micaceous quartzites also in part phosphatic; micaschists and quartzites passing laterally into graphitic schists with apatitic intercalations, micaceous and sericitic schists and quartzites, transgressively overlying the Sapa Series.

Kalmykov *et al.* (1959 unpublished report, 1962) divided the Coc San Series into two parts–a lower unit, the Da Dinh Series, of ?Archaean age, and an upper unit, considered to be Sinian and known as the Coc San Series. Subsequently, Dovzhikov *et al.* (1965) divided the Coc San Series into three parts – a lower ?Proterozoic unit, a middle phosphatic unit of pre-Devonian age, and an upper unit of possible Devonian age. In their discussion on the phosphatic unit Dovzhikov *et al.* (1965) stated that 'using the comparative method of study, we find that the age of the Cam Duong Formation could be equivalent to the Cambrian phosphatic sediments of Kunming, Yunnan Province (China) which is located along strike of the Can Duong Formation and contains trilobites of Early Cambrian age. Therefore the Cam Duong Formation could represent the littoral facies of the western portion of the Cambrian seas'.

Hoang Thai Son discovered oncolites in the phosphatic interval which were determined by Z.A. Zhuravleva (*personal communication*) as *Medularites lineolatus, Vermiculites tortuosus, Ambigolamellatus horridus* and *Osagia nimia*. According to Zhuravleva, this assemblage is characteristic of the Yudomian (Vendian) of the southern Urals and Siberia. However, Tran Van Tri *et al.* (1979) considered that similar specimens they collected at Muong Khuong, about 50 km north of Lao Cai, lie immediately below Upper Cambrian trilobites with no stratigraphic break. On these grounds, Tran Van Tri *et al.* (1979) consider the Cam Duong Formation to be of Middle Cambrian to Lower Ordovician age. Conversely, Phan Kim Ngan (1976), pointing out that there is no break between the Cam Duong Formation and the Sa Pa Series, supports a Late Precambrian Late Sinian) age for the Cam Duong Formation. Work by Bui Phu My (1977) on a rich algal, pollen and spore flora enabled him to compare dolomites of the Da Dinh Series (the Da Dinh Member here) with limestones of the Upper Sinian Dengying Formation of Yunnan, China. He also proposed that rather than being Vendian as suggested earlier, the oncolites could be of Early Cambrian age. If this was the case, then the base of the Lower Cambrian in the Lao Cai area could coincide with the contact between the Cam Duong Formation and the Sa Pa Series. Phan Truong Thi (1980) and Nguyen Dang Khoa (1981) agree with this interpretation, considering on the basis of the oncolites that the Cam Duong Formation is of Cambrian age. Tran Huu Dan has recently obtained a new collection of acritarchs from the main phosphatic unit (CS-5) which may help to resolve this problem, but results are not yet available.

In conclusion, one of the important questions awaiting resolution is the age of the phosphorites of Vietnam. On the basis of comparison with the nearby Kunyang deposits of China or the more distant deposits of Karatau many geologists regard them as Early Cambrian. However, rare microfossils, microphytolites and oncolites, together with detailed stratigraphic studies have led others to suggest a Vendian to Early Cambrian age. Clarification of the age of the Cam Duong Formation, along with a better resolution of the depositional environments and palaeogeography of the Lao Cai is essential if we are to fully understand Late Proterozoic–Cambrian phosphogenesis in south-east Asia.

References

Bui Phu My (1977). Cambrian sediments of Northwest Vietnam. In *Geological problems of Northwest Vietnam*, pp. 65–81. Science and Technology Publishing House, Hanoi (in Vietnamese).

Do Hoang Luong (1974). Phosphorite formation of the southeastern area of the Lao Cai apatite deposit. *Geological Review*, **111**, 18–20 (in Vietnamese).

Do Hoang Luong (1976). On the homogeneity and heterogeneity of metamorphism of the Lao Cai apatite-bearing formation. *Geological Review*, **128**, 10–12 (in Vietnamese).

Dovzhikov, A.E. *et al.* (1965). *Geology of North Vietnam.* Science & Technology Publishing House, Hanoi, pp. 112–114 (in Vietnamese and Russian).

Fromaget, J. (1941). Étude géologique des gisements de phosphates (apatite) de Cam Duong (Tonkin) et des formations qui les contiennent. *Bulletin du Service géologique de l'Indochine*, **26(1)**, 1–30.

Głazek, J. & Juskowiak, O. (1964). Sur la stratigraphie et la métamorphose des roches de la zone de Bao-Ha–Yen-Bai (Vietnam). *Bulletin de l'Académie polonaise des Sciences*, **12(3)**, 195–204.

Hoang Hoa Cuong & Do Hoang Luong (1977). Discovery of Riphean fossils in the Cam Duong Suite of Lao Cai. *Geological Review*, **123**, 23–25 (in Vietnamese).

Hoang Thai Son (1973). Metamorphic features of the Cam Duong Suite. *Geological Review*, **106**, 23–25 (in Vietnamese).

Hoang Thai Son & Le Phung Le (1977). On the stratigraphy and the age of marine phosphorite-bearing sediments of Hoang Lien Son province. *Geological Review*, **130**, 1–5 (in Vietnamese).

Jacob, C. (1921). Études géologiques dans le Nord Annam et le Tonkin. *Bulletin du Service géologique de l'Indochine*, **10(1)**, 1–204.

Kalmykov, A.F., Sokolov, A.S. & Tushina, A.M. (1962). Mo Kok apatite deposits of Vietnam. In *Geology of Phosphorite deposits*. ed. B.M. Gimmelfarb. *Trudy Gosudarstvennogo Nauchno-issledovatelskogo Instituta Gornokhimicheskogo Syrya*, **7**, 139–91.

Nguyen Dang Khoa (1981). On the stratigraphic position of the Cam Duong metaphosphorite-bearing formation. In *Scientific Seminar of the Institute of Geology and Mining, Hanoi*, Abstract, p. 22.

Onokhin, F.M. (1961). Unpublished report on the structural setting of the Lao Cai phosphate deposit. Department of General Geology, Hanoi.

Phan Kim Ngan (1976). Biostratigraphy of the Cambrian in North Viet Nam. *Geological Review*, **125**, 1–3 (in Vietnamese).

Phan Truong Thi (1980). Geology of metamorphic complexes of southeast Asia. Doctoral Thesis, University of Moscow, 364 pp. (unpublished).

Tran Van Tri (1967). On the stratigraphy of metaphosphorite (apatite bearing sediments of Cam Duong, Lao Cai). *Geological Review*, **69**, 12–9 (in Vietnamese).

Tran Van Tri *et al.* (1979). *Geology of Vietnam – Northern part*, pp. 29–30 Science & Technology Publishing House, Hanoi (in Vietnamese).

Zukov, S.I. *et al.* (1979). Absolute dating of Precambrian and other rocks of Vietnam by the Uranium–Thorium–Lead method. *Review of Earth Sciences*, **1(2)**, 57–62 (in Vietnamese).

14

Proterozoic and Cambrian phosphorites – deposits: Khubsugul, Mongolian People's Republic

A.V. ILYIN, N.S. ZAITSEV AND Z. BJAMBA

Introduction

The Khubsugul Basin of southern Siberia and northern Mongolia contains predominantly calcareous sediments of latest Precambrian–Early Cambrian age. Many phosphorite occurrences are known in southern Siberia notably in the Eastern Sayan, Gornaya Shoriya and Minusinsk areas, but they are generally small low-grade deposits. However, in northern Mongolia, in the Khubsugul Basin, a total of eight large deposits have so far been discovered (Ilyin & Bjamba, 1980), the largest of which is the Khubsugul deposit (Fig. 14.1). The Khubsugul deposit is located in a mountainous area on the west side of Lake Khubsugul within a deep rift valley (Ilyin, 1973). It is this deposit which will be considered in some detail.

Stratigraphy, age and structure

Two principal rock assemblages are encountered in the vicinity of the Khubsugul deposit, notably Precambrian crystalline basement and Riphean (latest Precambrian) to Cambrian sediments (Fig. 14.2). The Precambrian basement consists of gneisses, calcsilicates and greenschists with the latter often progressively matamorphosed into high-grade crystalline rocks. Conversely, the gneisses are sometimes retrogressively metamorphosed into rocks similar to greenschists.

The basement is cut by numerous granites, the youngest of which is 823 m.y. old (by K/Ar whole rock). An Archaean–Proterozoic age has been proposed for the basement, but others consider the basement rocks no older than Riphean, basing their views on the fact that subsequent geological events have reworked the basement during Baikalian (the earliest Cambrian) and Caledonian orogenies.

The Darkhat Group lies unconformably on the basement. Its section commences with basal conglomerates passing upward into arkoses. Red-coloured shale and siltstone appear in the upper part of the Group with some dolomite layers. Near the top of the sequence, tillite-like rocks appear with numerous angular dolomite clasts. Volcanics of basaltic or rhyolitic composition are present in the group with comagmatic alkaline intrusions dated at 752 m.y. (K/Ar, biotite). The Khubsugul Group, which is 3000 m thick and dominated by carbonates, rests conformably on the Darkhat clastics, although with a depositional break. The Khubsugul Group includes both limestones and dolomites, the latter clearly confined to nearshore facies and the former in more seaward environments. A phosphatic interval occurs near the base of the thick carbonate succession which comprises the bulk of the Khubsugul Group.

The Khubsugul Group covers a broad time interval, from Late Riphean or Vendian at the base to Middle Cambrian at the top. Korobov (1980) has studied the sequence in the Khesen Valley and described two trilobite horizons; the lower with the genera *Elganellus* and *Malykania* and a slightly younger one with *Fallatospidella* and *Bulaiaspis*, both indicating a late Aldanian (Atdabanian)–Early Cambrian age. The phosphatic series is located some 500 m below the lowest known trilobite horizon and has been attributed by Korobov (1980) to the initial, so-called non-trilobite part of the Cambrian (the Tommotian Stage). However a latest Precambrian (Vendian) age of the phosphate series cannot be excluded since there are medusoid fossils in some deposits to the north of Khubsugul, correlated by B.S. Sokolov (personal communication) with the late Proterozoic Ediacarian System (Cloud & Glaessner, 1982) of southern Australia. In the Boxon area of the Eastern Sayan Ridge, rocks thought to be equivalent, with K/Ar ages of about 600 m.y. contain a bauxite layer (Ilyin, 1973). Throughout the Eastern Sayan Ridge the phosphatic sequence lies some 700 m below the earliest Cambrian fossil occurrences.

The structural pattern of the deposit area is dominated by the Khesen Syncline on the eastern side of the Basin striking meridionally along the shore of Lake Khubsugul (Fig. 14.3). The Khesen fold resembles to some extent a small basinal syncline with clearly defined northern closure, a more complicated southern closure and steeply dipping limbs. Dislocations within it are rather marked but probably of local extent caused by tectonic movements associated with the fault along the west side of the Lake – the oldest tectonically active feature, which now marks the western border of the Khubsugul rift. Some 5–7 km to the west of Lake Khubsugul is the Mongolian Fault, which cuts the core of the adjacent Mongolian Anticline, and still further to the west is the dome-like Arasan Anticline. All these structures control the outcrops of the phosphatic series, the most important of which are the

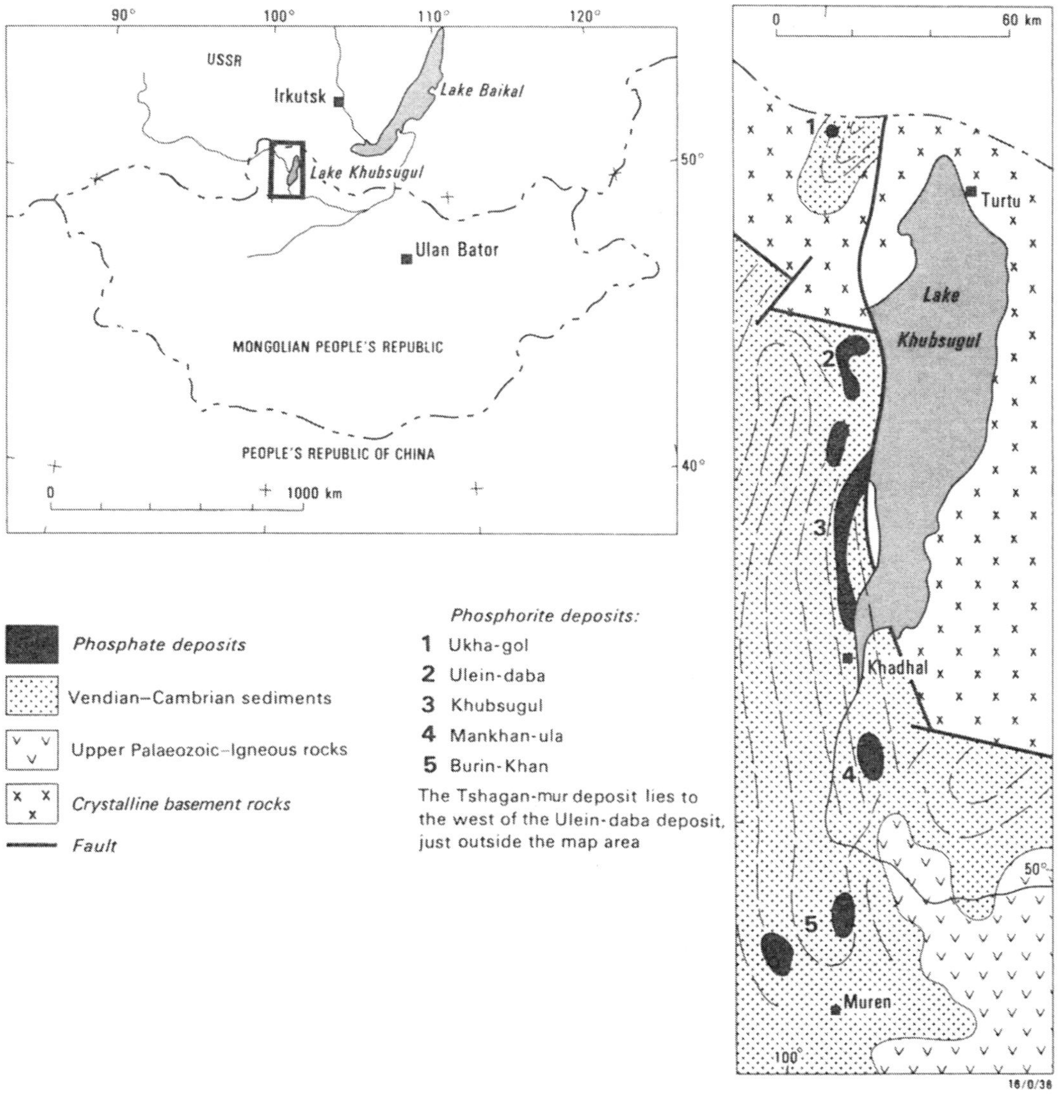

Fig. 14.1. Location of the Khubsugul Basin.

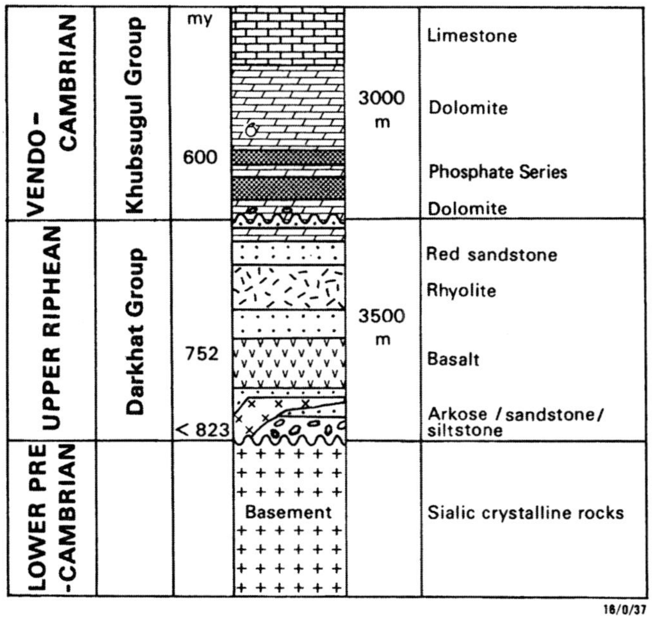

Fig. 14.2. Generalised stratigraphic sequence of the Khubsugul basin.

phosphorite beds flanking the Khesen Syncline in which the main phosphate reserves of the basin are concentrated.

The phosphatic interval (Fig. 14.4) rests either on a lower cherty dolomite or black, thin-bedded platy limestone, with a characteristic red-coloured 'underphosphate unit'. The phosphatic interval contains two productive units of which the lower one is the main ore horizon (Fig. 14.4). A chert bed caps the main phosphorite. The bed contains black, mostly non-phosphatic cherts, best exposed and most persistent along the strike with a thickness of 20–30 m. The chert is overlain by a unit 70–80 m thick composed of dark thin-bedded dolomites, with limestone in the middle part.

This is followed by the upper phosphatic unit exhibiting sharp facies changes. In general, the phosphatic series commences and finishes with coarse clastics, with pebbles derived from adjacent beds. These clastics grade upwards into finer-grained sediments and then into siltstone and marls and finally into phosphorites. This succession suggests a transgression, with a maximum water depth corresponding to the occurrence of the limestone overlying the black chert unit. The rest of the sequence appears to be regressive with a minimum water depth corresponding to the

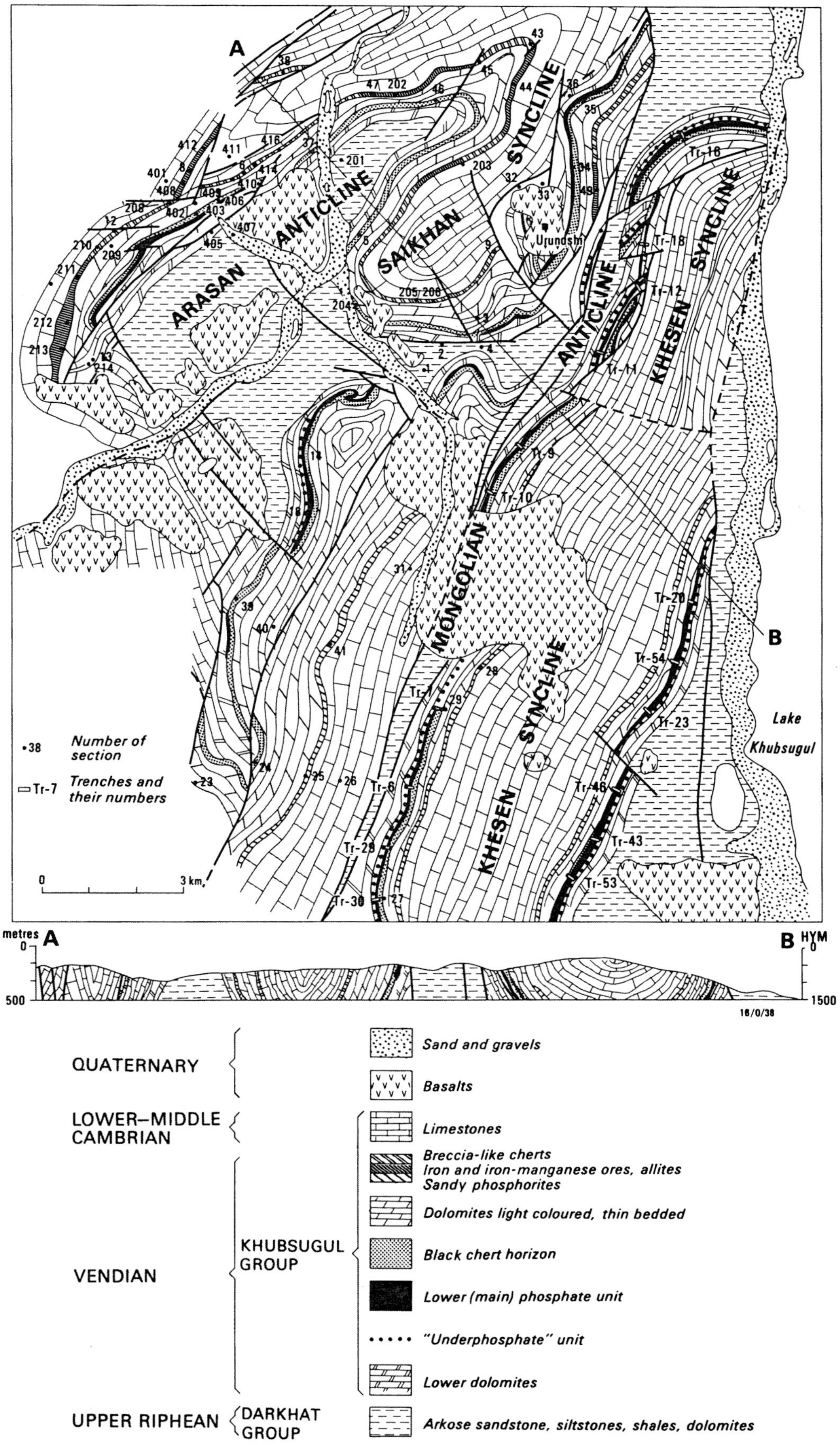

Fig. 14.3. Geological map of the northern portion of the Khubsugul deposit.

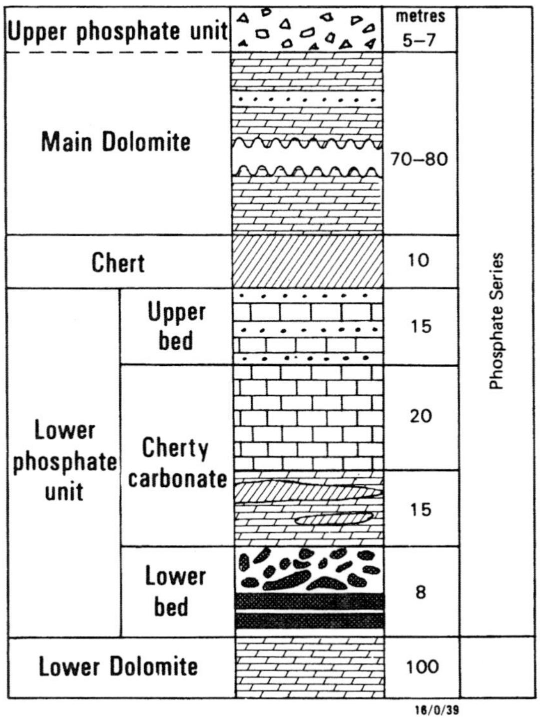

Fig. 14.4. Stratigraphic section of the productive phosphorite interval. Symbols: ,varigrained phosphorites; ,phosphatic sandstones and dolomites. Other symbols as in Fig. 14.2 and 14.3.

upper phosphorite. So in the broader sense, the duration of the whole of the phosphatic part of the Khubsugul Group correlates with the initial phase of a major transgression which climaxed during the later Early Cambrian with the accumulation of a monotonous limestone series. Carbonate sedimentation then continued into Middle Cambrian time.

The lower phosphatic unit shows many facies changes and consequently it is not uniform in its phosphate distribution. There are three basic types of sequences:

(1) Sequences where phosphate concentrates (22–28% P_2O_5) at the base and the top, forming lower and the upper horizons or beds each about 10 m thick. The remainder of the sequence is composed of limestone and dolomite, with no appreciable quantity of phosphate. Trench 21 provides an example of such a sequence (Fig. 14.5).

(2) Sequences where phosphorite beds are separated by phosphatic dolomites and cherty dolomites. The average phosphate content of these types of phosphorites in trench 23 (Fig. 14.5) is about 22% P_2O_5. Dolomites show a considerable range in P_2O_5 content ranging from several per cent up to 15%. The contacts between the phosphorites and the dolomites are generally gradational. In most places there are five phosphorite layers though less commonly up to seven. With such a pattern of phosphate distribution in the phosphatic series it is difficult to correlate from one trench to the next (Fig. 14.5).

(3) Sequences which are essentially phosphatic throughout. Non-phosphatic rocks such as cherts and 'diatomite'-like siliceous rocks are present only as thin beds and intercalations within the phosphorite, Trench 16 (Fig. 14.5) providing an example of this type of sequence. The lower phosphate unit has been traced along the eastern limb of the Khesen Syncline for about 25 km.

The western part of the deposit, within the Arasan Anticline, is of considerably lower grade. The productivity falls to 1–3 tonnes P_2O_5/m^2. Only thin phosphorites are present and even these are low in P_2O_5. In this same area the upper phosphate unit changes to manganese and iron–manganese ores and low-grade bauxites. Some of the Fe and Fe-Mn ores are phosphorus-poor, but generally the rock types included in the phosphate series are geochemically related. The manganese ores and allites contain up to 3% P_2O_5 and phosphorites and some associated sediments locally contain thin lens-like intercalations of manganese minerals. In general the phosphatic series shows frequent facies changes but the overall rock assemblage remains constant, consisting everywhere of phosphorite, chert, dolomite and limestone although the relative proportions of the various lithologies changes considerably. The amount of phosphate in the cherts and dolomites also varies. The productivity of the series ranges broadly from 5–8 to 40–45 t/m^2 P_2O_5*. The productivity of both of the phosphatic horizons diminishes basinwards, as their lithology varies.

Conglomerates and red-coloured beds wedge out completely to the west and dolomite grades into limestone. The facies changes show a source area located immediately east of the Khubsugul deposit. They also indicate phosphorites formed in a nearshore location.

Deposits of the Khubsugul Basin

There are two phosphogenic zones in the Basin – eastern and western, some 80 km apart. The Khubsugul deposit itself lies in the middle part of the eastern zone and will be discussed in detail below. The Ukhagol deposit occupies the northern flank of the zone and the Burin-Khan deposit the southernmost position.

The *Ukhagol deposit*, located partly in Mongolia and partly in the USSR, occurs in a high mountainous area. The phosphatic series crops out extensively and is notable for its generally black colour which contrasts sharply against the white carbonate rocks composing the adjacent parts of the sequence. The Ukhagol phosphorites are mostly grainstone phosphorite with a considerable admixture of fine, poorly rounded quartz grains. Unlike the other deposits, argillaceous phosphorites are significant here. Chemically the Ukhagol phosphorites are siliceous rather than calcareous with a SiO_2 content of about 15% and a P_2O_5 content of 20–25%. The organic carbon content is relatively high.

The *Burin-Khan deposit* has a chemical composition that differs markedly from that of the Khubsugul deposit. The most common phosphorite is composed of coarse (1–2 mm), rounded to angular phosphate grains, which commonly have a microcrystalline quartz core. The phosphate grains are cemented by a microcrystalline quartz matrix. In hand specimen the Burin-Khan phosphorites are massive unbedded black rocks, indistinguishable from cherts. There is every gradation along

*The amount of P_2O_5 present in 1 m^2 of the total thickness of the phosphorite unit.

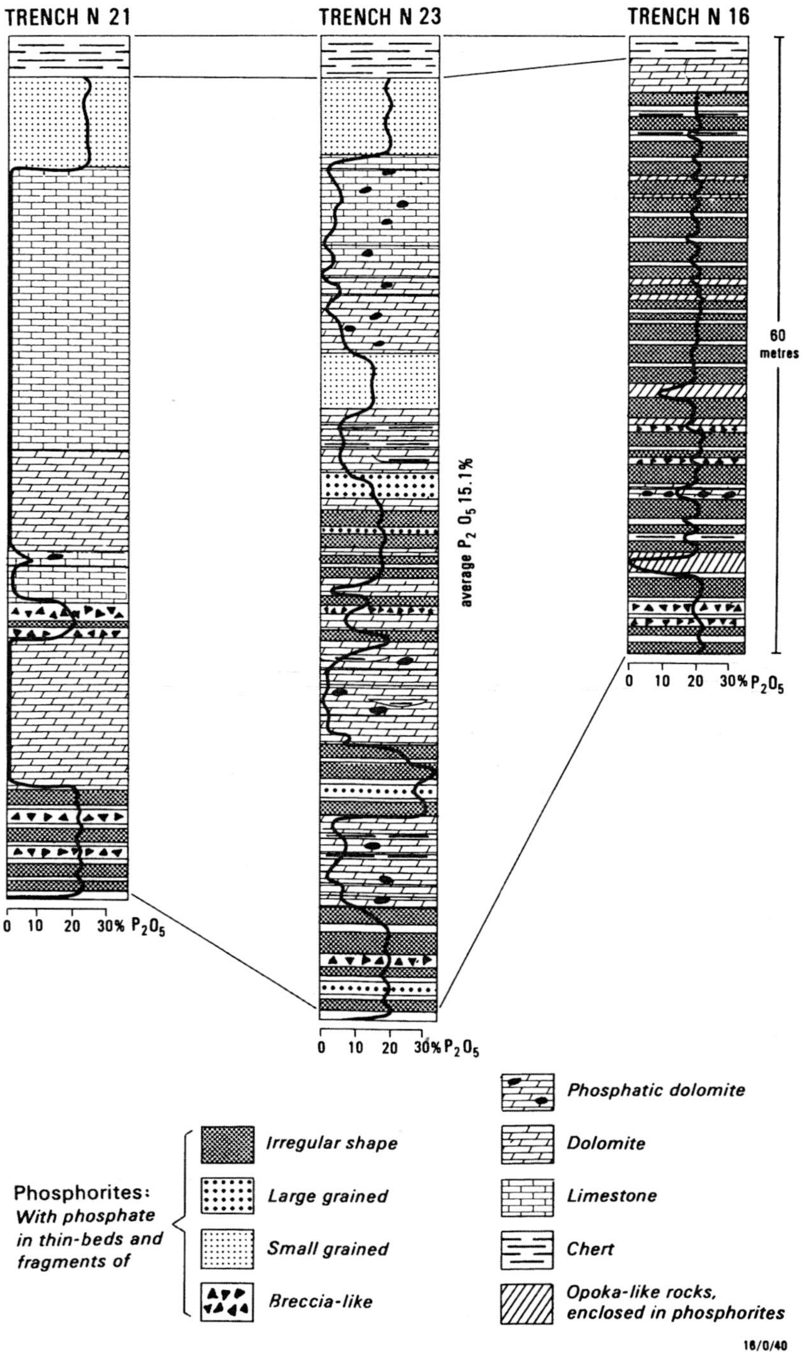

Fig. 14.5. Lithological variation in the lower phosphate unit in Trenches 16, 21, and 23, Khubsugul deposit.

strike (and across it) from weakly phosphatic cherts to siliceous phosphorites with 30% P_2O_5 and more. The Burin-Khan deposit has not been studied in any great detail but there is strong evidence to suggest that there has been extensive weathering, resulting in carbonate leaching and replacement with silica. The net result of this weathering is considerable relative enrichment of phosphate in the near-surface zone of the deposit. In sub-surface, fresh unweathered phosphorites have been encountered in a number of boreholes. Phosphatic uncemented sands, and 'karst phosphorites', have also been found in the sub-surface.

The *Tshangan-nur deposit* occurs in the western phosphogenic zone of the Basin. Five phosphate beds are present in the deposit, which is composed mostly of mudstone phosphorite. Chemically these are calcareous phosphorites with 6–8% SiO_2 and 25–27% P_2O_5. Coarse phosphatic sandstone, gravel, and conglomerate are developed in the upper part of the Upper Unit of the Tshangan-nur deposit.

Lithological framework of the Khubsugul deposit

The phosphorites of the Khubsugul deposit are clearly associated with dolomite, chert and manganese or iron–manganese ores. The association with dolomite is particularly close and distinct, and it is evident on a broad palaeogeographical scale that both kinds of rocks are associated with a nearshore facies. It is also apparent in the close association of apatite and dolomite in phosphorites. The dolomites underlying phosphorite beds are usually brecciated. Chert incrustations, stringers and irregular intercalations are also common within or protruding from dolomites. Algal structures and burrows are also fairly common in the dolomites. Within the phosphatic unit itself, dolomites contain chert intercalations and show breccia-like structures, whereas dolomites overlying the productive unit are light-coloured and thin-bedded. About 300 samples of carbonate rocks were analysed from the area of the Khubsugul deposit (Bliskovsky, personal communication). It was found that only 10% of the samples approximated a dolomite composition. The bulk of the samples showed a range of MgO content from 1% to 20%. The strontium content of the dolomite ranged from 80 ppm to 145 ppm, the average being about 100 ppm. However, more stoichiometric dolomites were richer in Sr, and a well defined inverse relation exists between Mg and Sr in dolomites and magnesium limestones. The sodium content of the dolomites averaged 370 ppm. On the whole, all the dolomites analysed are considerably lower in Sr and Na than would be expected for phases precipitated from sea water of normal salinity.

The limestones overlying the phosphorites of the Khubsugul deposit area are 10–20 times richer in strontium than the dolomites. These geochemical considerations, coupled with the preferred nearshore location for dolomites, suggest that the dolomites formed due to diagenetic replacement of primary disordered carbonate muds following the mixing of sea water and phreatic fresh water derived from subaerially exposed areas. The plausibility of this suggestion is strengthened by the latest Precambrian palaeogeographic reconstructions which indicate that land was located to the east of Lake Khubsugul. The Early Cambrian transgression resulted in the source area lying further to the east of the Khubsugul deposits and as a consequence fresh water could no longer reach the Khubsugul area, thus preventing the formation of dolomite. The association of phosphorites with chert is very marked. The whole sequence of the thick Khubsugul Group contains one prominent chert layer, black in colour and 20–40 m thick, occurring immediately above the upper phosphatic unit. A similar black chert layer occasionally occurs at the base of the lower (main) phosphatic unit, although it is thinner and rather less persistent than the upper chert. Within the phosphatic unit itself cherts are relatively common, occurring mostly as clasts up to 50 cm in diameter within a phosphatic dolomite matrix. In the southern part of the basin numerous fine chert interbeds, 3–5 mm thick, are common within phosphate beds, producing a series of phosphate–chert–phosphate laminae. Fine silica is often found included within phosphate grains, comprising in most cases about 20% of the grains, whereas the whole rock contains 6–8% SiO_2. The black chert member which intermittently overlies the lower (main) phosphatic unit is generally non-phosphatic, but large clasts of chert included in some of the dolomites are always slightly phosphatic.

Cross-sections through the Khubsugul deposit area (Fig. 14.6) show that cherts mostly accumulated to the west (seaward) of the areas where phosphatic sedimentation was most active. West of the Khubsugul deposit the phosphatic unit does not crop out and spatial relations are unknown between cherts and phosphorites in what is inferred to be the more offshore region. The southern portion of the eastern phosphogenic zone exhibiting the more offshore phosphatic facies is represented by siliceous phosphorites. In the case of the Mankhan-Ula deposit it is suggested that the phosphatic facies are replaced basinward by cherty rocks. Thus, cherts are closely related to phosphorites, comprising, together with dolomites, the productive phosphatic unit. On a broader scale they show some inverse relationships in spatial distribution since the zone of maximum chert accumulation lies outside of, or on the periphery of, the phosphogenic zone. Compared with other rock types of the phosphatic unit, cherts are enriched in Ni (200–500 ppm), Co (500 ppm) and Zn (600–800 ppm) but are relatively poor in manganese.

The association of phosphorites with iron and manganese ores is best developed in the western part of the Khubsugul area where they are strictly confined to the so called 'red chert horizon', a breccia-like multi-coloured chert. The iron and

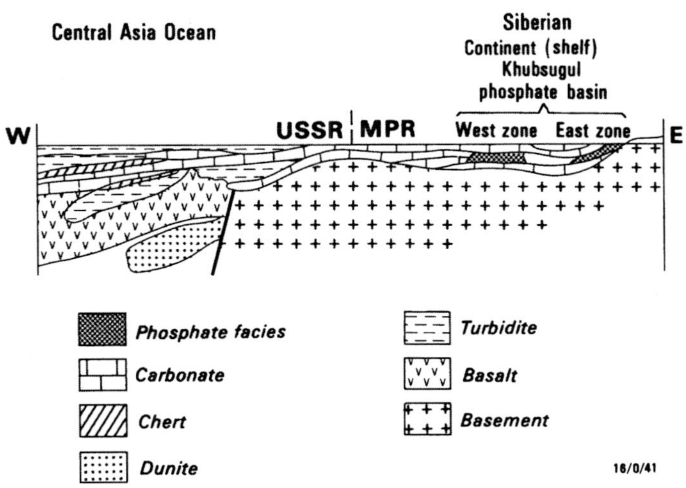

Fig. 14.6. Schematic section through Siberian continental shelf at about latitude 50° N (not to scale).

Table 14.1. *Major (%) and minor (ppm) element analysis of Khubsugul phosphorites and associated rocks*

	1	2	3	4	5	6	7	8	9	10	11
P_2O_5	26.3	12.0	15.05	11.9	8.21	22.1	20.1	29.9	5.5	2.3	—
CaO	43.3	37.5	40.3	37.2	33.6	36.8	39.2	49.1	10.7	5.6	2.20
F	2.1	1.1	1.4	1.2	0.8	2.1	1.9	1.5	0.5	—	—
CO_2	10.3	31.1	28.3	29.9	31.2	10.5	21.2	12.5	5.6	20.4	—
MgO	3.8	11.9	11.2	11.6	12.3	4.5	8.3	4.4	72.1	1.0	—
SiO_2	9.5	6.6	9.2	7.4	13.0	18.7	6.2	0.1	—	2.1	30.1
Al_2O_3	1.3	0.2	0.4	0.2	0.2	1.7	0.9	0.1	0.9	2.6	30.1
Fe_2O_3	1.2	0.3	0.5	0.4	0.4	0.7	0.8	0.4	0.9	6.7	23.2
FeO	0.8	—	0.3	0.3	0.3	0.1	0.2	0.1	0.5	0.8	—
K_2O	0.3	0.5	0.1	0.1	0.1	1.2	1.1	0.1	0.1	0.1	—
Na_2O	0.2	0.1	0.1	0.1	0.1	2.0	1.2	0.1	0.1	0.1	—
MnO	—	—		—	—	—	—	—	—	47.1	5.8
MnO_2	—	—	—	—	—	—	—	—	—	10.7	—
SO_3	—	—	—	—	—	—	—	—	—	—	—
C_{org}	0.5	0.2	0.4	0.3	0.1	0.9	—	—	—	—	—
Sr	1400	510	1300	480	400	1200	1000	700	—	75	120
As	1	13	20	5	18	2	2	10	—	20	30
Zn	6	10	5	8	15	8	10	18	—	35	8
Pb	1	1	3	4	3	8	7	8	—	20	48
Ag	1	1	1	2	2	1	1	4	—	70	20
Cu	8	8	10	7	5	10	8	15	—	20	10
Co	7	10	4	8	6	5	8	4	—	6	4
Ni	66	68	14	40	70	50	80	87	—	200	300
Cr	53	57	8	32	41	14	45	40	—	16	10
Mo	2	1	1	3	2	3	2	7	—	14	—
V	22	15	50	40	32	28	48	45	—	100	80
Mn	75	1000	340	600	340	410	800	600	—	1000	1000
U	11	7	9	5	3	2	5	13	—	2	8

Samples 1–9 are from the lower phosphate unit as follows:

1. The lowest mudstone phosphorite bed, 7 m thick.

2–5. From the middle part of the lower phosphate unit corresponding to beds which are 3, 2, 2.5 and 3 m thick respectively. These are phosphorites with irregular grain shape and size. (Samples 1–5 are from Trench 25.)

6. The lowest mudstone phosphorite bed, 13 m thick, in Trench 20.
7. The lowest fragmental mudstone phosphorite, 4 m thick, in Trench 43. (Samples 6 and 7 are channel samples.)
8. Oolitic grainstone phosphorite from the upper bed of the lower phosphate unit in Trench 25.
9. Phosphatic sandstone in lower phosphate unit. (Samples 8 and 9 are hand specimens.)
10. Manganese carbonate–oxide ore.
11. Bauxite (allite).

manganese ores occur at a stratigraphic position which can be confidently correlated with the same level as the upper phosphate unit to the east. Further west from Lake Khubsugul, there is a well-defined lateral succession commencing with phosphatic calcareous sandstone, followed by siliceous phosphatic sandstone, gritstone and multi-coloured breccia-like cherts. The iron–manganese and manganese ores are found only within distinctive multi-coloured chert occurring as lens-like bodies, 1–2 km long and up to 15 km thick. The chert horizon itself has been traced for about 15–20 km, and ten ore bodies have been found within this distance, the largest being about 3 km long, and 15 m thick. In places cherts are totally replaced by Fe–Mn ores. Manganese ore bodies vary in composition: some are entirely composed of manganese, whilst others are low-grade manganiferous bauxites or iron–manganese ores.

Megascopically two kinds of manganese ores can be recognised: one dense, heavy, crystalline and associated with red-brown carbonate or carbonate–oxide; and the other earthy, porous, black and oxidised (Table 14.1). The first type consists of rhodochrosite (rarely ferro-rhodochrosite), manganite and brownite. The other type has no carbonate minerals, and is composed primarily of pyrolusite and psilomelane, along with brownite and manganite. The most oxidised ores are composed entirely of pyrolusite and psilomelane.

Petrography of the phosphorites

Each of the major phosphorites of the three main phosphatic units differ in dimensions, colour and in their textural features (Ilyin & Ratnikova, 1981).

(1) The basal bed of the lower unit (Fig. 14.4) is the most

persistent phosphorite. It includes several varieties of phosphorite which alternate on a variety of scales, ranging from microscopic upwards. The most conspicuous, though relatively uncommon type, is thin-bedded mudstone phosphorite 1–3 mm thick, interbedded with dolomite, 1–2 mm thick (Fig. 14.7A). Granular mudstone phosphorite, grainstone phosphorite and intraclastic grainstone phosphorite occur more commonly. Mudstone phosphorite is composed of submicroscopic apatite, and generally devoid of any internal structures, though rarely the phosphate shows a laminated and fibrous structure. It is dark brown or dark yellow-brown in colour. Some rocks are 'hieroglyphically' marked and irregularly mottled with black carbonaceous matter and micritic–dolomitic intercalations (Fig. 14.7B). Very rare argillaceous laminae are found alternating with some phosphorite laminae. Field observations show convincingly that mudstone phosphorite is the parent material for many other morphological types of fragments, grains and ovules. The grain types most commonly found in association with mudstone phosphorite are interbeds composed of elongated platy fragments of mudstone phosphorite in a dolomite matrix and oriented either parallel or oblique to the general direction of bedding. Without too much imagination, such fragments can be joined together to form a continuous phosphate layer which has undergone limited *in situ* brecciation, but little or no transport. With further brecciation, the phosphorite acquires the appearance of an intraformational breccia, with larger, equant and elongated fragments surrounded by smaller phosphate particles and all in a dolomite matrix.

Many mudstone phosphorites lose their textural uniformity in thin sections and become indistinctly 'grained' with darker-coloured and ill-defined 'grains' 0.5–1.0 mm across. These grains probably contain more carbonaceous matter relative to the lighter-coloured phosphate comprising the matrix. Microfractures developed within the phosphate matrix help to emphasise the grain borders. At a more advanced stage of granulation, dolomite penetrates along fractures and progressively replaces intergranular phosphate by intrusion from the periphery of the interbedded phosphate (Fig. 14.7C). Gradually dolomite fills the space, pushing apatite grains aside to produce a new carbonate matrix. During this diagenetic process, phosphate grains retain spherical or elliptical outlines. Carbonaceous matter seems to be mainly concentrated within large, black phosphate grains, 0.5–0.7 mm across (Fig. 14.7D) which also always contain euhedral rhomboidal dolomite crystals, 0.2–0.3 mm across. Phosphate grains become more clearly defined when micro-crystalline quartz or chalcedony are concentrated within them.

The most common variety of Khubsugul phosphorite consists of phosphate fragments, 0.5–3 mm across, in a dolomite matrix. In thin-section these phosphate fragments are most frequently black, and irregularly shaped, with projections and re-entrants along their margins (Fig. 14.7E), their shape defined by the dolomite which is replacing them. They are devoid of any internal structure and are believed to originate from the fragmentation and dolomitisation of beds of mudstone phosphorite. They are also devoid of any clastic admixture or signs of transportation. The phosphatic fragments contain 1–3% carbonaceous matter. In several areas the lower bed is a grainstone phosphorite (Fig. 14.7F) with spherical or ovular grains usually 0.5 mm across, composed of light-coloured phosphate. Several kinds of grains are discernible (i) ellipsoidal or spherical grains composed of uniform brown phosphate, internally structureless and devoid of inclusions; (ii) zoned grains with a black core and light rim, the outer ring occasionally composed of radially oriented apatite crystallites; (iii) phosphate grains with inclusions of quartz grains and dolomite rhombs; (iv) compound grains consisting of several simple ovules inside and accretionary, light-coloured phosphate outside. Compound ovules are accompanied by coarser quartz grains and reflect a complex depositional history. Grainstone phosphorites bear a strong resemblance to 'granulated' mudstone phosphorites which have undergone replacement of phosphate matrix by dolomite. Frequently, these two types of phosphorite are interbedded. The principle difference is that clastic quartz grains are always associated with the grainstone phosphorites but never observed in the mudstone phosphorites.

(2) The main phosphorite of the upper bed of the lower unit is a massive, uniform rock, with a chert-like appearance. Thin-sections reveal that the phosphate occurs as grains 0.08–0.1 mm in diameter. Three kinds of grains are recognisable:- pellets, ovules and ooids. (i) Grains (pellets) with quadrangular to hexagonal elliptical or circular outlines which are packed so densely that no intergranular space is left. The phosphate making up the grains is quite 'clean' and coloured in shades of grey in ordinary light. The outer portion of some grains consists of finely crystalline apatite, which gives the grains a characteristic appearance in thin-section. Sometimes the outer rims show a concentric pattern. Such distinctive rocks are known as 'oolitic grained' phosphorite in the Karatau Basin and as 'international-type' phosphorite in the Khubsugul Basin, because it has been described from so many phosphate basins in the USSR (Tushina, 1979, Fig. 2b, p. 113) and southern China (Bushinsky, 1966, Table XV, 4; Table XVI, 1). (ii) Grains (ovules), 0.08–0.1 mm in diameter, with a simple or compound form and one or more nuclei. The grains are composed of light-brown phosphate and are somewhat similar to the grains of the first group, though they are smaller, better sorted and always accompanied by phosphate ooids and fine angular quartz grains (Fig. 14.7G). In some areas quartz grains are coarser and more abundant, giving rise to sandy grainstone phosphorites, similar in appearance to those illustrated by de Keyser & Cook (1972, Table XX, Fig. 1). (iii) Ooids formed about a nucleus of quartz or apatite. Generally phosphorites in the upper bed of the lower unit are characterised by clastic quartz grains, lighter-coloured and finer phosphate grains, and by the presence of ooids.

(3) The upper unit is composed of phosphatic dolomite, sandstone and conglomerate containing detrital phosphate grains. Phosphorites contain angular or rounded phosphate grains or both. The phosphate grains are generally poorly sorted and rounded. Zoned grains are rarely encountered. Phosphatic dolomite is composed of thin lenticular laminae containing small well-rounded phosphate grains, and less well rounded quartz grains of approximately the same diameter.

The texture of the phosphorites in the three main units was found to be variable when examined by transmission and scanning electron microscope. Optically amorphous mudstone

A

B

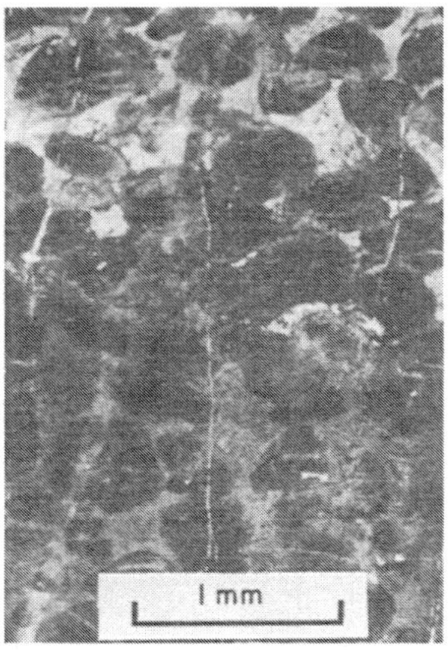

C

D

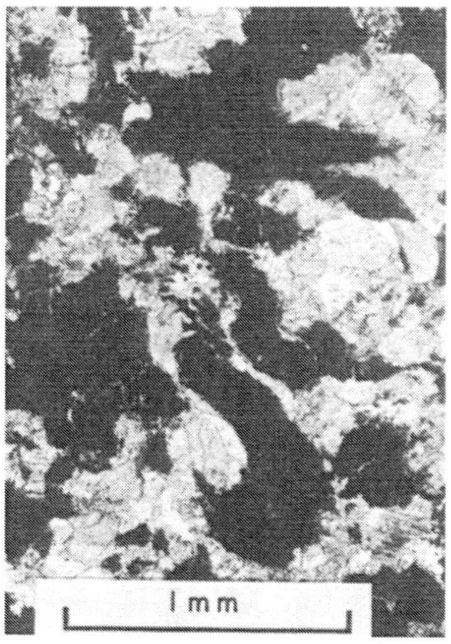

E

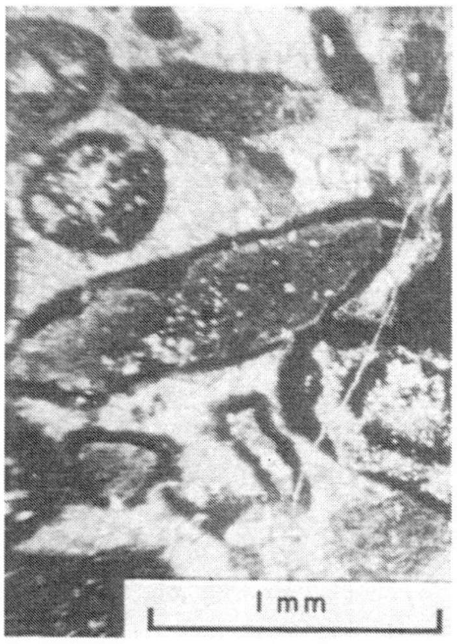

F

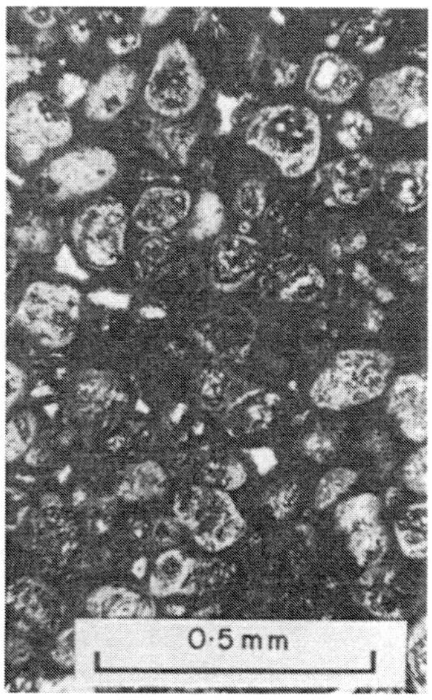

G

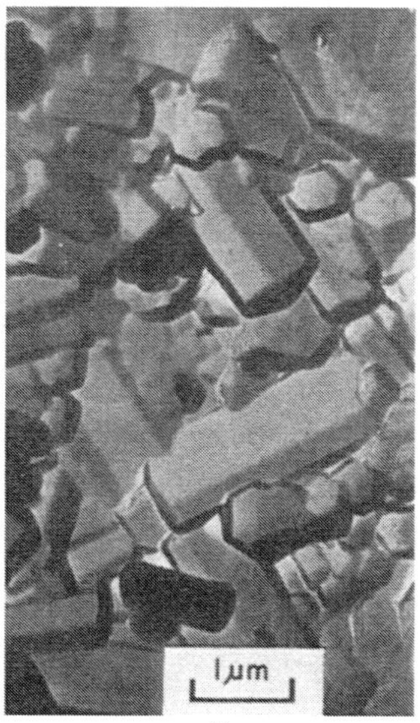

H

I

phosphorite consists of apatite crystallites, 0.5–1.0 μm in size, with typical hexagonal cross-section and tabular axial section. The degree of crystallinity is rather uniform. The crystallites have a random orientation (Fig. 14.7H), although sometimes they show a radiating pattern. Electron microscopy reveals the similarity in microstructure between phosphate taken from mudstone phosphorite interbeds and that from irregularly shaped fragments, supporting the genetic link between the two forms. The apatite making up the grains (pellets) is rather different. It tends to be in the form of fibrous radiating aggregates, with the crystal size increasing towards the outer rim of the grain (Fig. 14.7 I). The degree of crystallinity of apatite is notably less in grains than it is in beds and irregularly shaped fragments of mudstone phosphorite.

Petrological interpretation

Phosphorites formed by lithification of a phosphate mud are the source for many of the varieties of grain found in the lower phosphorite of the lower unit (Fig. 14.7). In a stable, low-energy environment the undisturbed phosphate mud was compacted and lithified to form structureless mudstone phosphorite. In an environment of somewhat higher energy the partly lithified mud suffered some disruption causing both granulation and disintegration of phosphatic beds. Besides being affected by the energy of the environment, the process of granulation was also governed by the composition of primary mud, which was far from uniform. In the case of a purely phosphatic unit, the process of granulation probably did not occur but with a higher percentage of silica and carbonate in the primary mud, very early diagenetic modification commenced and was subsequently modified by high-energy conditions. This led to the formation of microconcretions with silica concentrated in the inner portion and phosphate in the outer rim. Granulation and disintegration were closely followed by dolomitisation which probably started before the mud had been lithified. Superimposed on an isotropic (i.e. previously granulated) phosphate, the dolomitisation led to replacement or removal of intergranular phosphate, while the grains were not affected to any great extent. In the case of mudstone phosphorite, irregularly shaped and variably sized phosphate fragments or remnants of very varied grain size were left after dolomite had replaced the 'interfragmental' spaces. All the phosphorites composed of mudstone phosphorite fragments or remnants are believed to have formed *in situ*, for although many hundreds of thin-sections have been examined, no clastic quartz grains have been noted in any of them.

Clastic quartz is found in grainstone phosphorites thought to have developed in higher-energy environments. Structureless phosphate grains are essentially clastic grains which have suffered some transportation and rounding before being deposited. Grains with nuclei are probably authigenic grains with an accretionary origin.

The finer phosphate grains of the upper bed of the Lower Unit are thought to be formed in two ways: (1) by *in situ* granulation of previous mudstone phosphorite; and (2) as authigenic grains of an accretionary nature. In the latter case they may have been transported towards the shore of the basin to be deposited there along with ooids. During transportation the phosphate was considerably altered and much of the organic matter oxidised and lost. The grainstone phosphorites were also partially replaced by dolomite.

Phosphorites of the lower unit originated as primary chemical sediments precipitated at the 'sediment–water' interface. Orthochemical sedimentation seems to have been an important process in the formation of the Khubsugul phosphorites. The precipitation was a response to apatite saturation of the deep waters of the Late Precambrian–Early Cambrian Central Asia Ocean (Ilyin, 1980). The high content of dissolved phosphate was built up gradually starting during the Late Riphean and reaching its peak by earliest Cambrian time. Increased residence time of deep waters may have led to an accumulation of dissolved phosphate as proposed by Fischer & Arthur (1977).

Geochemical features of phosphorites

The most common type of phosphorite in the Khubsugul Basin is characterised by that in the lower bed of the lower unit. This contains 20–22% P_2O_5, 5–7% MgO and 6–8% SiO_2 (Table 14.1). However, in some areas, this lower unit contains sediments averaging 26–28% P_2O_5, or even higher. This is found in some parts of the main phosphorite unit on the western limb of the Khesen Syncline (Fig. 14.3) where phosphorites contain up to 39.3% P_2O_5.

Chemical analyses and thin-sections show that there is every gradation from 'pure' phosphate rocks practically devoid of carbonate, through phosphatic dolomite, to non-phosphatic

Fig. 14.7. Phosphorites from the lower bed of the Lower Phosphate Unit. (A) Thin-bedded mudstone phosphorite in hand specimen. Black phosphate interbedded with grey dolomite. Note fragmented phosphate layer in dolomitic matrix in upper half of photograph. (B) Mudstone phosphorite with mottled black carbonaceous matrix and thin lenticular dolomite intercalations, under crossed nicols. (C) Grainstone phosphorite, crossed nicols. (D) Large black phosphorite grains in dolomite matrix, under crossed nicols. (E) Dolomite replacing phosphate (black), producing irregular and variable sized phosphate relict fragments in dolomitic matrix, under crossed nicols. (F) Phosphate grains in a dolomitic matrix, under crossed nicols. (G) Small phosphate grains, quartz and carbonate nuclei, compound grains and phosphate matrix, under crossed nicols. (H) Electron micrograph of carbon replica showing micro-structure of apatite crystallites from a mudstone phosphorite layer. (I) Electron micrograph of carbon replica showing micro-structure of apatite from small pellets.

Table 14.2. *Major element analyses of phosphate concentrates*

Sample Number	P_2O_5	CaO	F	CO_2	SO_3	Fe_2O_3	Al_2O_3	MgO	SiO_2	Na_2O	K_2O	Loss on ignition	Σ	Cell dimension *a* (nm)
1	39.9	53.4	1.9	1.57	0.21	0.39	0.3	0.25	0.9	0.99	0.05	1.5	101.36	9.368
2	39.4	52.8	1.75	1.2	0.41	0.39	0.25	0.2	1.1	1.0	0.1	2.3	100.9	9.377
3	35.8	47.0	2.66	0.19	2.45	1.79	0.89	0.25	8.05	0.2	0.14	1.07	100.49	9.363
4	39.1	53.0	2.92	0.6	0.31	0.34	0.32	0.11	2.1	0.23	0.06	1.0	100.1	9.371
5	37.2	50.0	2.8	1.48	0.3	2.39	1.02	0.25	1.48	0.9	0.1	3.3	101.27	9.361

Samples 1, 2, 4 from Khubsugul; 3, from Ukha-gol; and 5, from Burin-khan.
1, mudstone phosphorite, sample K-3/7, lower bed of Lower Phosphate Unit, trench 16;
2, mudstone phosphorite, sample K-10/23, lower bed of Lower Phosphate Unit, trench 10;
3, grainstone phosphorite, sample T-2/3;
4, mudstone phosphorite, lower bed of Lower Phosphate Unit, trench 25;
5, coarse grainstone phosphorite, sample 62/1, trench 60.

carbonates. Two chemical types are clearly present: the generally dominant carbonate and the siliceous phosphorites. The latter can be divided into two groups depending on the amount of silica present: higher-grade phosphorites have 18–24% SiO_2 and poorer-grade phosphorites about 30% SiO_2.

The upper phosphorite unit includes two chemically different types, both of which contain up to 31% P_2O_5. In one type, silica varies from 6% SiO_2 to 10% SiO_2 while in the second it may fall to as low as 0.1% SiO_2. The first type corresponds to oolitic grainstone phosphorite where some thin concentric bands of the ooids are made up of silica. The second type is a fine-grained variety devoid of quartz debris. Sandy phosphorites or phosphatic sandstones and gravels of the Upper Unit are low grade, containing only 5–10% P_2O_5, and have a siliceous matrix.

Mineralogically the phosphate was found to correspond to carbonate–fluorapatite with probable deviation to carbonate hydroxy-fluorapatite in places. The chemical composition of phosphate concentrates together with their cell parameters are shown in Table 14.2.

Minor element contents in the phosphorites and associated rocks of the Khubsugul phosphatic series are shown in Table 14.1.

Organic carbon: No definite macroscopic or microscopic organic remains have been found in the Khubsugul phosphorites and other rocks of the phosphate series. The most common variety of phosphorite, with phosphate fragments of various sizes and shapes, contains about 0.5% C_{org}. Interbeds of mudstone phosphorite may contain as much as 1.65% C_{org}, while coarser grainstone phosphorites contain only 0.5–0.7% C_{org}. The fine-grained phosphorites of the upper productive unit are characterised by a low organic carbon content (average 0.1% C_{org}). On the whole, the Khubsugul phosphorites contain more C_{org} than the phosphorites of the Karatau Basin, where the average organic carbon content is 0.5–0.15%. However, it should be noted that petrographically similar varieties of phosphorites in both basins have approximately the same organic carbon content. In a lateral section the more seaward facies are richer in C_{org} than the nearshore facies, although the most seaward succession in the Khubsugul deposit is in fact low in both P_2O_5 and C_{org}. Considering the complete phosphatic series, the highest C_{org} content (2.4%) is found in the black shale. No correlation exists between the P_2O_5 and C_{org} contents nor does the organic carbon content of the phosphate series stand out compared to the Khubsugul sequence as a whole.

Strontium: Phosphorites are noticably enriched in strontium compared to other rocks. The strontium content in phosphorites (average 500–600 ppm Sr) is generally proportional to the phosphate content. High-grade mudstone phosphorites of the western phosphogenic zone (the Tsagan-nur deposit) contain about 1200–1400 ppm Sr, while low-grade carbonate phosphorites contain not more than 200–300 ppm Sr. The Mongolian phosphorites are considerably lower in strontium than the Mesozoic phosphorites of central Russia for example.

Uranium: A detailed study of uranium geochemistry has been undertaken in the Tsagan-nur deposit where it was found that typical mudstone phosphorites contain only 4 ppm of uranium. Even lower concentrations of uranium (1–2 ppm) were found in non-phosphatic rocks within the productive series. Considering the complete range of rocks, from those having 1–2% P_2O_5 or less to almost pure phosphate, a direct proportional dependence between P_2O_5 and U is evident. However the relationship becomes much less apparent in the interval 15–30% P_2O_5. If the P_2O_5 content is constant then the uranium content decreases with increasing CO_2 content. In other words carbonatisation of phosphorites leads to a loss of uranium. All varieties of grainstone phosphorite are found to be somewhat richer in uranium than mudstone phosphorite, a feature also noted in the Georgina Basin by Cook (1972).

Rare-earth elements: Five samples were analysed for REE (Table 14.3). They were found to vary from the detection limit to 300 ppm. The highest amount of REE is found in the high-grade mudstone phosphorites while the lowest REE content is found in phosphatic dolomites. As is the case with uranium, REE also tend to be more concentrated in grainstone phosphorites than in mudstone phosphorites.

Depositional environment and tectonic history

Three main lithological assemblages exist in the uppermost Precambrian and Lower Cambrian sequences of southern Siberia and Mongolian People's Republic. The first is domi-

Table 14.3. *Rare earth element (REE) analyses of phosphorites (recalculated to 100% REE)*

No.	TR%	La	Ce	Pr	Nd	Sm	Eu	Gd	Tb	Dy	Ho	Er	Tu	Yb	Lu
1	0.03	18.8	26.7	6.0	20.4	6.8	2.5	5.3	0.6	4.9	1.2	2.5	0.6	2.4	0.3
2	0.009	21.5	21.5	—	24.6		2.0	9.7	—	7.5	1.8	3.9	1.20	4.9	0.3
3	0.027	17.4	28.5	8.4	22.8	7.0	1.6	5.4	0.5	3.2	1.2	1.9	0.5	1.9	0.2
4	0.012	13.7	30.3	5.8	23.8	9.0	1.8	5.5	1.0	2.9	1.0	1.8	0.4	1.6	0.2
5	0.011	20.7	33.1	4.7	23.5	5.4	1.2	4.5	0.6	3.5	0.7	1.4	0.3	0.8	0.2

1, Mudstone phosphorite, trench 30; 2 and 4, mudstone phosphorite with irregular phosphate fragments, trench 29; 3, grainstone phosphorite, from the upper bed of the main unit, trench 54; 5, phosphatic shale, Ukha-gol deposit.

nated by various basaltoides, and the second by variegated clastic rocks, carbonates and evaporites. The third assemblage is of special interest here as it is represented by thick and monotonous carbonate sequences covering wide areas of eastern Tuva, northern Mongolia, the Eastern Sayan Basin, near Lake Baikal, and other areas on the southern, south-eastern, south-western and western margins of the Siberian Platform. The composition and distribution of this third assemblage suggests shelf deposition. The phosphorite of the Khubsugul Basin formed on an earlier shelf bordering the Siberian continent.

From the Khubsugul area, the uppermost Precambrian and Lower Cambrian carbonate sediments extend north-eastwards into the Lake Baikal area, north-westwards into the Eastern Sayan Basin, and further north towards the Yenisei Ridge. These ancient carbonates have not been preserved everywhere but occur as relatively small patches or inliers among extensive Caledonian batholiths as, for example, in the area to the east of Lake Baikal. From their spatial distribution, it appears that a broad belt of carbonate sediments, 300–400 km wide, rimmed the Siberian continent during latest Proterozoic and Early Cambrian time.

Latest Proterozoic–Cambrian phosphorites are widespread on an older shelf which covered the areas of Tuva, Buryatia, northern Mongolia, eastern Sayan and adjacent areas. In some places the phosphorites are considered to have a late Riphean age, whilst others are thought to be Vendian or Early Cambrian. In other words phosphogenesis may be diachronous across this shelf region. The oldest phosphorite occurrences tend to be smaller in size than the younger ones, and are found closer to the shelf edge. The younger deposits are larger and higher-grade and are located in inner shelf environments.

A possible explanation for these relationships may be that at the initial stage of transgression the outer part of the subsiding continental margin was narrow, but with subsequent transgression and increasing subsidence of the continental margin the shelf broadened, permitting the deposition of extensive carbonate sheets. By the end of the Proterozoic and the beginning of the Cambrian it was 3–400 km wide (Fig. 14.6), and favourable for phosphate deposition and accumulation following invasion by cold deep upwelled oceanic water. As these waters flooded the shallow shelf areas they were warmed and degassed causing the precipitation of both phosphate and carbonate.

Carbonate sedimentation was especially common on and around the Siberian palaeo-continent in the latest Precambrian–Cambrian. Not only were the marginal shelf seas bordering the continent loci of carbonate accumulation but also intracontinental basins formed additional loci, though the thickness of accumulated carbonates and lithological associations are somewhat dissimilar in the two areas. Also, the intracontinental basins were filled mostly with limestones and dolomites but never with phosphorites.

Modern carbonate sedimentation is especially abundant in the equatorial belt. Consequently, the widespread development of these ancient carbonate sediments may indicate a low-latitude position for the Siberian continent in the latest Precambrian–Cambrian. This supposition is strengthened by some other lithological peculiarities such as the occurrence of bauxites in the latest Precambrian of the area. Palaeomagnetic data, though scarce, also support a low-latitude location.

Turning to the global palaeotectonic reconstruction of Scotese *et al.* (1979), in the latest Precambrian–Early Cambrian all the continents and continental blocks were located within the equatorial belt, while oceans occupied both polar areas. Quite large oceans also occurred in the equatorial belt between continents, such as the Iapetus (Palaeo-Atlantic) and Central Asian Oceans. The Siberian continent, situated in the equatorial zone, had an opposite orientation to that of the present-day (i.e. its present southern margin faced north). Further to the north lay the smaller central Mongolian continent, with the Central Asian Ocean in between, elongated latitudinally in that area.

According to Yanshin (1964), the oldest period of massive accumulation of phosphate coincided with a period of intense lithosphere rifting accompanied by sialic magmatism dated at 820 m.y. A further development of ancient rifts associated with younger magmatism at 752 m.y. is expressed by bimodal volcanics, and comagmatic alkaline granite. The Darkhat–Khubsugul rift was a part of this old rift system developed around Baikal and Khubsugul. It was filled with clastic rocks of arkosic composition, with some rhyolites and minor basalts (i.e. the Darkhat Group). The source rocks for these sediments were undoubtedly granite–gneissic terrains of an older Precambrian basement. The tectonic relief of the neighbouring land masses was rather marked initially but was subsequently subdued due to peneplanation. This seems to be illustrated in the Darkhat Group sections by the upwards-fining sequence from conglomerate to sandstone, mudstone, and finally shale with some carbonate interlayers.

A general subsidence of the continental margin commenced at

about 700–750 m.y. This was accompanied by the accumulation of shelf carbonate sediments as the Siberian palaeo-continent drifted northwards across the equator. This situation not only favoured the formation of carbonates but, as sustained drifting moved the region into appropriate latitudes for upwelling of oceanic waters to occur, also favoured the development of phosphogenesis and the formation of phosphorites.

References

Bushinsky, G.I. (1966). Old phosphorites of Asia and their genesis. *Akademiya Nauk SSSR, Trudy Geologicheskogo Instituta*, **149**, 195pp. (Israel Programme for Scientific Translations, Jerusalem, 1969).

Cloud, P.E. & Glaessner, M.F. (1982). The Ediacarian period and system: Metazoa inherit the earth. *Science*, **217(4562)**, 783–92.

Cook, P.J. (1972). Petrology and geochemistry of the phosphate deposits of northwest Queensland, Australia. *Economic Geology*, **67**, 1193–213.

de Keyser, F. & Cook, P.J. (1972). Geology of the Middle Cambrian phosphorites and associated sediments of northwestern Queensland. *Bureau of Mineral Resources Geology and Geophysics, Bulletin*, **138**, 79pp, 28pls.

Fischer, A. & Arthur, A.M. (1977). Secular variation in the pelagic realm. *Society of Economic Paleontologists and Mineralogists, Special Publication*, **25**, 19–50.

Ilyin, A.V. (1973). Khubsugul phosphorite-bearing basin. *Trudy Sovmestnaya Sovetsko-Mongolskaya Nauchno-issledovatelskaya Geologicheskaya Ekspeditsiya*, **6**, 168pp. Geologicheskiy Institut, Akademiya Nauk SSSR; Nilzarubezhgeologiya Ministerstvo Geologii SSSR; Geologicheskiy Institut, Akademiya Nauk MNR; Izdatelystvo 'Nauka', Moscow.

Ilyin, A.V. (1980). A reevaluation of the spatial and temporal distribution of sedimentary phosphorite deposits in the light of Plate Tectonics – a discussion. *Economic Geology*, **75**, 771–3.

Ilyin, A.V. & Bjamba, Z. (1980). *Guidebook for excursion, phosphorites of the Khubsugul basin, Mongolian People's Republic. Field Conference of IGCP project 156*, 135pp. Geologicheskiy Institut, Akademiya Nauk SSSR, Moscow.

Ilyin, A.V. & Ratnikova, G.I. (1981). Primary, bedded, structureless phosphorite of the Khubsugul Basin, Mongolia. *Journal of Sedimentary Petrology*, **51(4)**, 1215–22.

Korobov, M.N. (1980). Lower Cambrian trilobites and the age of the phosphorites of western Prekhubsugul (Mongolia). *Izvestiya Akademiya Nauk SSSR, Seriya Geologicheskaya*, **1980(9)**, 97–107 (in Russian).

Scotese, C., Bambach, K., Barton, C., van der Voo, R. & Ziegler, A. (1979). Paleozoic Base Maps. *Journal of Geology*, **87(3)**, 217–77.

Tushina, A.M. (1979). Petrography of the Karatau phosphorites. *Composition of Phosphorites*, ed. Yu. N. Zanin, pp. 90–9. Akademiya Nauk SSSR, 'Nauka' Moscow.

Yanshin, L.A. (1964). Perspectives and scientific problems in the search for potassium salts and phosphorites in Siberia. *Byulleten, Moskovskoye Obshchestvo Ispytateley Prirody, Otdelenie Geologicheskiy*, **39(5)**, 1–17.

15

Proterozoic and Cambrian phosphorites – deposits: Karatau, southern Kazakhstan, USSR

E.A. EGANOV, YU K. SOVETOV AND A.L. YANSHIN

Introduction

Phosphorites were first discovered in the Karatau Mountains (north-west Tien Shan) in 1937 (Mashkara, 1938). By 1939, almost all of the deposits now known in the region had been found, and it had been established that they constituted a large phosphorite-bearing province (Bezrukov, 1938) (Fig. 15.1). In 1946 the first phosphate mine and beneficiation plant came into operation near the town of Karatau. Subsequently, larger deposits were found near the town of Dzhanytas in the northwest part of the Karatau Basin. Since that time, mining operations have continued to expand to the point where the Karatau mines provide a significant percentage of the total phosphate rock production of the USSR.

The Lower Cambrian phosphorites occur near the base of a thick (up to 3000 m) sequence of Cambrian–Ordovician carbonates. These carbonates outcrop as a series of prominent north-west-trending strike ridges, rising 200–300 m above the valley floor making up the Maly (Lesser) Karatau Mountains

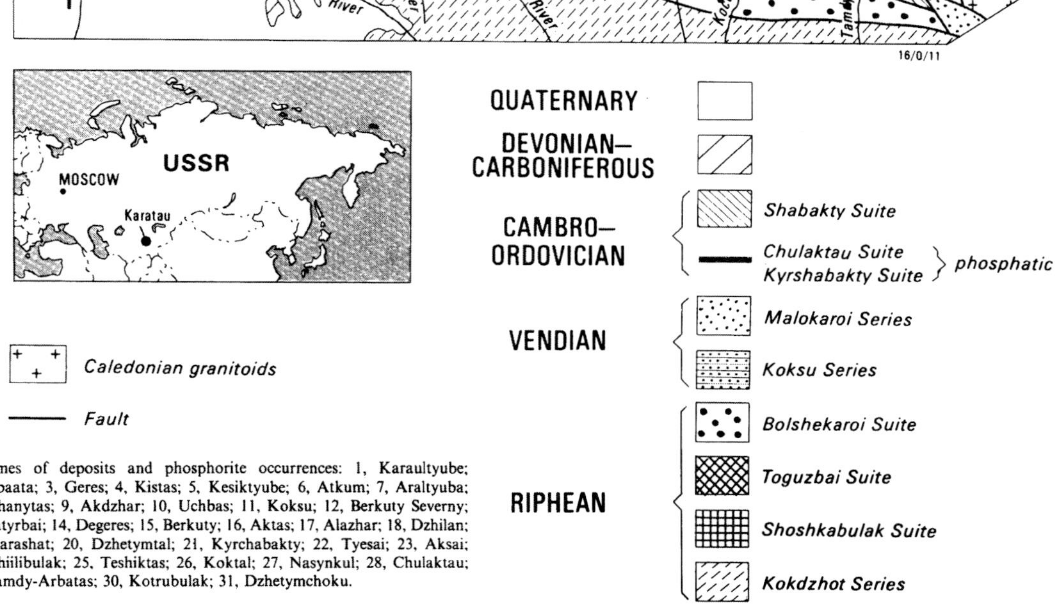

Names of deposits and phosphorite occurrences: 1, Karaultyube; 2, Babaata; 3, Geres; 4, Kistas; 5, Kesiktyube; 6, Atkum; 7, Araltyuba; 8, Dzhanytas; 9, Akdzhar; 10, Uchbas; 11, Koksu; 12, Berkuty Severny; 13, Batyrbai; 14, Degeres; 15, Berkuty; 16, Aktas; 17, Alazhar; 18, Dzhilan; 19, Karashat; 20, Dzhetymtal; 21, Kyrchabakty; 22, Tyesai; 23, Aksai; 24, Chiilibulak; 25, Teshiktas; 26, Koktal; 27, Nasynkul; 28, Chulaktau; 29, Tamdy-Arbatas; 30, Kotrubulak; 31, Dzhetymchoku.

Fig. 15.1. Schematic geological map of the phosphorite-bearing Karatau Basin, showing named phosphate deposits.

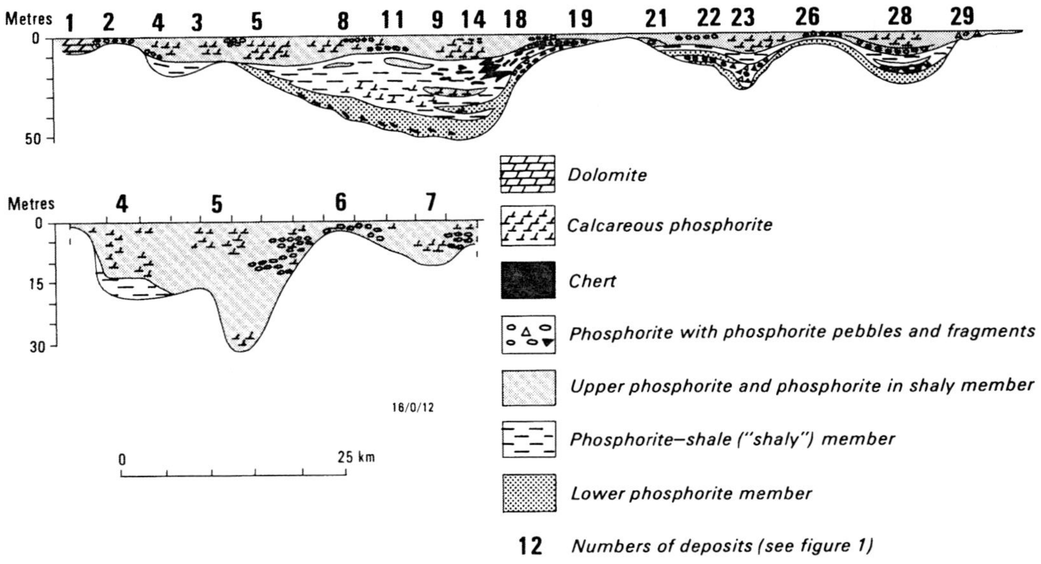

Fig. 15.2. Schematic cross-section of the productive phosphatic interval of the Karatau Basin.

which extend in a 20-km wide belt for about 120 km along strike. The Cambrian–Ordovician calcareous sequence is underlain by less well exposed Precambrian terrigenous and tuffaceous deposits. To the south-west the Lesser Karatau is terminated by a fault trending north-east which separates the folded area of the Bolshoi (High) Karatau, extending 200 km further to the north-west from the Lesser Karatau. In the High Karatau, Middle Palaeozoic rocks are exposed, with Mesozoic and Palaeogene sequences present in places.

Regional structure

The Karatau Fold Belt refers to the north-western branch of the geosynclinal Caledonian fold belt of the northern Tien Shan. The Karatau Basin is confined to the main Karatau fault zone. The northern part of the Tien Shan 'Main Fault' is a large lineament which divides the Karatau phosphatic basin into two parts. South-west from this fault, in the central part of the Karatau Basin (High Karatau), there is abundant evidence of Palaeozoic and some Precambrian folding. Here, Devonian and Carboniferous folded terrigenous and calcareous formations infill a large trough, which overlaps Caledonian structures. The Devonian and Carboniferous sediments in turn overlie sediments of the Chatkala–Naryn structural zone which comprises Upper Ordovician flysch-like deposits, Middle to Lower Ordovician and Cambrian siliceous–carbonaceous deposits, and the terrigenous Ulutau Series of Late Riphean to Vendian age. Below this are the Middle to Upper Riphean Bakyrlin and Kainer Suites of carbonate and volcanic rocks. The gneissic–amphibolitic Beszas Group, which may represent oceanic crust (Makarychev, 1978), is the oldest unit in the area.

The Karatau Phosphatic Basin proper is situated within the Talaso–Karatau structural belt, which lies to the north-east of the Tien Shan 'Main Fault'. It is made up of phosphorite-bearing carbonaceous sequence with flysch and molasse deposits. All Lower Palaeozoic and older formations are overlapped by Devonian to Carboniferous deposits. A cross-section of the productive phosphatic interval of the Karatau Basin is shown in Fig. 15.2.

The Lesser Karatau sequence represents a zone trending north-west with multiple thrust repetition of the sequence, dipping at 30–80° to the north-west. The thrusting is associated with the formation of a large flexure during the Caledonian Orogeny. In places there have been later transverse displacements.

Precambrian and Cambro-Ordovician formations are separated from each other and from Devonian–Carboniferous sediments by unconformities. There is an especially well defined unconformity at the base of the Middle Palaeozoic. Below this, the sediments are strongly deformed. The phosphorite-bearing Cambrian–Ordovician series of the Lesser Karatau in the south-east of the Karatau Basin is intruded by granites, and overlain by Middle Palaeozoic deposits (Fig. 15.1).

Stratigraphy

Riphean and Vendian

Precambrian formations make up the lower folded portion of the Karatau sequence. The Cambro-Ordovician phosphorite-bearing series overlies Precambrian basement transgressively and with angular unconformity in places. The Precambrian sequence is given in Table 15.1 and Figure 15.3.

The Kokdzhot Series is the oldest sequence, confined to the horst flanking the Main Karatau Fault situated in the south-western part of the Lesser Karatau. The south-western side of the horst has a displacement of 5–6 km (Nikolayev, 1933; Peive, 1945) and a lateral north-west displacement of about 200 km (Burtman, 1964). The Kokdzhot Series is a tightly folded sequence of green-grey sericitic–chloritic shales with quartz, ankerite and pyrite, and abundant quartz veins. About 30–50 km to the south, the Kokdzhot Series is composed of unmetamorphosed or slightly metamorphosed terrigenous sediments. Here the Series is divided into two parts (Sovetov,

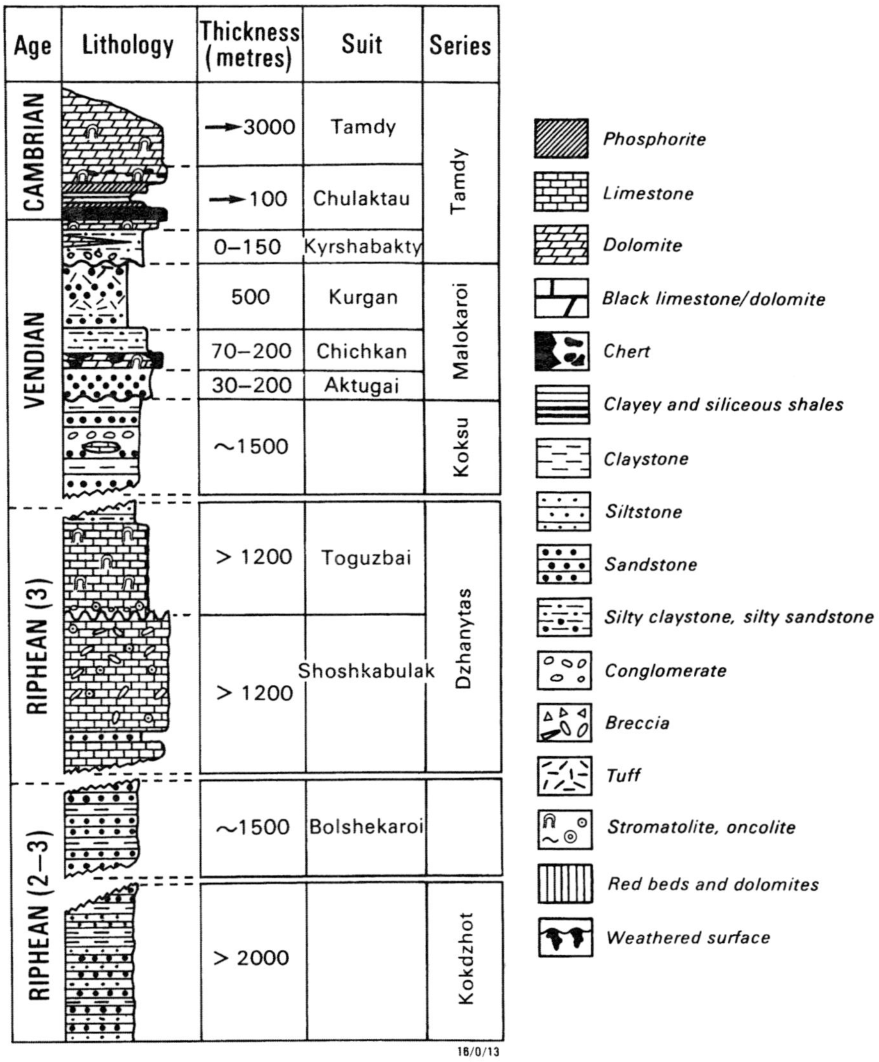

Fig. 15.3. Generalised stratigraphy of the Karatau Basin.

1981a); a lower terrigenous series 1300–1500 m thick and an upper terrigenous and carbonaceous series 500–700 m thick. The lower series consists of silty to sandy turbidites, silty to clayey pelagites and, less commonly, traction-load siltstones. In the upper series cross-bedded clastic limestones predominate, with rare andesites and andesitic tuffs.

The Bolshekaroi Series is separated from the Kokdzhot Series by the Karoi Fault. The Series is 1500 m thick, only slightly metamorphosed, and consists of alternations of silty-sandy turbidites and silty-clayey pelagites (Eganov & Sovetov, 1979).

The Dzhanytas Series is approximately equivalent in age to the Bolshekaroi Series (Sovetov, 1981a): it is 2500 m thick and consists of stromatolitic limestones (containing microphytolites) intercalated with micritic and fragmental (mud-chip breccia) limestones and less commonly dolomites. The Series is made up of two suites. The lower unit, the Shoshkabulak Suite (1200 m) is a part of a large 'reef-like' massif (Sovetov, 1981a), made up of microphytolitic limestones and calcareous breccias deposited in a 'reef' lagoon complex. Similar rocks have been noted in the Upper Riphean deposits of Siberia (Eganov & Sovetov, 1979). The Toguzbai Suite, which unconformably overlies the Shoshkabulak Suite, consists chiefly of 1200 m of stromatolitic limestones believed to represent a biostromal massif of Late Riphean age (Eganov & Sovetov, 1979).

The Koksu Series reflects the molasse stage of the northern Tien Shan Geosyncline. It rests unconformably on the Bolshekaroi Series. Conglomerates developed at the base of the Series contain greywacke and limestone pebbles and boulders from the Toguzbai Suite. The maximum thickness of the Koksu Series is 1500 m. A particular feature of the Koksu Series is the alternation of red mudstones and limestones containing 'phytogenic onkoids' (Sovetov, 1975). The 'phytogenic onkoids' contain an association of microphytolites comparable with those from the Yudoma Suite of the Siberian Platform (Eganov & Sovetov, 1979), suggesting a Vendian age for the Koksu Series.

The Malokaroi Series extends throughout much of the eastern part of the Karatau Basin and into the Talassky Ridge area of the Tien Shan. The Malokaroi Series transgressively, but apparently conformably, overlies the Koksu Series, representing the beginning of the formation of the quasiplatform mantle on the folded Riphean basement. The Malokaroi Series is divided into three suites – the Akatugai, Chichkan and the Kurgan.

The Aktugai Suite ranges from 30 m to 200 m in thickness and comprises quartzose and arkosic cross-laminated shallow-marine gritstones and sandstones. There is a widespread basal conglomerate, 3–4 m thick, throughout much of the Lesser Karatau. The source of the arkosic material was probably leucocratic granites, situated to the south-west of the Lesser Karatau (Sovetov, 1981b). The Chichkan Suite (70–200 m,) is a shallow shelf sequence made up of partly glauconitic green-grey siltstones, silicified argillites and silicified stromatolitic dolomites of Vendian age (Krylov, 1967) containing remains of blue-green algae in the cherts after dolomites (Schopf & Sovetov, 1976). The Kurgan Suite (about 500 m) consists of alternations of tuffaceous siltstones, acid tuffs, arkoses, volcanic sandstones and green, pink and red gritstones.

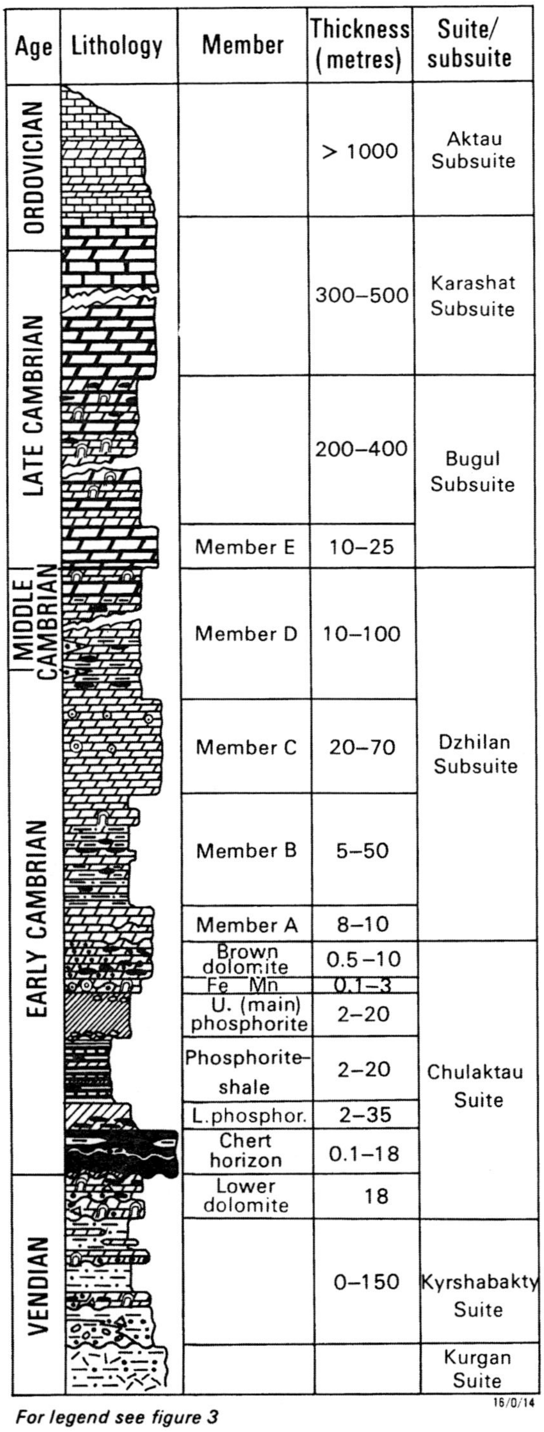

Fig. 15.4. Stratigraphic section of the Tamdy Series.

Cambrian and Ordovician

The Tamdy Series The Vendian–Ordovician Tamdy carbonaceous series, which contains phosphorites in places, unconformably overlies Proterozoic units. The unconformity is marked by an erosion surface and a light-coloured non-resistant zone of eluvium where it overlies with marked angular unconformity the Bolshekaroi and Koksu Suites (Fig. 15.4).

In many parts of the Lesser Karatau, thin phosphorites occur within the basal unit of the Tamdy Series, represented by red fine-grained silty, clayey and dolomitic units of the Kryshabakty Suite (Korolev & Maksumova, 1976; Eganov & Sovetov, 1979) in the form of phosphate fragments, nodules, and scattered grains.

Because the Kryshabakty Suite infills erosional pockets in the underlying bedrock topography, it has a very variable thickness ranging from 0 m to 150 m. The sequence commences with a conglomerate containing poorly rounded clasts of tuffaceous material from the underlying Kurgan Suite, in a mudstone or sandstone matrix. This is overlain by a very widespread pink dolomite 10–12 m thick containing in places thin (1–5 cm) interbeds of black mudstone, black aphanitic phosphorite, and chert nodules and lenses. In turn, this is overlain by a mixed

Table 15.1. *General stratigraphy of the Karatau Basin*

	Devonian and Carboniferous	
	--- Unconformity ---	
Cambro-Ordovician	Tamdy Series	
	Shabakty Carbonate Suite	
	Aktau Subsuite	(700 m)
	Karashat Subsuite	(5–700 m)
	Bugul Subsuite	(300 m)
	Dzhilan Subsuite*	(200–250 m)
	Chulaktau Suite*	(up to 70 m)
	Kyrshabakty Suite	(0–150 m)
	--- Unconformity ---	
Vendian	Malokaroi Series	
	Kurgan Suite	(500 m)
	Chichkan Suite	(200 m)
	Aktugai Suite	(200 m)
	Koksu Series	(1500 m)
Riphean	--- Unconformity ---	
	Dzhanytas Series	
	Toguzbai Suite	(1200 m)
	--- Unconformity ---	
	Shoshkabulak Suite	(1200 m)
	Bolshekaroi Series	(1500 m)
	Kokdzhot Series	(2000 m)
	Upper Terrigenous Suite	
	Lower Terrigenous Suite	

*Phosphatic portions of the sequence.

carbonate–terrigenous unit. The terrigenous component, which is up to 50 m thick, is composed of green siltstone, quartz sandstone, and claystone. This intertongues with, or in places is interbedded with, dolomitic units up to 20 m thick containing varying proportions of glauconite, chert and terrigenous sand. In places the dolomite is stromatolitic. Thin laminae and beds of mudstone phosphorite are present in places. In subsurface, black siltstones and claystones of possible lagoonal origin have been encountered. The age of the Kryshabakty Suite is somewhat uncertain, but is generally regarded as Late Vendian possibly extending into the earliest Cambrian (Table 15.1).

The Chulaktau Suite has been divided by Bezrukov (1954) into seven units (Table 15.2). Each of these units is laterally very persistent, though their thickness is variable. The age of the suite is somewhat uncertain. The only known fossils are hyoliths, hyolithelminths, and other similar forms which Mambetov & Missarzhevsky (1971), Mambetov & Repina (1979) and Missarzhevsky & Mambetov (1981) compare with fossils found in the Tommotian of the Siberian Platform. Mambetov & Repina (1979) and Missarzhevsky & Mambetov (1981) have distinguished biostratigraphic zones (Table 15.2), and they suggest that most of the sequence is of Tommotian (early Aldanian) age.

The Lower Dolomite is a thin (2–15 m) but very persistent unit of the Chulaktau suite. It conformably overlies the red-beds of the Kyrshabatky Suite in many areas, with a stromatolitic dolomite (in which the stromatolites form bioherms up to 1 m high) at the base. Elsewhere, it disconformably overlies tuffaceous units of the Kurgan Suite with a well-defined erosional contact. In these areas, the basal unit is a sandy and pebbly dolomite with quartz sand lenses and thin layers and lenses of mudstone phosphorite. Glauconite is abundant. Towards the top of the Lower Dolomite massive columnar stromatolites with lenses and nodules of chert and stromatolitic phosphorite characteristically occur.

The Chert Horizon is persistent, but may be locally absent. The contact with the underlying Lower Dolomite is irregular and erosional. The unit commences with a thin (0.1–0.3 m) siliceous mudstone phosphorite, with massive lenticular black chert. This is overlain by a finely layered unit comprising interlaminae of chert and mudstone ranging in thickness from 1 cm to 10 cm. The chert commonly consists of microcrystalline aggregates of brown-stained quartz and chalcedony, commonly with a spongiolith structure. Where the suite is condensed, the massive cherts give way to dolomite with irregular chert nodules and patches. In places the chert contains scattered phosphate grains or thin layers of grainstone or mudstone phosphorite. Because of this, the phosphate content of the chert ranges from 1% P_2O_5 to 10% P_2O_5 (see Fig. 15.6A).

The Lower Phosphorite Member is not present everywhere or alternatively it cannot always be distinguished from the Upper Phosphorite Member. The contact with the underlying Chert Horizon is gradational. The unit consists of siliceous mudstone and grainstone phosphorites with a dolomitic cement or matrix. Some of the phosphorites are cross-bedded and conglomeratic.

The Phosphorite–Shale Member has a gradational contact with the underlying Lower Phosphorite Member. It consists of thinly laminated 'muddy siltstones and black (when fresh) siliceous shaly mudstones. Thin interbeds of argillaceous mudstone phosphorite and grainstone phosphorite a few centimetres thick occur within the sequence. In places the phosphorite is up to 2 m thick. Along strike this unit is rather variable with the typical lithology of the Phosphorite–Shale Member giving way to a predominantly siliceous unit or a richly carbonaceous

Table 15.2. *Nomenclature and biostratigraphy of the phosphatic portion of the Karatau Basin*

Shabakty Suite (Dzhilan Subsuite)			Lenian
Member D	(80–90 m)	*Probowania asiatica* Zone	
Member C	(15–80 m)	Late *Redlichia chinensis* Zone	
Member B	(20–25 m)	Early *Redlichia chinensis* and *Ushbaspis limbata* Zones	
Member A	(3–8 m)	*Hebediscus orientalis* and *Ushbaspis* sp. 1 Zones	
Chulaktau Suite			Atdabanian
Brown Dolomite	**(1–10 m)**	*Rhombocorniculum cancellatum* Zone	
Ferromanganese Horizon	(0.1–3.0 m)	*Bercutia cristata* Zone	
Upper Phosphorite Member	(0.5–15 m)	*Pseudorthotheca costata* Zone	Tommotian
Phosphorite–Shale Member	(2–35 m)		
Lower Phosphorite Member	(0–10 m)		
Chert Horizon	**(0–20 m)**		
Lower Dolomite	(0–15 m)	*Protohertzina anabarica* Zone	Vendian
Kyrshabakty Suite			

interval; in places it may pinch out almost completely to a thin compact phosphorite.

The Upper (Main) Phosphorite Member overlies the Lower Phosphorite or the Phosphorite–Shale Member. It consists primarily of high-grade grainstone (ovular and oolitic) and mudstone phosphorites with a carbonaceous or phosphatic matrix. In the lower part of the Member, a conglomeratic unit in which the clasts are composed either of phosphorite or chert, is fairly common. In the grainstone phosphorites quartz grains are relatively common. Conglomeratic pebbly phosphorites are also relatively common in the upper part of the unit. The maximum development of this unit occurs near the margin of the Karatau Basin. The facies changes within this unit are also illustrated in Figure 15.2.

The Ferromanganese Horizon is a thin (0.1–3 m) but extensive interval composed of Fe and Mn-stained dolomite with abundant stromatolites, oncolites, hyoliths and brachiopods. Thin phosphorites, phosphate pebbles and chert nodules are common. The contact with the underlying Upper Phosphorite Member is erosional in places. There is also an erosional surface in the middle of the unit in places.

The 'Brown' Dolomite Member is 1 m to 10 m thick and is composed of brown, slightly phosphatic (0.5–5% P_2O_5, rarely up to 10% P_2O_5) dolomite. Chert nodules are very abundant in this unit. Detrital quartz grains are common and in the south-west part of the Karatau Basin the Brown Dolomite grades laterally into a quartz sandstone. Towards the top of the unit there is a very widespread thin (0.1–0.2 m) dolomitic layer containing abundant patches of black phosphorite composed of bioclastic (oncolitic, biohermal and hyolithid) material.

The Shabakty Carbonate Suite overlies the Chulaktau Suite (Table 15.2). This unit was considered to be of Middle Cambrian age (Bezrukov, 1954; Gimmelfarb, 1950). However, on the basis of structure, correlations with the Chinese sequence (Shatsky, 1955) and from its trilobite faunas (Ergaliev & Pokrovskaya, 1977) it is now generally accepted that the lower part of the Shabakty Suite has an Early Cambrian (Lenian) age. The Shabakty Suite has been sub-divided by Eganov & Sovetov (1979) into four Subsuites (Table 15.1). The basal subsuite, the Dzhilan Subsuite, has in turn been divided into four members (Table 15.2).

Member A is a massive spotted grey dolomite within rare thin crusts and films of black aphanitic phosphorite concentrated along the bedding planes. It overlies the Chulaktau Suite with an erosional contact in places. Ergaliev & Pokrovskaya (1977) consider that Member A occurs within the *Hebediscus orientalis* Zone which is believed to be equivalent to the late Early Cambrian Lenian Stage of the Siberian Platform.

Member B is predominantly a thin-bedded dolomite with varying amounts of terrigenous sand, silt and clay. Glauconite grains, chert nodules, and scattered phosphatic material (grains and shell debris) are found within the member, which has an average P_2O_5 content of around 1–2% P_2O_5. In the south-west part of the Karatau Basin (the Geres deposit) the fossil assemblage indicates the *Ushbaspis limbata* Zone, which is indicative of a Lenian age when compared to the Siberian Platform, and to the lower part of the *Redlichia chinensis* Zone of China (Ergaliev & Pokrovskaya, 1977; Mambetov & Repina, 1979). In the north-east part of the Basin (the Koksu deposit) the faunal assemblage belongs to the *Redlichia chinensis–Kootenia gimmelfarbia* Zone which Mambetov & Repina (1979) believe to be comparable to the latest part of the Early Cambrian of Korea, China, India, Iran and Australia (but see Shergold & Brasier, this volume).

Member C is composed of light-grey, thickly bedded to massive dolomite with a 'clotted' appearance. Stromatolites and oncolites are present in places. This member belongs to the youngest part of the *Redlichia chinensis* Zone.

Member D consists of interbeds of massive and laminated grey to black dolomite with some yellow, red and green argillaceous dolomites and marls. The lower part of the member is placed in the Early Cambrian *Probowmania asiatica* Zone (Ergaliev & Pokrovskaya, 1977), whereas the upper part (which includes the Akdzhar deposit) contains Middle Cambrian trilobites.

The Bugul Subsuite, which overlies the Dzhilan Subsuite with a gradational contact, (Fig. 15.4) consists of dark, thin-bedded stromatolitic dolomites, with dolarenites and dolomite breccias. This is in turn conformably and gradationally overlain by the Karashat Subsuite which consists of fine-grained dolomites, in part stromatolitic or brecciated. According to Ergaliev (1980) the Bugul and Karashat Subsuites are of Middle and Late Cambrian age, possibly extending into the Early Ordovician. The Shabakty Carbonate Suite is terminated by the Aktau Subsuite which is made up of grey dolomite and limestone, with bedding ranging from thin to massive. In the north-east part of the Karatau Basin, the carbonates grade upwards into mudstones and siltstones. Fossil hash, composed of fragments of brachiopods, gastropods, trilobites and other fossils, is fairly common in the carbonates. This unit is Early to Middle Ordovician in age.

Phosphate deposits of the Karatau Basin

There are 45 phosphate deposits and occurrences in the Karatau Basin (Fig. 15.1), occurring in five distinct regions (Gimmelfarb & Tushina, 1969) (Table15.3). Most of the phosphate reserves of the basin (more than 75%) are concentrated in five deposits – Chulaktau, Aksai, Koksu, Dzhanytas, and Kokdzhon. In the Chulaktau, Aksai and Koksu deposits, in the north-east part of the Karatau Basin, there is a well-defined tripartite sub-division of the productive sequence. In the north-west, the lower phosphorite was not deposited nor was it eroded prior to deposition of the upper part of the sequence. Consequently, in this area, the Upper Phosphorite or Shale Member rests directly on the Chert Horizon of the Lower Dolomite Member. The productive portion of the sequence reaches a maximum thickness of 40–50 m in the north-west part of the basin. To the south-east and extreme north-west the sequences thin out. The phosphorites appear to have accumulated in downwarps: on uplifts and towards the basin margins they thin markedly and also become coarser-grained. In some places phosphate pisoliths and phosphate pebbles (some of them with phosphatic shells) are present (Eganov & Sovetov, 1979).

In the Dzhilan and Chulaktau areas (regarded as 'distal uplifts'), the argillaceous units of the middle member of the

Table 15.3. *Phosphate deposits and occurrences in the Karatau Basin*

Area	Deposits/occurrences	% of total basinal reserves
1. Karatau	*Chulaktau*. Tamdy, Arbatas. Suleymansai, Koturbulak, Dzhetymchoku	3.5%
2. Aksai Malokaroi Valleys	*Aksai*, Chiilibulak, Tyesai, Koktal, Kyrshabakty	13%
3. NE and central part of basin	Dzhetymtal, Karashat, Dzhilan, Berkuty, Berkuty Severny	3.5%
4. N part of basin	*Koksu*, Ushbas, Akdzhar, Arkharlysai, Dzhartas	24%
5. NW and central part of basin	*Dzhanytas*, *Kokdzhon*, Gimmelfarbovskoye, Geres, Babaata, Karaltyube	56%

Major deposits are italicised

productive sequence are replaced by spongiolites.* In the northwest part of the Karatau Basin, and in areas regarded as 'proximal uplifts', the productive part of the sequence is replaced by dolomites.

Each of the major deposits will now be described briefly. A more detailed description is provided for the Dzhanytas deposit.

Chulaktau

The phosphorites of the Chulaktau deposit extend for 5 km, before being faulted out. The western part of the deposit has been metamorphosed by adjacent granitic intrusions. The productive portion of the deposit consists of three members, from the top downwards:

3. Upper Phosphorite Member (6–12 m): grainstone and mudstone phosphorites, up to 36% P_2O_5
2. Phosphorite–Chert Member (1.5–7 m): thin, interbedded siliceous phosphorite and chert, 2–22% P_2O_5
1. Lower Phosphorite Member (1–6 m): 26–29% P_2O_5 in near-surface locations

The upper unit, economically the most important interval, comprises a conglomeratic phosphorite, 0.3–0.5 m thick at the base of the interval, then a series of grainstone (partly oolitic) and mudstone phosphorites. The proportion of detrital quartz, chert and feldspar increases towards the top.

Aksai

The productive interval of the Aksai deposit extends for 8 km and comprises four members, from the top:

4. Carbonate–Phosphorite Member (20–22% P_2O_5): becomes a phosphorite conglomerate and a phosphatic dolomite along strike.
3. Chert–Carbonate Phosphorite Member (3–12 m; 25–26% P_2O_5): grainstone phosphorite composed of phosphate pellets, ooids, and some intraclasts, together with detrital quartz grains. Interbeds of siliceous claystone and chert.
2. Carbonaceous–Phosphorite Member (2–10 m; 20–27% P_2O_5): Predominantly phosphorite with a carbonaceous matrix. These grade laterally into grainstone and conglomeratic phosphorites.
1. Cherty-Phosphorite Member (5–30% P_2O_5): interbedded phosphorites, siliceous phosphorites, phosphatic cherts, siliceous dolomites, and dolomitic breccias.

Kokdzhon

The Kokdzhon deposit, which is subdivided into several smaller deposits – Kistas, Kesictube and Atkum – is located in the western part of the Karatau Basin. These deposits extend for 40 km. The productive part of the sequences is divided into two members. From the top down these are:

2. Phosphorite Member (8–10 m; 26–30% P_2O_5): In the northwest the deposits consist of black mudstone phosphorite with thin (up to 1 cm) dolomite interbeds. In the centre the deposits are composed of grainstone phosphorite, carbonaceous clayey phosphorite and interbeds of conglomeratic phosphorite and black mudstone phosphorite. In the southeast they consist of grainstone (oolitic) phosphorite with a phosphatic matrix, or less commonly, a calcareous–siliceous matrix.
1. Clayey – Siliceous Phosphorite (or Phosphorite–Shale) Member (2.4–10.0 m; 2–20% P_2O_5): siliceous claystones and shales with minor thin phosphorite interbeds.

Koksu

The productive sequence of the Koksu deposit in the northeastern part of the Karatau Basin extends for 13 km, and is divided into 6 members. From the top these are:

6. Upper Phosphorite Member (up to 5 m; 25–31% P_2O_5): partly oolitic grainstone phosphorite with some interbeds of conglomeratic phosphorite
5. Shaly Phosphorite Member (up to 10 m; 10% P_2O_5): phosphorite with thin interbeds of sericitic and siliceous shales
4. Siliceous–Clayey (shaly) Member (4–22 m; 4–5% P_2O_5): sericitic and siliceous shales with thin interbeds (ranging from a few millimetres to 1 m) of siliceous phosphorite
3. Argillaceous–Siliceous Phosphorite Member (3–6 m; 7–15% P_2O_5): interbedded phosphorite and siliceous shale
2. Lower Phosphorite Member (4–12 m; 26–28% P_2O_5): cherty phosphorite
1. Cherty Phosphorite Member (2–10 m; 11–15% P_2O_5)

Dzhanytas

Because the Dzhanytas deposit is the largest of the Karatau deposits it will be discussed in rather more detail than the others. It is situated in the central north-west part of the Karatau Basin, where the deposits extend for about 23 km along strike; on the flanks, the deposits are faulted out. At the present time the Dzhanytas deposits are being intensively exploited by open-cut mining methods. The general stratigraphy and the

*A sediment almost entirely made up of sponge remains.

more detailed stratigraphy of the phosphatic portion of the sequence, which is summarised in Table 15.4, is typical of most of the other economic deposits of the Karatau Basin.

Table 15.4. *Stratigraphy of the Dzhanytas Deposit*

Tamdy Series (Cambrian)

Shabakty Carbonate Suite
Thin bedded, clayey varigated dolomites and marks.

Chulaktau Suite

16. Brown dolomite – siliceous (with patches of chert) sandy and phosphatic, with thin (0.02–0.1 m) beds of black conglomeratic and grainstone phosphorite (8.2 m).
15. Ferromagnesian Horizon including phosphatic and stromatolitic dolomites (up to 1 m).
14. Productive Member
    d. Upper Phosphorite Sub-member. Grainstone phosphorite; Conglomeratic in part (5.2 m).
    c. Phosphatic Shale Sub-member. Siliceous and argillaceous shales with dispersed phosphate grains; average P_2O_5 content of 13–20% P_2O_5 (15–20 m).
    b. Lower Phosphorite Sub-member. Grainstone phosphorite predominate; average P_2O_5 content of 21–30% P_2O_5 (16–17 m).
    a. Phosphate Chert Sub-member. Phosphatic cherts and siliceous grainstone phosphorites. P_2O_5 content of 8–26% P_2O_5.
13. Chert Member. Black to grey dense massive 'spongiolite' with conchoidal fracture. P_2O_5 content of 3–10% P_2O_5 (17–19 m).

Kyrshabakty Suite

12. Dolomite – grey and pink, stromatolitic (undulatory in lower part, columnar in upper part). Thin (1–3 cm) phosphorites in middle (2.5–3).
11. Intercalated marls, siltstones and fine grained sandstones with abundant glauconite. Grey to black when fresh; blue-grey to green when weathered (5–6 m).
10. Dolomite – pink ferruginous (possibly after pyrite) stromatolitic At the top are thin (1–10 cm) layers of black aphanitic phosphorite (28% P_2O_5) or phosphatic breccia (0.3–1 m).
9. Dolomite – massive grey and pink, partly brecciated laminar stromatolitic (1–2.5 m).
8. Phosphorite breccia with intraclasts in a dolomite matrix (0.1–0.4 m).
7. Clayey dolomite (0.3 m).
6. Sandstones and marls; green (0.3 m).
5. Breccia with angular clasts of aphanitic phosphorite (0.1 m).
4. Sandstone and marls; phosphatic, fine grained, grey and tan; laminate to thin bedded (5–5.5 m).
3. Sandstone – quartzose with dolomite fragments; fine to coarse grained (0.8–1 m).
2. Dolomite – sandy, pyritic; grey and green (0.3 m).
1. Gritstone – quartz pebbles up to 1 cm; flat pebbles of the Toguzbai marls and siltstones; red to grey (0.7–12 m).

------------------------ Angular unconformity ------------------------

Dzhanytas Series (Riphean)

Toguzbai Suite

Shoshkabulak Suite

Dzhanytas Series The phosphatic Tamdy Series overlies the Riphean Dzhanytas Series with marked angular unconformity. The basal Tamdy Series rests on progressively older rocks towards the margin of the Tamdy basin of deposition. The oldest unit, the Shoshkabulak Suite, was deposited in the form of a 'reef' overlying the Toguzbai 'biostromal' massif. Facies changes have been delineated by Sovetov & Eganov (1976) and Sovetov (1981c) and show that on the south-west flank of the deposit the Dzhanytas Series consists of stromatolitic limestones of a 'reef plateau', whereas on the north-west flank it comprises calcarenites and dolomites of lagoonal origin. The uppermost units of the Dzhanytas Series are red-beds, thinly bedded silty marls, dolomites, and siltstones. These were originally thought to be of Vendian (Malokaroi Series) age but Eganov & Sovetov (1979) subsequently regarded them as Riphean on the basis of the associated stromatolites and microphytolites, and the structure of the sequence.

Kyrshabakty Suite The Kyrshabakty Suite of the Tamdy Series is composed of 12 units (Table 15.4) of which several are phosphatic. The phosphate has various forms. In Unit 4 it is in the form of phosphatic cement in sandstone and marl. In Units 5, 8 and 10 the phosphate is in the form of angular clasts of aphanitic phosphorite (mudstone phosphorites). Unit 10 also has thin laminae of mudstone phosphorite towards the top of the unit. Thin (1–3 cm layers of mudstone phosphorite also occur within the stromatolitic dolomites of Unit 12, but in addition it also occurs as siliceous phosphate between some of the stromatolitic bioherms. Unit 12 probably corresponds with the Lower Dolomite of the Chulaktau Series elsewhere in the basin (Table 15.2). Therefore in the Kyrshabakty Suite the phosphate occurs as grainstone phosphorite together with glauconite in the silty-clayey units. In the stromatolitic unit it occurs as thin layers of mudstone phosphorite or as phosphate-bearing dolomites or as oolitic grainstone phosphorites on the bedding planes of bioherms.

Chulaktau Suite In the vicinity of the Dzhanytas deposits there are four major units included in the Chulaktau Suite; one of these units, the phosphatic (Productive) member, is in turn divided into four sub-units. Therefore the stratigraphic sequence is similar to that elsewhere in the Karatau Basin (Table 15.4); although as pointed out earlier, the Lower Dolomite which is generally included in the Chulaktau Series is here included in the Kyrshabakty Series.

Immediately overlying the Lower Dolomite is a layer, 0.2 m to 0.3 m thick, of black mudstone or grainstone phosphorite which 'drapes' over the relief of the underlying stromatolitic 'bioherms'. This layer is included within the Chert Member because of the lack of any evidence of a break in sedimentation. There are other thin phosphorite layers within the Chert Member containing up to 10% P_2O_5. However, most of the phosphate occurs in the Productive Member (Unit 14, Table 15.4). The basal phosphate chert sub-member is represented in the south-east part of the deposit by a black high-grade oolitic grainstone phosphorite with layers of siliceous phosphorite and black chert. In the central part of the deposit, this unit is represented by massive black phosphatised chert and siliceous phosphorite, and to the north-west by siliceous grainstone phosphorites with intercalations of phosphatised cherts. The lower phosphorite sub-member also shows some facies variations. In the south-east part of the deposit the unit is composed of high-grade homogeneous, dark-grey or black, partly oolitic grainstone phosphorites. In the central part of the deposits, siliceous phosphorites and phosphatic cherts predominate in the lower part of the unit but these give way to grainstone phosphorites in the upper part. To the north-west, oolitic grainstone and mudstone phosphorites and siliceous and carbonaceous phosphorites predominate, though in the middle part of the unit phosphatic siliceous clayey shales are present in places. Locally this unit completely pinches out. The phosphatic shale sub-member (Unit 14, Table 15.4) extends from the south-west throughout much of the deposit before pinching out to the north-west. The unit comprises grey sandy chert and chalcedony with microcrystals of apatite, muscovite and sericite, together with a few phosphate grains (some of them oolitic). The upper phosphorite sub-member consists in the south-east part of the deposit of dark-grey to black dense, massive, partly oolitic grainstone phosphorites, with a phosphorite conglomerate up 3.4 m thick in places. There are also some lenses of carbonaceous phosphorites within the high-grade grainstone phosphorites. Three kilometres to the north-west, siliceous grainstone phosphorites with thin beds of phosphatic siliceous claystone predominate in the lower part of the unit, whilst conglomeratic phosphorite (up to 0.6 m thick), sandy phosphorites and calcareous and siliceous phosphorites become more common in the upper part, with the uppermost unit becoming a quartzose grainstone phosphorite. Therefore the upper phosphorite sub-member shows extensive intertonguing of various phosphatic facies including conglomeratic phosphorites, grainstone phosphorites, cherty phosphorites, phosphatic cherty mudstones and calcareous phosphorites over a distance of a few kilometres.

The overlying Ferromagnesian Horizon also contains phosphatic stromatolitic dolomites, generally overlying conglomeratic phosphorites of the upper unit (Table 15.4, 14d) of the Productive Member. The top unit of the Chulaktau Suite (the Brown Dolomite Member) contains some phosphatic dolomite and a few thin beds of conglomeratic phosphorite.

Composition of the Dzhanytas phosphorites Two types of phosphatic material have been recognised in the Dzhanytas deposits – cryptocrystalline apatite (commonly termed collophane or francolite) making up grains and cement (analyses 2 and 3, Table 15.5), which has a composition (based on its refractive index and cell parameters) approximating to carbonate fluorapatite. The second variety, termed crystalline apatite (Table 15.5, analysis 1) is close to a fluorapatite. Petrographically, several types of phosphate grains and cements can be recognised, including those which are composed of phosphate alone, and those which are siliceous–phosphatic, siliceous–carbonaceous–phosphatic or carbonaceous–phosphatic grains and cements. Dolomite and calcite are present, the latter being especially abundant in zones of tectonic dislocation and weathering. Acid-insoluble minerals include chalcedony, quartz, feldspar, and hydromica. These components form particular assemblages including: (a) detrital grains of quartz and feldspar; (b) siliceous (chalcedonic) cement; (c) microcrystalline and aphanitic chalcedony within phosphatic grains; (d) crystals and veins of secondary quartz; and (e) hydromicas in association with siliceous–argillaceous units. Variations in the nature of the phosphorites and the phosphatic sequence in different parts the Dzhanytas deposit have been shown by Gimmelfarb *et al.* (1962) to be quite marked. Some of these changes are summarised in Table 15.5.

Table 15.5. *Chemical analyses of phosphorites from the Dzhanytas deposit*

	1 %	2 %	3 %
P_2O_5	37.75	40.00	38.1
SiO_2	1.34	2.09	1.1
Al_2O_3	0.47	trace	0.75
Fe_2O_3	1.13	2.29	0.1
MgO	trace	trace	—
CaO	53.71	53.56	53.13
CO_2	2.48	0.84	2.13
Na_2O	—	—	0.65
K_2O	—	—	0.04
MnO	—	—	—
TiO_2	—	—	0.12
H_2O^+	—	—	1.58
H_2O	—	—	0.0
F	2.93	3.34	3.0
Total	100.37	102.12	100.76
(O—F)	1.26	1.41	1.26
recalculated Total	99.11	100.71	99.5

1, microcrystalline phosphorite; 2, grainstone phosphorite (Gimmelfarb *et al.*, 1969); 3, grainstone phosphorite (Yu.N. Zanin, *personal communication*).

—, not determined.

Phosphorite types

The phosphorite types of the Karatau Basin have been considered at length by Bezrukov (1954), Smirnov & Tushina (1962), Trukhacheva (1962), Tushina, Usanova & Maimistrova (1969), Smirnov (1972), Krasilnikova & Shmelkova (1975) and many others. They have shown that within the basin, mudstone phosphorites are rare and that the overwhelming majority of the deposits are grainstone phosphorites. On the basis of their size and structure a number of grain types can be recognised. These include: (1) homogeneous phosphatic grains with a rounded or oval form ranging from 0.02 mm to 0.6 mm in diameter; inclusions range in abundance from 0–50% of the total grains and comprise particles of chalcedony, clay, carbonates, and iron hydroxides; (2) oolitic grains variously consisting of a core of microcrystalline phosphate, a mixture of phosphate and chalcedony, fragments of pre-existing ooids and fossil fragments, with an oolitic fringe made up of thin concentric layers of dark-brown isotropic phosphate–ooid aggregrates and spherulitic textures covered by a single oolitic margin; (3) rounded and oval grains ranging from 1.5 mm to 2 mm in diameter, composed of fragments of ooids and mudstone phosphorite; (4) grains composed of phosphatic

Table 15.6. *Variation of phosphorite type within the Dzhanytas deposit*

	SE part of deposit	NW part of deposit	Central part of deposit
Type	Phosphate alone and carbonaceous-phosphatic	Siliceous–phosphatic	All varieties
Grainsize	0.07–0.25 mm	0.1–0.2 mm	0.03–0.1 mm
P_2O_5 content	26–31%	25–28%	24–26%
Acid insoluble fraction	15%	15–17%	high variable
Cement	siliceous		
Total thickness of phosphatic series	30 m	25–30 m	60–70 m

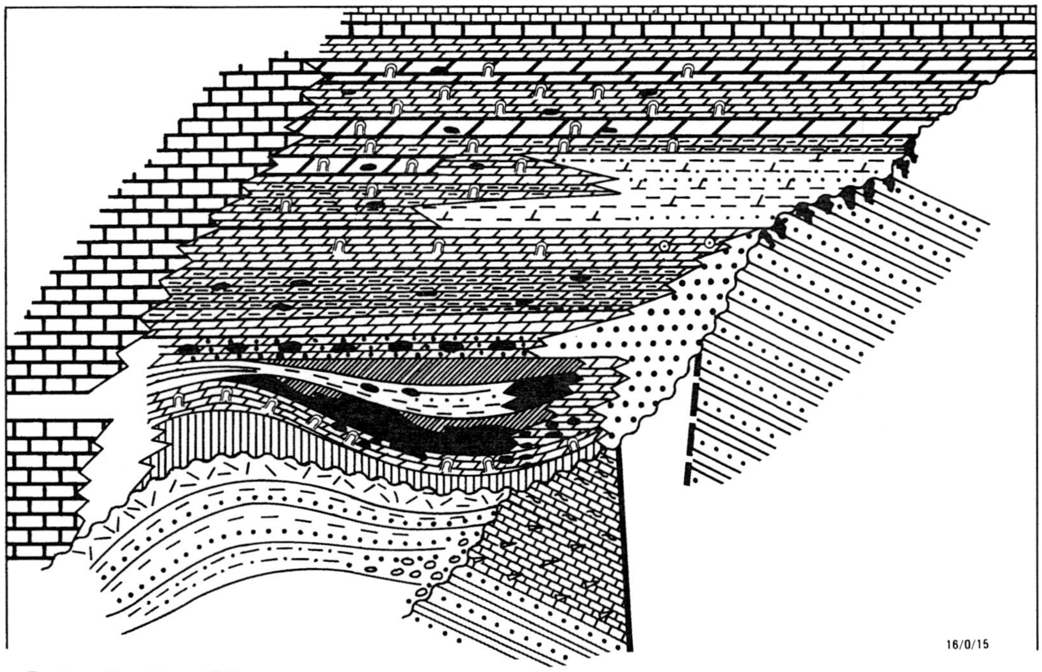

For legend see figure 15.3

Fig. 15.5. Schematic representation of facies relationships from SW to NE across the Karatau basin.

fragments of hyolithid shells ranging in size from 0.02 mm to 0.2 mm.

The phosphorites also contain rare grains of detrital quartz, glauconite, dolomite, and iron hydroxides.

On the basis of grain size and textures Trukhacheva (1962) concluded that the phosphate grains making up the phosphorites were not detrital but represented 'micro-concretions' which formed diagenetically; subsequently, concentric overgrowths formed during a reworking phase. Cements in the phosphorites include isotropic microcrystalline phosphate, fine- to coarse-grained dolomite, microcrystalline quartz, or a mixture of these minerals.

Both grainstone and mudstone phosphorites are present in the Karatau Basin deposits. On the basis of grain types and size, a number of types of phosphorite have been distinguished (Tushina *et al.*, 1969; and the present authors), though all gradations have been observed between these various types as summarised in Table 15.6 and Fig. 15.5. The mudstone phosphorites or 'poorly grained phosphorites' are composed of dense, massive dark-grey to black aphanitic phosphorite. This has a characteristic spherulitic or clotted structure and a well-developed conchoidal fracture. It also includes rare feldspar grains, quartz grains, iron hydroxides (maximum of 5%), and isolated phosphate grains (including simple and compound ooids). Typical chemical compositions for this phosphorite type are given in Table 15.7.

There are several types of grainstone phosphorite. The 'grained (micro) concretionary phosphorite' is a dense massive, grey to black laminate to thin-bedded grainstone phosphorite (Fig. 15.6B, C, D). It is composed of grains of phosphate or a mixture of phosphate and chert, with grains of quartz, hydromica, pyrite and iron hydroxides. The matrix is phosphatic, siliceous or carbonaceous. Typical chemical compositions are given in Table 15.7.

Table 15.7. *Chemical composition of phosphorites from the Chulaktau Suite*

	Mudstone phosphorite			Grainstone phosphorite			Oolitic grainstone phosphorite		
	1	2	3	4	5	6	7	8	9
P_2O_5	37.75	33.62	29.83	32.41	24.58	15.83	35.77	22.13	20.21
SiO_2	1.34	5.62	—	10.88	25.79	—	4.24	—	4.89
Al_2O_3	0.47	0.74	1.26	2.39	0.57	0.16	0.16	1.17	0.09
Fe_2O_3	0.91	0.33	1.15	0.68	0.75	0.82	1.05	1.24	0.43
FeO	0.22	0.95	0.73	1.56	2.06	0.59	0.78	0.36	0.21
R_2O_3	—	0.37	0.65	trace	1.07	0.56	trace	1.41	trace
MgO	trace	1.30	2.93	0.12	0.09	8.54	0.10	2.13	7.98
CaO	53.71	50.44	46.27	46.30	35.93	36.99	50.98	36.12	43.09
CO_2	2.48	3.18	7.00	2.47	3.12	19.60	2.54	6.53	20.09
SO_3	0.50	0.56	1.49	0.54	0.58	1.92	0.12	1.15	0.60
F	2.93	3.34	2.35	3.04	2.34	1.58	3.41	2.27	2.08
Total	100.53	100.92	100.15	100.72	101.49	100.15	99.36	99.50	99.90

1, 'spherulitic phosphorites'; 2, clotted phosphorite; 3, 'microsphorite', 4, 'high-grade monophosphatic phosphorite'; 5, 'siliceous phosphorite'; 6, carbonaceous phosphorite; 7, 'monophosphatic phosphorite'; 8, siliceous phosphorite; 9, carbonaceous phosphorite.
—, not determined.

A second type of grainstone phosphorite (termed a 'grained oolitic phosphorite') has the same texture and composition except that the grains are predominantly oolitic. The composition of some oolitic grainstone phosphorites is given in Table 15.7. In a third type of grainstone phosphorite the phosphate grains are of silt and clay size, well sorted but poorly rounded. Rounded grains and clasts (interclasts?) of earlier generations of grainstone phosphorites are present. Admixed with the phosphate grains are grains of quartz, chert and muscovite.

All of these grainstone phosphorites show bedding (including cross-bedding), and signs of reworking. The chemical composition of these phosphorites is rather variable, due to the varying proportions of phosphate grains, non-phosphatic grains, detritus, and the matrix.

Bioclastic phosphorites composed of phosphatic hyolith shells, together with terrigenous grains and glauconite do occur, but are rare. Conglomeratic phosphorites composed of phosphorite pebbles up to 5 cm in diameter are similarly rare. A further type of coarse grainstone phosphorite is one in which the grains are pisolitic, ranging in diameter from 2 mm to 12 mm. The nucleus of the pellets is commonly a small fragment of oolitic grainstone phosphorite; this is fringed by concentric layers of radial crystals of phosphate, or layers of cryptocrystalline phosphate.

Gimmelfarb *et al.* (1969) have distinguished five types of phosphate ores on the basis of average chemical composition, and petrological criteria. These, together with estimated phosphate reserves in the Karatau Basin (following Krasilnikova & Shmelkova, 1975), are summarised in Table 15.8.

Geochemistry of the Chulaktau Suite

Earlier work (Tabyldiev, 1966) on the geochemistry of the Chulaktau suite showed that Be, Na, Cr, Ni, Y, Mo, La, Cd, Yb, and Pb are concentrated in the phosphorites; V, Ti, Cu and Ba in siliceous rocks; and Mn, Co, Ce, La, Er and Yb in the carbonaceous rocks. Kholodov (1973) has similarly divided trace elements studied by him into three groups: (1) those associated with the phosphorites such as Th which occurs in the crystal structure of the apatite, Sr which is in the form of celestite, and possibly Be; (2) those which preferentially accumulated in siliceous rocks (usually within carbonaceous-argillaceous units) such as V (which is adsorbed by organic matter, clay minerals and silicates), and Mo and Cu which are concentrated in the sulfides; (3) those present in separate mineral phases, disseminated throughout the rock or in fractures, geodes and pore spaces of phosphorites, cherts and other rock types. These include Pb, Ag, Zn and Ba. The average elemental composition of phosphorites and cherts is given in Table 15.9. The organic carbon content of the phosphorites ranges from 0.5% to 1% in unweathered material (Kholodov, 1973) and from 0.05% to 0.1% in weathered phosphorites (Bezrukov, 1954). V_2O_5 is found to be enriched in the more carbonaceous and argillaceous cherts. The major element composition of siliceous rocks of the Chulaktau Suite is given in Table 15.10.

Palaeogeography of the Karatau Basin

By Late Vendian time the land surface in the vicinity of the Lesser Karatau had been peneplaned, though with a slope dipping slightly to the north-east. The south-east part of the region, the foldbelt, had similarly been peneplaned in pre-Vendian time. In the north-east, the syntectonic foredeep deposits were exposed and weathered during the Precambrian. Therefore, throughout the region there had been widespread sedimentation, although there were in places minor interruptions, immediately prior to the Precambrian–Cambrian boundary interval.

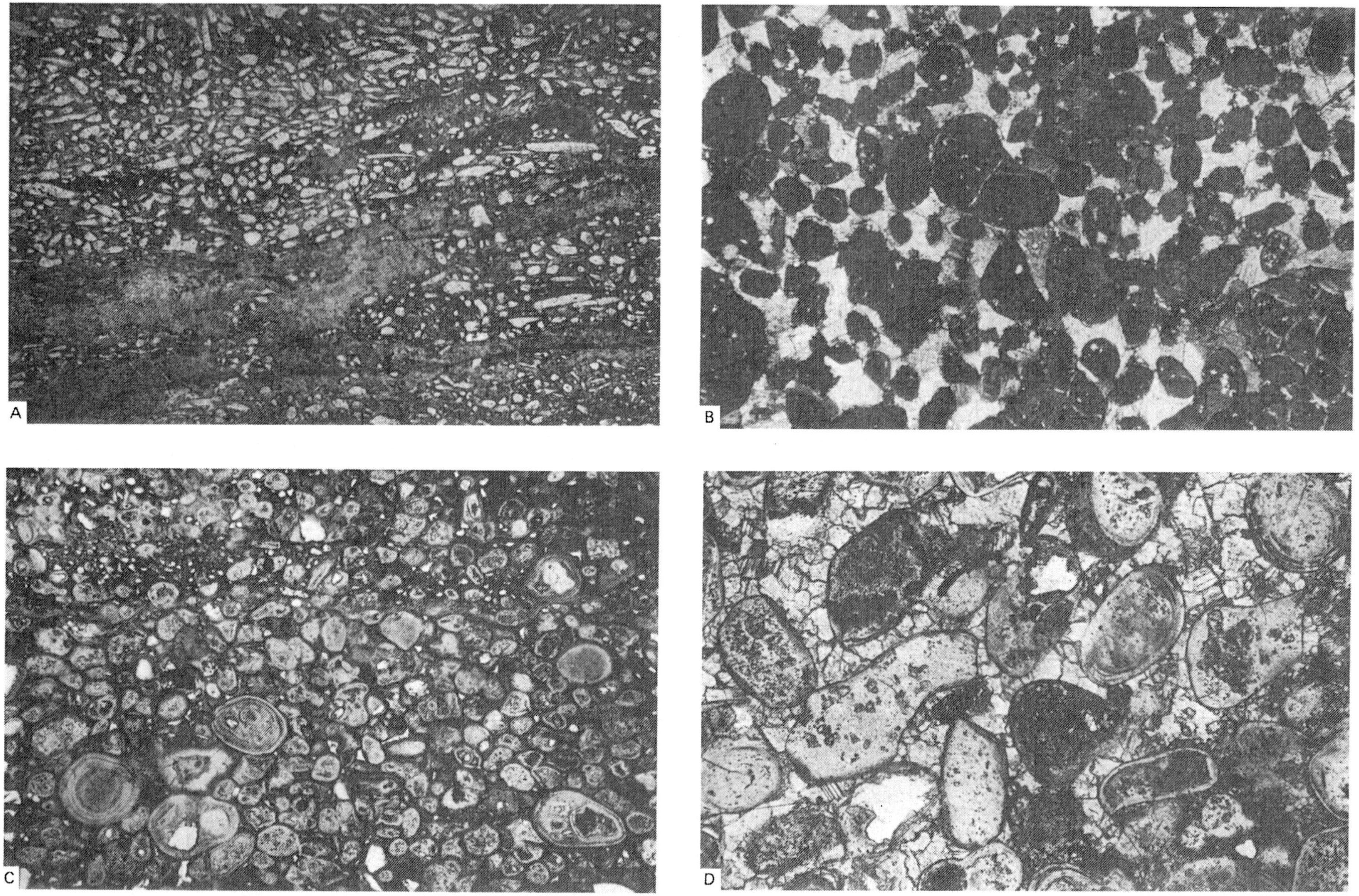

Fig. 15.6. (A) Spongiolite; ordinary light, × 50. (B) Grainstone phosphorite; lower phosphorite unit, Dzhanytas deposit; crossed nicols, × 40. (C) Oolitic grainstone phosphorite; upper phosphorite unit, Dzhanytas deposit; ordinary light, × 30. (D) Oolitic grainstone phosphorite with dolomite cement; upper phosphorite unit, Dzhanytas deposit; ordinary light × 80.

Table 15.8 *Main phosphate ores, phosphate contents and proportion of reserves of the Karatau Basin*

Ore type	P_2O_5 content	% of total reserves of basin
Monomineralic	> 28%	3.0%
Carbonaceous	25–27%	42.2%
Siliceous	22–25%	23.2%
Carbonaceous–argillaceous	16–18%	5.6%
Metamorphosed	25–26%	9.6%

Total phosphate reserves of the Karatau Basin–3 billion tonnes with an average phosphate content of 23% P_2O_5.

Adapted from Gimmelfarb *et al.*, (1969). Total reserve figures after Krasilnikova & Shmelkova (1975).

Table 15.9. *Trace element contents of phosphorites and associated siliceous rocks of the Chulaktau Suite, adapted from Kholodov (1973)*

	Phosphorites (ppm)	'Spongiolites' (ppm)	Cherts (ppm)
ΣREE	900	20	17
V_2O_5	70	60	2900
Ga	2	3	4
Be	3.69	1.08	1
Pb	2500	1100	1300
Ag	210	240	145
Mo	1	1	20
Li	4	100	700
Sr	780	4	150
Cr_2O_3	—	—	220

—, not determined

Table 15.10. *Composition of siliceous sediments of the Chulaktau Suite, adapted from Kholodov (1973)*

	'Spongiolites'		Chert	
	1 %	2 %	3 %	4 %
P_2O_5	0.01	0.59	1.49	1.49
SiO_2	96.90	94.92	67.25	74.87
Al_2O_3	0.09	0.52	9.36	7.08
Fe_2O_3	1.71	1.52	3.72	3.30
FeO	0.43	1.00	0.72	0.57
TiO_2	0.09	0.09	0.72	0.51
MnO	0.02	0.04	0.07	0.02
MgO	0.02	0.07	1.80	0.70
CaO	0.77	1.32	4.33	2.88
Na_2O	0.07	0.14	0.73	0.71
K_2O	0.17	0.17	3.78	2.71
CO_2	0.12	0.06	2.66	0.34
S	—	—	2.14	2.17
H_2O^+	0.11	0.03	2.12	1.77
H_2O^-	0.14	0.18	0.48	0.36
C_{org}	—	—	0.26	1.19
Total	100.35	100.65	100.63	100.67

—, not determined.

At the beginning of the Cambrian there was a well-defined transgression over the Vendian and older surfaces which resulted in the deposition of various sediment types in the basal unit of the Tamdy Series, including red and variegated sandy dolomites, and quartzose and feldspathic conglomerate with rock fragments from the underlying units (tuffs, greywacke, marl, siltstone, dolomite, conglomerate). The region in the vicinity of the present-day Lesser Karatau became an area of shallow-marine sedimentation during Kyrshabakty time, when sedimentation was predominantly as stromatolitic dolomites and mixed terrigenous–calcareous sediments with glauconite and phosphate. Differential rates of subsidence resulted in changes in thickness and styles of sedimentation within the basin. North-east across the basin the Kyrshabakty sediments increase in thickness from 15–20 m to 120–150 m. There were also three cycles of sedimentation at this time related to transgressions, which resulted in the progradation of various facies, such as the stromatolitic dolomites, across the basin.

At the close of Kyrshabakty time the Lower Dolomite was eroded and a fairly rugged topography, in part related to the distribution of bioherms, was developed on its upper surface. This was followed by deposition of the Chert Member, commonly with phosphorite intercalations at the base. There was a shift in the locus of sedimentation at the conclusion of deposition of the Chert Member with an elongate depression forming along the present axis of the Lesser Karatau, flanked to the north-east and south-east by zones of uplift. To the south-west, towards the uplifted area, and also towards the basin margin, in the vicinity of Geres, Kokdzhon, Karashat and Tamdy, the Chert Member is replaced by dolomites, some of them stromatolitic. Conversely, in the north-east part of the Karatau Basin, on the palaeoslope in the vicinity of Berkuty Severny, Batyrbai and Aktas, the Chert Member pinches out. The general palaeoslope was to the north-west. This is consistent with the palaeocurrent direction in the clayey spongiolite ooze, the thickness of the Chert Member, the increase in the phosphate content of the sediments and the relative abundance of admixed silt and clay. At the start of Chulaktau time, sedimentation was dominated by spongiolites. This was followed by spongiolites and cherts together with organic matter and phosphate, probably as a result of the accumulation of abundant phytoplankton remains, for which there is supporting microscopic evidence. Initially, at the commencement of deposition of the Lower Dolomite, phosphorite comprised a relatively minor component of the total sediment, but subsequently became progressively more important. Phosphorites appear to have preferentially accumulated in pre-existing depressions. However, their area of deposition does extend beyond the basins and on to some of the flanking highs. Within the phosphatic part of the sequence there are some clear facies changes. From the south-west edge, north-eastwards across the basin, the changes include: (1) the appearance of a thicker and more complete sequence (Bezrukov, 1954); (2) a decrease in the size of phosphate grains (Bushinsky, 1963); (3) an increase in the relative abundance of siliceous shales in the sequence; and (4) a decrease in the abundance of organic carbon in the phosphorites. In the marginal south-western facies of the phosphorite suite only granular phosphatic sandy dolomites are present (Eganov & Sovetov, 1979). It is likely that to the south-west, near the coastline flanking the peneplaned land, these granular dolomites are replaced by quartz sandstones in much the same way that at Kokdzhon and Karaultyube the Ferromanganese Horizon and the Brown Dolomite are represented by fine-grained quartz sandstones.

Thus, the facies distribution evident in the phosphatic units indicates that accumulation of phosphorites occurred under conditions ranging from extremely shallow water (Tamdy, Karashat, Babaata) to water tens of metres deep (Dzhanytas, Koksu, Ushbas). Tabyldiev & Cherbyanova (1976) have shown that there was considerable reworking, movement and mixing of phosphate grains. Smirnov & Tushina (1962) have proposed several stages in the formation of the Karatau phosphorites:

1. the formation of a phosphatic-siliceous gel, preserved in places as a cryptocrystalline phosphorite
2. the formation of microconcretions, or grains, during diagenesis
3. the development of micro-ooids
4. the formation of detrital phosphate grains and compound ooids, pisolites and pebbles

Smirnov & Tushina (1962) found that grainstone phosphorites also show a well-developed trend to the north-west, from grains with no banding to grains with a single external concentric band and finally grains with multiple concentric bands. At the same time the average grain size ranges from 0.109 mm to 0.123 mm to 0.135 mm. There is no correlation between grain types and the phosphate or silica content.

The mode of formation of the primary siliceous-phosphatic gel is uncertain. A volcanogenic hypothesis has been proposed by Shatsky (1955), Korolev *et al.* (1973) and Tabyldiev (1966) and a modified upwelling hypothesis by Bezrukov (1954) and Smirnov (1972). It has been shown by Eganov & Sovetov (1979) that the phosphatic sequence of the Chulaktau Suite is the result of cyclic sedimentation with the basic sequence repeated up to five times. As a result of these cycles, sedimentation zones migrated across the shallow shelf as far as the uplifed region on the north-eastern edge of the Karatau Basin. The deepening of the Basin resulted in phosphatic sedimentation being succeeded by predominantly calcareous sedimentation (black deep-water limestones with trilobites) with clear evidence of slump breccias in these deeper water facies. However, landwards (to the south-west) the carbonates became more dolomitic (dolarenites) and in places with a greater abundance of terrigenous sand.

Previously, Ankinovich (1961) and Tushina (1973) have proposed the existence of a 'phosphatic sound', however there is little supporting evidence for this. The deep-water areas, from which the phosphate-rich waters of the Karatau Basin were presumably derived, were situated to the north-east of the Karatau Basin, i.e. beyond the Muyunkum Basement Block (now the site of the Chui Depression). This deep sea area, in the far north-east (in central Kazakhstan), was in turn connected to deeper water areas, probably an ocean where cherts, 'phtanites' and other products of submarine volcanism were accumulating (Shatsky, 1955; Korolev *et al.*, 1973).

Therefore the main factors responsible for phosphate accumulation in the Karatau Basin include the following palaeogeographic and sedimentological factors:

1. a low continental hinterland and an adjacent broad shelf dipping towards a deep sea basin
2. an embayment or partly enclosed sound through which ocean currents did not extend
3. open connection to the ocean which provided a source for the P, Si, and minor elements which in turn resulted in the development of a high organic-productivity system on the shelf
4. sulfate reduction and the development of anoxic bottom waters during periods of phosphate accumulation precluded the development of carbonate bioherms and biostromes, and microphytolitic banks
5. There was periodic reworking of siliceous–phosphatic ooze, including the winnowing-out of fine grained components

The nature of the underlying basement appears to have been irrelevant to the accumulation of phosphorites apart from its influence on the topography, such as the development of the foredeep. In general the Cambrian sediments mantled the basement with a new geosynclinal and platform pattern of sedimentation unrelated to the Precambrian.

References

Ankinovich, S.G. (1961). *Lower Palaeozoic of the vanadium-bearing basin of the Northern Tien-Shan and western outlying districts of central Kazakhstan. Part 1.* Akademiya Nauk Kazakhskoy SSR, 272 pp., Alma-Ata.

Bezrukov, P.L. (1938). Lower Silurian geosynclinal phosphorites in the northeastern branch of the Karatau Range. *Trudy Akademiya Nauk SSSR*, **18(7)**, 453–6.

Bezrukov, P.L. (1954). Geological structure of the phosphorite-bearing Karatau basin and principal results of geological prospecting. In *Phosphorites of Karatau*, pp. 3–66. Akademiya Nauk Kazakhskoy SSR, Alma-Ata.

Burtman, V.S. (1964). Talaso-Fergansky wrench-fault (Tien-Shan). *Trudy Geologicheskiy Institut, Akademiya Nauk SSSR*, **104**, 142pp.

Bushinsky, G.I. (1963). On shallow water phosphorite deposits and their genesis. In *Deltaic and shallow marine deposits*, ed. P.L. Bezrukov, M.S. Shvetsov & V.S. Yablokov, pp. 102–7. Akademiya Nauk SSSR, Moscow.

Eganov, E.A. & Sovetov, Yu.K. (1979). Karatau – a model for phosphorite deposition. *Trudy Instituta Geologii i Geofiziki, Akademiya Nauk SSSR, Sibirskoe Otdelenie*, **427**, 192 pp.

Ergaliev, G.Kh. (1980). *Trilobites of the Middle and Upper Cambrian of the Lesser Karatau.* Institut Geologicheskikh Nauk, Akademiya Nauk Kazakhskoy SSR, 212 pp.

Ergaliev, G.Kh. & Pokrovskaya, N.V. (1977). *Lower Cambrian Trilobites of the Lesser Karatau (Southern Kazakhstan).* Geologicheskiy Institut, Akademiya Nauk SSSR, Institut Geologicheskikh Nauk, Akademiya Nauk, Kazakhskoy SSR, 138pp. Alma-Ata.

Gimmelfarb, B.M. (1950). The question of the age of phosphorites of Karatau. *Doklady Akademii Nauk*, **71(1)**, 127–8.

Gimmelfarb, B.M. & Tushina, A.M. (1969). Characteristics of phosphorite deposits of the Karatau Basin. In *Phosphorites of Karatau*, ed. B.M. Gimmelfarb, A.S. Sokolov & A.M. Tushina, *Trudy Gosudarstvennogo Nauchno-issledovatelskogo Instituta Gornokhimicheskogo Syrya*, **13**, 26–36.

Gimmelfarb, B.M., Tushina, A.M., Smirnov, A.I. & Maimistrova, R.I. (1962). Geological structure and ore types of the phosphorite deposit of Dzhanytas. In *Geology of phosphorite deposits*, ed. B.M. Gimmelfarb, *Trudy Gosudarstvennogo Nauchno-issledovatelskogo Instituta Gornokhimicheskogo Syrya*, **7**, 71–131.

Gimmelfarb, B.M., Tushina, A.M., Smirnov, A.I., Maimistrova, R.I. & Usanova, V.G. (1969). Material composition and industrial ore types of the main deposits of the Karatau phosphorite basin. In *Phosphorites of Karatau*, ed. B.M. Gimmelfarb, A.S. Sokolov & A.M. Tushina, *Trudy Gosudarstvennogo Nauchno-issledovatelskogo Instituta Gornokhimicheskogo Syrya*, **13**, 37–124.

Kholodov, V.N. (1973). Sedimentary ore-genesis and vanadium metallogeny. *Trudy Geologicheskiy Institut, Akademiya Nauk SSSR*, **251**, 262 pp.

Korolev, V.G., Dzhumaliev, T., Kiselev, V.V. & Maksumova R.A. (1973). Peculiarities of the geological history and environment of phosphate accumulation in the Talaso-Karatau zone. In *Conditions of formation of geosynclinal phosphorites*, ed. A.S. Sokolov, pp. 62–70. Akademiya Nauk SSSR, 'Nauka' Moscow.

Korolev, V.G. & Maksumova, R.A. (1976). Kyrshabakty Suite and the correlation of the Karoy and Tamdy Series of the Lesser Karatau Range. *Izvestiya Akademii Nauk SSSR, Seriya Geologicheskaya*, **1976(7)**, 84–9.

Krasilnikova, N.A. & Shmelkova, Yu.F. (1975). Petrographic types of phosphorites and economic deposits. *Trudy Gosudarstvennogo Nauchno-issledovatelskogo Instituta Gornokhimicheskogo Syrya*, **31**, 5–27.

Krylov, I.N. (1967). Riphean and Lower Cambrian stromatolites of the Tien Shan and Karatau. *Akademiya Nauk SSSR, Trudy Geologicheskogo Instituta*, **171**, 76 pp.

Makarychev, G.I. (1978). *Geosynclinal process and continental crust formation in the Tien-Shan*. Izdatelstvo 'Nauka', Moscow.

Mambetov, A.M. & Missarzhevsky, V.V. (1971). New data on fossils from the phosphorite-bearing sequence of the Lesser Karatau. In *Stratigraphy of the Precambrian of Kazakhstan and Tien-Shan*, ed. Yu.A. Zaitsev, B.M. Keller & M.A. Semikhatov, pp. 217–21. Materialy Karagandskogo Stratigraficheskogo Soveshchaniia, Karaganda 1969. Izdatelstvo Moscow University.

Mambetov, A.M. & Repina, L.N. (1979). The Lower Cambrian of the Talass Ala-Too and its correlation with sections of the Lesser Karatau and Siberian Platform. In *Biostratigraphy and paleontology of the Lower Cambrian of Siberia*, ed. I.T. Zhuravleva & N.P. Meshkova. *Trudy Instituta Geologii i Geofiziki, Akademiya Nauk SSSR, Sibirskoe Otdelenie*, **406**, 98–138. Novosibirsk.

Mashkara, I.I. (1938). The stratigraphy and tectonics of the north-eastern branch of the Karatau (Southern Kazakhstan). *Problemy Sovetskoy Geologii*, **8(1)**, 49–59.

Missarzhevsky, V.V. & Mambetov, A.M. (1981). Stratigraphy and fauna of the Cambrian-Precambrian boundary beds of Lesser Karatau. *Akademiya Nauk SSSR, Trudy Geologicheskogo Instituta*, **326**, 90 pp.

Nikolayev, V.A. (1933). About the most important structural line of the Tien-Shan. *Mineralogical Society Notes*, **62(2)**, 347–54.

Peive, V.A. (1945). Deep faults in geosynclinal areas. *Izvestiya Akademii Nauk SSSR, Seriya Geoligicheskaya*. **1945(5)**, 23–46.

Shatsky, N.S. (1955). Phosphorite-bearing formations and the classification of phosphorite deposits. In *Conference on sedimentary rocks, Fasc. 2*, ed. G.D. Afanasiev, pp. 7–100. Izdatelstvo Akademii Nauk SSSR, Moscow.

Shopf, J.W. & Sovetov, Yu.K. (1976). Microfossils in *Conophyton* from the Soviet Union and their bearing on Precambrian biostratigraphy. *Science*, **193**, 143–6.

Smirnov, A.I. (1972). Composition and conditions of formation of the principal types of phosphorite. *Trudy Gosudarstvennogo Nauchno-issledovatelskogo Instituta Gornokhimicheskogo Syrya*, **14**, 196pp.

Smirnov, A.I. & Tushina, A.M. (1962). Composition and genesis of phosphorites of the Ak-Say deposit. In *Geology of phosphorite deposits*, ed. B.M., Gimmelfarb. *Trudy Gosudarstvennogo Nauchno-issledovatelskogo Instituta Gornokhimicheskogo Syrya*, **7**, 41–70.

Sovetov, Yu.K. (1975). Phytogenic onkoids in Upper Precambrian of the Lesser Karatau (Southern Kazakhstan). *Doklady Akademii Nauk*, **224(3)**, 673–6.

Sovetov, Yu.K. (1981a). Terrigenous formations of Later Precambrian in the North Tien-Shan geosyncline. In *Evolution of the Lithosphere in Earth History*. ed. A.L. Yanshin, M.A. Zharkov & Yu.P. Kazansky, pp. 51–61. Akademiya Nauk SSSR, Institut Geologii i Geofiziki; Siberskoe Otdelenie, Novosibirsk.

Sovetov, Yu.K. (1981b). Arkose formation of Vendian of the North Tien-Shan Riphean Geosyncline. *Doklady Akademii Nauk*, **257(2)**, 449–52.

Sovetov, Yu.K. (1981c). Upper Riphean reef complex of the Lesser Karatau (Southern Kazakhstan). *Doklady Akademii Nauk*, **257(4)**, 975–9.

Sovetov, Yu.K. & Eganov, E.A. (1976). Late Precambrian disconformities in the Lesser Karatau (Kazakhstan). *Doklady Akademii Nauk*, **229**, 702–5.

Tabyldiev, K.T. (1966). Lithologic-geochemical peculiarities of phosphorite-bearing series and the questions of genesis of phosphorites in the Lesser Karatau. Pp. 81–90. In *Lithologic investigations in Kazakhstan*, ed. P.T. Tazhibaeva, pp. 81–90. Akademiya Nauk Kazakhskoy SSR, Alma-Ata.

Tabyldiev, K.T. & Cherbyanova, L.F. (1976). Environmental facies of sedimentation of phosphorite-bearing series of the Karatau basin. In *Lithology of phosphorite-bearing deposits*, ed. A.S. Sokolov, pp. 76–9. Izdatelstvo 'Nauka' Moscow.

Trukhacheva, A.G. (1962). Petrography of bedded phosphorite deposits of Chulak-Tau. In *Geology of phosphorite deposits*, ed. B.M. Gimmelfarb. *Trudy Gosudarstvennogo Nauchno-issledovatelskogo Instituta Gornokhimicheskogo Syrya*, **7**, 5–40.

Tushina, A.M. (1973). Environments of formation of Karatau phosphorites. In *Environments of formation of geosynclinal phosphorites*, ed. A.S. Sokolov, pp. 71–9. Akademiya Nauk SSSR, 'Nauka' Moscow.

Tushina, A.M., Usanova, V.G. & Maimistrova, R.I. (1969). Lithology and facies of phosphorite-bearing deposits of the Lesser Karatau. In *Phosphorites of Karatau*, ed. B.M. Gimmelfarb, A.S. Sokolov & A.M. Tushina. *Trudy Gosudarstvennogo Nauchno-issledovatelskogo Instituta Gornokhimicheskogo Syrya*, **13**, 125–86.

16

Proterozoic and Cambrian phosphorites – deposits: Hazara, Pakistan

MIRZA TALIB HASAN

Introduction

Marine phosphorite deposits of Cambrian age are located in the north-west of Pakistan on the flanks of the structurally complex Hazara–Kashmir Syntaxis. Phosphorite deposits are developed at two horizons within the bedded cherty dolomite of the Abbottabad Formation and a horizon within the shale and siltstone of the overlying Hazira Formation. The phosphorite is partly siliceous, generally granular, but also dense and laminate, and dark in colour. The phosphate content ranges from 10% P_2O_5 to 30% P_2O_5 and reserves of phosphate rock are estimated at 22.9 million tonnes. However, the deposits will be difficult to mine because of extensive faulting and folding.

Geological setting

The rocks of the Hazara region range in age from Precambrian to Quaternary and include sedimentary, igneous and metamorphic rocks. Clastic units of Precambrian age, and extending in places into the middle Palaeozoic (Hazara and Tanol Formations) are present. However, the sequence extending from Cambrian to Eocene is predominantly calcareous. The Himalayan Orogeny which began in the early and middle Tertiary is reflected in the deformation of all rocks in the region. The major structural feature in the Hazara area is the bend of 180° in the geological structures. Wadia (1931), who first studied this feature, called it the 'Syntaxis of North West Himalaya'. Two phases of the Tertiary Himalayan Orogeny are evident, and the extension of the syntaxis development into the post-Miocene period is evident from the deformation of the Murree Formation, of Oligocene to Miocene age.

The main episode of igneous activity took place during Cretaceous to Tertiary times when synorogenic intrusion of the Mansera Granite occurred. Palaeozoic and older rocks are metamorphosed. In the vicinity of the Hazara phosphate deposits there are argillites, phyllites, slates and quartzose schists, together with unaltered shale, quartzose siltstone, dolomite, limestone and quartzite. To the north, the degree of regional metamorphism increases along the syntaxial bend leading to the development of garnet, kyanite, staurolite and sillimanite-bearing schists and tremolite-bearing marble.

Table 16.1 *Generalised stratigraphic setting of rocks in the Hazara region*

Lockhar Limestone	Palaeocene
Kawagarh Formation	Late Cretaceous
Lumshiwal Formation	Early Cretaceous
Chichali Formation	Late Jurassic
Samanasuk Formation	Mid Jurassic
------- Unconformity	-------
Hazira Formation	Early Cambrian
Abbottabad Formation	Early Cambrian
Tanawal Formation	Precambrian
Hazara Formation	Precambrian

Stratigraphy

The rock units exposed in the Hazara area range in age from Precambrian to Recent (Table 16.1).

Hazara Formation

The oldest formation exposed in the region of the phosphorites consists of a thick and widespread sequence of slate, phyllite, greywacke sandstone, algal limestone and shale for which Calkins & Matin (1968) used the name Hazara Formation, and Latif (1970) the Hazara Group. A major fault, the Tarnawal Fault (Fig. 16.1), extends through the area on the eastern margin of Hazara Formation. East of the fault the Hazara Formation is unconformably overlain by the basal conglomerates of the Abbottabad Formation. West of the fault the Hazara Formation apparently has a conformable relationship with the Tanawal Formation. The contact is poorly exposed but is probably gradational. In this area the Hazara Formation can be broadly divided into two major units, one predominantly of slate and phyllite, the other of gypsum and limestone.

No fossil evidence is available to assign a definite age to the Hazara Formation. It is thought to be of Precambrian age on the basis of correlation with the Dogra Slate of Kashmir which is overlain by fossiliferous Cambrian rocks (Shah, 1977).

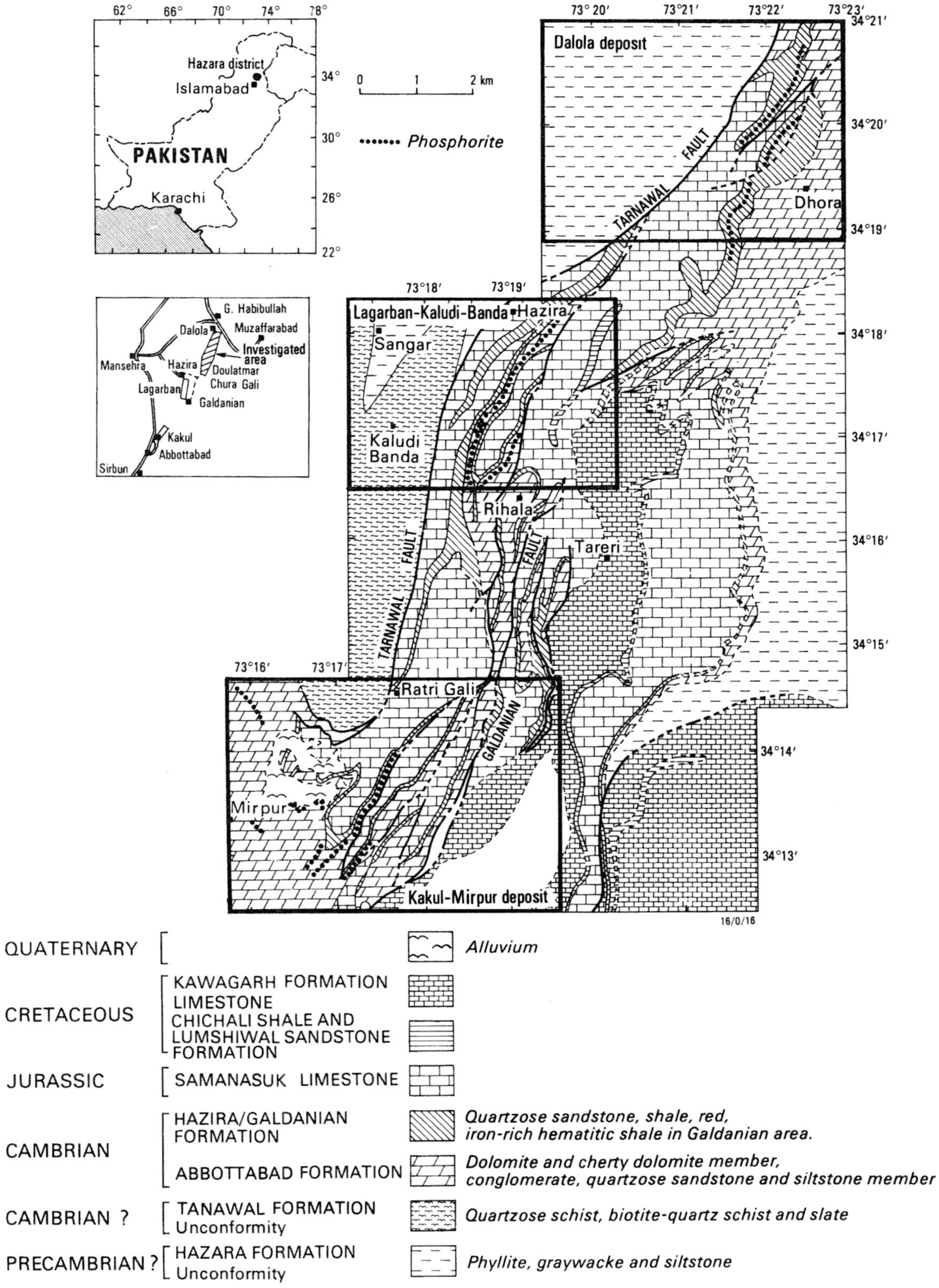

Fig. 16.1. Location and distribution of the Hazara phosphate deposits.

Tanawal Formation

This sequence, of low grade medium- to fine-grained quartz–mica schist, micaceous quartzite, and micaceous quartzose siltstone, is only found west of the Tarnawal Fault. Because of the tight folding its thickness cannot be ascertained. In places it is intruded by the Mansera Granite. The Tanawal Formation is unconformably overlain by the Abbottabad Formation. The age of the formation is uncertain as it is devoid of fossils. Calkins & Matin (1968) have correlated it with the Muth Quartzite on the basis of lithology and assigned a Siluro-Devonian age. However, according to Shah (1977), the formation is Early Cambrian. Here it is regarded as Precambrian, because of gradational relationships with the Hazara Formation, and unconformity with the overlying Abbottabad Formation.

Abbottabad Formation

Overlying the Tanawal Formation is the conglomerate, shale, sandstone, siltstone and dolomitic limestone of the Abbottabad Formation (Calkins *et al*., 1969). It is best developed near the town of Abbottabad where it consists of an upper unit of dolomite, limestone and shale; a middle unit of quartzose sandstone, shale and limestone; and a lower unit of sandstone, variegated shale, quartzite and dolomitic limestone. The Tanaki Conglomerate Member occurs at the base of the Abbottabad Formation (Fig. 16.2).

Gardezi & Ghazanfar (1965) and Latif (1970) have divided the Abbottabad Formation into two units. The lower (Kakul) unit, represented by the Tanaki Conglomerate, lower Dolomite and middle Quartzose Sandstone on Fig. 16.2, consists of basal conglomerate followed by a sequence of orthoquartzite, dolomite, sandstone and shale in the middle part; and variegated

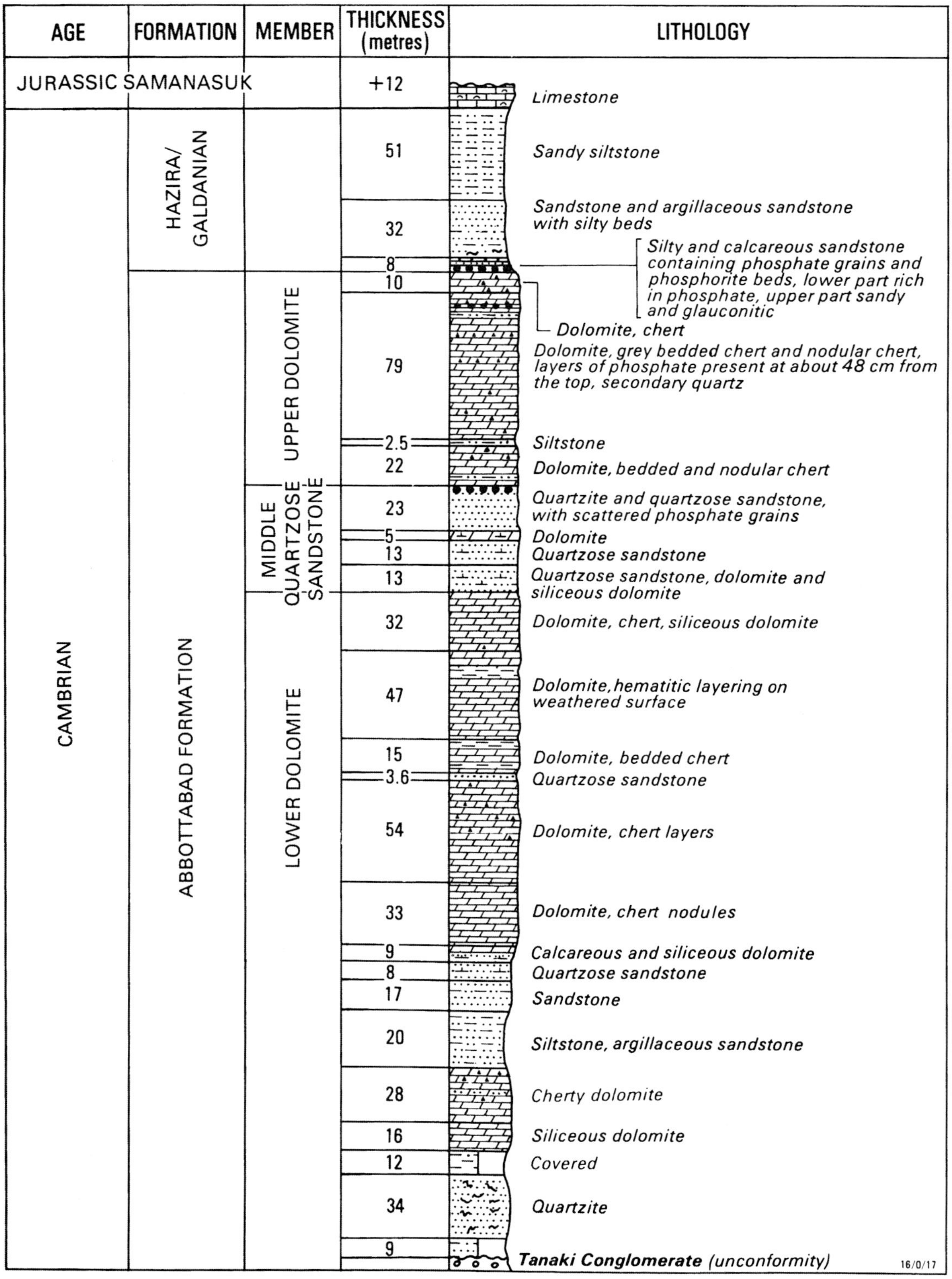

Fig. 16.2. Generalised stratigraphy of the phosphatic intervals.

shale and siltstone in the upper part. The lower dolomitic part is medium- to thick-bedded, buff in colour and nodular in part; synsedimentary breccias are present in the basal carbonate units. Quartzose sandstones in the middle and upper parts of this unit are hard, massive, grey in colour, and contain well-sorted, sub-rounded quartz grains in siliceous or calcareous cement.

The upper cherty dolomitic part of the Abbottabad Formation (Upper Dolomite) consists of dolomite with bedded chert, dolomitic limestone with a few silty beds and nodular chert in places. Phosphorite is developed within this upper unit at Kakul, Sirban, Lagarban, Kaludi Banda and Bagla Gali. Its contact with the overlying Hazira Formation seems to be gradational at most places. The Abbottabad Formation is assigned an Early Cambrian age by Shah (1977).

Hazira Formation

This formation occupies the stratigraphic position between the Samanasuk Limestone and Abbottabad Formation, and is equivalent to the Galdanian Formation of Calkins &

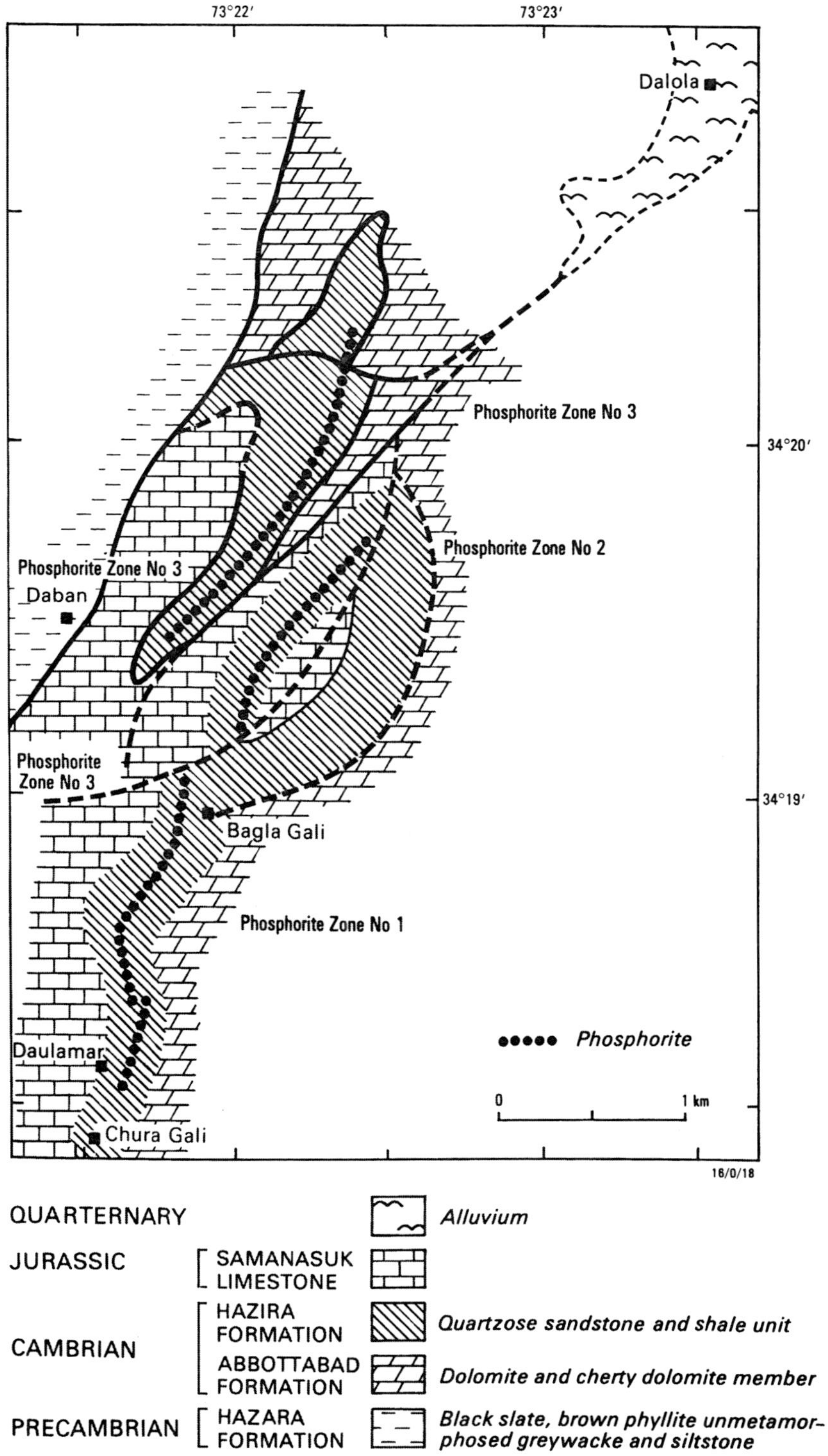

Fig. 16.3 Geology of the Dalola Deposit.

Matin (1968). The formation consists of beds of granular and partly oolitic, hematitic siltstone, claystone with minor ferruginous and siliceous quartzose sandstone, and siltstone. In places, calcareous sandstone, glauconitic sandstone, and siltstone with some hematitic intercalations are present, and there is some interbedded rhyolite. The Hazira Formation has two facies – the 'Galdanian facies', consisting of hematitic quartzite, volcanics, and claystone, and the 'Hazira facies' in which siltstone, claystone and glauconitic sandstone are dominant. Phosphorite is also developed in this formation at many places at Dalola, and Sirbun Hill (see later). An Early Cambrian age is assigned to this formation by Shah (1977), on the basis of hyolithid shells found at Sirbun Hill. A major unconformity separates the Hazira Formation from the overlying Mesozoic

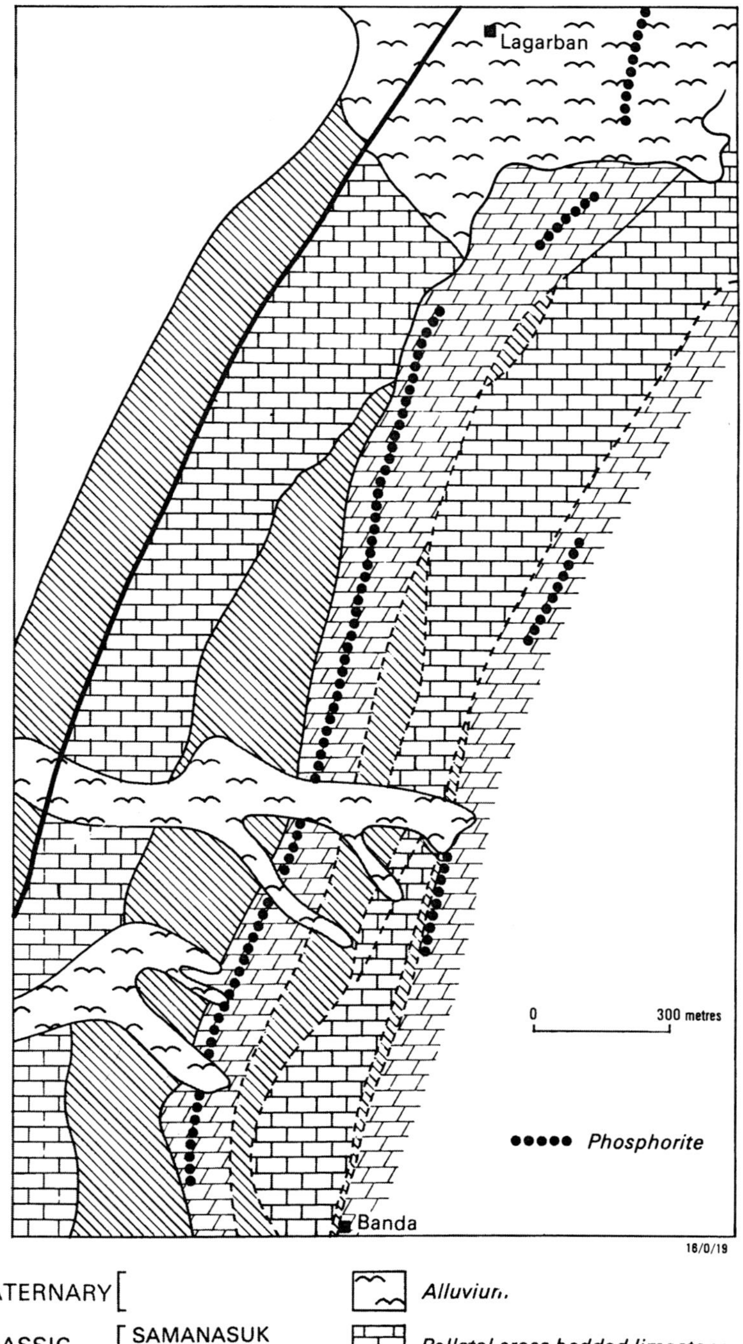

QUATERNARY — Alluviun.

JURASSIC — SAMANASUK FORMATION — Pelletal cross bedded limestone

CAMBRIAN — HAZIRA/GALDANIAN FORMATION — Sandy and silty quartzite, shale

ABBOTTABAD FORMATION — Dolomite and cherty dolomite, phosphorite with dolomite intercalations, cherty dolomite and sandy dolomite

Fig. 16.4. Geology of the Lagarban–Kaludi–Banda Deposit.

and Tertiary sediments (Table 16.1) which are essentially unrelated to the depositional pattern of the Hazira Formation and therefore are not discussed here.

General structure

The Hazara area lies on the western flank of the Hazara–Kashmir Syntaxis (Wadia, 1931). The phosphorite deposits of the Dalola–Doulatmar and Lagarban areas are located in a south-plunging synclinal structure called the Garhi Habibullah Syncline (Figs. 16.3 and 16.4). This syncline is bounded by the Jhelum Fault in the east and the Tarnawal Fault in the west. At Garhi Habibullah the syncline is generally overturned to the west but is relatively simple: to the south it becomes complex in the Galdanian–Kakul area (Fig. 16.5) where the Cambrian to Eocene sediments have been deformed into doubly plunging isoclinal folds commonly overturned westward on the flanks of the regional synclinal structure. In the Kakul area, the folding is disrupted by numerous large strike faults (Fig. 16.1), some of which show lateral movement closely related to folds. Mostly these faults dip eastwards, although a few dip to the west. The structural complexity of the area has adversely affected the phosphorite deposits.

Phosphate deposits

Collophane, dahllite and minor francolite are the common phosphatic constituents at Hazara. Glauconite, dolomite,

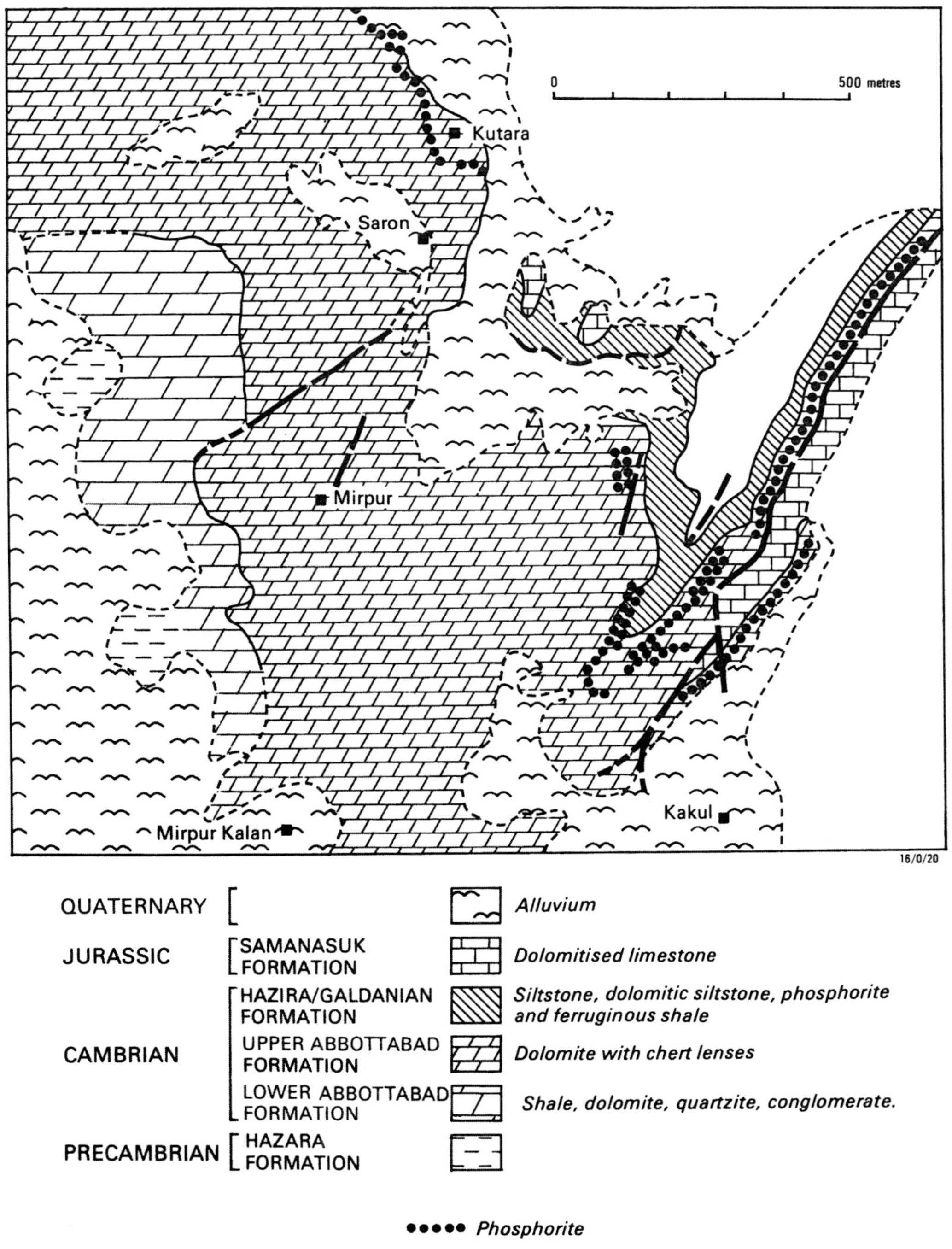

Fig. 16.5. Geology of the Kakul–Mirpur Deposit.

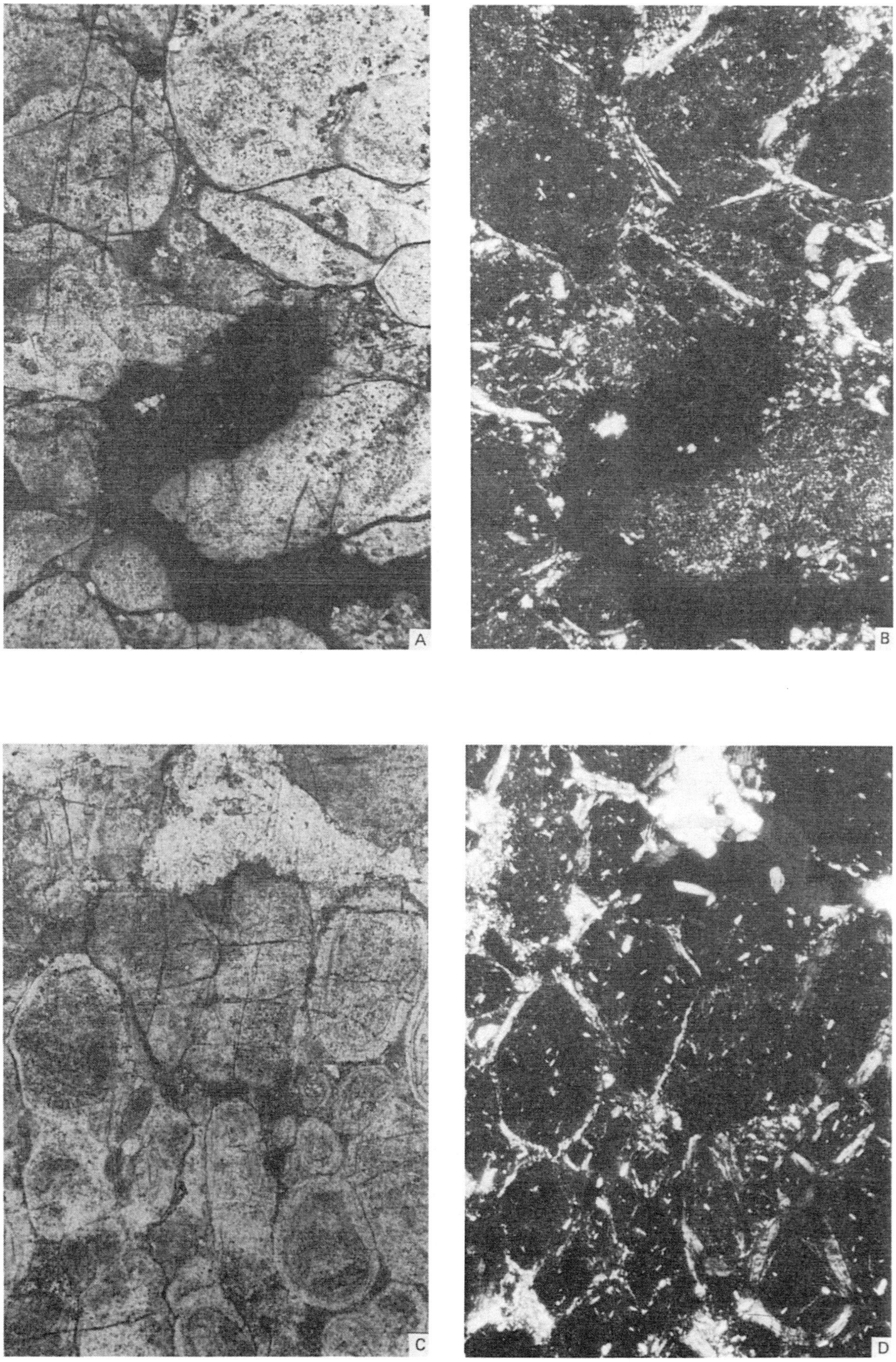
A
B
C
D

iron oxide and pyrite are common impurities; their relative abundances differ markedly from place to place.

There are rapid changes in lithology within the phosphatic interval which comprises the upper cherty dolomite of the Abbottabad Formation and lower part of the Hazira Formation. The phosphate content ranges from 25% P_2O_5 in the phosphorites, to as little as 1% P_2O_5 in the associated sediments. Generally the phosphorite is granular; in places, however, such as Dolola, the phosphorites of the Hazira Formation are dense, dark, and organic-rich. Additionally, in some areas the outlines of the phosphate grains are ill-defined due to diagenesis. At Sirbun Hill elongated tubes are present. A few of them are still siliceous or calcareous in composition, but most have been completely phosphatised. These tubes are interpreted as hyolithids. The cement is commonly dolomitic or cherty, or less commonly, secondary apatite.

A few of the phosphate grains show a concentric internal banding, however the majority show no internal structure, although inclusions of quartz, calcite, dolomite, clay, pyrite and glauconite are found within many grains. Photomicrographs of selected samples are illustrated in Figs 16.6 and 16.7.

Description of deposits

In the Hazara area, phosphate deposits extend from Kakul to Dolola (Fig. 16.1). There are two main types, the cherty phosphorites of the Abbottabad Formation, and the silty, calcareous phosphorites of the Hazira Formation. However, grade and thickness vary greatly from one area to another. The geochemistry of each deposit is summarised in Table 16.2.

Kakul–Mirpur area

The deposits of the Kakul–Mirpur area are located about 9.6 km from the city of Abbottabad, and are spread over an area of about 13 sq. km. The main and most promising deposit is located in the southern part of the map area near Kakul (Fig. 16.5). Other deposits of less economic interest are located to the north, at Rashian, Nari-de Gali and Lambi-Dogi (Fig. 16.1).

The phosphorite deposits are found at two well-defined horizons in the upper part of the cherty dolomite of the Abbottabad Formation, one at the contact with the Hazira Formation, and the other about 30 m below the contact. The main phosphorite horizon at Kakul is the uppermost one, developed in the Abbottabad–Hazira transition zone. It extends for more than 500 m along strike, and has an average thickness of 4.5 m. Elsewhere, the deposits occur as irregular lenses, ranging in thickness from 0.04 m to 1.4 m. The lower phosphorite horizon, in the upper Abbottabad Formation, occurs

Table 16.2. *Geochemistry of the Hazara deposits (mean percentage)*

	Kakul	Sirbun Hill	Dalola	Lagarban–Kaludi–Banda
SiO_2	40	30	40	20
Fe_2O_3	2	6	6	5
Al_2O_3	3	2	2	2
P_2O_5	27	15	20	28
CaO	25	30	26	45
MgO	1	2	3	1
F	1	1	1	0.8

sporadically from Rashian and Nara-de-Gali to Lambi-Dogi (Fig. 16.1). It is in the form of numerous lenses from a few centimetres to several metres thick and extends along strike for up to 70 m.

Tight folding and extensive faulting is the normal feature of the Kakul–Mirpur area and this has adversely affected the phosphorite deposits. Reserves of phosphate rock in the Kakul–Mirpur area total about one million tonnes of phosphate rock with an average phosphate content of 24% P_2O_5.

Lagarban–Kaludi–Banda area

The phosphate deposits of the Lagarban and Kaludi–Banda areas (Fig. 16.1, 16.4) are situated about 40 km northeast of Abbottabad (Ghaznavi & Tahir, 1978). They occur along the two overturned limbs of the syncline (the Garhi–Habibullah Trough) extending for up to 3 km along strike. The deposits of the western limb of the syncline occur 60–75 m below the top of the upper cherty dolomite of the Abbottabad Formation in the form of irregular lenses with maximum thickness ranging from 3 m to 8 m. In places the phosphorite is intercalated with dolomite layers. A second lenticular phosphatic horizon is developed about 30 m below the main horizon near Kaludi–Banda close to the faulted contact between the Hazira and Abbottabad Formations. The phosphorite deposits of the eastern limb of the syncline are not as well developed, or as high in grade as those of the western limb, although they are located at the same stratigraphic horizon (the upper cherty dolomite of the Abbottabad Formation). The lateral extension of this deposit is up to 1 km, with a maximum thickness of 6 m. They occur as sporadically developed lensoid bodies of phosphorite intercalated with layers of dolomite. In the eastern limb of the Garhi Habibullah Trough, a phosphorite is also developed in the Hazira Formation but it is not of economic significance. These deposits are located in areas of

Fig. 16.6. (A, B) Phosphorite from the Kakul area as seen under plane polarised light (A) and crossed nicols (B). Phosphate is seen in the matrix as well as in the form of irregular grains. Silica and carbonate grains are associated with the phosphate grains.
(C,D) Phosphorite from Kakul Mine under plane polarised light and crossed nicols (D). Phosphate occurs both in the matrix and as discrete grains, associated with silica and carbonate grains. Some phosphate grains have a well developed concentric banding. Scale not provided by author but average diameter of phosphate grains in the order of 0.3–0.5 mm.

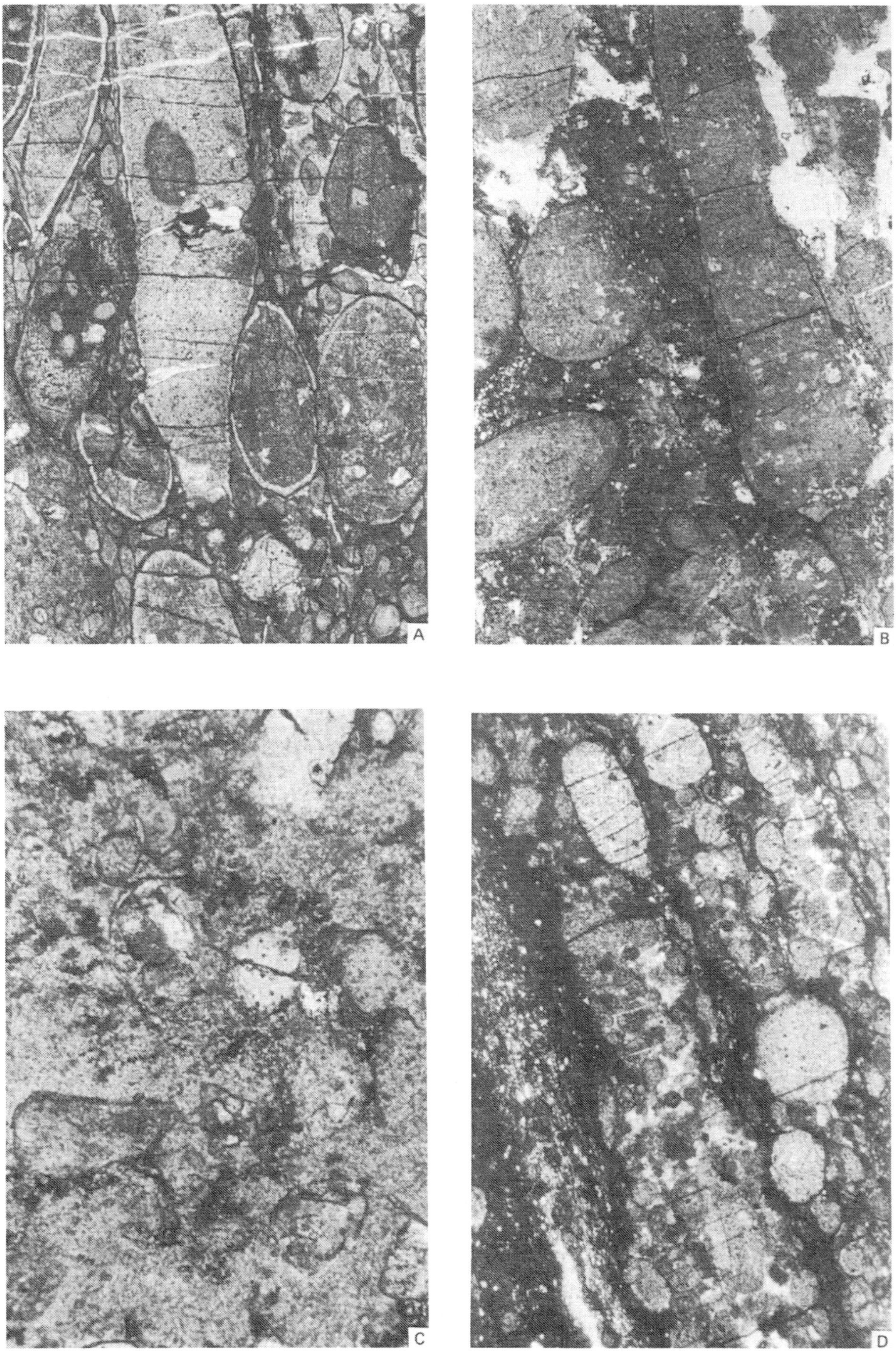
A
B
C
D

severe folding and faulting which has markedly affected all of these deposits.

The reserves of the Kaludi–Banda deposits are estimated at 1.9 million tonnes of phosphate rock with an average P_2O_5 content of 34% in the eastern limb of the syncline, and 6.7 million tonnes of 25% P_2O_5 in the western limb. The reserves of the Lagarban area are estimated at 0.8 million tonnes of approximately 30% P_2O_5, and 1.07 million tonnes with 1.5–27% P_2O_5.

In the Lagarban and Kaludi–Banda areas the phosphorite is generally a grainstone phosphorite (oolitic or ovulitic) or less commonly a mudstone phosphorite, and is mainly associated with chert and dolomite. The main phosphorite mineral is recognised as collophane. Some dahllite occurs in the outer shells of the grains. These grains are compressed, dispersed and at times binucleated. The gangue minerals associated with collophane are quartz, chert, pyrite and iron oxide. In places calcite and dolomite grains are also visible.

Dalola area

The deposits of the Dalola area are located about 6 km south of Garhi Habibullah (Fig. 16.1), between Dalola and Chura Gali (Fig. 16.3), (Ghaznavi & Tahir, 1979). Three phosphorite zones have been recognised in the silty beds of Hazira Formation extending up to 2 km along strike. In this area the Abbottabad Formation consists of hard, medium to massive, grey-coloured dolomite where phosphorites are absent. The Hazira Formation in this area mainly comprises a quartzose sandstone in the lower and upper parts and siltstone in the middle parts. Phosphorite is developed in the middle silty part of this formation. Phosphorite is absent south of Daultmar. In the Chura Gali area (Fig. 16.3), manganese lenses are developed at the same horizon within the Hazira Formation.

This area is structurally very complicated with severe folding and faulting; the phosphorite zone has been terminated by a fault. The uppermost phosphorite zone extends for 1700 m along strike, varying in thickness from 13 m in the south to only 3 m in the north. The middle phosphorite zone is thicker and higher in grade than the other two. It extends for about 900 m along strike and is lenticular in shape with a maximum thickness of 23 m. The lowest phosphorite zone mainly comprises phosphatic intervals with silty beds although at the top there is a hard calcareous friable bed. This phosphorite is different from that at Kakul in that it is not typically granular but black, dense and structureless. It is commonly silty, calcareous and cherty with a small amount of manganese which gives it a shiny black appearance. Total reserves of phosphate rock at Dalola are estimated at 9.20 million tonnes with a P_2O_5 content ranging from 15% to 25%.

Sirbun Hill

The deposits of the Sirbun Hill area are located on the eastern side of Havalian–Abbottabad road (Fig. 16.8). Low grade phosphorite extends for up to about 500 m along strike with an average thickness of about 4 m. The total reserves are estimated at 0.8 million tonnes of phosphate rock. There are two other non-economic phosphorite horizons, one in the cherty dolomite and other in the quartzose sandstone of the Abbottabad Formation. The total reserves of these two horizons are small.

The upper phosphatic horizon of the Abbottabad Formation is located near the passage into the Hazira Formation, and is a cherty phosphorite. This type of phosphorite is not well developed in the Sirbun area and it has a limited thickness and lateral extension. The middle phosphorite horizon of the Abbottabad Formation, having an average P_2O_5 content of 10%, is developed in cherty dolomite, 15–18 m below the contact with the Hazira Formation. This occurs as recurrent thin phosphorite veins and layers and dispersed phosphate grains in dolomite. The third horizon, which is not of any economic significance, is developed in quartzose sandstone near the middle of the Abbottabad Formation. All of these phosphorite deposits vary laterally and vertically in mineralogy, lithology and P_2O_5 content. They are generally lenticular in nature and irregularly developed.

The main phosphorite horizon in the area is located in the 'passage beds' of the Abbottabad and Hazira Formations. The phosphorite is low grade with a lateral extension of about 1 km. The area is structurally very complicated and all the phosphorite beds are affected by faulting. Thrust faults at the contact of the Abbottabad and Hazira Formations terminate the Hazira phosphorite.

Glauconite, microcrystalline calcite, and some quartz grains and clay minerals are commonly associated with the phosphate rock in this formation, and a few pyrite grains are also present.

Fig. 16.7. (A) Phosphorite from Sirbun Hill seen under crossed nicols. Collophane is present in the form of subrounded or elongated grains of variable size. Pale rims surrounding some grains are due to secondary alteration. The fine-grained matrix is mostly siliceous and calcareous. Veinlets of iron oxide are developed due to weathering. (B) Rounded and subrounded collophane grains from Sirbun Hill, interpreted as phosphatised carbonate grains. An elongate calcareous tube has also been replaced. Secondary quartz and calcite grains are visible within the phosphatised grains, and veinlets of calcite can also be seen. The matrix is mainly of fine silt, with some phosphate and carbonate. (C) Phosphorite from Sirbun Hill showing secondary alteration around the grain margins. Secondary silica, calcite and iron oxides are developed. The matrix is mainly siliceous. (D) Clusters of siliceous and calcareous grains within elongate phosphate grain. The grains are welded by a silica or calcitic matrix. Scale not provided by author but average diameter of phosphate grains in the order of 0.3–0.5 mm.

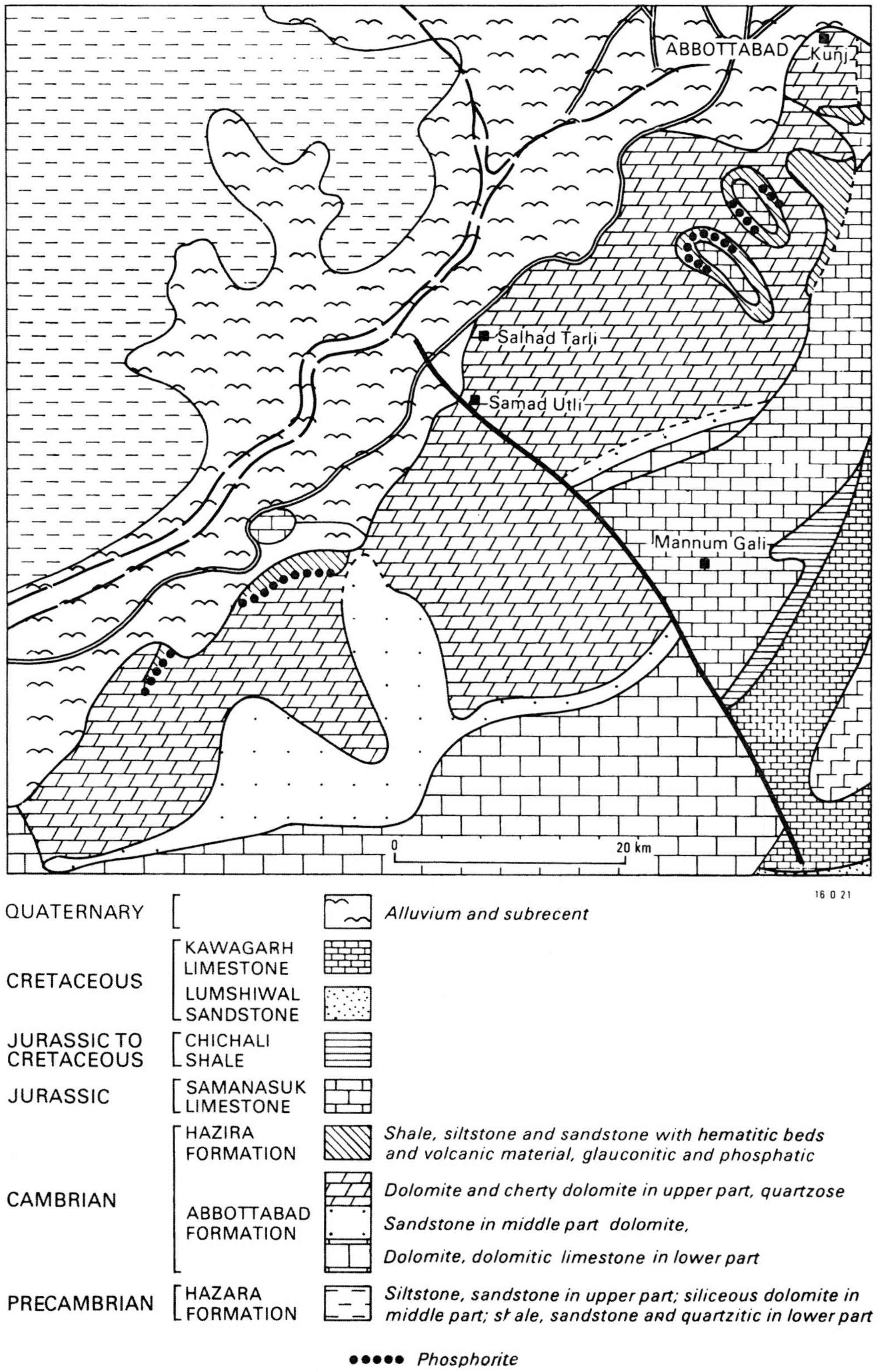

Fig. 16.8. Geology of the Sirbun Hill area.

Collophane is the dominant variety of apatite; and dahllite occurs as an alteration product around the margin of colophane grains. These phosphate grains are dark in colour, subrounded to round or elongate. Calcite and some iron oxide is also present in the form of secondary veins, matrix and cementing material.

The total reserves of the Sirbun Hill area, including both high and low grade ore, comprise 1.9 million tonnes of phosphate rock with a phosphate content of 10–25% P_2O_5.

Mining and utilisation

Detailed exploration and evaluation of the Hazara phosphorite deposits commenced in 1975 and the first phase was

Table 16.3. *Probable and proven reserves of phosphorite in the Hazara district (in million tonnes)*

	Dip depth	Proved	Probable
1. Kakul–Mirpur			
(a) High to medium grade	91 m	0.4	
(b) Low grade			0.68
2. Lagarban (North)			
(a) High to medium grade		0.8	
(b) Low grade			1.7
3. Kalu-de-Bandi (Lagarban South)			
(a) 1. Eastern phosphorite	152 m		1.0
2. Western phosphorite			6.4
4. Dalola			
(a) Low to medium grade	91 m		2.3
(b) Low grade			6.9
5. Sirbun			
Low grade			1.9
	Total	1.2	21.78

(a) High to medium grade between 25% P_2O_5 and 34% P_2O_5;
(b) Low grade below 25% P_2O_5.

completed in 1978. A feasibility study was prepared based on an annual production of 300 000 tonnes of phosphate rock per annum with an average grade of approximately 26% P_2O_5. A mine was developed at Kakul with two levels of mining. So far, only 15 300 tonnes of phosphate rock have been mined from these two levels. A test mine at Lagarban is also in progress. However, the mine development work depends greatly on the grade, size and quality of the phosphorite, and further detailed investigations of these variables are required. Beneficiation tests still have to be completed. The rock is hard to grind and at times contains impurities which are detrimental to the normal sulphuric acid process thus raising the cost of treatment. Moreover, the phosphorite is less reactive and less amenable to treatment processes, and thus upgrading will be expensive. Thirty-one exploratory holes have been drilled in the test area, totalling 2091 m of section. Of these, 20 holes were drilled at Kakul (1415 m), six at Lagarban (401 m) and five at Dalola (275 m). Exploration to date suggests that reserves of phosphate rock in all the deposits of the Hazara area are estimated at 23 million tonnes (Table 16.3).

References

Calkins, J.A. & Matin, A.S. (1968). The geology and mineral resources of the Garhi Habibullah Quadrangle and the Kakul area, Hazara district. *US Department of State, Project Report*, **PK 3p**, 7–22.

Calkins, J.A., Offield, T.W. & Ali, S.T. (1969). Geology and mineral resources of the southern Hazara district, west Pakistan and part of western Kashmir. *US Department of State and Government of Pakistan, Project Report*, **PK-43**, 17–43 and 50–9.

Gardezi, A.H. & Ghazanfar, M. (1965). A change of facies at the base of the Jurassic in district Hazara, west Pakistan. *Geological Bulletin of the Punjab University*, **5**, 53–4, geological map 1:45 000.

Ghaznavi, M.I. & Karim Tahir (1978). Phosphorite deposits of Kalu-de-Bandi and Lagarban area, district Hazara, NWFP. *Geological Survey of Pakistan, Information Release*, **103**, 31pp.

Ghaznavi, M.I. & Karim Tahir (1979). Phosphorite deposits of the Dalola-Daulatmar area, Hazara division, NWFP, Pakistan. *Geological Survey of Pakistan, Information Release*, **104**, 29pp.

Latif, M.A. (1970). Explanatory notes on the geology of southeastern Hazara to accompany the revised geological map. *Geologisches Bundesanstalt, Jahrbuch, Sonderband*, **15**, 5–20.

Shah, S.M.J. (Ed.) (1977). Stratigraphy of Pakistan. *Geological Survey of Pakistan, Memoir*, **12**, 138pp.

Wadia, D.N. (1931). The syntaxis of the northwest Himalaya; its rocks, tectonics and orogeny. *Records of the Geological Survey of India*, **65(2)**, 189–220.

17

Proterozoic and Cambrian phosphorites – deposits: Jhamarkotra, Rajasthan, India

R. CHOUDHURI AND A.B. ROY

Introduction

The phosphorite deposit of Jhamarkotra, situated about 26 km south-east of Udaipur, in Rajasthan (India), is of considerable academic and commercial interest. Not only does it constitute the largest known phosphate repository amongst the Precambrian Aravalli deposits, with proven reserves of 65 million tonnes of 'ore', but the occurrence of phosphate as one of the main constituents within a stromatolitic framework makes the deposit an extremely uncommon one. The occurrence of such a deposit is likely to encourage the search for phosphorite in Precambrian rocks of similar lithology.

This paper gives an account of the regional framework of the stromatolitic Proterozoic phosphorite occurrence of Jhamarkotra. It is not merely because of its relatively large size that the Jhamarkotra deposit has been singled out for special attention, but because it occurs in an area where the stratigraphic succession and sedimentation history is well documented, and phosphatic stromatolites show their most prolific and diverse morphological development in a single continuous marker horizon, indicating a conspicuous biological role in phosphate accumulation. Besides being an active mining area, the grade behaviour of the deposit and the factors controlling the grade can be studied to a considerable extent.

Regional geological setting

The Aravalli rocks, of which stromatolitic phosphorite forms a part, constitute an important element in the Precambrian mosaic of north-western India. Lying unconformably on basement granites, gneisses, amphibolites, mica-chlorite schist etc., generally referred to as the Banded Gneissic Complex and the Berach Granite, the Aravalli Group comprises a thick pile of shale, sand and carbonate with basal volcanics. The other stratigraphic formations, which together with the Aravalli and the pre-Aravalli rocks constitute the protracted Archaean–Proterozoic sequence of Rajasthan and its neighbourhood, include the Delhi, Champaner and Vindhyan Groups, besides a host of granitic, basic, ultrabasic and alkaline intrusive bodies (Heron, 1935, 1953).

The type region of the Aravalli rocks is the belt that extends from Jhabua in the south to Nathdwara in the north, through Zawar and Udaipur (Fig. 17.1). In this terrain of generally low- to medium-grade metamorphism, the Aravalli rocks are sharply divided into two contrasting lithological associations (Fig. 17.2). The eastern belt, which fringes the Banded Gneissic Complex, shows such diverse rock types as greenschist (altered lava flows and tuffs), carbonate, orthoquartzite, carbonaceous phyllite and mica schist, arkose–conglomerate and greywacke–shale–phyllite–diamictite. By contrast, the western belt is a carbonate-free sequence of phyllite–mica schist and quartzite (Heron, 1953). Sedimentary attributes of these two belts indicate nearshore shelf and deep sea environments (Poddar, 1966; Roy & Paliwal, 1981).

The rock phosphate deposits in the Aravalli sequence show a restricted distribution pattern both in space and time. As far as our present knowledge goes, the majority of the deposits are located either around Amlimal–Sallopat (Jhabua neighbourhood) or around Udaipur. Several large occurrences are known near Udaipur besides that of Jhamarkotra which is the biggest commercially exploitable deposit (Fig. 17.2). Stratigraphically, the Udaipur cluster of deposits belong to a single rock formation, whose distribution in space suggests a complex strandline pattern (Roy & Paliwal, 1981).

Isotope dating by Crawford (1970) and Choudhary *et al.*, (1981) indicates that the basement to Aravalli sedimentation is a pre-2500 m.y. crystalline basement and that granites coeval with the first folding of the Aravalli rocks were emplaced at around 2000 m.y. (Choudhuri, 1979).

Regional correlation

A recent study by Roy & Paliwal (1981) proposed a new stratigraphic succession of the phosphate-bearing Precambrian rocks of the Udaipur region (Table 17.1). These authors have recognised four formations, although one of these (Formation A) is not developed at Jhamarkotra, where the Aravalli succession starts with a clean-washed quartz arenite passing upwards as well as laterally into an extensive carbonate sequence which includes the phosphorite bed (Figs. 17.3 & 17.4). Correlation of this formation, including the basal arenite wedge, with Formation B is based on physical continuity and gross lithological similarity.

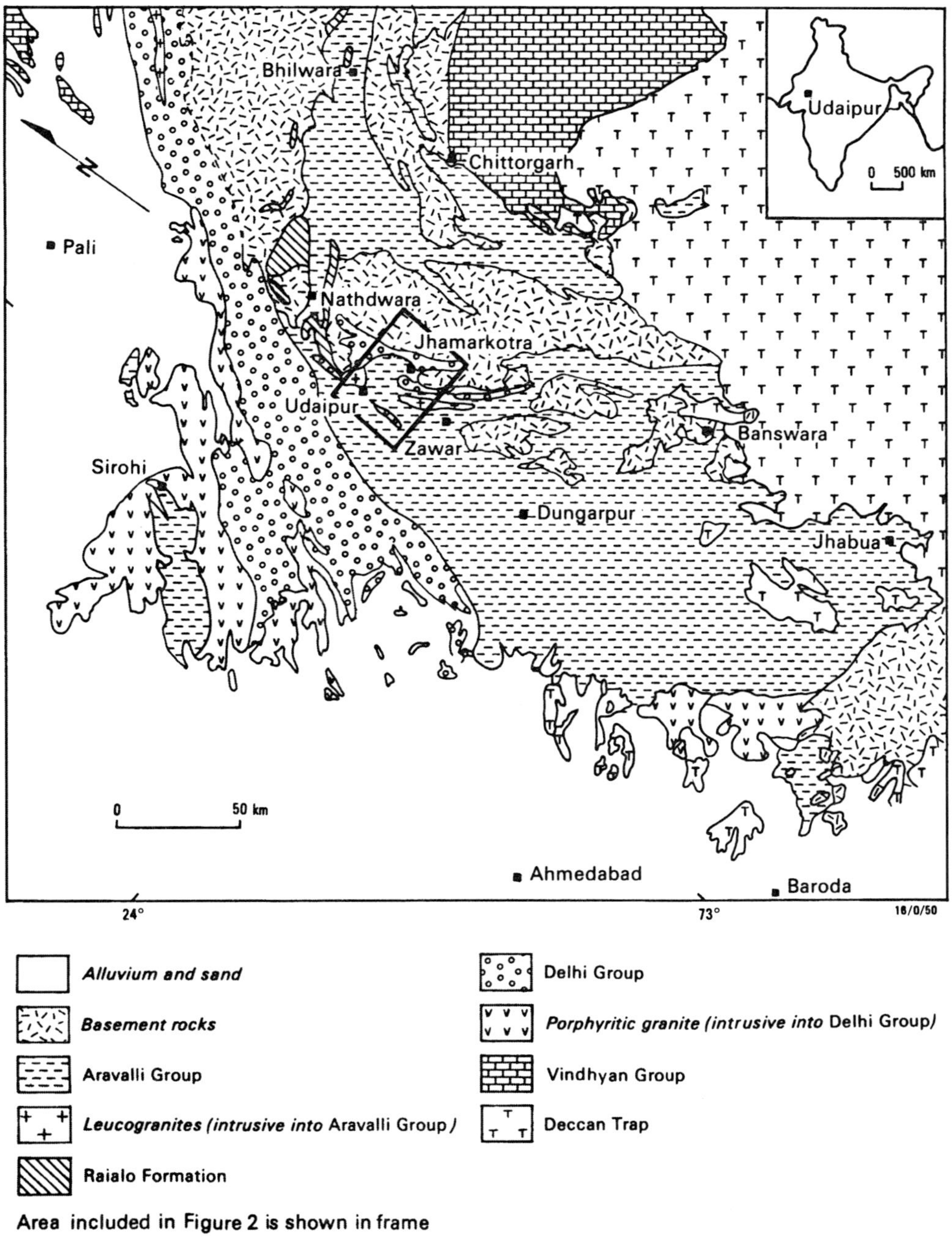

Fig. 17.1. Map showing distribution of the phosphate bearing Aravalli Group in southern Rajasthan and neighbourhood. (Based on the map of the Geological Survey of India, 1969.)

The passage from carbonates to the shales (phyllites and mica schists), and locally black shales (carbonaceous phyllites and graphite schists of the eastern part of the area), is gradational (Fig. 17.5). In the central part of the area, lithic arenite succeeds the shale sequence, whereas to the east orthoquartzite has developed in its place. The two lithological bodies, lithic arenite and orthoquartzite, which show similar stratigraphic relationship to the underlying units, are correlated as Formation C. The mutual stratigraphic relationship between the rock formations developed at Jhamarkotra, Debari–Dabok (east of Udaipur city), and in the Udaipur neighbourhood, is shown diagrammatically in Fig. 17.6.

Deformational history

Four phases of folding have shaped the rocks of the Jhamarkotra region (Roy *et al.*, 1980), and the structural architecture noted here is comparable in all details with that in the rocks of Udaipur (Roy, Paliwal & Goel, 1971). Axial traces of the major folds are shown on the map (Fig.17.3). The third generation folds have apparently not changed the gross outcrop pattern and have produced only minor folds with horizontal axial planes. As highlighted by the rock phosphate horizon, the outcrop pattern shows strong effects of second and fourth generation folds superimposed on the limbs of a large first generation isoclinal fold with co-axially refolded axial planes.

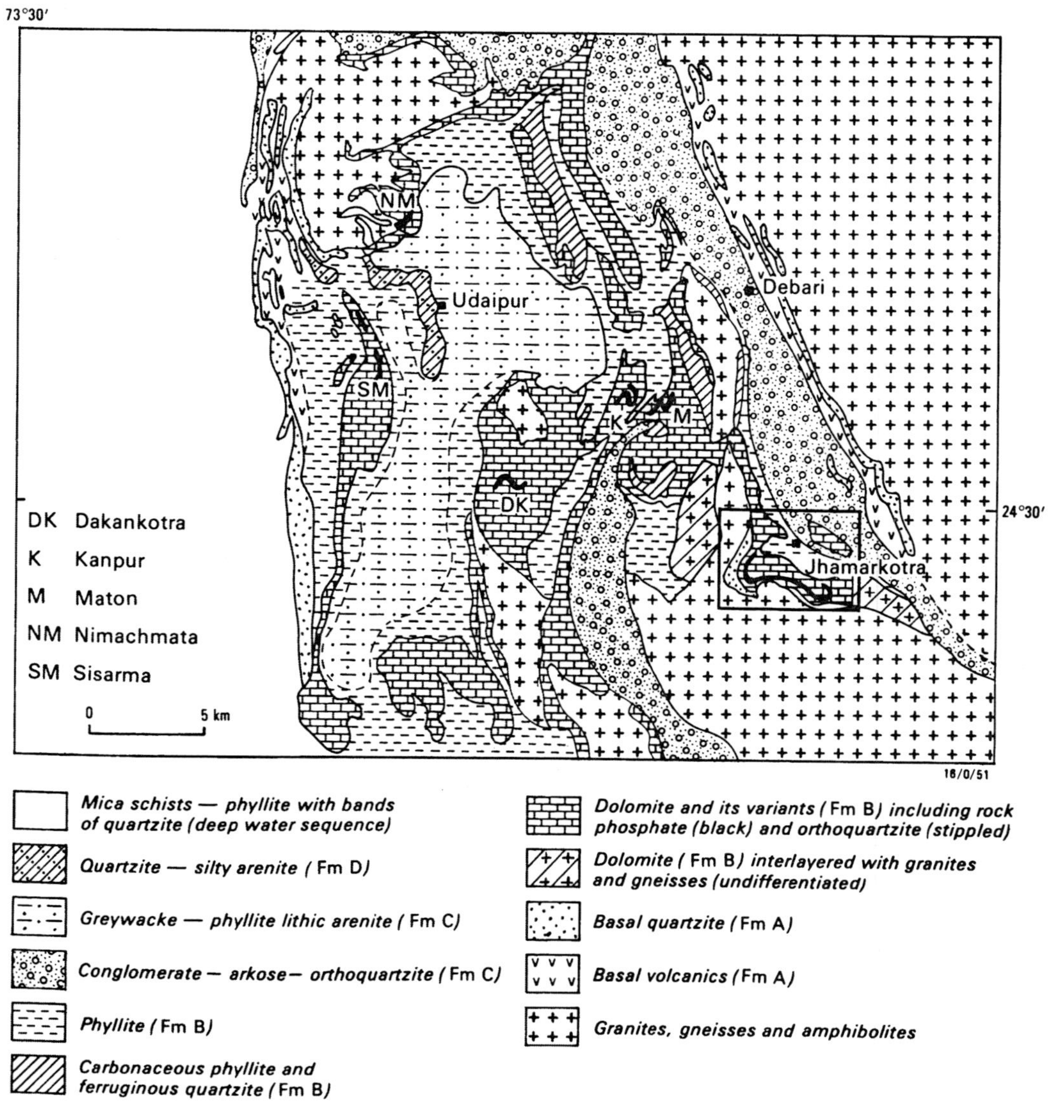

Fig. 17.2. Lithological distribution of the Aravalli rocks showing location of the major rock phosphate deposits around Udaipur (Based on Roy & Paliwal, 1981, and modified after Heron, 1953.)

Table 17.1 *Stratigraphic succession of Aravalli rocks around Udaipur*

	Deep water facies		Shallow epicontinental shelf facies
Aravalli Formation	Mica-schists, phyllites and quartzites	Formation D:	Quartzites of Udaipur
		Formation C:	Greywackes, lithic arenites, phyllites, pebbly and bouldery phyllites of Udaipur valley and Jhamarkotra. Conglomerate-arkose-quartzite of Debari.
		Formation B:	Dolomites, orthoquartzites, stromatolitic rock phosphates, carbonaceous phyllites, silicified rocks, phyllites with carbonate lenses.
		Formation A:	Quartzite (locally arkose) conglomerates, amphibolites and greenschists (altered basic volcanic flows and ash falls)
		Unconformity	
	Pre-Aravalli basement rocks (> 2500 m.y.)		

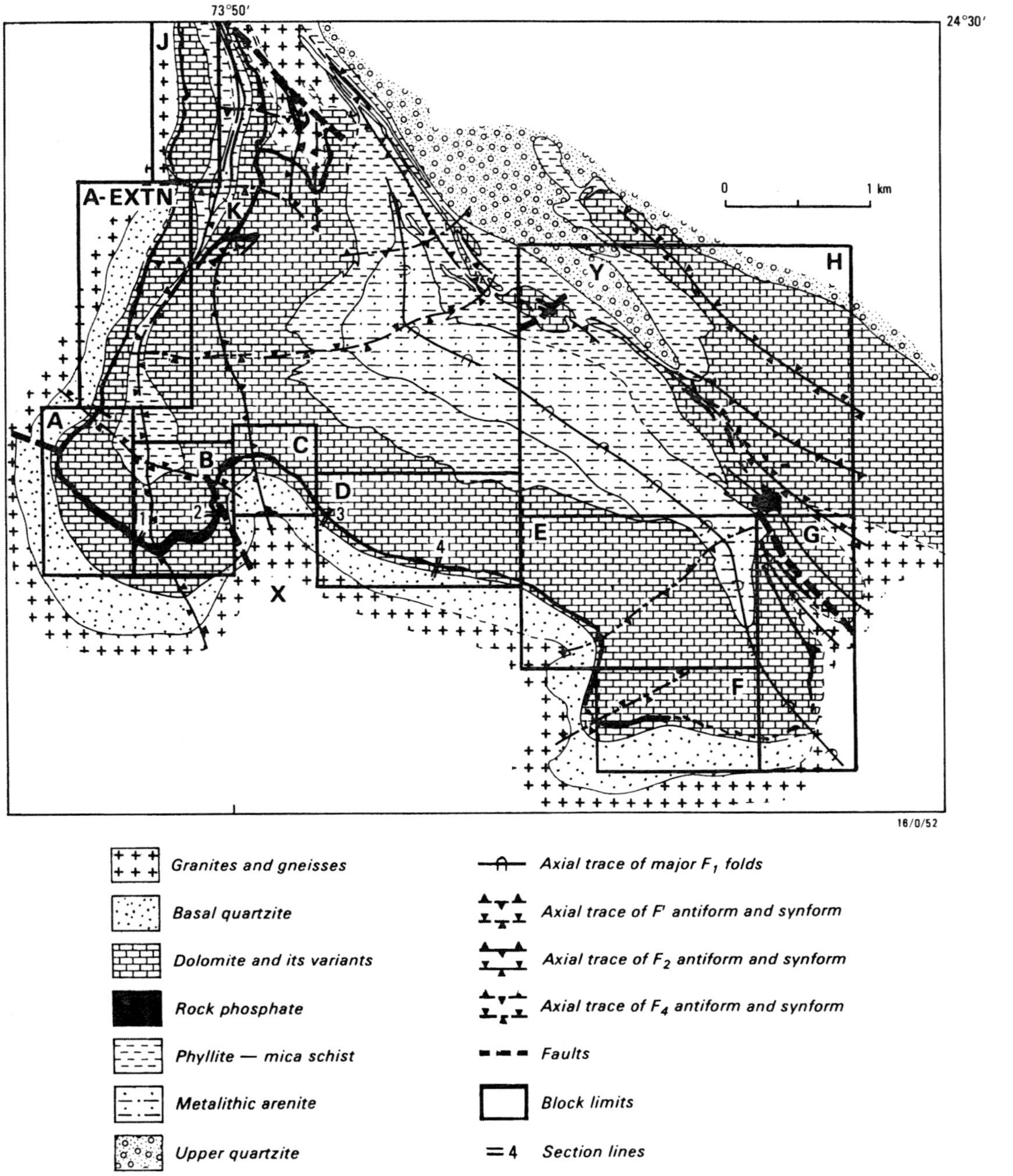

Fig.17.3. Geological map of the Jhamarkotra rock phosphate deposit. (Based on Roy, *et al.*, 1980; phosphorite outcrop based on the Rajasthan Directorate of Mines & Geology, 1977.)

The second generation folds have variable interlimb angles and vary in trend from roughly north–south in the western part to about north-west to south-east in the east. Besides folding, the rocks have also been affected by a number of faults, and suffered metamorphism and migmatisation along certain zones. The basement configuration seems to have controlled not only the sedimentary geometry but also the tectonic pattern.

Stratigraphy of the Jhamarkotra area

The Aravalli rocks in the neighbourhood of Jhamarkotra lie unconformably on basement rocks which comprise banded and migmatitic gneiss, porphyritic granite, amphibolite and chlorite–mica schist. Evidence of angularity between the basement gneisses and the covering Aravalli rocks is clear in the southern and western part of the area.

Basal quartzite

The basal Aravalli beds in the Jhamarkotra area include a wedge of arenite below the dolomite phosphorite unit. At places the arenite is conglomeratic, containing small-sized pebbles of vein quartz. The clastic character of this basal unit is not clear everywhere, and at several localities the rock appears as massive vitreous quartzite. Field relations suggest that at least a part of this is a replacement product (silicified dolomite). The quartzite grades into dolomite both upwards and laterally, and there are also some thin lenses of dolomite within the arenite body. At

UNIT	THICKNESS (metres)		LITHOLOGY
UPPER QUARTZITE / LITHIC ARENITE			Orthoquartzite / Lithic arenite and pebbly quartzite
PHYLLITE – MICASCHIST	250		Phyllite – micaschist, marly phyllite, locally carbonaceous phyllite at the base
UPPER DOLOMITE	450 – 50		Dolomite (marble) ferruginous and manganiferous dolomite, quartzite, brecciated and silicified dolomite
ROCK PHOSPHATE	20 – 0		Stromatolitic rock phosphate
LOWER DOLOMITE	350 – 0		Dolomite (at places stromatolitic) silicified ferruginous and argillaceous dolomite
BASAL QUARTZITE	250 – 0		Orthoquartzite (highly ferruginous) locally pebbly at the base
PRE-ARAVALLI BASEMENT ROCKS			Granite – gneiss, amphibolite, porphyritic granite, chlorite actinolite schist, etc

16/0/53

Fig. 17.4. Stratigraphic sequence of the rock phosphate-bearing sedimentary package of the Jhamarkotra area.

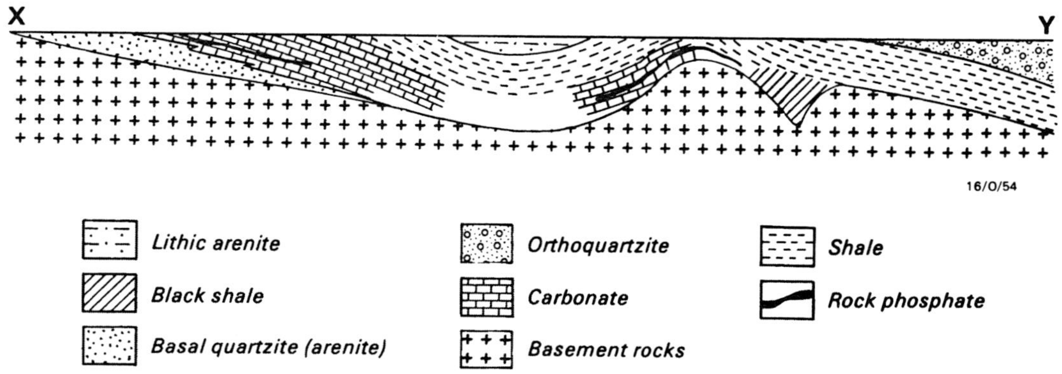

Fig. 17.5. Restored generalised section through X–Y line (see Fig. 17.3) showing sedimentary facies and environment. Vertical scale exaggerated.

some places, however, a sharp contact between the quartzite–arenite and the overlying carbonate has been recorded (Fig. 17.3).

Carbonates and their variants

Carbonates cover a large surface area in the Jhamarkotra region, and include several facies variants and alteration products. Among the most commonly occurring lithological types are: (1) grey and greyish-white dolomite grading to marble; (2) brownish-coloured ferruginous dolomite; (3) dark grey coloured carbonaceous dolomite grading to carbonaceous phyllite; (4) argillaceous dolomite and marly phyllite; (5) brownish-black manganese-bearing dolomite; (6) brecciated dolomite; and (7) silicified rock.

The first mentioned type is by far the most dominant member of the carbonate formation and appears as an extremely monotonous poorly bedded unit. Excellently preserved bedding is present in the lower dolomite unit underlying the rock phosphate in the B block area (Fig. 17.3). It is, however, possible that many of the laminations present in this part of the sequence may be of algal origin (Fig. 17.7A). Stromatolites are common here, showing divergent morphological forms which include undulating laminations, domes (convex upwards), columnar structures, etc. Some more complex types may be the result of tectonic overprinting (Paliwal, 1980).

The thickness of the dolomite unit underlying the rock phosphate is extremely variable. In certain cases only a ribbon of carbonate (in places silicified) intervenes between the phosphate and the underlying rock (either basal quartzite or gneiss–amphibolite), whereas in the B block at Jhamarkotra the maximum thickness of the dolomite unit is about 100–120 m.

The dolomites are generally micritic, but coarsely recrystallised sparry dolomites occur at the contact of granitic intrusions. Mineralogically, the pure dolomites are composed of 90% dolomite (MgO 16–20%), a little quartz, and micas. The proportion of mica minerals increases as the dolomite grades to more argillaceous types. Manganese-bearing and carbonaceous dolomites have comparatively restricted occurrences, the latter

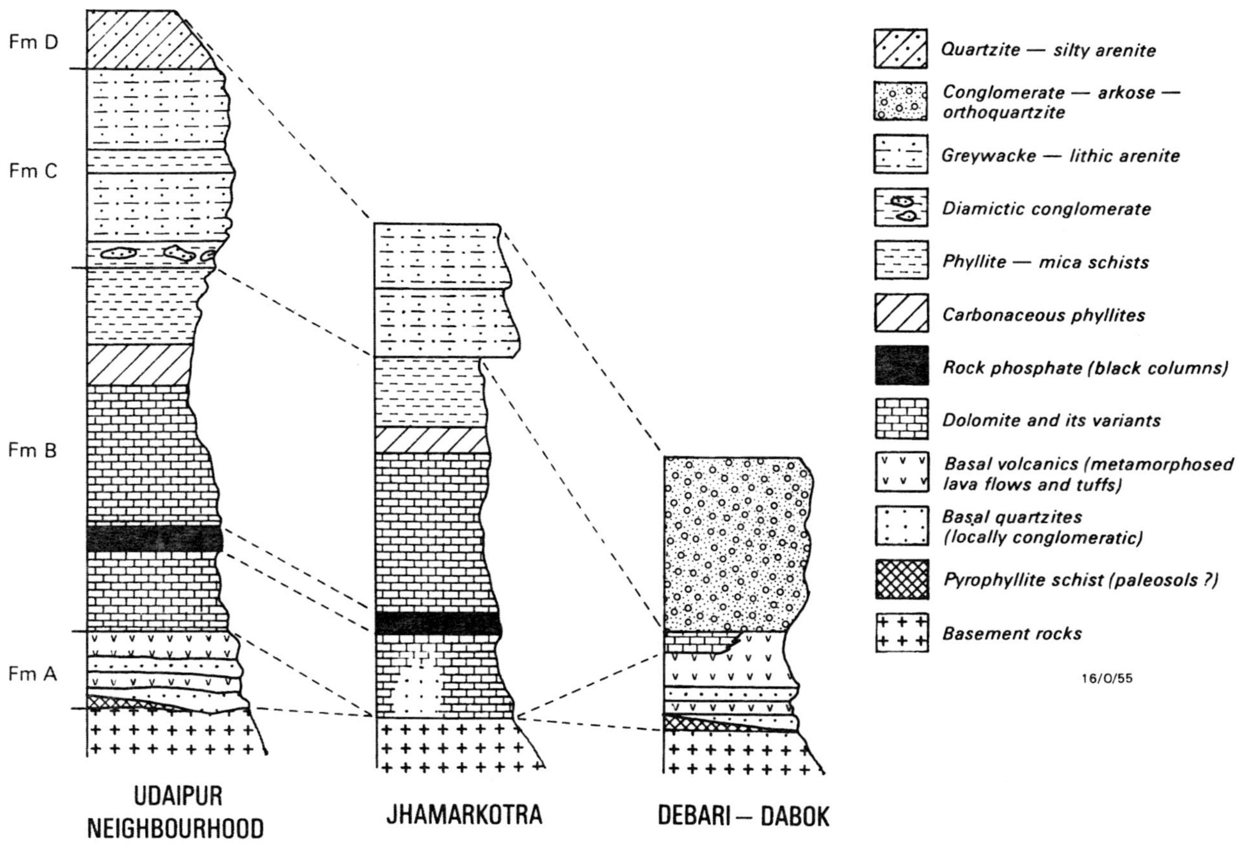

Fig. 17.6. Schematic stratigraphic correlation of the rocks of Jhamarkotra, Udaipur neighbourhood and Debari–Dabok sections.

mostly cropping out in the eastern part of the area. The manganese minerals are of secondary nature and are concentrated along different structural planes.

Brecciated and silicified dolomites are commonly encountered, but their distribution is rather erratic. It is possible that the brecciation and silicification are a pre-tectonic phenomenon related to slumping, caving and karstification during diagenesis.

Stromatolitic phosphorite

Taking into consideration the sporadic outcrops, the phosphorite bed is exposed continuously over a strike length of 16 km in the Jhamarkotra area. The phosphorite bed in this area has a variable thickness, ranging from 1 m to 30 m, wedging out in places. It has an average thickness of 15 m along the continuous stretch of 5 km strike length, which in turn accounts for more than 80% of the reserve. However, because of repeated folding in some areas, the phosphorite bed has a thickness of up to 110 m. Taken as a whole, the phosphorite horizon with its conspicuous strike continuity, relatively uniform thickness, and sharply defined contacts with the dolomite, forms an excellent marker bed for both structural and stratigraphic reconstruction. The petrography of the stromatolitic phosphorite bed is discussed separately.

Phyllites and mica schists

Phyllites and mica schists occupy an intervening position between the carbonate sequence and metalithic arenite. With the exception of a few coarsely crystalline mica schists containing large porphyroblasts of garnet, the majority of the rocks are finely schistose, in places crenulated and intersected by discrete planes of crenulation cleavage. Phyllites grade into highly ferruginous marly phyllites at the contact with the dolomitic limestone. To the east, normal argillaceous phyllite passes into carbonaceous phyllite towards the base.

Metalithic arenite

The main outcrop of metalithic arenite (sub-greywacke) is in the central part of the area. There is also a narrow strip of this unit in the north-western part, around Dhamdhar. Petrological variants of metalithic arenite include psammopelite and actinolite schist. There are also some pebbly beds containing small quartz pebbles about a centimetre or so in diameter. Besides these pebbles, the rocks are seen to be studded with large angular to smooth ellipsoidal pebbles and boulders (up to 0.5 m in diameter) of phyllite, rock phosphate, dolomite, quartzite and granite.

Upper quartzite (outlier of Delhi System of Heron, 1953)

The entire north-eastern and eastern part of the map area is covered by a ferruginous orthoquartzite belonging to a thick sequence of conglomeratic arkose, arkose and orthoquartzite mapped earlier by Heron (1953) as outliers of the Delhi System. Although recent studies indicate that this unit is a part of the Aravalli sequence (Poddar, 1966; Roy & Paliwal, 1981), opinions differ on its precise position within the Aravalli Group (Banerjee, 1971a, b; Roy *et al.*, 1980).

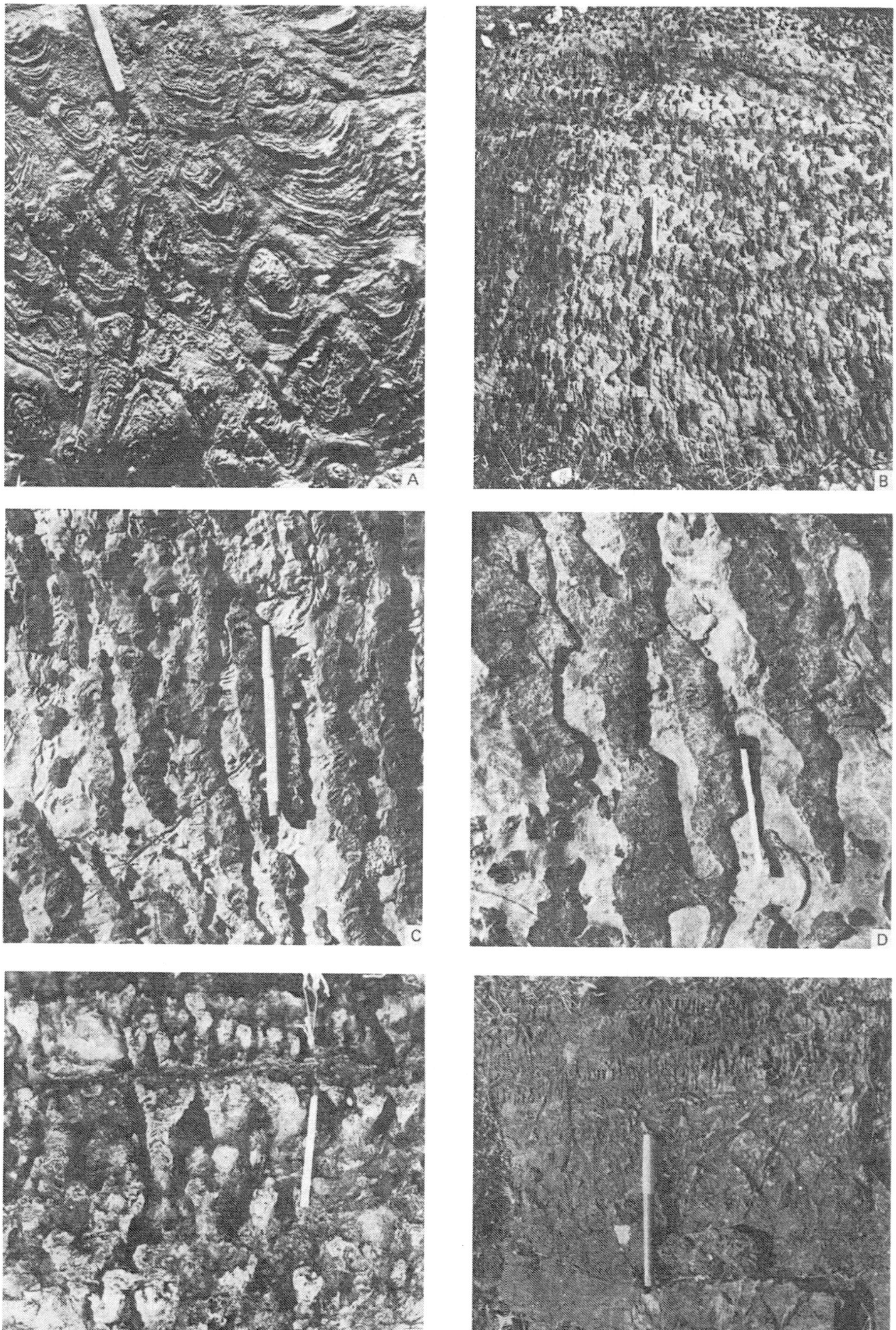
A
B
C
D
E
F

Petrography and mineralogy of stromatolitic phosphate

The Jhamarkotra phosphorite bed is characterised by an intimate association of columnar (Fig. 17.7B) and laminated stromatolites (algal mats) which are invariably the preferential locales of phosphatic constituents (now represented by microcrystalline apatite). It is dominantly a biohermal (biostromal) type of deposit and is confined to the carbonate unit. The phosphorite occurrences of Jhamarkotra can be described under the following headings: primary stromatolitic phosphorite; weathered/reworked variants of the primary stromatolitic phosphorite; and secondary calcareous phosphorite.

Primary stromatolitic phosphorite

The stromatolitic phosphorites are characterised by a spectacular bluish bloom, although in patches they show various shades of white, red or snuff brown. The columnar stromatolites are discrete or branched (Fig. 17.7C, D), the branches usually parallel, ramifying into minor columns of the same form and growing parallel upwards without any increase in the overall diameter (the alpha parallel type of Walter, 1972). Where the branched growths are arrested, the columns assume a somewhat irregular configuration along the margins in the form of knotty protuberances (Fig. 17.7D). Notwithstanding the intense deformation the area has been subjected to in certain areas, the columnar stromatolites bear ample evidence of their growth normal to the bedding. In the Jhamarkotra area, the phosphatic stromatolite columns show differential growth patterns and their length ranges from about a centimetre in the dwarf variety (Fig. 17.7E) to as high as 60 cm in narrow and erect discrete or branched types. The diameters of the individual columns also show a corresponding range of variation from microscopic to as high as 3–4 cm.

In some weathered zones (e.g. in D and E blocks) phosphorite occurs as sharply defined laminated bands (1 cm to about 1 m thick) at the top and bottom of the phosphate horizon. This is possibly a morphological variant of flat low algal mat type (Golubic, 1976). Some of these laminated bands are fragmental in nature and give an impression of a thin brecciated zone. The laminated or banded zone from the bottom upwards shows imperceptible transition into the columnar variety which in turn grades in some instances to the laminated type. The stromatolite of these types may be referred to as non-columnar stromatolite, following Preiss (1976).

Besides the tall columnar variety, the occurrence of phosphate as laminated or stratiform algal mat in association with the dwarf columnar variety is quite common in K block (Fig. 17.7F). It appears that this dwarf columnar stromatolite, overlying the phosphatic planar algal mat and characterised by fine laminoid fenestrae, is distinct from the cumulate non-columnar stromatolite type (Fig. 17.7E). A distinction is necessary between the type described above and those produced by tectonic deformation (Fig. 17.8A, B). In the stratified stromatolite the laminations can be described as either repetitive or alternating in character; in the latter type, phosphate laminae alternate with carbonate.

The tall columnar stromatolites of the Jhamarkotra region are usually characterised by the presence of bluish-grey convex phosphate laminae alternating with carbonate laminae. It has been observed that in the Jhamarkotra area the phosphate laminae are comparatively thicker than those of carbonate, thus constituting the dominant framework of the stromatolite. By volume these occupy on an average more than 90% of the bulk of the stromatolite and not infrequently they are almost completely phosphatic. The individual phosphate lamina in the columnar stromatolite has a thickness from a few microns to about 2 mm. Instances are common where several laminae coalesce along the stromatolitic margin to form a sheath-like structure.

The intercolumnar spaces and the stromatolites are usually made up of predominantly carbonate constituents and in some instances incorporate a few stromatolite (phosphate) intraclasts (as in the K block area).

Curiously, almost all the morphological variants of phosphatic stromatolite that can be seen in the different phosphorite outcrops of the Udaipur area are also found at Jhamarkotra, although the relative scale of development of particular types of stromatolite may vary between the two areas. In this respect the Jhamarkotra Basin perhaps embodies the entire spectrum of environmental conditions that supported the growth of the various types of phosphatic stromatolites of the Udaipur area.

In the Jhamarkotra area, Banerjee (1971b) and Chauhan (1973, 1979) described *Collenia columnaris* Fenton & Fenton, *Baicalia prima* Semikhatov, and *Mingeria calceolata* Korylyak as the dominant stromatolitic form species. The age of the phosphorite bed on the basis of the above stromatolite forms has been regarded as middle to late Riphean (Banerjee, 1971b).

Fig. 17.7 (A) Development of non-phosphatic stromatolite in the lower dolomite showing divergent morphological forms, e.g. undulating lamination and irregular domes. (B) Columnar stromatolite (dark grey) within dolomitic matrix (light grey) attributing a characteristic 'crocodile skin' appearance to the rock. (C) Discrete, tall, columnar phosphatic stromatolite showing sharp contact with the dolomitic matrix (light grey). Differential weathering of dolomite and phosphate is evident. (D) Columnar phosphatic stromatolite (dark grey) showing parallel branching. The sharp contact of stromatolite framework and the intercolumnar carbonate is noticeable. Clasts of a few phosphatic stromatolite framework can be seen in the matrix (matchstick measures 4 cm). (E) Dwarf vertical phosphatic columnar stromatolite alternating with the planar algal mat characterised by fine laminoid fenestrae. (F) Development of thick-bedded algal mats (in the lower part) along with the dwarf columnar phosphatic stromatolite (above the pen).

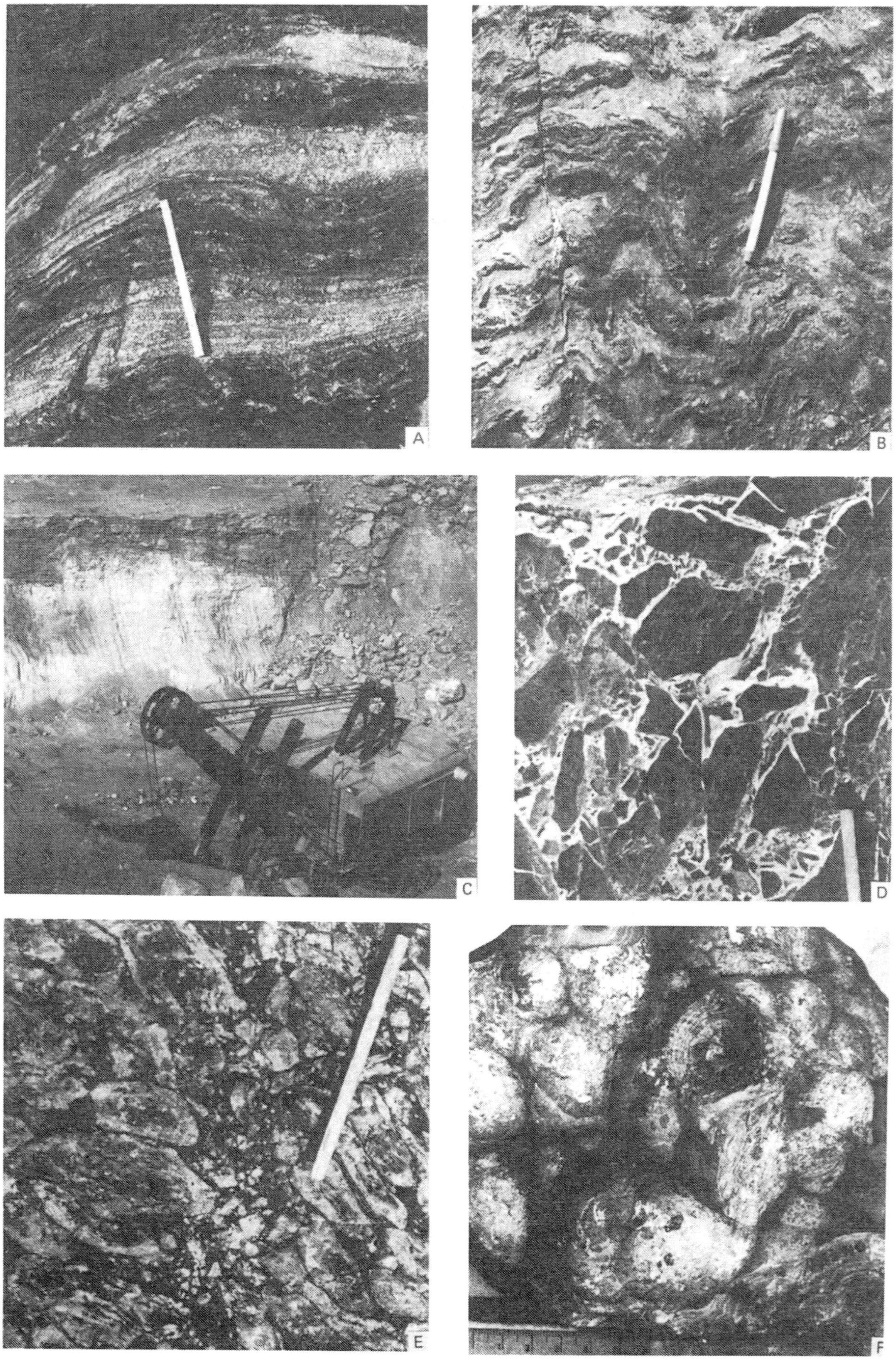
A
B
C
D
E
F

Weathered reworked phosphorite

Along a considerable stretch of the phosphorite bed, particularly in D, E, F and in parts of A, A-extension, B and H blocks, broken phosphatic columnar and laminated stromatolitic fragments occur as a non-cohesive, but tightly packed mass (Fig. 17.8D). This type of phosphate occurrence has been variously described as weathered, reworked, brecciated, fragmented or slumped (Fig. 17.7D, E). A characteristic feature of this type of weathered fragmented phosphorite is the absence of dolomite matrix and bedding structures. At many places the fragments are cemented by secondary silica (chalcedony) to produce a breccia.

Secondary calcareous phosphate

In addition to the two abovementioned dominant primary types of phosphorite, secondary calcareous apatite (Choudhuri, 1981), occurring as a fine network of veins controlled by the fracture pattern of the host rock, and as colloform and globular accretionary bands presenting a deceptive appearance of banded agate, has been found in some of the blocks, particularly in G and E blocks. The secondary apatites show oolitic, pisolitic, reniform and botryoidal forms (Fig. 17.8F). The external form of the botryoids is invariably dependent on the shape of the core (Fig. 17.9A). The occurrence of angular fragments of argillaceous dolomite floating in creamy white secondary phosphate and giving a collapse breccia-like structure (karst) is quite common.

Microscopic features of Jhamarkotra phosphate

The phosphatic constituents forming the stromatolitic framework consist predominantly of clusters of microcrystalline apatite, mostly in the form of laminae which in deformed columns show a fairly high degree of preferred dimensional orientation, particularly along the direction of flattening. When the grain size is extremely fine, the cluster appears as a fine mat without any visible grain boundaries, forming an isotropic mass under crossed nicols. Crystallinity of the apatite grains in the case of the Jhamarkotra phosphorite ranges from 1 μm to as high as 50 μm, with an average of 10–12 μm. By contrast the constituent microsphorites (Riggs & Freas, 1965) in many phosphorite deposits, including Florida, Morocco, Queensland and Khubsugul, are very fine (0.82–1 μm) (Ilyin, 1979). The coarse crystallinity of apatite in the Jhamarkotra phosphorite can be attributed to the metamorphism.

In the unweathered phosphorite, sparry dolomite and calcite occur along the sharply defined margin of the stromatolitic columns, and there is a marked contrast in the degree of crystallinity of the apatite constituting the stromatolite framework and that of carbonates which occupy the intercolumnar spaces (Fig. 17.9B). The intercolumnar carbonates are free from apatite except for a few dispersed grains. This is clearly seen in the X-ray mapping photograph (Fig. 17.9C) and scanning electron photomicrographs (cf. Banerjee, Basu & Srivastava, 1980). Quite commonly, discrete dolomicrite occurs in the compact microcrystalline apatite aggregate in such a way as to give a sieve-like texture. Silt-sized grains of quartz are also found enclosed in the microcrystalline apatite aggregates, as well as in the carbonates. Evidence of recrystallisation in these quartz grains is evident from their granoblastic texture. Trails of micritic carbonate microlaminae alternating with the thick phosphate laminae in the columnar stromatolite are commonly seen. These are at places traversed by fine anastomosing veins of carbonate.

Mica minerals such as sericite, muscovite and, rarely, lepidolite (Duchatelle, 1976) and specks of cubic pyrite crystals often invisible to naked eye, are the associated minerals found in the Jhamarkotra phosphorite. Brown-coloured goethite, frequently altered to limonite, is found to be associated with the phosphorite in the A-extension block, where weathered phosphorite is strongly stained with limonitic coatings. The occurrence of secondary pyrolusite veins and fracture fillings in the weathered phosphorite has also been observed.

Under the microscope the secondary calcareous apatite encrustations, occurring in the form of concentric bands, show the development of needle-shaped crystallites.

X-ray and infrared studies

Phosphorite samples collected from Jhamarkotra, as well as from neighbouring phosphorite occurrences in the Udaipur area, have been subjected to X-ray diffraction studies (Choudhuri *et al.*, 1968; Mukherjee & Chatterjee, 1972; Rao & Rao, 1971; Choudhuri, 1979; and Banerjee *et al.*, 1980). The samples chosen in most of the cases were either the bulk samples or the concentrate.

The cell dimensions determined for the various Udaipur phosphorite occurrences, including Jhamarkotra, are given in the Table 17.2. The variation in the *a* dimension is from 0.94 to 0.9376, while that of *c* dimension is from 0.688 to 0.6901.

Fig. 17.8. (A) Flattened and stretched phosphatic stromatolite (black) assuming a laminated appearance; the stretching is parallel to the bedding plane of the dolomite (white) host. (B) Pseudo-laminated phosphatic stromatolite (black) showing conspicuous folding. (C) A quarry face in E block showing a non-cohesive but highly compacted mass of powdery (fragmental) phosphatic stromatolitic (white) capped by dolomite (grey). The highly weathered character of the rock is evident by the 'shovel bite' mark in the phosphorite bed. (D) Brecciated phosphorite comprising internally laminated angular phosphatic fragments (stromatolitic) embedded in siliceous (white) matrix. (E) Moderately deformed ellipsoidal phosphatic stromatolites and their finer broken aggregate (possibly slumped) as found in E block area. (F) Secondary apatite showing typical botryoidal form. Note the development of concentric banding in the sliced patch on the top and also in the centre.

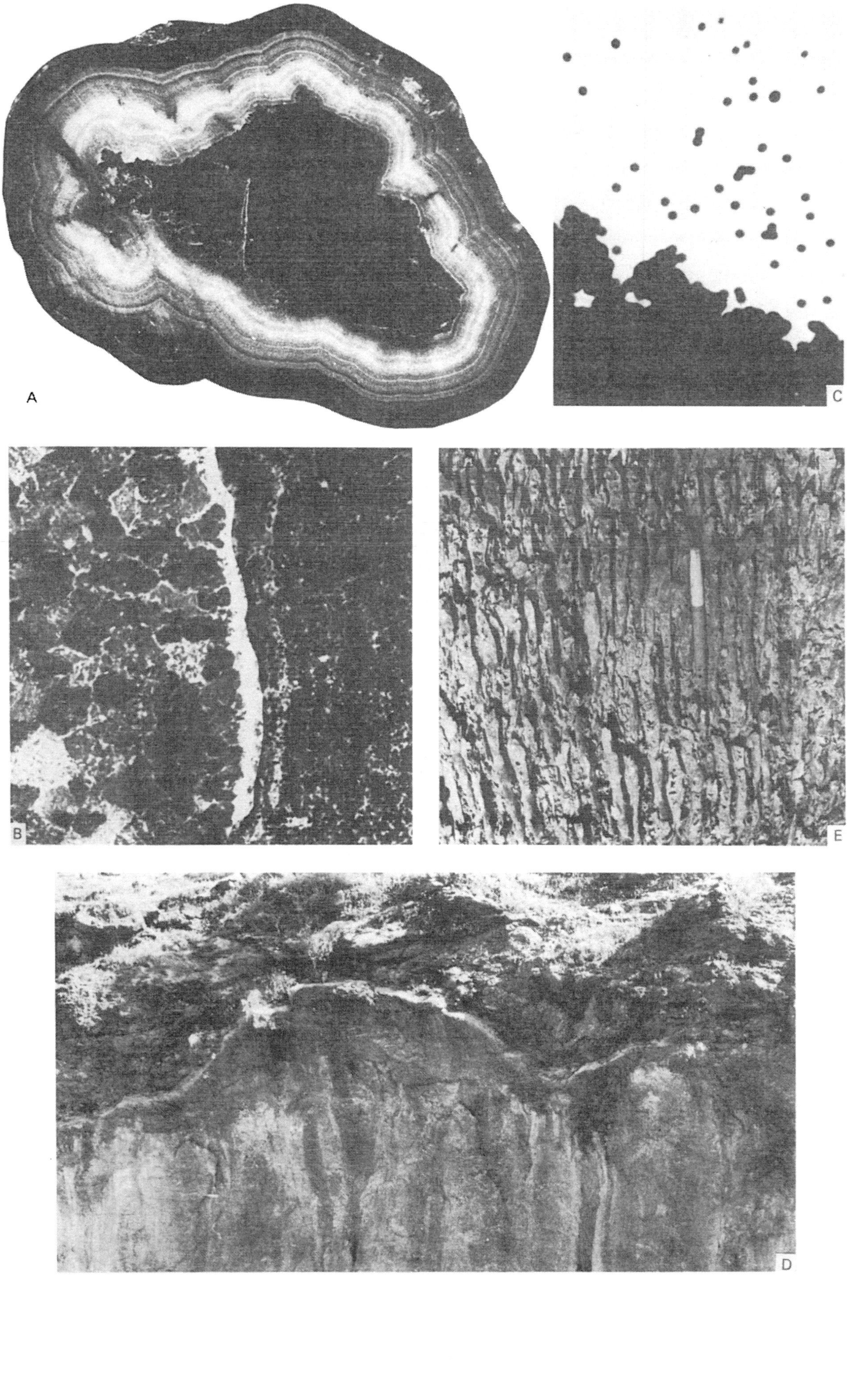
A
C
B
E
D

Banerjee *et al.* (1980), attempted to quantify the degree of $(CO_3)^{-2}$ substitution for $(PO_4)^{-2}$ and indicated that the range at CO_2 entrapped in the apatite lattice was of the order of 1–1.15%. On the other hand, Rao & Rao (1971), on the basis of the chemical analysis of the almost pure apatite, estimated the amount of lattice-substituted CO_2 at 0.5–0.6%. Thus, it will appear that the degree of carbonate substitution in the apatite lattice is low. However, the relative high degree of crystallinity, coupled with the lower amount of lattice-substituted $(CO_3)^{-2}$ and a low fluorine content in the constituent apatite, support the view that the sedimentary apatite of the Jhamarkotra and other deposits around Udaipur has a composition close to a pure fluorapatite (Lehr, 1969; Rao & Rao, 1971; Choudhuri, 1979; Banerjee *et al.*, 1980).

The progressive post-depositional alteration of a carbonate apatite to fluorapatite, with the simultaneous loss in structural CO_2 and the regular change in the *a* cell dimension due to metamorphism, is now a well-observed phenomenon (Lehr *et al.*, 1967; Ilyin, 1979; Ando, 1980; Lucas *et al.*, 1980; McClellan, 1980). However, the control of metamorphism on the composition and the crystal size of the apatite is not yet precisely known. Lehr *et al.* (1967), on heating apatite to as high as 1000 °C, observed a progressive alteration of francolite to fluorapatite. Curiously, in the case of the Lao Cai phosphorite deposit in Vietnam (Ilyin, 1979; Ando, 1980), the constituent carbonate–apatite is reported to have undergone significant transformation to almost pure fluorapatite with an attenuated size increase of the individual crystal of exceptionally high magnitude up to 1000 μm (Ando, 1980). These rocks have been subjected to regional metamorphism of greenschist facies (Ilyin, 1979). In the Jhamarkotra area the grade of metamorphism observed ranges from lower greenschist facies to epidote–amphibolite facies, but the corresponding crystal size of apatite in the phosphorite ranges on average between 10 μm and 20 μm; compositionally, the loss of CO_2 is perhaps not as marked as in the case of the Lao Cai apatite.

The study of infrared spectra of fresh and weathered phosphorite samples collected from the Jhamarkotra, Maton and Kanpur areas by Choudhuri *et al.* (1968), Rao & Rao (1971) and Banerjee *et al.* (1980) indicated P–O absorption around 568–580 cm^{-1} and also at 590–600 cm^{-1}. The other P–O bands are found in the 1040–1045 cm^{-1} and 1090–1100 cm^{-1} ranges. For the $(CO_3)^2$ ion, strong doublets in the region of 1430 cm^{-1} and 1455 cm^{-1} have been observed in all three cases. On the basis of the bidentate absorption bands in the 7 μm region, shown by all the infrared spectra, the samples have been identified as carbonate fluorapatite, with partial $(CO_3)^{-2}$ substitution. Interestingly, the infrared spectra of all the abovementioned types of weathered and fragmental group of phosphorite samples show a striking similarity. Rao & Rao (1971), on the basis of determining 0.58% CO_2 and 0.39% CO_2 and following McConnell (1958), considered the samples as fluorapatite.

Chemical composition of Jhamarkotra phosphorite

Table 17.3 shows the results of analysis of some of the trench and borehole samples typical of individual blocks. A composite sample of low-grade ore representing dominantly the unweathered part of the phosphorite bed (prepared out of 60 tonnes of composite samples drawn from several trenches) has also been analysed. In addition, a typical analysis of secondary carbonated apatite is given. When analysed separately, the columnar stromatolite framework in the Jhamarkotra area shows a P_2O_5 concentration broadly ranging from 25% to 30% (cf. Choudhuri *et al.*, 1980); the higher concentration of P_2O_5 is invariably found in columns where the intracolumnar carbonates are almost absent. In the intercolumnar carbonate matrix, the P_2O_5 is usually present in an insignificant amount: the maximum content of P_2O_5 percentage recorded is about 4% in rare cases. This pattern of P_2O_5 distribution is particularly common in A, B, C, and in some of the parts of D and E blocks where the intercolumnar matrix is largely free of stromatolite intraclasts.

The Jhamarkotra phosphorite, like many of the Precambrian-Cambrian phosphorite deposits, is characterised by the presence of a very low percentage of uranium, the value

Fig. 17.9. Secondary apatite showing accretionary concentric band (dark grey to whitish grey) over the carbonate fragment (black). The growth of the accretionary bands has been controlled by the shape of the carbonate core. (B) Negative print showing sharply defined contact between the columnar phosphatic stromatolite marked by tiny but perfectly crystalline equant apatite grains (to the right) as well as the coarsely recrystallised carbonate grains (to the left). The laminated nature of the phosphatic sheath of the columnar stromatolite at its contact with the carbonate contact is clearly visible. (C) X-ray scanning photograph of the distribution of phosphate along the contact of stromatolite column (dark) and the intercolumnar carbonate contact (white). The selective distribution of phosphate in the stromatolite framework is also conspicuous. (D) The contact between the weathered phosphorite and the overlying dolomite showing the undulating character resulting from local sagging. The voids created in the stromatolitic phosphorite bed due to intense leaching of carbonate were responsible for the sagging of phosphorite bed under compaction. (E) Densely populated cylindrical phosphatic stromatolite marking the preferential zone of higher concentration of phosphate. The conspicuous ridge and furrow structures characterise the intense preferential leaching of carbonate which is a recent phenomenon.

Table 17.2. *Cell dimensions in the Udaipur cluster of phosphorites*

Deposit	Cell parameter	Cell dimensions (nm) 1	2	3	4	5
Jhamarkotra	*a*	0.9368××	—	0.9376×	0.9366××	0.936–0.9368+
	c	0.6886	—	0.6890	—	0.6880
Maton	*a*	0.9340×	0.9358××	—	—	0.9369–0.9371
	c	0.6890×	0.6884	—	—	0.6880
Kanpur	*a*	0.9340×	0.9357××	—	—	0.935–0.936
	c	0.6890×	0.6901	—	—	0.690

1. Choudhuri *et al.*, 1968; Choudhuri, 1979; 2. Rao & Rao, 1971; 3. Centre d'Etudes et de Recherche des phosphates Mineraux, Paris, unpublished report, 1976; 4. McClellan, 1980; 5. Banerjee *et al.*, 1980. ×, Bulk sample. × ×, Concentrate sample. +, Average apatite sample —, not determined.

Table 17.3. *Chemical Composition of Jhamarkotra phosphorite*

Locality and sample type	Loss on ignition	P_2O_5%	CaO%	MgO%	SiO_2%	R_2O_3%	F%
Jhamarkotra D block (Run of mine weathered sample. Mean of 50 analyses)	n.d.	33.58	49.03	0.63	8.87	1.83	2.32
Jhamarkotra low grade (composite sample prepared from 60 tonnes of ore from trenches covering A, B, C, D & E blocks)	27.04	15.80	41.23	9.72	3.81	1.81*	1.49
Jhamarkotra A block (Mean of 10 trench samples)	25.42	16.43	39.89	11.81	3.48	1.72	1.51
Jhamarkotra B block (Mean of 9 trench samples)	26.60	18.03	39.27	11.70	2.30	1.63	1.49
Jhamarkotra C block (Mean of 13 trench samples)	22.60	19.65	41.03	8.67	3.54	2.92	1.62
Jhamarkotra E block (Mean of 20 weathered run of mine samples)	n.d.	36.18	52.36	0.22	4.60	—	2.68

*Value expressed in terms of Fe_2O_3 only.
n.d.—not determined

seldom exceeding 100 ppm. Surprisingly, a significant percentage of uranium has been observed to be present in the carbonaceous phyllite bed of the neighbouring Umra region, about 8 km north-west of Jhamarkotra.

Grade distribution of phosphorite and the controlling factors

The Jhamarkotra phosphorite shows a significant variation in its composition in terms of P_2O_5, MgO and SiO_2. The range of variation noted in the phosphate content is from 8% P_2O_5 to 39% P_2O_5. Choudhuri & Balasubramanian (1980) have suggested a statistical model which helps in understanding the grade distribution pattern of the Jhamarkotra phosphorites. Their study reveals a broad but distinct bimodal distribution of P_2O_5 percentage (Fig. 17.10). A general bimodal character of the grade is also indicated in the selective association of relatively high silica and extremely low MgO content in the high grade 'ore', and in the inverse relation in the low grade 'ore'.

Different mechanisms have been proposed to account for the enrichment of phosphorite. Sheldon (1981) has invoked three main processes which cause concentration of ore: winnowing, reworking, and weathering.

In the case of the Jhamarkotra ores, weathering appears to be the single most important factor causing enrichment of grade. Located in the warm semiarid climatic belt of north-western India, the rocks are constantly under the action of solution, leaching and oxidation and the effect of weathering is conspicuous everywhere.

The entire so-called high grade belts in the D, E, F, and in part of other blocks at Jhamarkotra probably represent a strongly weathered zone marked by the almost total absence of carbonate in the interstitial spaces of the broken angular fragments comprising the columnar phosphatic stromatolitic mass. The action of leaching seems to have preferentially affected also

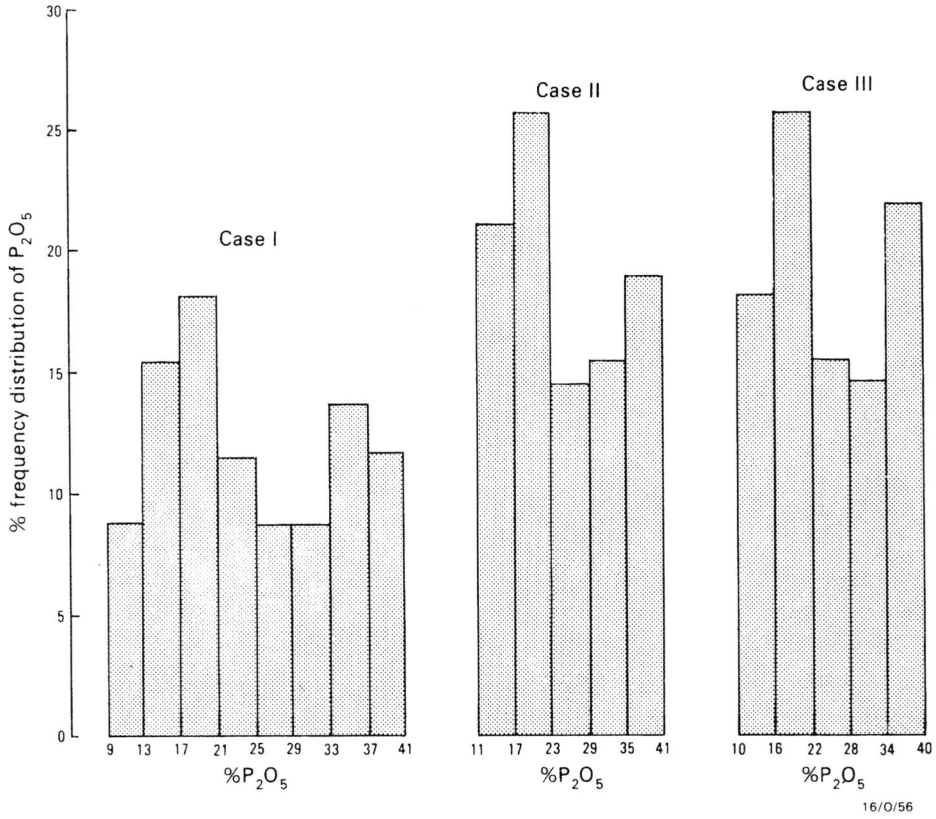

Fig. 17.10. Three cases with variable class limits of $P_2O_5\%$ showing frequency distribution of percentage of P_2O_5 (of the unit length analysis of drill core).

the dolomite underlying the phosphorite bed all along the weathered zone. Thus, the basal dolomite is either totally absent or too thin, or is marked by extremely weathered argillaceous dolomitic limestone probably representing a dissolution surface (cf. Southgate, 1980). The leaching of the intercolumnar carbonate material, as well as of the carbonate underlying the phosphorite bed, seems to have induced *in situ* collapse of stromatolite columns as well as the algal mat. As has been observed by Altschuler (1973), and Gulbrandsen & Krier (1980), the leaching effect of this otherwise primary rock component from the phosphorite bed must have been instrumental in the creation of porosity equivalent to the volume of the component lost. The voids thus created in the weathered phosphorite portion at Jhamarkotra perhaps were responsible for the local sagging and caving and for the rather rolled or undulatory character of the contact zone between the phosphorite bed and the overlying carbonate (Fig. 17.9D). Consequently, the overlying carbonates have been brecciated at their contact with the phosphorite bed.

However, besides weathering, the density population of the stromatolite per unit area also has a bearing on the grade variation at Jhamarkotra. It has been mentioned earlier that the columnar stromatolites are the preferential locale of phosphate and the range of phosphate in the individual columns when analysed varied from 25% P_2O_5 to 39% P_2O_5 (Choudhuri *et al.*, 1980). It follows that the higher the column population density the higher will be the P_2O_5 content per unit area. Accordingly, depending on the density of population of columnar stromatolite, even relatively mild weathering may also produce high grade ore.

Although it has not yet been possible to establish any generalised relationship between the effect of structural deformation *vis-à-vis* the concentration of phosphorite, in certain instances a marked enrichment in grade has been observed along the folded trough zones (Fig. 17.11).

It appears that weathering and the relative growth density of the stromatolites were the primary factors responsible for concentration of apatite in the phosphorite bed at Jhamarkotra, with structural disturbance being of importance locally.

The abundance of the phosphatic stromatolite intraclasts in the intercolumnar spaces of the stromatolite in the Neemuch Mata area, and the presence of scour channels cutting across the biolithites, led Chauhan (1979) to believe that the strong action of tidal waves caused erosion of the phosphatic sheaths of the columnar structure as well as the fragmentation of developing stratiform structures. The occurrence of such types of phosphorite with bioclastic fragments in the intercolumnar carbonate is limited in the Jhamarkotra area and is confined solely to the K block. It may further be noted that while wave actions were responsible for the mechanical fragmentation of the stromatolite columns, they were not effective enough to effect enrichment through either mechanical reworking and winnowing (Cook, 1976) or winnowing action alone (Sheldon, 1981).

Depositional environment and phosphogenesis

The lithofacies distribution of the Aravalli rocks in

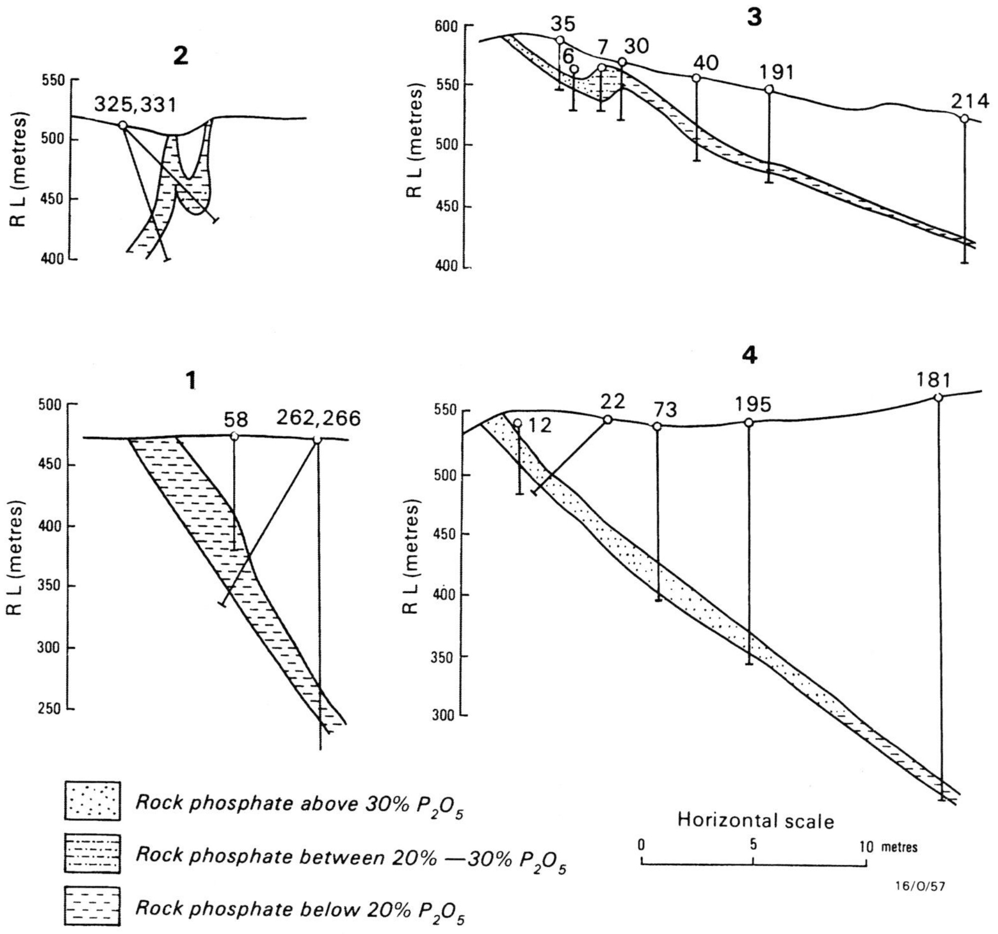

Fig. 17.11. Four section lines across the phosphorite bed at Jhamarkotra showing the interpolated grade behaviour of the phosphorite bed at depth. (1, 2, 3, 4 in Fig. 17.3). The selective increment of P_2O_5 grade at the folded trough is noticeable. The unrelated nature of the grade as a function of depth of burial is also evident.

southern Rajasthan and neighbouring areas shows the development of an extensive carbonate sand shelf along continental margins covered by shallow water (Poddar, 1966; Banerjee, 1971a, b; Roy & Paliwal, 1981). Within this long stretch of shelf deposits evidence of algal activity in the form of stromatolites can be seen at several localities, but surprisingly the occurrence of phosphatic stromatolites is limited to well-defined geographic confines. Out of the two main groups of localities where phosphorites are reportedly present, the one around Udaipur (including Jhamarkotra) is considered to be the most important commercially, and it is in this area that the growth of stromatolites has been most prolific. A close study of the Udaipur Basin, of which Jhamarkotra constitutes a part, is likely to yield information which might help in the understanding of the intriguing problem of phosphogenesis.

Recently, Roy & Paliwal (1981) have elucidated the palaeogeographic framework of the deposition of the phosphorite-bearing Aravalli rocks of Udaipur. The authors suggested the presence of a land-locked epicontinental (epeiric) sea within a shallow 'continent dominate' shelf (cf. Banerjee *et al.*, 1980), and considered it to be the most important tectonic sedimentary factor controlling the growth of phosphatic stromatolites. The depositional basin was presumed to have remained starved of terrigenous detritus and the circulation of water during the entire period of phosphate formation was restricted.

Certain features emerge out of the palaeogeographic study of the Udaipur Basin which are relevant to phosphogenesis. Although the non-phosphatic stromatolites are present in the carbonate formations both within and outside the realm of the 'epicontinental sea', the phosphatic stromatolites are restricted within it. Stromatolitic columns and algal mats are the exclusive locales of phosphate. Intercolumnar areas do not show any evidence of apatite except for some 'reworked intraclasts'. The phosphorite–dolomite interface is extremely sharp both at the top and bottom and there is no repetition of primary phosphorite in the stratigraphic column. There is a marked difference in the morphology of stromatolites of phosphatic and non-phosphatic types, the latter being more varied in form and association than the former.

All the abovementioned features unequivocally point towards a particular type of alga which developed and flourished under the very special situation existing in the epicontinental sea (cf. Howard, 1972; Riggs, 1980) and the critical role played by algae in fixing apatite in the edifice, directly or indirectly. The onset of phosphate deposition, therefore, marks a major event in biological evolution. The abruptness with which the phosphorite starts and ends in the stratigraphic column indicates a delicate and ephemeral ecosystem which could sustain a special type of alga ('phosphate horders' of Murthy, 1980).

Assuming the role of algae in the deposition of phosphorite, two questions are raised which are relevant to the problem of phosphogenesis in the Udaipur area in general and the Jhamarkotra area in particular: (1) what was the source of phosphorite?; and (2) how did the phosphate become fixed in the stromatolitic edifice?

Regarding the first question, three possible sources are generally suggested: (i) marine upwelling; (ii) continental weathering; and (iii) volcanogenic products. A major difficulty in assuming that the upwelling of oceanic water was responsible for the concentration of phosphorus in the Udaipur Basin was its apparently land-locked situation. Banerjee *et al.* (1980) have also strongly argued against the upwelling theory for the genesis of the Udaipur phosphorites on more or less the same grounds (see also Cook, 1972 and Howard, 1972, for similar studies on the Georgina Basin phosphate). Sheldon (unpublished discussion, IVth International Seminar on phosphorites, Udaipur, 1981), however, feels that the Udaipur Basin might have had some channels linking it with the open sea and that phosphorus in the epicontinental sea became recharged through upwelling currents. It is now extremely difficult to prove or disprove the existence of any such channel linking the epicontinental sea with the open sea. The Jhamarkotra basin was very much a part of the Udaipur epicontinental sea. Palaeogeographic studies, however, indicate that it was a sheltered bay cut off from the Udaipur sea by a basement high, now represented by the Udaisagar Granite. The only connection between the Udaipur sea and the Jhamarkotra bay, if any, was to the north, and it is possible to conceive the Jhamarkotra situation in terms of a modern analogue like Hamelin Pool in Shark Bay, Australia, where it is also difficult to apply an upwelling model.

Absence of evaporite beds in the stratigraphic sequence of Udaipur and Jhamarkotra precludes the possibility of phosphate precipitation through mixing of phosphorus-rich continental brines with cold oceanic water (Hite, 1978). Furthermore, the high degree of concentration of phosphate per unit area in the phosphorite deposit (about 5.2 tonnes/m^3, assuming 80 million tonnes of phosphate reserves of all categories in an area of 6 km long, 150 m wide, and 15 m average thickness), makes it difficult to imagine a situation in which the water of the Jhamarkotra sea could have built up the necessary concentration of phosphate from a continentally derived source (cf. Pevear, 1966). In this respect it is relevant that, except for the thin clastic wedge at the base, the pre-phosphorite terrigenous contribution in the sedimentary package is insignificant.

In the absence of any compelling evidence in favour of either marine upwelling or a continental contribution for augmenting phosphate in the Udaipur sea, it may be pertinent to discuss the possibility of a volcanic source. Roy & Paliwal (1981) indicated that sedimentation in the shelf and epicontinental sea was probably triggered by fault-controlled troughs along with contemporaneous volcanicity. The terrigenous debris was shed from areas of high relief. Deposition of carbonate and subsequently rock phosphate followed after erosion of ephemeral relief in the provenance area. Therefore, notwithstanding the time-lag between the period of volcanism and the deposition of carbonate, and the absence of tuffaceous material in the phosphorite (cf. Cathcart, 1968), the possibility of a volcanogenic supply of phosphate to the shelf sea cannot be altogether ruled out (cf. Riggs, 1980).

It is uncertain whether the fixation of apatite in the stromatolite framework was chemogenically precipitated with the algae acting as catalysts favouring an inorganic phosphate precipitation, or whether the phosphatic stromatolites represented the preferential site for the burial of algal debris along with their organically precipitated phosphate.

Precipitation of phosphate is intimately linked with concentration of phosphorus in the aqueous system in which it is dissolved, and precipitation in turn is brought out by change in the pH of the system. The role of organisms, particularly of autotrophic plants, in modifying the CO_2-bicarbonate equilibrium and thereby controlling phosphate deposition is a well-known phenomenon (McConnell, 1965; Trudinger, 1979; Yanshin, 1981). Depletion of CO_2 in shallow water conditions would occur at a high rate because of the gradual warming up of the water in the euphotic zone and low hydrostatic pressure. Volcanic exhalations might also have been responsible for increasing the temperature of the epicontinental sea water. Chauhan (1979) and Banerjee *et al.* (1980) have discussed in detail the role of organisms in concentrating and precipitating phosphate. However, it would appear from a critical review of the literature that in spite of overwhelming evidence for the biological accumulation of phosphate, there is no unanimity as regards the precise role that the blue-green algae might have played in the process of concentration of inorganic phosphorus in the sediment (cf. Azad & Borchardt, 1970; Banerjee, 1978; Trudinger, 1979; Chauhan, 1979; Riggs, 1979; Banerjee *et al.*, 1980; Verma & Barman, 1980). Verma & Barman (1980) postulated that the stromatolite framework of the Udaipur phosphorite occurrences accumulated phosphate in their intercellular organic-rich laminae for the purpose of photosynthesis, thereby espousing the view of 'accumulation of phosphate within the algal life, now represented by stromatolite and oncolites and not due to chemical precipitation from sea water or trapping of phosphate in the thalli of stromatolite-forming algae'. Banerjee (1978) however maintained that the process of phosphorus uptake by marine blue-green algae and its cycle in the algal life process alone cannot explain such large-scale concentration of phosphate. He was of the opinion that the dissolved organic phosphorus (DOP) present in the marine water was used by the alga for its nutrition and growth, subsequently forming a solid phase and accumulating in the stromatolites of the Udaipur Basin. The localisation of phosphate in the stromatolite framework according to Banerjee (1978) and Chauhan (1979) resulted from microbiological activity involving diurnal photosynthetic and respiratory processes affecting the water chemistry at the microenvironmental level. The rhythmic layering of phosphate alternating with carbonate in the columnar stromatolite framework was regarded by Banerjee (1978) and Banerjee *et al.* (1980) as evidence of periodicity in biochemical precipitation.

The only difficulty in accepting this interesting hypothesis of Banerjee (1978; Banerjee *et al.*, 1980) is the virtual absence of phosphate in the intercolumnar carbonate. The occurrence of macroscopically discernible carbonate phosphate layers alternating within the stromatolite columns (which is also a microscopic phenomenon) according to Patwardhan (1980) could not have been brought about by a 'well tuned system operating between a buffered pH range' since this would have required different pH-Eh systems for the columnar and intercolumnar areas.

Therefore, in the absence of any positive evidence regarding the precise role that the blue-green algae played in selective (but fairly uniform) localisation of the phosphatic constituent in their framework, the mechanism involved could be either: (a) that the stromatolites are the selective locales of burials of the algae with organically precipitated apatite forming an intrinsic constituent; or (b) that algae acted as a catalyst in 'siphoning' the dissolved phosphate out of the aqueous system leading to inorganic precipitation in the algal mat zones which are either continuous (laminated non-columnar) or discontinuous (columnar) stromatolite. One particular feature which militates against the concept of phosphatisation of carbonate as proposed by Chauhan (1979), is the report of sharp contact between the apatite and carbonate with their distinctive crystalline form (cf. Banerjee *et al.*, 1980, Fig. 8). Perhaps a more convincing answer to the problem can be found through a critical study of the organic content of the stromatolites and their associated sediments. This might help in establishing the effectiveness of the mechanism of biological reduction of phosphate to soluble hypophosphites followed by oxidation of the latter in an aerobic environment (Trudinger, 1979).

References

Altschuler, Z.S. (1973). The weathering of phosphate deposits – geochemical and environmental aspects. In *Environmental Phosphorus Handbook*, ed. E.J. Griffith & A. Beston, pp. 33–96. John Wiley & Sons, New York.

Ando, J. (1980). Conventional fertilizer manufacture – problems and trends. In *Proceedings of the Fertilizer Flow Conference on Fertilizer Raw Material Resources, Needs and Commerce in Asia and the Pacific*, ed. R. Sheldon, S. Ahmed & Yang Yuch-Hing, p. 46. East–West Resource Systems Institute, East–West Center, Honolulu.

Azad, H.S. & Borchardt, J.A. (1970). Variations in phosphorus uptake by algae. *Environmental Science Technology*, **4**, 737–43.

Banerjee, D.M. (1971a). Precambrian stromatolitic phosphorite of Udaipur, Rajasthan, India. *Geological Society of America Bulletin*, **82**, 2319–30.

Banerjee, D.M. (1971b). Aravallian stromatolites from Udaipur, Rajasthan. *Journal of the Geological Society of India*, **12**, 349–55.

Banerjee, D.M. (1978). Chemical rhythmicity in the Precambrian laminated phosphatic stromatolite and its bearing on the origin of algal phosphorite. *Indian Journal of Earth Science*, **5**, 102–10.

Banerjee, D.M., Basu, P.C. & Srivastava, N. (1980). Petrology, mineralogy, geochemistry and origin of the Precambrian Aravallian phosphorite deposits of Udaipur and Jhabua, India. *Economic Geology*, **75**, 1181–99.

Cathcart, J.B. (1968). Phosphate in the Atlantic and Gulf Coastal Plains. In *Proceedings of the 4th Forum on the Geology of the Industrial Minerals*, ed. L.F. Brown, Jr pp. 23–43. University of Texas Press, Austin.

Chauhan, D.S. (1973). Stromatolites from the Precambrian phosphorite bearing stratum of the Aravalli region of Udaipur, Rajasthan (India). *Doklady Akademiya Nauk SSSR*, **208**, 1429–31.

Chauhan, D.S. (1979). Phosphate bearing stromatolites of the Precambrian Aravalli phosphorite deposits of the Udaipur region, their environmental significance and genesis of phosphorite. *Precambrian Research*, **8**, 95–126.

Choudhary, A.K., Gopalan, K., Bose, U., Gupta, S.N., Ravindra, R. & Sastry, C.A. (1981). Rb–Sr chronology of granites from the Delhi Formation, Rajasthan. In *Symposium on three decades of developments in petrology, mineralogy and geochemistry in India May 1981*, p. 52. Geological Survey of India, Jaipur.

Choudhuri, R., Chakravarty, A.K., Ghosh, S.K. & Banerjee, K.C. (1968). Studies on Indian Phosphates – Part I. *Technology*, **5**, 285–93.

Choudhuri, R. (1979). The characteristics of the Jhamarkotra phosphorite deposit in Udaipur District, Rajasthan, India. In *Proterozoic–Cambrian Phosphorites*, ed. P.J. Cook & J.H. Shergold, pp. 40–2. ANU Press, Canberra.

Choudhuri, R. (1981). Botryoidal apatite (staffelite) in the Jhamarkotra area of Udaipur district, Rajasthan. In *Symposium on three decades of developments in petrology, mineralogy and petrochemistry in India, May 1981*, p. 106. Geological Survey of India, Jaipur.

Choudhuri, R. & Balasubramanian, N. (1980). Grade distribution pattern and its bearing on the formation of high and low grade in Jhamarkotra phosphorite deposit, Udaipur, Rajasthan. *Indian Journal of Earth Sciences*, **7(1)**, 89–93.

Choudhuri, R., Balasubramanian, N., Kapoor, G.P.S., Sharma, D.S. & Dhabai, D.R. (1980). Stromatolites as parameters in grade control on Jhamarkotra phosphorites. *Geological Survey of India, Miscellaneous Publication*, **44**, 330–9.

Cook, P.J. (1972). Petrology and geochemistry of the phosphate deposits of northwest Queensland, Australia. *Economic Geology*, **67**, 1193–213.

Cook, P.J. (1976). Sedimentary phosphate deposits. In *Handbook of Strata Bound and Stratiform Ore Deposits*, ed. K.H. Wolf, pp. 505–35. Elsevier, Amsterdam.

Crawford, A.R. (1970). The Precambrian geochronology of Rajasthan and Bundelkhand, Northern India. *Canadian Journal of Earth Sciences*, **7**, 91–110.

Duchatelle, G. (1976). Indian phosphate, Jhamarkotra, mineralogical composition and enrichability of apatite rock. Centre of Study and Research on Phosphate Minerals, Unpublished Report, 925.

Golubic, S. (1976). Organisms that build stromatolites. In *Stromatolites*, ed. M.R. Walter, pp. 113–48. Elsevier, Amsterdam.

Gulbrandsen, R.A. & Krier, D.J. (1980). Large and rich phosphorus resources in the Phosphoria Formation in the Soda Spring area, south-eastern Idaho, United States, *Geological Survey Bulletin*, **1496**, 1–25.

Heron, A.M. (1935). Synopsis of the Pre-Vindhyan geology of Rajputana. *Transactions of the National Institute of Sciences of India*, **1(2)**, 17–33.

Heron, A.M. (1953). The geology of central Rajputana. *Memoir of the Geological Survey of India*, **70**, 1–385.

Hite, R.J. (1978). Possible genetic relationship between evaporites, phosphorites and iron-rich sediments. *Mountain Geologist*, **14(3)**, 97–107.

Howard, P.F. (1972). Exploration for phosphorite in Australia – a case history. *Economic Geology*, **67**, 1180–292.

Ilyin, A.V. (1979). Late Precambrian–Cambrian phosphorites. In *Interregional Seminar on the Development Potential of Precambrian Mineral Deposits*, pp. 1–24. Proceedings of IGCP, GKES and the USSR Academy of Science, Moscow, 1–15 August.

Lehr, J.R., McClellan, G.H., Smith, J.P. & Frazier, A.W. (1967). Characterisation of apatites in phosphate rocks. *International Colloquium on Solid Inorganic Phosphates, Toulouse, 16–20 May*,

2, pp. 29, 44. Société Chemique de France, Paris.

Lehr, J.R. (1969). Characterisation of phosphate rocks from India. Unpublished report communicated to the State Directorate of Mines & Geology, Government of Rajasthan (July, 1969).

Lucas, J., Felicoteaux, R., Nathan, Y., Prévôt, L. & Shahar, Y. (1980). Different aspects of phosphate weathering. *Society of Economic Paleontologists and Mineralogists, Special Publication*, **29**, 41–51.

McClellan, G.H. (1980). Mineralogy of carbonate apatites. *Journal of the Geological Society, London*, **137(6)**, 675–81.

McConnell, D. (1958). The apatite like mineral of sediments. *Economic Geology*, **53**, 110–11.

McConnell, D. (1965). Precipitation of phosphates in sea water. *Economic Geology*, **60**, 1059–62.

Mukherjee, B. & Chatterjee, P.K. (1972). Mineralogy of phosphorites from Udaipur, Rajasthan. *Indian Minerals*, **44**, 101–3.

Murthy, M.V.N. (1980). Welcome address in the workshop on stromatolites, held in January, 1978, Udaipur, India. *Geological Survey of India, Miscellaneous Publication*, **44**, iii–v.

Paliwal, B.S. (1980). Systematics of the Precambrian stromatolites of Rajasthan – a review in the light of tectonic deformation. *Geological Survey of Indian Miscellaneous Publication*, **44**, 68–74.

Pathwardhan, A.M. (1980). Phosphorite-pyrite association and the genesis of stromatolitic and pelletal phosphorites. In *Proceedings of the 3rd Indian Geological Congress, Poona*, pp. 347–65.

Pevear, D.R. (1966). The estuarine formation of United States Atlantic coastal plain phosphorite. *Economic Geology*, **61(2)**, 251–6.

Poddar, B.C. (1966). An example of contrasted tectonic regimes from Precambrian of Udaipur, Rajasthan. *Indian Minerals*, **20**, 192–4.

Preiss, W.V. (1976). Basic field and laboratory methods for the study of stromatolites. In *Stromatolites*, ed. M.R. Walter, pp. 4–13. Elsevier, Amsterdam.

Rao, N. Krishna & Rao, G.V. (1971). Petrographic study of some Indian phosphorites. *Journal of the Geological Society of India*, **12(4)**, 356–67.

Riggs, S.R. (1979). Phosphorite sedimentation in Florida – a model phosphogenic system. *Economic Geology*, **74**, 285–314.

Riggs, S.R. (1980). Tectonic model of phosphate genesis. In *Fertilizer Mineral Potential in Asia and the Pacific*, ed. R.P. Sheldon & W.C. Burnett, pp. 159–90. East–West Resource Systems Institute, East–West Center, Honolulu.

Riggs, S.R. & Freas, D.H. (1965). Stratigraphy and sedimentation of phosphorite in the central Florida phosphate district. *American Institute of Mining Engineers, Preprint*, **65-H-84**, p. 13.

Roy, A.B., Nagori, D.K., Golani, P.R., Dhakar, S.P. & Choudhuri, R. (1980). Structural geometry of the rock phosphate bearing Aravalli rocks around Jhamarkotra Mines area, Udaipur district, Rajasthan. *Indian Journal of Earth Science*, **7(2)**, 191–202.

Roy, A.B. & Paliwal, B.S. (1981). Evolution of Lower Proterozoic epicontinental deposits. Stromatolite bearing Aravalli Rocks of Udaipur, Rajasthan, India. *Precambrian Research*, **14**, 49–74.

Roy, A.B., Paliwal, B.S. & Goel, O.P. (1971). Superposed folding in the Aravalli rocks around Udaipur, Rajasthan. *Journal of the Geological Society of India*, **12**, 342–8.

Sheldon, R.P. (1981). Ancient marine phosphorite. *Annual Review Earth and Planetary Science*, **9**, 251–84.

Southgate, P.N. (1980). Cambrian stromatolitic phosphorites from the Georgina Basin, Australia. *Nature*, **285(5764)**, 395–7.

Trudinger, P.A. (1979). Microbiological controls on phosphate accumulation. In *Proterozoic-Cambrian Phosphorites*, ed. P.J. Cook & J.H. Shergold, pp. 87–92. ANU Press, Canberra.

Verma, K.K. & Barman, G. (1980). Relationship between algal stromatolites and phosphate localisation within the carbonate sequence of the Aravalli Super Group in Udaipur and Banswara districts, Rajasthan. *Geological Survey of India, Miscellaneous Publication*, **44**, 355–63.

Walter, M.R. (1972). Stromatolites and the biostratigraphy of Australian Precambrian and Cambrian. *Palaeontological Association, Special Paper*, **11**, 1–190.

Yanshin, A.L. (1981). On possible causes of abundance of Late Precambrian and Early Cambrian deposition of phosphorites. In *4th International Field Workshop and Seminar on Phosphorite*. Udaipur, 26–29 November, 1981.

18

Proterozoic and Cambrian phosphorites – deposits: Fontanarejo, Spain

E. PERCONIG, F. VAZQUEZ, F. VELANDO AND F. LEYVA

Introduction

Most phosphate exploitation undertaken so far in Spain has been for deposits of igneous origin related to Variscan granites. Mines in such deposits were operational from 1866 until 1956 and provided a total production of 1.2×10^6 tonnes of phosphate rock with a $Ca_3(PO_4)_2$ content ranging from 35% to 80%. At the present time no mines are in operation and Spain relies largely on North African ore for domestic supplies, at a cost in 1982 of well over 100 million US dollars.

Spanish phosphate consumption rates have remained stable during the last five years, at about 2.9×10^6 tonnes per year. Requirements of phosphate rock for 1987 are estimated at 4.1×10^6 tonnes. In order to reduce dependency on foreign supplies of phosphate the Spanish administration encouraged research programmes developed by the Geological & Mining Survey of Spain (IGME) as well as financial incentives to private enterprise, which resulted in a marked increase in activities related to the search for likely phosphate-bearing formations.

Joint programmes by the private and public sectors included research in older dyke-type apatite mines. IGME focused its research on the potential existence of sedimentary phosphate deposits. The phosphate nodule-bearing shales of Tournaisian age in the Pyrenees and Cantabrian Cordillera were investigated as far back as 1978, but with negative results. In 1980 IGME entered into a joint venture with 'Empresa Nacional Adaro de Investigaciones Mineras, S.A.' (ENADIMSA) consisting of a systematic survey aimed at delineating areas where sedimentary phosphate deposits might be present. Many regional geological features were considered, including stratigraphy, tectonics and sedimentology, with a critical analysis of known Spanish phosphate occurrences in the light of current hypotheses of phosphogenesis. This resulted in the identification of a number of potentially interesting prospects, with exploration of the Proterozoic and Cambrian being given the top priority.

During recent years the Cambrian and Precambrian intervals have been the subject of considerable interest following the discovery of a number of phosphate occurrences of this age. Phosphate resources of Proterozoic and Cambrian age have been estimated by Notholt (1979) to comprise over 10^{10} tonnes of phosphate rock with a P_2O_5 content greater than 4%.

Both the Proterozoic and Cambrian intervals found in Spain show facies and palaeogeographic conditions very much like those in some of the important phosphogenic provinces in other parts of the world. The similarity extends to their palaeolatitude (about 30° S) which is within the normal latitudinal limits of phosphorites. A number of features therefore supported the initiation of a phosphate exploration programme over 60 000 km^2 of the country (almost one-eighth of Spain) known to have outcrops of Proterozoic and Cambrian rocks (Fig. 18.1). This exploration programme has benefited from recent publications originating from IGCP Project 156, project seminars and field-trip activities, and information exchange with international experts that make up the Working Groups of Project 156. The authors wish to mention especially the kind help provided by R.P. Sheldon and A.J.G. Notholt. A combination of geological mapping efforts by IGME (MAGNA Project) and the phosphate research team has resulted in the discovery of phosphorites in Upper Precambrian formations near Fontanarejo (Ciudad Real Province) where an important orebody is currently being evaluated.

Regional stratigraphy and tectonics

A summary of the main regional geological features, shown in Fig. 18.2, places the Fontanarejo deposits in their stratigraphic and tectonic framework. Given the divergent opinions existing among authors who have dealt with the pre-Ordovician stratigraphy of the region, the description below provides only a broad-scale summary. The divergence of views stems from the uniform lithology, poor outcrop and rather complex structure.

Precambrian–Lower Cambrian

The sequence believed to be of Proterozoic to Early Cambrian age is grouped within the 'Schistose–greywacke Complex', named because its predominant lithology consists of shales and greywackes with coarser-grained detrital intercalations (microconglomerates, conglomerates, breccias), and carbonate layers. It extends through the provinces of Ciudad Real, Toledo, Salamanca, Caceres, Avila and Badajoz in Spain, and north of the Tagus river into Portugal, and it has been

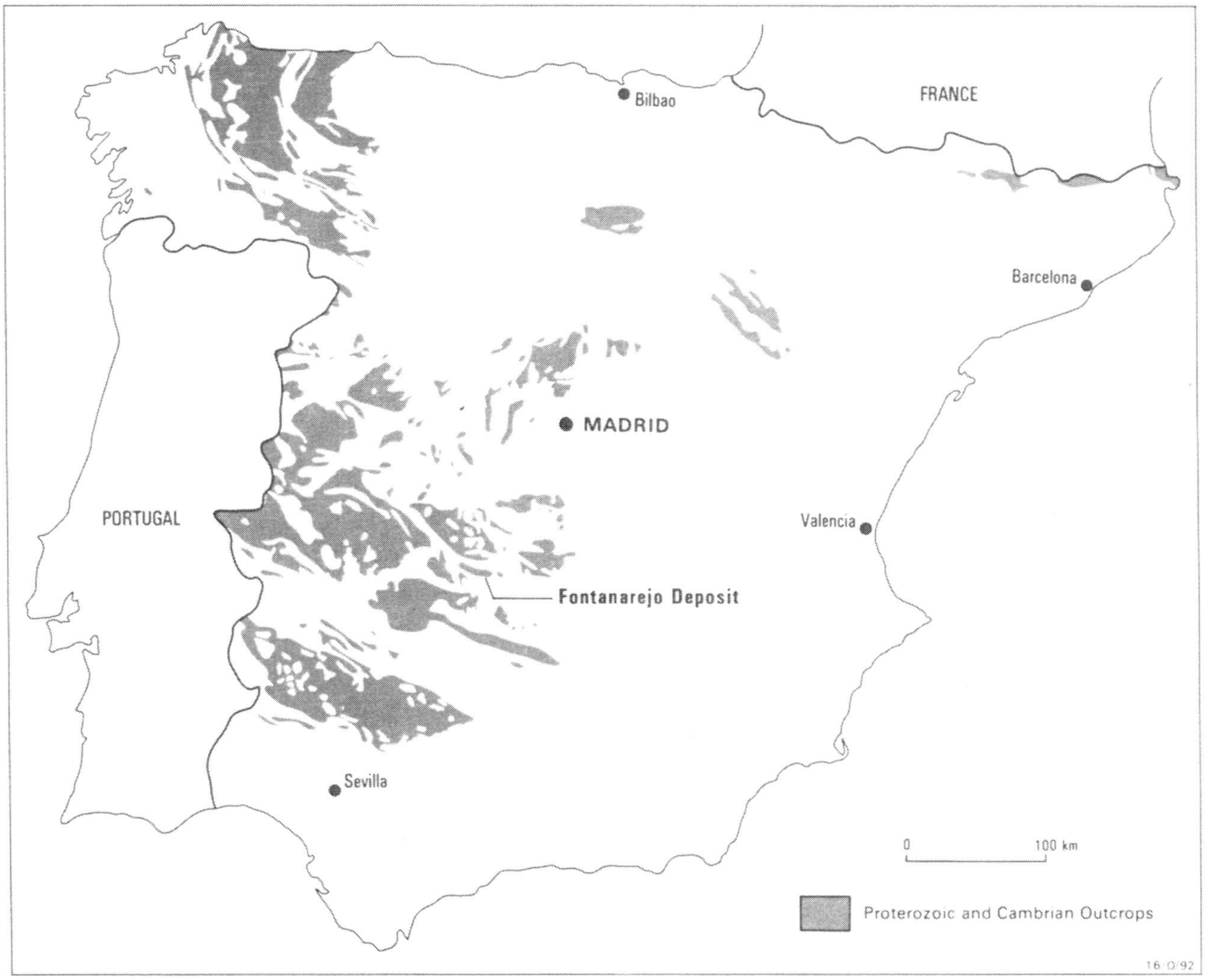

Fig. 18.1. Distribution of Proterozoic and Cambrian rocks in the Iberian Peninsula and location of the Fontanarejo deposit.

extensively studied by Teixeira (1954, 1968, 1979) who describes it as a 'flysch-type formation where shales and greywackes alternate'.

Lotze (1956) named the lower, conspicuously shaly, part of the Complex the Valdelacasa Series; the upper part, consisting of carbonates as well as greywackes and conglomerate, was named the Transitional Series, and it is this part of the sequence which is thought to have a Late Proterozoic–Early Cambrian age. More recently, Herranz, San Jose & Vilas (1977) have investigated the Schistose–greywacke Complex in the western Toledo Mountain Range, examining in detail a 5000 m thick sequence of shale and greywacke with alternations of pebbly horizons which become increasingly common up the section. There is one carbonate marker-horizon, the Villarta Limestone. The stratigraphic sequence proposed by Herranz *et al.* (1977) (Table 18.1) is as follows:

Top 6. *Azorejo Sandstone* – detrital shelf sediments with ichnofossils (*Astropolithon, Scolicia*) indicating an Early Cambrian age.

5. *Pusa Shale* – turbidites, gradational upwards
4. *Fuentes Level* – olistostrome containing carbonate pebbles
3. *Torilejo Shale* – turbidites up to 2000 m thick
2. *Villarta Limestone* – calcareous sequence up to 250 m thick
1. *Guadiana Shale* – turbidites with shaly clays predominating, but with minor lenses of pebbly mudstone, approximately 800 m thick.

Vegas, Roiz & Moreno (1977) agree in general with the stratigraphic scheme of Lotze (1956). However, within the same turbidite facies they distinguish between a basal, mainly pelitic sequence, almost 1200 m thick, and a greywacke sequence, over 1500 m thick in which pelitic and conglomeratic intercalations increase upwards. These formations are overlain by the Transitional Series, a rather heterogeneous group of lithologies which comprise shale, quartzite, limestone, calcareous and even feldspathic conglomerate, fanglomerate and olistostromes, representative of shelf environments.

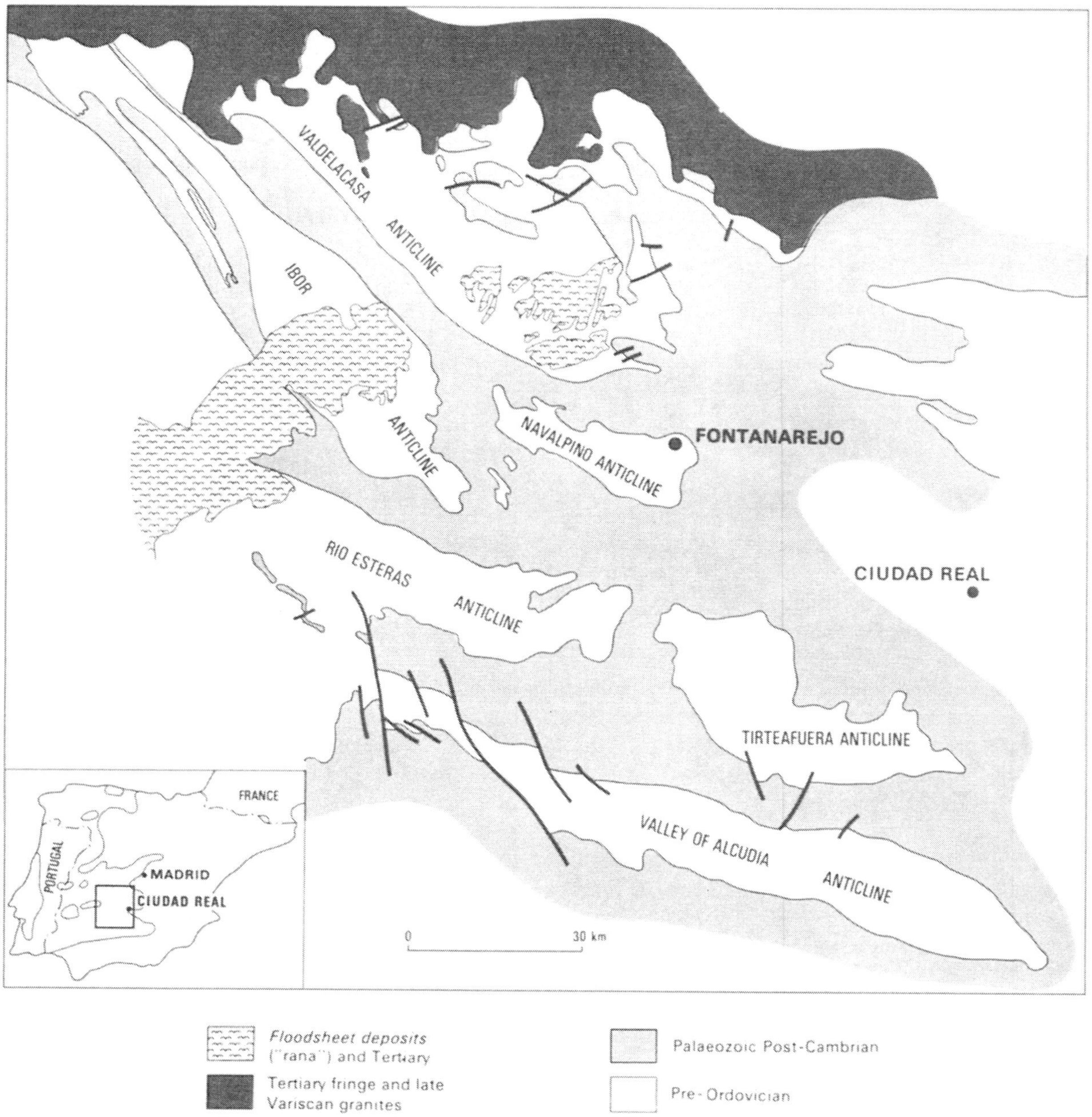

Fig. 18.2. Regional geology of the western Toledo Mountains, central Spain.

Cambrian–Ordovician

Within the Valdelacasa Anticlinorium, which is located to the north of the Navalpino Anticline and includes the Fontanarejo deposit, Moreno, Vegas & Marcos (1976) have distinguished three formations below the quartzites of the Armorican facies:

Top 3. An upper formation composed of shallow-water facies represented by sandstone, shale, conglomerate and quartzite, with red and purple-hued horizons dominant; dated as Tremadocian on the basis of ichnofossils (Fig. 18.3A).

2. A detrital facies unconformably overlying a pelitic sequence which contains a middle interval made up of intertidal limestones, 140 m thick, containing an archaeocyathid fauna dated as terminal Early Cambrian.

1. A lower formation consisting of 550 m of quartz sandstone, containing ichnofossils which indicate an Early Cambrian age. This unit overlies the turbidities of the Pusa Shale.

This relationship, where the Tremadoc unconformably overlies Lower Cambrian formations, exists in the area lying to the north of the axis of the Valdelacasa Anticlinorium. South of the axis no Lower Cambrian outcrops are known, and the Tremadoc lies unconformably on a variety of Precambrian horizons as a result of the intense erosion that occurred during or after the Sardinian orogenic phase.

Plio-Pleistocene

Recent sediments cover much of the outcrop area of the phosphate ore, which accordingly tends to be poorly exposed,

Table 18.1 *Stratigraphy of the Precambrian to Ordovician sequence in the western Toledo Mountains, central Spain*

	According to Moreno *et al.* (1976)		According to Herranz, San José & Vilas (1977)
Lower Ordovician	Upper Sequence	Quartzite shale, sandstone, and conglomerate	
Lower Cambrian	Cortijos de Malagon & Soleras Formations	Pelite, quartzite and sandstone (400 m)	
	Navalucillos Formation	Carbonate Series (140 m)	Navalucillos Limestone
	Gevalo Formation	Greywacke, shale and quartzite (> 500 m)	Azorejo Sandstone
	Transitional Series	Shale, conglomerate, limestone and quartzite (*c.* 1700 m)	Pusa Shale Fuentes Level
Precambrian	Valdelacasa Beds	Upper Sequence: greywacke and pelite (1500 m) Lower Sequence: mostly pelitic (1200 m)	Torilejo Shale Villarta Limestone Guadiana Shale

and makes correlation between outcrops difficult to establish with any certainty. Recent sediments mainly include flood-sheet deposits ('rañas') which formed under flash-flood conditions (Fig. 18.4C).

Tectonic phases

The structure of the region is a monotonous alternating series of broad Precambrian-cored anticlines with narrow, occasionally Devonian-cored synclines. There has been a series of tectonic events in the region. According to currently prevailing theory, no significant deformation took place during the Precambrian. The earliest structures are attributable to the Sardinian Movements which produced large wavelength, steep-sided, flat-crested folds consistent with tensional, basement-controlled tectonics. This Sardinian folding phase was responsible for first-order angular unconformities between the Lower Ordovician and underlying rocks. The main deformation, however, is attributed to the Variscan Orogeny that manifests itself in four first-order tectonic phases on which later deformation has had hardly any effect in the west Toledo Mountain Range (Moreno, 1977). Phases 1 and 4 did not have a major effect on the area and it is difficult to document these structural features with any confidence anywhere in the Navalpino Anticline.

Therefore, out of the four tectonic phases described by Moreno (1977), only those corresponding to his second and third-order folds have been recorded in the Navalpino Anticline:- (a) the main Variscan phase; and (b) the Transverse phase. The main Variscan phase accounts for the vast folds trending NW–SE as well as for their axial plane schistosity, whereas the Transverse phase fits the strike-slip model and is mainly observed in pelitic material. The Transverse phase similarly produces broad folds with axial trends ranging between N and N 20° W. The associated schistosity shows a poorly developed fracturing.

When the two phases interfere, the interaction of the two right-angled fold systems with vertical axial planes results in a dome-and-basin configuration. Regional metamorphism related to the Variscan Orogeny is low-grade, and never greater than anchimetamorphism. The sediments of the Navalpino Anticline have also been subjected to post-folding right-lateral strike-slip faulting trending between N 40° E and N 60° E. There is also another major N 100–120° E fault system.

The Fontanarejo deposit

The Fontanarejo phosphate deposit lies in the northwestern part of Cíudad Real Province, in the western part of the 1:50000 scale National Topographic Chart, sheet No. 735,

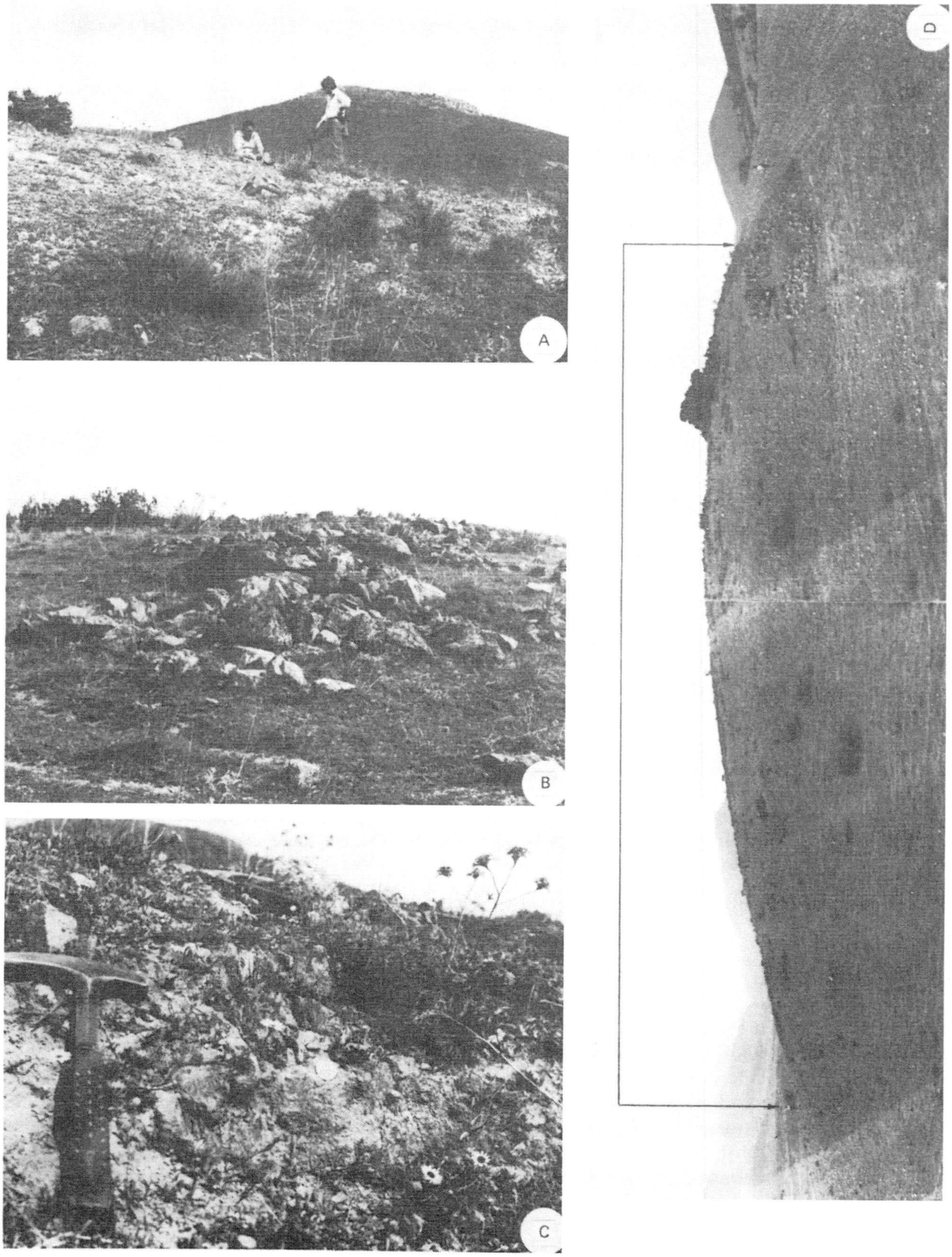

Fig. 18.3 (A) E. Hill–Foreground: phosphorites. Background: Arenig–Tremadoc series with ferruginous red sandstones and shales topped by 'Armorican' quartzites. (B) E. Hill–Exposure of phosphorite with a phosphate content of over 24% P_2O_5. (C) D. Hill–Close up of weathered phosphorite with a phosphate content of over 30% P_2O_5. (D) E. Hill – Southern scarp. Transverse section of phosphorite over 60 m thick at 70° E. The phosphate content is 20–24% P_2O_5.

Fig. 18.4. (A) D. Hill – Abundant quartzose microconglomerate showing resistance to weathering. (B) D. Hill–Exposure of quartz-poor phosphorite with a phosphate content of 20–25% P_2O_5. (C) F. Hill – Phosphorite outcrop partly covered by Plio-Pleistocene ('raña') deposits. (D) Arroyo del Puerto 15–20 cm thick algal phosphorite within a sequence. (E) Arroyo del Puerto – Very low angle channel cross-bedding.

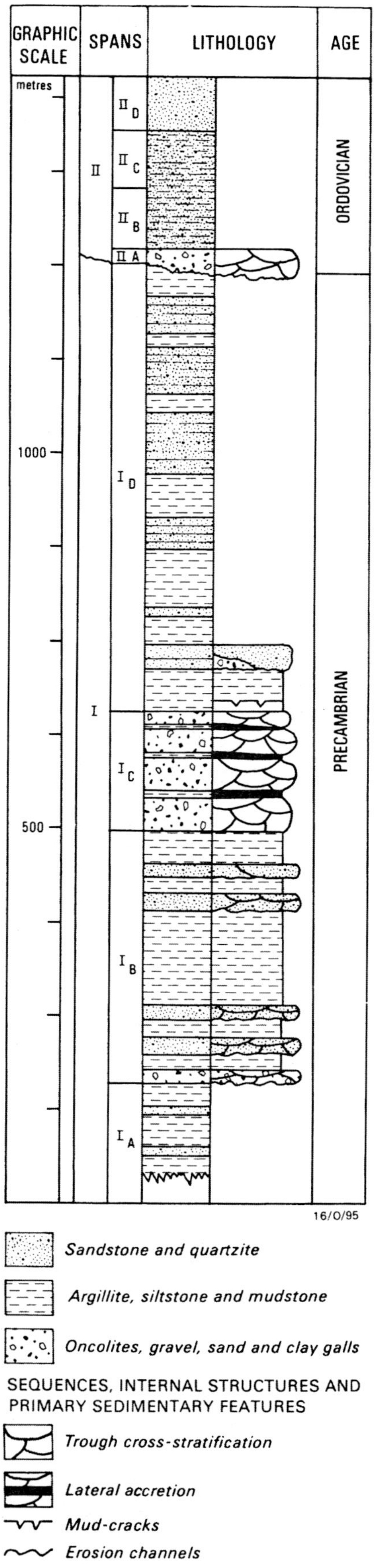

Fig. 18.5. Stratigraphic section in the Fontanarejo area.

about 2 km from Fontanarejo township. The area where the deposit is located is the Los Montes district, an economically depressed and sparsely populated region.

Three main elements characterise the landscape: (1) vast open plains which correspond to both the erosional and depositional late Tertiary peneplain, as well as lesser flood-sheet deposits ('rañas'); (2) hummocky relief in the areas of Precambrian outcrop where differential erosion between softer units (pelitic) and harder units (sandstones, greywackes, conglomerates) has resulted in well-defined differential topography; and (3) rugged areas controlled by lithology (quartzites and sandstones), weathering, and geological structure which also controls drainage.

Geological setting of the deposit

Precambrian, Ordovician and Recent deposits crop out in the vicinity of the phosphate deposit. The Ordovician deposits are moderately well exposed but the Precambrian deposits are commonly concealed by Recent sediments.

The Phanerozoic units have been previously described and consequently only the Fontanarejo Precambrian will be described here. A generalised stratigraphic column is shown in Fig. 18.5 and consists of the following units (from the base):

Interval 1a, at the base, spans about 250 m of section with no exposed footwall. It is made up of silty argillite and greyish-green siltstone, with occasional strongly silicified units as well as sandstone and microconglomeratic units. These intercalations consist of 70% quartz, rock fragments (quartzite, chert) and 5% accessory minerals, including zircon and muscovite. The matrix, which comprises 18% of the rock, is sericitic with some glauconite, while the cement (5%) is siliceous. The sandstones are clearly bimodal with the main mode medium to coarse sand and a secondary mode ranging from fine to very fine sand. Graded and rhythmic sediments are present.

Interval 1b is about 250 m thick (total thickness is hard to establish because of the schistosity), consisting of argillite, mudstone, siltstone and shale cut by channels infilled either by sandy-lutaceous gravel or sand and gravel. There are also litharenite to sub-litharenite units where glauconite abounds. As is evident in Fig. 18.5, pelitic material predominates.

Interval 1c is approximately 130 m thick and corresponds to a channel complex containing oncolites and sand–gravel with a variable quartz content. Both upward-coarsening and upward-fining sequences have been observed, as well as clasts in the centimetre range in the lag and rubified surfaces. There are conspicuous lateral pinch-outs. The channel material is phosphatic, consisting of orientated oncoforms (2–15 mm range) and a small proportion of quartz grains (2–5 mm range) with a ferruginous cement. Some samples contain large clasts (clay galls) which are phosphatised and include oncolites in places (Fig. 18.7E, Fig. 18.8A, B). The channels are separated by lutaceous intervals. A number of sedimentary features have been observed (Fig. 18.9) including: planar cross-stratification, trough cross-stratification resulting from mega-ripple translation and abundant clay galls in the lag. The morphology shown by these deposits, as well as the sedimentary petrography (see later), support their interpretation as tidal channel fillings. Reworking of both subtidal and intertidal deposits while

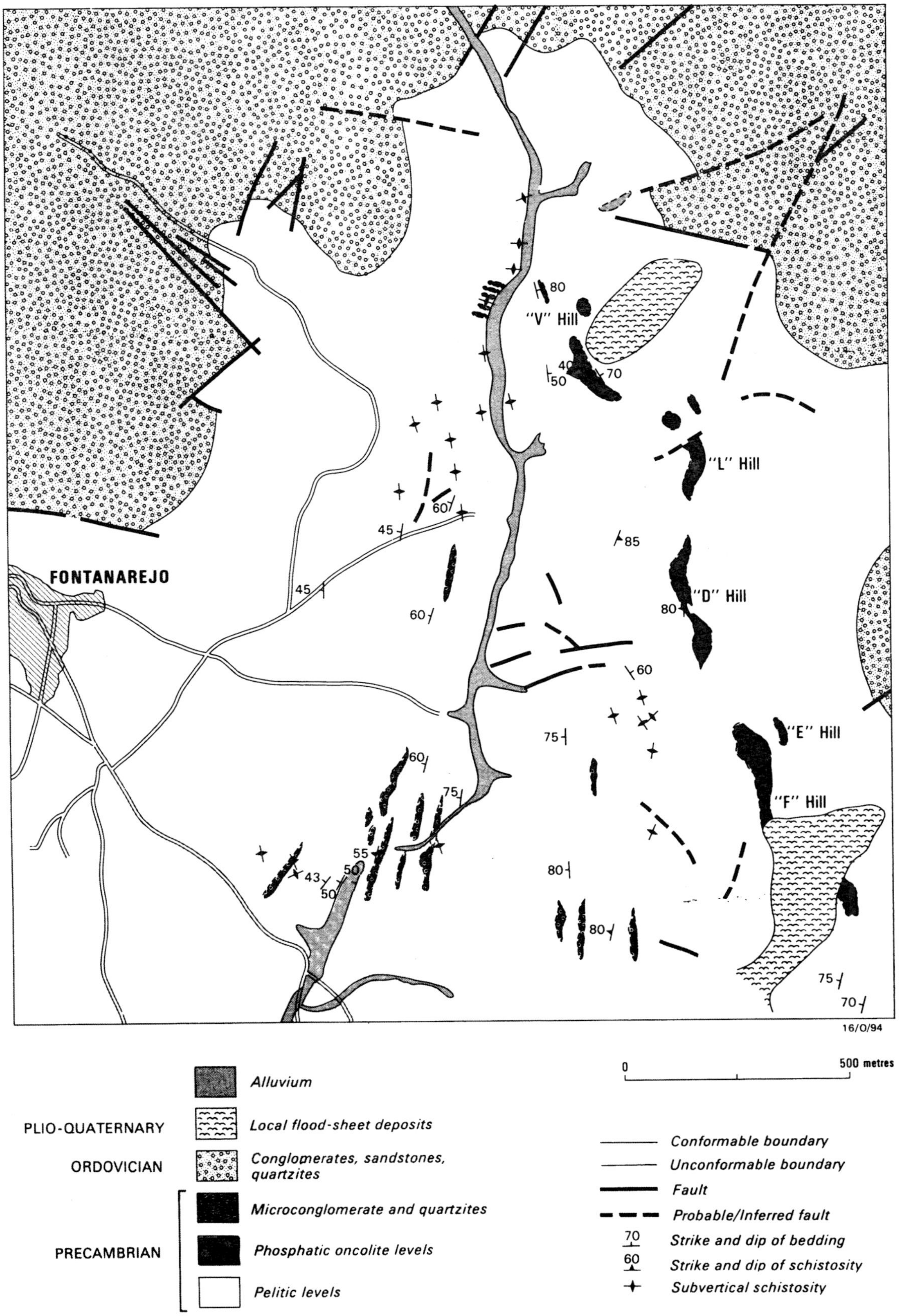

Fig. 18.6. Geology of the Fontanarejo deposit.

Fig. 18.7. (A, B) Bedding – schistosity relationship. Bedding is evident from the larger scale detrital layers. Conspicuous schistosity in the pelitic intervals. (C) Sample EH 1. Phosphatic rock made up by orientated oncolites which differ both in size (2–15 mm) and internal structure. Ferruginous cement. Quartz grains (2–15 mm) are rare. 30% P_2O_5. (D) Sample EH 2. Phosphatic rock made up of large oncolites with about 10% quartzite clasts. Ferruginised clasts are abundant. 26% P_2O_5. (E) Sample EH 4. Large phosphatised clast (phosphatic clay-galls). Random clay gall-enclosed oncolites (arrow). Phosphate content of both oncolites and matrix is 25% P_2O_5. The clasts range in phosphate content from 1% to 24% P_2O_5.

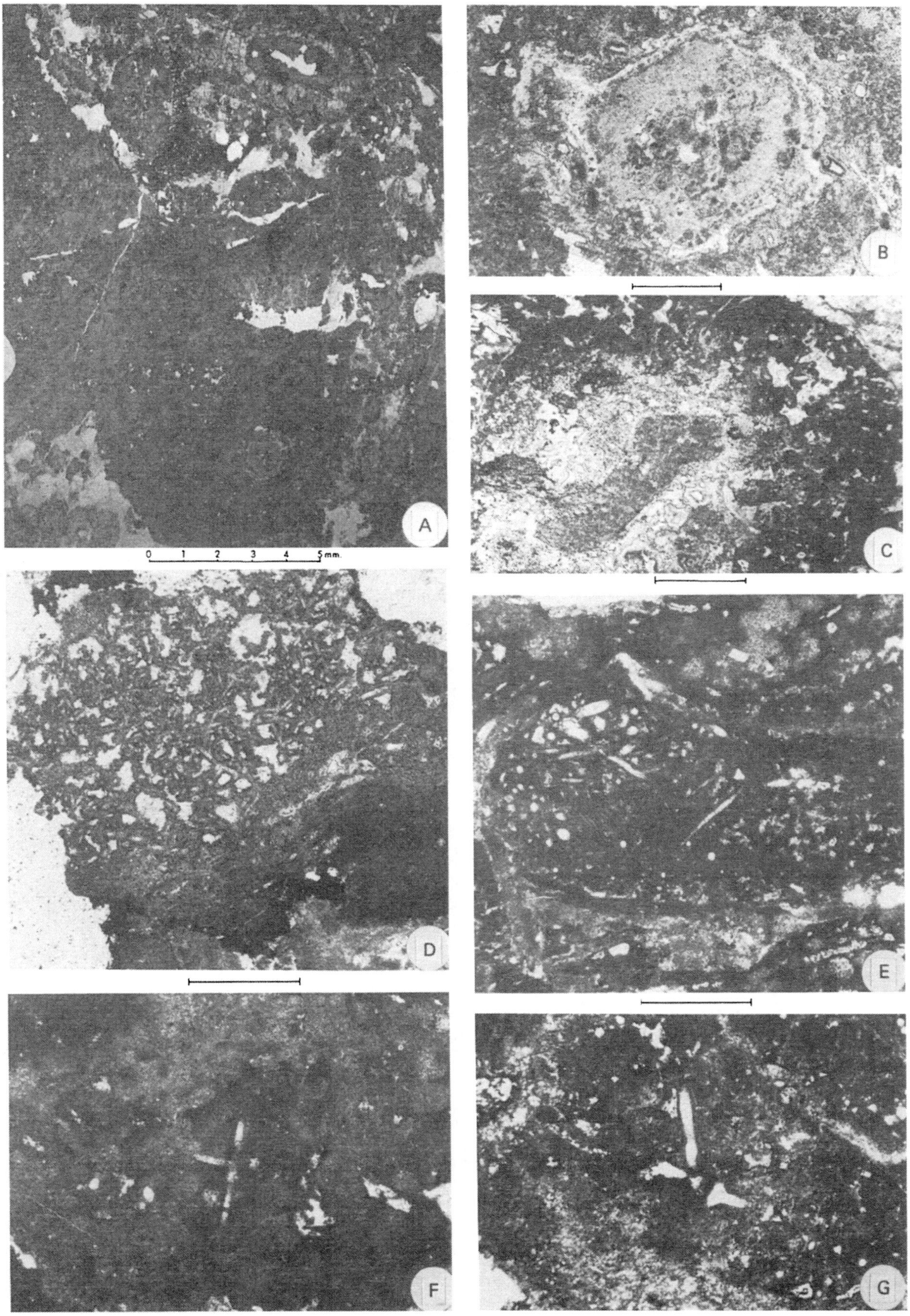

Fig. 18.8. (A) Sample 3–1. General view of phosphatic rock with oncolite-enclosing clay-gall (right lower corner). (B) Sample 3–1. Detail of previous oncolite. (C) Sample AP. Microcrystalline-nucleated oncolite made up of silicified cyanophyceous algal network (Chroococcales). The black outer nucleus area is collophane. (D). Sample LH (A). Sponge remains (E). Sample LH (B). Spicules in phosphatic groundmass. (F) Sample LH & SN Spicule details. *Bar scale indicates 0.5 mm.

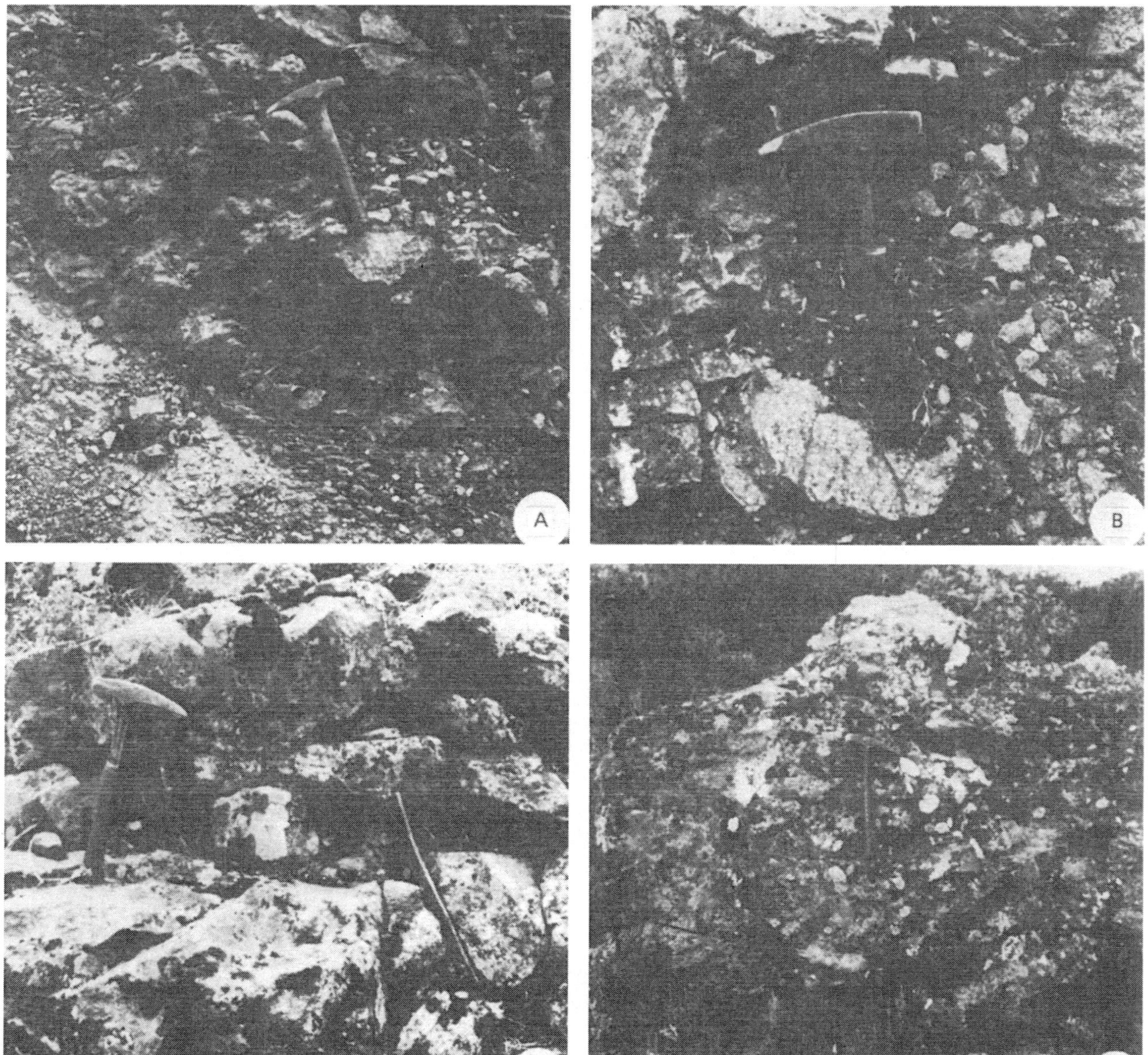

Fig. 18.9. (A) Channel cut zone. (B) Graded trough cross-bedding. (C) Tabular cross-bedding. (D) Channel trough cross-bedding. (E) Side view of mud-cracks with underlying shales. (F) Detail of mudcracks.

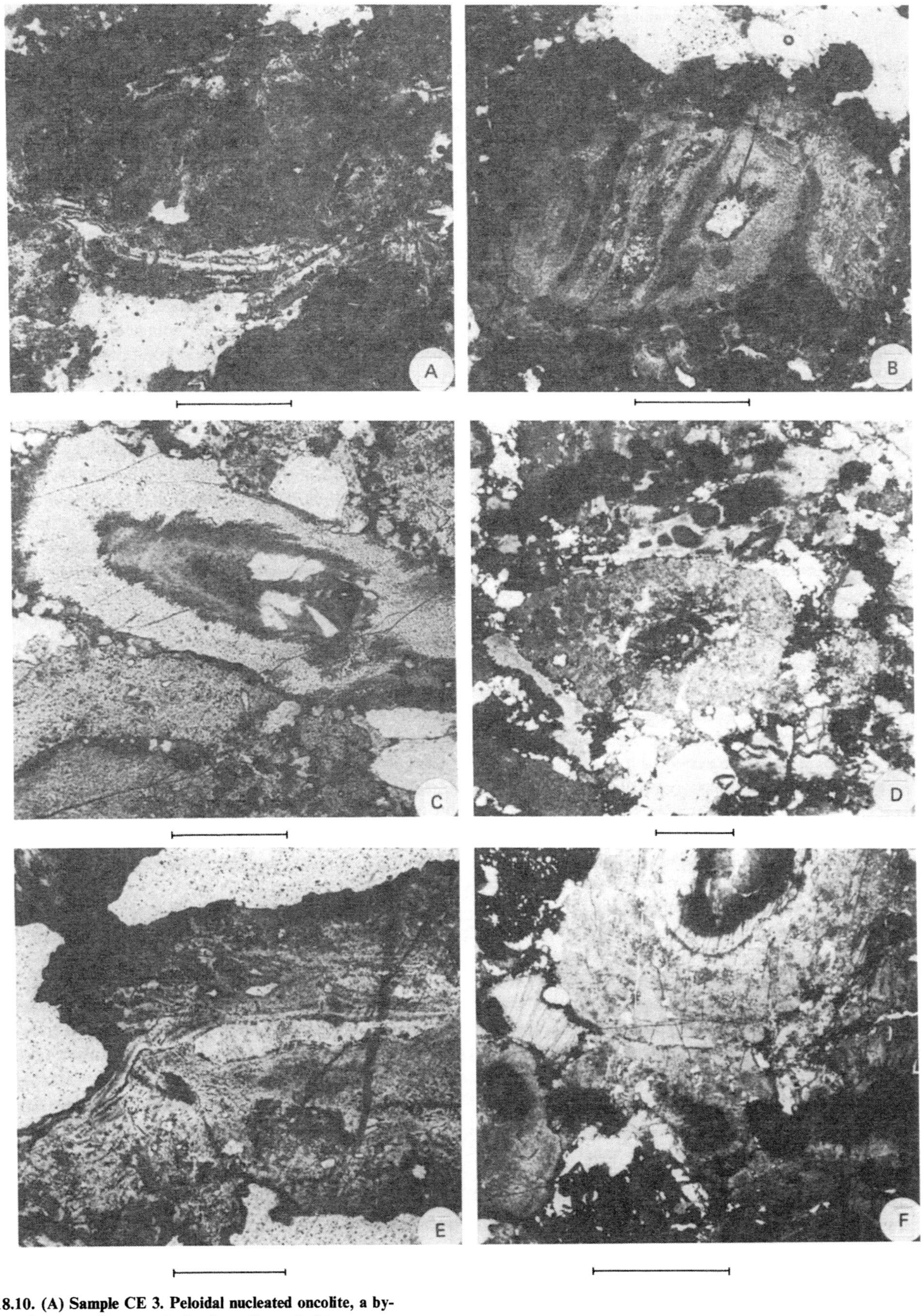

Fig. 18.10. (A) Sample CE 3. Peloidal nucleated oncolite, a by-product of a cyanophyceous algal colony. Cryptalgal-type outer margin. (B) Sample CE 4. Cryptalgal nucleated oncolite with a more complex outer margin. (C) Sample VH. Microcrystalline nucleated oncolite made up of silicified cyanophyceous algal colonies. (D) Sample VH. Phosphatised detrital nucleated (dark central zone) oncolite. Light patches correspond to fine-grained quartz. (E) Sample CV 2. Detail of complex cryptalga. (F) Sample EH 2. Collophane-nucleated oncolite (dark zone) with a phosphatic outer cover. *Bar scale is 0.5 mm.

channels were migrating appears to have produced lateral accretion and the observed petrographic variations. Similarly, lateral and vertical migration of the channel system implies a changing geography for the various sub-environments.

Interval 1d includes approximately 450 m of claystone, mudstone and siltstone, with sandstone and quartzite intercalations. Rapid lateral pinching-out and a channel complex up to 10 m thick occurs 50 m above the footwall. Oncolites occur in some units, while at the base light claystones (white to pinkish) and mud-cracks are prominent (Fig. 18.9E, F).

Folded and fractured Ordovician sediments define the eastern closure of the Navalpino Anticline (Fig. 18.6). The Variscan tectonic phases previously described have affected all lithologies in the area. The main phase produced fracture schistosity in the pelitic rocks and upright axial planes, steep-dipping flanks and moderately tight folds. The second phase affected previous structures producing interference tectonics as a result of the intersection of folds with axes that are perpendicular to each other. The schistosity is also locally folded. The Precambrian seldom crops out as it is mainly covered by Recent deposits. Plasticity within and erosion without have jointly masked folding.

Sedimentary petrography

Some of the petrographic features of the phosphorite are shown in Figs. 18.9 and 18.10. In thin-section, the phosphorites are seen to be composed of oncolites, clasts, terrigenous components, matrix, cements and other diagenetic components. Most of the samples are homogeneous.

Organo-sedimentary structures, especially of the oncolitic type, represent a major component, comprising 55–70% of the total sediment. Several types of oncolites ('oncoforms' as they are referred to in Spanish literature) can be distinguished, including those which are predominantly microcrystalline, pellet-shaped, cryptalgal or have other algal components. Biogenic builds-up other than these are also present. In a number of instances zoned oncolites occur showing well-defined concentric banding. Microcrystalline or pellet-shaped nucleated oncolites as a rule have silicified nuclei, whereas the algal layers are phosphatic. Phosphate is also present as a cement, probably as a result of diagenetic mobilisation of phosphate which replaced the original matrix.

Clasts are an important component, comprising up to 7% of the total rock. The grain size ranges between coarse sand and gravel, and the sediment is well sorted. Usually present are rounded to subrounded clasts (clay galls) of phosphatic mudstone. Development of oncolitic envelopes around clasts supports a syndepositional origin, i.e. prior to lithification. Phosphatisation and the development of phosphatic mobilisation were also diagenetic processes. Original terrigenous clasts which have been phosphatised have been mentioned elswhere. The terrigenous sediment is made up of 5–15% quartz and quartzite clasts, well sorted, subrounded to subangular, and within the coarse sand–gravel size interval. Compared to oncolites, terrigenous clasts are rarely heterometric grains. Apart from a few fine-grained zircons, heavy minerals are very scarce. The matrix which comprises 5–10% of the total rock is sericitic. Diagenetic phosphatisation of the groundmass has occurred locally. The cement which comprises 5–10% of the total rock is predominantly siliceous, but has been partly replaced by diagenetic phosphate. Other types of cement (e.g. ferruginous) may have existed in some channels or horizons, but later diagenetic processes have obscured their original form, and have made their genesis hard to identify. There is variable metallic mineralisation, which may range in places from 3% to 10% metal content. The bulk of the metalliferous ore seems to be of hydrothermal diagenetic origin and is characteristically accompanied by feathery chalcedony.

A poorly represented group of sediments, which have pronounced petrographic differences from the samples so far described, correspond to microconglomeratic intercalations and palaeochannel deposits stratigraphically overlying the main phosphatic interval. They comprise coarse sandstones (quartzarenites) and gravels with subrounded to subangular grains. The matrix makes up 4–6% of the rock and the cement, which is a mixed siliceous-phosphatic material and constitutes less than 10% of the total rock.

The remaining sediments of the phosphatic sequence comprise 50–55% terrigenous components of medium sand- to gravel-size, rare oncolites and phosphatic clasts, as well as a high percentage of sericitic matrix and mixed siliceous-phosphatic cement. Hydrothermal mineralisation producing metalliferous ores and feathery chalcedony is relatively important in places.

Overall, the sediments studied seem to have resulted from the interaction of biogenic processes, mainly related to algal activity (cryptalgal mats, etc.) and clastic processes mainly related to energy currents. The degree of sorting and rounding, the almost complete lack of matrix, plus the oncolites and terrigenous features, indicate a well-washed, mature sediment deposited within an intertidal to subtidal environment but with supratidal conditions occurring from time to time.

Economic geology

The exposed phosphatic intervals range in thickness from 10 m to 60 m and dip at about 70°. They extend over a strike length of nearly 1.5 km forming low hills trending N–S (Fig. 18.6). Vegetation, recent deposits and local flood-sheet deposits, together with tidal palaeochannels with complex geometry and lithological variability, hinder conclusive evaluation of these phosphorites. A programme of drilling and excavation of test pits has been devised in order to determine the location and extent of other phosphatic levels under the 'raña'. Current ore estimates of about 2×10^6 tonnes are based on existing data and may be considerably increased in the future. X-ray fluorescence analysis of several samples from oncolite levels with a low concentration of detrital quartz gave the mean analytical values shown in Table 18.2. X-ray diffraction indicates the main phosphate mineral to be fluorapatite.

Age of the phosphorites

As previously mentioned, the Fontanarejo deposit is placed in the upper part of the Schistose-greywacke Complex, below the Ordovician (Tremadocian) transgression. Intense erosion related to the Sardinian tectonic phase accounts for the local absence of the Cambrian as well as the phosphate-bearing deposits in parts of the Navalpino Anticline. At the Valdelacasa

Table 18.2. *Selected X-ray fluorescence analyses of samples from the Fontanarejo deposit, Spain*

	Value (mean)
SiO_2	33.2%
P_2O_5	24.0%
CaO	33.9%
Al_2O_3	2.86%
Fe_2O_3	1.9%
MgO	1.14%
F	1.8%
PF	1.5%
S	25 ppm
Cl	43 ppm

Anticlinorium north of the Navalpino Anticline, Brasier, Perejon & San Jose (1979) have observed the occurrence of *Planolites* in the Fuentes olistostromic horizon, near the Uso river. Further up the sequence, in the Pusa Shale, they identified *Beltanelloides sorichevae* and *Chuaria circularis*, both forms being apparently restricted to rocks of Riphean and Vendian age. However, they have also found arthropod traces below the *Chuaria* levels, and attributed them to the species *Monomorphichnus lineatus*, which is considered to be no older than Early Cambrian, as well as other Early Cambrian ichnofossil traces including *Phycodes* and *Treptichnus*. This apparent overlapping of Precambrian and Cambrian forms cannot be resolved by the authors, and it is uncertain whether we are dealing with the survival of Precambrian fossils into the Cambrian or the extension down into the Precambrian of fossils previously considered to be solely Cambrian.

Within a stratigraphic level close to the phosphatic unit at Fontanarejo (the 'Casas del Membrillar Olistostrome', which is probably contemporaneous with the 'Fuentes Olistostrome'), T. Palacios (personal communication) has reported the presence of organic-walled microfossils, including *Bavlinella favoelata*, a species prevailing during the Varanginian glaciation, and which is found worldwide in the range Early Vendian to terminal Early Cambrian. Associated occurrences of *Trachysphaeridium laufeldi* pursuaded Palacios (personal communication) to date the olistostrome as roughly early–middle Vendian. The remaining organic remains are too dark for morphological feature identification. Cyanophyceous algae of the Order Chroococcales, which occasionally occupy the inner part of oncolitic structures, and thread-like forms that could be attributed to the Order Oscilatoriales (Golubic, 1976; Boureau & Dars, 1979) have also been observed. Further south, in the Valdemanco del Esteras and Abenojar geological sheet areas at levels which could be stratigraphically equivalent to those at Fontanarejo, floral remains of the Vendotaenida group, of late Vendian age, have also been found.

Within the Sevilleja de la Jara geological sheet area, at levels which seem to be stratigraphically lower than the 'Fuentes Level' and probably correspond to the shales, greywackes and sandstones of the 'Valdelacasa Series', the following acritarchs have been identified: *Bavlinella favoelata*, *Pterosphermopsimorpha*, *Protosphaeridium*, *Leiosphaeridia*, *Trachisphaeridium*, *T.* cf. *laminaritium*, *Uniporata*, and *Faroeosphaeridium*. These are taken to indicate a late Riphean–Vendian age, with the exception of the genus *Uniporata* which is dated as Tommotian (Early Cambrian on the Siberian Platform). In the same geological sheet area, in the 'Pusa Shales', which are stratigraphically higher than the 'Fuentes Level', identifiable fossils include *Beltanelloides sorichevae*, a typical Vendian species in the USSR, and a host of acritarchs represented by *Polytorama*, *Bavlinella faveolata*, *Protosphaeridum*, *Pterosphermopsimorpha*, *Kildinella sinica* and *Leiosphaeridia* among the vegetal remains, all of which point to a late Vendian age according to the authors of the geological maps, and *Planolites* and *Phycodes pedum* among the ichnofossils.

During the symposium 'Chronology of the Geological Record', held in London on May 11, 1982, the difficulty in defining the base of the Cambrian was discussed. The decision of the IUGS Working Group on the Precambrian–Cambrian Boundary (and IGCP Project 29) was to place the boundary stratigraphically near the first appearance of Tommotian-type fossil assemblages (Snelling, 1982). More recently, Cloud & Glaessner (1982) have formally proposed a major revision to this part of the geological time scale, one which extends the Phanerozoic beyond the Cambrian to include a new interval of geological time, the Ediacarian. In so doing, they group together rocks from many countries which contain the first occurrence of the Metazoa, the oldest known multicellular animals, and their fossil traces. These appeared worldwide at low palaeolatitudes after the terminal Proterozoic glaciation. Rocks of the Ediacarian System extend to the 'massive onset of shelly fauna and distinctive trace fossils that mark the initial Cambrian at many places'. The Ediacarian is overlain by the initial Cambrian Tommotian Stage. The authors consider that only the Vendian warrants consideration and comparison. The term 'Vendian' is generally taken to include the later Proterozoic glaciogenic deposits, which Cloud & Glaessner, place below the Ediacarian.

In a tentative attempt to fit the Fontanarejo deposit with the generally accepted chronostratigraphy, all the data which were discussed earlier in this chapter as being consistent with a Middle to Late Vendian age point towards a probable Ediacarian attribution in the sense given to this term by Cloud & Glaessner (1982). However the finding of sponge spicules in thin-sections of some Fontanarejo phosphorites (Fig. 18.8D–G) is worth noting, since sponges are generally considered to originate at the end of Precambrian time (de Laubenfels, 1955; Reid, 1979) although there is no proof of this. De Laubenfels (1955), in summarising doubtful Precambrian sponge occurrences, proposed the new genus and species *Eospicula cayeuxi*, which he placed among the Class Hyalospongea on the basis of some specimens found by Cayeux (1895) in the Precambrian of Brittany. More recently, Sepkoski (1978, p. 226, fig. 2) has placed the first appearance of Porifera at the beginning of the 'Lower' Tommotian (Cambrian). It is only during the Early Cambrian that true spongiolites accumulated, such as those in the Karatau phosphorites (Bushinsky, 1969). Sponges are multicellular and thus fall within the Metazoa although the almost independent cells and absence of distinct individuality have led to their placement among the Parazoa branch i.e. 'Eumetazoa' (many-celled animals *sensu stricto*).

In dating the Fontanarejo deposits, we are therefore confronted by the fascinating dilemma: do the sponge spicules indicate the presence of a multicellular megascopic organised evolutionary level in the latest Proterozoic, or are the deposits of initial Cambrian age? The data at our disposal are presently insufficient to resolve this question. But whatever the final decision on the precise age of the Fontanarejo deposit, there is no doubt that its discovery has greatly increased the prospects for finding sedimentary phosphate deposits in Spain in rocks of late Precambrian and Cambrian age.

References

Boureau, E. & Dars, R. (1979). Sur les organismes précambriens récoltés au Mali. *Bulletin des Centres de Recherches Exploration-Production Elf-Aquitaine*, **3**, 481–7.

Brasier, M.D., Perejon, A. & de San Jose, M.A. (1979). Discovery of an important fossiliferous Precambrian–Cambrian sequence in Spain. *Estudios Geológicos*, **35**, 379–83.

Bushinsky, G.I. (1969). *Old phosphorites of Asia and their genesis*, 266pp. Israel Program for Scientific Translations, Jerusalem IPST Press.

Cayeux L. (1895). De l' existence de nombreux débris de spongiaires dans le précambrien de Bretagne. *Annales de la Société Géologique du Nord*, **23**, 54–65.

Cloud, P., Glaessner, M.F. (1982). The Ediacarian Period and System: Metazoa inherit the Earth. *Science*, **217(4562)**, 783–92.

de Laubenfels, M.W. (1955). Porifera. In *Treatise on Invertebrate Paleontology, Part E. Archaeocyatha and Porifera*, ed. R.C. Moore, pp. E21–112, Geological Society of America and University of Kansas Press, Lawrence, Kansas.

Golubic, S. (1976). Taxonomy of extant stromatolite-building Cyanophytes. In *Stromatolites*. ed. M.R. Walter, pp. 127–40. Elsevier, Amsterdam.

Herranz, P., de San Jose M.A. & Vilas, L. (1977). Ensayo de correlación del Precámbrico entre los Montes de Toledo occidentales y el Valle del Matachel. *Estudios geológicos*, **33**, 327–42.

Lotze, F. (1956). Das Präkambriums Spaniens. *Neues Jahrbuch für Geologie, Paläontologie, Monatshefte*, **8**, 373–80.

Moreno F. (1977). Estudio geológico de los Montes de Toledo occidentales. Ph.D. Thesis, Madrid University (Unpublished).

Moreno, F., Vegas, R. & Marcos, A. (1976). Sobre la edad delas series ordovícicas y cámbricas relacionadas con la discordancia 'Sardica' en el anticlinal de Valdecasa (Montes de Toledo). *Breviora Geologica Asturica*, **20(1)**, 8–16.

Notholt, A.J.G. (1979). Resources of Precambrian and Cambrian sedimentary phosphate rock. In *Proterozoic – Cambrian Phosphorites*, ed. P.J. Cook & J.H. Shergold, pp. 71–86. ANU Press, Canberra.

Reid, R.E.H. (1979). Porifera. In *The Encyclopaedia of Paleontology*, ed. R.W. Fairbridge & D. Jablonski, pp. 635–41. Dowden, Hutchinson & Ross, Stroudsburg, Pennsylvania.

Sepkoski, J.J. (1978). A kinetic model of Phanerozoic taxonomic diversity. I. Analysis of marine orders. *Paleobiology*, **4(3)**, 223–51.

Snelling, N.J. (1982). Chronology of the geological record. Conference Rept. *Episodes*, **1982(2)**, 26–7.

Teixeira, C. (1954). Os conglomerads de complexo xisto-grauváquico ante-silurico. Sua importancia geologica e paleogeográfica. *Comminicacoes Survicos Geologicos, Portugal*, **35**, 33—49.

Teixeira, C. (1968). Quelques problèmes de la géologie du Portugal. *23rd International Geological Congress, Prague, Proceedings of Section 13, Other Subjects*, 233–42.

Teixeira, C. (1979). Le Precambrien portugais. *Compte Rendu Sommaire des Séances de la Société Géologique de France*, **5/6**, 228–30.

Vegas, R., Roiz, J.M. Moreno, F. (1977). Significado del complejo esquisto-grauváquico en relación con otras series 'pre Arenig' de España Central. *Studia Geologica*, **12**, 207–15.

19

Proterozoic and Cambrian phosphorites – deposits: Volta Basin, West Africa

J. LUCAS, A.V. ILYIN AND A. KÜHN

Introduction

A late Proterozoic phosphogenic province is located along the margins of the West African craton (Ilyin *et al.*, 1979). Three phosphogenic regions or basins have been identified so far: the north-eastern part of the Volta Basin; the western part of the Taoudeni Basin; and the Anti-Atlas region (see Slansky, this volume). Of these, the largest deposits occur in the Volta Basin.

The Precambrian phosphate deposits of the northern Volta Basin, in Upper Volta, Niger and Benin, have been described by Trompette *et al.* (1980). These authors have also discussed the stratigraphy and structure. Five main phosphatic areas can be distinguished: Kodjari and Arli, occurring in the flat-lying sediments of the Basin; Tapoa and Mekrou in the folded portion of the Basin; and Aloub Djouana, in the structurally complex, folded and slightly metamorphosed Buem structural unit (Fig. 19.1). The Tapoa–Mekrou area is the most important area for phosphate occurrences and deposits. Research carried out in this area in the period 1974–1976 has led to an evaluation of those deposits and a determination of their sedimentation pattern, and has provided a spatial picture of the distribution of the phosphorite-bearing series.

General geology of the Volta Basin

The Volta Basin occupies the periphery of the West African craton, and its formation was probably associated with ancient fractures rimming it. The sediments filling the basin rest disconformably on the igneous and metamorphic Birrim Formation and in places probably on the Tarkvay Molasse. The total sedimentary thickness is poorly known: according to the recent data of Bessoles & Trompette (1980) it may exceed 1500 m in the central part of the basin in Ghana and may be even thicker towards the east. The western and central parts of the basin have a platform character whereas the eastern part is deeper, and is regarded by some authors as a geosynclinal trough.

The stratigraphic sequence of the Volta Basin comprises three main sedimentary units: the Voltaian, Buem and Atakor Series. Initially, the Atakor and Buem were considered to be the oldest (Early Proterozoic) sedimentary cycles, and the Voltaian to be mainly Cambrian. However, further investigations by Grant (1969), Bozhko (1969), Pascal (1972), Affaton, Sougy & Trompette (1980) and Bessoles & Trompette (1980) have shown the Atakor to be a mildly metamorphosed and folded equivalent of the Lower Voltaian Series as a result of the Pan-African Orogeny, and the Buem to be wholly or partially a lateral equivalent of the Middle Voltaian which has been strongly tectonised and cut by basaltic extrusions.

The Voltaian Series is best developed in Ghana where three lithological groups can be recognised. In this part of the Volta Basin, the Lower Voltaian Series consists of sandstones and conglomerates (regarded by some authors as tillite); the Middle Voltaian Series consists mainly of claystones referred to as the Oti Group; and the Upper Voltaian Series is represented by the predominantly arenaceous Obosum Group (which is restricted to this part of the Basin). In the northern Volta Basin, in Togo and Upper Volta, where only the Lower and Middle Voltaian Series are known, the lower part of the Series is represented by sandstones of the Dapango–Bombouaka Group which are considered to be the equivalents of the basal clastic group of Ghana. In Niger and Benin, the basal Voltaian Series is represented by the Kirtachi sandstone which includes hematitic claystones and conglomerates. In general this complex of sedimentary rocks is of continental character resulting from slow infilling of a relatively shallow basin by material brought in by rivers. Conglomeratic beds occurring in the upper parts of the Kirtachi Sandstone resulted probably from alluvial fan deposition that terminated the deposition of the Lower Voltaian Series.

North from the Togo–Benin border, the Volta Basin thins and becomes a narrow trough between two crystalline massifs. The southern part of the trough in the area of Benin is called the Pendjari Basin which in turn has a prolongation into the Tapoa–Mekrou area of Niger. The sediments of the Tapoa–Mekrou–Pendjari area comprise the Pendjari Group, predominantly shales, silty shales, sandstones and greywackes, equivalent to the Oti Group of Ghana, and represent a new Middle Voltaian cycle resting discordantly on the Lower Voltaian continental sediments. The commencement of this phase of deposition can be dated at *c.* 615 m.y. on the basis of

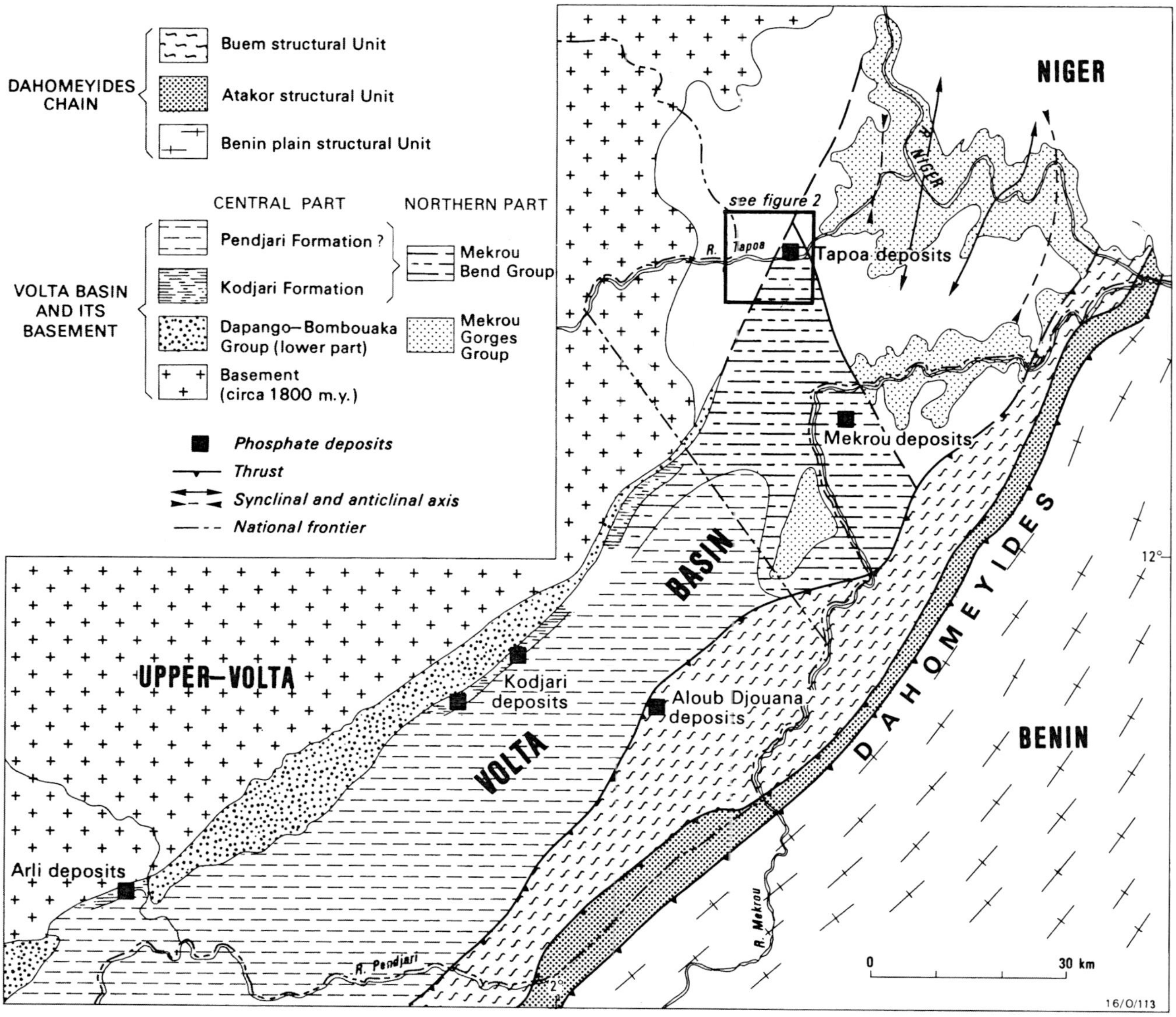

Fig. 19.1. Schematic structural map of the northern part of the Volta Basin and the Dahomeyides Chain in Benin, Upper Volta and Niger, with location of the main phosphate deposits (after Trompette *et al.*, 1980). (For broader context see Slansky, this volume, Fig. 9.2.)

radiometric age determinations on the upper part of the Oti Group by Clauer (1976). The Middle Voltaian Pendjari Group contains the phosphorite deposits of the northern Volta Basin. Stratigraphic relationships are summarised in Table 19.1.

Deposits of the undeformed Volta Basin

Stratigraphy

The main deposits of this part of the northern Volta Basin are at Arli and Kodjari. The stratigraphic position of these phosphate deposits has been fairly well established by Trompette *et al.* (1980) who assign them to the Pendjari Group, previously dated (Bozhko, 1969; Clauer, 1976) as late Proterozoic (660–615 m.y.). In this part of the Basin, the Pendjari Group consists of a basal Kodjari Formation, which contains about 50 m of tillite, dolomitic limestone, chert and phosphorite, and an upper Pendjari Formation comprising 1000–2000 m of clastic sediments. At Kodjari, the sequence consists of a basal breccia (0.1 m), followed upwards by tillite (0.15 m), dolomitic limestone (0.5–3 m), bedded chert (25–30 m) and an uppermost unit (0–30 m) which varies in lithology from argillaceous siltstone to phosphorite. In his detailed study of this part of the sequence at Kodjari, Ouedraogo (1982) distinguished three main facies consisting of: phosphatic siltstones overlain by fine grainstone phosphorites, which again are overlain by a thin, irregular, channelled bed of coarse grainstone phosphorite containing numerous rounded grains ('phosphoncolites'). This sequence corresponds to an environment of increasing energy. At the Arli deposit, similar in its general stratigraphy to Kodjari, the phosphatic horizon is poorly bedded and only about 10 m thick.

Petrology

From the petrological descriptions of the phosphatic material from these areas given by Trompette *et al.* (1980), Lucas, Prévôt & Trompette (1980), Lucas & Prévôt (1981) and

Table 19.1 *Correlation of stratigraphic units of the Volta Basin*

Series	Ghana/Togo	Togo/Upper Volta	Niger/Benin	Age
Upper Voltaian	Obosum Group (615–550 m.y.)			Cambrian
	local			
Middle Voltaian	Oti Group	Pendjari Group (675–615 m.y.): Pendjari Formation; Kodjari Formation*	Moussa Kwara Formation; Phosphorite formation*; Clayey-silty formation	Vendian; Late Proterozoic
	angular			
Lower Voltaian	Dapango Bombouaka Geoup: Panabako Formation; Poubogou Formation; Tosseigou Formation	Dapango-Bombouaka Group (1000–650 m.y.)	Kirtachi Sandstone	Late Riphean; Late Proterozoic
	major			
	Basement (2030–1700 m.y.)			Early Proterozoic

*Position of phosphatic sequences
Based on Affaton *et al.*, 1980; Trompette *et al.*, 1980.

Ouedraogo (1982), it is possible to point to a number of features; for example the phosphate grains are variable in size and appearance. At Kodjari, where they seldom exceed 100 μm in size, they are rounded, without visible internal structure, and are transparent to brown. The smallest of them (50 μm) are spheroidal with an internal 'brain' structure. In the very highest levels at Kodjari and at Arli, the grains are coarser (10–100 μm) and more complex. In addition to simple coarse ovular grains, there are numerous grains which have variably thick aureoles. These vary from a simple coating to a thick accumulation of fine concentric envelopes. The aureoles may encase a single grain or several which already have aureoles. Thus, several successive stages of aureolation may form aggregates or lumps. These aggregates, as well as their enclosed grains, are either 'fresh' or abraded and fractured. They are evidence of an important reworking process during which periods of agitation (abrasion and fracturing) are succeeded by periods of quiescence (aureolation). Ouedraogo, following Champetier & Joussemet (1979), considers that these aureolated grains are oncolites.

The primary cement is difficult to discern, although it is probably represented by the calcite of the non-weathered or slightly weathered beds found in the Kodjari boreholes. The

Table 19.2. *Mineralogical composition of the main beds (after Ouedraogo, 1982)*

	Upper horizon (altered)	Upper horizon (unaltered) and lower horizon (unweathered)
Major minerals	Quartz and particularly microquartz	Apatite
		Abundant microquartz
	Abundant oxides and hydroxides	Oxides and hydroxides
	Apatite, strontium phosphate, crandallite, wavellite	Calcium carbonate (thick horizon)
	Clays	Clays
Minor minerals	Turquoise, vivianite	Pyrite
	Barium sulfate	Abundant organic material
	Rare pyrite	
	Traces of organic material	No diagenetic phosphate

rock is most affected by diagenetic silicification which fills the intergranular spaces with a mixture of silica and phosphate, or as at Arli, replaces in part the grains and their aureoles. The phosphate, which is also mobile, may be replaced by silica, or displaced into the matrix. The results of meteoric weathering superimposed on the effects of diagenesis led to the transformation of apatite to aluminium or iron phosphates at the surface (Table 19.2). This is a complicated process: deposition, reworking, diagenesis and weathering have probably modified the primary phosphate mineral, and are responsible for its present abnormal characteristics. X-ray diffraction of the Kodjari and Arli apatites indicates a rather uniform composition consistent with a non-carbonate fluorapatite. However, chemical analyses of rocks from different localities indicate a greater complexity. They confirm the almost complete lack of lattice CO_2, but they reveal that the Ca/P_2O_5 ratios are closer to those of carbonate fluorapatite than to non-carbonate fluorapatite. A similar result was given by whole rock chemical analyses (Table 19.3) and microprobe analyses (Table 19.4). These features suggest that the primary mineral was indeed a carbonate fluorapatite similar to that found in most of the non-weathered phosphate deposits. Transformation was caused by diagenesis and weathering during a long and complex post-depositional history.

Deposits of the deformed Volta Basin

The Tapoa deposit (Fig. 19.2) is the largest and most important phosphate deposit in the Volta Basin with reserves, according to Kühn & Zajaczkowski (1975) of about 100 million tonnes, and subsequent detailed exploration indicating up to 470 million tonnes (Mining Annual Review, 1977). However, the stratigraphic position of the productive sediments is not clear, their relationship to the basement rocks is uncertain, and their correlation with the sediments of the flat-lying part of the Basin is in some doubt. These uncertainties are caused by the location of the deposit in an isolated graben. Nevertheless, it is thought that the Tapoa phosphorites can be correlated with the Kodjari Formation of the unfolded part of the Basin (and that all deposits in the Basin are of the same age), or with the lower part of the younger Pendjari Formation. In either case, however, a late Vendian age is indicated.

Stratigraphy

The Tapoa deposit has been extensively drilled, and the following stratigraphy established (from base upwards):

Table 19.3. *Average chemical analysis of phosphorites from the Volta Basin (after Lucas et al., 1980)*

	SiO_2	Al_2O_3	MgO	CaO	Fe_2O_3	Na_2O	K_2O	P_2O_5	F	C	$CaO:P_2O_5$	$F:P_2O_5$	Sr	Ba	Mn
(1)	24.07	2.42	0.19	40.0	1.3	0.17	0.2	27.83	2.7	0.48	1.40*	0.097	1661	860	403
(2)	20.9	3.18	0.39	41.45	1.38	0.16	0.5	28.35	2.88	0.5	1.46	0.101	1443	641	173
(3)	29.24	3.31	2.11	34.56	1.7	0.23	0.6	19.7	1.98	1.76	1.75**	0.1	842	423	622
(4)	21.72	1.06	0.13	43.38	0.9	0.22	0.1	28.54	3.1	0.37	1.52	0.108	1145	785	350

Major elements in wt. %; Sr, Ba, Mn in ppm.
(1) Kodjari, surface rocks, 6 samples.
(2) Kodjari, drill core, carbonate-free, 4 samples.
(3) Kodjari, drill core, carbonate-bearing, 8 samples.
(4) Arli, 5 samples.
* Average of 8 samples strongly influenced by 2 samples at 1.50; other values are 1.25, 1.32, 1.36, 1.38, 1.43, 1.44.
** Value excessive due to presence of calcite; 2 samples without calcite yield ratios of 1.49 and 1.50.

Table 19.4. *Comparison of CaO/P_2O_5 and F/P_2O_5 ratios from various parts of the Upper Volta phosphorites (after Lucas & Prévôt, 1981)*

Origin and number of analyses		CaO/P_2O_5 Maximum value	CaO/P_2O_5 Minimum value	CaO/P_2O_5 Average	F/P_2O_5 Maximum value	F/P_2O_5 Minimum value	F/P_2O_5 Average
Arli							
Apatite cortex	(8)	1.38	1.24	1.33	0.164	0.115	0.149
Recrystallised cortex	(6)	1.50*	1.26	1.35	0.141	0.08	0.106
Pure nucleus	(3)	1.36	1.31	1.33	0.187	0.13	0.155
Cement	(1)			1.38			0.121
Phosphate grain	(3)	1.37	1.23	1.30	0.155	0.093	0 125
Kodjari							
Phosphate grain	(3)	1.44	1.43	1.44	0.153	0.133	0.146

* These values were obtained by microprobe analysis and F is probably an overestimate.

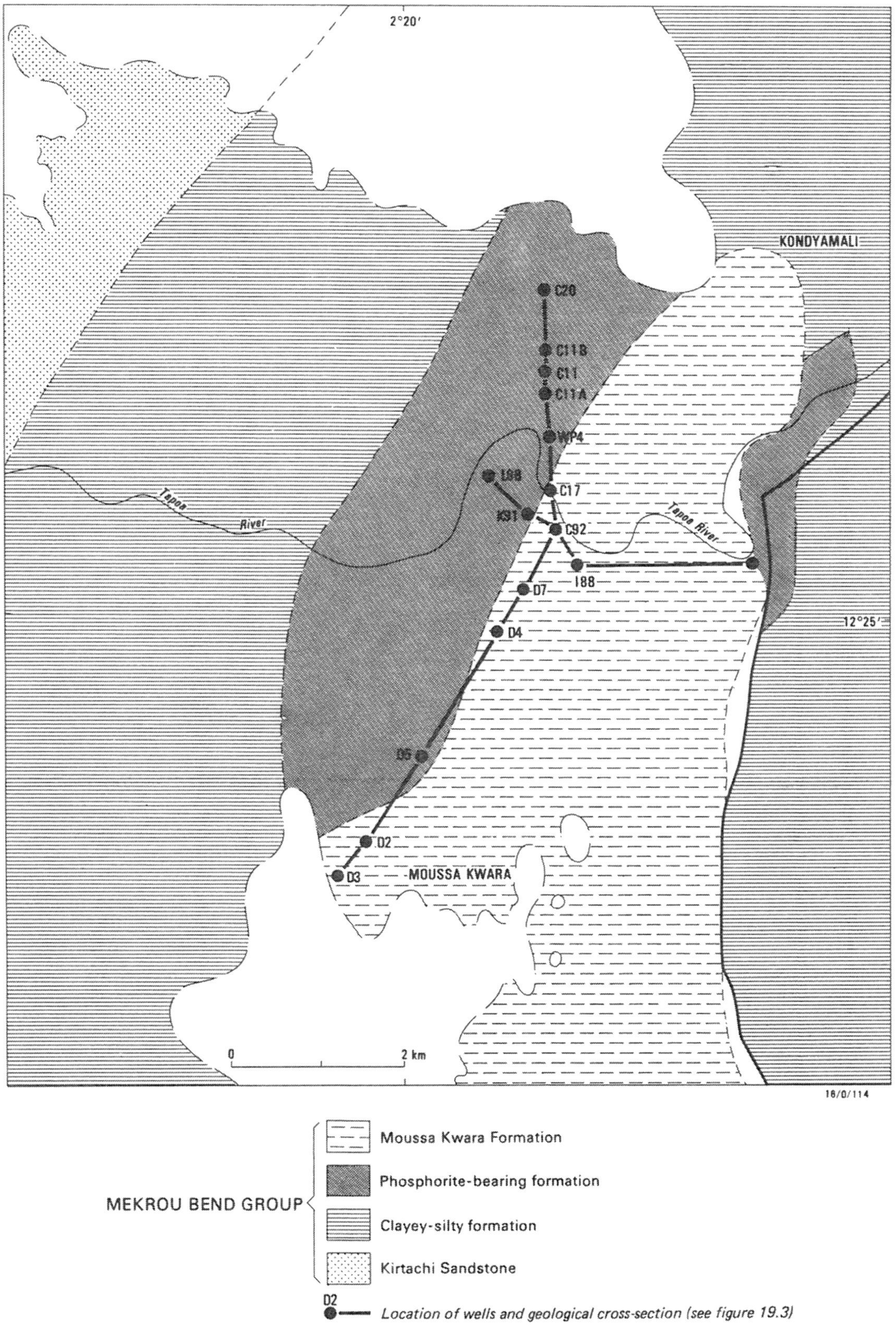

Fig. 19.2. Schematic map of the Tapoa region.

(1) At the base, a clayey silty interval with pyrite underlies the phosphatic sequence and consists of a series of black mudstones and siltstones alternating with sandstones and intraformational breccia. These thinly laminated, sometimes silicified sediments with horizons of calcareous-clayey shale, thin pyrite bands, and some laminae containing phosphatic grains, look like the basal part of the formation at Kodjari. These sediments contain about 3% of organic carbon (Ilyin *et al.*, 1979).

(2) The phosphate-bearing formation begins with medium-

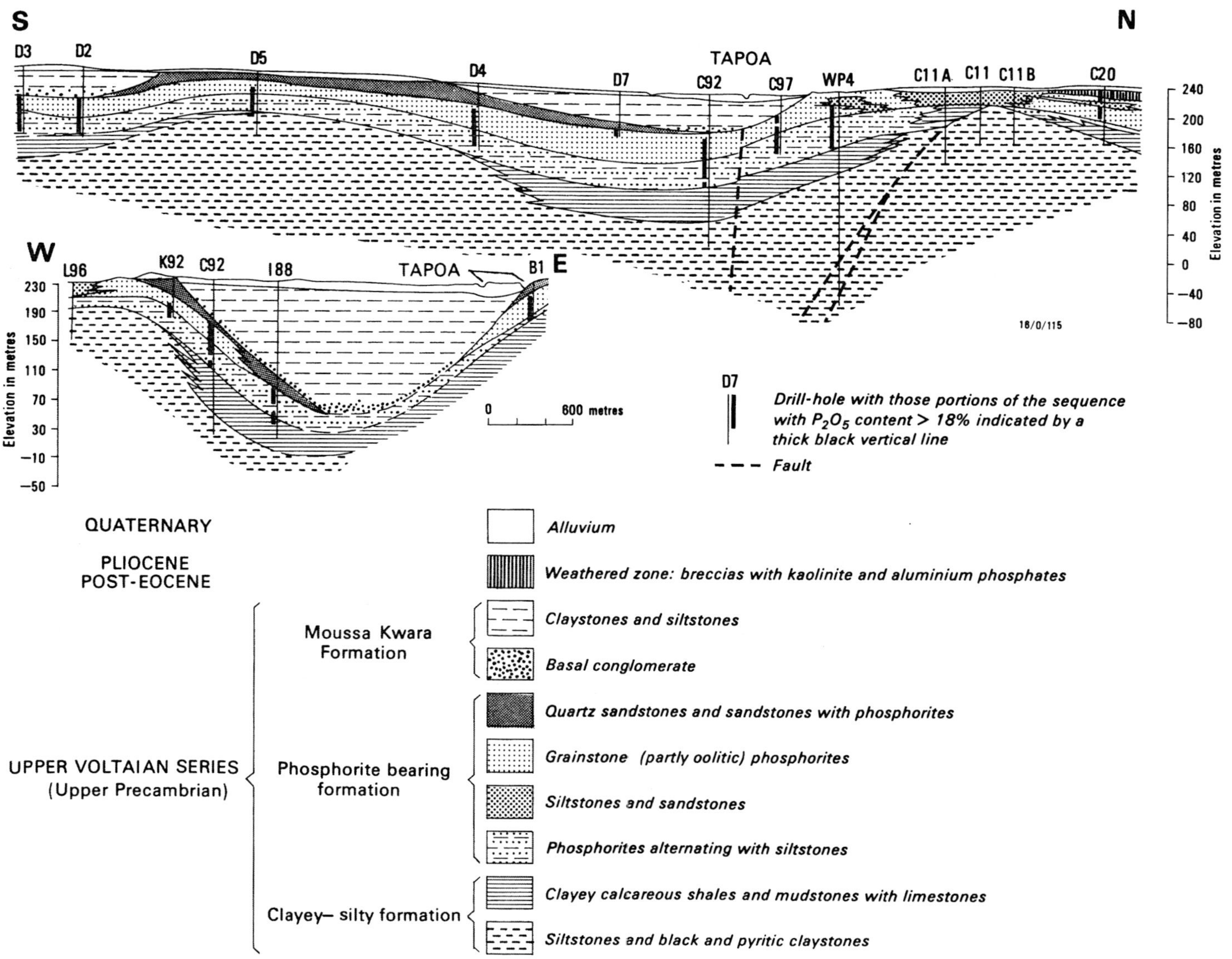

Fig. 19.3. Cross-sections through the Volta Basin at Tapoa.

to coarse-grained phosphatic cross-bedded sandstone (10–15% P_2O_5), with mudstone and siltstone interbeds. The phosphate grains in these beds are contact packed and consist of peloids and subordinate ooids. Phosphatic matrix makes up to 16% of the rock.

(3) The next unit consists of phosphorite and fine- to medium-grained phosphatic sandstone or sandy phosphorite (18–30% P_2O_5), alternating with mudstone and siltstone interbeds. These phosphorites exhibit poor sorting and the grains are peloid, ooids, ooid envelopes and lithoclasts of packstone phosphorite.

(4) The upper unit of the formation shows lateral changes of facies, with relatively uniform phosphorite (20–30% P_2O_5), and black and grey pyritic siltstone. Phosphorites are characterised by packstone and grainstone components of which about 50% are ooids, aggregates, black intraclasts; cross-bedding is poorly developed in places, laminae and lenticular bedding are present in other places.

(5) The Moussa Kwara Formation, which overlies the phosphate-bearing formation, consists of black shale, clayey siltstone and pyritiferous mudstone, and completes the Series.

Kühn (1980) has been able to determine from borehole information the stratigraphic relationships in the Tapoa deposits (Fig. 19.3).

Petrology

The Tapoa phosphorites can be grouped into two distinct varieties: a 'non-bedded' variety and a 'thin-bedded' variety.

The non-bedded to poorly bedded, massive, black phosphorite is granular in thin-section. Simple, complex and compound phosphate grains are present (Fig. 19.4A). The simple grains are structureless or rarely concentric. They vary in size from 0.07 mm to 0.15 mm, averaging about 0.10 mm, and are composed of yellow-green or brown phosphate. Compound phosphate grains are an aggregation of simple grains, and vary considerably in size and form (Fig. 19.4B). These are second generation grains composed of two or three simple grains. Even more complex third generation phosphate grains embracing first and second generation products also occur, the grain size increasing with complexity. The outer phosphatic rim of a complex grain is always lighter than the phosphate of a simple grain. There are two kinds of complex or compound phosphate

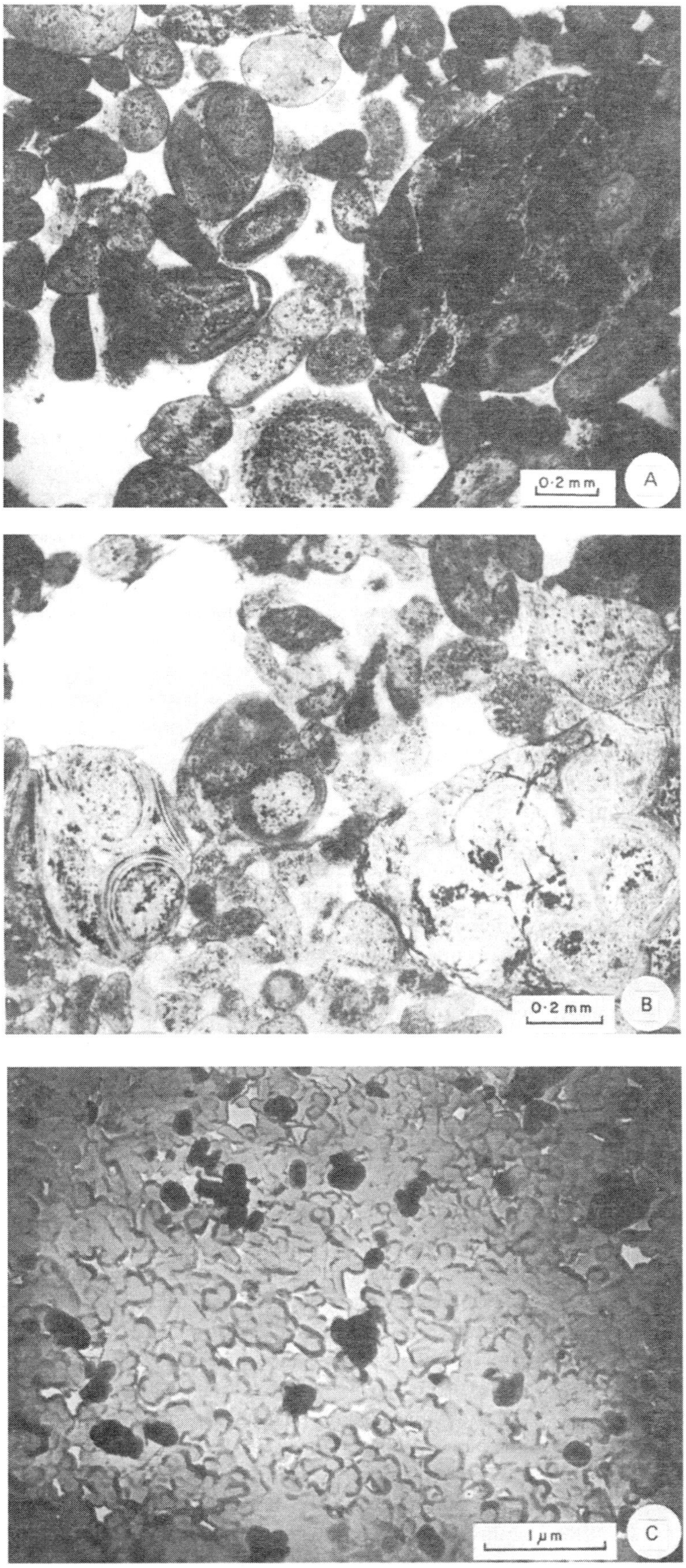

Fig. 19.4. Tapoa phosphorites. (A) Simple and compound phosphate grains in a clayey siliceous matrix. (B) Compound grains with abraded simple grains. (C) Electron micrographs of Tapoa phosphorite.

Table 19.5. *Major elements, organic carbon and uranium (%) analysis of Tapoa phosphorites (samples 1–4), and apatite concentrate (samples 5–6), all from boreholes*

Sample Number	P_2O_5	CaO	CO_2	F	SiO_2	MgO	Fe_2O_3	Al_2O_3	K_2O	Na_2O	C_{org}	SO_2	U	Σ1	O–F	Σ2
1	31.3	44.4	1.2	3.1	24.7	0.3	1.0	0.8	0.4	0.5	1.7	—	8×10^{-4}	99.6	1.23	98.37
2	27.1	42.8	1.8	2.7	20.1	0.1	1.3	1.3	0.8	0.8	1.4	—	7×10^{-4}	100.2	0.9	99.3
3	21.7	32.2	1.2	1.6	40.7	0.2	0.3	0.4	0.2	0.3	0.8	—	6×10^{-4}	99.6	0.85	98.75
4	8.6	12.1	3.4	0.4	61.1	1.7	3.2	5.6	3.4	0.7	3.1	—	4×10^{-3}	100.2	0.35	99.85
5	35.0	47.6	1.48	3.04	3.85	0.4	1.99	1.53	0.3	0.4	1.66	2.36	9×10^{-4}	99.61	1.21	98.41
6	35.9	48.8	1.84	3.23	2.8	0.3	2.39	1.15	0.2	0.6	1.4	2.09	9×10^{-4}	100.7	1.24	99.46

grains depending on the form of the simpler grains. In the first case, the simple grains preserve their primary rounded shape, and they are closely agglomerated inside an outer rim which is somewhat like a coating. In the second case the outer phosphatic rim appears to truncate simple grains which can be either rounded or angular. In the first case the compound grains were formed by *in situ* agglomeration, and in the second the simple phosphate grains were mixed with quartz debris and broken prior to agglomeration. Much larger elongated phosphate grains oriented parallel to each other are sometimes present. They are composed of structureless laminated phosphate. Some angular phosphate grains are also present.

The other main constituent of the rock is detrital quartz grains. These are small, well sorted and poorly rounded if found in the company of simple phosphate grains. Quartz grains are either equal in size to simple phosphate grains or a little smaller. As soon as larger complex phosphate grains are present, quartz grains became much larger and more rounded. In some instances, the quartz grains become so numerous that the phosphorite must be classified as a phosphatic sandstone.

Thin-bedded phosphorite is composed of alternating black phosphatic and grey non-phosphatic laminae. The phosphatic laminae comprise simple phosphate grains, 0.08–0.10 mm in size, admixed with finer quartz grains. Non-phosphatic laminae are either silty mudstone or shale.

When viewed under the electron microscope the phosphate was found to consist of small hexagonal apatite crystallites, 0.3 μm long and 0.1 μm across. No particular orientation of crystallites was observed, but rarely they show an obscure radial pattern (Fig. 19.4C). Phosphorite of the Tapoa deposit is siliceous (Table 19.5); carbonate is only a minor constituent, especially in the higher grade varieties.

The iron and alumina content are also low, while C_{org} is moderately high. Pyrite is a conspicuous component. (This can be of great value for direct application of the ground phosphorite to certain soil types). The phosphate mineral is carbonate fluorapatite with a high degree of substitution of SO_3 for PO_4. The CO_2/P_2O_5 ratio is 0.47–0.52, and F/P_2O_5 is 0.086–0.089. Cell dimensions (in nm) for samples 5 and 6 on Table 19.5 are as follows: (5) $a = 0.9354$, $c = 0.6889$, (6) $a = 0.9350$, $c = 0.6887$. The refraction index is 1.627.

Origin of the Volta Basin deposits

The Kodjari deposit, with its fine-grained, well-bedded material, and its calcareous and ferruginous argillaceous cement, is a primary deposit, which has undergone little reworking and formed in a shallow marine environment during the early part of a transgression (Trompette *et al.*, 1980). This possibly corresponded to a transitional transgression, possibly of tectonic origin. The Arli deposit and the upper part of the Kodjari deposit contain coarser material, with aggregates, broken phosphate grains and more abundant detrital material. They can be interpreted as reworked material from a pre-existing formation lying in very shallow and agitated water favourable for oncolite development. After deposition, these formations underwent diagenesis (dissolution, silicification, etc.) and weathering (aluminisation of apatite).

The interpretation of the depositional conditions for the Tapoa deposit is more difficult, because of its location in a tectonised area. Nevertheless, for one of the authors (J. Lucas) the morphological similarity of the phosphate grains with those of Arli suggests a similar reworking history, perhaps more important because of the slightly finer size of the phosphate grains and the greater abundance of detrital material. This suggestion is similar to that of Kühn (1980) who proposes a comprehensive model of formation on a continental shelf in very shallow marine conditions, with reworking and accumulation of phosphate particles within a barrier zone by the combined action of tidal currents and wave action (Fig. 19.5). The similarity of the formations underlying the phosphate-bearing formation at the Arli–Kodjari and Tapoa deposits supports a similar origin. However, the stratigraphic correlation between Kodjari and Tapoa is not certain, and A. Ilyin believes in a separate history for this deposit where some of the thin-sections show a predominance of structureless, equidimensional and densely-packed phosphate grains. These show no signs of transportation, but only a disturbance of a supposed primarily pure phosphatic sediment. He suggests that phosphate grains are formed by the granulation of phosphatic sediment *in situ*. Prior to lithification, the phosphatic sediment may have undergone considerable disturbance. In a nearshore environment clastic quartz grains were added to the phosphate grains, while further offshore the clastic admixture was less abundant. Quartz grains were mixed with phosphate grains resulting in a secondary sediment composed of phosphate and quartz grains. In the lower-energy conditions which followed, deeper water impinged upon the shelf environments and phosphate precipitated using as 'centres of crystallisation' the pre-existing simple phosphate

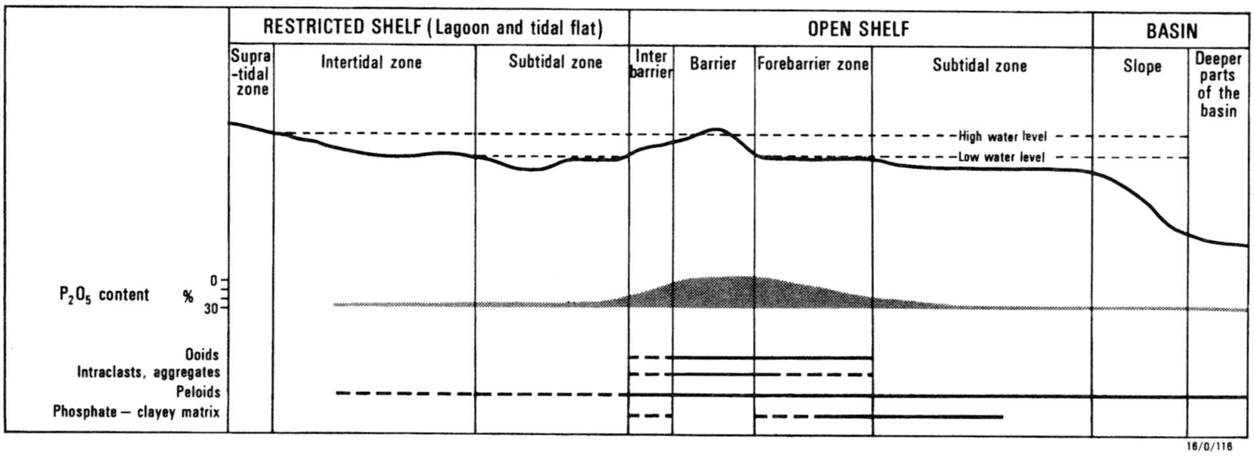

Fig. 19.5. Explanatory model for the Tapoa–Mekrou deposit with the various observed phosphatic subfacies and accumulations within the basin. (After Kühn, 1980.)

grains and quartz grains to produce more complex compound grains. In any event transportation and redeposition of phosphatic grains, and disturbance and fragmentation of the phosphatic sediment, played a major role in the formation of these grainstone phosphorites.

More general considerations on the origin of phosphorites have to be made in order to explain the formation of deposits at this geological epoch, and to compare them with other deposits. During the late Proterozoic, the Volta Basin probably had a connection from the north, through the Gourma zone, with the Taoudeni Basin. This is supported by the similarity of the Ydouban Series of the Gourma Basin with the clayey-silty Oti Series of the Volta Basin. In the present state of geological knowledge it may reasonably be assumed that by the end of the Proterozoic the basins which developed at the margins of, or partly within, the West African craton were interconnected during the early stages of their development. This is supported by a palaeogeographic reconstruction for the Late Proterozoic at the terminal period of the Pan-African Orogeny (Black, 1978). In this palaeogeographic scheme the Tapoa–Mekrou–Pendjari part of the Volta Basin is located within the equatorial zone.

In summary, the general palaeogeographical and genetic conditions required to produce these (and other) phosphorite deposits include a continental margin with very shallow water zones, an equatorial latitude, and development of biological activity producing an abundance of organic matter. These conditions are the same throughout geological time and space.

References

Affaton, P., Sougy, J. & Trompette, R. (1980). The tectono-stratigraphic relationships between the Upper Precambrian and Lower Paleozoic Volta Basin and the Pan-African Dahomeyides orogenic belt, West Africa. *American Journal of Science*, **280(3)**, 224–48.

Bessoles, B. & Trompette, R. (1980). Géologie de l'Afrique. La chaîne panafricaine 'zone mobile d'Afrique Centrale (partie sud) et zone mobile soudanaise'. *Bureau de Recherches Géologiques et Minières, Mémorie*, **92**, 397pp.

Black, R. (1978). Propos sur le Pan-Africain. *Bulletin de la Société Géologique de France*, **20**, 843–850.

Bozhko, N.A. (1969). Stratigraphy and tectonics of the Voltaian Basin. *Annales de la Faculté des Sciences, Université Clermont-Ferrand*, **41(19)**, 3–4.

Champetier, Y. & Joussemet, R. (1979). Découverte de nubéculaires et d'oncolithes en tant qu'éléments phosphatés. *Compte Rendu Académie des Sciences Paris, Série* D, **288(7)**, 673–5.

Clauer, N. (1976). Géochimie isotopique du strontium des milieux sédimentaires. Application à la géochronologie de la couverture du craton ouest-africain. *Sciences Géologiques, Strasbourg, Mémoire*, **45**, 256pp.

Grant, N.K. (1969). The late Precambrian to early Paleozoic pan-african orogeny in Ghana, Togo, Dahomey and Nigeria. *Geological Society of America, Bulletin*, **80**, 45–56.

Ilyin, A.V., Volkov, R.I. Vydzin, V.N. & Zabrodin, V.E. (1979). West African late Precambrian-Cambrian phosphate province. *Lithology and Mineral Resources*, **14(3)**, 344–57.

Kühn, A. (1980). Les phosphorites du Bassin des Volta République du Niger, Ph.D. Thesis, Institute of Geology, University of Warsaw (in Polish).

Kühn, A. & Zajaczkowski, W. (1975). Evaluation du gisement de phosphates, secteur Tapoa Nord. Rapport final: Niamey, Buremi, Ministry Mines and Hydraulics, 41pp. (unpublished).

Lucas, J. & Prévôt, L. (1981). Caractères pétrographiques et microchimiques de phosphorites précambriennes du Bassin des Volta (Afrique de l'Ouest). Considérations génétiques. *Bulletin de la Société Géologique de France*, [7] **23(5)**, 515–24.

Lucas, J., Prévôt, L. & Trompette, R. (1980). Petrology, mineralogy and geochemistry of the late Precambrian phosphate deposits of Upper Volta (W. Africa). *Journal of the Geological Society*, **137**, 787–92.

Mining Annual Review (1977). Mining Journal, London, p. 467.

Ouedraogo, Z.R. (1982). Contribution à l'étude des gisements de phosphates d'âge Precambrian du Sud-Est voltaïque Geologie – valorisation. Thèse Dr. 3è cycle. Institut National Polytechnique de Lorraine. 205pp. (unpublished).

Pascal, M. (1972). Réinterprétation de la géologie et examen des possibilités minières de l'ensemble Voltaïen-Buem-Atacorien du Togo, Dahomey, Haute-volta et Niger. *Bureau de Recherches Géologiques et Minières, Rapport*, **72 RME 014AF**, 77pp. (unpublished).

Trompette, R., Affaton, P., Joulia, F. & Marchand, I. (1980). Stratigraphic and structural controls of late Precambrian phosphate deposits of the northern Volta Basin in Upper Volta, Niger and Benin, West Africa. *Economic Geology*, **75**, 62–70.

20

Proterozoic and Cambrian phosphorites – specialist studies: chemical and mineral characteristics of some Cambrian and Precambrian phosphorites

G.H. McCLELLAN AND F.N. SAAVEDRA

Introduction

As recently as the early 1960s, significant deposits of Cambrian and Precambrian phosphorites were thought to be absent from the geological column (Geijer, 1962). However, Davidson (1963) and Bushinsky (1969) pointed out that this was not the case and indicated that a number of large and small deposits were known that could be assigned to these periods. In 1979, Cook & Shergold edited a collection of papers that covered a broad spectrum of deposits including many examples from the Cambrian and Precambrian. A number of factors combined to stimulate interest in studying these ancient phosphorites. Part of the stimulus was the need to develop indigenous phosphate resources to meet the food requirements of growing populations. Interestingly, this need became apparent at a time when geological skills and economic pressures were increasing. The result has been the discovery of large tonnages of phosphate resources at various geographic locations on different continents as well as an expansion of information on many previously known occurrences. Such disparate locations as Australia, Brazil, China, India, Mauritania, Mongolia, Pakistan, USSR, United States, and West Africa contain phosphorites of Cambrian and Precambrian ages that are of present or future commercial interest.

Cook & McElhinny (1979) estimate that world phosphate deposits in the Phanerozoic contain 3×10^{11} tonnes of P_2O_5. Of this total, about 10% occurs in Cambrian rocks. The data for Precambrian deposits are less clear, and only general orders of magnitude on their quantity have been estimated. Notholt (1979) compiled data showing 600×10^6 tonnes of P_2O_5 in Precambrian occurrences. While the total resources appear to be large, only a small proportion is commercially exploited at present. Unexploited deposits may lack sufficient tonnages, bed thicknesses, or rock quality to be considered economic. As countries continue to make inventories of their mineral resources, it is likely that additional discoveries of ancient phosphatic sediments will be made. Some comparatively recent examples of such discoveries are known in Brazil, Niger, Senegal, Upper Volta, and the United States.

Since 1965, considerable work has been done by the authors and their colleagues on the mineralogy, petrography, and chemistry of phosphorite samples from a large number of geographic locations. Emphasis has been placed on developing petrographic data to support mineral dressing investigations and chemical process treatments. One purpose of the petrographic studies was to develop textural descriptions that could be used to classify phosphate ores and to serve as a guide in the evaluation of mineral dressing processes. This is particularly helpful in identifying grinding conditions which would allow optimum liberation without overgrinding. To accomplish this, an effort has been made to develop quantitative petrographic procedures for comparing the prominent mineralogical and textural features of the various ores. A careful mineralogical study was made of the francolites (carbonate fluorapatites) which were the major, and usually the only significant, phosphate constituent of these ores.

Although considerable work has been done on the stratigraphy of a number of Cambrian and Precambrian phosphorites, few data have been published on the mineralogy, petrography, and chemistry of these ores. Some notable exceptions include the descriptions of Queensland deposits (Cook, 1972; Howard & Hough, 1979), the Indian deposits (Banerjee, Basu, & Srivastava, 1980), and the Upper Voltan deposits (Lucas, Prévôt, & Trompette, 1980; Trompette *et al.*, 1980).

We have reviewed much of the previously published information and present new data on samples of Cambrian and Precambrian phosphorites from 15 locations in Africa, Asia, Australia, North America, and South America. All of these samples are analysed and treated in the same manner in order to establish a limited data base for identifying differences and similarities in these ancient phosphorites. Methods used are a combination of optical and scanning electron microscopy (SEM), X-ray diffraction, and analytical chemical procedures that have been developed over more than a decade of such studies.

The principal phosphate mineral species in these samples will be described as francolites, where appropriate, rather than 'collophane'. Collophane is a textural description for cryptocrystalline carbonate apatite for which the fluorine is not known (McConnell, 1973). The *Glossary of Geology*, 2nd edition, (Bates & Jackson, 1980) contains a very imprecise definition of collophane but agrees that it is mainly a textural term for carbonate apatites. Francolites have been defined by Sandell, Hey, & McConnell (1939) as carbonate apatites that contain more than 1% F and appreciable amounts of CO_2 in which the carbonate for phosphate substitutions is balanced by an additional substitution. As the data will show, the fluorine requirement of the definition is easily fulfilled, but the CO_2 content of the apatites may be small. This interpretation will be supported by detailed X-ray diffraction and refractive index data. In one special case, infrared spectroscopy was required to clarify a particular point concerning the apatite in a sample. Energy dispersive X-ray analysis was routinely used in combination with the SEM to confirm details of the francolite and accessory mineral morphologies and textures.

X-ray diffraction data were collected using modifications of the procedures described in McClellan & Lehr (1969). The X-ray and chemical data were interpreted using the statistical methods described in McClellan & Clayton (1980) and related publications.

Sources of phosphate ore samples

Most of the samples studied have come to the authors in the normal course of work on the mineralogical and chemical characterisation of phosphorites. However, several samples were provided by interested parties who are recognised in the acknowledgement at the beginning of the volume. The study is based on 31 samples from 15 geographic locations (see Table 20.1). In some cases, data were collected for one of several

Table 20.1. *Mineralogical content (%) of Precambrian phosphorites*

	Apatite	Quartz	Hematite	Kaolinite	Goethite	Muscovite -sericite	Secondary phosphate	Calcite	Phlogopite
Australia, Rum Jungle									
Sample 1	14	79	1	2	2	1	1 (W, M, D)		
Sample 2[a]	61	16	10	9	2	1	1 (W, M, D)		
Sample 3[a]	70	10	7	8	1	3	1 (W, M, D)		
Sample 4	49	24	9	12	2	3	1 (W, M, D)		
US–Michigan									
Dead River[b,c]	30	58		4		6	TR (D)		
Huron River[b,d]	28	25			15	4		18	
US–Minnesota									
Sample 1[c]	33	58			3	5	TR (D)		
Sample 2[c]	41	52			1	4	TR (D)		
India									
Sample 1	69	8			1	1		20	1
Sample 2	90	5			2	1			2
Sample 3[e]	40	7			TR	TR			2
Sample 4	48	8			1	1		40	2
India–Madhya Pradesh									
Sample 6	51	45			1	3			
Sample 7	84	15			1	TR	1		
Brazil									
Patos de Minas[f]	60	15		1	2	16	3 (W, C)		
Lagamar[a,g]	66	13		5	3	9			
United States									
Tennessee[h]	73	10			1	1	2 (L, C, R)		
Upper Volta									
Pit 1[i]	65	24		1	2	6	1 (W)		
Pit 2[i]	43	40		1	3	11	1 (W)		
Niger, Parc W									
Sample A	65	30			2		2 (V)		
Sample B	70	25			2		2 (V)		

a. Trace fluorite.
b. 1% graphite.
c. 1% pyrite.
d. 10% barite.
e. 50% dolomite.
f. 4% montmorillonite.
g. Trace psilomelane.
h. 13% albite.
i. 1% glauconite.

TR, trace amount (less than 1%); W, wavellite; M, millisite; D, dufrenite; C, crandallite; L, lacroixite; R, rocherite; V, variscite.

samples from a location that was considered 'typical'. While there is a certain hazard associated with this approach to sample collection, it does permit persons familiar with a deposit to select the sample as being typical. The sample locations are presented in the episodic sequence given in Table 1 of Cook & McElhinny (1979).

Rum Jungle, Northern Territory, Australia

Preliminary work on these deposits began in the early 1960s. Phosphate has been found in 15 or more deposits which are about 65 km south of Darwin. These deposits are associated with hematitic silt and sandstones and a hematitic quartz breccia in intensely folded and shattered Lower Proterozoic sediments (Pritchard & Cook, 1965). These deposits may contain 4 million tonnes of phosphate rock averaging 10% P_2O_5. The high Fe_2O_3 and Al_2O_3 content makes this rock unsuitable for conversion to chemical fertilizers. Data are reported on four samples of ore from this deposit.

Huron River and Dead River Basin, Michigan, United States of America

This deposit consists of phosphatic pebbles near the base of the metasedimentary Marquette Range Supergroup in the central part of the Upper Peninsula of Michigan (Cannon & Klasner, 1976). These deposits are Early Proterozoic (Precambrian X) in age. Overall, the phosphates are a minor component of the rock; however, at one locality a conglomeratic bed 15 m thick averaging about 15% P_2O_5 occurs. The resources of this deposit have not been estimated because of inadequate information, but they are likely to be large. Two similar phosphate occurrences were reported earlier from this area by Manusco, Longheed, & Shaw (1975). They describe a carbonate apatite that occurs in a lean carbonate iron formation near the base of the Baraga Group in the Baraga Basin. Data are presented on a sample from the Huron River and a sample from Dead River Basin (Cannon & Klasner's Section 15 locality). Brown, Rule, & Dallenbach (1981) have presented some detailed data on the mineralogy and beneficiation.

Carleton Country, east-central Minnesota, United States of America

In November 1978, the discovery of an Early Proterozoic phosphate occurrence was made during a routine exploration for uranium in Carleton County, Minnesota (Ullmer, 1981). The phosphatic units and surrounding rocks are part of the Thomson Formation (Animikie Group) consisting of metasedimentary clastics and iron formation which occur in eastern Minnesota, northern Wisconsin, and Michigan's Upper Peninsula. The phosphate occurs in quartzite units associated with graphitic phyllites. The phosphate units pinch and swell and vary from a thickness of 2–4 m to absence over small areas. The francolite content of the rock may be 50% (about 20% P_2O_5). Although no quantitative estimates were made for this deposit, it is likely that the resources will be large. Data are presented for two samples from Carleton County.

Jhamarkotra, Udaipur District, south-central Rajasthan, India

These stromatolitic phosphorites are the major indigenous deposit presently exploited in India. The deposit contains more than 50 million tonnes of rock including 17 million tonnes of ore containing more than 30% P_2O_5 and thus may contain half of the total known reserves of India. A production of 0.5 million tonnes per year of rock represents up to 80% of the indigenous phosphate produced (Choudhuri, 1979). The phosphate occurs in the Aravalli rocks of Early to Middle Proterozoic age (Sant, 1979) in a typical upwelling suite of rocks – chert, black shale, and phosphorite. The rocks have been altered by regional metamorphism to the greenschist or amphibolite facies, so the upwelling suite has been replaced by their metamorphic equivalents, marble, phyllite, and quartzite (Sant, 1980; Pant, 1980). There are several different types of ore in this deposit. Data will be presented for five types including brecciated stromatolitic ore, columnar stromatolitic ore, Block D ore, Block G closely spaced stromatolitic ore, and massive siliceous ore from Block G.

Jhabua, Madhya Pradesh, India

This deposit was discovered in 1973 and represents a southern extension of the main Early to Middle Proterozoic Aravalli belt from Rajasthan into western Madhya Pradesh (Khan, 1979). The rocks are folded and metamorphosed Precambrian dolomitic limestones and quartzites. There are two distinct phosphate horizons in this area. One is an older carbonate type similar to the Jhamarkotra deposit which contains stromatolites. The younger horizon is siliceous and also contains stromatolites. These deposits contain about 7.5 million tonnes of reserves and resources that average about 15% P_2O_5. Data are presented for both the stromatolitic and massive ore types.

Patos de Minas/Lagamar, Minas Gerais, Brazil

The Bambuí Group is reported to be of Late Precambrian or Early Cambrian age and contains phosphate in the Indaia Member of the Paraopeba Formation in Minas Gerais (Cathcart, 1978). It may, however, be older (see Dardenne *et al.*, this volume). The phosphorite is laminated and isoclinally folded by slumping between two competent beds. The phosphate occurs in thickened areas of the folds and is usually bounded by faults. The resources of this deposit are at least 500 million tonnes with an average grade of 11% P_2O_5. Cook & McElhinny (1979) give this deposit an older age than Cathcart (1978). Most of the ore will require beneficiation prior to use in chemical fertilizer although a small tonnage of high-grade ore (more than 32% P_2O_5) is known. Some limited exploitation started at Lagamar in 1980 (0.2 million tonnes per year). Data are presented for typical samples from the Patos de Minas and Lagamar locations.

East Fork District, Sevier County, Tennessee, United States of America

Phosphorite has been reported in the Ocoee Series (Late Proterozoic II) of the East Fork manganese district, Sevier County, Tennessee, since 1946 (Hamilton, 1961; Wedow, Carpenter, & Lehr, 1968; Brower, 1973). The local stratigraphic unit containing the phosphate is part of the Wilhite Formation. The rock types include a lower unit of silty shale, with beds of limestone and dolomite overlain by a massive manganese-bearing dolomite and carbonaceous shale, and an upper unit of coarse-grained phosphatic arkose and associated fine-grained clastics. The phosphorite, containing 25%–35% P_2O_5, occurs at the base of the upper unit in a bed up to 1.5 m thick and 400 m along the apparent strike.

Fluorapatite occurs as crystalline aggregates in, and adjacent to, veins in the phosphorite and probably formed by alteration of the cryptocrystalline francolite. No resource estimates are available for this deposit, but phosphates have been reported in an intraformational conglomerate in the Wilhite Formation up to 120 km from the principal study area. This suggests that the deposition of phosphate may have been widespread in this area, and the total phosphate resources may be large. This deposit is an example of the possible association of manganese and phosphate in ancient deposits. Data are reported for one example of high-grade rock.

Kodjari, Upper Volta

This deposit is part of the recently discovered phosphates in the north-western part of the Volta Basin (Trompette, *et al.*, 1980; Lucas, *et al.*, 1980). The geochronological date of the phosphate is 660 ± 8 m.y., indicating that they are Late Proterozoic in age. The lithological sequence begins with tillite, limestone, mudstone, and chert associated with the phosphorite, which are overlain by sandstone and shales (up to 2000 m thick) and continental clastics. The main phosphate bed is up to 15 m thick and contains 20%–25% P_2O_5. The deposit is reported to contain at least 60 million tonnes of phosphate rock. Present production is about 10 000 tonnes per year which is crushed for use as direct application. Data are presented from two locations (pit 2 and pit 3) in the area being mined.

Parc W, Tapoa area, Niger

These deposits occur on the northern perimeters of the Volta Basin and are the largest known in the region. The phosphate unit contains 18%–35% P_2O_5 and is 35–40 m thick. Estimated resources are at least 250 million tonnes (Ilyin, 1978; Kühn & Radlicz, 1979). The phosphate occurs in a well-bedded siltstone, with a fine to very fine texture. These rocks can be correlated in part with the Upper Proterozoic Kodjari Formation and perhaps with the basal portions of the overlying Pendjari Formation. In general, the phosphate in these rocks is reported to be slightly coarser in particle size than the phosphate at Kodjari, Upper Volta. Data are reported on two samples from the Parc W deposit.

Namel, Senegal

These phosphorites are located on the western side of the West African craton. The phosphatic sequence is considered to have an Infracambrian age (650 m.y.) according to Slansky (1979), making it Late Proterozoic II in the scheme of Cook & McElhinny (1979). The rock sequence above the Precambrian basement (Birrimian) comprises 3 m of tillite, 18 m of fine-grained siliceous rocks with some calcite and dolomite (pelite), 9 m of pelite with phosphate ranging from 1.5% P_2O_5 at the base to 8% P_2O_5 at the top, 2 m of calcareous phosphate (15% P_2O_5), 8 m of phospharenite (22%–32% P_2O_5), and 6 m of coarse surficial dolomite blocks and laterite (Slansky, personal communication). Lateral variations of the formation combined with erosion, weathering, faults, and doleritic sills limit the resources which have not been estimated. Data are presented on a surface sample and a subsurface sample (8 m depth) of the phosphorite.

Nouedgui-Bou Naga area, D'Ijibiten region, Mauritania

These samples are from the north-western sector of the Mauritanides along the West African craton south-west of the town at Iriji (Lecorche, 1980). These phosphatic sands occur near an angular disconformity thought to be caused by glaciation (Trompette, 1973) in the uppermost Proterozoic (Infracambrian). These late Precambrian phosphorites belong to Super-Group 2 of Atar, which transgressively overlies the basement. The usual sequence includes a basal conglomerate followed by sandstone and limestone–phosphorite. The phosphate occurs as a slightly metamorphosed, fine-grained, ferruginous sand with some sericite and detrital mica. These deposits are not well known, so their reserve/resource potential has not been defined although their areal distribution is large.

Georgina Basin, Queensland, Australia

Since its discovery in 1966, this has been perhaps the most thoroughly studied Cambrian phosphate deposit. Studies by the Australian Bureau of Mineral Resources and Mines Exploration Proprietary Limited, a subsidiary of BH South Ltd. now part of the Western Mining Corporation, showed that phosphatic sediments of regional significance occurred in Middle Cambrian formations (see Rogers & Crase, 1980 for a detailed review of the geology and literature). The rock suite containing the phosphate is the usual siltstone–chert–carbonate sequence. The only commercial exploitation of this deposit was at the Duchess deposit where from 1976 to 1978 the high-grade part of the Monastery Creek Phosphorite Member of the Beetle Creek Formation was mined as a direct shipping grade ore. The ore is predominantly grainstone phosphorite with minor laminae of mudstone phosphorite. Mining at the site was scheduled to resume during 1981. Reserves and resources in the Duchess–Ardmore area are greater than 1000 million tonnes; total Georgina Basin resources are more than twice that amount. Rock with a grade of greater than 31% P_2O_5, so-called direct shipping grade, is about 20 million tonnes. Data are reported for samples of both the siliceous and calcareous Beetle Creek Formation phosphate.

Mirpur/Kakul area, Hazara District, Pakistan

These deposits are located in the northern region of Pakistan at the foot of the Himalaya. The phosphorites are associated with dolomite and chert in the top 100 m of the Abbottabad Formation (Bhatti, 1979). Fossil evidence indicates that the phosphate beds are of Cambrian age and may be related to Russian and Chinese deposits. The phosphate horizons are in a complex structural setting and vary in thickness from a few millimetres to a few metres. Reserves for this deposit are 3–5 million tonnes with up to 20 million tonnes of resources known in the area. The phosphorites have a variable composition and appearance but are usually dark coloured and granular. The grains are francolites (up to 5 mm) and occasionally show secondary overgrowths. There is some commercial exploitation of these deposits, with plans for expansion in the near future. Data are presented for a sample of the siliceous ore.

Burin Khan, Khubsugul District, Mongolian People's Republic

Since its discovery in 1963, the phosphorites of the Khubsugul Basin have been extensively studied. In 1967, phosphorites were traced to a promising new area north of the city of Muren on the south side of the basin. Although the correlations are not precise, the Burin Khan deposits are thought to have an Early Cambrian age, the same as those of the main Khubsugul deposit (Ilyin & Bjamba, 1980). Phosphorite horizons at Burin Khan are siliceous and occur in black cherts. The middle units have the most potential with seven phosphorite horizons attaining a thickness of 250 m; the thickest single horizon is 55 m. The phosphorites are rich at some localities (28%–30% P_2O_5) and resources are large, at least several hundred million tonnes. Data are presented for a sample of the ore that will be mined in the future.

Kaiyang area, Guizhou Province, People's Republic of China

The three phosphorite beds in this district represent portions of the Doushantuo Formation (690 m.y.) of Late Proterozoic (Late Sinian) age. The sedimentary sequence is described as consisting of dolomite, phosphorite, limestone, and siliceous rock (Liang & Chang, 1981). These ores vary in texture from microcrystalline phosphorite to clastic phosphorite to a banded or striped dolomitic phosphorite. Littoral sedimentary structures are common, and phosphatic stromatolites occur locally. The beds vary from 0 m to 10 m in thickness. Reserves are thought to be large but are not defined at this time. Two mines are reported to be operating in the area at Kaiyang and Fuchan. The grade of the ore is reported to be quite variable, but some ores contain up to 35% P_2O_5.

Mineralogy/petrography

The mineralogy of the phosphorites is rather simple in that only a few species are required to describe the bulk phases present (Tables 20.1 and 20.2). In these samples nearly all of the apatites are either francolites with low degrees of carbonate-for-phosphate substitution or essentially fluorapatites (Tables 20.3 and 20.4). The unit-cell dimensions and mean refractive indices showed a limited range of variation compared with the general data collected on francolites in the past (McClellan, 1980). In addition to phosphate, quartz is a ubiquitous constituent in all of the sediments. After these major components, it becomes difficult to generalise, for individual rocks may contain varying amounts of iron-minerals, micas, clays, carbonates, and a number of minor constituents (5% or less by weight). The clay minerals in these samples are also limited in types and quantities. In general the muscovite–sericite–illite series is the most common, particularly in rocks showing little or no weathering.

Table 20.2. *Mineralogical content (%) of Cambrian phosphorites*

	Apatite	Quartz	Goethite	Muscovite-sericite	Calcite	Fluorite	Heavy minerals
Senegal – Namel							
surface[a,b,c]	59	26	1				
8-Metre depth[b,d,e]	59	9	1	TR	15		
Mauritania							
505[f,g]	22	57	6	11		TR	2 (H, T)
506[f]	20	55	6	16		1	2 (H, T)
522[f]	26	59	4	8		1	2 (H, T)
Australia – Queensland							
calcareous	24	4	1	1	68	1	
siliceous	68	27	1	2		2	
Pakistan							
Hazara[i]	71	22	3	3			TR (B, R)
Mongolia							
Burin Khan	90	6	TR	1	2	TR	
China							
Guizhou[j]	54	2	TR			TR	TR

a. 1% clintonite.
b. Trace psilomelane.
c. 10% talc, 2% tremolite, 1% chlorite.
d. 1% phlogopite, 4% albite.
e. 6% talc, 5% chlorite.
f. Trace hematite, pyrite.
g. Trace graphite.
h. 1% glauconite.
i. 1% dufrenite.
j. 44% dolomite.

CL, clintonite; H, hornblende; T, tourmaline; B, biotite; R, rutile, Z, zircon; TR, trace.

Table 20.3. *Chemical and crystallographic data for Precambrian samples*

	P_2O_5	CaO	Na_2O	MgO	CO_2	F (wt %)	Al_2O_3	Fe_2O_3	SiO_2	K_2O	Total S	Loss on ignition
Australia, Rum Jungle												
Sample A	5.8	8.4	0.02	<0.01	0.16	0.53	0.43	1.3	78.7	0.02	0.01	2.7
Sample B	25.7	33.3	0.14	0.06	0.68	2.1	4.7	8.7	17.3	0.08	0.03	7.3
Sample C	29.5	39.1	0.14	0.09	0.49	2.6	4.9	5.9	11.4	0.55	0.04	5.1
Sample D	20.7	25.8	0.14	0.08	0.63	1.6	6.1	7.7	35.2	0.53	0.02	4.8
United States, Michigan												
Huron River	11.1	24.2	0.04	0.13	8.0	1.0	1.4	9.3	25.2	0.27	0.64	7.2
Dead River Basin (Section 15)	12.6	16.6	0.04	0.11	0.32	1.1	1.6	2.8	58.3	0.46	0.06	3.5
United States, Minnesota												
Sample A	13.4	17.5	0.04	0.05	0.20	1.1	1.0	2.0	59.4	0.33	0.23	3.0
Sample B	16.9	22.4	0.03	0.05	0.28	1.4	0.79	0.69	53.0	0.23	0.02	1.6
India, Rajasthan												
Sample 1	32.2	50.6	0.01	0.15	6.0	2.9	0.13	0.45	9.2	0.01	0.01	7.7
Sample 2	37.3	50.6	0.07	0.33	1.5	3.6	0.16	0.30	4.8	0.07	<0.01	7.0
Sample 3	16.2	42.5	0.05	12.7	26.1	1.5	0.12	0.14	2.1	0.07	0.01	27.0
Sample 4	29.4	51.7	0.02	0.22	9.9	1.8	0.40	0.46	5.9	0.02	0.02	12.0
India–Madhya Pradesh, Jhabua												
Sample 6	20.4	27.4	0.04	0.35	0.42	2.2	0.66	0.30	46.7	0.12	<0.01	1.2
Sample 7	34.7	46.2	0.41	0.16	1.0	3.7	0.04	0.18	14.1	0.01	<0.01	5.2
Brazil												
Patos de Minas	18.1	24.1	0.08	0.88	0.65	1.6	8.4	3.5	29.5	1.9	0.30	8.5
Lagamar	27.4	39.6	0.17	0.37	1.3	2.6	4.6	2.4	19.2	0.84	0.02	6.2
United States												
Tennessee, Sevier County	31.6	42.2	0.65	0.03	0.81	2.9	2.1	0.48	19.2	0.02	0.04	4.0
Upper Volta, Kodjari												
Pit 2	26.3	36.1	0.15	0.40	1.2	2.6	2.0	1.6	26.0	0.13	0.03	5.4
Pit 3	17.0	23.2	0.10	0.13	0.78	1.7	2.2	2.8	47.1	0.22	0.03	4.6
Niger, Parc W												
Sample A	35.2	49.0	0.20	0.13	2.1	3.6	1.6	0.81	4.7	0.09	0.14	5.1
Sample B	29.0	40.6	0.28	0.21	1.1	2.7	1.6	9.4	4.1	0.08	0.33	9.4
Tapoa A	28.8	37.0	0.21	0.04	1.0	2.4	1.3	0.63	24.1	0.04	0.04	4.8
Tapoa B	29.6	39.5	0.21	0.03	1.0	2.4	1.8	1.4	25.0	0.02	0.03	3.1

	SrO	ZnO	CaO	NiO	MnO (parts per million)	Cr_2O_3	TiO_2	BaO	Unit cell *a* (nm)	Unit cell *c* (nm)	Mean Refractive index
Australia, Rum Jungle											
Sample A	<50	120	60	2700	<30	210	1200	350	0.9377	0.6889	1.631
Sample B	105	140	<20	300	370	840	7500	340	0.9375	0.6887	1.631
Sample C	<50	210	<20	780	50	590	7200	340	0.9373	0.6889	1.631
Sample D	80	190	30	1300	110	1500	8100	280	0.9378	0.6889	1.631
United States, Michigan											
Huron River	840	150	45	1100	10 000+	520	960	68 000	0.9374	0.6886	1.638
Dead River Basin (Section 15)	180	120	50	3400	100	440	1400	340	0.9372	0.6883	1.638
United States, Minnesota											
Sample A	310	140	50	3700	100	880	850	370	0.9383	0.6889	1.638
Sample B	810	130	50	3700	<30	1500	440	350	0.9385	0.6889	1.638
India, Rajasthan											
Sample 1	300	100	<20	840	480	80	100	430	0.9372	0.6891	1.631
Sample 2	150	100	50	460	450	<30	100	460	0.9365	0.6889	1.632
Sample 3	450	110	40	270	440	<30	100	360	0.9363	0.6890	1.632
Sample 4	310	90	50	90	135	<30	100	590	0.9372	0.6891	1.631

Table 20.3 (*Contd.*)

	SrO	ZnO	CaO	NiO	MnO	Cr_2O_3	TiO_2	BaO	Unit cell *a* (nm)	Unit cell *c* (nm)	Mean Refractive index
				(parts per million)							
India–Madhya Pradesh, Jhabua											
Sample 6	230	120	40	2400	140	160	400	370	0.937	0.6883	1.630
Sample 7	900	130	30	1100	120	200	90	350	0.937	0.6888	1.630
Brazil											
Patos de Minas	550	250	150	920	100	<30	330	370	0.9371	0.689	1.626
Lagamar	1100	<50	40	360	470	40	3600	180	0.9366	0.6892	1.624
United States											
Tennessee, Sevier County	1600	250	40	1600	80	300	830	280	0.9366	0.6888	1.634
Upper Volta, Kodjari											
Pit 2	960	120	40	450	330	30	1200	470	0.9355	0.689	1.622
Pit 3	540	120	30	450	280	<30	2100	530	0.9354	0.6889	1.622
Niger, Parc W											
Sample A	1500	140	<20	2700	220	50	910	540	0.9364	0.6889	1.624
Sample B	2100	130	<20	1200	320	340	920	2400	0.9364	0.6889	1.624
Tapoa A	1400	80	30	80	530	<30	1100	340	0.9357	0.6891	1.627
Tapoa B	260	340	<20	310	530	470	2000	720	0.9368	0.6885	1.634

Table 20.4. *Chemical and crystallographic data for Cambrian samples*

	P_2O_5	CaO	Na_2O	MgO	CO_2	F	Al_2O_3	Fe_2O_3	SiO_2	K_2O	Total S	Loss on ignition
					(wt %)							
Senegal, Namel												
surface	24.2	34.0	0.01	1.9	0.45	2.0	0.79	0.82	29.8	0.04	0.01	3.5
8-metre depth	19.0	30.4	0.09	1.1	4.8	1.9	0.39	0.74	40.0	0.05	0.06	7.7
Mauritania, Nouedgui												
Bou Naga–505	8.8	12.7	0.07	0.05	0.30	0.7	4.4	4.3	68.3	0.7	0.02	2.4
506	7.5	10.6	0.15	0.22	0.29	0.7	6.8	4.2	66.7	1.5	0.02	1.7
522	10.0	13.9	0.07	0.19	0.69	0.9	3.1	3.3	69.0	0.61	0.02	1.2
Australia, Queensland												
calcareous	9.7	52.0	0.16	0.19	31.2	1.0	0.31	0.31	4.2	0.10	0.12	35.0
siliceous	27.0	37.4	0.4	0.13	0.93	2.7	0.66	0.41	26.5	0.05	0.25	5.9
Pakistan, Hazara area	27.8	38.6	0.23	0.13	1.8	2.7	1.1	1.8	23.0	0.28	0.13	3.8
Mongolia, Burin Khan	38.3	52.7	0.54	0.11	1.7	3.0	0.12	0.28	3.2	0.02	0.06	4.9
China, Guizhou region	22.0	47.9	0.2	9.8	20.7	2.8	0.23	0.20	1.0	0.05	0.14	21.4

	SrO	ZnO	CuO	NiO	MnO	Cr_2O_3	TiO_2	BaO	Unit cell *a* (nm)	Unit cell *c* (nm)	Mean refractive index
				(parts per million)							
Senegal, Namel											
surface	1500	80	<20	2200	380	330	530	2000	0.9373	0.6891	1.630
8-metre depth	2500	90	<20	3200	200	<30	330	8800	0.9360	0.6893	1.627
Mauritania, Nouedgui											
Bou Naga–505	995	130	30	50	290	150	3600	390	0.9382	0.6882	1.630
506	620	160	20	60	260	140	7200	560	0.9370	0.6880	1.630
522	540	90	20	70	440	180	1900	510	0.9370	0.6880	1.630
Australia, Queensland											
calcareous	470	130	40	220	340	<30	210	460	0.9356	0.6884	1.628
siliceous	540	260	180	920	80	<30	330	370	0.9359	0.6889	1.628
Pakistan, Hazara area	670	120	40	1400	290	160	620	440	0.9360	0.6889	1.627
Mongolia, Burin Khan	1900	160	70	400	30	70	200	790	0.9372	0.6888	1.629
China, Guizhou region	620	160	30	40	250	80	170	390	0.9362	0.6888	1.626

The other clays, particularly kaolinite, are found in the weathered samples. The distribution of secondary phosphate minerals is also affected by weathering.

The textures of the samples are quite variable. Some retain very characteristic features of their sedimentary origins (bedding and fossils among others), while the others show various degrees of alteration to metasediments with foliar textures and low-grade metamorphic mineralogies (talc-tremolite for example). However, there are few generalities that can apply to samples from such diverse locations. The main emphasis has been on a systematic study in which the same techniques are applied to all the samples. The following individual descriptions will provide details of materials from the various locations.

Rum Jungle, Australia

These samples are weathered portions of a metasedimentary, siliceous phosphorite (Table 20.1). Metamorphism has caused a recrystallisation of the phosphate to fluorapatite and the development of specular hematite (Fig. 20.1). Much of the recrystallised apatite occurs as short prismatic crystals (Fig. 20.2), although some occurs as fibrous vein fillings and goethite-stained, subrounded masses that resemble distorted francolite grains. The apatite crystals show zonal growth bands that contain scalar hematite which sometimes is concentrated along the banding. Occasionally the vein fillings of apatite are coarsely crystalline and contain occluded plates of specular hematite. Some veins are filled with quartz and are rimmed by apatite crystallites which range in size from 20 μm to 100 μm. They are randomly oriented, and their interpenetrating growth appears to be the primary cementation in the rock.

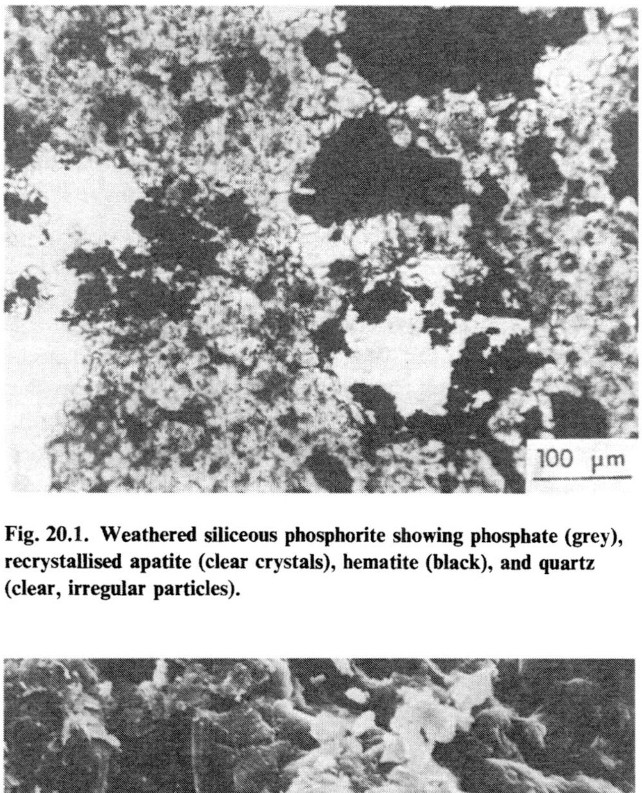

Fig. 20.1. Weathered siliceous phosphorite showing phosphate (grey), recrystallised apatite (clear crystals), hematite (black), and quartz (clear, irregular particles).

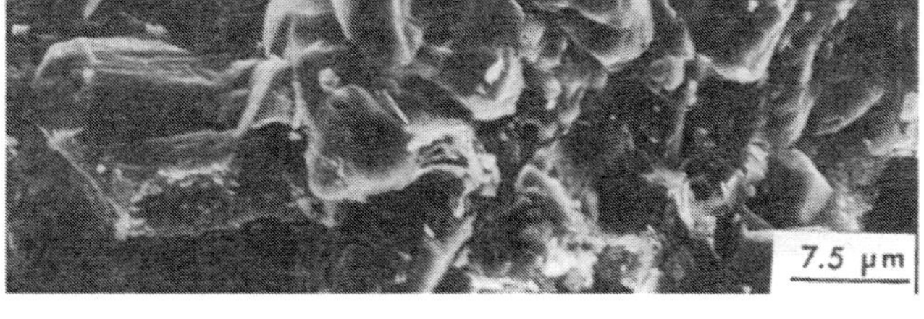

Fig. 20.2. Subhedral recrystallised apatite crystals.

Fig. 20.3. Hematite rosettes and tabular apatite crystals (right).

Hematite occurs in intimate association with the apatite as inclusions, filling the interstices of phosphate masses, and as plates, scales, and rosette aggregates (Fig. 20.3). Despite its varied occurrences, the hematite does not appear to contribute significantly to the cementation of the rock.

Authigenic quartz is found as veins, lenses, and discrete particles generally with inclusions of apatite crystals, hematite, and possibly fluorite. Some patchy silica masses show wavy extinction. Detrital quartz is also present as skeletal grains that have been severely pitted by leaching.

Kaolinite occurs as void and fracture fillings and sometimes is strongly stained by goethite. It may occur as an interstitial phase of curled scales or flakes between apatite grains. In some cases, the kaolinite ground mass occludes apatite crystallites.

Secondary phosphates (wavellite, millisite, dufrenite, and perhaps bassetite) occur as void fillings and often are intimately mixed or coated with goethite. These alteration products are the result of weathering.

Some samples of this phosphorite are formed of fragments of a phosphatic quartzite breccia. Angular quartzite clasts and some phosphate fragments are cemented by a matrix of sugary quartz, kaolinite, millisite, and goethite.

All the apatites in these four samples are essentially flourapatite in composition ($F/P_2O_5 = 0.084$) with little or no carbonate substitution for the phosphate. The unit-cell data ($a = 0.9373$–0.9378 nm, $c = 0.6887$–0.6889 nm) and mean refractive index ($N_d^{25} = 1.631$) support this interpretation.

Dead River Basin, Michigan

This rock consists of apatite plus quartz, chert, and sericite with other accessories (Table 20.1). It is a metasedimentary conglomerate with a poorly developed lepidoblas-

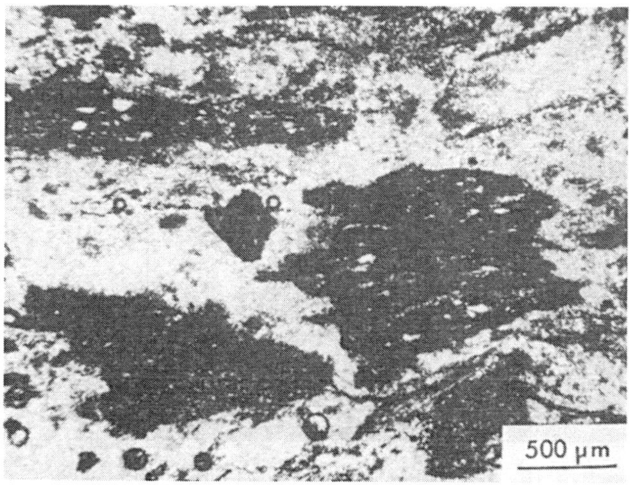

Fig. 20.4. Phosphatic sand and slate clasts in a quartz matrix (crossed polars).

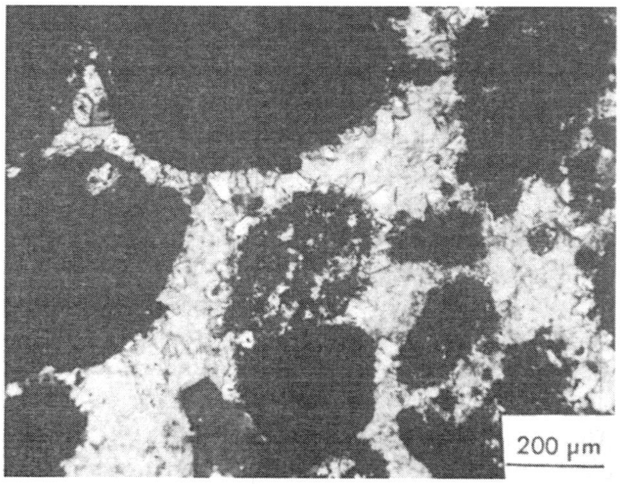

Fig. 20.5. Phosphate grains with overgrowths of quartz crystals.

tic texture and slaty cleavage (Fig. 20.4). The clasts are phosphate sand and slate pebbles in a matrix of fine-grained quartz with disseminated pelitic silicates and skeletal pyrite crystals. The phosphate occurs as irregular, partially recrystallised masses, as individual crystals disseminated in the matrix and as flat, black, aphanic pebbles. The apatite crystals disseminated in the matrix usually are hexagonal prisms up to 100 μm, wide in their basal sections. The black aphanic phosphate cements sand-sized quartz into pebbles.

The matrix contains irregular, local enrichments of sericite. The grey to black colour of the matrix is due to flakes of microcrystalline graphite and disseminated remnants of pyrite crystals that have been altered to goethite. Fractures and voids on the slaty cleavage are filled by a secondary phosphate identified as dufrenite.

The apatite in this sample has a composition very close to that of fluorapatite as indicated by the unit-cell data ($a = 0.9372$ nm, $c = 0.6883$ nm) and mean refractive index ($N_d^{25} = 1.638$).

Huron River, Michigan

The bulk mineralogy of this sample (Table 20.1) shows that it is predominantly a mixture of phosphate plus accessory silica and carbonates. The rock is a metasedimentary conglomerate containing clastic pebbles and sand-sized grains of subrounded and distorted apatite grains, goethite, and various rock debris. The phosphate is present as fine recrystallised grains which are mixed in varying proportions with quartz, goethite, sericite, and, less frequently, with calcite and barite. Occasionally the phosphate is cryptocrystalline to amorphous in appearance, stained by iron oxides, and contains inclusions of goethite as an alteration product of pyrite. The clasts frequently are rimmed or coated by secondary silica that consists of authigenic quartz crystals extending from the surfaces of the particles (Fig. 20.5). This drusy silica occasionally contains needle-like aggregates of goethite and/or pyrolusite. The phosphate crystallites are smaller than 10 μm and are subhedral to anhedral in morphology. The brown colour of the phosphate clasts results from staining by microcrystalline earthy goethite.

The interstitial materials in this sample are principally sparry calcite, chert, and sometimes barite. The sparry calcite and barite probably have a post-depositional hydrothermal origin. Energy dispersive X-ray fluorescence studies show that the calcite usually contains small amounts of Fe and Mn.

The sericite–muscovite in this sample occurs as small flakes mixed with the apatite crystallites or mixed with the quartz crystallites in the slate. Authigenic quartz crystals are also formed on the detrital sand grains in this sample.

The apatite in this sample has a composition quite close to that of fluorapatite. The unit-cell data ($a = 0.9374$ nm, $c = 0.6886$ nm) and mean refractive index ($N_d^{25} = 1.638$) are indicative of fluorapatite.

East-central Minnesota

Two samples were studied, and their bulk mineralogies are reported in Table 20.1. These samples are both metasedimentary phyllites, probably in the low-grade greenschist facies, and have lepidoblastic textures (foliated or schistose).

The apatite occurs as fine-grained crystallites forming discontinuous, oriented, parallel bands, irregular patches, or ragged masses associated with foliar sericite and alternating quartz bands (Fig. 20.6). Metamorphism has altered the rock texture

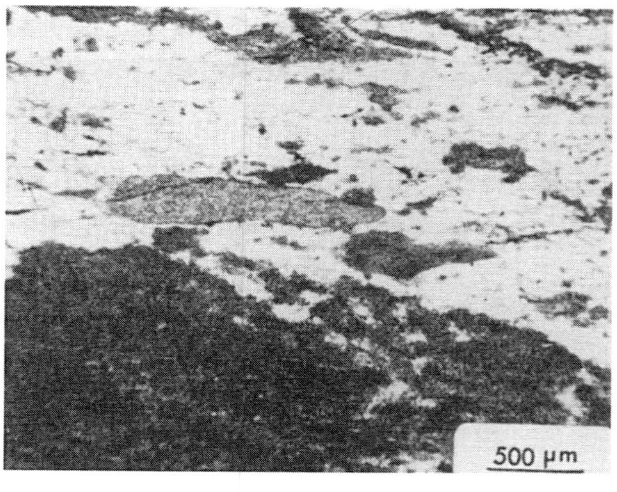

Fig. 20.6. Fine-grained, banded phosphate showing lepidoblastic texture with quartz cement.

significantly, making it difficult to know the original phosphate form; however, some of the less recrystallised masses resemble distorted grains. The recrystallised phosphate is subhedral in appearance in both optical and SEM studies and usually smaller than 10 μm. The apatite–sericite bands usually have a variable brownish coloration that is due to goethite staining and the presence of very fine, disseminated graphite and pyrite.

The quartz in these rocks is usually fine grained and occurs in bands alternating with the apatite–sericite. The crystallite size of the silica is about 50 μm. These bands often develop a pink colour similar to potassium feldspars because of inclusions and staining by goethite after pyrite. Some of the voids in the silica are filled by goethite and secondary phosphate (dufrenite).

The sericite occurs as flakes and platelets in parallel orientation between the apatite or quartz crystallites. Some of the goethite and most of the secondary phosphate occurs as fracture fillings or in spaces between the schist planes.

The apatite in these samples has a unit-cell *a* value that is slightly greater than that of fluorapatite ($a = 0.9386$ nm) and may contain some OH^{-1} substituting for F^{-1}. The mean refractive index (N_d^{25}) was measured as 1.638.

Udaipur, Rajasthan, India

The bulk mineralogy of these four samples (Table 20.1) is similar although the quantities of constituents vary. Because the textures of the various samples are very different, they will be discussed under separate headings.

Calcareous phosphorites One of these samples is a stromatolitic phosphorite from Block G (Banerjee *et al.*, 1980) and the other is a columnar stromatolite from Block G. Both are metamorphosed, brecciated, calcareous phosphates. The metamorphism has caused extensive fracturing, crushing, and recrystallisation of the original francolite with the formation of a secondary, microcrystalline fluorapatite (Fig. 20.7). Clasts and fragments of this phosphate are cemented in a micritic, calcite matrix. Veins and networks are filled by calcite that is occasionally sparry. Deposition of the calcite cement has corroded and replaced parts of the phosphorite. Replacement features are more extensive in the columnar stromatolitic sample where only a few patches or relicts of the phosphorite remain. Calcite is pseudomorphic after phosphorite in crystallite size and shape. In some places a honeycomb texture develops where calcite has replaced the edges of a particle, whereas the core remains apatitic. Most of the calcite is stained by goethite. The replacement was selective, with the phosphate being most affected. Other minerals only show minor effects of dissolution, reprecipitation, and redistribution.

In individual fragments the apatite shows polyhedral forms with a subparallel to parallel alignment that suggests that banding formed before fragmentation. Some fragments have been rotated from their original position after breakage. The particle size of the clasts varies from a few micrometres to centimetres, while the particle size of the secondary apatite averages 5–10 μm.

The quartz present is usually within the phosphate clasts and shows a crude preferred orientation parallel to that of the apatite crystallites. Some granoblastic quartz is present in the intraclasts and it is often associated with phlogopite.

Block D high-grade ore and stromatolitic phosphorite The high-grade ore (Table 20.1) is a brecciated metasediment. The colourless clasts are concentrically banded (stromatolitic) units of polyhedral, fine-grained, secondary fluorapatite crystallites with inclusions of recrystallised quartz and colourless phlogopite crystals. The clasts are cemented by a micritic mixture of phlogopite/apatite matrix which is stained by goethite. The apatite in this matrix may represent a third generation of phosphate, with fluorapatite crystallites that are smaller and more poorly developed than in the clasts. The phlogopite and goethite in this matrix are also less well developed than they are in other areas of the sample.

Occasionally the matrix is cut by veins and veinlets filled by divergent, radially fibrous to coarsely crystalline apatite that may represent a fourth generation of phosphate formation (Fig. 20.8). In many places it is possible to see relicts of the matrix with this generation of apatite growing from the surface. This last apatite phase is colourless. The associated phlogopite

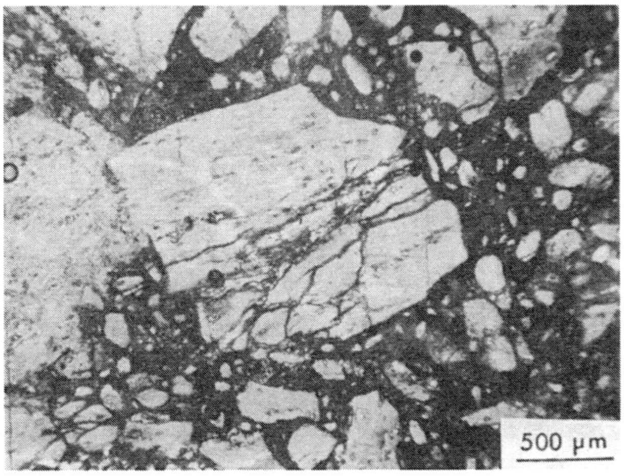

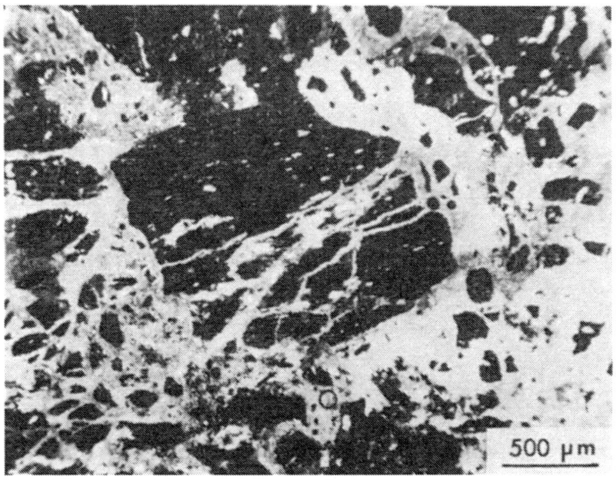

Fig. 20.7. Fractured phosphate clasts in a dolomite matrix (left, polarised light, right, crossed polars).

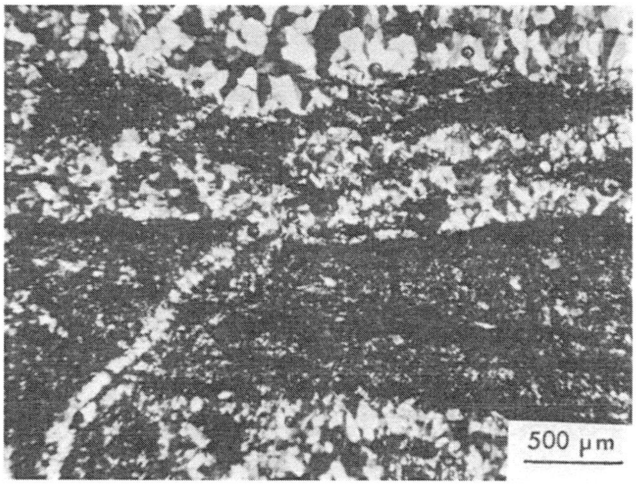

Fig. 20.8. Several stages of recrystallised, birefringent apatite surrounding and cross-cutting isotropic phosphate clasts (crossed polars, inclined 10°).

has been recrystallised into coarse grains. Voids are rimmed by coarse, hexagonal prismatic crystals of apatite which may be stained by goethite.

As stated above, the phlogopite in the matrix is associated with the apatite as small, preferentially aligned scales. When the apatite is coarse and colourless, the phlogopite is also coarse and is associated with quartz.

The apatite in the clasts is between 5 μm and 15 μm; it is about 5 μm in matrix, and up to 60 μm in veins and void fillings.

The columnar stromatolitic phosphorite is a dolomitic metasedimentary ore with a lepidoblastic texture. The columns (or dark bands) are principally phosphate, whereas the light-coloured interstitial material is dolomite. The columns may be discontinuous and pass into lenses whose constituents show preferential alignment parallel to the banding, thus giving rise to the lepidoblastic texture (Fig. 20.9).

The apatite occurs as fine-grained, polyhedral, secondary fluorapatite bands and lenses showing preferred orientation parallel to the banding. Recrystallised quartz and phlogopite occur in the phosphate as particles and scalar crystals growing between and in alignment with the apatite crystals. Some fine bands of the apatite are occluded in the dolomite. Generally, the apatite crystal size is about 10 μm except in some secondary veins where 50-μm crystals occur.

The dolomite occurs as a granular mosaic showing preferred orientation parallel to the banding. Generally the crystallite sizes are larger than those of the apatite; the crystallite size varies from 50 μm to 100 μm. Sometimes there are small patches and discontinuous lenses occluded in the phosphate.

Some of the last secondary veins are filled by sparry calcite, and voids are rimmed by coarse prismatic apatite crystals. Occasionally the quartz grains show apatite crystallite inclusions.

The apatites in the sample are francolites with a low degree of carbonate substitution approaching fluorapatite composition. The unit-cell values ($a = 0.9365–0.9372$ nm, $c = 0.6889–0.6891$ nm) and the mean refractive indices ($N_d^{25} = 1.631–1.632$) support this observation.

Jhabua, Madhya Pradesh, India

Two samples from this area were studied. One is a stromatolitic ore and the other is a banded ore. These samples are from siliceous metasedimentary rocks (Table 20.1). The stromatolitic ore contains discontinuous, alternating bands of an apparently authigenic, microcrystalline apatite and fine-grained, recrystallised quartz with pelitic sericite–muscovite inclusions (Fig. 20.10). Small intersecting veins and veinlets filled with quartz are common. The quartz in the bands has a brown colour and consists of fine-grained masses with disseminated inclusions of apatite crystallites and iron-oxide-stained pelitic sericite which is parallel to subparallel with the banding. The iron oxide stain is the source of the brown colour. The quartz often contains a cloudy core surrounded by clear silica that may include extremely fine, pigmented inclusions or larger grain sizes formed from a fibrous opaline or chalcedonic nucleus. In the SEM, these grains show a porous, fibrous core rimmed by a better crystallised material resulting from metamorphism. Energy dispersive X-ray data show a small amount of Nb in these areas. Some quartz contains small (less than 10 μm)

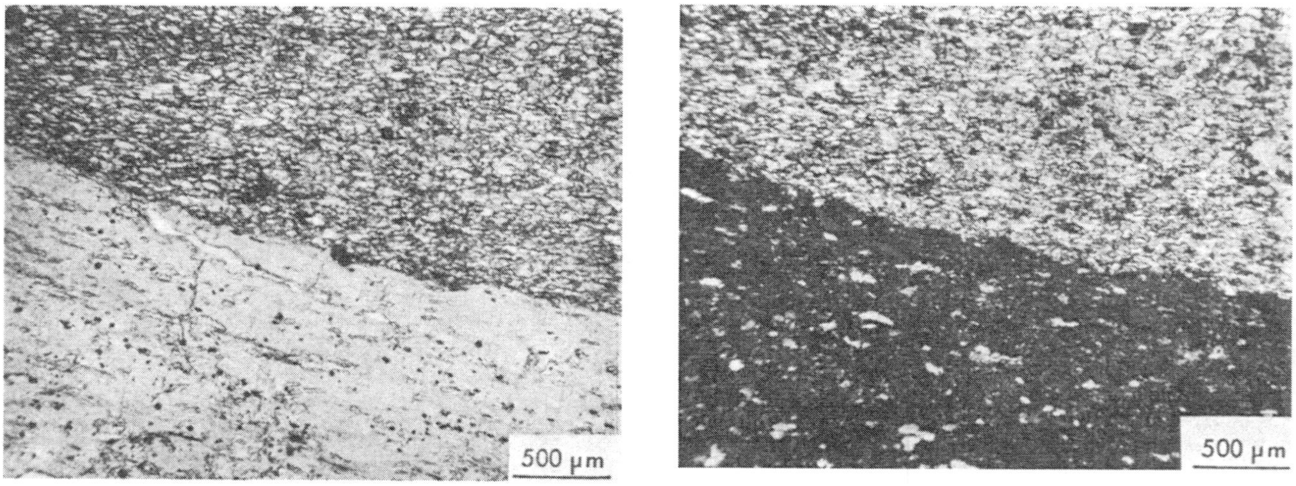

Fig. 20.9. Phosphate (lower half) and dolomite (upper half) showing parallel alignment of crystals and dolomite disseminated in the phosphate (left, polarised light; right, crossed polars).

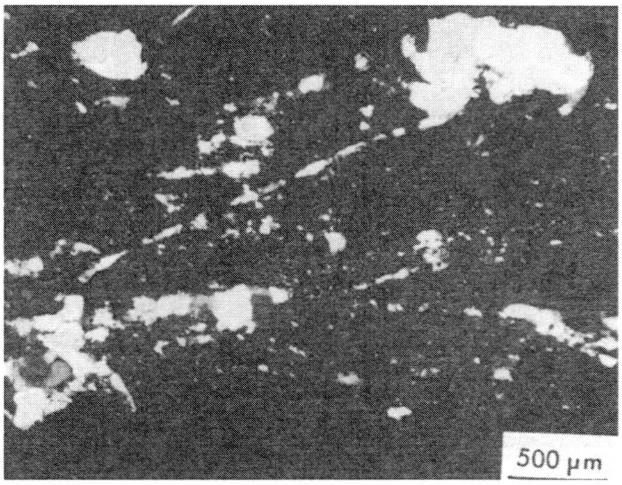

Fig. 20.10. Birefringent quartz disseminated in a fine-grained matrix (crossed polars).

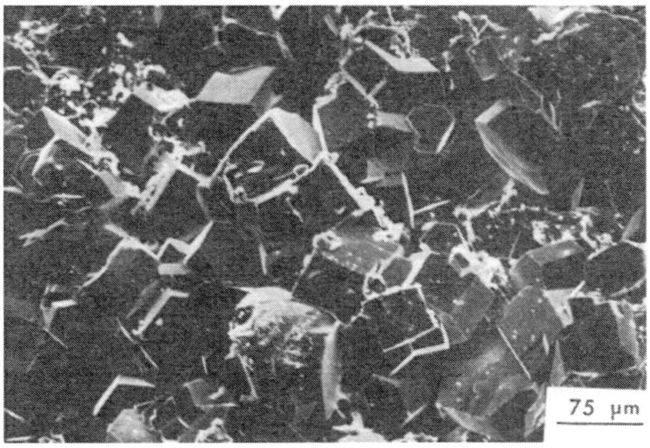

Fig. 20.11. Coarse, recrystallised, tabular apatite crystals filling a fracture.

birefringent carbonate inclusions. Other grains are eroded and iron stained.

In the banded phosphate the siliceous matrix is extensively fractured, and the veins are filled with quartz and fibrous, radial chalcedony rimmed by coarsely crystalline apatite (up to 100 μm) with some iron staining (Fig. 20.11). As in the stromatolitic ore, the phosphate is a fine-grained, microcrystalline material formed by the partial recrystallisation of the original francolite. Much of the phosphate is extensively fractured, and the veins are filled by quartz. The deposition of the quartz has caused a recrystallisation of the phosphate along the vein contact and thus has resulted in larger apatite crystallites and some embayment of the phosphate by quartz. Quartz is also present as inclusions in the phosphate ore in the form of granoblastic particles or in small lenses aligned with the banding.

Voids in the banded ore are rimmed by iron or manganese oxides. Some contain secondary phosphates as needle-like aggregates and appear to be mixture of fairfieldite and a complex, unidentified phosphate.

The francolite in these samples now has a composition very close to that of fluorapatite. The stromatolitic phosphate (unit-cell a value $= 0.9373$ nm, mean refractive index $N_d^{25} = 1.630$) and the banded phosphate ($a = 0.9370$ nm, $N_d^{25} = 1.630$) contain essentially the same apatites.

Patos de Minas–Lagamar, Minas Gerais, Brazil

These samples represent a medium-grade phosphate rock consisting of a metasedimentary, folded shale. The shale contains decimetre-thick bands of sand-sized particles which are rich in phosphate (Table 20.1) and show folding on macro- and micro-scale (Fig. 20.12). The phosphate occurs as discontinuous bands, lenses, and distorted grains consisting of aphanic to fine-grained francolite. The rock is a cherty quartz-sericite matrix that has been extensively fractured and weathered. The fractures, voids, and schist planes are filled with secondary, euhedral fluorapatite ($3 \times 50\,\mu$m) and wavellite crystals (Fig. 20.13). In some phases, the secondary fluorapatite is growing on the relict surface of the original francolite. The primary francolite shows evidence of dissolution and reprecipitation to form this secondary fluorapatite. The secondary apatite represents about 10%–15% of the phosphate, and the wavellite is about 1%–2% of it.

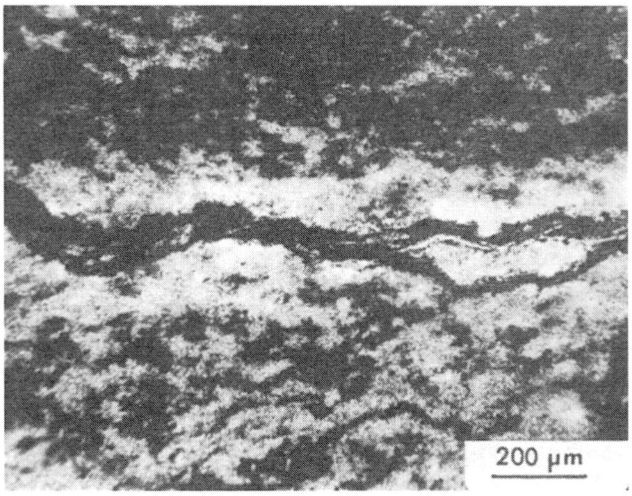

Fig. 20.12. Bands of phosphate and sericite showing microfolds.

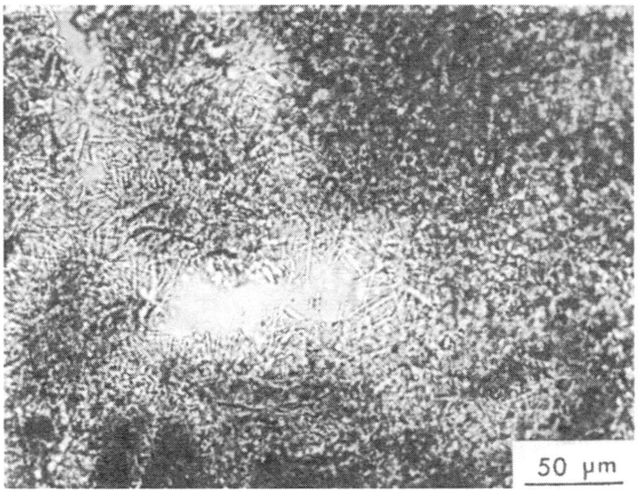

Fig. 20.13. Secondary apatite crystallites filling voids (crossed polars).

The bands of sand-sized materials are highly porous and extensively leached. They are rich in phosphate with a small amount of pelitic quartz–sericite. The phosphate occurs as distorted, metamorphosed grains fused into banded masses. The grains and voids of this porous zone are partially cemented and filled with an amorphous calcium, aluminium phosphate having a mean refractive index higher than that of crandallite. The energy dispersive X-ray data show Ca, Al, P with minor peaks for Ti and Fe.

The interstitial mineral phases are illite–sericite, montmorillonite, kaolinite, and goethite. The latter is usually present as a surface stain on the other components. Most of the detrital quartz and recrystallised chalcedony grains are coated by aggregates of radial wavellite crystals.

The francolite in this sample has a low degree of carbonate for phosphate substitution and is essentially a fluorapatite ($a = 0.9371$ nm, $c = 0.6890$ nm, $N_{\mathrm{d}}^{25} = 1.626$).

Sevier County, Tennessee

This sample is a mixture of phosphate and silicates (Table 20.1) with some secondary phosphate minerals. This rock is a metasedimentary, arkosic phosphorite. The metamorphism has caused extensive fracturing of the rock and thus has produced a mylonitic texture for the detrital grains and recrystallisation of a large part of the original francolite (Fig. 20.14). The recrystallised phosphate occurs as secondary, thin, elongated, euhedral crystals as well as coatings on quartz grains; it embays the recrystallised albite crystals and cleavage fragments. In some areas the apatite is present as inclusions in the albite. These are low-carbonate francolites with a unit-cell *a*-value (0.9366 nm) slightly lower than that of fluorapatite as is the mean refractive index ($N_{\mathrm{d}}^{25} = 1.633$).

The aphanic apatite in this sample has an anomalously low index of refraction ($N_{\mathrm{d}}^{25} = 1.600$). Energy dispersive X-ray fluorescence indicated that this material contains Al^{+3} and has a low Ca/P ratio. This may be an aluminium-rich apatite as described by Fisher & McConnell (1969), but the unit-cell and optical data do not agree with their values.

The energy dispersive X-ray fluorescence did confirm the presence of small amounts of Zn^{+2} in the lacroixite and rocherite.

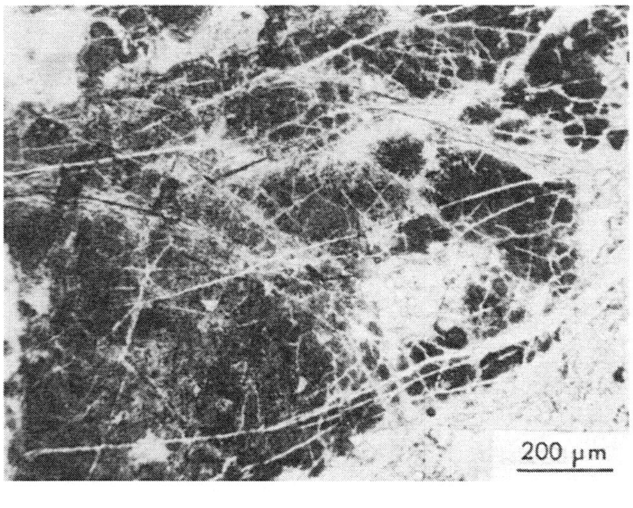

Fig. 20.14. Mylonitic texture with apatite vein fillings a phosphatic matrix.

Kodjari, Upper Volta

The bulk mineralogical data (Table 20.1) confirm that these are siliceous phosphorites of variable grade. The samples are from two different sites in the area that is presently being mined. The samples are bedded, metasedimentary phosphorites that appear to have undergone significant diagenetic alteration and very low-grade metamorphism resulting in a phosphatic shale. The phosphate occurs as angular fragments, subrounded grains, distinct grains (including composite grains and pseudo-ooliths) (Fig. 20.15). The grainstone phosphorites are mostly isotropic, aphanic francolites. The secondary phosphatic overgrowths are fibrous, concentrically layered francolites with weak birefringence. Some of the grains were deformed during metamorphism, and a few have welded textures (Fig. 20.16); however, there has not been extensive recrystallisation of the francolite. The grains have a dark colour, usually brown, that results from iron staining; they contain fine quartz particles, scalar sericite, and patches of fine-grained goethite. The size of the grains varies from 30 μm to 100 μm and averages about 90 μm.

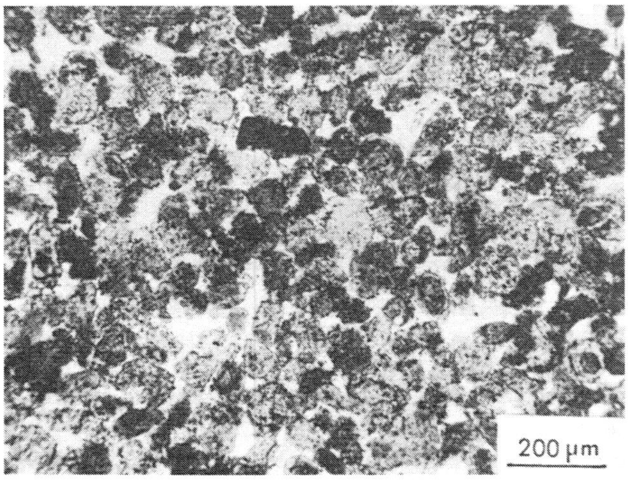

Fig. 20.15. Phosphate grains in a siliceous matrix.

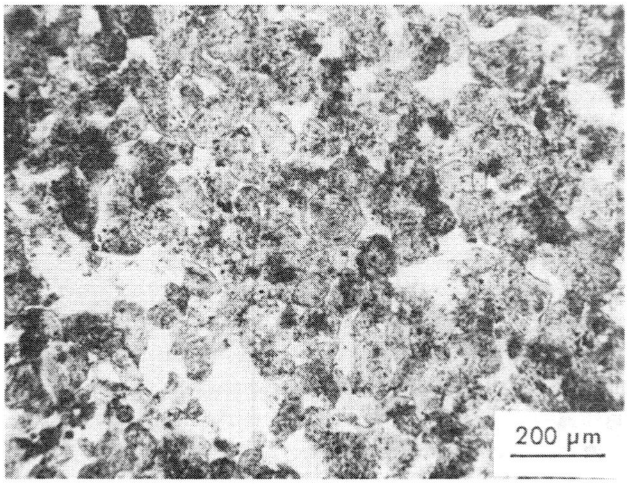

Fig. 20.16. Welded and fractured grains in a siliceous matrix.

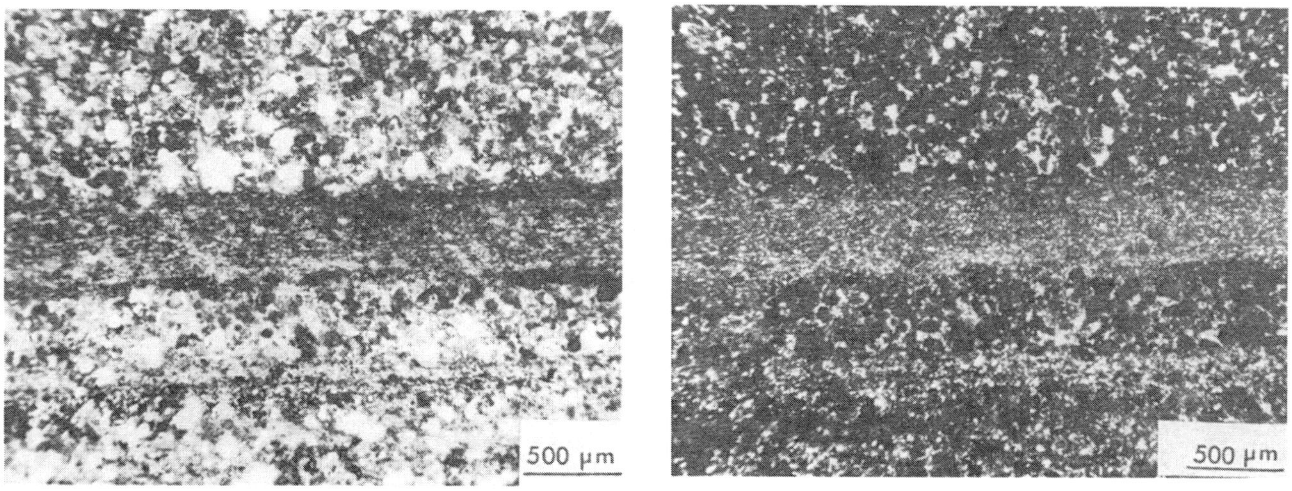

Fig. 20.17. Sericitic shale lenses in the granular, siliceous phosphate matrix (left, polarised light; right, crossed polars).

The phosphate is tightly cemented by a cherty-quartz mixture with some interstitial cementing by sericite and secondary francolite. Some detrital quartz and partially altered glauconite grains are also present; fibrous wavellite occasionally fills the voids as radial clusters, and rarely some kaolinite is seen.

The two samples are very similar mineralogically, but the Pit 3 sample is more impure and contains thin, interbedded lenses (2–3 mm) of sericitic shale (Fig. 20.17). This shale is heavily stained by iron oxide, and the sericite in the shale shows a good alignment parallel to the bedding. There is little or no phosphate in these micaceous lenses.

The apatites in these samples are francolites with a low degree of carbonate for phosphate substitution as shown by the unit-cell data ($a = 0.9355$ nm, $c = 0.6890$ nm) and mean refractive index ($N_d^{25} = 1.622$).

Parc W, Tapoa area, Niger

These are medium- to high-grade ores from the siliceous phosphates. These two samples were separated initially because of the significant differences in colour. Sample A is slightly mottled, with a dark to light grey colour and with occasional brown to orange streaks caused by goethite. Sample B is highly mottled in appearance and is dark brown to light grey in colour. Mineralogically the samples are similar (Table 20.1). These rocks are metasediments containing sand-sized phosphate which usually occurs as rounded to subrounded grains varying in size from 40 µm to 250 µm (Fig. 20.18). In some samples there are pseudo-ooliths or grains with secondary overgrowths of apatite. Some of the grains are fractured. In thin section the phosphate varies from light to dark brown with iron staining. Generally the phosphate is aphanic, but some shows partial recrystallisation of material disseminated in the matrix and embayment of the original detrital quartz grains. The phosphate grains also contain inclusions of fine-grained (less than 10 µm) quartz, goethite, clay, and anhedral to subhedral fluorite crystals.

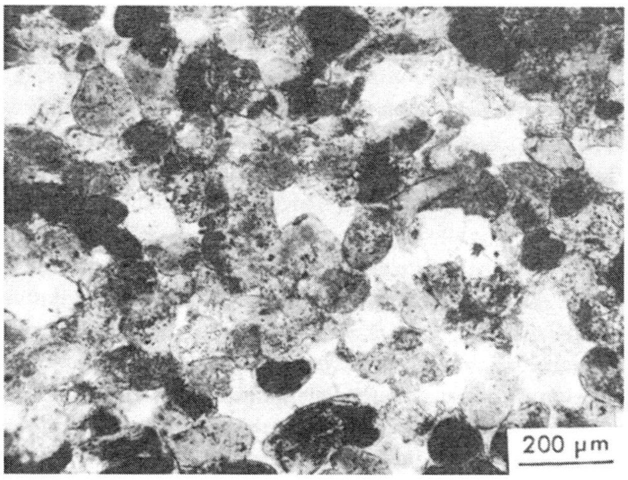

Fig. 20.18. Sand-sized phosphatic grains in a siliceous matrix.

The matrix is principally quartz that appears to be recrystallised chert. This silica is tightly bonded, both to the phosphate grains and to the detrital quartz. Liberation of the phosphate by grinding would be difficult. Variscite occurs in the interstitial voids as fine-grained, radial fibrous clusters and partly cements the rock when silica is absent.

The brown colour of Sample B seems to result from local concentrations of goethite–variscite cement, an absence of silica, and a decreased frequency of phosphate grains in the matrix. These factors combine to intensify the mottled appearance of this sample.

The francolite in these samples shows a low degree of carbonate for phosphate substitution, as indicated by the unit-cell data ($a = 0.9364$ nm, $c = 0.6889$ nm) and mean refractive index ($N_d^{25} = 1.624$).

Namel, Senegal

Two samples were examined from this area. One is a surface sample, and the other represents a phosphorite from 8 m below the subsurface (Slansky, personal communication). The bulk mineralogies (Table 20.2) show significant differences in these samples. The superficial sample is high in mafic silicates, and the subsurface sample is high in calcite.

The surface sample contains sand-sized phosphate in a

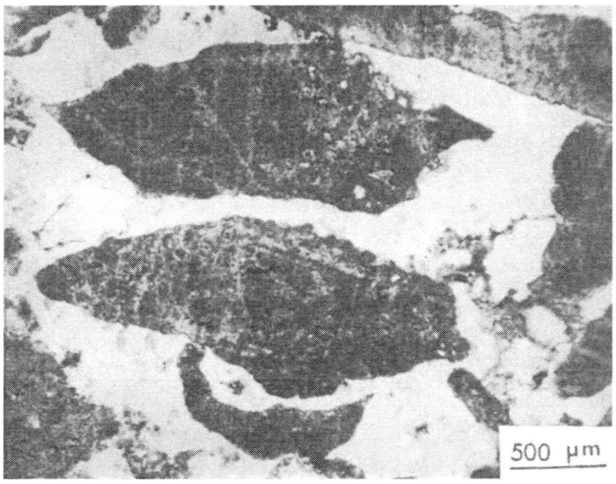

Fig. 20.19. Deformed, fine-grained phosphate grains in a siliceous matrix.

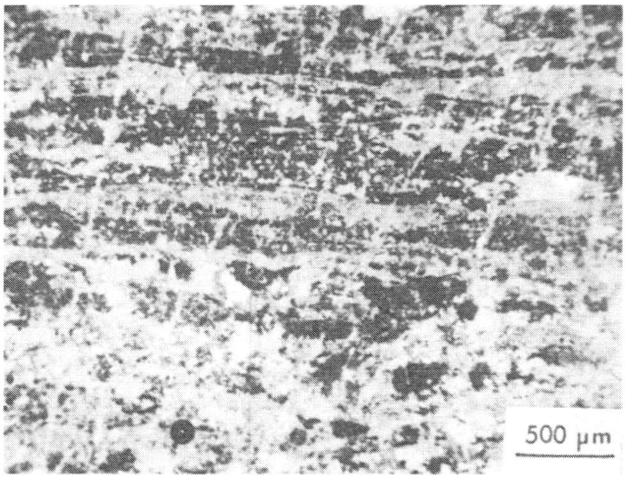

Fig. 20.20. Banded, calcareous phosphate showing carbonate-filled fractures and replacement textures.

metasediment that may originally have been a dolomitic chert. The magnesian minerals have been altered to tremolite/talc and the chert to a grainy quartz. The phosphate occurs as elongate, deformed grains sometimes rod-like, in which the francolite has been recrystallised into a fine-grained, microcrystalline texture (Fig. 20.19). These crystals are smaller than 10 μm and usually are polyhedral to hexagonal in shape. Individual well-formed crystals are also occasionally disseminated in the matrix. The particle size of the phosphate grains varies but is as large as 3.0 mm. In a hand specimen, the grains are nearly white, but in thin section, they appear dark grey because of inclusions of iron and manganese oxides and organic matter.

The cementing matrix is a fine-grained, quartz mosaic. The polyhedral quartz crystals range in size from 10 μm to 150 μm with a mean size of about 50 μm.

Talc usually occurs as an interstitial phase as patches and as elongate, irregular masses of flaky to fibrous aggregates embayed by quartz. The talc varies in size from 50 μm to 250 μm and sometimes shows a yellow-brown colour due to iron oxide staining. Opaque inclusions which are high in titanium also occur in the talc occasionally.

This rock contains some secondary fractures which are filled with quartz, chert, chlorite, and psilomelane. The psilomelane always has some barium associated with it.

The subsurface sample is a banded, calcareous, metasedimentary phosphorite with postdepositional fracturing (Fig. 20.20). The phosphate occurs as brown, discontinuous bands and lenses, subrounded pelletal grains, and secondary overgrowths. The bands and lenses consist of aphanic to fine-grained francolite that shows extensive fracturing. Some fractures are filled with high magnesian calcites; these form a vein network which has caused extensive replacement and embayment textures. The francolite grains have been deformed into subrounded, elongated forms. Secondary francolite rims most of the minerals but also occurs as thin lenses and spherulites.

The secondary magnesian calcite occurs as discontinuous white bands, lenses, and cross-cutting veins with micritic, fibrous, or, less commonly, sparitic textures. Some rhombohedral grains are zoned.

Quartz occurs as detrital grains which sometimes are cracked or welded and which show wavy extinction. Calcite replacement has eroded the edges of many grains.

Albite occurs as detrital grains, as tabular, authigenic crystals, and as a pelitic groundmass mixed with chlorite and phosphate. Sometimes the detrital grains show eroded edges and relict polysynthetic twinned grains in a pelitic groundmass of fine-grained quartz or, less frequently, sericite.

Talc, chlorite, and phlogopite occur as void and fracture fillings caused by leaching. Sometimes they replace primary minerals as flaky, interlocking crystal aggregates, fibrous aggregates, and foliated crystal masses. Goethite usually occurs as staining on the other mineral components.

The apatite in the surface sample is nearly pure fluorapatite (unit cell $a = 0.9373$ nm, $c = 0.6981$ nm, mean refractive index $N_{\rm d}^{25} = 1.630$), while the subsurface apatite is a francolite with a low degree of carbonate substitution (unit cell $a = 0.9360$ nm, $c = 0.6893$ nm, mean refractive index $N_{\rm d}^{25} = 1.627$).

Nouedgui-Bou Naga area, Mauritania

These rock samples are light yellow-brown, fine- to medium-grained, bedded quartzites. The samples contain abundant, thin grey-to-black bands (0.5–1.0 cm). Low-grade metamorphism has caused some recrystallisation of the quartz and phosphate.

The rock is a metasedimentary sandstone (Table 20.2) in which the phosphate occurs in a fine-grained quartz–sericite matrix as fine-grained, recrystallised masses and strongly distorted, partially recrystallised grains (Fig. 20.21). The phosphate masses contain subhedral crystallites ranging from 10 μm to 100 μm in size, with recrystallisation masses approaching 200 μm in diameter. In areas of strong metamorphic alteration, the apatite occurs as colourless crystals disseminated in the siliceous matrix. The distorted phosphate grains appear as patches, irregular masses, and lenses bordered by metaquartz and sericite. The megascopic banded appearance of the rock is the result of graphite inclusions, some of which occur in the apatite. In less altered portions of the samples, the phosphate

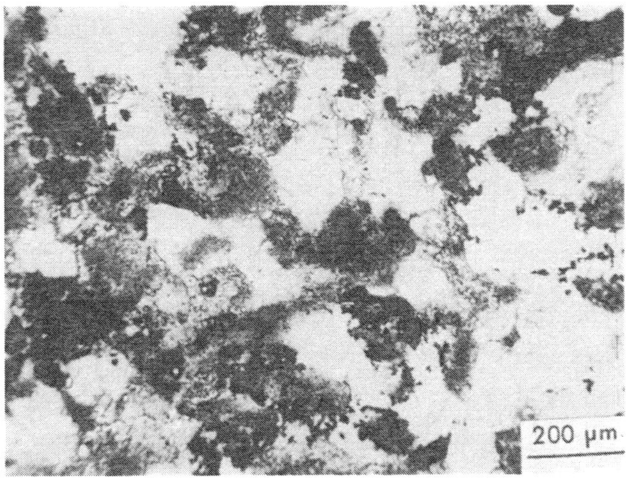

Fig. 20.21. Phosphatic sandstone containing fine-grained phosphate (grey to black) as deformed grains and recrystallised masses in a quartz–sericite matrix.

occurs as irregular, lenticular grains with an aphanic to fine-grained texture.

The matrix cementing the rock is an intergrowth of fine-grained, recrystallised quartz–sericite. The sericite usually occurs as an interstitial filling between the quartz and phosphate grains. Quartz also occurs as secondary, cross-cutting veins. Sometimes the quartz occurs as a granular mosaic.

The apatite composition in these three samples varies with the amount of metamorphism. In the rocks showing the highest grade of metamorphic change, the apatite is a fluorhydroxy variety (unit cell $a = 0.9382$ nm, $c = 0.6882$ nm and a mean refractive index $N_d^{25} = 1.634$). The lower grade rocks contain a mixture of fluorapatite and francolites with near-fluorapatite compositions (unit cell $a = 0.9370$ nm, $c = 0.6880$–0.6888 nm, mean refractive index $N_d^{25} = 1.630$).

Georgina Basin, Queensland, Australia

The bulk mineralogy of the calcareous phosphate sample (Table 20.2) shows that this is a low-grade phosphatic limestone. The rock is composed of sand-sized grains, including irregularly shaped ovular phosphate grains, phosphatised shells, and fragments in a matrix of sparry calcite (Fig. 20.22). The irregularly shaped phosphate usually contains inclusions of calcite, quartz, flaky sericite, and goethite. The inclusions are all smaller than 15 μm, and calcite is the most abundant. The phosphate grains consist of aphanic apatite. The grain boundaries are often replaced by calcite or secondary apatite that is weakly birefringent. The fossils and fragments present are well preserved and usually have phosphatised organic structures, while the voids are filled with calcite. Other voids in the samples are lined by banded, secondary apatite. The phosphate particles vary in size from 50 μm to 250 μm. The phosphate seems aligned parallel to the bedding plane of the rock.

The calcite matrix that contains the phosphate is sparry and often contains crystals showing polysynthetic twinning. Glauconite grains also occur as an interstitial accessory mineral as both fresh green particles and brown, iron-stained particles partially converted to oxides. Goethite and fluorite also occur as interstitial accessory minerals. Occasionally cherty quartz is present in the matrix and seems to be replacing and embaying parts of the calcite.

The siliceous sample from the Beetle Creek Formation is mainly a mixture of apatite and silica (Table 20.2) with lesser amounts of other accessory minerals. This sample contains sand-sized phosphate which occurs as distorted grains, composite grains, and occasional pseudo-ooliths with concentrically banded secondary apatite in a cherty quartz matrix (Fig. 20.23). The apatite in the grains is usually microcrystalline to aphanic in appearance and often has a spherulitic texture. Void fillings of secondary apatite occur as small aggregates of hexagonal prisms. The phosphate particles range in size from 30 μm to 500 μm, but most are between 100 μm and 250 μm.

The phosphate grains contain fine-grained inclusions of quartz, goethite, and, occasionally, euhedral fluorite. These inclusions usually are less than 10 μm in size. Fractured grains often contain cherty quartz as void fillings. Some samples show evidence of strong weathering, where leaching has produced

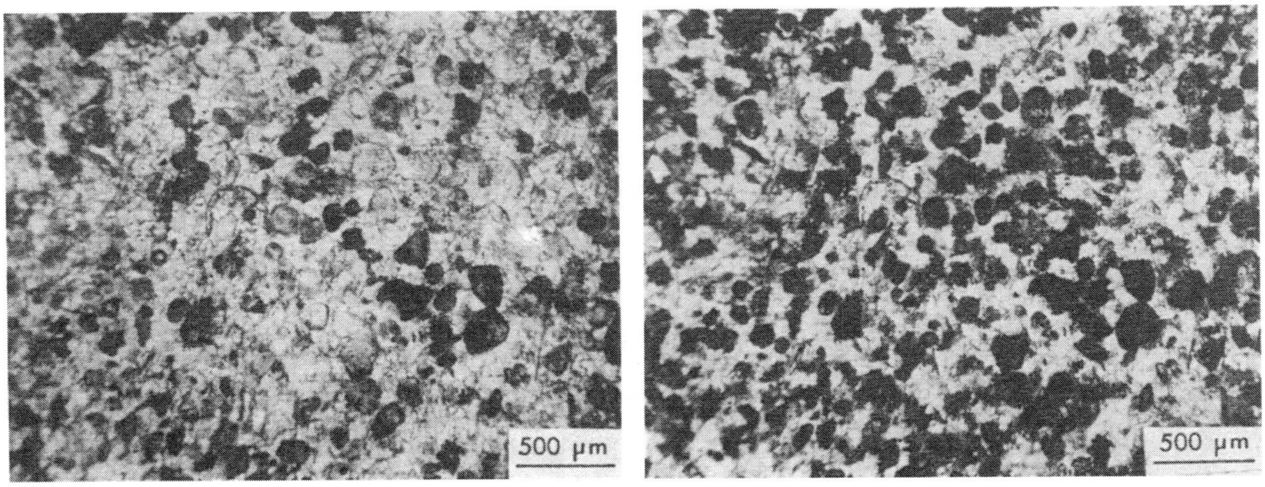

Fig.20.22. Phosphatic limestone containing phosphate grains, phosphatised fossils, and sparry calcite (left, polarised light; right, crossed polars).

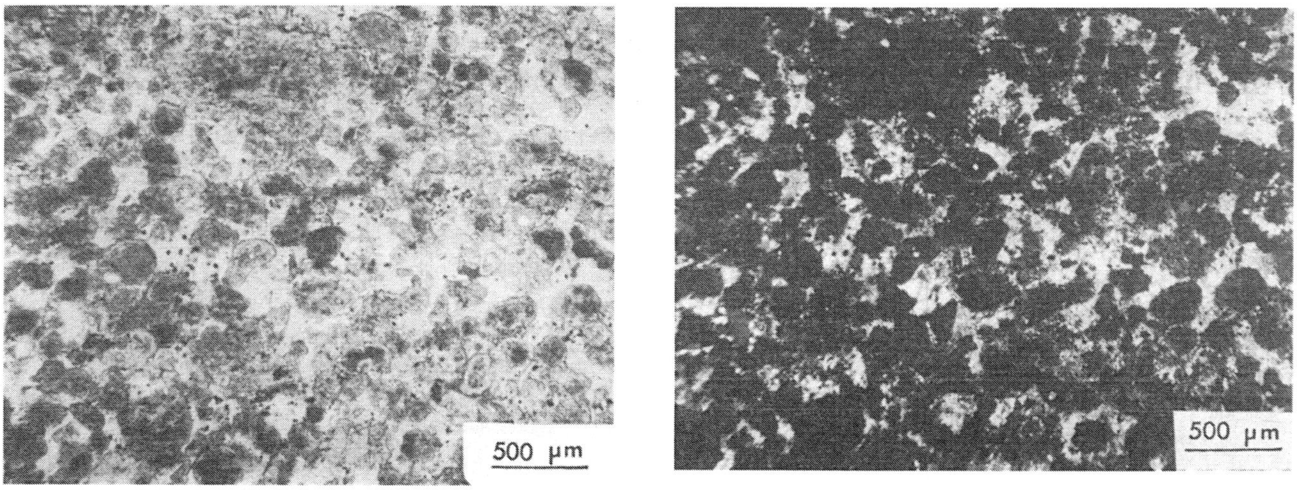

Fig. 20.23. Sand-sized, distorted, phosphate grains in a cherty quartz matrix (left, polarised light; right crossed polars).

voids that may have contained carbonates. The fossils in the sample are often fragmented and cemented by the cherty quartz matrix.

The accessory interstitial minerals include sericite which occurs as small plates or scalar crystals, fluorite occurring as small cubes or penetration twins, goethite pseudomorphic after pyrite, and spheroidal mixtures of metallic oxides that may have been formed by the weathering of glauconite.

The apatite in these samples occurs as francolite with a low degree of carbonate for phosphate substitution as shown by the unit-cell and optical data ($a = 0.9363$ nm, $c = 0.6890$ nm, mean refractive index $N_{\rm d}^{25} = 1.628$).

Kakul area, Hazara, Pakistan

This rock has undergone significant post-depositional diagenetic change as well as slight metamorphism. It is a siliceous phosphorite (Table 20.2) in which metamorphism has caused extensive fracturing (Fig. 20.24). The phosphate occurs as a variety of nodular clasts, coarse, fine fragmented and composite grains, ooliths, and pseudo-ooliths. The grains often show secondary concentric overgrowths of layered, fibrous phosphate that cause them to appear oolitic. Many are cracked or have welded edges as a result of metamorphism. The phosphate shows an aphanic to microcrystalline texture.

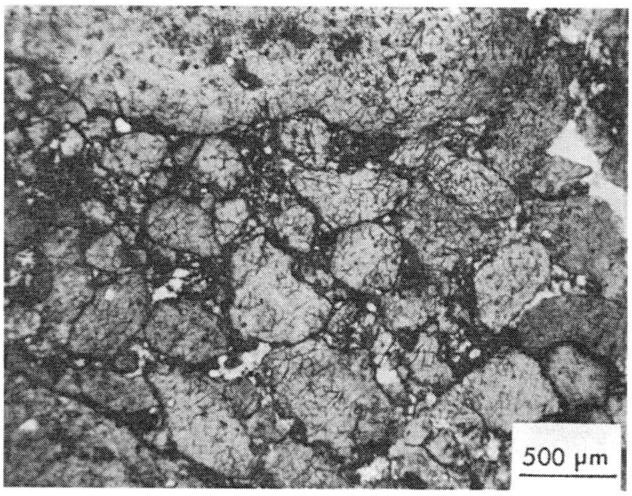

Fig. 20.24. Fractured siliceous grainstone phosphorite showing a variety of clast shapes and secondary textures.

Most phosphate grains contain occluded detrital quartz, authigenic quartz, sericite–muscovite, and some goethite. Some grains contain well-preserved spherulitic, organic structures. Occasionally voids are filled with aggregates of sericite–muscovite crystals, and chert–quartz mixtures fill some fractures and veins.

The phosphate is cemented by a chert–quartz matrix forming an interconnecting network throughout the rock. The matrix sometimes embays relicts of recrystallised, fine-grained phosphate. The sericite–muscovite occurs in this siliceous matrix and helps to cement the rock. Detrital quartz and biotite occur as intraclasts. Secondary phosphates (dufrenite) and goethite fill fractures and voids and cause local staining.

Generally the phosphate occurs as sand-sized grains (500 μm), but some local bands of silt-sized grains and fragments occur intimately mixed with the silica matrix. The silt-sized bands are often aligned parallel to bedding in the rock.

The apatite in this rock is francolite with a low degree of carbonate for phosphate substitution (unit cell $a = 0.9364$ nm, $c = 0.6890$ nm, mean refractive index, $N_{\rm d}^{25} = 1.627$).

Burin Khan, Mongolia

This rock is a black, slightly metamorphosed, siliceous phosphorite. The phosphate occurs as aphanic to microcrystalline masses which are brown in colour with disseminated inclusions of cherty quartz, organic matter, and occasional sericite and iron oxide (Table 20.2). Metamorphism has caused extensive fracturing of the rock and partial recrystallisation of the primary francolite (Fig. 20.25). These fractures are filled with a colourless mixture of fine-grained quartz and microcrystalline to coarse-grained fluorapatite formed by recrystallisation of the francolite. This can clearly be seen in some veins where the recrystallised fluorapatite surrounds and contains the brown francolite, showing gradational changes in texture from the edge to the centre of the relicts. Apatite crystals are always found in

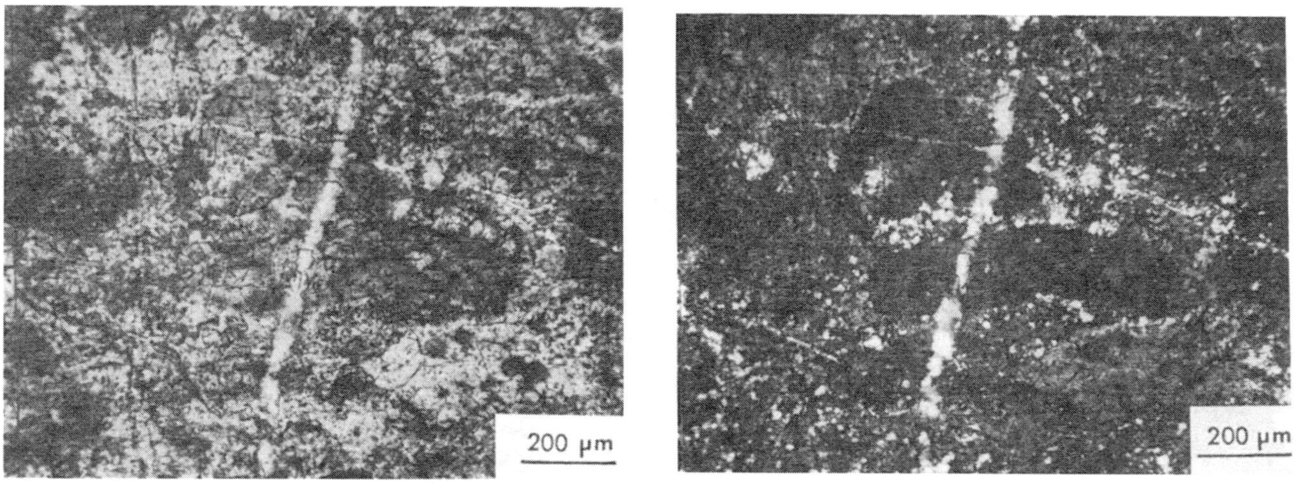

Fig. 20.25. Fractured, siliceous grainstone phosphorite with recrystallised fluorapatite (birefringent) and fine-grained quartz (left, polarised light; right, crossed polars).

Fig. 20.26. Well rounded to angular phosphate grains in a dolomitic matrix (left, polarised light; right, crossed polars).

voids along the edges of fractures. Veins filled with quartz are rimmed by apatite crystallite, while those veins filled with apatite show occasional inclusions of authigenic quartz crystals.

Minor mineral phases that occur as inclusions are rutile, calcite, sericite, muscovite, and fluorite. These inclusions generally are less than 10 μm in size.

The apatite in this rock is essentially fluorapatite (unit-cell $a = 0.9372$ nm, $c = 0.6888$ nm, mean refractive index $N_d^{25} = 1.629$–1.632).

Guizhou, China

The ore sample is a metamorphosed carbonate. The bulk mineralogy of this sample (Table 20.2) shows that this ore is mainly a mixture of phosphate and dolomite. The phosphate occurs as light brown, isotropic grains (Fig. 20.26) and as massive irregular bands in the dolomitic matrix. The grains range in size from 100 μm to 350 μm in the maximum dimension, and are circular, ovate, or irregular in shape. Small authigenic quartz crystals (30–120 μm) are common as inclusions in the grains. The spherical grains commonly exhibit a texture characterised by an anastomosing network of small tubules like that of algal stromatolites (Fig. 20.27). These grains may be surrounded by concentric, wavy bands of phosphate and appear to be pseudo-ooliths. Some smaller grains inside the larger 'algal' grains display a spherulitic texture. Other grains lack any distinct internal structures. Phosphate grains are often corroded around their margins and are partially replaced by dolomite.

The massive areas of apatite are also light brown and isotropic, aphanic to fine-grained in texture, with some dark areas of iron oxide staining and goethite grains. These massive areas may have been formed by the welding of several smaller grains. The matrix dolomite is commonly very coarsely crystalline in the proximity of massive apatite, perhaps because of metamorphic recrystallisation (Fig. 20.28).

Dolomite is present as subhedral, coarsely crystalline mosaics. Crystal sizes range from 50 μm to 350 μm. Euhedral, rhombic dolomite is common. Some of these rhombic crystals appear to be zoned and in optical continuity with the surrounding dolomite.

Primary quartz is present as endogangue or inclusions in the phosphate. Zircon is also present in small quantities in the phosphate. Minor goethite and fluorite occur throughout the

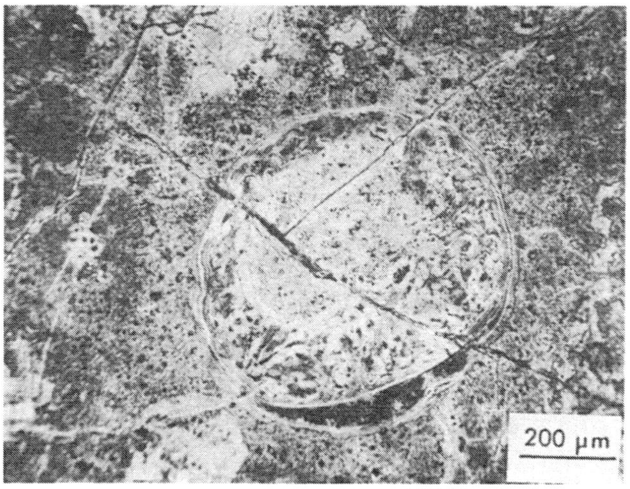

Fig. 20.27. Well rounded phosphate grains with algal stromatolite structure.

Fig. 20.28. Zoned dolomite crystals developed near the massive apatite.

sample. Fluorite may be present owing to the loss of fluorine from the francolite during metamorphism.

The apatite in this sample is a francolite with a small amount of carbonate-for-phosphate substitution (unit-cell $a = 0.9363$ nm, $c = 0.6888$ nm mean refractive index $N_d^{25} = 1.626$).

Analytical methods

X-ray diffraction analyses of whole rock powders and optical microscopy were used to identify the major mineral phases for all samples. The X-ray diffraction interplanar spacings were measured using computer controlled step-scan and peak search methods. The unit-cell dimensions were determined using linear least squares regression methods (McClellan & Lehr, 1969). Clay mineralogy was identified using the oriented specimens on glass slides. Scanning electron microscopy and energy dispersive X-ray fluorescence analysis were used to confirm minor mineral components and to clarify certain textural features.

Major and minor element analyses were determined by X-ray fluorescence and supplemented by specific ion electrode, gravimetric gas analysis, and atomic absorption analysis as needed. The sums of the analytical data for these samples are all close to 100% and confirm that no major phases are omitted. With the exception of rare earths and uranium, only those trace elements most frequently found in phosphorites (Prévôt & Lucas, 1980) were included in these analyses.

The data obtained (Tables 20.3 and 20.4) show that 20 analyses were carried out on each of the 33 samples and included the major and minor components. Because of their diverse nature, it is difficult to establish even qualitative relationships among these samples and their compositional variations. The raw data were normalised by dividing each determination by the mean of all determinations for that element. To facilitate the interpretation of these data a simple statistical factor analysis was used (Reeves & Saadi, 1971; Nathan *et al.*, 1979; Jarvis, 1980). Factor analysis can reduce complex relationships among large numbers of variables to a few simplified relationships that explain most of the variation in the data. Using the Statistical Package for the Social Sciences (SPSS), variables (in this case chemical data) are grouped into factors on the basis of their intercorrelations (Davis, 1973). These factors, which are associations of highly correlated variables, are derived from a Pearson correlated matrix. The principal factor matrix was rotated to produce a varimax orthogonal solution in which each factor represents a dimension of the data. A rotation to an oblique matrix was also tried; however, it did not give any improvement or simplify the geological interpretation. Factor analysis assembles the data into a readily interpretable form. The most significant results for the whole rock analytical data were obtained with only five factors that explained more than 94% of the variance (Table 20.5) The factor analysis results are here interpreted as interelemental correlations within particular minerals. The result is that one factor may be linked to the same mineral assemblage. Such factors may indicate different processes of mineral genesis or different stages of genesis. Some factors cannot be clearly related to a specific mineral or mineral assemblage. In this case, where the data set is relatively small, outliers (specific species that occur infrequently) can influence that data set. In any case, apart from the induced correlations, there should be no strong correlation of factors. The failure of the oblique matrix to improve the correlations is a confirmation that the factors are not correlated.

Only loadings greater than 0.25 are reported in the varimax rotated factor matrix (Table 20.5), and lower values are considered insignificant. The factors have been correlated to minerals according to the loadings of the major elements. For example, the factor with a P_2O_5 loading is called apatite, CO_2 loading is called carbonate, Fe_2O_3 with Al_2O_3 is called oxides–hydroxides, Al_2O_3 with K_2O is called clay minerals (illite–sericite–muscovite), Fe_2O_3 with S and BaO is called sulfide–sulfate (pyrite, barite) and an unnamed residual factor.

It is interesting to note that a silica factor does not occur independently, although quartz is the most abundant mineral after apatite in these samples. Silica has a significant negative loading in the apatite and carbonate factors showing an antipathetic relationship with these major components.

In the apatite factor, there is a strong positive loading for CaO, showing its common association with apatite and the

Table 20.5. *Varimax rotated factor matrix for composite chemical data*

	Apatite	Carbonate	Oxides–hydroxides	Clays	Sulfide–sulfate	
Na_2O	0.69					0.52
Fe_2O_3			0.83		0.52	
S					0.78	
F	0.90	−0.33				
CaO	0.88	0.36				
SiO_2	−0.86	−0.44				
MgO		0.52				
K_2O				0.72		
P_2O_5	0.89	−0.38				
Al_2O_3			0.53	0.75		
CO_2		0.96				
SrO	0.36	−0.27	−0.42	−0.30		0.61
ZnO			−0.32	0.68		0.39
CuO			−0.43	−0.31		
NiO	−0.68	−0.44				
MnO	0.41					0.71
Cr_2O_3	−0.35		0.59			
TiO_2			0.85			
BaO					0.91	
Eigen value	4.92	3.55	3.35	2.05	1.54	1.01
% of variation	30.00	21.60	20.40	12.50	9.40	6.20

Blank spaces indicate varimax factors less than 0.25.

general absence of calcite and gypsum in these samples. Fluorine and Na_2O are also positive, showing they usually occur in the apatite. The SrO and MnO are the only trace elements with positive loadings in the apatite factor, while NiO and Cr_2O_3 have negative loadings. The positive loading of Na_2O, MnO, and SrO clearly indicates that all of these elements are associated with the apatite, although their mode of occurrence is not known. The NiO and Cr_2O_3 are not associated with the phosphate and must occur in some undefined accessory phase. The absence of CuO and BaO as significant loadings in this factor is of interest. These elements may not be associated with the apatite found in these ancient phosphorites because of various alteration processes. The apatite factor accounts for 30% of the variation in the data.

The carbonate factor shows a positive loading for the expected elements (CaO, MgO, CO_2, LOI (loss on ignition)) and expected negative loadings for SiO_2 and P_2O_5 (antipathetic relationships). Of the trace elements, NiO and SrO show significant negative loadings and may represent the group of metals associated with organic matter (Prévôt & Lucas, 1980). The carbonate factor accounts for nearly 22% of the variation in the data.

The oxide–hydroxide factor contains only Fe_2O_3 and Al_2O_3 as positive loadings of major constituents. The trace elements with positive loadings are TiO_2, Cr_2O_3, and ZnO. These elements generally are associated with the goethite–hematite assemblage reported in several samples as well as some weathering products. The CuO and SrO are trace elements that have negative loadings in this factor. The oxide–hydroxide factor accounts for more than 20% of the variation in the data.

The clay factor represents the combined effects of several minerals; illite, muscovite, sericite, and kaolinite are the most important contributors. The K_2O and Al_2O_3 are the major positive loadings contributing to the clay factor. Post-depositional alteration has limited the clay mineralogy so that the 2 : 1 mica-related types are dominant. Kaolinite and montmorillonite are restricted to those samples that show evidences of weathering. The CuO is the only trace element with a positive loading in the clay factor and may be associated with rather than substituted in these minerals. The NiO and SrO show negative loadings indicating that they are not part of the clay mineral compositions. The clay factor accounts for 12.5% of the variation in the data.

The sulfide–sulfate factor contains positive loading for Fe_2O_3, S, and BaO. This factor collected the pyrite–barite mineralogy with a single factor and accounts for 9.4% of the variation of the data. It is interesting that this factor is so easily separated considering the sparseness of the data for these components.

The remaining factor seems to contain the residual values and cannot be associated with identifiable mineral assemblages. The Na_2O is the only major component associated with this factor and might be associated with the occurrence of albite in two samples. The absence of other major factors makes this interpretation rather speculative. The loadings of trace elements in this factor also fail to indicate any specific crystalline phases. The small amount of variation included in this factor makes interpretation difficult and perhaps unimportant.

Comparison with other phosphorites

This study of ancient phosphorites and their francolites has led to several observations concerning their general mineralogy and the compositions of their apatites. Some of the samples are rather similar to more recent phosphates, but most show only remnants of their original textures and compositions. It is remarkable to see the residual granular structures in the Lower Precambrian phosphorites from Michigan and Minnesota and speculate on the processes that formed the mineralogy and texture of these and other phosphatic rocks.

The bulk mineralogy of the samples studied is rather simple considering their diverse geological origins. Silica, carbonate, and clays are the main ingredients combined with the phosphate. The abundance of silica with its diverse modes of occurrence is an important feature in these samples and seems to contrast with the carbonate–phosphate association of irregular deposits. Many textural features in these rocks indicate that the silica cementation was a post-depositional phenomenon and may have displaced carbonates and other primary minerals. The silicification processes caused significant alteration of the rocks and often allowed the development of well-formed apatite crystals in the matrix or as cavity linings and void fillings.

The clay mineral compositions are rather uniform throughout the samples with the illite–sericite–muscovite assemblage being most common. No doubt these minerals are the alteration products of the original clays. Other clay minerals (kaolinite, montmorillonite) that occur in these samples are the result of weathering.

About 25% of the samples studied contain carbonates, with calcite occurring more frequently than dolomite. This is surprising considering that up to 75% of the world's known phosphate resources occur in carbonate rocks. With the exception of the Georgina Basin phosphates, the carbonate samples are characterised by an absence of organic remains and often show significant effects of metamorphism or replacement textures. One important role of the carbonates is to minimise the alteration of the francolite in the rocks by acting as a buffer against thermal and chemical processes.

The absence of feldspars in these samples is noteworthy. Most younger phosphates contain potassium feldspars (microcline being a minor, but common, accessory in Phanerozoic deposits). The only feldspar encountered was albite, which is so unusual in phosphorites that its occurrence is significant. It is possible that the potassium feldspars were altered to clay minerals and only their equivalents persist.

The apatites in ancient phosphorites show only a limited range of composition between low carbonate francolites and fluorapatite. This is a predictable result according to Chien & Black (1976) who showed that the free energy of formation ($\Delta G^{\circ}_{F,Ca}$) and solubility product constant (K_{CA}) of francolite increase systematically with the increase in moles of CO_3 per mole of francolite. The result establishes that francolites are metastable with respect to fluorapatite, the stable end member of the series. The metastability predicts a systematic alteration of francolites towards fluorapatite compositions as a result of various geological processes such as metamorphism, weathering, and diagenesis (McClellan, 1980). The observed and predicted systematic alteration seems to indicate that all francolites start off with nearly the same composition and then are always changed by various processes through a series of intermediate steps towards a single final composition. The observation seems to confirm the 'original constant composition' hypothesis of McArthur (1978) and explains much of the variability observed in francolite compositions. These unaltered francolites occur in select geological environments (sea floor and shelf deposits and some onshore deposits) and are characterised by high CO_3 content and poorly developed crystallinity. Previous work (McClellan, 1980, and others) has shown that these 'primary' francolites (i.e., high CO_3 varieties) will have unit-cell *a*-values of 0.9325 nm and low mean indices of refraction (1.600 ± 0.005). These crystallographic properties vary systematically with francolite composition (McClellan & Gremillion, 1980). Thus the francolites with high degrees of carbonate for phosphate substitution represent the closest known materials to the original francolite.

Several examples of phosphorite weathering (Morocco, Senegal, Israel, Florida, and Togo) have shown that the francolites lose CO_2 and change composition towards fluorapatite (Lucas *et al.*, 1980; McClellan, 1980). Slight weathering of the type seen in portions of Senegal, Morocco, Israel, and Florida can cause an enrichment of the deposit by removal of carbonate from the matrix and the francolite. In many of the ores, the less altered francolites occur at depth in the deposits or in intimate association with calcite or dolomite. When the accessory carbonates are completely removed by weathering, the francolites can be altered more rapidly and may form iron–aluminium phosphates. This removal of carbonates has occurred in the phosphate ores of Christmas Island, Senegal, and Florida.

Thermal alterations of francolites can also produce fluorapatite from francolites. Lehr *et al.* (1967) showed that heating francolites stepwise to 1000 °C caused a progressive decarbonation of the francolites. As alteration progressed, intermediate compositions and, eventually, fluorapatite formed, with the attendant systematic changes in unit-cell *a*-dimensions. These changes caused a large decrease in reactivity, as measured by solubility, when temperatures increased and compositional changes and recrystallisation took place. The progressive thermal changes in francolites have been used to interpret the results of thermal metamorphism in the 'Mottled Zone' of Israel (Mathews & Nathan, 1977). Low-grade, regional metamorphism may also account for some of the variations observed in samples from various areas of the Phosphoria Formation (Gulbrandsen, 1970). If the hypothesis concerning the original composition of francolites is correct, the observed variations of francolites from the same deposit are most probably the result of post-depositional alteration rather than original environments of deposition. There are no data on the effect of pressure alone or in combination with temperature on the alteration of francolites. The presence of the gaseous constituents (F and CO_2) in francolites would imply that pressure could increase the field of stability and might offset the effects of temperature over a given range. Thus, mixtures of carbonates with francolite could protect the apatites from thermal as well as chemical change.

Temporal models have been developed to explain the distribution of phosphates in sediments over the last 3400 m.y. (Cook & McElhinny, 1979). The compilation shows that major phosphogenic episodes have occurred from Early Proterozoic to Miocene–Pliocene times. The variation in francolite composition (using the *a*-value as an indicator) for a number of deposits has been plotted according to Cook & McElhinny's time classification scheme, and it clearly shows the expected time trend of the metastable francolites toward fluorapatite composition (Fig. 20.29). The systematic variation of francolite composition with time is interpreted as clear support for the hypothesis of an 'original constant composition' because of the simple relationship that seems to exist between the starting and final francolite composition. This simple relationship is restricted to those apatites which are identified as belonging to the francolite–fluorapatite series. It is interesting that the deposit ages compiled by Cook & McElhinny (1979) provide a near perfect rank-order relationship for these apatite compositions. The highly variable geological histories and sampling errors included in the data make this correlation even more remarkable.

The data collected on the Precambrian–Cambrian samples show that their apatites are the near-end and end member compositions in the francolite–fluorapatite series. The variation in apatite composition within this group covers the last 40% of the total variation of francolites. This is somewhat surprising since compositional variation in these apatites is usually thought to be small. According to the models of McClellan & Clayton (1980), the francolite compositions range from $Ca_{9.82}Na_{0.13}Mg_{0.05}(PO_4)_{5.46}(CO_3)_{0.54}F_{2.21}$ with 39.1% P_2O_5, to pure fluorapatite $Ca_{10}(PO_4)_6F_2$ containing 42.2% P_2O_5.

The $F:P_2O_5$ ratios in these francolites are low and approach or equal the 0.089 value of fluorapatite. In general, these low $F:P_2O_5$ ratios can be interpreted as good indicators of francolite compositions, and experience has shown their applicability to most occurrences. However, work on recent phosphorites (Price & Calvert, 1978; Burnett, 1977) has shown this ratio may be low when some OH^{-1} replaces F^{-1}. Cheney, McClellan & Montgomery (1979) reported similar results for some low F francolites from Peru. There the $F:P_2O_5$ ratios

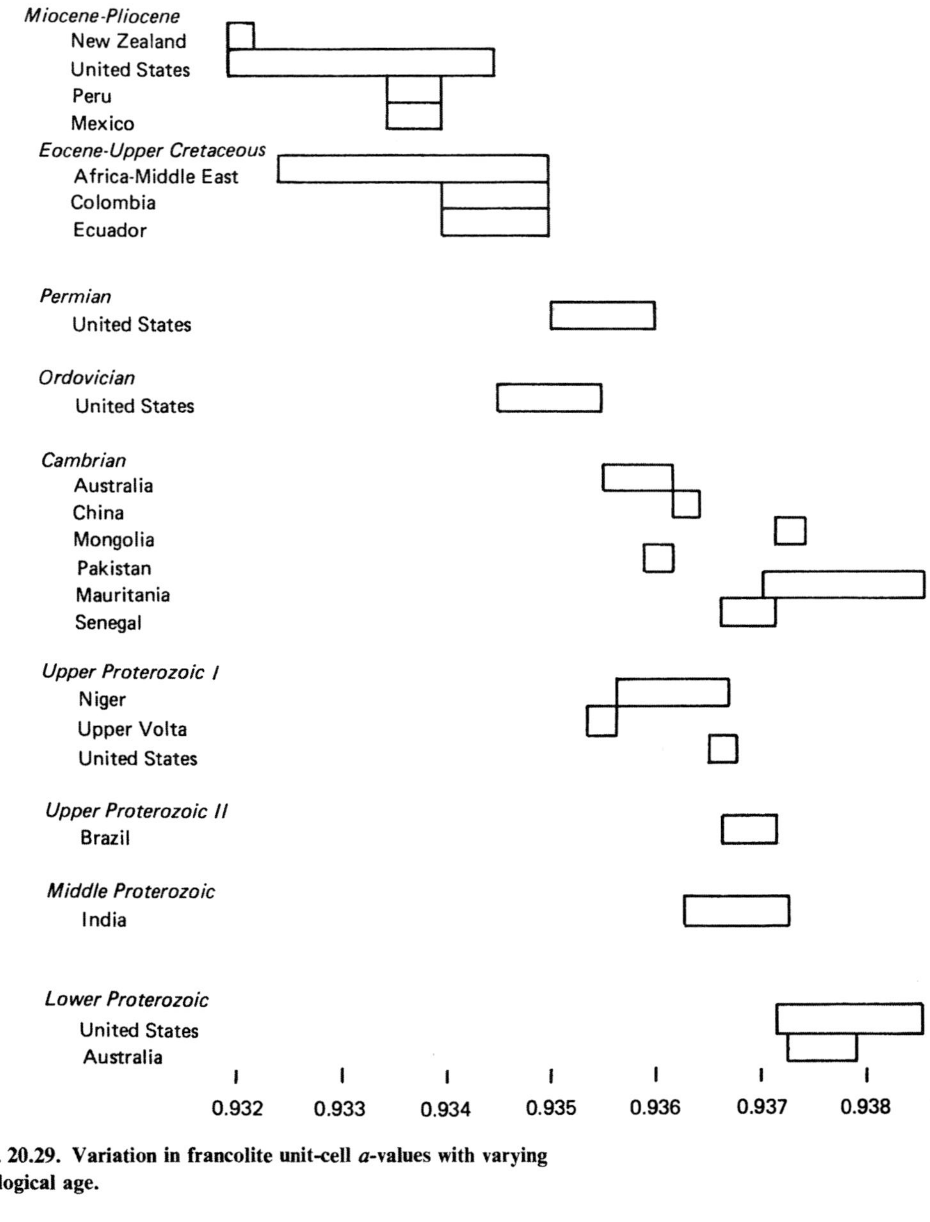

Fig. 20.29. Variation in francolite unit-cell *a*-values with varying geological age.

seems to fit most but not all francolite examples.

In conclusion the mineralogy of the Precambrian–Cambrian francolites can provide very useful information if the data are collected carefully and interpreted reasonably. The differences between samples largely result from post-depositional alteration by any of a variety of processes. The results of the alteration seem to follow a simple, predictable course that systematically changes the crystallographic properties of francolites originating from diverse samples. As more data are collected on ancient phosphorites, then perhaps new and more refined relationships can be established that will enhance the understanding of a complex mineral species with great economic and scientific importance. The corollary that ignorance of these changes and of their geological importance can lead to misinterpretations and potentially serious errors is equally important.

References

Banerjee, D.J., Basu, P.C. & Srivastava, N. (1980). Petrology, mineralogy, geochemistry, and origin of the Precambrian Aravallian phosphorite deposits of Udaipur and Jhabua, India. *Economic Geology*, **75**, 1181–99.

Bates, R.L. & Jackson, J.A. (1980). *Glossary of Geology*, 2nd ed, American Geological Institute Publishers, Falls Church, Virginia.

Bhatti, N. (1979). Hazara Phosphate Deposits – Pakistan. In *Proterozoic–Cambrian Phosphorites*, ed. P.J. Cook & J.H. Shergold, p. 44. ANU Press, Canberra.

Brower, J.C. (1973). Geology of the East Fork mine and vicinity, Sevier County, Tennessee. Unpublished M.Sc. Thesis, University of Tennessee, Knoxville, Tennessee.

Brown, L.L., Rule, A.R. & Dallenbach, C.B. (1981). Characterization and beneficiation of phosphate-bearing rocks from Northern Michigan. *United States Department of Interior, Bureau of Mines Report of Investigations*, **8562**, 16 pp.

Burnett, W.C. (1977). Geochemistry and origin of phosphorite deposits from off Chile and Peru. *Geological Society of America Bulletin*, **88**, 813–23.

Bushinsky, G.I. (1969). *Old Phosphorites of Asia and Their Genesis*, 266pp. Israel Program for Scientific Translations, Jerusalem.

Cannon, W.F. & Klasner, J.S. (1976). Phosphorite and other apatite-bearing sedimentary rocks in the Precambrian of northern Michigan. *United States Geological Survey Circular*, **746**, 6 pp.

Cathcart, J.B. (1978). Phosphorite deposits near Patos de Minas, Minas Gerais, Brazil. *Transactions American Institute of Mining and Metallurgical Engineers*, **264**, 1703–8.

Cheney, T.M., McClellan, G.H. & Montgomery, E.S. (1979). Sechura phosphate deposits, their stratigraphy, origin, and composition. *Economic Geology*, **74**, 232–59.

Chien, S.H. & Black, C.A. (1976). Free energy of formation of carbonate apatites in some phosphate rocks. *Journal of the Soil Science Society of America*, **40**, 234–9.

Choudhuri, R. (1979). The characteristics of the Jhamarkotra phosphorite deposit in Udaipur district, Rajasthan, India. In *Proterozoic–Cambrian Phosphorites*. ed. P.J. Cook & J.H. Shergold, pp. 40–2. ANU Press, Canberra.

Cook, P.J. (1972). Petrology and geochemistry of the phosphate deposits of Northwest Queensland, Australia. *Economic Geology*, **67**, 1193–213.

Cook, P.J. & McElhinny, M.W. (1979). A reevaluation of the spatial and temporal distribution of sedimentary phosphate deposits in the light of plate tectonics. *Economic Geology*, **74(2)**, 315–30.

Cook, P.J. & Shergold, J.H. (Eds.) (1979). *Proterozoic–Cambrian Phosphorites*, 106 pp. ANU Press, Canberra.

Davidson, C.F. (1963). Phosphate deposits of Precambrian age. *Mining Magazine*, **109**, 205–8.

Davis, J.C. (1973). *Statistics and Data Analysis in Geology*. John Wiley, New York, 550pp.

Fisher, D.J. & McConnell, D. (1969). Aluminium-rich apatite. *Science*, **164**, 551–3.

Geijer, P. (1962). Some aspects of phosphorus in Precambrian sedimentation. *Arkiv for Mineralogi och Geologi*, **3(9)**, 165–86.

Gulbrandsen, R.A. (1970). Relation of carbon dioxide content of apatite of the Phosphoria Formation to the regional facies. *United States Geological Survey Professional Paper*, **700B**, 9–13.

Hamilton, W. (1961). Geology of the Richardson Cove and Jones Cove Quadrangles, Tennessee. *United States Geological Survey Professional Paper*, **349-A**, 55pp.

Howard, P.E. & Hough, M.J. (1979). On the geochemistry and origin of D Tree, Wonarah, and Sherrin Creek phosphorite deposits of the Georgina Basin, Northern Australia. *Economic Geology*, **74**, 260–84.

Ilyin, A.V. (1978). Late Precambrian–Early Cambrian phosphorites in Africa. *Newsletter of International Geological Correlation Programme, Project 156-Phosphorites*, **2**, April 1978, pp. 8–9.

Ilyin, A.V. & Bjamba, Z. (1980). *Guidebook for Excursion: Phosphorites of the Khubsugul Basin, Mongolian People's Republic, 17–28 June, 1980*. USSR National Committee for IGCP, Moscow, pp. 1–61 (in Russian), 62–112 (in English), 113–35 (English and Russian).

Jarvis, I. (1980). Geochemistry of the phosphatic chalks and hardgrounds of the Santonian to early Campanian (Cretaceous) of northern France, *Journal of the Geological Society*, **137**, 705–22.

Khan, H.H. (1979). An appraisal of Precambrian phosphorites of Jhabua M.P., India. In *Proterozoic–Cambrian Phosphorites*, ed. P.J. Cook & J.H. Shergold, pp. 42–3. ANU Press. Canberra.

Kühn, A. & Radlicz, K. (1979). Development of phosphate bearing deposits of Tapoa-Mekrou Basin, southwestern Niger. *Newsletter of International Geological Correlation Programme, Project 156-Phosphorites*, **5**, November 1979, 11–13.

Lecorche, J.P. (1980). The Mauritanides in front of the West African craton. Structure of a key-sector: The region of D'Ijibiten. D.Sc. Thesis, St Jerome, Marsielle.

Lehr, J.R., McClellan, G.H., Smith, J.P. & Frazier, A.W. (1967). Characterization of apatites in commercial phosphate rocks. In *Colloque International sur les Phosphates Mineraux Solides, Toulouse, 16–20 May 1967*, **2**, 29–44. Société Chemique de France, Paris.

Liang, T.Y. & Chang, A.C. (1981). On the characteristic and genesis of late Precambrian phosphorites associated with Gondwana. *Abstract, 4th International Field Workshop and Seminar on Phosphorite*, pp. 36–7. International Geological Correlations Programme, Project 156, Udaipur, India.

Lucas, J., Flicoteaux, R., Nathan, Y., Prévôt, L. & Shahar, Y. (1980). Different aspects of phosphorite weathering. *Society of Economic Paleontologists and Mineralogists, Special Publication*, **29**, 41–51.

Lucas, J., Prévôt, L. & Trompette, R. (1980). Petrology, mineralogy, and geochemistry of the late Precambrian phosphate deposits of Upper Volta (West Africa). *Journal of the Geological Society*, **137(6)**, 787–92.

Manusco, J.J., Longheed, M.S. & Shaw, R. (1975). Carbonate-apatite in Precambrian cherty iron-formation, Barga County, Michigan. *Economic Geology*, **70**, 583–6.

Mathews, A. & Nathan, Y. (1977). The decarbonation of carbonate fluorapatite (francolite). *American Mineralogist*, **62**, 565–73.

McArthur, J.M. (1978). Systematic variations in the contents of Na, Sr, CO_3, and SO_4 in marine carbonate fluorapatite and their relation to weathering. *Chemical Geology*, **21**, 89–112.

McClellan, G.H. (1980). Mineralogy of carbonate fluorapatites. *Journal of the Geological Society*, **137(6)**, 675–81.

McClellan, G.H. & Clayton, W.R. (1980). Francolite: the commercial phosphate mineral. In *Proceedings of 2nd International Congress on Phosphorus Compounds*, pp. 131–44. IMPHOS, Paris.

McClellan, G.H. & Gremillion, L.R. (1980). Evaluation of phosphatic raw materials. In *The Role of Phosphorus in Agriculture*, ed. F.E. Khasawneh, E.C. Sample & E.J. Kamprath, pp. 43–80. Madison, Wisconsin.

McClellan, G.H. & Lehr, J.R. (1969). Crystal chemical investigation of natural apatites. *American Mineralogist*, **54**, 1379–91.

McConnell, D. (1973). *Apatite: Its Crystal Chemistry, Mineralogy, Utilisation, and Geologic and Biologic Occurrences*, 111pp. Springer-Verlag, New York.

Nathan, Y., Shiloni, Y., Roded, R., Gal, I. & Deutsch, Y. (1979). The geochemistry of the northern and central Negev phosphorites. *Geological Survey of Israel, Bulletin*, **73**, 41pp.

Notholt, A.J.G. (1979). Resources of Cambrian and Precambrian sedimentary phosphate rocks. In *Proterozoic-Cambrian Phosphorites*, ed. P.J. Cook & J.H. Shergold, pp. 71–86. ANU Press. Canberra.

Pant, A. (1980). Resource status of rock phosphate deposits in India and areas of future potential. In *Fertilizer Mineral Potential in Asia and the Pacific*, ed. R.P. Sheldon & W.C. Burnett, pp. 331–58. East-West Resource Systems Institute, Honolulu.

Prévôt, L. & Lucas, J. (1980). Behavior of some trace elements in phosphatic sedimentary formations. *Society of Economic Paleontologists and Mineralogists, Special Publication*, **29**, 31–9.

Price, N.B. & Calvert, S.E. (1978). The geochemistry of phosphorites from the Nambian shelf. *Chemical Geology*, **23**, 151–70.

Pritchard, P.W. & Cook, P.J. (1965). Phosphate deposits of the Northern Territory. In *Geology of Australian Ore Deposits*. Volume 1, 2nd edn, ed. J. McAndrew, pp. 219–28. Australasian Institute of Mining Metallurgy, Melbourne.

Reeves, M.J. & Saadi, T.A.K. (1971). Factors controlling the deposition of some phosphate-bearing strata from Jordan. *Economic Geology*, **66**, 451–65.

Rogers, J.K. & Crase, N.J. (1980). The Phosphate Hill rock phosphate deposit. North-West Queensland, Australia – an outline of geological development. In *Fertilizer Mineral Potential in Asia and the Pacific*, ed. R.P. Sheldon & W.C. Burnett, pp. 307–31. East–West Resource Systems Institute, Honolulu.

Sandell, E.B., Hey, M.H. & McConnell, D. (1939). The composition of francolite. *Mineralogical Magazine*, **25**, 395–401.

Sant, V.N. (1979). Precambrian phosphorite of Rajasthan, India: a case history. In *Proterozoic–Cambrian Phosphorites*, ed. P.J. Cook & J.H. Shergold, pp. 39–40. ANU Press, Canberra.

Sant, V.N. (1980). Geology of the Indian Platform and its Phosphate Occurrences. In *Fertilizer Mineral Potential in Asia and the Pacific*, ed. R.P. Sheldon & W.C. Burnett, pp. 29–48. East-West Resource Systems Institute, Honolulu.

Slansky, M. (1979). West African Infracambrian phosphorites. In *Proterozoic–Cambrian Phosphorites*, ed. P.J. Cook & J.H. Shergold, p. 51. ANU Press, Canberra.

Trompette, R. (1973). The Upper Precambrian and Lower Paleozoic of the 'Adrar of Mauritania' western part of the Taoudeni Basin, western Africa. Thesis summary, Works of the Laboratory of Earth Sciences, Saint Jerome, Marseille.

Trompette, R., Affaton, P., Joulia, F. & Marchand, J. (1980). Stratigraphic and structural controls of late Precambrian phosphate deposits of the northern Volta Basin in Upper Volta, Niger, and Benin, West Africa. *Economic Geology*, **75**, 62–70.

Ullmer, E. (1981). A mid-Proterozoic phosphate occurrence in east-central Minnesota. *Newsletter of International Geological Correlation Programme, Project 156-Phosphorites*, **8**, April 1981, 22–5.

Wedow, H. Jr, Carpenter, R.H. & Lehr, J.R. (1968). Phosphorite in the Precambrian Ocoee Series of the East Fork manganese district, Sevier County, Tennessee. *Geological Society of America, Special Paper*, **101**, 380 (Abstract).

21

Proterozoic and Cambrian phosphorites – specialist studies: geochemistry of organic matter in Middle Cambrian phosphorites from the Georgina Basin, north-eastern Australia

M.W. SANDSTROM

Introduction

The organic matter in sediments that have experienced little thermal stress is the result of contributing source organisms and the biological and chemical diagenetic processes that have occurred after deposition of the sediments. Characterisation of the nature and distribution of this organic matter in terms of its source and transformations has proved to be a useful approach to the interpretation of sedimentary biogeochemical processes and depositional environments. Biological marker analysis is one such approach widely used in environmental and petroleum source-rock analysis in which specific hydrocarbons in organic extracts are related to contributing organisms and diagenetic processes (Eglinton, 1973; Seifert & Moldowan, 1979). Another, less specific technique is the chemical and isotopic characterisation of complex macromolecular organic compounds (e.g. humic acids and kerogen) isolated from sediments (Stuermer, Peters & Kaplan, 1978). Sedimentary phosphorites are particularly suitable for this type of study because the organic matter associated with the phosphorites may be directly or indirectly related to their origin, possibly serving as the immediate source of the phosphorus ultimately fixed in a mineral phase in the sediments.

Investigations of modern and ancient phosphorites have recognised the importance of biological productivity as a mechanism for concentrating the phosphorus, although a detailed understanding of the nature of this association is lacking (Sheldon, 1981). Indirect processes for concentrating the phosphorus include regeneration of plankton-derived organic matter (Baturin & Bezrukov, 1979) or fish debris (Suess, 1981) during early diagenesis, while a more direct role for concentrating and/or precipitation of phosphorus as apatite by micro-organisms within the sediments has also been suggested (Riggs, 1979; O'Brien & Veeh, 1980). Analysis of biological marker compounds derived from such organisms may provide a better understanding of the importance of these processes in different sedimentary phosphorite deposits. In addition, the preservation and distribution of organic matter in the phosphorites may reflect the winnowing and sorting processes that are proposed to explain the formation of high-grade phosphorite deposits.

This paper examines the relationship between organic matter and phosphorus in some Middle Cambrian phosphorites from the Georgina Basin, northern Australia. These phosphorites are part of a major phosphogenic province that extends from Australia through south-east Asia to eastern USSR (Cook & Shergold, 1979). Previous studies have identified a number of local and global palaeoceanographic and palaeoclimatic factors which may have been important for the formation of these phosphorites. Characterisation of the organic matter associated with the phosphorites may provide a better understanding of the relative importance of these factors in existing genetic models. Moreover, a comparison of the characteristic features of the organic matter in these phosphorites with that in other phosphogenic provinces may ultimately aid in understanding the genesis of phosphorites.

Unweathered phosphorites and associated sediments from the Duchess and Ardmore phosphorite deposits were selected for organic geochemical analysis. Capillary gas chromatography (GC) and computerised gas chromatography–mass spectrometry (GC–MS) were used for biomarker analysis of the hydrocarbons in organic extracts. Humic acids and kerogen were characterised by chemical and isotopic analysis.

Geological setting

The Georgina Basin consists of a sequence of mainly Middle Cambrian to Lower Ordovician marine sediments. Major sedimentary phosphorite deposits occur within the Middle Cambrian Beetle Creek Formation, and its equivalents, and are located near the present margins of the basin (Figure 21.1). One of the largest is the Duchess phosphorite deposit, which occurs on the edge of a small, fault-bounded outlier near the south-eastern margin of the basin. The stratigraphy, sedimentology and depositional environments of these sediments have been summarised by Cook & Shergold (1979).

The Beetle Creek Formation is divided into the Monastery Creek Phosphorite and Lower Siltstone Members. The former consists of interbedded grainstone phosphorite, chert, and phosphatic and siliceous siltstone and dark phosphatic chert. Sedimentary features and fossils of the phosphorites indicate a very shallow marine environment. The Inca Formation discon-

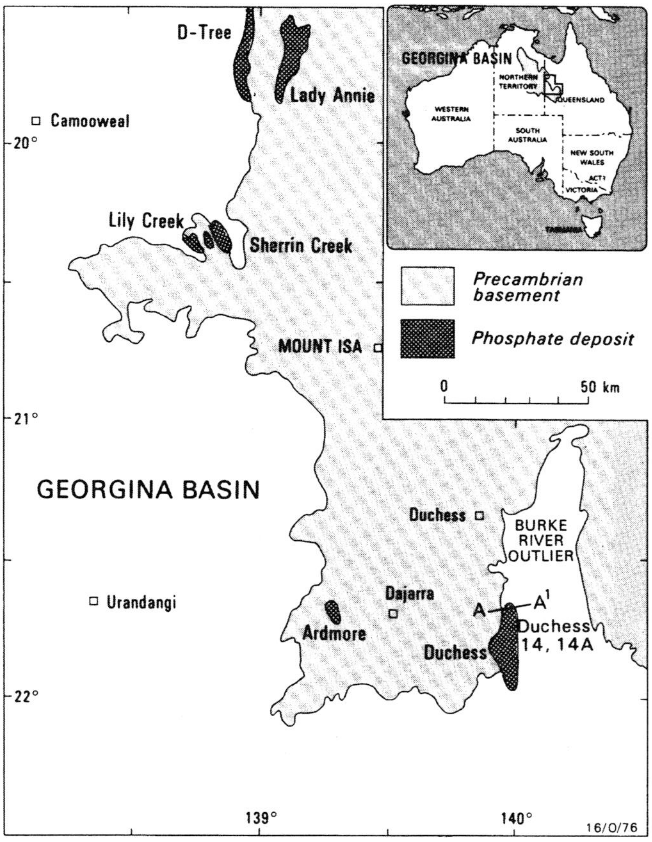

Fig. 21.1. Location of Duchess and Ardmore phosphorite deposits in the Georgina Basin.

formably overlies the Beetle Creek Formation in this area of the Georgina Basin, and consists of thinly laminated siliceous shale and siltstone. Sediments from both formations are black and pyrite-rich below the zone of weathering.

Deposition of Middle Cambrian sediments in the Georgina Basin was associated with a major marine transgression that resulted in the establishment of shallow submergent and epeiric seas over much of the craton. Extension of this narrow seaway from the inferred edge of the continental shelf on the present south-eastern margin of the basin may have led to the establishment of sea current systems, and the associated upwelling was related to the deposition of phosphorites and organic-rich sediments which characterise this time interval (Cook, 1982).

The Georgina Basin phosphorites are attractive for organic geochemical study because the Middle Cambrian sediments have experienced little thermal stress. The sediments are mainly flat-lying or gently dipping, and have not been metamorphosed or intruded by igneous rocks (Draper, Shergold & Heighway, 1978). The maximum depth of burial of phosphorites from the Duchess deposit is estimated to be less than 750 m (Powell, Cook & McKirdy, 1975).

Methods

Drill core samples were selected from Bureau of Mineral Resources (BMR) stratigraphic drill cores Duchess 14 and 14A located near Monastery Creek in the Duchess phosphorite deposit (Fig. 21.1). These holes penetrated the Inca Formation and Monastery Creek Phosphorite Member of the Beetle Creek Formation below the zone of weathering (Shergold & Walter, 1979). Unweathered samples from the Inca Formation and Lower Siltstone Member of the Beetle Creek Formation equivalents in the Ardmore phosphorite deposit were obtained from Broken Hill South drill cores A2 and A3. A complete list of samples is given in Table 21.1.

Core samples were prepared by removal of the outer surfaces with a diamond saw, crushed to small chips which were cleaned ultrasonically in dichloromethane, and pulverised to a fine powder in a stainless steel ring mill. The powdered rock samples were extracted in a Soxhlet apparatus with dichloromethane-methanol (9:1) for 100 hrs, with one solvent change after 24 hrs. The total extractable organic matter (EOM) was chromatographed on a column packed with 1:2 alumina over silica gel, successively eluting with two column-volumes each of hexane (aliphatic hydrocarbons), benzene (aromatic hydrocarbons) and methanol (heteroatomic hydrocarbons).

Gas chromatographic (GC) analyses were conducted using a Varian model 2440 instrument having a flame ionization detector, splitless injector and low volume detector fittings, and a fused silica capillary column (Hewlett-Packard, 0.2 mm × 25 m) coated with SP-2100. Aliquot samples (0.1–0.4 μl) were injected at ambient temperature, and after two minutes, the oven temperature was increased to 70 °C and then at 4 °C/min to 265 °C. Helium was used as carrier gas, at a flow rate of 22 cm/sec.

Combined gas-chromatographic–mass spectrometric (GC–MS) analyses were carried out on a Finnegan model 4023 quadrupole mass spectrometer interfaced directly with a Finnegan model 9610 gas chromatograph, equipped with a glass WCOT column (30 m × 0.25 mm i.d.) coated with SP-2100 and temperature programmed from 100 °C to 270 °C at 4 °C/min. The separator temperature was 250°C, ion source temperature was 260 °C and ionization voltage was 70 eV. The mass spectrometric data were acquired and processed using a Finnigan-Incos Model 2300 data system.

Humic acids and kerogen were isolated by the methods described by Stuermer *et al.* (1978). The humic acids were purified by gel permeation chromatography on a column (1.5 × 40 cm) of Bio-gel P-2 (Bio-Rad, Inc.). Kerogen was purified by treatment with 25% $AlCl_3$ $6H_2O$ to remove fluoride minerals. Carbon, hydrogen, nitrogen, sulfur, and ash analyses were determined by the analytical facility of the Australian National University. Iron was determined by atomic absorption spectrometry and used to calculate sulfide sulfur for correction of organic sulfur values (Durand & Monin, 1980). Stable carbon isotope analyses were determined by the method of Sofer (1980).

Results

Carbon and phosphorus distribution

There are major differences in the amount and type of organic matter in the Beetle Creek Formation phosphorites compared to other Middle Cambrian sediments. The distribution of organic carbon, sulfur, and phosphorus in samples from Inca and Beetle Creek Formations from the Duchess deposit, and stratigraphic equivalents from the Ardmore deposit, are listed in table 21.2. Organic carbon concentrations in

Table 21.1. *Description of samples (drillcore, subsurface depth, stratigraphic unit, and lithology) from Duchess and Ardmore phosphorite deposits*

Sample	Subsurface depth (m)	Stratigraphic unit	Lithology
BMR drill core Duchess 14 (21°40'S; 139°50'E)			
5677	75.2	Inca Formation	Laminated, black siltstone
5678	76.9	Inca Formation	Laminated, black siltstone
5671	79.1	Inca Formation	Laminated, black siltstone
5679	81.2	Inca Formation	Laminated, black siltstone
5672	82.5	Beetle Creek Formation; Monastery Creek Member	Black grainstone phosphorite
5673	82.9	Beetle Creek Formation; Monastery Creek Member	Black grainstone phosphorite
5680	84.5	Beetle Creek Formation; Monastery Creek Member	Black grainstone phosphorite
5681	88.2	Beetle Creek Formation; Monastery Creek Member	Black grainstone phosphorite
5674	89.1	Beetle Creek Formation; Monastery Creek Member	Black grainstone phosphorite
BMR drill core Duchess 15 (50 m west of Duchess 14)			
5675	85.1	Beetle Creek Formation, Monastery Creek Member	Black grainstone phosphorite
5676	85.5	Beetle Creek Formation, Monastery Creek Member	Grey, friable phosphorite
Ardmore A2			
5685	46.3	Blazan Shale Formation (Inca equivalent)	Laminated, black siltstone
5686	52.0	Simpson Creek Formation; Lower Siltstone Member	Laminated, black siltstone
5687	58.2	Thorntonia Limestone	
Ardmore A3			
5688	58.5	Simpson Creek Formation; Lower Siltstone Member	Laminated, black siltstone

Table 21.2. *Results of organic carbon, sulfur, phosphorus, and semi-quantitative X-ray diffraction mineralogical analysis of drillcore samples from Duchess and Ardmore phosphorite deposits. All values in weight %.*

Sample	Organic carbon	Sulfur	Phosphorus	Quartz	Apatite	Pyrite	Apatite–CO_2^1
Inca Formation							
5677	4.76	1.88	1.16	85	11	4	—
5678	3.52	1.75	0.29	81	9	4	—
5671	2.82	0.22	0.29	100	—	—	—
5679	2.52	0.39	0.37	91	2	5	—
Monastery Creek, Beetle Creek Formation							
5672	1.16	0.51	14.41	9	89	2	2.1
5673	0.23	0.45	14.67	14	84	2	1.2
5680	1.16	0.60	10.24	28	69	3	1.8
5681	0.51	0.49	15.24	8	88	4	1.8
5674	0.40	0.29	6.19	59	39	2	2.4
5675	1.51	0.76	13.49	24	74	2	2.1
5676	0.19	0.41	14.92	13	85	2	1.2
Blazan Shale							
5685	4.11	2.27	0.35	82	—	5	—
Lower Siltstone; Simpson Creek Formation							
5686	8.87	2.62	0.89	95	—	5	—
5688	8.55	1.87	0.18	94	—	4	—
Thorntonia Limestone							
5687	1.11	4.11	4.91	39	43	8	—

—, not detected.

[1]Calculated indirectly using peak-pair method of Gulbrandsen (1970).

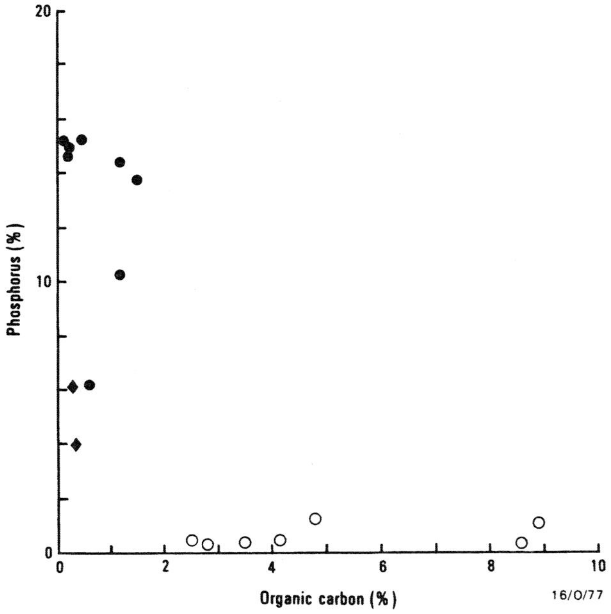

Fig. 21.2. Phosphorus and organic carbon concentrations in Middle Cambrian phosphorites (●) and siltstones (○) from the Georgina Basin, Queensland, Australia and phosphorites (◆) from Yunnan, China.

the Monastery Creek phosphorites are low, from 0.4% to 1.5%, compared to the overlying Inca siltstones, which have values from 2.5% to 4.8%. The Lower Siltstone samples from Ardmore, which occur stratigraphically below the phosphorites, have very high organic carbon contents, up to 8.9%. Sulfur concentrations are also lower in the Monastery Creek phosphorites compared to the associated siltstones. Sulfur occurs in the form of fine-grained, framboidal pyrite in all samples. The distribution of phosphorus in the samples mainly reflects variations in the abundance of carbonate–fluorapatite, the main phosphorus-bearing mineral phase which occurs as fine-grained ovules and skeletal fragments in the samples.

Phosphorus and organic carbon concentrations are shown in Fig. 21.2, together with values for samples from the Cambrian Yunnan phosphorite deposit of China. All Cambrian phosphorite samples are characterised by lower amounts of organic carbon than the associated nonphosphatic siltstones. Similar results for other phosphogenic provinces have also been reported by Powell *et al.* (1975) and Maughan (1980).

Petrographic examination of polished thin-sections of drill-core samples indicates that, although the amounts of organic matter are different, morphologically similar types of organic matter occur in all stratigraphic units. The predominant form is very fine lamellar filaments compressed between mineral grains. In reflected white light some of these filaments are dark grey and exhibit a dull orange or no fluorescence. Additional light-grey lamellar filaments display variable fluorescence intensities and colours ranging from yellow-green to dull orange. Other types of organic matter include discrete irregularly shaped particles which have a mottled or spongy appearance, are dark grey in reflected white light, and have little fluorescence. Small spheri-

Table 21.3. *Gravimetric data for extractable organic matter (EOM) and hydrocarbon compound class distribution in Georgina Basin drillcores*

Sample	Organic carbon (%)	EOM	Hydrocarbon compound class*: Aliphatic	Aromatic	Heteroatomic	S/A[†]	Yield[‡]	HC/OC[§]
		μg/g dry sediment					mg/g	
Duchess 14; Inca Formation								
5677	4.76	1898	183	206	431	0.89	39.9	8.2
5671	2.82	2007	360	236	713	1.53	71.2	21.1
Duchess 14; Monastery Creek Member, Beetle Creek Formation								
5672	1.16	83	14	20	49	0.69	7.2	3.0
5673	0.23	(19)	1	2	30	0.63	8.2	1.1
5680	1.16	171	27	29	79	0.94	14.7	4.9
5674	0.40	57	57	8	24	0.66	14.1	3.3
Duchess 15; Monastery Creek Member, Beetle Creek Formation								
5675	1.51	487	23	55	248	0.41	32.2	5.2
5676	0.19	(39)	2	3	49	0.81	20.7	2.5
Ardmore A2								
5685	4.11	2646	159	1097	688	0.14	64.4	30.6
5686	8.87	3935	262	1073	1792	0.24	44.4	15.1
5687	1.11	248	57	104	299	0.55	67.4	14.5
Ardmore A3								
5688	8.55	5034	986	2016	1756	0.49	58.9	35.1

*Gravimetric quantitation of column chromatography fractions.
†Ratio of aliphatic to aromatic hydrocarbon fractions.
‡Yield of EOM, normalised to organic carbon.
§Sum of aliphatic and aromatic hydrocarbons relative to organic carbon.

cal, discrete particles which may be acritarchs or other microfossils display bright yellow green fluorescence. In the Monastery Creek phosphorites, the same types of organic matter occur both within and interstitial to the phosphate grains.

The reflectance measurements of these discrete dark-grey filaments and particles, which are from 0.2% to 0.5%, and the bright fluorescence intensities of the other particles confirm that the samples have experienced little thermal stress.

Hydrocarbon distribution

The amount of extractable organic matter (EOM) and bulk hydrocarbon compound class distibution determined by gravimetric analysis of selected samples from the Duchess and Ardmore phosphorite deposits are given in Table 21.3. In general, Monastery Creek phosphorite samples have much lower amounts of EOM and hydrocarbons than samples from associated nonphosphatic siltstones. These differences are smaller, although still apparent, when the values are normalised to organic carbon content. The yields of EOM and hydrocarbons (i.e. proportion of organic matter or hydrocarbons soluble in organic solvents) are plotted in Fig. 21.3. The Monastery Creek phosphorites are characterised by relatively low EOM and hydrocarbon yields compared to other Middle Cambrian sediments sampled, or shales of low maturity in general. However, the hydrocarbon compound class distribution is similar for samples from all stratigraphic units, with the bulk of the EOM consisting of sulfur, nitrogen, and oxygen-containing (heteroatomic) hydrocarbons.

These results differ from the very high EOM yields reported by Powell *et al.* (1975) as a characteristic feature of the phosphorites they examined from different phosphogenic provinces. Since differences in analytical techniques (e.g. extraction solvents) may explain part of this difference, additional sampling and analysis by the same technique is required to determine if the Cambrian phosphorites have different EOM yields compared with other phosphorites.

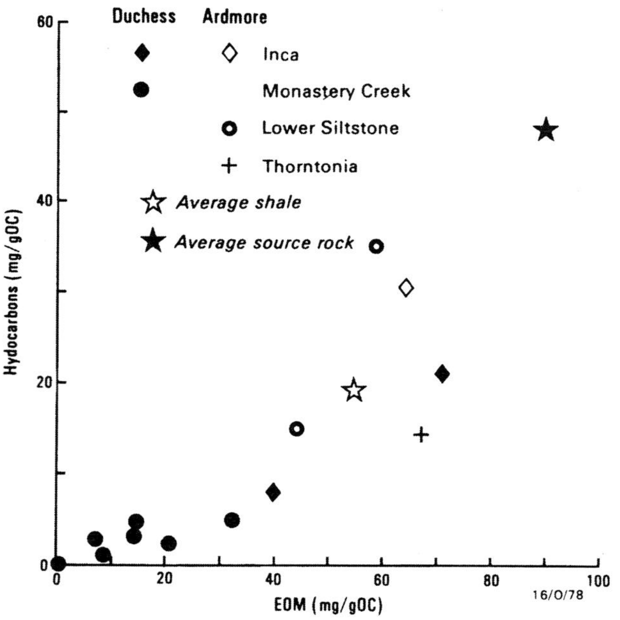

Fig. 21.3. Yield of extractable organic matter (EOM) and hydrocarbon in Middle Cambrian sediments from Duchess and Ardmore phosphorite deposits. Values for average immature shale and source rock from Tissot & Welte (1978).

Biomarker analysis

The molecular composition of the aliphatic hydrocarbon fractions demonstrate a variety of hydrocarbon distributions within the Middle Cambrian sediments. Quantitative amounts of individual components are lowest in Monastery Creek phosphorites, although qualitatively similar hydrocarbon distributions occur in all stratigraphic units. Examples of capillary gas chromatograms of samples from Monastery Creek phosphorites are shown in Fig. 21.4. Resolved components consist predominantly of *n*-alkanes, acyclic isoprenoids, and *n*-alkylcyclohexanes, which vary in abundance relative to the unresolved complex mixture (UCM).

A homologous series of *n*-alkanes with carbon numbers from 13 to 28, maximising near *n*-C_{17} or *n*-C_{18}, and exhibiting no carbon number predominance, are the most abundant resolved components in all samples. These hydrocarbons are attributed to either direct contribution from bacteria (Han & Calvin, 1969), or to the indirect result of microbial diagenesis of algal biolipids (Johnson & Calder, 1973; Hatcher, Simoneit & Gerchakov, 1977). The *n*-alkane distributions are generally comparable to those characteristic of the phosphorites reported by Powell *et al.* (1975) although the carbon number maxima are lower in the Monastery Creek phosphorites compared to the predominance of *n*-C_{22} observed in other phosphorites. Most of the phosphorites analysed by Powell *et al.* (1975) were from outcrop, and so these differences may be merely due to evaporative or weathering loss.

The acyclic isoprenoids pristane, phytane, and the C_{16} and C_{18} homologues occur in each sample, with minor differences in distributions within the stratigraphic units. Pristane is the most abundant isoprenoid, with more variable but approximately equal amounts of phytane and the C_{18} isoprenoid (Fig. 21.5) occurring in all samples. Pristane to phytane ratios for all samples are within a narrow range of 1.2–2.4, and demonstrate no correlation with stratigraphic unit, organic carbon, sulfur or phosphorus content. Acyclic isoprenoids with carbon numbers greater than 20 were not detected.

The distribution of acyclic isoprenoids in these samples is also attributed to direct or indirect derivation from microorganisms. The neutral lipids of Archaebacteria and certain other species of anaerobic photosynthetic bacteria contain minor amounts of C_{15}–C_{20} acyclic isoprenoids (Han & Calvin, 1969; Holzer, Oro & Tornaberne, 1979). The methanogenic and thermoacidophilic micro-organisms, however, also contain relatively large amounts of C_{25} and C_{30} isoprenoids, which were not detected in the Middle Cambrian sediments. Phytol from algal-derived chlorophyll may also be a precursor to the isoprenoids in the sediments. In Recent sediments, the geological diagenesis of phytol results in the formation of pristane or phytane, depending on the oxidation state of the sediments (e.g. Didyk *et al.*, 1978; Brooks, Gould & Smith, 1969). Hence the pristane/phytane ratios between 1.2 and 2.4 for the Middle Cambrian sediments would indicate mildly oxic conditions during sedimentation. Yet this contrasts with the relatively high

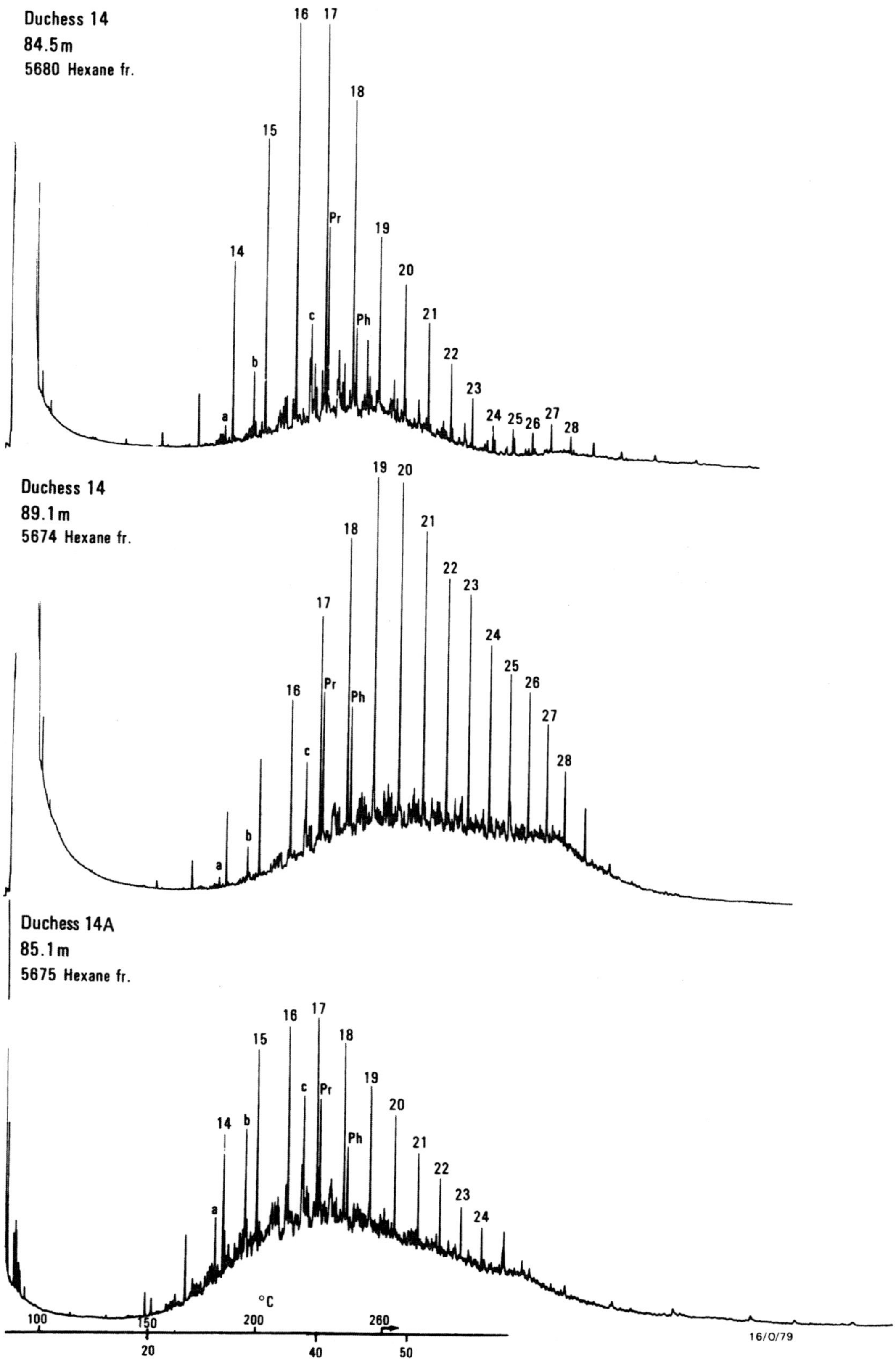

Fig. 21.4. Gas chromatograms of aliphatic hydrocarbons extracted from Monastery Creek phosphorites, BMR drillcores Duchess 14 and 14A. Designated peaks refer to *n*-alkane carbon numbers and acyclic isoprenoids: Pr = pristane; Ph = phytane; a = C_{15} isoprenoid; b = C_{16} isoprenoid; c = C_{18} isoprenoid.

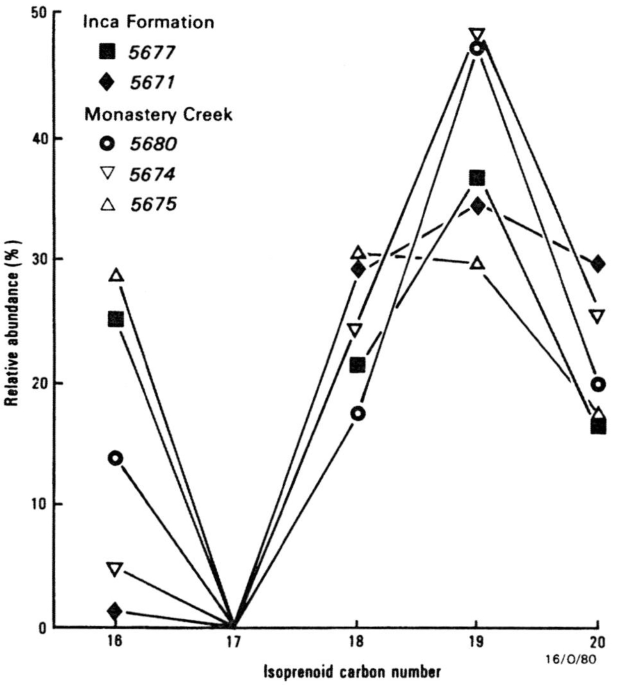

Fig. 21.5. Relative abundance of acyclic isoprenoids in aliphatic hydrocarbons extracted from Middle Cambrian sediments, Duchess phosphorite deposit, BMR drillcores Duchess 14 and 14A.

sulfur content, and abundance and preservation of pyrite that suggests a more anaerobic depositional environment. This geological diagenesis of phytol also does not explain the presence of the C_{16} and C_{18} isoprenoids. Alternatively, biological diagenesis of chlorophyll may provide a better explanation for the distribution of acyclic isoprenoids in these sediments. For example, a comparable distribution of C_{18}–C_{20} isoprenoids occurs in coorongite, and is attributed to microbial degradation of chlorophyll from the *Botryococcus* alga which is the predominant source of organic matter in the coorongite (Douglas, Eglington & Maxwell, 1969).

Other minor resolved components tentatively identified by GC–MS are a homologous series of *n*-alkylcyclohexanes with carbon numbers from 17 to 22, maximising at 18. An example of the m/z 83 mass fragmentogram, characteristic for this compound class, and the corresponding reconstructed ion chromatogram are shown in Fig. 21.6.

The *n*-alkylcyclohexanes do not occur in living organisms and thus represent products of biological or geological diagenesis. Similar compounds have been identified in a Precambrian seep oil and Devonian shale extract and attributed to the cyclisation of monounsaturated fatty acids (Johns *et al.*, 1966). Another possible precursor is the cyclisation of alk-l-enes. These unsaturated hydrocarbons have been identified in coorongite (Douglas *et al.*, 1969), which also has *n*-alkane and isoprenoid distributions similar to that of the Monastery Creek phosphorites. The *n*-alkylcyclohexanes in the phosphorites may be derived from microbial degradation processes similar to those that have occurred to the algal organic matter in the formation of coorongite.

A characteristic feature of the molecular composition of the Monastery Creek phosphorites is the lack of, or very low concentration of steroid and triterpenoid hydrocarbons. These compound classes are ubiquitous biological marker hydrocarbons, occurring in many Recent and ancient sediments and fossil fuel samples, predominantly derived from algae and bacteria. Characteristic mass fragmentograms for steranes (m/z 217), rearranged steranes (m/z 259), and triterpenoids (m/z 191) indicated the absence of these compounds in Monastery Creek phosphorites. In contrast, both steranes and triterpanes occur in sample 5671 from the Inca Formation siltstones.

In these siltstones, the predominant steroid hydrocarbons are C_{27}–C_{29} diasteranes with the 13β(H), 17α(H) stereochemistry, and the 20 S compound more abundant than the 20 R epimer. These compounds are formed by clay-catalysed rearrangement of sterenes during geological diagenesis (Rubinstein, Sieskind & Albrecht, 1975). Also present are minor amounts of C_{27}–C_{29} regular steranes. The stereochemical configuration of the C_{29}-steranes, with an abundance of the geologically stable 5α, 14α, 17α 20 S and 5α, 14β, 17β 20 R,S isomers compared to the 5α, 14α, 17α 20 R compound indicate an advanced stage of thermodynamic equilibrium for the organic matter in the sediments (Seifert & Moldowan, 1981). However, recent studies have shown that extensive biodegradation results in preferential removal of regular steranes compared to diasteranes, and of the 5α, 14α, 17α 20 R compound compared to the 5α, 14α, 17α 20 S epimer (Seifert & Moldowan, 1979), as observed in the Inca Formation. hence the relative sterane abundances may be better explained by extensive biodegradation rather than an advanced stage of thermodynamic equilibrium.

The triterpanes identified in the Inca Formation include a series of tricyclic alkylated diterpanes which range from C_{20} to C_{27}, excluding the C_{22} compound, and maximise at C_{23}. Minor amounts of C_{27}, C_{29}, and C_{30} 17α(H), 21β(H)-hopanes, and 18α(H),22, 29, 30-trisnorhopane also occur in this sample.

Humic acids and kerogen

Humic acids and kerogen were isolated from selected samples of Monastery Creek phosphorites and Inca siltstones from the Duchess deposit. Chemical composition and stable carbon isotope data for these fractions are given in Tables 21.4 and 21.5. The amount of humic acids relative to organic carbon and phosphorus content of the samples is shown in Fig. 21.7.

The Monastery Creek phosphorites are characterised by relatively high amounts of humic acids, which represent 5–57% of the total organic carbon. The proportion of kerogen carbon varies accordingly, comprising 3–80% of the total organic carbon, with EOM and water- and acid-soluble components representing minor portions of the total organic carbon. In contrast, Inca Formation samples contain very small amounts of humic acids while kerogen represents more than 90% of the total organic carbon. Samples from the Lower Siltstone Member from Ardmore also contained negligible amounts of humic acids.

In Recent marine sediments humic acids may comprise a substantial part of the organic matter, representing between 10% and 70% of the total organic matter (Nissenbaum & Kaplan, 1972; Stuermer *et al.*, 1978; Huc & Durand, 1977). During burial diagenesis humic substances are converted to kerogen through loss of functional groups and increased

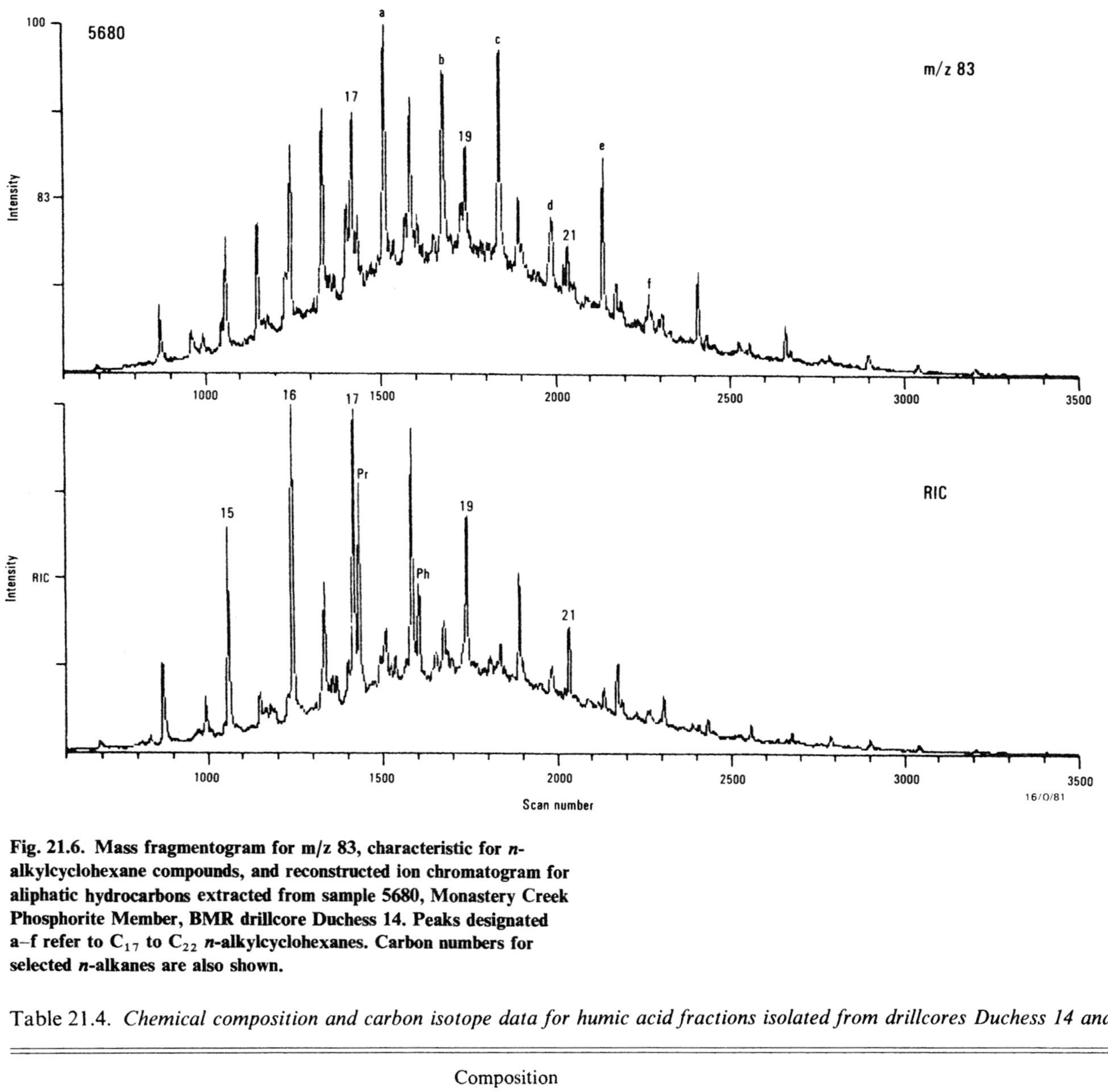

Fig. 21.6. Mass fragmentogram for m/z 83, characteristic for *n*-alkylcyclohexane compounds, and reconstructed ion chromatogram for aliphatic hydrocarbons extracted from sample 5680, Monastery Creek Phosphorite Member, BMR drillcore Duchess 14. Peaks designated a–f refer to C_{17} to C_{22} *n*-alkylcyclohexanes. Carbon numbers for selected *n*-alkanes are also shown.

Table 21.4. *Chemical composition and carbon isotope data for humic acid fractions isolated from drillcores Duchess 14 and 14A*

Sample	Sample* yield (mg/g DS)	Composition (% ash-free) C	H	N	S	O†	P	Ash	Atomic ratio H/C	N/C	S/C	O/C	C/P	$\delta^{13}C_{PDB}$ (ppt)
Inca Formation														
5677	0.56	51.24	6.08	1.31	0.07	41.3	—	44.87	1.41	0.022	—	0.61	—	—
Beetle Creek Formation														
5672	3.66	49.73	3.89	1.17	2.38	42.84	—	23.66	0.93	0.02	0.017	0.65	—	—
5673	2.82	53.13	3.34	1.83	2.85	38.81	0.054	35.36	0.75	0.029	0.019	0.55	2537	−30.5
5680	3.43	37.05	5.13	1.24	1.31	55.27	—	56.57	1.65	0.029	0.013	1.12	—	−30.7
5681	5.86	47.31	4.8	1.49	2.13	44.19	0.079	42.13	1.21	0.027	0.016	0.7	1545	−32.3
5674	1.44	37.82	3.33	0.65	1.35	56.85	0.013	32.41	1.05	0.015	0.013	1.13	7502	—
5675	1.69	59.17	4.38	2.09	2.99	31.36	0.016	18.3	0.88	0.03	0.018	0.4	9537	—
5676	2.94	57.76	3.36	2.33	3.01	33.48	0.055	36.6	0.69	0.035	0.018	0.44	2708	—

*Dry weight of humic acid fraction, mg/g dry sediment.
†Determined by difference.
—, not determined.
PDB in $\delta^{13}C_{PDB}$ refers to a belemnite, *Belemnitella americana* from the Cretaceous Peedee Formation of South Carolina. The ^{13}C to ^{12}C ratio measured on this fossil has become an internationally recognised standard reference point.

Table 21.5. *Chemical composition and carbon isotope data for kerogen fractions isolated from drillcores Duchess 14 and 14A*

Sample	Sample* yield (mg/g DS)	Composition (% ash-free) C	H	N	S†	O‡	P	Fe	(%) Ash	Atomic ratio H/C	N/C	S/C	O/C	C/P	$\delta^{13}C_{PDB}$ (ppt)
Inca Formation															
5677	71.88	87.62	8.25	2.23	4.1	<1.0	0.041	4.93	23.8	1.12	0.022	0.017	<0.01	5512	−32.0
5678	63.75	84.68	8.8	2.01	3.01	1.46	0.047	6.27	27.0	1.24	0.02	0.013	0.013	4647	−31.6
5671	36.2	79.6	8.26	1.64	2.98	7.5	0.026	0.78	10.21	1.24	0.018	0.013	0.071	7876	−32.2
Beetle Creek Formation															
5672	10.36	76.98	6.15	1.37	2.4	13.08	0.025	4.51	10.64	0.95	0.015	0.011	0.13	7941	−32.0
5673	2.62	44.73	2.73	1.28	<0.1	51.08	0.18	1.06	43.88	0.73	0.024	—	0.86	641	−32.0
5680	7.81	75.05	6.84	1.59	4.73	11.75	0.038	5.16	16.93	1.09	0.018	0.022	0.12	5094	−32.6
5681	7.02	67.77	5.28	1.85	10.05′	15.04	—	—	14.82	0.93	0.023	0.052′	0.17	—	
5674	3.41	69.04	5.26	1.49	16.08′	8.14	—	—	20.87	0.91	0.019	0.082′	0.09	—	−31.6
5675	18.75	85.86	7.91	2.16	5.57	<1.0	0.031	7.49	16.73	1.1	0.022	0.023	<0.01	7143	−32.2

*Dry weight of kerogen fraction, mg/g dry sediment.
‡Determined by difference.
†Determined by difference of total sulfur and equivalent sulfide sulfur from iron analysis.
Includes sulfide sulfur.
—, not determined.

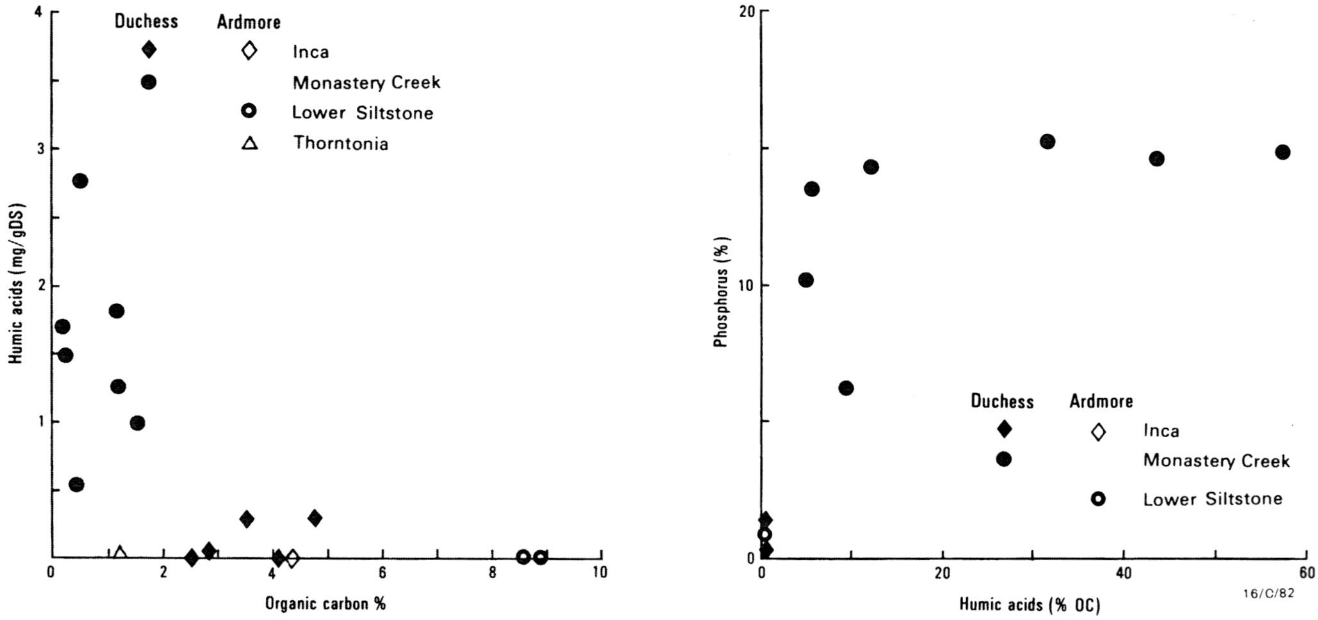

Fig. 21.7. Amount of humic acids (mg/g dry sediment) compared to percentage of organic carbon, and expressed as percentage of total organic carbon compared to total phosphorus content of Middle Cambrian phosphorites (●,◆) and siltstones (O,◇) from Georgina Basin.

polycondensation and polymerisation reactions (Welte, 1974; Nissenbaum, 1974) so that both ancient and thermally mature sediments generally contain few humic substances. The presence of humic acids in the Cambrian phosphorites, therefore, is unusual compared to other ancient sediments.

The differences in humic acid content between the Monastery Creek phosphorites and associated nonphosphatic siltstones may be the result of early diagenetic processes such as anaerobic microbial degradation of organic matter associated with the phosphorites. Although the long term effect of burial diagenesis is the conversion of humic acids to kerogen, during early diagenesis reactive portions of sedimentary organic matter (protokerogen) may undergo oxidative biological degradation to form humic substances (Stuermer *et al.*, 1978). A significantly higher amount of aerobic degradation of the organic matter associated with the phosphorites may explain an initial higher proportion of humic acids. Lithological differences may have been equally important. For example, the inclusion of early formed humic material within phosphate grains and the adsorption of humic acids on the surfaces of carbonate or apatite grains may have prevented subsequent polymerisation of humic acids to form kerogen. The limited data available indicate some correlation of humic acid content with phosphorus (i.e. apatite) content (Fig. 21.7).

The chemical composition of the humic acid and kerogen fractions isolated from Monastery Creek phosphorites and Inca siltstones also indicates a different degree or type of diagenesis of organic matter rather than different source. Kerogen isolated from Inca siltstones has relatively high H/C (1.1–1.2) and low O/C (0.01–0.07) ratios. This is similar to kerogen derived from lipid-rich algal material, such as coorongite or algal mats, and is classified as Type I in the scheme of Tissot *et al.* (1974). Kerogen isolated from Monastery Creek phosphorites has slightly lower H/C (0.9–1.1) and higher O/C (0.01–0.17) ratios than the Inca kerogens, and is similar to that of Type II kerogen, which in pre-Devonian sediments is thought to be derived from partially degraded algal and microbial organic matter (McKirdy, 1977). The humic acids are characterised by higher O/C ratios compared to the kerogen fractions and is a reflection of the higher content of oxygenated functional groups in the humic acids, and suggests that aerobic degradation processes and incorporation of oxygen into the organic matter were important during formation of the humic acids.

The sulfur content of the kerogen isolated from the Monastery Creek phosphorites is higher than kerogen from the Inca siltstones, in spite of the much higher total sulfur and pyrite content of the Inca sediments. This suggests that diagenetic reactions between organic matter and microbially-produced sulfide were favoured in the depositional environment of the phosphorites, while in the Inca siltstones sufficient iron was available for pyrite formation and minimal organic matter–sulfide reactions. Relatively high S/C ratios of kerogen isolated from other phosphorites was previously described as a characteristic feature of the organic matter (Powell *et al.*, 1975).

The stable carbon isotope values for kerogen and humic acid fractions isolated from Inca siltstones and Monastery Creek phosphorites from drillcore Duchess 14 are given in Tables 21.4 and 21.5. All kerogen samples fall within the range -31.6 to -32.6 ppt, with no significant difference in carbon isotope values between the stratigraphic units. These results suggest a common source for the organic matter in both stratigraphic units.

The carbon isotope values of the humic acids are enriched in carbon-13 compared to the kerogen fractions, as a result of the higher content of oxygenated functional groups (Galimov, 1980). These results also demonstrate that the humic acids are syngenetic and have not been introduced as a result of postdepositional groundwater movement, since modern soil humic acids have carbon isotope values in the range -25.0 ppt to -29.0 ppt (Nissenbaum, 1974).

Discussion

Depositional environment

The distribution and type of organic matter associated with the Middle Cambrian phosphorites from the Georgina Basin are best explained by a more extensive and perhaps aerobic, microbial degradation of the organic matter compared to that in associated organic-rich shales and siltstones. This is consistent with the assumption that organic matter, with an atomic C/P ratio of approximately 106:1, served as the immediate source of the phosphorus now fixed in the sediment in the form of sedimentary minerals, since it implies degradation and removal of a significant amount of organic carbon.

The results reported here demonstrate that humic acids represent 5–60% of the total organic carbon in the Monastery Creek phosphorites, whereas the associated organic-rich shales and siltstones contain negligible amounts of humic acids. In addition, the kerogen isolated from the phosphorites has lower H/C and higher O/C ratios than the kerogen from the Inca Formation samples, although the carbon isotopic composition of the kerogen from both stratigraphic units is similar. These data suggest that the humic acids in the Middle Cambrian phosphorites were formed by microbial degradation of protokerogen during the early stages of biological diagenesis. Because the humic acids have higher oxygen content than the kerogen, this transformation of the organic matter was probably the result of aerobic microbial processes.

The molecular composition of the aliphatic hydrocarbons isolated from the Monastery Creek phosphorites are also attributed to extensive degradation and resynthesis of precursor algal material by micro-organisms. The smooth distribution of *n*-alkanes maximising near *n*-C_{18}, the occurrence of C_{16}–C_{20} acyclic isoprenoids, and presence of the *n*-alkylcyclohexanes are comparable to hydrocarbon distributions observed in some Recent sediments and coorongite which have been attributed to microbial degradation of algal biolipids. For example, the series of *n*-alkylcyclohexanes may be derived from cyclisation of alk-l-enes which occur in algae and coorongite. Although the hydrocarbon distributions are non-specific as to the type of microbial organism responsible for the degradation of the algal material, the absence of C_{25} and C_{30} acyclic isoprenoids in the phosphorites suggests that methanogens and extreme halophiles were not predominant micro-organisms, at least at the final stages of biological diagenesis of the phosphorites.

At the same time, the complete degradation and resynthesis of algal biolipids by micro-organisms means that interpretation of the original phytoplankton source organism is not possible. Consequently the molecular composition of the aliphatic hydrocarbons provides no information or constraints on the type of algae (planktonic or benthic) responsible for the initial accumulation of phosphorus now contained in the sediment in the form of apatite. These results are consistent with the petrographic examination of the phosphorites, which indicates that there are no well-preserved algal bodies, with the exception of minor amounts of acritarchs.

The distribution of aliphatic hydrocarbons in the Inca Formation and the Lower Siltstone Member shales and siltstones is qualitatively similar to the Monastery Creek phosphorites, although quantitatively the hydrocarbons are at least an order of magnitude more abundant in the shales and siltstones. This indicates that if a different (aerobic?) community of micro-organisms is responsible for the lower organic carbon and higher phosphorus and humic acid content which characterise the phosphorites, these micro-organisms did not contain a unique assemblage of hydrocarbons which were fossilised in the EOM of the sediments.

Although microbial degradation is the preferred explanation for the lower amount of organic matter contained in the phosphorites, physical processes such as winnowing may also have been important. Baturin (1971) attributed the decrease in organic carbon in progressively more lithified Recent phosphorites to such a mechanism, while Maughan (1980) similarly explained the negative correlation between organic carbon and phosphorus in Phosphoria Formation sediments to winnowing by waves and currents. Although the grain texture and sedimentary structures in the Monastery Creek phosphorites indicate moderately high energy conditions at times (de Keyser & Cook, 1972), the occurrence of fine-grained organic matter between the phosphate grains indicates that winnowing was limited and can not entirely account for the low organic carbon content of the phosphorites. Similarly, the low amount of steranes and triterpanes contained in the phosphorites is better explained by additional, or more extensive, microbial degradation of the organic matter associated with the phosphorites than removal by waves and currents.

In summary, the nature and content of organic matter contained in Monastery Creek phosphorites suggest that the organic matter has been subjected to more extensive aerobic degradation than the organic matter associated with shales and siltstones of the Inca and the Lower Siltstone Member of the Beetle Creek Formation. The source of the organic material now contained in all of the stratigraphic units was probably a similar type of marine alga. This degradation of the organic matter in the phosphorites may have occurred entirely within the water column of the Duchess embayment prior to burial, or may have occurred as a result of subsequent winnowing and erosion of the sediments or as a combination of both processes. In either case, this more efficient aerobic degradation resulted in lower total organic carbon content, and higher amounts of humic acids which characterise the Monastery Creek phosphorites. The phosphorus released from this degradation may have been fixed in the sediments as a direct result of this process (e.g. humic acid–phosphorus adsorption) or because of the presence of suitable inorganic substrate for the diagenetic or authigenic precipitation of apatite. The accumulation of phosphorus and removal of carbon in the phosphorites, in contrast to the associated organic-rich shales, indicates that local sedimentological and oceanographic factors, as well as global oceanographic factors, are important in determining whether the sediments will ultimately be considered potential sources of phosphorus or petroleum.

Comparison with other phosphogenic provinces

Although data available on the organic geochemistry of sediments from other phosphogenic provinces is limited, the organic matter associated with the Cambrian phosphorites appears to share features common to other phosphorites. Powell *et al.* (1975) and Maughan (1980) noted the lower content of total organic carbon in phosphorites compared with associated organic-rich shales in the Permian Phosphoria province. The *n*-alkane distribution patterns in phosphorites from Cretaceous to Eocene Tethys province (Amit & Bein, 1982) and the Permian Phosphoria province (Powell *et al.*, 1975) are similar to those of the Cambrian phosphorites and are attributed to microbial degradation of precursor algal biolipids. This hydrocarbon distribution is not unique to phosphorites, however, since it also occurs in coorongite (Douglas *et al.*, 1969) and Recent carbonate sediments (Johnson & Calder, 1973).

Both the Cambrian and Cretaceous–Eocene phosphorites are characterised by relatively high amounts of humic acids as a percentage of total organic carbon. The Cambrian phosphorites are also similar to modern submarine phosphorites from Peru–Chile in the absence of steranes and triterpanes in the aliphatic hydrocarbons (Sandstrom, 1982), which are ubiquitous components of nonphosphatic sediments. These may be general features of the organic matter in phosphorites related to oxidation of the organic matter during a winnowing stage in the formation of the phosphorites.

Summary

Phosphorites from the Georgina Basin are characterised by relatively low amounts of EOM, high concentration of humic acids, and an aliphatic hydrocarbon distribution indicative of extensive microbial transformation of plankton-derived organic matter. Humic acids represent 5–60% of the total organic carbon in the phosphorites, while the associated organic-rich shales and siltstones contain negligible amounts of humic acids. These results, and the lower H/C and higher O/C ratios of the phosphorite kerogen, suggest that the humic acids were formed by aerobic microbial degradation of the phosphorite protokerogen during the early stages of diagenesis. The molecular composition of the aliphatic hydrocarbons is also attributed to extensive microbial degradation and resynthesis of precursor algal material. The smooth distribution of *n*-alkanes maximising near n-C_{19}, the C_{16-20} acyclic isoprenoids, the *n*-alkylcyclohexanes, and low amounts of steranes and triterpanes are evidence of this biological diagenesis.

The diagenetic sequence leading to the formation of the phosphorites may have involved a reworking or winnowing stage as proposed in models for the origin of Recent phosphorites, at which time the organic matter was exposed to more extensive microbial diagenesis. Since phosphorus is released preferentially during early diagenesis, and the geopolymers (i.e. kerogen and humic acids) do not represent a siginficant sediment phosphorus reservoir (Sandstrom, 1982), it is unlikely that this reworking and subsequent stage of diagenesis provided a source of phosphorus that was fixed in the sediment in the form of apatite. Moreover, the absence of steranes and triterpanes, and occurrence of fine-grained intergranular organic matter suggest that the low amount of organic matter in the phosphorites is the result of biological diagenesis rather than simple winnowing.

References

Amit, O. & Bein, A. (1982). Organic matter in Senonian phosphorites from Israel – origin and diagenesis. *Chemical Geology*, **37**, 277–87.

Baturin, G.N. (1971). Formation of phosphate sediments and water dynamics. *Oceanology*, **11**, 372–6.

Baturin, G.N. & Bezrukov, P.L. (1979). Phosphorites on the sea floor and their origin. *Marine Geology*, **31(314)**, 137–332.

Brooks, J.D., Gould, K. & Smith, J.W. (1969). Isoprenoid hydrocarbons in coal and petroleum. *Nature*, **222**, 257–9.

Cook, P.J. (1982). The Cambrian palaeogeography of Australia and opportunities for petroleum exploration. *Australian Petroleum Exploration Association Journal*, **22(1)**, 42–64.

Cook, P.J. & Shergold, J.H. (Eds.) (1979). *Proterozoic-Cambrian Phosphorites*, 106pp. ANU Press, Canberra.

de Keyser, F. & Cook, P.J. (1972). Geology of the Middle Cambrian phosphorites and associated sediments in northwestern Queensland. *Bulletin, Bureau of Mineral Resources, Geology and Geophysics of Australia*, **138**, 79pp.

Didyk, B.M., Simoneit, B.R.T., Brassell, S.C. & Eglinton, G. (1978). Organic geochemical indicators of paleoenvironmental conditions of sedimentation. *Nature*, **272**, 216–22.

Douglas, A.G., Eglinton, G. & Maxwell, J.R. (1969). The hydrocarbons of coorongite. *Geochimica et Cosmochimica Acta*, **33**, 569–77.

Draper, J.J., Shergold, J.H., & Heighway, K.A. (1978). A review of petroleum exploration and prospects in the Georgina Basin. Record, Bureau of Mineral Resources, Geology and Geophysics of Australia, **1978/10** (Unpublished).

Durand, B. & Monin, J.C. (1980). Elemental analysis of kerogens. In *Kerogen*, ed. B. Durand, pp. 114–42. Editions Technip, Paris.

Eglinton, G. (1973). Chemical fossils: a combined organic geochemical and environmental approach. *Pure and Applied Chemistry*, **34**, 611–32.

Galimov, E.M. (1980). $^{13}C/^{12}C$ in kerogen. In *Kerogen*, ed. B. Durand, pp. 271–99. Editions Technip, Paris.

Gulbrandsen, R.A. (1970). Relation for carbon dioxide content of apatite of the Phosphoria Formation to regional facies. *United States Geological Survey Professional Paper*, **700-B**, B913.

Han, J. & Calvin, M. (1969). Hydrocarbon distribution of algae and bacteria and microbiological activity in sediments. *Proceedings of the National Academy of Sciences of the USA*, **64**, 436–43.

Hatcher, P.G., Simoneit, B.R.T. & Gerchakov, S.M. (1977). The organic geochemistry of a Recent sapropelic environment: Mangrove Lake Bermuda. In *Advances in Organic Geochemistry 1975*, ed. R. Campos & J. Goni, pp. 469–84. Enadimsa, Madrid.

Holzer, G., Oro, J. & Tornabene, T.G. (1979). Gas chromatographic-mass spectrometric analysis of neutral lipids from methanogenic and thermoacidophilic bacteria. *Journal of Chromatography*, **186**, 795–809.

Huc, A.Y. & Durand, B. (1977). Occurrence and significance of humic acids in ancient sediments. *Fuel*, **56**, 73–80.

Johns, R.B., Belsky, T., McCarthy, E.D., Burlingame, A.L., Haug, P., Schones, H.K., Richter, W. & Calvin, M. (1966). The organic geochemistry of ancient sediments – Part II. *Geochimica et Cosmochimica Acta*, **30**, 1191–222.

Johnson, R.W. & Calder, J.A. (1973). Early diagenesis of fatty acids and hydrocarbons in a salt marsh environment. *Geochimica et Cosmochimica Acta*, **37**, 1943–55.

McKirdy, D.M. (1977). Diagenesis of microbial organic matter: a geochemical classification and its use in evaluating hydrocarbon-generating potential of Proterozoic and Lower Paleozoic sediments, Amadeus Basin, central Australia. Ph.D. Thesis, Australian National University, Canberra (unpublished).

Maughan, E.K. (1980). Relation of phosphorite, organic carbon and hydrocarbons in the Phosphoria Formation, western U.S.A. In: *Géologie Comparée des Gisements des Phosphates et de Pétrole, Colloque International*, pp. 63–91. Bureau de Recherches Géologiques et Minières Document 24.

Nissenbaum, A. (1974). The organic geochemistry of marine and terrestrial humic substances: implications of carbon and hydrogen isotope studies. In *Advances in Organic Geochemistry 1973*, ed. B. Tissot & F. Bienner, pp. 39–51. Editions Technip, Paris.

Nissenbaum, A. & Kaplan, I.R. (1972). Chemical and isotopic evidence for the *in situ* origin of marine humic substances. *Limnology and Oceanography*, **17**, 570–82.

O'Brien, G.W. & Veeh, H.H. (1980). Holocene phosphorite on the East Australian continental margin. *Nature*, **288**, 690–2.

Powell, T.G., Cook, P.J. & McKirdy, D.M. (1975). Organic geochemistry of phosphorites: relevance to petroleum geochemistry. *American Association of Petroleum Geologists, Bulletin*, **59**, 618–32.

Riggs, S.R. (1979). Phosphorite sedimentation in Florida – a model phosphogenic system. *Economic Geology*, **74**, 285–314.

Rubinstein, I., Sieskind, O. & Albrecht, P. (1975). Rearranged sterenes in shale: occurrence and simulated formation. *Journal of the Chemical Society, Perkin Transactions*, **1**, 1833–6.

Sandstrom, M.W. (1982). Organic geochemistry of phosphorites and associated sediments. Ph.D. Thesis, Australian National University, Canberra (unpublished).

Seifert, W.K. & Moldowan, J.M. (1979). The effect of biodegradation on steranes and terpanes in crude oils. *Geochimica et Cosmochimica Acta*, **43**, 111–26.

Seifert, W.K. & Moldowan, J.M. (1981). Paleoreconstruction by biological markers. *Geochimica et Cosmochimica Acta*, **45**, 783–94.

Sheldon, R.P. (1981). Ancient marine phosphorites. *Annual Review of Earth and Planetary Science*, **9**, 251–84.

Shergold, J.H. & Walter, M.R. (1979). BMR stratigraphic drilling in the Georgina Basin, 1977 and 1978. *Record, Bureau of Mineral Resources, Geology and Geophysics of Australia*, **1979/36**.

Sofer, Z. (1980). Preparation of carbon dioxide for stable carbon isotope analysis of petroleum fractions. *Analytical Chemistry*, **52**, 1389–91.

Stuermer, D.H., Peters, K.E. & Kaplan, I.R. (1978). Source indicators of humic substances and proto-kerogen. *Geochimica et Cosmochimica Acta*, **42**, 989–98.

Suess, E. (1981). Phosphate regeneration from sediments of the Peru continental margin by dissolution of fish debris. *Geochimica et Cosmochimica Acta*, **45**, 577–88.

Tissot, B.P. & Welte, D.H. (1978). *Petroleum Formation and Occurrence*, 538pp. Springer-Verlag, Berlin.

Tissot, B., Durand, B., Espitale, J. & Combaz, A. (1974). Influence of nature and diagenesis of organic matter in formation of petroleum. *American Association of Petroleum Geologists, Bulletin*, **58**, 499–506.

Welte, D. (1974). Recent advances in organic geochemistry of humic substances and kerogen. A review. *Advances in Organic Geochemistry 1973*, ed. B. Tissot & F. Bienner, pp. 1–13. Editions Technip, Paris.

22

Proterozoic and Cambrian phosphorites – specialist studies: Early Cambrian palaeogeography, palaeoceanography and phosphorites

J.T. PARRISH, A.M. ZIEGLER, C.R. SCOTESE, R.G. HUMPHREVILLE AND J.L. KIRSCHVINK

Introduction

Phosphorites are not distributed uniformly in time and space (Sheldon, 1964; McKelvey, 1967; Cook & McElhinny, 1979; Sheldon & Burnett, 1981a, b; Arthur & Jenkyns, 1981). The Early Palaeozoic and the Late Cretaceous–Tertiary were the most important phosphogenic times, as expressed in number of deposits and estimated reserves. Jurassic and Permian deposits, though fewer in number, also contain large reserves (Cook & McElhinny, 1979). The most significant phosphogenic episode of the Early Palaeozoic was during the Early Cambrian. The subject of this paper is the distribution of Early Cambrian phosphorites and the purpose is to put these deposits into a global perspective. In order to accomplish this goal, we have reconstructed Early Cambrian global palaeogeography and palaeoclimatology. In addition, we present Late Cambrian palaeogeographic and palaeoclimatic reconstructions representing revisions of those presented in Scotese *et al.* (1979) and Parrish (1982).

Reconstructing Cambrian global palaeogeography has proved difficult (Palmer, 1974; Ziegler *et al.*, 1977, 1979; Scotese *et al.*, 1979) because the positions of many continents at that time are not well known. No undeformed oceanic crust of Palaeozoic age exists, so the magnetic lineations formed during sea-floor spreading cannot be used to orient continents. Consequently, the palaeomagnetic data used in the reconstructions presented here are wholly from continental rocks. Palaeomagnetic information must be selected carefully to avoid relying on spurious poles that may have resulted from subsequent magnetisations and tectonic disturbance. In addition, global correlation of Cambrian rocks is difficult because radiometric dates are less reliable for older rocks and Cambrian biostratigraphic relationships are not well known on a global scale (Palmer, 1972). The uncertainties in dating and correlation make it difficult to reconstruct the palaeogeography for any given time. Especially difficult is the reconstruction of such transient palaeogeographic features as shorelines and highlands.

Despite these problems, we believe that new attempts to reconstruct the Cambrian world are worthwhile. Global reconstructions limited to one age are preferable, because it is not necessary to average a wide range of tectonic and sea-level changes. Elsewhere, we (Scotese *et al.*, 1979) have presented a global palaeogeographic reconstruction for the Franconian Age (Late Cambrian), which is here revised. In the Early Cambrian reconstruction, palaeogeographic features such as shorelines are based on lithological data for the entire epoch, which encompasses approximately 30 million years. Worldwide biostratigraphy for the Early Cambrian is not sufficiently detailed to permit choosing a shorter time span. However, radiometric dating does allow us to select palaeomagnetic data limited to the earliest Early Cambrian (roughly equivalent to the Tommotian).

Palaeogeography

Methods used to produce global palaeogeographic maps are discussed fully in Ziegler *et al.* (1977, 1979) and in Scotese *et al.* (1979). Our approach is to integrate data from a variety of sources, including palaeomagnetic, biogeographic, palaeoclimatic, and tectonic. The primary data for features such as shorelines consisted of published descriptions of stratigraphic sections. We relied on the palaeogeographic reconstructions of other workers as guides only, preferring to make our own interpretations of local palaeogeography in the context of the regional palaeogeographic setting. We have not presented here the lithological data used in constructing the palaeogeographic maps. Sources for Late Cambrian lithological data are found in Ziegler *et al.* (1979); sources for the Early Cambrian data are included in the references in this paper. The scarcity of known Cambrian rocks necessitated extrapolating features such as shorelines between control points from widely separated geographical areas, especially in Gondwanaland (see also Palmer, 1974).

Cambrian continents

The Cambrian reconstructions presented here are similar to those of Smith, Hurley & Briden (1981, based on palaeomagnetic data), Zonenshayn & Gorodnitsky (1977, based on tectonic data), and Burrett & Richardson (1980, based on faunal data). The Early Cambrian continents are outlined in Fig. 22.1, with present geographic features indicated for orientation. We have also included the modern 5° latitude-longitude

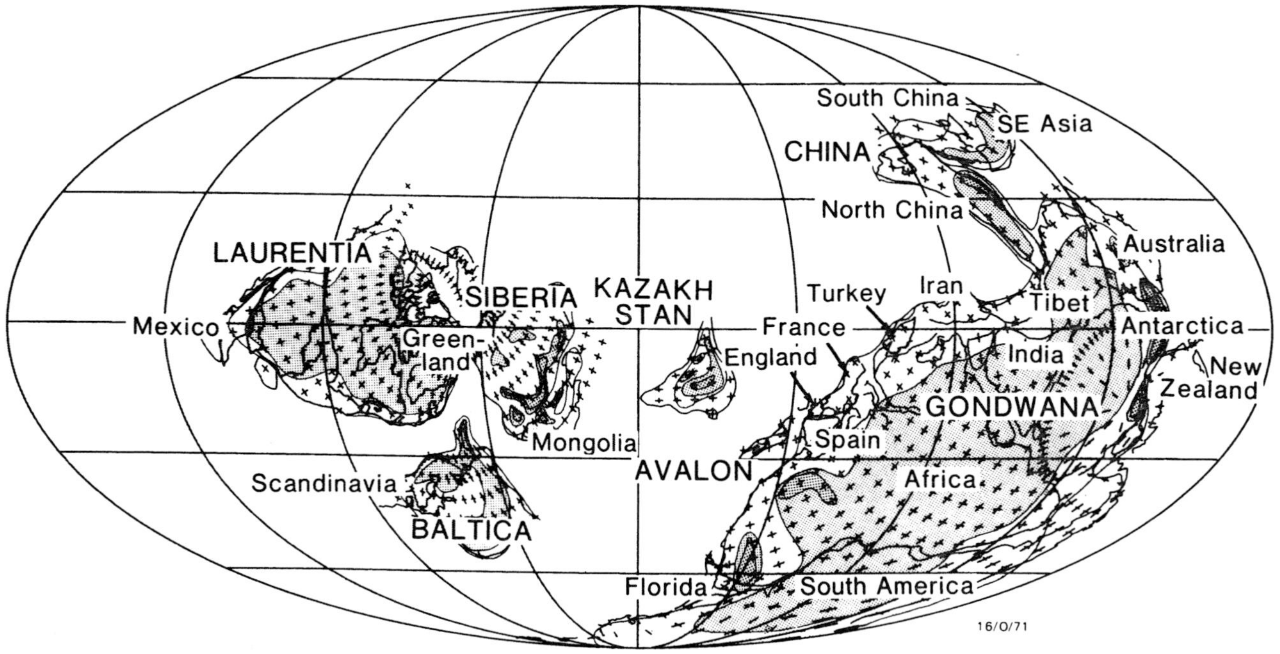

Fig. 22.1. The Early Cambrian continents. Bold letters indicate palaeocontinents (see text for discussion) and the light letters indicate modern geographic features. Palaeogeography: dark shading = highlands, medium shading = lowlands, light shading = shelf.

intersections. Gondwanaland consisted of present-day Africa, South America, Australia, Antarctica, India, Tibet, southern Europe, Iran, and Turkey. Avalon included Nova Scotia, south-eastern Newfoundland, and probably also England. Laurentia consisted of most of present-day North America. Siberia included northern Mongolia. Baltica consisted of northern Europe west of the Urals. We believe that Kazakhstan was a separate unit (see discussion following); China lay near Gondwanaland and may have been part of it.

We have kept north China, south China, and south-east Asia together in these reconstructions and in the position indicated by the Cambrian pole from north China reported by Khramov (1975). These land masses certainly were separated (McElhinny *et al.*, 1981), but we do not know what their relative positions were. In the Early Cambrian, and indeed throughout the Palaeozoic, faunas in north and south China, though distinctive, were never very different. This faunal evidence suggests that the two continental fragments were in proximity throughout their pre-collision histories. On the other hand, the Lower Cambrian phosphorite deposits in north China are commonly associated with gypsum, which is not seen in the deposits of south China (Yeh, Sun & Chen, 1982). Moreover, Cambrian lithofacies in south China indicate that it might have been part of another land area to the west, although western South China subsequently has been so deformed that a detailed palaeogeographic analysis of the region may now be impossible. We will return to a discussion of China's palaeogeographic position later in this paper.

Kazakhstan is now generally accepted to have been a separate land mass at some time in its history (e.g. Burke, Dewey & Kidd, 1977; Ziegler *et al.*, 1977; McElhinny *et al.*, 1981); when and for how long it was separate is still unknown. For several reasons we reconstruct Kazakhstan as separate in the Cambrian. First, it contains ophiolites that increase in age, from Devonian to Precambrian, toward the centre (Zonenshayn, 1973; Antonyuk *et al.*, 1977), indicating a long history of accretion of smaller blocks. Kazakhstan consists of at least two microcontinents, the Yerementau–Balkhash block and the Kokchetav–Chuya block. The latter block is itself composed of smaller fragments formed during the early and middle Precambrian, including the Ulu Tau Massif, site of the major Kazakhstanian phosphorites. That the Yerementau–Balkhash and Kokchetav–Chuya blocks were joined by Cambrian time is not clear, a collision having been inferred from the circumstantial evidence that both blocks underwent a severe 'destructive' phase (Antonyuk *et al.*, 1977), beginning at the Precambrian–Cambrian boundary. Secondly, Kazakhstan has abundant latest Precambrian to Cambrian tholeiitic basalts and andesites, which are indicative of subduction in an island-arc setting (Antonyuk *et al.*, 1977; Vinogradov, 1968). Indeed, the geochemical evolution of the volcanic suites suggested to Antonyuk *et al.* (1977) that the Yerementau–Balkhash block was initially an island arc, becoming more truly continental in the Palaeozoic, as evidenced by the increase in alkalinity and silica. Thirdly, no evidence exists that Kazakhstan was part of a larger continent prior to or during the Cambrian. Cambrian Kazakhstan can be envisaged properly as an agglomeration of island arcs and older island arc terrains, similar to the present East Indies.

Continental positions

Palaeomagnetic data for the Cambrian are scarce, poorly dated, of uneven quality, and often give conflicting results. Consequently, the reconstructions presented here must be regarded as preliminary; work in progress (Van der Voo,

Scotese & McCable, 1982) provides a more complete documentation of Cambrian palaeomagnetic data. Our earliest Early Cambrian reconstruction of continental positions is based on palaeomagnetic data from the Upper Arumbera Sandstone of Australia (Kirschvink, 1978), unnamed Tommotian rocks of the Lena River Valley, USSR (Kirschvink & Rozanov, 1979; Kirschvink, 1980), intrusives of latest Precambrian–Early Cambrian age in Colorado (French, Alexander & Van der Voo, 1977), and unnamed lowermost Cambrian rocks of the north China Platform (Khramov, 1975). The results from the Arumbera Sandstone are in good agreement with latest Precambrian (Vendian) data from other Gondwanaland continents (Hagstrum *et al.*, 1980; McWilliams, 1981). The Arumbera pole, in African coordinates, plots at 16° S, 25° W; the mean pole for all Gondwanaland continents is at 10° S, 30° W. The new palaeomagnetic determination from Siberia puts that continent on the equator and requires a 90° counterclockwise rotation through the Cambrian. The position of Kazakhstan in our reconstructions is derived from two lines of evidence. First, we extrapolated Kazakhstan's position along a trajectory defined by later Palaeozoic palaeomagnetic data, which show Kazakhstan moving northward. Second, Kazakhstan has Cambrian carbonates that generally are indicative of low latitudes, and late Precambrian tillites that are suggestive of higher latitudes. The simplest explanation of these data is that Kazakhstan maintained its northward movement for a long time, beginning in the southern hemisphere in the late Precambrian (Ziegler *et al.*, 1977). The orientation of Kazakhstan is rather arbitrary and for lack of better data is retained from that indicated by the Devonian pole (Scotese *et al.*, 1979).

The revision of Scotese *et al.*'s (1979) Late Cambrian palaeogeography is different from the earlier reconstruction in two major ways. Gondwanaland is rotated slightly clockwise and is placed farther south. Siberia also is placed farther south. We have repositioned these continents based on new palaeomagnetic data from Kirschvink (1980) and Kirschvink & Rozanov (1979).

As in our previous reconstructions (Scotese *et al.*, 1979), palaeomagnetic data are preferred for determining the palaeolatitudes of the continents. Supplementary techniques can provide additional information about the palaeolatitudinal and palaeolongitudinal positions of the continents, but many of these techniques were helpful only marginally for the Cambrian because the necessary data were not available or were exceptionally ambiguous. The supplementary techniques included using the distribution of climatically significant sedimentary rocks, such as evaporites and coals; using biogeographic patterns; and extrapolating the positions between times of continental rifting and collision.

Climatically significant sediments

Evaporites tend to be deposited in low mid-latitudes where rainfall is low, and coals tend to form in the equatorial and high mid-latitude rainy belts (Drewry, Ramsay & Smith, 1974). Coal and evaporite deposits are most helpful for positioning a continent when they are present together in time or space. A moderate temporal separation with, say, evaporites overlying coals, suggests movement of the continent through a wet belt and then through a dry belt. A spatial separation of coeval evaporites and coals on the same continent is particularly helpful for determining the exact orientation of the continent with respect to the wet and dry belts. Such distributions of coals and evaporites in time and space are found on many continents – for example, in China through the Triassic and Jurassic (Parrish, Ziegler & Scotese, 1982). However, single deposits of coals or evaporites may not be helpful. First, a single deposit suggests a broad latitudinal zone but does not give any clues to the orientation of the plate or its direction of movement, both of which can be important in locating a continent at any given time. Secondly, certain geographic configurations, such as those promoting monsoonal circulation (e.g. Permian – Triassic Pangaea: Parrish, Parrish & Ziegler, in press; Parrish, Ziegler & Scotese, 1982) can distort the latitudinal positions of wet and dry belts. Oceanographic conditions adjacent to the continents, such as upwelling currents, can also distort wet and dry belts from a strictly latitudinal arrangement. For these reasons, the Early Cambrian evaporites of Siberia and Australia, although extensive, might not be particularly helpful by themselves for positioning the continents. Both are the only such deposits on their respective continents and both may have been deposited in regions influenced by upwelling (Cook, 1982). Moreover, Siberia and Australia are two regions for which the palaeomagnetic control is particularly good. The positions of the evaporites away from the equatorial rainy belt at about 15° S palaeolatitude (Siberia) and 12° N palaeolatitude (Australia), confirm the palaeomagnetic data. The new, palaeomagnetically determined positions of Gondwanaland and Siberia in the Late Cambrian (Fig. 22.5) are also more consistent with the distribution of Late Cambrian evaporites than were the positions in the previous reconstruction (Ziegler *et al.*, 1979).

Coals are generally formed in rainy regions by the accumulation of vascular land-plant material. To date, no Cambrian vascular land plants have been described. Lu & Zhu (1981) report coals from China (the so-called 'stone coals' of the Qiongzhusi Formation of late Meishucun age), but these probably are marine and are the offshore equivalents of the Meishucun age phosphorites (Organizing Committee, 1982).

Biogeography

Longitudinal as well as latitudinal positions of continents can commonly be inferred from biogeographic relationships. Cambrian biogeographic patterns were similar throughout the period (Palmer, 1979; Burrett & Richardson, 1980), and we have taken some clues from biogeography in our placement of continents. We recognise that faunal similarity does not necessarily imply geographic proximity (Jell, 1974; Palmer, 1977). For example, in the Early Cambrian, deeper-water faunas are demonstrably more cosmopolitan (Taylor, 1976; Taylor & Forester, 1979), whereas shallow-water faunas tend to be endemic (Palmer, 1977). Nevertheless, the faunas are not entirely endemic nor are they entirely facies controlled, as Jell (1974) demonstrated. Palmer (1979) implied that the provinces were entirely facies controlled, pointing out that the environments, as indicated by lithology, were somewhat different on each continent. However, it seems to us unlikely that the biogeographic units, if they were only facies controlled, would coincide exactly with distinct continental units.

Early Cambrian biogeographic divisions of trilobite faunas are rather diffuse (Palmer, 1979). Similarities exist among faunas in Australia, Antarctica, and China, and among faunas in North America, South America, and Baltica (Palmer, 1979). Siberia, north-western Africa, and south-western Europe contain elements of both biogeographic divisions (Palmer, 1974, 1979). Burrett & Richardson's (1980) quantitative analysis of Early Cambrian trilobite faunas revealed stronger similarities between Siberia and Australia/Antarctica than between Siberia and the other regions. Their analysis also isolated south China as a separate province.

By the Middle Cambrian, the biogeographic divisions were clearer (Jell, 1974). The trilobite faunas in Australia and China, especially in southern China and Antarctica, are quite similar to each other and also share elements with the faunas of Siberia and Kazakhstan. Jell (1974) placed the faunas from all these regions in one province. The fauna of North America warrants its own provincial designation (Jell, 1974); it contains trilobites also found in the medial Middle Cambrian fauna of Argentina, but is otherwise distinct. Jell's (1974) third province included Baltica, southwestern Europe, Avalon, and north-western Africa. In the late Middle Cambrian, the northern Siberian faunas and those of Baltica were somewhat similar. This similarity may be a clue to the timing of the rotation of Siberia throughout the Cambrian. Burrett & Richardson's (1980) analysis largely corroborated Jell's (1974) results.

Late Cambrian biogeographic divisions include four shallow-water and two deep-water provinces (Palmer, 1979). The shallow-water provinces are identical to the Middle Cambrian provinces recognised by Jell (1974) except that Siberia is distinct from China/Australia/Antarctica (Palmer, 1979; Burrett & Richardson, 1980). The deep-water provinces correspond to the positions of the northern and southern subtropical gyres (Ziegler *et al.*, 1980). In summary, the observed Cambrian biogeographic divisions are consistent with the Cambrian palaeomagnetic data that are available.

Rifting and collisional events

Major rifting and continental collisional events were few just before, during, or just after the Cambrian, although evidence of Andean-style convergence is ample (Ziegler *et al.*, 1979). Precambrian rifts are so poorly dated and were so much earlier than the Early Cambrian that they are of little help in constraining the Early Cambrian positions of the continents. The Taconic Orogeny was too late to provide more than moderate constraints on the Late Cambrian continental positions.

Palaeoceanography

Many phosphorite deposits are considered to be related to ancient upwelling currents – areas of high biological productivity in the oceans. Indeed, Sheldon (1964), McKelvey (1967), and Freas & Eckstrom (1968) successfully projected modern upwelling zones on palaeogeographic reconstructions as a method of phosphorite exploration. However, Early Palaeozoic land areas were quite different from the present, and the effects of the palaeogeography on circulation must be taken into account. The most persistent and productive oceanic upwelling currents are driven by winds associated with the major features of atmospheric circulation, which are determined by the rotation of the earth, land–sea contrasts, and seasons. Therefore, atmospheric circulation must be modelled first in order to predict the locations of upwelling currents. For the purpose of this paper, we have followed the lead of Hansen in Ziegler *et al.* (1977), of Parrish (1982), and of Parrish & Curtis (1982) and have constructed generalised, qualitative models of Early and Late Cambrian atmospheric circulation. From these models, we have predicted the locations of Cambrian upwelling zones (Figs 22.2 and 22.3). The Late Cambrian circulation is modified and generalised from the maps presented by Parrish (1982), to take into account the revision in palaeogeography. Following is a brief outline of a more detailed and extensive discussion of how atmospheric circulation and upwelling are predicted (Parrish, 1982).

Atmospheric circulation is driven primarily by the equator-to-pole temperature contrast. On a non-rotating earth, the heat exchange between warm equatorial air and cold polar air would be effected by two vertical circulation cells, one in each hemisphere. Air rising at the equator would flow poleward in the upper atmosphere and sink at the poles, to return equatorward along the earth's surface. The rotation of the earth modifies this simple heat exchange by deflecting the moving air from a straight path; this deflection is called the Coriolis effect. Schematically, the heat exchange is effected by three vertical cells in each hemisphere. At the surface, the flow in these cells is expressed in three zonal wind belts in both south and north latitudes: (1) equatorial easterlies converging on the equator between about 20° S and 20° N; (2) the westerlies, between about 35° and 50°; and (3) the polar easterlies, at latitudes higher than about 65°. Barometric pressure patterns also have a zonal configuration, with high pressure (sinking air) at the poles and at about 30° north and south latitude and low pressure (rising air) at the equator and at about 60° north and south latitude.

The zonal pattern that results from the equator-to-pole temperature contrast and the rotation of the earth is modified by the thermal contrast between land and sea. The degree of modification depends upon the intensity of the contrast, that is, how much heat must be exchanged from one area to another. The zonal pattern is broken most strongly into major atmospheric circulation cells at mid-latitudes (between about 25° and 70° north and south latitude) because the land–sea contrasts are greatest there. This effect can be enhanced by seasonal variations. For example, during the northern winter, the water around the southern tip of Greenland is much warmer than is the adjacent land area. During that season, an intense low-pressure (rising air) cell forms over the region in response to this contrast. At lower mid-latitudes, the ocean tends to be cooler than the adjacent land and the high-pressure belt predicted for those latitudes by the zonal model is intensified into high-pressure cells over the oceans, particularly during the summer when the land is warmest.

Using these general principles as guides and taking analogies from modern atmospheric circulation for special situations such as the Asian monsoon, ancient wind direction (and, by exten-

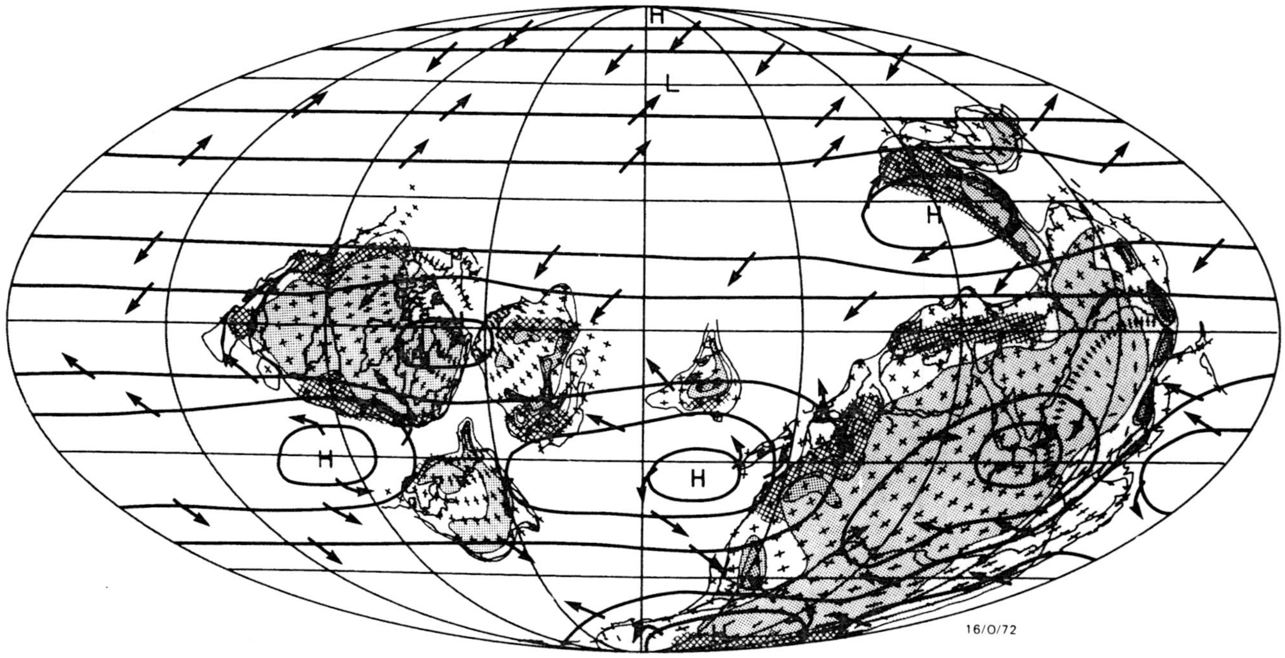

Fig. 22.2. Early Cambrian palaeogeography, atmospheric circulation, and upwelling. Circulation is generalised for the northern winter. The major difference between northern winter and northern summer would have been the circulation over southern Gondwanaland, which would have been dominated by a high-pressure cell in the northern summer. This change probably would not have affected the circulation causing upwelling. Circulation: heavy lines = isobars, arrows = wind direction, L = low pressure, H = high pressure, crosshatch = upwelling. The isobar lines should be regarded as contours.

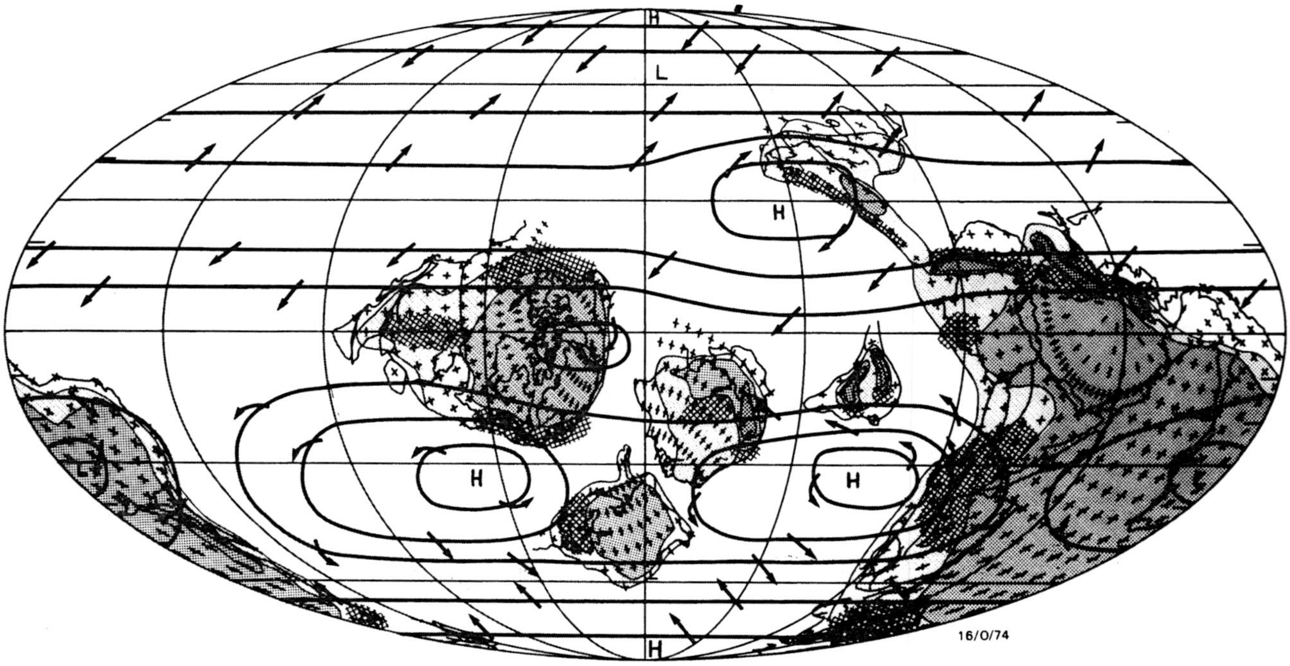

Fig. 22.3. Late Cambrian circulation and upwelling, generalised for northern winter. Circulation: heavy lines = isobars, arrows = wind direction, L = low pressure, H = high pressure, crosshatch = upwelling. The isobar lines should be regarded as contours.

sion, ancient upwelling currents) can be predicted. Two assumptions inherent in these models must be emphasised. First, we assumed that the difference in the earth's rotational speed between the present and the Early Cambrian, though apparently substantial (Mohr, 1975; Scrutton, 1978; Sündermann & Brosche, 1978), was not sufficient to require the abandonment of the six-cell zonal model. The effect of a higher rotational rate is not a question that has been addressed adequately in the atmospheric dynamics literature, although the numerical experimental work of Hunt (1979) suggests that a breakdown of the six-cell zonal circulation would require rotational speeds far higher than those predicted for the Cambrian. Second, we assume that any possible differences in the equator-to-pole temperature gradient between the Cambrian and the present would not have been sufficient to invalidate use of the principles of modern atmospheric circulation.

Two general types of wind-driven upwelling are recognised – coastal upwelling and oceanic divergence. Coastal upwelling is driven by persistent longshore winds that flow so that the net transport of water is offshore, that is, to the right of the wind in the northern hemisphere and to the left of the wind in the southern hemisphere. The most familiar settings for coastal upwelling are the west coasts of continents at mid-latitudes where upwelling is driven by winds associated with the subtropical high-pressure cells. A modern example is the Peru Current. The predominance of this type of upwelling current at present is a product of the predominantly north–south coastlines. Coastal upwelling also can occur along east–west coastlines that are situated favourably with respect to the zonal winds and along low-latitude, east-facing coasts that are influenced by monsoonal winds. At present, the former type is represented by upwelling off northern Venezuela and the latter type by upwelling during the summer off the coast of Somalia.

Oceanic water divergences occur under persistent atmospheric low pressure, along the equator and at about 60° latitude north and south. At the equator, winds converge from the north and south and flow westward. Because the Coriolis effect changes sign across the equator, with water flowing to the right of the wind in the northern hemisphere and to the left of the wind in the southern hemisphere, the net transport of water is away from the equator. Equatorial upwelling is dependent largely upon wind speed because the Coriolis effect is zero at the equator; exceptionally strong and persistent winds are required to create a significant upwelling current. The major equatorial upwelling today is in the Pacific Ocean.

Equatorial upwelling is most intense in the eastern part of the ocean where the tradewinds are strongest. Vigorous upwelling is also present in a band around Antarctica, where the polar easterlies and the westerlies converge into a belt of low pressure. The winds converge from opposite directions and the net transport of water is away from the line of convergence, resulting in upwelling. In the northern hemisphere, no continuous ocean exists at 60° latitude, and the low pressure predicted from the zonal model is manifested as low-pressure cells. In the northern hemisphere, the winds spiral counterclockwise into low-pressure cells and the net transport of water is away from a point in the centre of each cell. Cellular divergence upwelling exists around the southern tip of Greenland and along the Aleutian chain.

Phosphorites have also been deposited in zones of dynamic upwelling (McKelvey, 1967). This type of upwelling current is not wind-driven, but is related to the dynamics of fast oceanic currents, being most prominent in the strong western boundary currents such as the Gulf Stream and the Kuroshiro Current. Along the Gulf Stream, dynamic upwelling is manifested in a narrow band of cold water and high biological productivity on the western side of the current (Yentsch, 1974). At the present time, the upwelling is over relatively deep water, but in the past, the flow might have been over much shallower shelf environments (Pinet & Popenoe, 1980). The possibility of strong current flow in shallow water has also prompted the consideration of the significance of topographically-induced upwelling, which also can be augmented by winds (Pietrafesa & Janowitz, 1980; Blanton *et al.*, 1981). Distinguishing dynamic from topographically-induced upwelling might be difficult in the fossil record. Topographically induced upwelling is not easily predicted for the past because detailed palaeobathymetry must be known.

The regions indicated by a crosshatch pattern in Figs. 22.2 and 22.3 represent predicted Cambrian upwelling zones. For the Early Cambrian, we predict west-coast upwelling for China, south-western Gondwanaland (northern Africa, Avalon, and southern Europe), and north-western Baltica (Norway and Sweden, Fig. 22.2). Two belts of zonal upwelling would have existed off northern and southern Laurentia (now western and eastern North America), off southern Siberia (now south-western Siberia and Mongolia), in southernmost Gondwanaland (north-western South America) and, given its indicated orientation, in the southern (now western) regions of Kazakhstan (Fig. 22.2). Equatorial upwelling would have occurred over the shelves in western Laurentia (now Mexico), northern (now north-eastern) Siberia, and western Gondwanaland, in the vicinity of what are now Iran and Turkey, Tibet and northern India, and eastern Saudi Arabia (Fig. 22.2). Dynamic upwelling might have been present in eastern Baltica, north-eastern Gondwanaland (eastern Australia), and south-eastern Gondwanaland (Antarctica and southern South America, Fig. 22.2).

For the Late Cambrian, we predict west-coast upwelling in China, western Gondwanaland (Tibet, northern India, Iran, Saudi Arabia) and western Baltica (Scandinavia). Zonal upwelling would have been present in Kazakhstan and Laurentia as in the Early Cambrian, and also in southern (now western) and central Siberia, southernmost Gondwanaland (northern Africa, southern Europe, and Avalon), and northernmost Gondwanaland (eastern Australia and western Antarctica). Western Laurentia (now southern USA and Mexico) and north-western Gondwanaland (north-western Australia) would have had equatorial upwelling zones.

Many factors other than continental positions could change the individual predictions. Most important is bathymetry, which strongly controls the location of upwelling. A minimum depth of about 50 m is typically required for upwelling (Wooster, 1969; Smith *et al.*, 1981; O'Brien, 1971; Mooers,

Collins & Smith, 1976). Although upwelling can develop over a wide shelf that has a low bathymetric gradient, the upwelling will be at about the 50 m isobath, and not at the shoreline. If the 50 m isobath is considerably offshore, the upwelling zone also will be offshore. If the shelf gradient is such that the 50 m isobath is so far offshore as to underlie another wind regime, upwelling will not develop. On the wide shelves illustrated in Figs 22.2 and 22.3, we have shown for each prediction the entire area of possible upwelling, from the shoreline across the width of the shelf. An upwelling locus also can migrate with sea-level change (Fischer & Arthur, 1977) and, within the time intervals considered here, wide areas of phosphorite deposition could have resulted from such a migration.

Phosphorites

Descriptions of Cambrian phosphorite deposits and their distribution were assembled from the literature. Especially helpful was the compilation of references by Notholt (1979). We

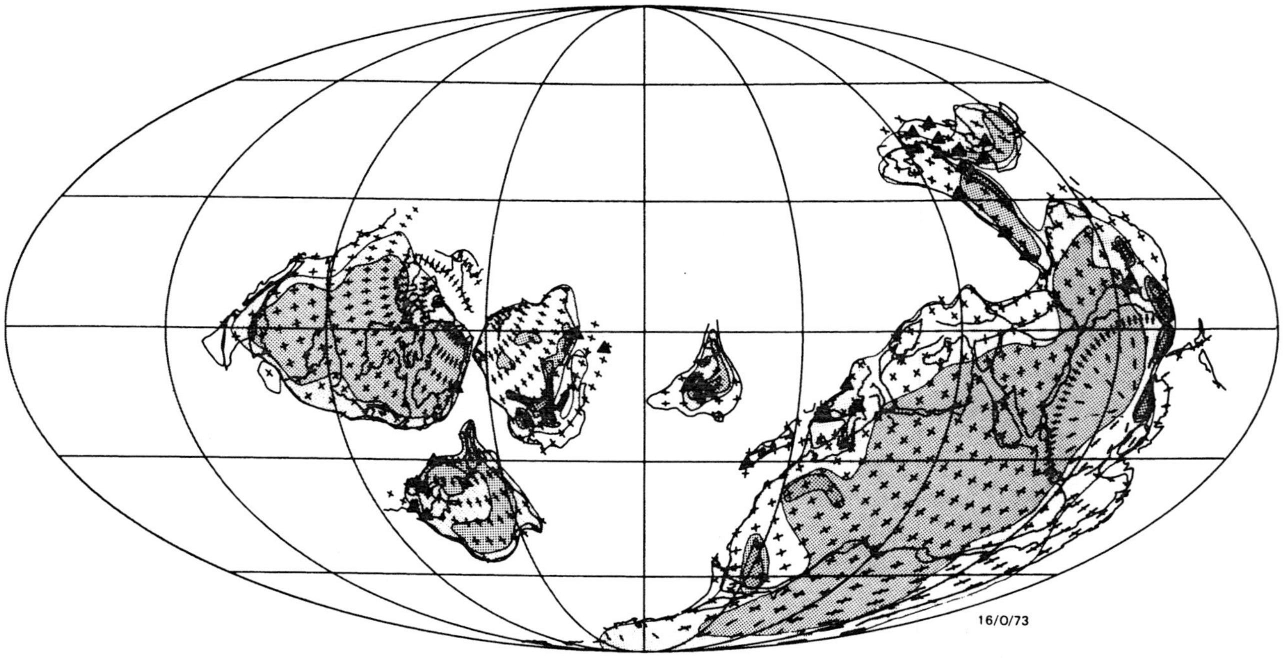

Fig. 22.4. Early Cambrian phosphorites. For data sources, see Table 22.1. Symbol: (▲), phosphorite deposits.

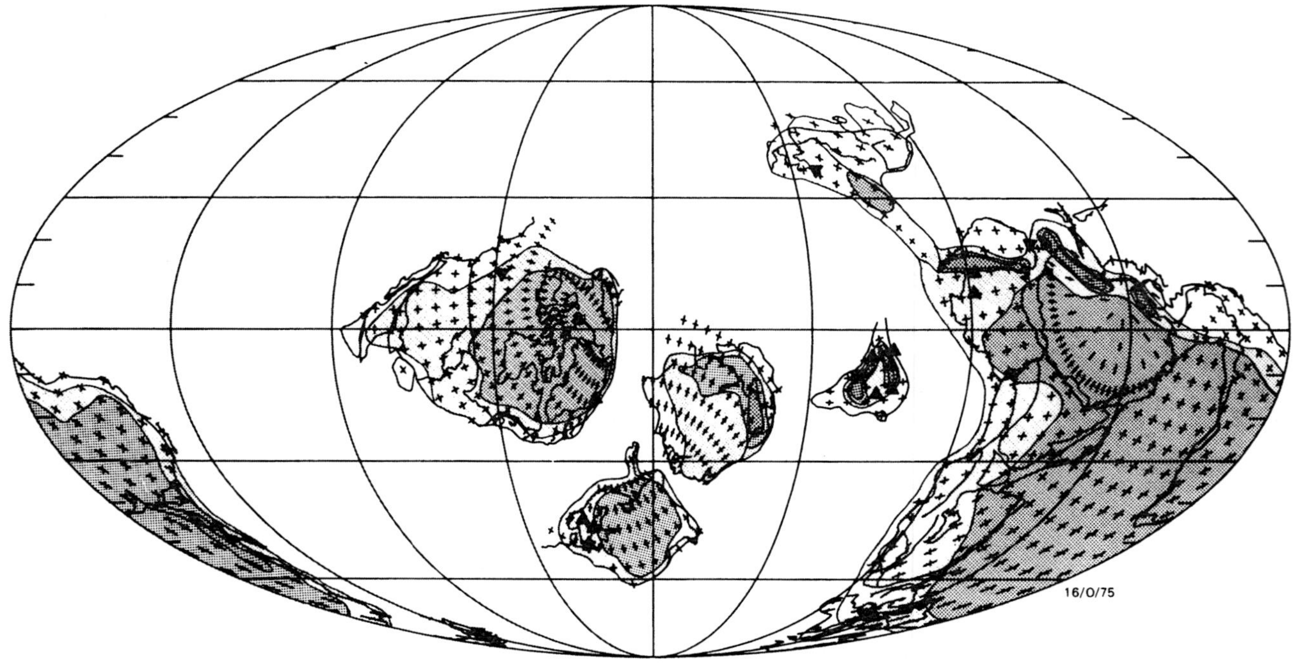

Fig. 22.5. Middle and Late Cambrian phosphorites. Symbols: (▲), Middle Cambrian phosphorites, (►), Late Cambrian phosphorites. Avalon is the distorted 'blip' in the palaeogeography at the bottom of the reconstruction. For data sources, see Table 22.2.

were able to obtain most of these references, and from them compiled detailed information on associated lithologies, types of phosphorites, and their ages. The data presented in Fig. 22.4 are for the Early Cambrian only and those in Fig. 22.5 are for the Middle and Late Cambrian.

Phosphorites are formed as sands, conglomerates, hardgrounds, nodules, pellets, and ooids, and are associated with a variety of sedimentary rock types. There may be no palaeogeographic significance to the type of phosphorite deposit, but the characteristics of the associated rocks are important in that two general types of phosphorite associations are recognised. First, phosphorites commonly occur with glauconite and chert, and with carbonaceous shale or marlstone that often contains pyrite or other indicators of reducing conditions. This phosphorite association is generally regarded as typical of upwelling deposits because siliceous sediments, anoxia, and glauconite are frequently found in the sediments of modern upwelling zones, especially off Peru and south-west Africa, and are all directly or indirectly related to the high biological productivity. More specifically, siliceous oozes and anoxic conditions are found beneath the most intense part of an upwelling current. The anoxic zone is rimmed by a zone of phosphorite nodules, also commonly associated with siliceous oozes. The phosphorite zone is in turn surrounded by a zone of glauconite (Burnett, 1980; Odin & Letolle, 1980; Bremner & Rogers, 1981; Waples, 1982). Secondly, phosphorites commonly are found in carbonates (Jarvis, 1980). The origin of this association has been the subject of much debate, and many of the carbonate phosphorites are reworked, further complicating the understanding of their genesis. Cambrian phosphorite deposits and some of their characteristics are listed in Tables 22.1 and 22.2.

The phosphorite data are plotted on Figs. 22.4 and 22.5. A comparison of the distributions of predicted upwelling and phosphorites (Figs. 22.2 to 22.5) shows that many of the Cambrian phosphorite deposits are grouped into predicted upwelling zones. For the Early Cambrian, the following correspondences are notable:

(1) Most of the Chinese phosphorite deposits are associated with chert and/or black shale or marlstone and many contain glauconite.

(2) Phosphorites in the outer shelf deposit in France are nodules in black shale containing pyrite, whereas phosphorites of the inner shelf are in limestone or sandstone with glauconite. This is precisely the pattern expected in and near an upwelling zone. According to Brasier (1980), the southern European phosphorites were deposited in what he calls an inner detrital belt setting, between the shoreline and a reef, although whether or not they were separated from the open ocean by a reef at the time of deposition was not made clear. Some of the English Lower Cambrian phosphorites are in black shale with pyrite (Hicks, 1875; Rushton, 1974), similar to the outer shelf deposits in France. The phosphorites in Nova Scotia (Avalon) occur in limestone and sandstone (Brasier, 1980).

(3) Phosphorite and glauconite are frequently associated in Scandinavia (Martinsson, 1974; Brasier, 1980) in sandstone, siltstone, and limestone. According to Brasier (1980), the Scandinavian deposits represent offshore, open-shelf environments.

(4) The phosphorite deposits in eastern (now south-eastern) Siberia contain 'coaly' siliceous shale and might have been deposited under an equatorial upwelling regime. Some of the deposits in southern Siberia (now south-western Siberia and Mongolia) are in limestone and dolomite, but others contain chert, bituminous carbonates, and black, siliceous shale.

Some of the Lower Cambrian phosphorite deposits in

Table 22.1. *Early Cambrian phosphorites*

Plate	Location	Type of phosphorite	Associated lithologies	References
Laurentia	E. Greenland	Nodules	Conglomerate, sandstone, limestone, shale, glauconite	Brasier, 1980
Baltica	67° N 23° E (Finnmark)	Phosphatic sediments	Conglomerate, sandstone	Martinsson, 1974
Baltica	S. Norway	Not reported	Shale, conglomeratic sandstone, glauconite	Brasier, 1980
Baltica	60° N 15°E (composite*)	Not reported	Sandstone, siltstone, glauconite (bituminous marl and chert reported by Andersson, 1895)	Brasier, 1980; Martinsson, 1974; Andersson, 1895
Baltica	Denmark	Not reported	Shale, sandy limestone	Brasier, 1980
Baltica	Scania, Sweden	Conglomerate, nodules	Limestone, sandstone, siltstone,	Brasier, 1980; Regnéll & Hede, 1960
Siberia	45° N 133° E (composite)	Not reported	Siliceous, 'coaly' shale	Vinogradov, 1968; Bushinsky, 1969, loc. 21
Siberia	54° N 145° E	Not reported	Limestone, volcanics, chert	Vinogradov, 1968
Siberia	52° N 101° E (composite)	Phosphatic sediments	Sandstone, siliceous rocks, bituminous carbonates	Bushinsky, 1969, loc. 16; Donov *et al.*, 1969; Gimmelfarb & Egorova, 1969; Zaitsev & Ilyin, 1970

Table 22.1. (*Contd.*)

Plate	Location	Type of phosphorite	Associated lithologies	References
Siberia	54° N 89° E (composite)	Nodules, breccia	Carbonates, calcareous and siliceous black shale	Bushinsky, 1969, loc. 12–14; Vinogradov, 1968; Sokolova, 1961
Kazakhstan	53° N 72° E	Beds, oolite	Conglomerate, sandstone, shale	Knipper, 1958
Kazakhstan	49° N 67° E	Oolite	Siliceous shale, tuffaceous sandstone	Bushinsky, 1969, loc. 1
Kazakhstan	47° N 69° E	Beds	Chert, carbonate	Notholt, 1972
Kazakhstan	49° N 71° E	Phosphatic sediments	Sandstone, siliceous shale	Bushinsky, 1969, loc. 3
Gondwanaland (Australia)	35° S 139°E (composite)	Lenses, nodules	Carbonates, shale, sandstone	Blisset & Callen, 1967; Callen 1969
Gondwanaland (Spain)	43° N 3° W	Nodules, hardground	Black shale with pyrite, limestone	Prian, 1980
Gondwanaland (France)	44° N 2°E	Not reported	Sandstone, glauconite	Brasier, 1980
Gondwanaland (England)	52°N 3°W (composite)	Phosphatic sediments	Black shale with pyrite, argillaceous limestone, glauconite	Hicks, 1875; Brasier, 1980; Rushton, 1974
Gondwanaland (Avalon)	45° N 66° W	Not reported	Sandstone, argillaceous limestone	Brasier, 1980
China	42°N 82°E	Nodules, beds, breccia	'Coaly', siliceous shale; dolomite; bituminous marlstone	Bushinsky, 1969, loc. 10
China	40° N 92° E	Nodules, beds	Dark, argillaceous limestone;	Bushinsky, 1969, loc. 11
China	39° N 107° E (composite)	Nodules, conglomerate, sand	Shale, cherty limestone, sandstone, glauconite	Bushinsky, 1969, loc. 24, 25; B.S.C., 1971, p. 160
China	30° N 105° E	Beds, lenses, grains, flat-pebble conglomerate, nodules	'Coaly', siliceous shale and marlstone; glauconite	Bushinsky, 1969, loc. 43, B.S.C., 1971, p. 161
China	33° N 112° E (composite)	Beds, nodules, lenses, cement	Limestone, shale, marlstone, glauconite	Bushinsky, 1969, loc. 26, 27, 41, 42; B.S.C.., 1971, p. 158
China	32° N 117° E	Nodules	Limestone, dolomite, shale, pyrite, glauconite	Bushinsky, 1969, loc. 28
China	27° N 103°E (composite)	Beds, nodules	Sandstone, black shale, marlstone, siliceous carbonates, chert, glauconite	Bushinsky, 1969, loc. 44, 45, 47; Wang, 1942; B.S.C., 1971, p. 161
China	27° N 107° E (composite)	Nodules, beds, oolite, pellets	Organic-rich, siliceous shale; chert	Chang Ta, 1959; Bushinsky, 1969, loc. 34–37
China	28° N 111° E (composite)	Beds, nodules, grains	'Coaly', pyritic, black shale; chert	Bushinsky, 1969, loc. 38–40
China	29° N 116° E (composite)	Beds, nodules	Siliceous dolomite; siliceous and carbonaceous shale; chert	Chang Ta, 1959; Bushinsky, 1969, loc. 29, 30
China	24° N 104° E (composite)	Nodules	Dolomitic, pyritic, 'coaly' shale; chert	Chiang *et al.*, 1968; Bushinsky, 1969, loc. 46, 48; B.S.C., 1971, p. 160–161
China	23° N 113° E (composite)	Nodules, pebbles	'Coaly' shale; sandstone, shale, conglomerate	Bushinsky, 1969, loc. 32, 33
China	40° N 116° E	Oolite, nodules, conglomerate	Shale, conglomerate, pyrite glauconite	B.S.C., 1971, p. 158–159

Locations of phosphorites are illustrated in Fig. 22.3. *Several data points on Fig. 22.3 are composites of closely spaced phosphorite deposits. Latitudes and longitudes are generalised. Bushinsky's (1969) locality numbers are given for data taken from his reference. B.S.C. = British Sulphur Corporation. (Note: the data presented in this table are slightly different from those presented in Parrish, Ziegler & Humphreville (1983), owing to slightly different compilation techniques).

Table 22.2. *Middle and Late Cambrian phosphorites*

Plate	Location	Age	Form in which the phosphorite occurs	Associated lithologies	References
China	40° N 116° E	l	Phosphatic sediments	Shale, conglomerate, pyrite, glauconite	Hsieh & Chao, 1948
Kazakhstan	48° N 69° E (composite*)	m	Beds, nodules, veins, oolite, lenses	Chert; carbonate; siliceous, pyritic shale; sandstone; siltstone	Bushinsky, 1969, loc. 1, 3
Kazakhstan	43° N 69° E (composite)	m	Beds, nodules, lenses, oolite, concretions, flat-pebble conglomerate	Dolomite; radiolarian chert; siliceous, 'coaly' shale	Bushinsky, 1969, loc. 6; Gimmelfarb & Tushina, 1966; Kholodov, 1969; Smirnov, 1959; Tushina, 1968; Meng, 1959
Kazakhstan	43° N 73° E (composite)	m	Concretions, oolite	Dolomite; siliceous, 'coaly' shale; black limestone; radiolarian chert; sandstone	Bushinsky, 1969, loc. 5, 7; Vinogradov, 1968
Kazakhstan	42° N 78° E (composite)	m	Nodules	Organic-rich, siliceous pyritic, limestone and shale	Bushinsky, 1969, loc. 8, 9
Kazakhstan	52° N 73° E (composite)	m, l	Oolite, lenses, beds, nodules	Sandstone, limestone, siltstone, dolomite, conglomerate, siliceous shale	Bushinsky, 1969, loc. 2; Dvortsova, 1958;
Kazakhstan	49° N 79° E	m	Not reported	Siliceous (radiolarians), carbonaceous sediments	Polyansky, 1977
Laurentia	71° N 22° W	m	Nodules	Conglomerate, limestone, shale, sandstone, glauconite	Brasier, 1980
Laurentia	58° N 122° W	l	Nodules	Shale	Christie, 1980
Baltica	61° N 11°E (composite)	m	Conglomerate	Sandstone, shale, glauconite	Brasier, 1980; Martinsson, 1974
Baltica	56° N 15°E (composite)	m, l	Nodules, conglomerate, phosphatic sand	Sandstone, conglomerate, black limestone, chert, pebbles, pyrite, glauconite	Brasier, 1980; Andersson, 1895; Martinsson, 1974
Baltica	61° N 17° E	m	Conglomerate	Sandstone	Martinsson, 1974
Gondwanaland (Australia)	21° S 135° E (composite)	m	Not reported	Limestone, sandstone, shale, dolomite	B.S.C., 1971, p. 168
Gondwanaland (Australia)	21° S 140° E (composite)	m	Pellets, stromatolitic laminae	Chert, limestone, siltstone, shale, dolomite	Cook, 1972; Howard, 1972; B.S.C., 1971, p. 168; Fleming, 1977; Russell & Trueman, 1971; Southgate, 1980
Gondwanaland (Australia)	37° S 146° E	l	Not reported	Not reported	B.S.C., 1971, p. 171
Gondwanaland (France)	44° N 3° W (composite)	m	Nodules, hardground	Sandstone, limestone, black shale, pyrite, glauconite	Prian, 1980; Brasier, 1980
Gondwanaland (Avalon)	47° N 54° W	m	Nodules	Siltstone, black slate, pyrite	Douglas, 1970; Hutchinson, 1962
Gondwanaland	45° N 66° W	l	Nodules	Sandstone, black shale, limestone	B.S.C., 1971, p. 3; Douglas, 1970

Locations of phosphorites are illustrated in Fig. 22.5. *Several data points in Fig. 22.5 are composites of closely spaced phosphorite deposits. Latitudes and longitudes are generalised. Bushinsky's (1969) locality numbers are given for data taken from his reference. B.S.C. = British Sulphur Corporation. m = Middle Cambrian, l = Late Cambrian.

Kazakhstan also correspond to a predicted upwelling zone. The phosphorite in what is now eastern Australia might have been deposited in a dynamic upwelling zone if the land areas to the east had been less continuous; the present level of reconstruction does not permit resolution of this possibility, although Cook (1982, p. 74) also suggested upwelling as a setting for these deposits.

Correspondence exists between predicted upwelling and Upper Cambrian phosphorites in south-eastern Australia, China, southern (now western) Kazakhstan, western Baltica (southern Scandinavia), southern Laurentia (eastern Greenland), and southernmost Gondwanaland (Avalon). (Middle Cambrian phosphorites are shown on Fig. 22.5 for the reader's information.)

Discussion

Unfortunately, many of the phosphorite deposits are in regions whose palaeogeographic positions are most poorly understood, that is, in Kazakhstan and China, and the correspondence between predicted upwelling and phosphorites in those regions perhaps ought to be regarded as fortuitous. As pointed out by Palmer (1974), a particularly large gap exists in the belt of Cambrian rocks surrounding Gondwanaland, in the area of what is now western Australia and northern India. The strong similarity of the faunas in Australia and southern China (Burrett & Richardson, 1980) might suggest that southern China filled that gap.* This position would solve the 'missing' land problem in the western part of southern China. This position of southern China also would account for the 'missing' phosphorites that are expected but not observed in Gondwanaland in that region. Cook & McElhinny (1979) offered a reasonable alternative explanation of the Australian and Chinese deposits. In their model (their Fig. 4B). China and Australia bounded an east–west seaway at 40° and 25° north latitude, respectively, permitting zonal upwelling currents predicted by the positions of the zonal components of the general atmospheric circulation.

Although Vendian (latest Precambrian) phosphorites are not illustrated here, they are, with one exception, in the same areas as some of the Lower Cambrian deposits, namely in southern Kazakhstan, southern Siberia, western Baltica, and China (Spjeldnaes, 1967; Vinogradov, 1968; Bushinsky 1969). The exception is in north-western Africa (Ilyin, 1977; Lucas, Prévôt & Trompette, 1980), where Vendian phosphorites are associated with black shale, pyrite, and chert. If Gondwanaland's clockwise rotation through the Cambrian had begun earlier, north-western Africa could have been influenced by upwelling in the Vendian.

In this study, we assumed that present principles of circulation dynamics can be applied to the past. Dissenting views have been expressed, however, mainly by Sheldon (1980, 1981) and Sheldon & Burnett (1981a, b), who suggested that the principles of circulation derived from present conditions cannot be applied to all times in the past. In their model, warm equable climates ('oligotrophic' times of Fischer & Arthur, 1977) led to the disruption of the zonal pattern recognised in the present world. According to the hypothesis of Sheldon (1980, 1981) and Sheldon & Burnett (1981a, b), the oceanic circulation assumed a simple configuration of downwelling at high latitudes and upwelling along the equator, and no other upwelling zones. During these times, the equatorial easterlies driving the upwelling spread the upwelled nutrients (including phosphorus) toward the continents, where phosphorite subsequently was deposited on the continental shelves. They did not discuss several difficulties with their model, namely (1) how the simple circulation they proposed could be possible on a rotating earth, and (2) how the nutrients could be carried hundreds or thousands of kilometres (Sheldon, 1981, Fig. 5) without being utilised by plants in the water column. The immediacy of nutrient utilisation in modern upwelling zones has been noted by Barber & Smith (1981). Nevertheless, the data as presented by Sheldon (1980, 1981) and Sheldon & Burnett (1981a, b) show an intriguing correlation between times of glaciation and the distribution of west-coast phosphorites and a corresponding correlation between times of no glaciation and a predominance of low-latitude (their equatorial) phosphorites. Low-latitude phosphorites, however, were not necessarily deposited in equatorial upwelling zones. Sheldon (1980, 1981) and Sheldon & Burnett (1981a, b) did not take into account types of coastal upwelling other than west-coast upwelling, for example, zonal coastal upwelling. In that sense, their model and ours really are not comparable. In their hypothesis, one of the times of no glaciation and corresponding low-latitude phosphorite deposition was the Early Cambrian. As our maps show, the distribution of phosphorites in low latitudes in the Early Cambrian may have been primarily a result of the distribution of the continents, almost all of which were in low latitudes, as noted also by Cook & McElhinny (1979). The Early Cambrian also had many potential west-coast upwelling zones, some of which correspond in the present reconstruction to phosphorite deposits (Figs. 22.2 and 22.4). Moreover, several low-latitude sites of coastal upwelling probably were zonal. Of course, our upwelling predictions may change as palaeomagnetic and palaeogeographic reconstructions improve. Until that time, however, we maintain that by using the principles of modern circulation and modifying the circulation patterns in accordance with the palaeogeography, one can predict the location of extensive zones of ancient upwelling. No radical changes in circulation are required to explain the distribution of phosphorites, at least for Early Cambrian time.

Lower Cambrian rocks have far more phosphorite than do Upper Cambrian ones (Figs. 22.4 and 22.5; Cook & McElhinny, 1979). Arthur & Jenkyns (1981) noted a number of factors that together appear to explain most variations in phosphogenesis through the Phanerozoic, namely (1) high and rising sea level, and (2) warm climate, which leads to increased chemical weathering on land and to widespread oxygen-depleted waters. According to the sea-level curves of Vail, Mitchum & Thompson (1977), Cambrian sea level rose rapidly through the Early Cambrian, peaking in the Late Cambrian. Evaluating Cambrian palaeotemperatures is almost impossible. Glaciation was extensive in the Late Precambrian (Frakes, 1979) and did not recur until the latest Ordovician. The extensive Cambrian archaeocyathid reefs have been interpreted as indicative of warm climate (Newell, 1972). Therefore, two of the requirements proposed by Arthur & Jenkyns (1981) for increased deposition of phosphorites are apparently satisfied for the Cambrian, although no climatic difference is apparent between the Early and Late Cambrian. Despite our contention that the principles of modern oceanic and atmospheric circulation can be applied to the past, our models cannot explain the vast difference in the amount of Lower and Upper Cambrian phosphorite. If the difference has a palaeogeographic or palaeoclimatic explanation, the discovery of that explanation awaits reconstructions that have finer resolution.

In conclusion, we note that upwelling zones may not be the only settings for the deposition of phosphorites. For example, some of the Australian phosphorites were deposited in very shallow water, if not intertidally (Cook, 1972; Southgate, 1980).

*Note added in proof: this position is now confirmed by Lin, L.J., Fuller, M. & Zhang, W.Y. (1985). Preliminary Phanerozoic polar wander paths for the North and South China Blocks. *Nature,* **313**, 444–9.

The problem of supplying phosphorus to such systems may force the postulation of a nearby upwelling zone (and, from our models, such a zone may have existed), but clearly no upwelling was possible in the immediate vicinity of the site of phosphorite deposition. Our models should be regarded as working hypotheses, to be tested with detailed studies of the local palaeogeography, of the phosphorites, and of the rocks associated with them.

References

Sources for the data used in the Early Cambrian palaeogeographic reconstruction are marked with an asterisk and are not cited in the text.

*Adie, R.J. (Ed.) (1972). *Antarctic Geology and Geophysics*, 876pp. Universitetsforlaget, Oslo.

Andersson, J.G. (1895). Über cambrische und silurische, phosphoritführende Gesteine aus Schweden. *Bulletin of the Geological Institution of Uppsala*, **2**, 133–238.

Antonyuk, R.M., Lyapichev, G.F. Markova, N.G., Pavlova, T.G., Rozen, O.M., Samygin, S.G., Tokmachev, S.G., Shuzhanov, V.I. & Shcherba, I.G. (1977). Structures and evolution of the crust of central Kazakhstan. *Geotectonics*, **11**, 365–72.

Arthur, M.A. & Jenkyns, H.C. (1981). Phosphorites and paleoceanography. In *Proceedings 26th Session International Geological Congress, Paris*, pp. 83–96. *Oceanologica Acta*, Montreuil.

Barber, R.T. & Smith, R.L. (1981). Coastal upwelling ecosystems. In *Analysis of Marine Ecosystems*, ed. A.R. Longhurst, pp. 31–68. Academic Press, New York.

*Belov, A.A. (1971). Paleozoic tectonics of the western and central Taurus (Turkey). *Geotectonics*, **5**, 31–8.

*Bender, F. (1971). Geologie von Jordanien. *Beitrage zur Regionalen Geologie der Erde*, **7**, 23pp. Gebruder Bohrtraeger, Berlin.

Blanton, J.O., Atkinson, L.P., Pietrafesa, L.J. & Lee, T.N. (1981). The intrusion of Gulf Stream water across the continental shelf due to topographically-induced upwelling. *Deep-Sea Research*, **28**, 393–405.

Blisset, A.H. & Callen, R.A. (1967). Myponga phosphate deposit. *South Australia Department of Mines, Mineral Resources Review*, **127**, 93–103.

*Borrello, A.V. (1971). The Cambrian of South America. In *Cambrian of the New World*, ed. C.H. Holland, pp. 385–438. John Wiley, London.

Brasier, M.D. (1980). The Lower Cambrian transgression and glauconite–phosphate facies in western Europe. *Journal of the Geological Society*, **137**, 695–703.

Bremner, J.M. & Rogers, J. (1981). Major lithofacies of the Namibian continental margin. In *Programme and Abstracts from the Conference on Coastal Upwelling: Its Sediment Record*, Sept. 1981, Portugal. Advanced Research Institute.

*Binkmann, R. (1976). *Geology of Turkey*, 158pp Elsevier, Amsterdam.

British Sulphur Corporation (1971). *World Survey of Phosphate Deposits*, 3rd edn, 180pp. British Sulphur Corporation, London.

*Brown, D.A., Campbell, K.S.W. & Crook, K.A.W. (1968). *The Geological Evolution of Australia and New Zealand*, 409pp. Pergamon Press, Oxford.

Burke, K., Dewey, J.F. & Kidd, W.S.F. (1977). World distribution of sutures – the sites for former oceans. *Tectonophysics*, **40**, 69–99.

Burnett, W.C. (1980). Apatite–glauconite associations off Peru and Chile: palaeo-oceanographic implications. *Journal of the Geological Society*, **137**, 757–64.

Burrett, C. & Richardson, R. (1980). Trilobite biogeography and Cambrian tectonic models. *Tectonophysics*, **63**, 155–92.

Bushinsky, G.I. (1969). Old phosphorites of Asia and their genesis. *Trudy Geolicheskii Institut, Akademiya Nauk SSSR*, **149**. (translated by Israel Program for Scientific Translations, Jerusalem), 266pp.

Callen, R.A. (1969). Sedimentary phosphate exploration. Part 1. The Cambrian of the Flinders Ranges. *South Australia Department of Mines. Mineral Resources Review*, **130**, 9–34.

Ch'ang Ta (1959). *The Geology of China*. Peking. Translated by US Department of Commerce, Off. Tech. Serv., Joint Publ. Res. Serv., Washington, DC, 623pp.

Chiang, N.-J., Wang, T.-C. & Chen, Y.-K. (1968). Cambrian stratigraphy of eastern Yunnan. *International Geology Review*, **10**, 1153–72.

Christie, R.L. (1980). Canadian occurrences of Proterozoic and Cambrian phosphorites. *International Geological Correlation Programme, Project 156, Newletter*, **5**, 28.

*Cielinski, S., Czerminske, J., Kozlowski, S., Pajchlowa, M., Ryka, W. & Slacka, A. (1971). *Lithostratigraphical table of geological deposits in Poland* 2pp. Instytut Geologiczny, Warsaw.

Cook, P.J. (1972). Petrology and geochemistry of the phosphate deposits of northwest Queensland, Australia. *Economic Geology*, **67**, 1193–213.

Cook, P.J. (1982). The Cambrian palaeogeography of Australia and opportunities for petroleum exploration. *Australian Petroleum Exploration Association Journal*, **22**, 42–64.

Cook, P.J. & McElhinny, M.W. (1979). A reevaluation of the spatial and temporal distribution of sedimentary phosphate deposits in the light of plate tectonics. *Economic Geology*, **74**, 315–30.

*Cowie, J.W. (1974). The Cambrian of Spitsbergen and Scotland. In *The Cambrian of the British Isles, Norden, and Spitsbergen*. ed. C.H. Holland, pp. 123–156. John Wiley, London.

*Dessila-Codarcea, M. (1968). Evolution préalpine des massifs crystallophylliens des Karpates roumaines. In *23rd Session International Geological Congress, Prague, Sec. 3*, pp. 121–7.

*Dhonae, N.B. & Holland, C.H. (1974). The Cambrian of Ireland. In *The Cambrian of the British Isles, Norden, and Spitsbergen*. ed. C.H. Holland, pp. 157–76. John Wiley, London.

Donov, N.A., Edemsky, E.V., Elyanov, A.A., Ilyin, A.V. & Muzalevsky M.M. (1969). Cambrian phosphorites in the Mongolian People's Republic. *Sovetskaya Geologiya*, **3**, 55–60.

*Douglas, R.J.W. (1970). Geology and economic minerals of Canada. *Canadian Geological Survey, Economic Geology Report*, **1**, 838pp.

Drewry, G.E., Ramsay, A.T.S. & Smith, A.G. (1974). Climatically controlled sediments, the geomagnetic field, and trade wind belts in Phanerozoic time. *Journal of Geology*, **82**, 531–53.

Dvortsova, K.I. (1958). Cambrian phosphorite deposits in the Kendyktas Mountains (South Kazakhstan). *Doklady Akademii Nauk SSSR*, **123**, 961–2.

*Elliot, D.H. (1975). Gondwana basins of Antarctica. In *Gondwana Geology*, ed. K.S.W. Campbell, pp. 493–536. ANU Press, Canberra.

Fischer, A.G. & Arthur, M.A. (1977). Secular variations in the pelagic realm. *Society of Economic Paleontologists and Mineralogists, Special Publication*, **25**, 19–50.

Fleming, P.J.G. (1977). Faunas, lithologies, and the origin of phosphorites in parts of the Middle Cambrian Beetle Creek Formation of NW Queensland. *Geological Survey of Queensland, Publication*, **364**, 21p.

Frakes, L.A. (1979). *Climates Throughout Geologic Time*, 310pp. Elsevier, New York

*Frakes, L.A., Kemp, E.M. & Crowell, J.C. (1975). Paleozoic glaciation. Part IV, Asia. *Geological Society of America, Bulletin*, **86**, 454–64.

Freas, D.H. & Eckstrom, C.L. (1968). Areas of potential upwelling and phosphorite deposition during Tertiary, Mesozoic, and late Paleozoic time. In *Proceedings of the Seminar on Sources of Mineral Raw materials for the Fertilizer Industry in Asia and the Far East*. pp. 228–38. ECAFE Mineral Resources Development Series 32, United Nations.

French, R.B., Alexander, D.H. & Van der Voo, R. (1977). Paleomagnetism of Upper Precambrian to lower Paleozoic intrusives from Colorado. *Geological Society of America, Bulletin*, **83**, 1785–92.

*Fullager, P.D. (1971). Age and origin of plutonic intrusions in the piedmont of the southeastern Appalachians. *Geological Society of America, Bulletin*, **82**, 2852–62.

*Furon, R. (1956). *Lexique Stratigraphique International. Afrique*,

Fascicule IV, Libye, 62pp. International Geological Congress.
*Furon, R. (1964). *Le Sahara: Géologie, Resources, Minérales.* 2nd edn, 313pp. Payot, Paris.
*Gardiner, P.R.R. (1975). Plate tectonics and the evolution of the southern Irish Caledonides. *Scientific Proceedings of the Royal Dublin Society Series A, Part III, Geodynamics Project*, **5**, 385–96.
*Gignoux, M. (1950). *Géologie Stratigraphique.* Masson, Paris, 735pp.
Gimmelfarb, B.M. & Egorova, O.P. (1969). Geology of the Khubsugul phosphorite deposit, Mongolian People's Republic. *Lithology and Mineral Resources*, **1969(2)**, 139–44.
Gimmelfarb, B.M. & Tushina, A.M. (1966). Principal phosphorite ore deposits of the Kara-Tau. *Lithology and Mineral Resources*, **4**, 88–102.
*Gonçalves, F. (1974). *Carta geológica de Portugal, Folha 36-B Estremoz*, 64pp. Servicos Geológicos Portugal, Lisbon.
*Gupta, V.J. (1973). *Indian Palaeozoic Stratigraphy.* Hindustan Publishing Company, Delhi.
Hagstrum, J. T., Van der Voo, R., Auvray, B. & Bonhommet, N. (1980). Eocambrian-Cambrian paleomagnetism of the Armorican Massif, France. *Geophysical Journal of the Royal Astronomical Society*, **60**, 489–517.
*Harland, W.B. (1975). Phanerozoic relative motions of North Atlantic Arctic lands. *Canadian Society of Petroleum Geologists, Memoir*, **4**, 235–56.
*Hatcher, R.D. Jr (1974). North American Paleozoic foldbelts and deformational histories: a plate tectonics anomaly? *American Journal of Science*, **274**, 135–47.
Hicks, H. (1875). On the occurrence of phosphates in the Cambrian rocks. *Quarterly Journal of the Geological Society of London*, **31(123)**, 368–83.
Howard, P.F. (1972). Exploration for phosphorite in Australia – a case history. *Economic Geology*, **67**, 1180–92.
Hsieh C.Y. & Chao C.H. (1948). Note on the phosphate deposits of China. *Bulletin of the Geological Society of China*, **28**, 71–4.
Hunt, B.G. (1979). The effects of past variations of the Earth's rotation rate on climate. *Nature*, **281**, 188–91.
Hutchinson, R.D. (1962). Cambrian stratigraphy and trilobite faunas of southeastern Newfoundland. *Geological Survey of Canada, Bulletin*, **88**, 156pp.
Ilyin, A.V. (1977). Late Precambrian–Cambrian phosphorites. In *Interregional Seminar on the Development Potential of Precambrian Mineral Deposits*, Moscow. United Nations, Adademiya Nauk SSSR, International Geological Correlation Programme.
Jarvis, I. (1980). The initiation of phosphatic chalk sedimentation – the Senonian (Cretaceous) of the Anglo-Paris Basin. *Society of Economic Paleontologists and Mineralogists, Special Publication*, **29**, 167–92.
Jell, P.A. (1974). Faunal provinces and possible planetary reconstruction of the Middle Cambrian. *Journal of Geology*, **82**, 319–50.
*Julivert, M. (1971). Decollement tectonics in the Hercynian Cordillera of Northwest Spain. *American Journal of Science*, **270**, 1–29.
*Julivert, M. (1971). L'évolution structurale de l'arc Asturien. In *Histoire Structurale de Golfe de Gascogne*, vol. 1. Institute Français du Pétrole, Collection Colloques et Seminaires, 22, 28pp. Éditions Technip, Paris.
Kholodov, V.N. (1969). Secondary alterations of bedded phosphorites of Lesser Karatau in the hypergene zone. *Lithology and Mineral Resources*, **1969(3)**, 176–288.
Khramov, A.N. (1975). Paleomagnetic directions and paleomagnetic poles, data from USSR In *Materials for World Data Center* B, Vol. 3, 4–43. Soviet Geophysical Committee, Academy of Sciences, Moscow.
Kirschvink, J.L. (1978). The Precambrian–Cambrian boundary problem: magneto-stratigraphy of the Amadeus Basin, central Australia. *Geological Magazine*, **115**, 139–50.
Kirschvink, J.L. (1980). The least-squared line and plan and the analyses of heliogenetic data. *Geophysical Journal of the Royal Astronomical Society*, **62**, 699–2718.
Kirschvink, J.L. & Rozanov, A. Yu. (1979). Paleomagnetism of fossiliferous Lower Cambrian sediments – new results from the Tommotian Stage of the Siberian Platform. *International Union of Geodesy and Geophysics, General Assembly, Abstracts*, **17**, 508.
Knipper, A.L. (1958). Geosynclinal phosphorites in southern Ulu Tau, central Kazakhstan. *Doklady Akademii Nauk SSSR*, **115**, 609–11.
Lu Y.H. & Zhu Z.L. (1981). Summary of the Cambrian biostratigraphy of China. *United States Department of the Interior, Geological Survey Open-File Report*, **81–743**, 121–2.
Lucas, L., Prévôt L. & Trompette, R. (1980). Petrology, mineralogy, and geochemistry of the late Precambrian phosphate deposits of Upper Volta (W Africa). *Journal of the Geological Society*, **137**, 787–92.
McElhinny, M.W., Embleton, B.J.J., Ma, X.H. & Zhang Z.K. (1981). Fragmentation of Asia in the Permian. *Nature*, **293**, 212–6.
McKelvey, V.E. (1967). Phosphate deposits. *United States Geological Survey, Bulletin*, **1252–D**, 1–21.
McWilliams, M.O. (1981). Paleomagnetism and Precambrian tectonic evolution of Gondwana. In *Precambrain Plate Tectonics*, ed. A. Kröner, pp. 649–88. Elsevier, Amsterdam.
*Martinsson, A. (1974). The Cambrian of Norden. In *The Cambrian of the British Isles, Norden, and Spitsbergen.* ed. C.H. Holland, pp. 185–283. John Wiley & Sons, London.
Meng, H.H. (1959). The petrography of phosphorites of the Karatau Basin. *Doklady Akademii Nauk SSSR*, **126**, 524–6.
Mohr, R.E. (1975). Measured periodicities of the Biwabik (Precambrian) stromatolites and their geophysical significance. In *Growth Rhythms and the History of the Earth's Rotation.* ed. G.D. Rosenberg & S.K. Runcorn, pp. 43–56. John Wiley, London.
Mooers, C.N.K., Collins C.A. & Smith, R.L. (1976). The dynamic structure of the frontal zone in the coastal upwelling region off Oregon. *Journal of Physical Oceanography*, **6**, 3–21.
Newell, N.D. (1972). The evolution of reefs. *Scientific American*, **226**, 54–65.
Notholt, A.J.G. (1972). Palaeozoic phosphate rocks in southwest Asia and the Near East. *Institute of Mining and Metallurgy Transaction, Section B*, **81(B), 789**, B158–63.
Notholt, A.J.G. (1979). Resources of Precambrian and Cambrian sedimentary phosphate rock. In *Proterozoic-Cambrian Phosphorites.* ed. P.J. Cook & J.H. Shergold, pp. 71–86. ANU Press, Canberra.
O'Brien, J.J. (1971). A two dimensional model of the wind-driven North Pacific. *Investigatión Pesquera*, **35**, 331–49.
Odin, G.S. & Letolle, R. (1980). Glauconitization and phosphatization environments: a tentative comparison. *Society of Economic Paleontologists and Mineralogists, Special Publication*, **29**, 227–38.
Organizing Committee, Fifth International Field Workshop and Seminar on Phosphorite, (1982). *Guidebook to Field Excursions*, 50pp. International Geological Correlation Programme, Project 156, Kunming, China.
Palmer, A.R. (1972). Problems of Cambrian biogeography. In *24th Session International Geological Congress, Montreal, Section 7, Paleontology*, pp. 310–5.
Palmer, A.R. (1974). Search for the Cambrian world. *American Scientist*, **62**, 216–24.
Palmer, A.R. (1977). Biostratigraphy of the Cambrian System – a progress report. *Annual Review of Earth and Planetary Science*, **5**, 13–34.
Palmer, A.R. (1979). Cambrian. In *Treatise on Invertebrate Paleontology*, Part A, ed. R.A. Robison & C. Teichert, pp. A119–35. Geological Society of America and University of Kansas, Lawrence.
Parrish, J.M., Parrish, J.T. & Ziegler A.M. (in press). Permian–Triassic paleogeography and paleoclimatology and implications for therapsid distributions. In *The Biology and Ecology of Mammal-like Reptiles*, ed. N. Hotton, III, p. McLean, J. Roth & C. Roth Smithsonian Press, Washington, DC.
Parrish, J.T. (1982). Upwelling and petroleum source beds, with reference to the Paleozoic. *American Association of Petroleum Geologists, Bulletin*, **66**, 750–74.

Parrish, J.T. & Curtis, R.L. (1982). Atmospheric circulation, upwelling, and organic-rich rocks in the Mesozoic and Cenozoic. *Palaeogeography, Palaeoclimatology and Palaeoecology*, **40**, 31–66.

Parrish, J.T., Ziegler, A.M. & Humphreville, R.G. (1983). Upwelling in the Paleozoic Era. In *Coastal Upwelling: Its Sediment Record*, vol. 8, eds J. Thiede & E. Suess, pp. 553–78. Plenum Press, New York.

Parrish, J.T., Ziegler, A.M. & Scotese, C.R. (1982). Rainfall patterns and the distribution of coals and evaporites in the Mesozoic and Cenozoic. *Palaeogeography, Palaeoclimatology and Palaeoecology*, **40**, 67–101.

Pietrafesa, L. J. & Janowitz, G.S. (1980). Upwelling in the region of the Charleston Bump and areas of the outer continental shelf. *Eos*, **61**, 986.

Pinet, P.R. & Popenoe, P. (1980). Cenozoic flow patterns of the Gulf Stream over the Blake Plateau. *Geological Society of America, Abstracts with Programs*, **2**, 500.

Polyansky, N.V. (1977). Discovery of phosphate shows in lower Paleozoic deposits of the Chingiz-Tarbagatai meganticlinorium in eastern Kazakhstan. *Lithology and Mineral Resources*, **12**, 108–10.

*Powers, R.W. (Ed.) (1968). *Lexique Stratigraphique International. Asie, Fascicule 10 b 1, Arabie Saoudite*, 177pp. International Geological Congress.

Prian, J.-P. (1980). Caractéristiques des paléoenvironnements des phosphorites cambriennes du versant septentrional de la Montagne Noire (sud du massif Central; France). In *Géologie Comparée des Gisements de Phosphates et de Pétrole*. Colloque International, Orléans 6–7 Novembre, 1979, *Documents du BRGM*, **24**, 93–111.

*Rankin, D.W. (1976). Appalachian salients and recesses: Late Precambrian continental breakup and the opening of the Iapetus Ocean. *Journal of Geophysical Research*, **81**, 5605–19.

Regnéll, G. & Hede, J.E. (1960). The lower Palaeozoic of Scania, the Silurian of Gotland. *21st Session International Geological Congress, Norden, Guide to Excursions A22, C17*, 89pp.

Rushton, A.W.A. (1974). The Cambrian of Wales and England. In *Cambrian of the British Isles, Norden and Spitsbergen*, ed. C.H. Holland, pp. 43–121. John Wiley, London.

Russell, R.T. & Trueman, N.A. (1971). The geology of the Duchess phosphate deposits, northwestern Queensland, Australia. *Economic Geology*, **66**, 1186–214.

Scotese, C.R., Bambach, R.K., Barton, C., Van der Voo, R. & Ziegler, A.M. (1979). Paleozoic base maps. *Journal of Geology*, **87**, 217–77.

Scrutton, C.T. (1978). Periodic growth features in fossil organisms and the length of the day and month. In *Tidal Friction and the Earth's Rotation*, ed. P. Brosche & J. Sündermann, pp. 154–96. Springer-Verlag, Berlin.

Sheldon, R.P. (1964). Paleolatitudinal and paleogeographic distribution of phosphorite. *United States Geological Survey, Professional Paper*, **501-C**, C106–13.

Sheldon, R.P. (1980). Episodicity of phosphate deposition and deep ocean circulation – an hypothesis. In *Society of Economic Paleontologists and Mineralogists, Special Publication*, **29**, 239–48.

Sheldon, R.P. (1981). Ancient marine phosphorites. *Annual Review of Earth and Planetary Science*, **9**, 251–84.

Sheldon, R.P. & Burnett, W.C. (1981a). Changing patterns of phosphogenesis in Mesozoic and Cenozoic. *American Association of Petroleum Geologists, Bulletin*, **65**, 991.

Sheldon, R.P. & Burnett, W.C. (1981b). Paleogeography of Phanerozoic phosphorites and extreme upwelling systems. In *Programme and Abstracts from the Conference on Coastal Upwelling: Its Sediment Record, September 1981, Portugal*. Advanced Research Institute.

*Shergold, J.H., Cooper, R.A., Mackinnon, D.I. & Yochelson, E.L. (1976). Late Cambrian Brachiopoda, Mollusca, and Trilobita from northern Victoria Land, Antarctica. *Palaeontology*, **19**, 247–91.

Smirnov, A.I. (1959). New data on elemental constitution of the phosphorites in the Karatau Basin. *Doklady Akademii Nauk SSSR*, **125**, 205–7.

Smith, A.G., Hurley, A.M. & Briden, J.C. (1981). *Phanerozoic Paleocontinental World Maps*, 102pp. Cambridge University Press, Cambridge.

Smith, R.L., Enfield, D.B., Hopkins, T.S. & Pillusbury, R.D. (1971). The circulation in an upwelling ecosystem: the *Pisco* cruise, *Investigación Pesquera*, **35**, 9–24.

Sokolova, Ye. A. (1961). The distribution of manganese and phosphorus in various types of rocks of the Usinsk Formation (Lower Cambrian of Kuznetsk Ala-Tau). *Doklady Akademii Nauk SSSR*, **135**, 1081–3.

*Sokolowski, S. (Ed.) (1976). *Geology of Poland. Volume 1: Stratigraphy, Part II: Mesozoic*. Wydawnictwa Geologiczne Publishing House, Warsaw.

Southgate, P.N. (1980). Cambrian stromatolitic phosphorites from the Georgina basin, Australia. *Nature*, **285**, 395–7.

Spjeldnaes, N. (1967). Fossils from pebbles in the Biskopasen Formation in southern Norway. *Norges Geologiske Underskelse*, **251**, 53–83.

*Stocklin, J. (1968). Structural history and tectonics of Iran: a review. *American Association of Petroleum Geology, Bulletin*, **52**, 1229–58.

*Stocklin, J. (1974). Possible ancient continental margins in Iran. In *The Geology of Continental Margins*, ed. C.A. Burk & C.L. Drake, pp. 873–88. Springer-Verlag, New York.

Sündermann, J. & Brosche, P. (1978). Numerical computations of tidal friction for present and ancient oceans. In *Tidal Friction and the Earth's Rotation*. ed. P. Brosche & J. Sündermann pp. 125–43. Springer-Verlag, Berlin.

Taylor, M.E. (1976). Indigenous and redeposited trilobites from Late Cambrian basinal environments of central Nevada. *Journal of Paleontology*, **50**, 668–700.

Taylor, M.E. & Forester, R.M. (1979). Distributional model for marine isopod crustaceans and its bearing on early Paleozoic paleozoogeography and continental drift. *Geological Society of America, Bulletin*, **90(I)**, 405–13.

Tushina, A.M. (1968). Lithofacies characteristics of the formation of the phosphorite stratum of Karatau. *Lithology and Mineral Resources*, **1968(4)**, 456–65.

Vail, P.R., Mitchum, R.M. Jr & Thompson, S. III, (1977). Seismic stratigraphy and global changes of sea level, Part 4: global cycles of relative changes of sea level. In *Seismic Stratigraphy – Applications to Hydrocarbon Exploration*, ed. C.E. Payton, pp. 83–97. American Association of Petroleum Geologists, Memoir 26, Tulsa.

Van der Voo, R., Scotese, C.R. & McCable, C. (1982). Was there an Eocambrian supercontinent before the Paleozoic Wilson Cycle? *Eos*, **63**, 308.

*Veevers, J.J. (1976). Early Phanerozoic events alongside the Australasian-Antarctic platform. *Journal of the Geological Society of Australia*, **23**, 183–206.

*Vinogradov, A.P. (Ed.) (1968). *Atlas of the Lithological-Paleogeographical Maps of USSR. Volume I, Precambrian, Cambrian, Ordovician, and Silurian*. Ministerstvo Geologiya SSSR, Akademiya Nauk SSSR, Moscow.

Wang, Y.L. (1942). Phosphate rocks of Chungyitsun, Kunyang, Yunnan. *Geological Bulletin, National Geological Survey of China*, **35**, 34–8.

Waples, D.W. (1982). Phosphate-rich sedimentary rocks: significance for organic facies and petroleum exploration. *Journal of geochemical Exploration*, **16**, 135–60.

*Wolfart, R. (1967). Zur entwicklung der paläozoischen Tethys in Voderasien, *Erdöl und Kohle, Erdgas, Petrochemie*, **20**, 168–80.

Wooster, W.S. (1969). Equatorial front between Peru and Galapagos. *Deep-Sea Research*, **16** (Suppl.), 407–19.

Yeh L., Sun S. & Chen Q. (1982). Facies Characteristics of the Chinese phosphorite deposits. In *Abstracts, 5th International Field Workshop and Seminar on Phosphorite*, pp. 79–82. International Geological Correlation Programme, Project 156 Kunming, China.

Yentsch, C.S. (1974). The influence of geostrophy on primary production. *Tethys*, **6**, 111–8.

*Zaitsev, N.S. & Ilyin, A.V. (1970). Tectonics of the Khubsugul Trough,

Mongolia. *Geotectonics*, **5**, 283–6.
Zaitsev, N.S. & Ilyin, A.N. (1970). The Khubsugul phosphate basin (Mongolia). *Doklady Akademii Nauk SSSR*, **192**, 23–6.
Ziegler, A.M., Barratt, S.F. & Scotese, C.R. (1980). Paleoclimate, sedimentation and continental accretion. In The origin and evolution of the Earth's continental crust. *Philosophical Transactions of the Royal Society of London, Series A, Mathematical and Physical Sciences*, **301(1461)**, 253–64.
Ziegler, A.M., Hansen, K.S., Johnson, M.E., Kelley, M.A., Scotese, C.R. & Van der Voo, R. (1977). Silurian continental distributions, paleogeography, climatology, and biogeography. *Tectonophysics*, **40**, 13–51.
Ziegler, A.M., Scotese, C.R., McKerrow, W.S., Johnson, M.E. & Bambach, R.K. (1979). Paleozoic paleogeography. *Annual Review of Earth and Planetary Science*, **7**, 473–502.
Zonenshayn, L.P. (1973). The evolution of central Asiatic geosynclines through sea-floor spreading. *Tectonophysics*, **19**, 213–32.
Zonenshayn, L.P. & Gorodnitsky A.M. (1977). Paleozoic and Mesozoic reconstructions of the continents and oceans. Article 1. Early and middle Paleozoic reconstruction. *Geotectonics*, **11**, 83–94.

23

Proterozoic and Cambrian phosphorites – specialist studies: biochronology of Proterozoic and Cambrian phosphorites

J.H. SHERGOLD AND M.D. BRASIER

Introduction

This paper presents a review of the palaeontological data required to establish a biochronological framework for the classification of Proterozoic and Cambrian phosphorites and occurrences of phosphatic sediments. It also attempts to resolve the question of continuity versus episodicity of phosphogenesis during this time interval.

Proterozoic and/or Cambrian phosphorites are known on all continents with the exception of Antarctica (see Notholt, 1979 for distribution). However, the quality and quantity of palaeontological data required to date these phosphorites accurately is enormously variable. For example, the phosphorites and stratigraphy of Europe are unusually well documented and permit us to formulate a detailed time scale for the elucidation of phosphogenic events. The complexity of the sedimentary events and highly refined biochronological framework used to date them is not easy to summarise, especially as the detail available gives us a potentially most useful calibration of these events which may be correlated elsewhere in the world. Since we have had to compress this part of the text substantially, we refer readers to Figs. 23.1 to 23.3 which attempt to assist comprehension and summarise the situation schematically. Additionally, a brief summary follows each major section.

In other parts of the world, particularly in Russia and China, there are equally numerous references to phosphorites and phosphatic sediments (e.g. Yanshin, 1979; Lu, 1979; Kazarinov & Krasilnikova, 1980), but it has proved difficult in some cases to establish the palaeontological aspects of many of these, and it is only possible to surmise a broad age. Accordingly, in these areas we have concentrated on those phosphatic sedimentary sequences for which we have the most adequate biological control. Inevitably, there is an unavoidable degree of overlap with parts of some regional reviews and details of some deposits described in other chapters of this book. Here, however, observations on a global scale are drawn together.

In general, an accepted global biochronological scheme for Cambrian time is lacking. Therefore, for the purpose of this review we regard Cambrian time as commencing with shelly fossil and archaeocyathid assemblages which define the beginning of the Tommotian Stage of south-eastern Siberia (Rozanov & Missarzhevsky, 1966; Rozanov *et al.*, 1969; Cowie, 1978), and its conclusion as coinciding with the beginning of the Tremadoc Series of North Wales, marked by the incidence of *Dictyonema* of the *flabelliforme* type. For the internal subdivision of Cambrian time we have used local scales based on biological events in Europe (plus eastern maritime North America), the Siberian Platform, China, and Australia. These scales are shown on the figures which accompany discussion of the major phosphogenic regions.

Europe

The classic Cambrian and Precambrian successions of Europe are the remnants of at least five formerly distinct platforms with contrasting lithofacies and biofacies: North American (Scotland and Greenland), Avalon (England, Wales, south-east Newfoundland to Massachusetts), Baltic (Scandinavia, Poland, Russia, i.e. including the East European Platform of other authors), Armorican (north France, Belgium, Germany) and southern European (south France, Sardinia, Spain). For a geological and palaeogeographic résumé see Brasier (1980) and Notholt & Brasier (this volume). Biostratigraphic correlation between these is particularly difficult in the Early Cambrian, where phosphorites are numerous, but a scheme is suggested here that attempts to use trilobites, archaeocyathans, acritarchs and small shelly fossils (Fig. 23.3). The Middle and Late Cambrian biostratigraphic schemes of the Avalon and Baltic Platforms are similar and of great importance for international correlation (Figs. 23.1, 23.2).

North-western Scotland and Eastern Greenland

These were both formerly part of the North American continent. The late Precambrian sediments of north-west Scotland comprise a varied suite of non-marine clastics of which the Torridonian Cailleach Head Formation contains phosphorite nodules (Peach *et al.*, 1907) with simple acritarchs of probable late Riphean age (Downie, 1975). The Cambrian rocks which unconformably overlie these rocks contain no recorded phosphorites, but there are several occurrences in the thicker, more complete sequence of East Greenland (Fig. 23.2). Lower parts of the succes-

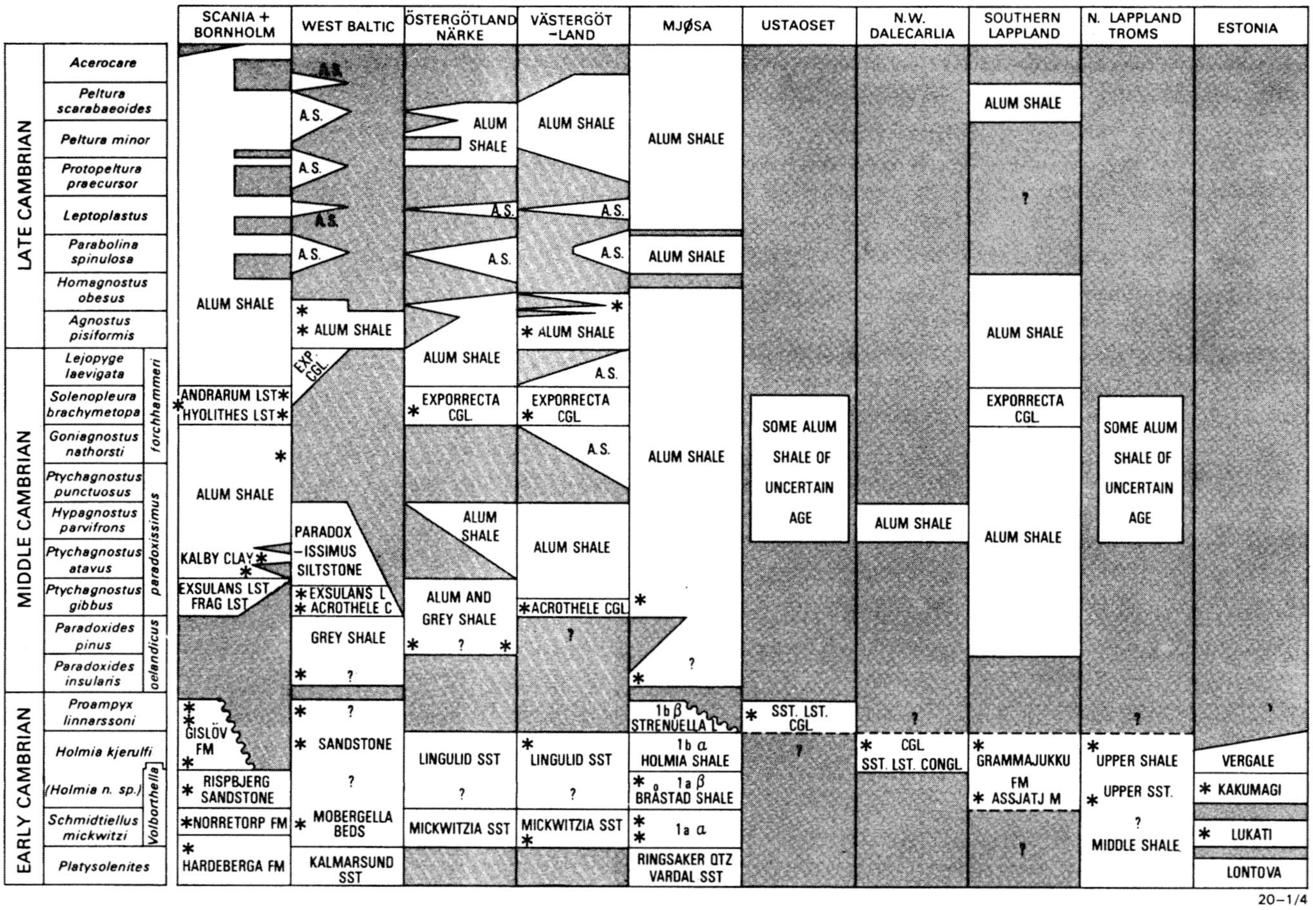

Fig. 23.1. Correlation of the phosphorites and phosphatic sediments of the Baltic Platform. Sources of information include Ahlberg & Bergström (1978), Bergström (1981), Bergström and Ahlberg (1981), Føyn & Glaessner (1979), Martinsson (1974) and Vidal (1981b, 1981c).

sion there contain Vendian acritarchs such as *Bavlinella faveolata* (Vidal, 1979) and these rocks were peneplained before deposition of the arenaceous Cambrian Kløftelv Formation, which is barren of acritarchs and shelly fossils but thought to contain the burrow *Skolithos linearis* (Cowie & Adams, 1957). Downie (1974) and Vidal (1979) correlate the unit with the Sub *Holmia* and Lomonosov Cambrian beds of eastern Europe (cf. Tommotian).

A disconformity separates the Kløftelv from the Lower Bastion Formation (Fig. 23.2), comprising glauconitic sandstones, siltstones and shales with a glauconitic, siliceous conglomerate at the base containing phosphate grains, pellets and nodules; a similar occurrence is found at the top of the formation (Cowie & Spencer, 1970; Henriksen & Higgins, 1976). The trace fossils *Scolicia, Skolithos, Phycodes, Arenicolites* and arthropod scratch marks (Cowie & Spencer, 1970) suggest an age not older than that of the Lontova Formation (cf. Fedonkin, 1977) (Fig. 23.1), and acritarchs like *Micrhystridium tornatum* found by Downie (1974) would not contradict this, although Vidal (1979) found only long-ranging sphaeromorphs. The glauconitic Upper Bastion Formation commences with the Lower Shell-Limestone, which is ferruginous and locally conglomeratic and phosphatic, suggesting a disconformity prior to deposition (Haller, 1971). The Lower Shell-Limestone and Lower Shales contain olenellid trilobites and hyoliths (Poulsen, 1932; Cowie & Adams, 1957) that suggest the late Lower Cambrian *Bonnia–Olenellus* Zone (Cowie, 1974; cf. Fritz, 1972) (Fig. 23.3). The overlying Upper Shell-Limestone and Upper Shales contain a rich fauna with olenellid and eodiscid trilobites (Poulsen, 1932; Cowie & Adams, 1957; Cowie, 1971) which also belong to this zone. *Calodiscus lobatus* suggests correlation with the *Holmia kjerulfi* Zone of Scandinavia and *Holmia* B beds of eastern Europe (Vidal, 1979) (Fig. 23.3). Acritarchs from the Upper Bastion Formation (Downie, 1974) confirm such a correlation, although Vidal (1979) reports only long-ranging forms. The conformably overlying limestones of the Ella Island Formation lack phosphorites; the shelly fauna suggests correlation with the *Bonnia–Olenellus* Zone (Cowie, 1974) while acritarchs compare with those of the *Protolenus* Zone of eastern Europe (Vidal, 1979). The overlying carbonates are less fossiliferous and may span high Lower Cambrian to Lower Ordovician (Cowie, 1971).

In summary, phosphatic sediments of the European segment of the North American Platform occur as nodules in the Vendian Cailleach Head Formation of north-western Scotland; and at two levels in the Bastion Formation of eastern Greenland. The latter is separated from underlying formations by disconformity, and the phosphorites levels within it are also separated by disconformity. They are both associated with

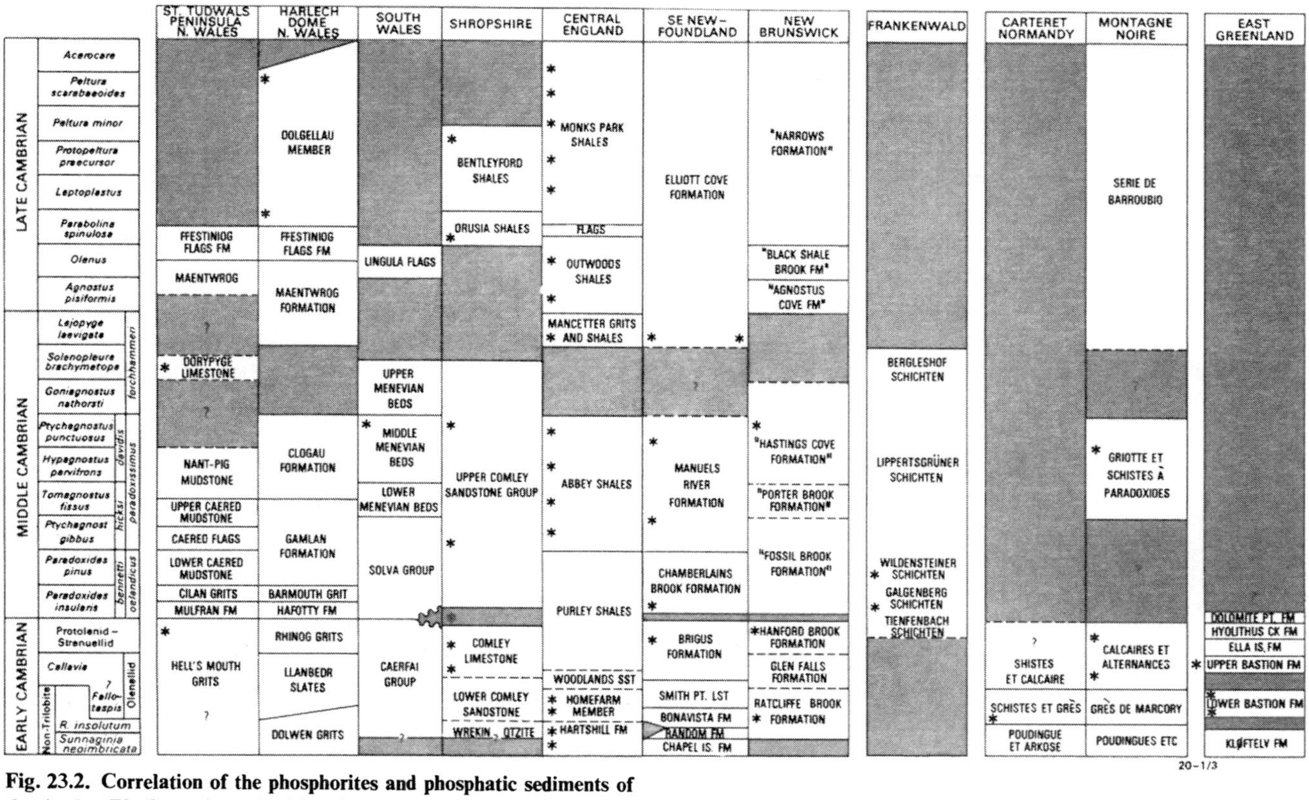

Fig. 23.2. Correlation of the phosphorites and phosphatic sediments of the Avalon Platform, from the following sources: Bengtson & Fletcher (1981), Brasier & Hewitt (1981), Cowie *et al.* (1972), Hutchinson (1962), Martinsson (1974), Rushton (1974) and Sdzuy (1964).

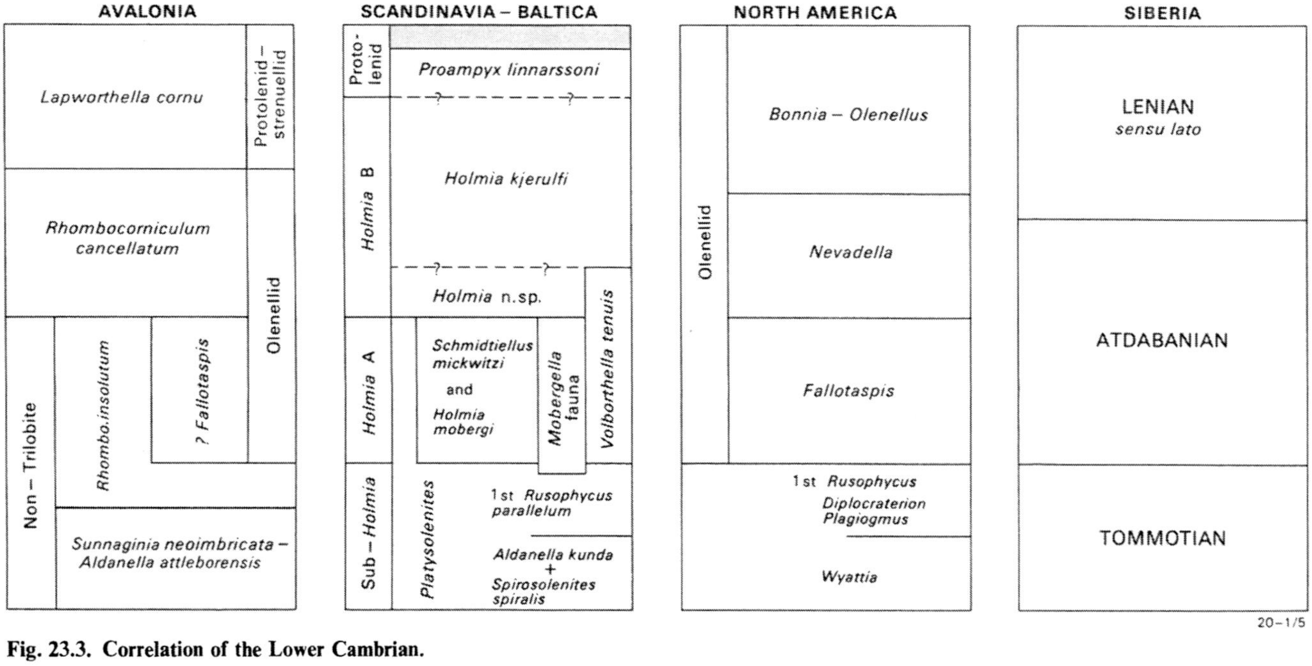

Fig. 23.3. Correlation of the Lower Cambrian.

trilobites and other fauna and flora indicative of the late Early Cambrian *Bonnia–Olenellus* Zone on the North American biochronological scale (see Fig. 23.3).

Baltic Platform

The thin Cambrian deposits of the Baltic Platform (Fig. 23.1) contain a classic biostratigraphic succession, extending down into Vendian or Riphean rocks in Finnmark, Lappland, around Lake Mjøsa (south Norway) and Lake Vättern (south-east Sweden) and several basins in Poland and the Moscow region. Minor phosphorites are many and widespread and the occurrences will be dealt with zonally.

(1) Vendian Precambrian phosphorites are scarce in the Baltic region. The sparagmite succession below the Moelv tillite of Lake Mjøsa contains phosphatic pebbles in the

Biskopås conglomerate, thought to have formed from the erosion of carbonates during a regional regression (Bjørlykke, Elvsborg & Høy, 1976). These phosphatic clasts contain acritarchs and the problematic microfossil *Papillomembrana* (Spjeldnaes, 1967; Manum, 1967); an early Vendian age is suggested by the acritarchs (Vidal, 1981a). The Riphean–Vendian sediments of the upper part of the Visingsö Series around Lake Vättern contain two beds of phosphatic nodules: abundant microfossils include protist loricae, *Chuaria circularis* and other acritarchs that indicate a late Riphean to early Vendian age (Vidal, 1976, 1981a; Knoll & Vidal, 1980).

The Vendian of the Baltic area generally lacks glauconite and phosphorite but these become common in the three transgressive–regressive cycles in the uppermost Vendian–Lower Cambrian, termed the Baltian, Liivian and Aisciaian cycles by Mens & Pirrus (1981). Phosphatic sediments are particularly conspicuous in the thinner and discontinuous platform sequences of Estonia and Scandinavia (Fig. 23.1).

The late Vendian was a period of relatively low sea level. Sediments bearing the alga *Vendotaenia* are confined to the cratonic basins of eastern Europe, Finnmark and Lappland (Vidal, 1979); trace fossils from these horizons comprise horizontal trails and burrows (e.g. Fedonkin, 1977).

(2) Platysolenites or Sub Holmia Zone A widespread regression marked by red sandstone deposition or erosion (Brangulis *et al.*, 1981) was followed by the initial deposits of the Baltian transgression, which includes the glauconitic sandstones of the Rovno horizon of the East European Platform. This contains the alga *Tyrasotaenia* and organic tubes of *Sabellidites cambriensis* in the lower part, plus agglutinated *Platysolenites antiquissimus* in the upper part (Aren & Lendzion, 1974; Rozanov, 1976; Føyn & Glaessner, 1979; Brangulis *et al.*, 1981); the trace fossils are simple *Planolites* and bilobite traces (Fedonkin, 1977). The most widespread facies of the Baltian cycle is the blue clay with glauconitic sandstones of the Lontova Formation (Estonia) or Mazoviecka Series (Poland) (Fig. 23.1), which yield *S. cambriensis, P. antiquissimus* and the small shells of *Aldanella kunda, Hyolithes mickwitzi* and *Hyolithellus* sp. (Fedonkin, 1977). Contrary to confused earlier reports, trilobites are lacking (Mardla *et al.*, 1968; Yankauskas & Posti, 1973), and the Lontova Formation is referred to the Sub *Holmia* Zone (e.g. Samsonowicz, 1960; Vidal, 1979). Trace fossils present are genera common in Cambrian and younger rocks and include *Rusophycus, Teichichnus, ?Rhizocorallium, Treptichnus* and *Skolithos* (Fedonkin, 1977). Acritarch assemblages, however, lack baltisphaerids and are dominated by sphaeromorphs plus a few *Micrhystridium tornatum* with long stratigraphic ranges (Yankauskas & Posti, 1973; Aren & Lendzion, 1974).

The Baltian cycle concluded with glauconitic sandstones and sparse clays (the Glebevo, Zawiszyn or *Mobergella* beds) which show desiccation cracks and other features of regression (e.g. Lendzion, 1968; Aren *et al.*, 1979) and are confined mainly to the Moscow Basin. *Mobergella holsti, M.* cf. *radiolata, M. turgida, M. braastadi* and *Hyolithellus* spp. are reported from Poland (Lendzion, 1968, 1972; Aren *et al.*, 1979; but see Bengtson, 1977).

In Estonia, the Sub *Holmia* Zone is represented by Lontova clays with a fauna that includes *P. antiquissimus, Spirosolenites* sp., *A. kunda* and *H. mickwitzi*, but the uppermost *Mobergella* facies and fauna is missing (Öpik, 1956; Mardla *et al.*, 1968; Mens & Pirrus, 1975; Bergström, 1981). Pebbles of phosphatic siltstone in conglomerates of the overlying Lükati Sandstone are probably from phosphorites formed during stillstand or regression following Lontova deposition, and reworked by the Lükati transgression of the Liivian cycle (Mens & Pirrus, 1975). No phosphorites of Sub *Holmia* age can be proven in Scandinavia; the presence of *Baltisphaeridium* in many basal Cambrian deposits implies a younger age for the major transgressive sediments with *Platysolenites, Mobergella* and *Camenella* (Vidal, 1979, 1981b; Bergström, 1981), although the *Platysolenites*-bearing beds of the Breivik Formation in Finnmark and the Middle Shale in Lappland may be of Sub *Holmia* age (Føyn & Glaessner, 1979; Bergström, 1981). The Middle Shale overlies sandstones with *Vendotaenia* and the medusoid *Kullingia*, of latest Vendian age, separated from it by basal conglomerates and breccias (*ibid*; Vidal, 1981a). The Middle Shale contains glauconitic sandstones and red shales with *P. antiquissimus, S. cambriensis, Tyrasotaenia* sp., *Aldanella kunda, ?Orthotheca, Torellella laevigata* and *Volborthella tenuis* (Føyn & Glaessner, 1979) although the last may be from a higher bed than the others. The rocks are barren of acritarchs but Vidal (1981a) infers a *Holmia* age.

(3) Schmidtiellus mickwitzi or Holmia A Zone The Liivian cycle and *Holmia* A Zone is clearly distinguished from the Baltian cycle below by the presence of trilobites and baltisphaerid acritarchs (Mens & Pirrus, 1981; Mens, 1981). The phosphatic and glauconitic sandstone at the base of the Lükati Series in Estonia (see above) was followed by shallow-water sandstones with *V. tenuis, Mickwitzia monilifera*, the olenellids *Schmidtiellus mickwitzi* and *Holmia mobergi*?, and the trace fossils *Diplocraterion* and *Eophyton; Platysolenites* is confined to the base (Öpik, 1956; Mardla *et al.* 1968; Bergström, 1981). The equivalent (Talsy) horizon elsewhere in eastern Europe contains *Mobergella* at the base (Rozanov, 1976; Volkova *et al.*, 1979) and *?Fallotaspis* (Brangulis *et al.*, 1981). There is also a distinctive assemblage of acanthomorph acritarchs (Volkova *et al.*, 1979).

Sediments of *Schmidtiellus mickwitzi* Zone (Bergström, 1981) age in Scandinavia overlap much of the crystalline basement. *Holmia* A Zone acritarchs are found in the basal sandstones or phosphatic and glauconitic siltstones of Scania, Bornholm, Kalmarsund (West Baltic) and Västergötland (Vidal, 1981b). On Bornholm, primary phosphate nodules (Poulsen, 1967) and reworked pebbles (Vidal, 1981b) occur in the glauconitic siltstones of the Green Shales, with *Coleoloides typicalis* and other shells also found in the Home Farm Member of central England (Poulsen, 1967; Brasier & Hewitt, 1981). Scarce trilobite exuviae and *Volborthella* occur in phosphatic limestone nodules and a phosphatised steinkern of *?Schmidtiellus* found in younger conglomerates may be from these beds (Poulsen, 1978). In Scania there are at least two phosphatic horizons of this age, separated by a member of Hardeberga Quartzite. The lower, glauconitic Brantevik Sandstone has a conglomerate with allochthonous phosphate pebbles in glauconitic quartz sandstone at the base (Hadding, 1927, 1929, 1932) without shelly fossils. The higher Norretorp Siltstone is glauconitic, with several beds of primary phosphatic nodules, tabular phosphorite and some allochthonous phosphorite pebbles near the top and base (Hadding, 1929, 1932); these beds contain *Holmia mobergi, S. mickwitzi* and *Wanneria lundgreni* (Bergström, 1973).

Glauconitic sandstones and drifted boulders of quartzose conglomerate and phosphatic conglomerate occur in Kalmarsund and Öland (West Baltic) with a fauna of *Mobergella holsti, M. radiolata, M. turgida, Camenella baltica, C.* cf. *plana, Hyolithellus* sp. and *Micromitra undosa* (Åhman & Martinsson, 1965; Bengtson, 1968, 1970, 1977). In borings on Öland and Gotland, glauconitic sandstones have phosphate nodules and quartzose conglomerates on the same level as *Mobergella*, while higher phosphorites occur near beds containing *M. monilifera, V. tenuis* and *T. laevigata* (Hadding, 1932; Thorslund & Westergard, 1938; Bengtson, 1977). The Mickwitzia Sandstone of Västergötland is only slightly glauconitic and Hadding (1927) reports small rounded phosphatic clasts in neptunean dykes; these are associated with *M. monilifera* and *T. laevigata*. But Lindström (1977) notes that shells of *T. laevigata* from the basal beds may be infilled with phosphate. *Holmiella* and *Mobergella* are rare elements found here; the trace fossils *Rusophycus dispar, Diplocraterion* and *Eophyton* (Bergström, 1980, 1981) are more typical.

The Vardal Sandstone and Ringsaker Quartzite of Mjøsa in south Norway are barren of acritarchs (Vidal, 1981b), although the latter beds contain *Skolithos* and *Diplocraterion* and cannot be older than the *Platysolenites* Zone. Both units lack glauconite and phosphorite but clasts of the latter are present in an overlying conglomerate (Vogt, 1924; Bjørlykke *et al.*, 1976). Transgressive deposits of alternating calcareous sandstone and shale followed locally (laα_1), bearing *Holmia mobergi* (?), hyoliths, inarticulate brachiopods; *P. antiquissimus* and *Mobergella* sp. indet. occur in sandy shales and glauconitic sandstones (laα_2), terminated in several areas by a conglomerate with phosphatic pebbles, occurring just below Bråstad shales with *V. tenuis* (Vogt, 1924; Skjeseth, 1963; Bengtson, 1977; Ahlberg & Bergström, 1978; Bergström, 1980). Phosphate nodules also occur through the lower part of the Upper Sandstone unit of the Dividal Group, Lappland (Hadding, 1929), which is tentatively assigned to this zone by Bergström (1981). Acritarchs are equivocal here, but Vidal (1981c) suggests a younger *Holmia* age.

(4) Holmia kjerulfi or Holmia B Zone The third transgressive cycle, the Aisciaian (Mens, 1981 and Mens & Pirrus, 1981) begins during or just before the Zone of *Holmia kjerulfi* (or *Holmia* B Zone). The initial deposits in eastern Europe (the Vergale Stage) are clastic sediments with *Strenuaeva primaeva, Holmia kjerulfi, Proampyx grandis, Ellipsostrenus* cf. *gripi, V. tenuis* and *Obolella* cf. *rotundata* (Lendzion, 1968; Aren *et al.* 1979; Brangulis *et al.* 1981; Mens, 1981) and many distinctive acritarch species (Volkova *et al.*, 1979; Vidal, 1981b). In Estonia, the Kakumagi beds have a cross-bedded basal conglomerate with phosphorite pebbles containing *M. monilifera, M. formosa, Scenella discinoides* and trilobite fragments (Öpik, 1956; Martinsson, 1974), suggesting a preceding hiatus.

Two zones are suspected in Scandinavia: a lower one with *Holmia* n. sp., poorly known from Mjøsa, and a higher one with *Holmia kjerulfi* in Norway or *H. sulcata* in Sweden plus solenopleurids (Ahlberg & Bergström, 1978; Ahlberg, 1981; Bergström, 1981). In Scania and Bornholm, the glauconitic Rispbjerg Sandstone may mark the relatively regressive base of the *Holmia* n. sp. Zone (e.g. Vidal, 1981c; Bergström, 1981). In both areas, the upper part contains phosphate nodules and the fossils *Rusophycus dispar* and *Hyolithellus micans* (Hadding, 1958; Poulsen, 1967; Bergström & Ahlberg, 1981). Samples of phosphatic pebbly conglomerate from the base and top of this sandstone at Brantevik are barren of acritarchs (Vidal, 1981b). The overlying shales, sandstones and thin limestones of the Gislöv Formation comprise a transgressive to regressive cycle with phosphate nodules near the base (*Holmia kjerulfi* Zone) and top (*Proampyx linnarssoni* Zone; e.g. de Marino, 1980; Bergström & Ahlberg 1981). The *H. kjerulfi* Zone is indicated by *H. sulcata, Ellipsocephalus nordenskioeldi, Proampyx rotundatus, P.* cf. *sularpensis, Strenuaeva* sp., *Calodiscus lobatus, Trapezotheca?* sp., *Helcionella?* sp., *Magnicanalis* sp., *Nisusia?, Lingulella nathorsti?* and *Hyolithellus* cf. *micans* (Bergström & Ahlberg, 1981) plus *Lapworthella cornu* (Bengtson, 1980); sparse acritarchs are of *Holmia* B age or younger (Vidal, 1981b). The Lingulid Sandstone of Västergötland also begins with a relatively regressive cross-bedded sandstone which bears *Skolithos* and desiccation cracks. Higher slightly glauconitic beds (Hadding, 1929) contain *Lingulella* sp. and *Obolella favosa*. The top is again regressive with a calcareous sandstone containing *Skolithos linearis*, and *Diplocraterion* and is overlain by a conglomerate with pebbles of phosphatic sandstone (containing *Holmia grandis?*) and basement rocks (Hadding, 1927; Martinsson, 1974; Bergström, 1981). Acritarchs could indicate a *Holmia* B age, comparable with the Vergale rocks (Vidal, 1981b).

In Mjøsa, the Bråstad Shale (la$_\beta$) follows phosphatic conglomerates at the top of laα_2 (e.g. Vogt, 1924; Skjeseth, 1963) and also contains small phosphate nodules plus *Holmia* n. sp. and *V. tenuis* (*ibid*; Ahlberg & Bergström, 1978; Bergström, 1981). The succeeding *Holmia* Shale (lbα) contains a classic fauna of *H. kjerulfi, Proampyx grandis, Kjerulfia lata, Strenuaeva primaeva, E. gripi? Weymouthia nobilis, Hyolithes* spp., *Obolella rotundata, T. laevigata, Scenella depressa* and *Helcionella rugosa* (Kiaer, 1916; Skjeseth, 1963; Ahlberg & Bergström, 1978; Bergström, 1981) with acritarchs of *Holmia* B type (Vidal, 1981b). The presence of *V. tenuis* is questioned, however, in bed lbα by Yochelson, Henningsmoen & Griffin (1977).

To the north and west of Mjøsa sediments of the *H. kjerulfi* Zone transgressed on to Precambrian rocks. The Lower Cambrian of north-west Dalecarlia, in central-western Sweden, contains beds of massive, phosphatised sandstone (Hadding, 1929); fossils from drift boulders include *H. kjerulfi, T. laevigata* and lingulids (Tegengren, 1962). In south Lappland phosphatic conglomerates (Hadding, 1929) are associated with shales and limestones that, near the top, contain *E. gripi* and *Fallotaspis ljungneri* (Bergström, 1973; Bergström & Ahlberg, 1981;

Ahlberg, 1981) and *Holmia* B type acritarchs (Vidal, 1981b) which are therefore referred to this zone. In Troms and north Lappland, this zone is represented by the Upper Shale with a thin sandy limestone at the base; the latter contains some phosphate and glauconite grains (Hadding, 1958) and locally contains *E.* cf. *gripi* and lingulids (Ahlberg, 1980). About 15 m of green shales separate these from an upper fragmented limestone with *Proampyx triangularis, Comluella? lapponica* and *Strenuaeva inflata*, which are provisionally referred to the *H. kjerulfi* Zone (Ahlberg, 1980, 1981; Bergström, 1981).

(5) Proampyx linnarssoni or Protolenus Zone The Aisciaian cycle that began in the *H. kjerulfi* Zone continued into the *Protolenus* Zone of eastern Europe, and its regressive phase continued into the *Paradoxides oelandicus* Zones (Mens, 1981; Mens & Pirrus, 1981). Rocks of the Rausve horizon have a poor fauna in Russia. They are referred to the *Protolenus* Zone of Poland on account of their acritarchs (Volkova *et al.*, 1979; Vidal, 1979, 1981b). The classic *Protolenus* fauna includes species of *Protolenus, Strenuella, Ellipsocephalus* and eodiscids (e.g. *Serrodiscus speciosus*, Samsonowicz, 1956) but in many places this fauna is missing, perhaps removed by erosion in late *Protolenus* or early Middle Cambrian times.

The *Proampyx linnarssoni* Zone of Scandinavia is equated with the *Protolenus* Zone (Bergström & Ahlberg, 1981; Bergström, 1981; Ahlberg, 1981; Vidal, 1981b), although there is no supporting evidence yet from the acritarchs (Vidal, 1981b). Diagnostic taxa include *P. linnarssoni, E. lunatus* and *Comluella? scanica*, although *E. gripi* and *H. kjerulfi* may also persist (Vidal, 1981b). A fauna of this type occurs in the top half of the Gislöv Formation of south-east Scania, which consists of thin, dark, silty calcilutites with partial phosphatisation (Hadding, 1958; de Marino, 1980; Bergström & Ahlberg, 1981). Other fossils include *Strenuaeva?* sp., *Lingulella* sp., *Magnicanalis* sp., *Botsfordia* sp., acrotretids, *Indiana*? sp., *Hyolithellus* cf. *micans, Contitheca* sp. and *Amphigeisina danica* (Bergström & Ahlberg, 1981). Further north, at Hardeberga and Andrarum, the zone may be represented by calcareous and glauconitic sandstones with allochthonous phosphate clasts and inarticulate brachiopods, or it may be missing. Similar facies which abruptly overlie the calcilutites of south-east Scania and sandstones of Öland may be attributable to the Lower or Middle Cambrian and represent the regressive phase of a cycle (Hadding, 1958; de Marino, 1980; Bergström & Ahlberg, 1981).

In Mjφsa the *P. linnarssoni* Zone is missing (probably eroded) from the eastern autochthon but present in the western allochthonous sequence in the form of thin sandy limestones (lb_β) which contains the zonal index fossil plus *E. gripi?, T. laevigata, Helcionella* cf. *rugosa, Scenella antiqua, Acrothele bellapunctata, Obolella favosa* and *O. rotundata* (Kiaer, 1916; Skjeseth, 1963; Ahlberg & Bergström, 1978; Ahlberg, 1981; Bergström, 1981). Further west, at Ustaoset, a thin sequence of basal conglomerates, sandstones and fossiliferous phosphatic conglomerates with limestones (Goldschmidt, 1925) may be of this age. They overstep Precambrian gneisses and contain *P. linnarssoni?, T. laevigata, H. micans*? and *Nisusia oriens*? (Stφrmer, 1925; Ahlberg & Bergström, 1978). Similar deposits with phosphorite nodules occur at Risbekken, south-west of Mjφsa, but only contain *T. laevigata* (Skjeseth, 1963). Both sequences are overlain by the basal conglomerates of Alum Shale of uncertain age.

(6) Oelandicus Zones The Middle Cambrian is not well represented in the eastern Baltic area but sedimentation is thought to have continued across the Lower-Middle Cambrian boundary, with the Kibartai beds representing the final regressive phase of the Aisciaian cycle. These are sandstones with phosphatic conglomerates containing fossils indicative of the *Paradoxides oelandicus* Zones: *Ellipsocephalus puschi, E. polytomus, Strenuella (Comluella) samsonowicki, S. (C.) insoluta*, and *Lingulella ferruginea* (Mens, 1981; Mens & Pirrus, 1981).

Regressive conditions are also indicated in Scandinavia at this time. The *Paradoxides insularis* Zone is largely unrecorded and even at Mjφsa, where deposition was nearly continuous, there is a marked erosional contact between Lower and Middle Cambrian rocks. The basal *Oelandicus* deposits there include local phosphatic conglomerates with *Paradoxides* fragments (Skjeseth, 1963). Correlated limestones and alum shales contain *P. oelandicus* and *Acrothele granulata* or *E. torelli* (Martinsson, 1974), and overlying alum shales contain agnostids of the basal *Paradoxissimus* Zones (Skjeseth, 1963). On Öland, in the west Baltic, glauconitic sandstones with phosphorites occur at about the Lower–Middle Cambrian transition (Westergård, 1929; Hadding, 1932) and are followed by *P. oelandicus* shales with some arenaceous and calcareous beds but no phosphorite (Hadding, 1958). Phosphatic sandy conglomerates form the base to the alum shale sequence in Ostergötland and have been assigned to the *Oelandicus* Zones, although none of the index fossils are present (Westergård, 1922, 1940, 1944). The presence in Närke of *P.* aff. *oelandicus, Peronopsis fallax, Triplagnostus praecurrens, Acrothele granulata* and *Acrotreta* cf. *socialis* in basal glauconitic and phosphatic sandstones suggests a *Paradoxides pinus* Zone age (Westergård, 1940; Martinsson, 1974).

(7) Paradoxissimus Zones In many areas of Scandinavia, sediments of the *Paradoxides paradoxissimus* Zones overlie Lower Cambrian rocks with a considerable hiatus (Fig. 23.1). In south Norway they are also transgressive on to Precambrian basement.

Phosphatic sediments are widespread in sandy calcareous or conglomeratic rocks at the base of the apparently global *Ptychagnostus gibbus* transgression. In Scania, the Lower Cambrian Gislöv Formation terminates at a glauconitic sandstone or limestone having an erosional base; these may contain allochthonous phosphate pebbles (Hadding, 1929; de Marino, 1980), plus *Acrothele granulata, Acrotreta schmalensei, Lingulella* and *Indiania*? sp. Although of uncertain age, they seem to form the base to an overlying sequence of intraclastic limestones or alum shales tentatively referred to the *P. gibbus* Zone (Bergström & Ahlberg, 1981). These limestones contain a similar fauna, plus *P. paradoxissimus, Hydrocephalus* sp., *Acrothele coriacea?, Lapworthella bornholmensis, Gapparodus bisulcata, Proconodontus?* sp., *Amphigeisina danica* and *Hyolithellus* cf. *micans* (Bengtson, 1976; Bergström & Ahlberg, 1981). They may also contain phosphate nodules and coprolites

(Hadding, 1958). The intraclastic limestones are followed by dark alum shales, and these by the *Ctenocephalus exsulans* Limestone, also referred to the *P. gibbus* Zone. This contains little quartz and no phosphate or glauconite (Hadding, 1958).

At Øleaa, on Bornholm, the Exsulans Limestone is the lowest Middle Cambrian deposit and differs from that in Scania in its content of phosphate and glauconite (Hadding, 1958). Pebbles of phosphatic sandstone at its base are probably derived from the Rispbjerg Sandstone below, although the phosphatic matrix is autochthonous (Poulsen, 1963). Other reworked Lower Cambrian material includes fossils from the *S. mickwitzi* and *P. linnarssoni* Zones (Poulsen, 1966, 1978). The trilobites *Jincella parva, C. exsulans* and *Holocephalina* sp., indicative of the *P. gibbus* Zone, are not phosphatised (Poulsen, 1966). At Laesaa, the Kalby Clay forms the base; this contains worn phosphatic sandstone moulds of the above species, other Lower Cambrian exotics, and *Ptychagnostus atavus* so that Poulsen (1963) refers it to the latter zone. Perhaps as many as five successive beds have had their phosphatic remains reworked into this clay. A pyritic conglomerate with pebbles and irregular lenses of phosphorite overlies the clay but contains no determinable fossils, although this has also been referred to the *P. atavus* Zone (Poulsen, 1963). It seems that the Exsulans Limestone facies may have developed very widely and as far north as north Jämtland (Martinsson, 1974).

On Öland, the Oelandicus shales are followed by *Acrothele granulata* conglomerates, with small phosphatic clasts derived from the beds below (Hadding, 1932, 1958); these are referred to the *P. gibbus* Zone (Martinsson, 1974). The overlying Exsulans Limestone contains glauconite impregnation, phosphate grains, inarticulate brachiopods and trilobite fragments including *C. exsulans* (Hadding, 1929, 1958) and is likewise referred to this zone. A similar sequence is developed in the eastern Holy Cross Mountains of Poland, where phosphatic conglomerates (Kozlowski, 1931; Wawryk, 1932) are followed by alum shales with a fauna spanning the *P. gibbus* to *H. parvifrons* Zones: *P. gibbus, Peronopsis fallax, Paradoxides tessini, Parasolenopleura linnarssoni, Lingulella vistuale* (Samsonowicz, 1956).

In Västergötland, the Lingulid Sandstone is locally overlain by a transgressive conglomerate with *A. granulata*, glauconite and pebbles of phosphatic sandstone, assigned together with succeeding shales to the basal *Paradoxissimus* Zones (Westergård, 1944).

(8) Forchhammeri Zones The *Paradoxides forchhammeri* Zones have their typical development in Scandinavia where they followed widespread alum shales of the later *Paradoxissimus* Zones. In many places this alum shale deposition continued into *Goniagnostus nathorsti* times but a widespread change in sedimentation occurred then or in the following *Solenopleura brachymetopa* Zone. The Andrarum Limestone of Scania is a bituminous micrite that lacks phosphorite, glauconite and quartz but bears *S. brachymetopa* and *Centropleura loveni* (Hadding, 1958). Just below it lies the Hyolithes Limestone, a micro-coquina with phosphatised shell fragments and small phosphate nodules; small brachiopods and *Hyolithes* spp. are present (Holm, 1893), but quartz and glauconite are lacking (Hadding, 1958). On Bornholm, the Andrarum Limestone is underlain by a thin limestone with phosphatic nodules, and phosphatic conglomerates in the lower and middle part, plus *Ptychagnostus punctuosus, Ciceragnostus cicer, Paradoxides davidis* and *Acrotreta* sp., suggesting a late *Paradoxissimus* age (Grönwall, 1902; Hadding, 1958); but the presence of *Triplagnostus lundgreni* and *Goniagnostus nathorsti* implies an age transitional with the lowest *Forchhammeri* beds (Poulsen, 1966). An anthraconite limestone overlying the conglomerate contains abundant *P. punctuosus* while the succeeding intraclastic limestone has *S. brachymetopa* (Poulsen, 1966) and authigenic phosphorite pellets (Hadding, 1958). The Andrarum Limestone which follows, differs from that in Scania in its content of glauconite, quartz, small phosphate nodules and coprolites (Hadding, 1958). It yields *S. brachymetopa, S. holometopa, C. loveni, P. forchhammeri, Anomocare laeve, Anomocarina excavata* and *Phalagnostus glandiformis* (Poulsen, 1966).

In Ostergötland, alum shales with *Lejopyge laevigata* are underlain by sandstones with glauconite and phosphorite which Westergård (1944) tentatively referred to the *S. brachymetopa* Zone. The Exporrecta Conglomerate of Västergötland is underlain by limestones with a *Goniagnostus nathorsti* fauna and followed by alum shales with *Lejopyge laevigata*; the conglomerate contains pebbles of limestone derived from the *Paradoxissimus* beds and phosphate pebbles with a matrix fauna of *Billingsella exporrecta, P. forchhammeri* and hyoliths, referrable to the *S. brachymetopa* Zone (Wallerius, 1895, 1930; Westergard, 1946; Hadding, 1958).

(9) Late Cambrian Zones Late Cambrian deposits are few in the eastern Baltic area but widespread in Scandinavia, where they consist almost entirely of black alum shales and anthraconite limestone concretions, the latter containing abundant agnostids and olenids which furnish the classic biostratigraphy of the region. Conditions are thought to have been largely euxinic on the seafloor (Martinsson, 1974; Bergström, 1980). Phosphate nodules or pebbles are lacking but reappear near the base and top of the Tremadocian (Hadding, 1932; Poulsen, 1966) in alum shales, glauconitic limestones and sandstones or conglomerates. Minor occurrences, with phosphatised microfossils and faecal pellets are present, however, in the *Agnostus pisiformis* and *Homagnostus obesus* Zone anthraconites of Västergötland and Öland (Müller, 1979), and probably occur at other horizons.

In summary, phosphate typically occurs on the Baltic Platform as phosphatic nodules within shale sequences, e.g the Early Cambrian Bråstad Shale of Mjøsa, Norway, Gislöv Formation of Scania and Bornholm, Upper Shale of Lappland; the Middle Cambrian Grey Shale of the west Baltic region, and Alum Shale of Ostergötland, Närke and Västergötland in Sweden and the Mjøsa district of Norway; or, more rarely, in the Late Cambrian Alum Shale of the west Baltic. Less commonly, phosphate occurs as nodules, pebbles or pellets, sometimes reworked in glauconitic sandstone and siltstone formations, mainly of Early Cambrian age, in Estonia, southern Sweden, and the Mjøsa district of Norway. It may also occur in Early Cambrian phosphatic conglomerates in Norway and southern Sweden; and in both Early and late Middle Cambrian conglomerates and intraclastic carbonates, throughout southern Sweden.

Avalon Platform

The Cambrian facies of this platform, which formerly stretched from Massachusetts to central England, are sometimes phosphatic like those of the Baltic Platform; again, oolitic, peritidal and archaeocyathan facies are absent. Precambrian sediments are not phosphatic in this area and mainly comprise thick volcanic, volcaniclastic or molasse facies. Correlation of the Middle and Late Cambrian with Scandinavian zones is possible, but the Early Cambrian faunas are more distinct and the scheme used here is a modification of those by Hutchinson (1962) and Bengtson & Fletcher (1981) for south-east Newfoundland and of Cowie, Rushton & Stubblefield (1972), Brasier & Hewitt (1981) and Brasier (1984) for England. During the Cambrian, the Avalon Platform seems to have consisted of many swells and basins with local histories, so that sedimentary cycles have not yet been as widely correlated as they have in the Baltic area.

(1) Aldanella attleborensis–Sunnaginia neoimbricata Zone This zone is tentatively erected here for those beds which contain either of the above species but lack trilobites. *Aldanella kunda* and *Platysolenites antiquissimus* occur together in north Scandinavia and the Lontova Formation of the southern Baltic Platform (see above). *A. attleborensis* and *S. imbricata* are also common in the early Tommotian of Siberia (Rozanov *et al.*, (1969).

A major change from volcanic to coastal clastic sedimentation took place at this time, as shown by *P. antiquissimus* in the top part of the Dolwen Grits which lie at the base of the north Wales sequence (Rushton in Allen & Jackson, 1978) and *A. attleborensis* assemblages both below and above the *Monomorphichnus* and *Phycodes pedum* bearing Random Quartzite of south-east Newfoundland (Bengtson & Fletcher, 1981; Anderson, 1981). Other basal conglomerates and feldspathic sandstones may be of this age, such as the *Arenicolites*- and *Psammichnites*?-bearing Hartshill Sandstone of Warwickshire (central England, Brasier, Hewitt & Brasier, 1978; Brasier & Hewitt, 1979) and the Ratcliffe Brooke Formation of New Brunswick (Patel, 1976) in which Matthew (1889) reported *Platysolenites*. Small, well-rounded phosphatic clasts occur in quartz arenites of this age in the Random Formation, south east Newfoundland (Anderson, personal communication), and as phosphatic clasts and hardgrounds in the lower Home Farm Member of Nuneaton in Central England (Brasier & Hewitt, 1979). The biostratigraphy of this member has been examined by Brasier & Hewitt (1981) and Brasier (1984) and further work shows that the member spans two microfossil zones. The lowest, with *Sunnaginia neoimbricata* n.sp., begins with a cross-bedded quartzose conglomerate of pea-sized quartz grains and sparse phosphatic clasts set in a phosphate-rich matrix (14% calcium phosphate; Lapworth, 1898) containing *'Obolella' groomi* (a lingulid?), *Sunnaginia neoimbricata* n.sp., *Micromitra phillipsi*, *Hyolithellus* cf. *micans* and *Turcutheca* sp. A slightly higher limestone hard-ground and conglomerate is extensively phosphatised and has a similar fauna plus *Coleoloides typicalis*, *Eccentrotheca grandis* n.sp., and *Camenella baltica*. These beds have a fauna that is older than the *Mobergella*-bearing Comely Sandstone in Shropshire and the *Schmidtiellus mickwitzi* Zone of the Baltic Platform (Brasier & Hewitt, 1981).

(2) Rhombocorniculum insolutum Zone The succeeding part of the Home Farm Member is a 2 m thick, nodular limestone with hardgrounds and phosphatic clasts near the top. A richer calcareous fauna appears within Bed 10, including the molluscs *Randomia aurorae*, *Bemella pauper*, *Prosinuites emarginatus*, *Fordilla troyensis* and hyoliths. The conoid phosphatic microfossils *Rhombocorniculum insolutum*, *Sunnaginia parva* n.sp. and *Eccentrotheca kanesia* are important for correlation and indicate a late Tommotian–early Atdabanian age. The top bears Fe–Mn stromatolites and oncoliths like those from the limestones of the Bonavista and Smith Point Formations in south-east Newfoundland (Edhorn & Anderson, 1977; 'Stromatoporoids' of Walcott, 1900; 'cabbage heads' of Hutchinson, 1962). The Home Farm Member is probably to be correlated with the Random and Bonavista Formations but is older than the Smith Point Formation.

(3) Rhombocorniculum cancellatum–Callavia Zone The abrut appearance of *Callavia* and *R. cancellatum* roughly coincided with transgressive facies changes in England (Brasier & Hewitt, 1981) and within the transgressive Smith Point Limestone of south-east Newfoundland (Hutchinson, 1962). The upper part of this zone contains *Dipharus attleborensis* and *Serrodiscus bellimarginatus* in the Brigus Formation, Newfoundland, and in bed Ac_3 of the Comley Limestone in Shropshire; these are both members of the cosmopolitan *D. attleborensis* assemblage (Robison *et al.*, 1977) and are associated with *Sunnaginia angulata* n.sp. throughout the area. Bed Ac_3 is phosphatic in Shropshire (Rushton, 1974).

(4) Lapworthella cornu–protolenid–strenuellid Zone The higher beds of the Comley Limestone (Ac_4,Ac_5,Ad) all contain some phosphate including intraclasts in the top bed. The Strenuella Limestone (Ac_4) contains *Strenuella strenua*, *S. (Comluella) platycephalia*, *Galodiscus lakei*, *Serrodiscus robustus*, *Obolella atlantica* (Rushton, 1974) and *Lapworthella cornu* (Bengtson, 1980), which suggests correlation with the higher *H. kjerulfi* beds of Scandinavia (Bergström & Ahlberg, 1981). The succeeding Protolenus Limestone (Ac_5) has yielded *P. latouchei*, *P. morpheus*, *Mohicana clavata*, *Strettonia comleyensis*, *Calodiscus lakei*, *Serrodiscus robustus*, *Cobboldites comleyensis* and *Runcinodiscus index* (Rushton, 1974). Correlation with the *Proampyx linnarssoni* Zone of Scandinavia has been suggested (Bergström & Ahlberg, 1981). Hiatus and erosion are evident above this bed in Shropshire, where the phosphatic and manganiferous Lapworthella Limestone overlies scoured limestones and older sandstones (Cobbold & Pocock, 1934; Rushton, 1974); this bed has ?phosphatised stromatolites and an impoverished fauna with *Lapworthella cornu* and *L. cornu* forma *nigra*, *Rhombocorniculum cancellatum* and *Acrothyra* cf. sera. It may belong in the Lower Cambrian on account of its unconformable relationship with overlying beds (Cowie *et al.*, 1972).

Rocks of protolenid-strenuellid Zone age in central England are thick and without recorded phosphorites, but in north

Wales, turbidites ceased briefly near the end of this zone and mudstones with phosphatic nodules were deposited (Bassett & Walton, 1960) with a fauna of *Serrodiscus ctenoa?, Hamatolenus douglasi* and *Kerberodiscus succinctus* (Bassett, Owens & Rushton, 1976). Phosphatic nodules also occur in the top of the *Protolenus*-bearing shales of the Hanford Brook Formation in New Brunswick (Walcott, 1900), and in the lower part of the *Protolenus*-bearing beds in the Brigus Formation of south-east Newfoundland (Howell, 1925).

(5) Oelandicus Zones A break between Lower and Middle Cambrian sedimentation in Shropshire is suggested by local erosion and the development of a basal phosphatic skin, preceding the conglomerates and coarse glauconitic grits of the Upper Comley Sandstone (Cobbold & Pocock, 1934). Fossils include *Paradoxides groomi, Kootenia lakei, Bailiella longifrons, Corynexochus illingi* and *Cobboldites simplex*; thus Cowie *et al.* (1972) suggest placement in the *Paradoxides insularis* Zone. Elsewhere on the Avalon Platform, sediments of the *Oelandicus* Zones seem to have been manganiferous. A disconformity is suspected in east Cape Breton Island (Hutchinson, 1952) and Mn-rich horizons occur in Welsh shales (Bassett *et al.*, 1976), and at the base of the *Paradoxides bennetti* Zone in south-east Newfoundland (Walcott, 1900; Howell, 1925; Hutchinson, 1962) where abundant phosphatic nodules occur in shales containing a fauna of *P. bennetti, P. eteminicus, Lingulella ferruginea* and *Acrothele matthewi*. This Mn bed may mark an important hiatus between the Lower and Middle Cambrian in the Avalon Peninsula of south-east Newfoundland (T.P. Fletcher, Ph.D. thesis, University of Cambridge, 1972).

(6) Paradoxissimus Zones The sedimentary cycle in Scandinavia that begins with *Acrothele granulata*-bearing conglomerates and Exsulans limestones (Fig. 23.1) is also traceable in Wales, where the shallow water Caered Flags or Upper Solva sandstones gave way to deeper water and dark Menevian shale facies by *Tomagnostus fissus* times. Middle Menevian shales from south Wales contain small flattened nodules with 1.5–4.0% P_2O_5 and *Paradoxides davidis* exoskeletons with 20% P_2O_5 (Hicks, 1875). Other elements of the fauna indicate the *Hypagnostus parvifrons* and *Ptychagnostus punctuosus* Zones (Cowie *et al.*, 1972; Rushton, 1974). In Shropshire, the Hill House Flag Member of the Upper Comley Sandstone is phosphatic in places and yields *Kootenia lakei* (Cobbold, 1911); it may be of *Ptychagnostus gibbus* age (Cowie *et al.*, 1972). Higher in the succession, the lower part of the Shoot Rough Road Flags contains nodules which are possibly phosphatic; these beds yield *P. davidis, Peronopsis fallax, Eodiscus punctatus* and *Parasolenopleura* sp., suggesting a *Ptychagnostus punctuosus* age (Cowie *et al.*, 1972). In central England, the change from red Purley Shales to blue grey Abbey Shales is accompanied by a change from Mn-rich nodules to concretions with Mn and about 1% P_2O_5 (Illing, 1916). Phosphatic concretions also occur in the *P. hicksi* and *P. davidis*-bearing shales of the Manuels River Formation, south-east Newfoundland (Howell, 1925).

(7) Forchhammeri Zones A hiatus corresponding to part of the *Forchhammeri* Zones is found in England, Wales and maritime Canada, probably owing to regression during *Goniagnostus nathorsti–Solenopleura brachymetopa* times (Taylor & Rushton, 1972; Poulsen & Anderson, 1975; Allen, Jackson & Rushton, 1981). On the St Tudwal's Peninsula, north Wales, the higher part of the dark Nant-Pig Mudstone yields trilobites of *Hypagnostus parvifrons* age (Rushton, 1974). At the top is a 1m thick bed, here called the Dorypyge Limestone, which also has volcanic clasts, coarse quartz sand, rounded phosphatic clasts and ?phosphatic columnar stromatolites (Nicholas, 1915; M.D. Brasier, unpublished). The calcareous beds are crowded with fossils amongst which are *Dorypyge* sp., *Paradoxides* sp., *Linguagnostus kjerulfi* and *Bailiaspis nicholasi* (Nicholas, 1915; Rushton, 1974), which may indicate the *Solenopleura brachymetopa* Zone. Lithological comparison may be made with the broadly contemporaneous Exporrecta Conglomerate and Hyolithes and Andrarum Limestones of Sweden.

In south Wales and Shropshire, the Upper Menevian and Upper Comley Sandstone are both glauconitic grits and conglomerates with *Billingsella*, also suggesting shallow waters. In Shropshire comparison with the Swedish facies is improved by the presence of *B. exporrecta, Paradoxides forchhammeri* and hyoliths (Cobbold & Pocock, 1934; Cowie *et al.*, 1972), although the correlation is uncertain (Rushton, 1974).

In central England, there is no evidence for the *Goniagnostus nathorsti* nor *Solenopleura brachymetopa* Zones, the base of the Mancetter Grits and Shales being a calcareous conglomerate with igneous pebbles and glauconite, followed by shales with phosphatic material and a *Leiopyge laevigata* fauna (Illing, 1916; Taylor & Rushton, 1972). A very similar sequence is found in south-east Newfoundland, at Random Island and Manuels River, but there the basal conglomerate consists largely of angular phosphatic clasts (Howell, 1925) presumably derived from the *Ptychagnostus punctuosus* Zone, and succeeded by shales with *Andrarina costata, 'L. laevigata', Peronopsis insignis* and *Paradoxides* sp. (Poulsen & Anderson, 1975). In the centre of the depositional basin, however, the Highland Cove beds are suggested to span the missing zones with regressive sandstones, limestones, algal beds and conglomerates (Poulsen & Anderson, 1975), although Rushton (1978) has shown that these beds are probably older than the *brachymetopa* Zone. The *L. laevigata* transgression seems to have been a widespread event, occurring on both the Avalon and Baltic Platforms (Poulsen & Anderson, 1975; Rushton, 1978; Allen *et al.*, 1981).

(8) Late Cambrian Zones The depositional cycle that began in *Forchhammeri* times continued through the late Cambrian and into the Tremadocian with local breaks. Many Upper Cambrian beds were eroded prior to the unconformable Arenigian transgression.

In central England, small black phosphatic nodules occur scattered through the Outwoods Shales and Monks Park Shales, referred to the *Agnostus pisiformis* to *Peltura scarabaeoides* + *Acerocare* Zones (Taylor & Rushton, 1972). In Shropshire, deposition of the Orusia Shales commenced in *Parabolina spinulosa* times (Cowie *et al.*, 1972), with phosphatic nodules at the base and a fauna of *Orusia lenticularis, Parabolinoides bucephalus?* and *Parabolinites* aff. *williamsoni* (Cobbold, 1927). Of similar age may be the small black phosphatic nodules which occur at the base and 70 m above in dark shales of the Dolgellau Member, Harlech Dome; the base is characterised by *Parabolina spinulosa* and *Orusia lenticularis* (Allen *et al.*, 1981). Phosphatic nodules also occur at the top of the member, associated with sponge spicules (Allen *et al.*, 1981), and locally at the base and throughout the overlying shaly Dolcyn-afon Member of early Tremadocian age. Stubblefield (1956) regarded the nodules as evidence for lack of deposition during *Acerocare* time, although the presence of this zone has now been proven locally (Allen *et al.*, 1981). Phosphatic nodules near the top of the Bentleyford Shales of Shropshire are associated with a fauna of *Leptoplastus* sp., *Ctenopyge modesta* and *C. similis*? (Stubblefield, 1930; Henningsmoen, 1957) from the zone of *Peltura minor* (Rushton, 1974).

As in the case of the Baltic Platform, there is no significant phosphogenesis recorded in rocks of Tommotian age on the Avalon Platform, and a significant lesser amount in Atdabanian equivalents. That occurring in the Home Farm Member of the Hartshill Formation of central England is associated with Fe–Mn mineralisation, stromatolites and hard grounds. During the Early Cambrian, phosphate is most commonly found in rocks of late Atdabanian and Lenian age occurring mainly as phosphatic nodules in mudstone or shale sequences as in north Wales, New Brunswick and south-east Newfoundland, or phosphatic carbonates, as in central England. There is little evidence for phosphogenesis during the initial Middle Cambrian, although some phosphatic nodules occur in shale sequences in south-east Newfoundland, until the *Ptychagnostus gibbus* Zone transgression. At that time, phosphate is widespread as nodules in shale sequences on the Avalon Platform, particularly in central England and south-east Newfoundland, and where this style of sedimentation persists into the late Cambrian, phosphatic nodules continue to occur, particularly in England. Phosphate clasts are associated with stromatolites in the late Middle Cambrian of north Wales. During Middle and Late Cambrian time, there appears to have been a distinct shift of phosphogenesis from the Baltic to the Avalon Platform.

The Armorican and South European Platforms

Phosphorites have a very limited occurrence in the Cambrian of southern Europe (Fig. 23.2). On the Armorican Platform, a sequence from Carteret in Normandy has some occurrences (e.g. Doré, 1972). There, conglomerates and arkoses are succeeded by shales and sandstones, several of which contain phosphorite and glauconite. Overlying sandstones with calcareous nodules contain *Allonia tripodophora, Chancelloria* sp., *Orthotheca* sp., *Circotheca* sp., *Hyolithes* sp. cf. *impar, H.* cf. *communis* and *H.* cf. *americanus* which suggests a correlation with the *Rhombocorniculum insolutum* Zone of the Avalon Platform. This age is supported by a fauna which occurs 200 m higher in limestones containing *Bigotina bivallata, Protopharetra* cf. *grandicaveata, P. bigoti, Ajacicyathus carterensis, Uralocyathus* cf. *kidrjassovensis, Coscinocyathus rojkovi* and algae (Doré, 1972; Debrenne, 1964). The presence of simple archaeocyathans now assigned to *Aldanocyathus, Retecoscinus* and *Sibiricyathus* implies an Atdabanian age, as does the presence of *Bigotina* (Rozanov & Debrenne, 1974). The Frankenwald succession of north-east Bavaria may have formed part of the same platform. Lower Cambrian flysch is there followed by shallow water sandstones with phosphatic nodules termed the Galgenberg and Wildenstein Formations (Ludwig, 1969). The former contains *Kingaspidoides frankenwaldensis, Bailiella.* Solenopleuridae and Ellipsocephalidae, favouring correlation with the *Paradoxides insularis* Zone of Scandinavia (Würm, 1925; Sdzuy, 1964). The overlying Wildenstein Formation has *Paradoxides rugulosus, Condylopyge carinata, Bailiella,* Solenopleuridae. *Micmacca, Protolenus* and *Dawsonia*, suggesting equivalence with the *Paradoxides pinus* Zone (Würm, 1925; Sdzuy, 1964). Boreholes through probable Cambrian in the Thüringischen Schiefergebirges have yielded some phosphatic microfossil steinkerns (Wucher, 1967), here regarded as chancelloriid rays, but no phosphorite as such.

The succession in the Montagne Noire, south France, may have been deposited on a separate South European Platform. Conglomerates are here followed by shallow water sandstones and capped with carbonates. The Cambrian inlier at Vigal contains karst-associated phosphorites near the base of the carbonates (Orgeval & Capus, 1978); archaeocyathans indicate an Early Cambrian, Lenian, age for these carbonates at Malines (Debrenne, Orgeval & Verraes, 1976). At Hérault, on the south-east flank of the Montagne Noire, there are Lower Cambrian black shales and lithistid sponge-bearing limestones with phosphorites, broadly correlated on the basis of lithology with the Lower Cambrian limestones to the south (Prian, 1979). In the southern Montagne Noire, north Spain and Sardinia, the carbonate bank facies is overlain by glauconitic and griotte limestone facies with *Paradoxides* and *Conocoryphe*; these probably represent a transgression bringing cooler, more oceanic conditions to the South European Platform. This transgression reached northern Spain in early *Paradoxides oelandicus* times but the Montagne Noire and Sardinian assemblages are of *P. paradoxissimus* age (Sdzuy, 1972). On the northern flank of the Montagne Noire, lithologies resembling the Schistes à *Paradoxides* contain thin beds of limestone, locally phosphatised, and volcanic detritus (Prian, 1979). The *Girvanella*-Reitlingerellidae biota suggests a shallow water carbonate environment but gives no clue as to the precise age.

Soviet Union and Mongolian People's Republic

Proterozoic and Cambrian phosphorites are numerous and stratigraphically well documented in the Soviet Union (Bushinsky, 1966; Notholt, 1979). With the exception of the Cambrian Karatau deposit in Kazakhstan, however, biological information directly related to phosphatic sediments is not readily obtainable. Late Proterozoic phosphorite occurrences are in general dated by microphytoliths and oncolites with a rather low degree of accuracy: in some instances it is not possible to decide whether occurrences are Proterozoic or Cambrian. Of the large number of localities indicated in the above publications, representative biochronological information is presented here only for the Kazakhstan, Altai–Sayan, Khubsugul and Lena–Tunguss Basins (Fig. 23.4).

Different biochronological scales for the Early Cambrian of the USSR have been summarised by Rozanov & Debrenne (1974, fig. 3), and reviewed by Öpik (1975b) and Suvorova (1980). In this paper we follow Rozanov & Missarzhevsky (1966) in recognising the division of the Aldanian into Tommotian and Atdabanian Stages (Fig. 23.3). The Lenian,

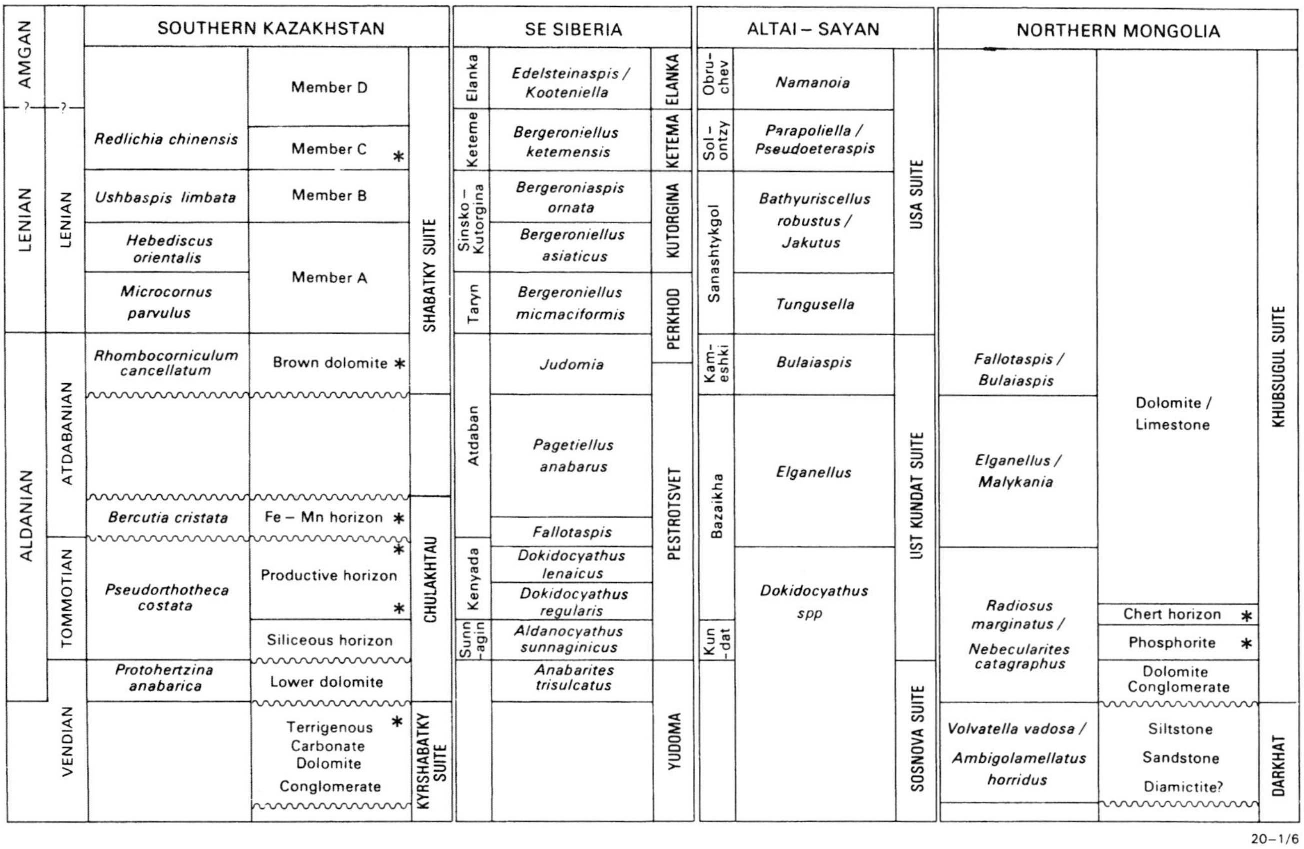

Fig. 23.4. Correlation of Lower Cambrian sections across the southern portion of the USSR and northern part of the Mongolian People's Republic.

however, is used here in a *sensu lato* manner to include the Botoman and Lenian Stages of Khomentovsky & Repina (1965) on the Siberian Platform (see also Öpik 1975b, Fig. 3), which are equivalent to the Lenian and Elankian as proposed by Rozanov & Debrenne (1974). The sequences of Gorizonts (Horizons) on the Siberian Platform and in the Altai-Sayan fold belt are those described by numerous authors (e.g. Öpik, 1975b, Fig. 3) and are well established.

Kazakhstan

According to Bushinsky (1966) phosphorites are found throughout the Ulutau–Tienshan region of central and southern Kazakhstan. There are no significant occurrences of Proterozoic phosphorites in this region, but commercial deposits of Cambrian age have been recorded at Shiderty, Atasu, in the Kendyktas Mountains and in both the Lesser and Greater Karatau, associated with platform deposits. The Lesser Karatau deposits are the most important and are the most thoroughly documented.

(1) Karatau Basin The Cambrian phosphorites of the Karatau Basin, which is located in the Lesser Karatau Range in south-central Kazakhstan, have been described in detail by Bushinsky (1966), Eganov & Sovetov (1979) and Eganov *et al.* (this volume). More biostratigraphic information is available for these phosphorites than for any other in the Soviet Union. They occur in the lower Tamdy Suite, in the upper part of the Kyrshabakty Suite, throughout the Chulaktau Suite, and in the lower part of the Shabakty Suite. The main commercial development occurs in the Chulaktau Suite (Bushinsky, 1966; British Sulphur Corporation, 1971: 146; Eganov, 1979: Fig. 10).

The Kyrshabakty Suite comprises a basal conglomerate overlain by dolomite, glauconitic limestone and red sandstone. Fine-grained phosphorite, not dated, is associated with the carbonate units. The Chulaktau Suite overlies the Kyrshabakty Suite with erosional contact. Its basal unit, the Lower Dolomite of previous authors, also referred to as the Berkutin Suite (Mambetov & Repina, 1979), contains the shelly fossils *Protohertzina unguliformis, P. anabarica, P. siciformis* and *Chancelloria* sp., associated with the microphytoliths *Vesicularites concretus, V. rectus*, and *V. nubecularites morulus*. This biota is referred to the *Protohertzina anabarica* Zone and is mostly regarded as having a late Yudomian (late Vendian) age (Mambetov & Missarzhevsky, 1971; Rozanov, 1976; Mambetov & Repina, 1979; Missarzhevsky & Mambetov, 1981). However, Ergaliev (1974) and Ergaliev & Pokrovskaya (1977), consider that it has an initial Cambrian age.

The Lower Dolomite is overlain with erosional contact by a Siliceous Horizon, which is in turn overlain by the Productive Horizon which contains two grainstone phosphorite layers separated by a shaly member. These units, representing the economic orebody, have yielded *Hyolithellus vladimirovae, Torellella biconvexa, Tiksitheca licis, Allatheca* sp., *Anabarites trisulcatus, Protohertzina siciformis, Chancelloria* sp., and Coreospiridae (Mambetov & Missarzhevsky, 1971; Rozanov, 1976; Mambetov & Repina, 1979; and Missarzhevsky & Mambetov, 1981). This shelly fauna is referred to the

Pseudorthotheca costata Zone (Ergaliev, 1974) which is considered to have an early Aldanian (Tommotian) age (Rozanov, 1976).

The phosphatic sequence is capped by a stromatolitic Iron–Manganese Carbonate Horizon containing *Hyolithellus vladimirovae, Torellella biconvexa, T. lentiformis, Camenella* sp., *Protohertzina siciformis, Conotheca mamillata, Uniformitheca* spp., *Allatheca* sp., *Sachites sacciformis, S. costulatus* and *S. amorphe* (Rozanov, 1976: Fig. 8; Mambetov & Repina, 1979; and Missarzhevsky & Mambetov, 1981). This fauna represents the *Uniformitheca* Zone of Ergaliev (1974) and the *Bercutia cristata* Zone of Mambetov & Repina (1979). Rozanov (1976) considered this as a late Tommotian fauna, but the other authorities quoted here consider it to compare with the early Atdabanian assemblage of the Siberian Platform. It also compares with the *Rhombocorniculum insolutum* Zone of the Avalon Platform and the *Schmitiellus mickwitzi* Zone of the Baltic Platform.

The uppermost unit of the Chulaktau Suite is a brown cherty dolomite, overlying the Iron–Manganese Horizon disconformably. According to Mambetov & Repina (1979) the fauna from this dolomite includes *Hyolithellus vladimirovae, Torellella biconvexa, Conotheca mamillata, Sachites sacciformis, S. amorphe, Cambrocalvus antis, C. clavus, C. undulatus,* and *Pseudoclavus singularis.* Ergaliev & Pokrovskaya (1977) have also recorded the late Atdabanian trilobites *Ushbaspis* sp., *Neopagetina* sp. and *Calodiscus korolevi* from this unit. By comparison with the Avalon Platform, they probably correlate with the late *Callavia* Zone (*Dipharus attleborensis* horizon).

The Shabakty Suite (Gilan Subsuite) disconformably overlies the Chulaktau Suite and contains several horizons of pelletal skeletal phosphorite associated with its basal dolomite units. The dolomites contain a sequence of trilobite assemblages which can be assigned late Atdabanian and Lenian ages. They also contain inarticulate brachiopods, hyoliths, hyolithelminthes, 'camenellids', monoplacophorans, and molluscs whose ranges are shown by Mambetov & Repina (1979: Fig. 5) and Missarzhevsky & Mambetov (1981: Figs. 7–8). Grainstone phosphorite occurs particularly during the latest Aldanian (latest Atdabanian), late *Rhombocorniculum cancellatum* Zone, just below the Atdabanian–Lenian contact according to Mambetov & Repina (1979). Lenian sediments are, however, transgressive in the Karatau Basin; according to Ergaliev & Pokrovskaya (1977) they rest on various older rocks, and phosphorites recorded at this level are presumed to have been reworked.

(2) Kendyktas Mountains Sediments with phosphate contents ranging from 2–20% P_2O_5 occur in the Narkyzyl Suite in the Kendyktas Mountain Range some 200 km west of Alma-Ata (Bushinsky, 1966). Previously considered to be of Middle Cambrian age, the faunas described by Borovikov & Kryskov (1963) indicate at least the presence of Idamean late Cambrian sediments when compared to the Australian biochronological scale, because they contain *Glyptagnostus reticulatus, Innitagnostus tchatertensis, Aspidagnostus laevis* and *Proceratopyge karaulensis,* amongst others.

Altai–Sayan Folded Belt

Phosphorites occur in rather similar lithofacies settings in south-west Siberia and northern Mongolia. In the Kuznetsk Alatau, Gornaya Schoriya, the Oka region of eastern Sayan, and the Khubsugul Basin of the northern part of the Mongolian People's Republic, Riphean, Vendian and Early Cambrian phosphorites have all been reported, generally in association with poorly fossiliferous dolomitic carbonates and siliceous sediments (Bushinsky, 1966; Kazarinov & Krasilnikova, 1980). Detailed palaeontological information is available, especially for the Rivers Bazaikha and Kiya in the Kuznetsk Alatau and the Batenev Range in the Oka River area, but, with the exception of the last is difficult to relate to the phosphatic sequences (Rozanov & Missarzhevsky, 1966; Rozanov *et al.*, 1969). Similarly, in the Gornaya Schoriya, palaeontological information published by Zhuravleva, Repina & Khomentovsky (1959) and Repina *et al.* (1964), cannot be directly related to the phosphatic horizons by the present authors.

(1) Kuznetsk Alatau/Gornaya Schoriya In the Gornaya Schoriya region of the Kuznetsk Alatau, low grade phosphorites are associated with dolomite and dolomitic limestone, often brecciated, of the Belka Suite, in the eastern Martayga and Mras Massifs. In these areas, the Belka Suite overlies weakly phosphatic dolomite and dolomitic breccia containing stromatolites and oncolites (Zapadnosibirsk Suite), and is itself overlain by algal (*Epiphyton*) and archaeocyathan carbonates containing intervals of dolomitic sandstone, chert and volcanogenic sediments considered to have a late Atdabanian (Kameshki) age by Zhuravleva *et al.* (1959). The Belka phosphorites themselves are considered to have a Vendian age on the basis of oncolites and their stratigraphic position.

Along the Usa and Belaya Usa Rivers phosphorite is associated with pyritic, manganiferous, carbonaceous shale immediately overlying the Kabyrza Suite of dark limestones, cherts, and shales with lenses of stromatolitic dolomite, of mid-Riphean age. Phosphate grains also occur in clastic interbeds within a dark grey dolomite and chert sequence referred to the Usa Suite (Bushinsky, 1966; Kazarinov & Krasilnikova, 1980). This suite contains Kameshkian archaeocyathans, suggesting a younger phase of phosphate reworking and deposition.

At Tamalyk Mountain, on the upper reaches of the Tom River on the eastern slopes of the Kuznetsk Alatau, low-grade phosphorites are associated with dolomitic limestone and chert in the Tamalyk Suite which rests unconformably on older dolomites (Bagzass Suite) containing *Osagia bothridioformis, Siberiphycus* sp., *Kabyrsinia* sp., *Vesicularites flexuosus, Collenia columnaris,* and stratiform stromatolites. The Tamalyk phosphorites have an Early Cambrian age (Ust-Kundatian?), containing trilobites and hyoliths, sponge spicules, radiolarians, the algal filament *Obruchevella,* and *Epiphyton* sp. These phosphorites, like those of the Belka Suite, are overlain by, and pass laterally into tuffaceous rock and porphyry and in turn by stromatolitic dolomites of the Usa Suite which contain late Atdabanian (Kameshki) and Botomian (Sanashtykgol) archaeocyathans (Kazarinov & Krasilnikova, 1980, Fig. 6).

According to Bushinsky (1966), siliceous, calcareous and brecciated Early Cambrian phosphorites also occur in the Batenev Range, on the eastern slopes of the Kuznetsk Alatau some 225 km east of Novokuznetsk, in the Martyukhina Suite (= ?Usian Formation of Rozanov, 1969). The stratigraphic situation is similar to that at Tamalyk Mountain in that near Bograd there is a conformable passage from alternating dark and pale grey dolomites and limestones containing *Vesicularites lobatus, Vermiculites irregularis* and stromatolites in the upper part of the Sosnovka Suite into dark algal limestones with *Epiphyton* and *Rasumovskia* and pale limestones and shales with archaeocyathans of Bazaikhan and Kameshkian (late Tommotian to Atdabanian) ages (Borisov & Rozanov, 1964; Rozanov, 1969, Fig. 20).

Elsewhere in the Kuznetsk Alatau, weak phosphatisation reportedly occurs throughout the Ust–Kundatian section along the Kiya River, below its intersection with the Kundat River. The Ust–Kundat sequence of dark detrital carbonates and shales contains *Scenella lanculiformis, Phyloxinella spiralis, Helcionella, Latouchella*, and species of *Chancelloria* possibly indicative of a mid-Tommotian age by comparison with the Siberian Platform (Rozanov, 1969; Rozanov & Missarzhevsky, 1966).

(2) Eastern Sayan Basin In the Oka region of the Eastern Sayan Basin, phosphorite occurs in the essentially carbonate Boxon Supersuite where it is associated with dolomite, limestone, sandstone, shale, and, in places (Boxon itself) bauxite, in the Zabit Suite. The age of phosphorites in this area is considered to be Riphean to Vendian on the basis of oncolites and catagraphs (Kazarinov & Krasilnikova, 1980). The Oka phosphorite-bearing region is separated from that of Khubsugul in northern Mongolia by the Shibit–Lambshitig anticlinorial structure.

(3) Khubsugul Basin In the Khubsugul Basin of northern Mongolia, which has been more intensively studied than other areas, commercial grade phosphorite occurs in the Khubsugul Series, a dominantly carbonate sequence of rocks, 3000–3500 m thick (Ilyin, 1973; Ilyin & Ratnikova, 1981; Ilyin *et al.*, this volume). Parts of three distinct sequences make up the Khubsugul Series in the Khesen Syncline adjacent to Lake Khubsugul. An initial sequence, comprising dolomite with breccia-like texture – possibly diamictite – overlain by red silts and shaly dolomite, is succeeded by pebble conglomerate, thin-bedded limestone, shale, phosphorites associated with shale, dolomite and tuffaceous rock, and black chert at the top of the lithosome. A third sequence consists of carbonates several hundred metres thick as in the Altai–Sayan Basin.

Only the upper carbonate sequence can be dated biochronologically with confidence as its lower beds contain hyoliths, trilobites and archaeocyathans, and its faunas extend into the Middle Cambrian. The earliest trilobite zone is that of *Elganellus–Malykania* (Korobov, 1980a). In the southern Altai, Repina & Romanenko (1978) correlate the *Elganellus* Zone with the upper Kenyada Horizon of the Lena River region to the north-east (the *Dokidocyathus lenaicus* Zone of Rozanov & Missarzhevsky 1966), of latest Tommotian age. Therefore, the underlying phosphatic sequence must predate the late Tommotian. Because it has a microphytolite assemblage (the *Radiosus marginalis–Nubecularites catagraphus* Zone of Korobov, 1980a) containing the typical Vendian *Volvatella* and *Ambigolamellatus* in association with *Nubecularites, Osagia, Radiosus* and *Vesicularites*, this phosphatic sequence has been assigned a Vendian (Yudomian) age (Ilyin, 1973, 1979; Bjamba, 1973; Ilyin & Bjamba, 1980; Ilyin & Ratnikova, 1981). However, an Early Cambrian age cannot be excluded, as demonstrated by the ranges of microphytolites, stromatolites and algae in the Vendian and Early Cambrian shown by Korobov (1980b) and Rozanov *et al.* (1969, table 15). Korobov (1980b) demonstrates their ranges against the trilobites and suggests an Early Cambrian (early Tommotian) age for the phosphatic sequence. In west Mongolia (Khasagt–Khairkhan Ridge) a similar microphytolite assemblage is associated with the alga *Renalcis*, generally regarded as a Cambrian form, and *Korilophyton*, and predates a Tommotian phosphatised shelly fauna (Korobov & Missarzhevsky, 1977).

Volcanics are more prevalent in west and south Mongolia, extending from probable Venedian to Early Cambrian time. At Burin Khan, in the southern-most part of the Khubsugul Basin, phosphorites, correlated with those in the Khesen Syncline, overlie a tuffaceous sequence, and are themselves overlain by carbonates with trilobites and archaeocyathans.

(4) Siberian Platform The general absence of a phosphatised shell/tube association in north Mongolian and south Siberian phosphatic sequences does not necessarily preclude an Early Cambrian age since it is precisely in these areas that algal sediments prevailed.

This situation is completely reversed in the Lena–Tunguss Basin of south-eastern Siberia where assemblages of phosphatic and phosphatised shells and tubes occur in the uppermost layers of the Vendian Yudoma Formation and throughout the Lower Cambrian Pestrotsvet Formation (Rozanov, 1976; Rozanov & Missarzhevsky, 1966; Rozanov *et al.*, 1969; Cowie & Rozanov, 1973; Matthews & Missarzhevsky, 1975). While these formations are often glauconitic, they are usually only weakly phosphatic, and their phosphate content seems to consist largely of phosphatic and phosphatised shells and their internal moulds, and pelletal grains derived from them.

Elsewhere in Siberia, phosphorite has been documented on the Gremyaka River, a tributary of the Yenisei River in its lower course (Dragunov *et al.*, 1966). The phosphorite occurs in a carbonate environment which contains *Pagetiellus lenaicus, Calodiscus* ex. gr. *schucherti, Serrodiscus, Judomia, Peronopsis, Pseudoamocarina, Solenopleura* and *Dawsonia*, these trilobites suggesting a mixture of Early and Middle Cambrian faunas. A late Atdabanian and Lenian age is indicated by part of the fauna, and a possible Amgan (Middle Cambrian) age by the remainder. Other phosphorites occur on the Yenisei River, downstream from Plakhin Island, between the mouths of the Rivers Tunguska and Kurieka, in two beds. The upper bed contains *Hebediscus attleborensis, H. ponderosus*, and

Neocobboldia dentata (Kazarinov & Krasilnikova, 1980; E.A. Eganov *in litt.*). It probably has an early Lenian (Botomian) age judging from the trilobite ranges published by Repina (1981).

In summary, phosphorites ranging in age from Riphean to Late Cambrian have been reported from the USSR. Throughout the Altai–Sayan to Eastern Sayan belts, they appear to have Riphean or Vendian ages on the basis of oncolites and microphytoliths, and they are pre-late Tommotian in the adjacent Khubsugul Basin in northern Mongolia. Elsewhere, in southern Kazakhstan, the Kuznetsk Alatau, and on the Siberian Platform, phosphorites and phosphatic sediments have a Tommotian to early Atdabanian (Early Cambrian) age. Possible Lenian or even Amgan occurrences have been reported from the Yenesei River, and those of the Kendyktas Mountains in central Kazakhstan are of Late Cambrian age.

Indian subcontinent

Proterozoic phosphorites are known from both the peninsular shield and Himalayan regions of the Indian subcontinent, specifically India and Nepal (see Banerjee, this volume). In the peninsular shield region phosphorite deposits have been well documented in the Udaipur Basin of Rajasthan (Choudhuri & Roy, this volume), but are also known in the Jhabua and Hirapur areas of Madhya Pradesh, and in the Cuddapah Basin of Andhra Pradesh. In Himalayan India, Proterozoic phosphorites have been studied in detail in the Pithoragarh and Almora districts of Uttar Pradesh (Valdiya, 1972), and these extend into Nepal.

In general, the Proterozoic phosphorites of India are associated with, and are dated by, stromatolites. The phosphorites of the Udaipur Basin, which includes Jhabua, occur in the Maton Formation of the Aravalli Group (Banerjee, 1971a; Chauhan, 1979), radiometrically dated at 2000–2500 m.y. (Crawford 1970) (see Banerjee, this volume, concerning doubts on this date). They are intimately associated with both columnar and stratiform stromatolites referred to *Collenia columnaris*, *C. kussiensis*, *C. baicalica*, *C. buriatica* (= *Minjaria* of Banerjee) and *Baicalia prima* (see Banerjee, 1971b; Chauhan, 1973) for which ages ranging from early to late Riphean on the Soviet morphochronological scale suggested by Cloud & Semikhatov (1969) have been deduced, i.e. within the range 600–1300 m.y.

In the Himalayan Pithoragarh–Almora belt, the phosphorites of the upper members of the Gangolihat Dolomite, are also stromatolitic (Valdiya, 1972), being characterised by the form *Baicalia*. Laterally, the phosphatic horizon passes into carbonates (Deoban Formation) with species of *Colonella*, *Kussiella*, *Minjaria* and '*Collenia symmetrica*' suggesting a mid to late Riphean age (Valdiya, 1980).

Cambrian sediments are widely distributed in the Himalayan region of the Indian subcontinent, occurring in the Salt Range of northwestern Pakistan, and extending in a long tectonically interrupted belt parallel to the Lesser Himalaya, from the Hazara district in northern Pakistan through Kashmir, Spiti, and Kumaun, to Nepal. Furthermore, a Cambrian age has been advocated for at least the upper part of the Vindhyan Supergroup which occurs in various peninsular Indian basins (Prasad, 1980).

In general, the ages of many Indian rock sequences thought to be Late Proterozoic or Cambrian are unclear. The earliest dateable trilobites so far found are the *Redlichia* faunas of the Salt Range (King, 1940), Kashmir (Shah, Raina & Razdan, 1980) and Spiti (Reed, 1910) which are presumed to have an Ordian age with reference to northern Australian biochronology. These are superseded by Tethyan Middle Cambrian assemblages containing conocoryphids, oryctocephalids, agnostids, hyoliths and inarticulate brachiopods, and later Middle and Late Cambrian faunas of Asian aspect (Sudan, 1981). Pre-Ordian Cambrian assemblages appear to be based largely on trace fossil associations. Mainly this interval is represented by slate, quartzite or red-bed sequences. However, phosphatic sediments are reported from the Hazara district of northern Pakistan and the Mussoorie district of Kumaun, Uttar Pradesh, India.

Hazara district

Phosphorites of the Hazara district are associated with a dolomitic facies developed at the top of the Abbottabad Formation and in the siltstones of the overlying Hazira Formation (see Hasan, this volume).

Dutro (in Shah, 1977) has identified the phosphatic tube *Hyolithellus*, associated with a hyolith, in the upper part of the Abbottabad Formation, and the Hazira Formation has yielded *Allonia tripodophora*, *Archiasterella pentactina* and 'stauractinid' sponge spicules, hyoliths, *Lapworthella* sp. and *Rushtonia* spp. (Fuchs & Mostler, 1972). Rushton (1973) compared the Hazira hyoliths to *Circotheca?* and *Linevitus?* and noted the occurrence of *Chancelloria*. All these determinations suggest an Early Cambrian age for the upper Abbottabad/Hazira interval.

Kumaun

A similar stratigraphic sequence occurs above the Blaini diamictite in the Mussoorie Syncline, western Kumaun. Overlying slate and sandstone is referred to the Infra-Krol Formation, and limestone, gypsiferous red shale, and upper dolomite, referred to the Krol Formation, suggests a sequence related to that of the Hazara district. The Krol Formation is overlain transitionally (according to some) or with disconformity by the Tal Formation whose lower black shale and chert unit contains commercial grade phosphorite. Phosphorite layers also occur in the basal beds of the Krol Formation near Durmala village (Kanwar & Ahluwalia, 1980) and in the brecciated transition zone from Krol to Tal (Shankar, 1970, 1971, 1976; Shrivastava *et al.*, 1981).

Previous authors have, in general, regarded the Blaini to Tal sequence as of Permian to Cretaceous age. The lower phosphatic Tal Formation, in particular, has yielded organisms regarded as annelids by Singh & Shukla (1981) who assigned them a late Cretaceous age. Specimens from the same locality, however, are interpreted as conodonts by Azmi, Joshi & Juyal (1981) who record: *Drepanodus homocurvatus*, *Prooneotodus tenuis*, *Oneotodus erectus*, *O. simplex*, *Scolopodus transitans*, *Paltodus* aff. *gracilis*, and *Cordylodus* sp. indicating an age close to the Cambrian–Ordovician transition. These conodonts have also been examined by Missarzhevsky & Dubrinina (in Azmi *et al.*, 1981) who prefer to interpret them as Early Cambrian forms (Atdabanian). The presence of *Cordylodus*, however, would seem not to support this interpretation. Also from the same locality are the problematical forms which Patwardhan (1978) called moravamminids, but which may be safely interpreted as *Obruchevella*-like algae. Moravamminids, however, are supposedly unknown from rocks younger than Permian (Patwardhan, 1978, 1979). In addition to conodonts and algae, the phosphatic portion of the Lower Tal Formation contains columnar and domal stromatolites and oncolites which have been compared with Proterozoic forms (Patwardhan, 1980). Thus there is a growing body of evidence to suggest that the phosphorites of the Mussoorie Syncline should at least be regarded as of Palaeozoic, and possibly of Cambrian age (see Banerjee, this volume, for further discussion). Preliminary studies by us suggest the assemblage is close to that of the phosphatic Zhongyicun Member of Yunnan, China.

In summary, the predominant form of phosphate found in the Proterozoic sediments of both the Himalaya and Indian subcontinent is associated with stromatolitic carbonates. Dates on these rocks are disputable, but a body of evidence now suggests a mid to late Riphean age, or maybe marginally younger. Early Cambrian phosphorite certainly occurs in north-west Pakistan associated with dolomite (Abbottabad Formation) or siltstone sequences (Hazira Formation) which pass laterally into ferruginous rocks (Galdanian facies). In the Kumaun Himalaya thin phosphatic layers in the carbonate Krol Formation may also have an Early Cambrian age. In the same area, phosphorite in the basal Tal Formation contains small shelly fossils which suggest an age close to the Precambrian–Cambrian boundary.

China

Late Proterozoic and Cambrian phosphorites are widespread in China, the former accounting for more than two-thirds of all Chinese phosphate reserves (Bushinsky, 1966; Notholt, 1979; Lu, 1979; Liang & Chang, 1981) (see Li, this volume).

Lu (1979) has recognised six periods of phosphogenesis during the Cambrian in China: three Early Cambrian events (chronologically the Yuhutzun, Xinji and Changping periods), two Middle Cambrian events (the Maochuang and Damao periods), and one late Middle Cambrian to early Late Cambrian event (Laoyeshan period). However, the Yuhutzun period appears the same as the phosphogenesis of the Meishucun Stage. The Xinji and Changping periods of the later Early Cambrian cannot be adequately discriminated biochronologically and appear to be both of mid-Canglangpuan (Tsanglangpuan) age. Similarly, the Middle Cambrian Damao and Maochuang periods seem to be synchronous. Accordingly, we recognise here phosphogenic events in the Meishucun, Canglangpu and Maochuang Stages, and the later Middle Cambrian event designated Laoyeshan by Lu (1979).

The Late Proterozoic and Cambrian periods of phosphogenesis are described here within the framework of the tectono-sedimentary provinces detailed by Chang (1980, Fig. 1). The quoted Cambrian trilobite biochronological scale also follows Chang (1980).

Yangzi Platform

Commercial grade phosphorites occur in the Upper Sinian Doushantuo and lower Dengying Formations, but are most important in the former which is considered the most productive horizon in China (Liang & Chang, 1981). The Doushantuo Formation, and its correlatives, extends throughout Guizhou, Hubei, Hunan, Jiangxi, Sichuan and Shaanxi Provinces in south-central China. Phosphorite occurs in both the dominantly carbonate upper and dominantly clastic lower members of the Doushantuo Formation in Guizhou, Hubei and Hunan Provinces, but in the other areas occurs only in the upper member. In Yunnan Province, the Doushantuo Formation is not phosphatic, in marked contrast to the occurrence of Cambrian phosphorite in this area. In the eastern part of the Yangzi Gorges district, Sin & Liu (1976) have recorded 21 genera and 37 species of microflora among which characteristic forms are *Margominuscula verrucosa, Lophosphaeridium ichangense, L. acietum, Nostocomorpha prisca, Paleamorpha punctulata* and elements of *Micrhystridium* and *Polyexdryxium*, associated with commonly occurring forms such as *Leiopsophosphaera infriata, Trachysphaeridium planum* and species of *Laminarites*. Additionally, Sin & Liu (1976) record sponge spicules. The forms indicate a Precambrian late Sinian (latest Vendian) age. The Doushantuo Formation also contains intervals of stromatolitic phosphorite (Zou & Li, 1965), *?Gymnosolen* f. (Cao Rui-ji *in litt.*), at Kaiyang, north of Guiyang, Guizhou, and north and south of Yichang, western Hubei (see the palaeogeographic maps in Li, this volume).

According to Liang & Chang (1981), thin seams of low-grade phosphorite also occur in the upper part of the overlying Dengying Formation. However, this formation has been divided into a number of members, both on a stratigraphic and regional basis (see Chang, 1980, Fig. 3; Xing, Ding & Luo, 1982, Table 5), and only the lower one (Baiyanshao Member of Yunnan, Baimatou Member of West Hubei, and Hungchunping Member of west Sichuan) can now be regarded as having a Proterozoic age. The pre-Cambrian part of the Dengying Formation is mainly non-phosphatic dolomite. From it, Sin & Liu (1976) have recorded some 27 genera and 50 species of microflora, predominantly acritarchs. Some 25 species of cyanobacteria, rhodophytes? and phaeophytes are recorded by Cao & Zhao (1978), as well as the calcareous stromatolites, *Baicalia* in Guizhou Province, and *Boxonia* f. at Shihmen, Hunan Province, microproblematica and vendotaenides. These fossils indicate terminal Sinian to initial Cambrian (Meishucunian) ages.

The distribution of Cambrian phosphorite deposits on the Yangzi Platform in Yunnan, Guizhou, southern Sichuan and Hubei Provinces, is shown by Lu (1979), and their sedimentological and palaeogeographic setting has been described in detail by Li (this volume). The richest and best known deposits occur on the southwestern margin of the Yangzi Platform in the upper part of the Dengying Formation. The lithostratigraphy of the upper Dengying Formation (equivalent to the Meishucun Formation of Chiang, Wang & Chen (1964) and Chang (1980) varies across south-west China and several members are recognised (Luo *et al.*, 1980, 1982; Xing *et al.*, 1982; Li, herein). These members are mainly composed of dolomite, phosphorite, chert, and sandstone or sandy chert, immediately and conformably overlying the basal dolomitic sequences of the lower Dengying Formation. They are themselves conformably overlain by the Qiongzhusi (Chiungchussu) Formation.

The Cambrian members of the Dengying Formation contain numerous small shelly fossils referred to Hyolitha, Brachiopoda, Gastropoda Monoplacophora, Conodontophorida, Rostroconchia, 'Camenellids' and Merismoconchia which predate the first Early Cambrian trilobites (Xing *et al.*, 1982; Luo *et al.*, 1982; Yin, Wang & Qian, 1983). Qian (1977) described 14 genera and 33 species from the upper Dengying Formation (Meishucun Formation) which included five hyolith genera and six genera of the problematical Cambroscleritida; and Liu (1979) has described 10 genera of brachiopods. Luo & Jiang (1982) summarise the faunas for each member of the upper Dengying Formation in Sichuan and Yunnan, and Yin *et al.*, (1983) have done the same for Guizhou. These faunas are referred to the initial Cambrian, Meishucun Stage, on the Yangzi Platform, which is considered approximately equivalent to the Tommotian Stage of Siberia. The earliest trilobites in China characterise the Qiongzhusi Stage, whose initial zone in eastern Yunnan is that of *Parabadiella*.

As the P_2O_5 content of the Meishucun Formation decreases as this formation is traced northwards across the Yangzi Platform, phosphatic intervals in the underlying Doushantuo Formation are correspondingly enriched, e.g. at Kaiyang in Guizhou Province. By western Hubei Province, thinly bedded phosphorites in the Huangshandong Formation have little commercial importance. Some 25 genera and 37 species of Hyolithida, Rostroconchia, Hyolithelminthes, Porifera and Cambroscleritida have been described from dark grey or black sandy siliceous phosphorites in the upper part of this formation (Qian, Chen & Chen, 1979); Yü (1979) has described 31 genera and 42 species of supposed Monoplacophora and Gastropoda in the Yichang district; and Chen, Chen & Zhang (1981) have added a further 10 species of shells and tubes. As in Yunnan, these phosphorites overlie dolomites of the lower Dengying Formation, and they have a similar Meishucunian age.

Chiangnan Belt

Small, low grade, nodular phosphorites occur commonly in the Cambrian of the Chiangnan Belt (Lu, 1979; Chang, 1980). Lu (1979) considers these to have a Meishucunian age, but palaeontological details are not available.

North China Platform

As indicated by Lu (1979, Fig. 1), Zhang *et al.* (1979, Fig. 1) and Li (this volume) Cambrian phosphorites also occur as an arcuate belt along the southern and western margins of the Ordos Basin, the southern margin of the Shanxi Platform, and the southern margin of the North China Basin in northern Anhui, south-western Henan, southern Shanxi and central Shaanxi Provinces, and the Ningxia Hui Autonomous Region. In this belt they occur in the Houjiashan Formation which is composed of 50–100 m of dolomite and dolomitic limestone. The basal 0.5–2 m of this formation is conglomeratic, containing pebbles of Sinian quartzite and dolomite and pre-Sinian metamorphic rocks all cemented by black or dark grey phosphorite.

Many species of protolenid trilobites, including several species of the genera *Hsuaspis (Hsuaspis)*, *Hsuaspis (Madainaspis)*, *Hsuaspis (Yinshanaspis)*, *Longxianaspis*, *Ningxiaspis*, and *Redlichia* have been described from the basal Houjiashan Formation by Zhang & Zhu (1979) who assigned this fauna a mid-Canglanpuan (*Drepanuroides* Zone) age (Fig. 23.5).

The Houjiashan Formation rests unconformably on various Sinian strata and is conformably overlain by the Manto Formation.

Tarim Platform

Cambrian phosphorites occur along the north-western and northern margins of the Tarim Platform. In the north-western region they occur in the Kelpin–Tagh (Bushinsky, 1966; Lu, 1979) where bedded and nodular phosphorite 0.2–1 m thick occurs with black siliceous shale and silicified limestone in the Xiaoerbulag Formation at the base of the Cambrian sequence. According to Chang (1980, p. 141) the Xiaoerbulag Formation contains species of *Kepingaspis*, *Metaredlichioides* and *Shizhudiscus*. In south-western China, the *Metaredlichioides–Changkouia* Zone is mid-Canglangpuan (Zhou & Yuan, 1980). As along the Ordos margins, therefore, phosphorites are apparently associated with a mid-Canglangpuan unconformity. Lu (1979) shows the further distribution of Canglangpuan phosphorites in the eastern Quruq–Tagh and the Beishan uplift (Fig. 23.5).

In the Beishan area of north-western Gansu Province, Chang (1980) has noted the occurrence of still younger (Middle Cambrian) black phosphorites associated with phosphatic chert, thin-bedded limestone and shale in the Shuangyingshan Formation. As in northern Australia, a lower phosphatic interval is associated with the trilobites *Galahetes* and *Pagetia*, and an upper one with *Centropleura* and *Ptychagnostus* species.

Qilian Geosyncline

Middle Cambrian phosphorites have also been reported from the Qilian Mountains (Lu, 1979) but we have no biochronological details of the locality. Lu (1979) also records a number of 'Middle to Upper Cambrian' phosphorites in the Datong region, Qinghai Province. These are phosphatic breccias and nodular or banded phosphorites interbedded with volcanogenic sediments and argillaceous limestones, and have little economic importance. Their age is uncertain: Chu (1960) described a late Middle Cambrian trilobite fauna from this area, but the relationship of the phosphatic sediments to the fauna is unclear to us.

South-east China Geosyncline

The most important and interesting phosphorites in this province occur at Yaxian, on Hainan Island, Guangdong Province. They occur at two levels in the Damao Group, where they are associated with siliceous shales in a predominantly quartzite/carbonate formation. The main phosphatic horizon lies 305 m from the base of the Damao Group from which *Xystridura yaxianensis*, *X. orientalis*, *Galahetes hainanensis*, *G. pyrus*, and species of *Kootenia* and *Pagetia* of Maochuangian age, have been described by Zhu and Lin (1978). This fauna is generically related to the Templetonian faunas of northwestern Queensland, as well as those in Beishan, northern Gansu Province.

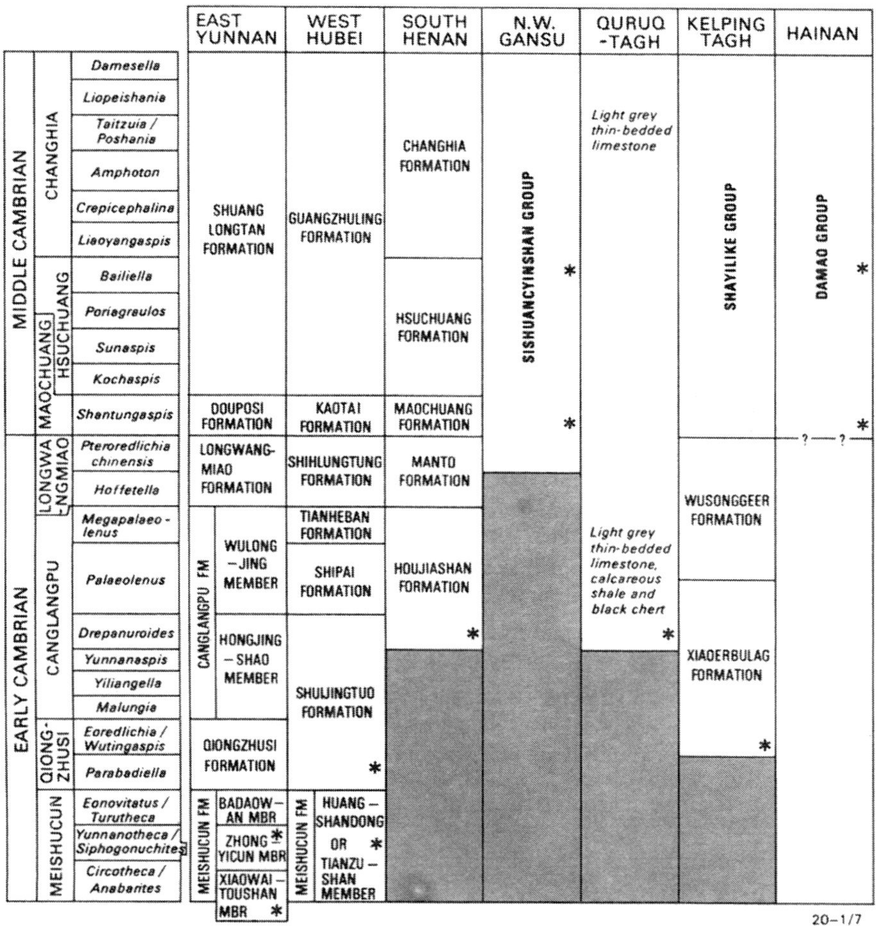

Fig. 23.5. Correlation of the Cambrian phosphorites of China, based on information supplied by Zhang Wen-tang (*in litt.*)

In summary, the most important late Proterozoic phosphorites occur in the Doushantuo Formation in south-central China where they contain palynomorphs, cyanobacteria and calcareous stromatolites indicative of a late Vendian age. The biochronological distribution of Cambrian phosphorites is shown in Fig. 23.5. They are best developed in the belt running from east Yunnan to west Hubei in which they have an initial Cambrian age (Meishucun Stage). On the Tsinling Rise (south Henan) and on the margins of the Ordos and Tarim Platforms where initial Cambrian is younger, phosphorites have an Early Cambrian Canglangpuan age. Two levels of Middle Cambrian phosphorites occur: in north-west Gansu Province, and on Hainan Island, whose associated faunas suggest correlation with the Templetonian and Undillan Stages of northern Australia.

Australia

Phosphatic sediments occur in the Early and Late Proterozoic and Early and Middle Cambrian of Australia (see Howard, this volume): there is no published evidence of Late Cambrian phosphorites. Early Proterozoic phosphatic sediments, associated with the Banded Iron Formations of the Hamersley Basin of Western Australia and the Willyama Inlier at Broken Hill in western New South Wales, are considered to lie outside the scope of this paper. Later Proterozoic phosphorite is recorded in the Adelaide Geosyncline, the Amadeus Basin, and the Pine Creek Geosyncline. Phosphorite of Early Cambrian age is mainly documented in the Adelaide Geosyncline, but phosphatic sediments of this age are also reported from the Amadeus Basin. Mostly, phosphogenesis occurred on the northern Australian Platform during the Middle Cambrian, though not all occurrences of phosphorite or phosphatic sediments can be accurately dated. That in the Depot Creek Sandstone of the Pine Creek Geosyncline, for example, has no biochronological control, though it is undoubtedly Proterozoic (see Howard, this volume).

There is no local biochronological scale for the Early Cambrian of Australia. South Australian phosphorites are dated with reference to the Siberian Early Cambrian biochronological scale because associated rocks contain Archaeocyatha and phosphatised shelly fauna which can be best dated with reference to Siberian sequences. A local biochronology is available for the Middle Cambrian of northern Australia, but there is still debate concerning the interpretation of Early and Middle Cambrian. The degree of overlap between what is considered equivalent to the Siberian Lenian Stage in South Australia and the Ordian Stage in northern Australia is not yet satisfactorily resolved. Both scales are used on our figures, in which the correlation proposed by Öpik (1968, 1975b) is used.

Adelaide Geosyncline

Low grade phosphorites occur in four areas of the Adelaide Geosyncline in South Australia: the Fleurieu Peninsula; the Mount Lofty Ranges; the Flinders Ranges; and Yorke Peninsula. Some 18 Late Proterozoic occurrences are listed by Howard (this volume), and 41 Cambrian occurrences have been plotted geographically by Callen (1969a, Fig. 1). Phosphate values and resources in the four areas are given by Callen (1969a, 1969b) and Johns (1976).

(1) Fleurieu Peninsula A small group of Proterozoic phosphorites in the vicinity of Noarlunga (Callen, 1968, 1969c), occur in the lower part of lateral equivalents of the Brighton Limestone (Preiss & Kinsman, 1978). This formation, together with the Tapley Hill Formation, which also contains lenses of phosphorite, lies between the Sturtian and Marinoan glacial sequences in the Umberatana Group. It contains stromatolites at various localities which indicate an age close to the late Riphean–Vendian boundary (Preiss & Forbes, 1981).

The most important phosphorite occurrence in the western Mount Lofty Ranges is in the Myponga–Sellick Hill district (Johns, 1961, 1976; Blissett & Callen, 1967; Callen, 1969b), in Cambrian siltstone and intraformational breccias in the uppermost part of the Mount Terrible Formation (Daily, 1963, 1972) at Myponga. According to Daily (1972, p. 25), member 3 of the Mount Terrible Formation consists of 'cavernously weathering calcareous quartz sandstone with some coarse siltstone interbeds', and is equivalent to the Hyolithes Sandstone of Abele & MacGowran (1959). Daily (1972, p. 25; 1976, Fig. 9) has recorded black phosphatic moulds of hyoliths, the monoplacophoran *Bemella*, and a pelagiellid mollusc, together with the calcareous actine *Chancelloria*, the phosphatic tube *Saarina* and problematica from this level. Earlier faunas, from the lower members of the Mount Terrible Formation, are also listed by Daily (1976, Fig. 9). These include: hyoliths, the rostroconch *Heraultipegma, ?Oneotodus*, cf. *Sachites, ?Turcutheca*, and *Chancelloria*, and are considered to have a Tommotian age on the Siberian biochronological scale (Daily, 1972, Fig. 5).

Callen (1969b, Table 2, Fig. 4) records a second weakly phosphatic interval associated with the contact between the Wangkonda and Sellick Hill Limestones, interpreted by Daily (1976, Fig. 8) as a hiatus, and a third within the Sellick Hill Formation. The Sellick Hill Formation contains hyoliths, *Helcionella* and *Tannuella* which are considered to indicate an early Atdabanian age (Daily 1972). The overlying Forktree Limestone contains archaeocyathans originally described by Taylor (1910) which Gravestock (in Daily, Jago & James, 1982) has redetermined as *Dokidocyathus* sp., *Picnoidocyathus* sp. and *Erugatocyathus sellicksi*. All of these are long-ranging forms which do not give a precise age within the Early Cambrian. Daily (1976) also records 'tommotiids' from the upper part of the Forktree Limestone.

Phosphatic nodules occur in the Black Shale Member of the overlying Heatherdale Shale (Campana & Wilson, 1954; Daily, 1963). *Helcionella*, hyoliths and sponges are recorded from this formation (Daily, 1963) but dating is imprecise. Daily (1976, Fig. 8) correlates the Heatherdale Shale with the uppermost part of the Ajax Limestone sequence, which in turn is laterally equivalent to the Oraparina Shale. Walter (1967) considers the last to have an early Lenian (Sanashtykgolian) age.

(2) Mount Lofty Ranges According to Howard (this volume), the oldest Proterozoic phosphorites of South Australia occur in the northern part of the Mount Lofty Range, around Burra. In this area, black shale and chert with phosphorite lenses is found at the transition between the Skillogalee Dolomite and the Kalachalpa Formation (of Ambrose, Flint & Webb, 1981). The Skillogalee contains stromatolites, including *Baicalia burra*, of Torrensian (late Riphean) age. Further north, in the general region of Orroroo, phosphorite lenses again occur in association with the Brighton Limestone, of late Riphean to Vendian age.

South of Burra, Callen (1969a, Fig. 1) indicates some 22 phosphorite occurrences in the area between Kapunda–Truro and Angaston, north-east of Adelaide, and an additional locality is shown at Tungkillo, east of Adelaide. In the Kapunda–Truro–Angaston area, the phosphatic interval is the Koonunga Phosphorite Member (Johns, 1976) which seems to lie below the Angaston Marble. There is no published evidence for the age of the phosphorites in this area. The Angaston Marble is traditionally regarded as a unit of the Cambrian Hawker Group, and Callen (1969a) shows it underlying the Heatherdale Shale. Thus the Koonunga Phosphorite Member is probably older than Atdabanian.

At Tungkillo, low grade phosphorite is associated with the Tungkillo Marble Member and Nairne Pyrite Member of the Inman Hill Formation, metasediments of the basal Kanmantoo Group (Callen, 1969a, 1969b; Horwitz, Thomson & Webb, 1959). The Kanmantoo Group overlies the Heatherdale Shale with significant lithological discontinuity, interpreted as a disconformity by Hortwitz *et al.* (1959) and as a fault contact by Daily & Milnes (1971, 1972). In the absence of recorded fauna, biochronological dating of this occurrence is not possible.

(3) Flinders Ranges In the southern Flinders, phosphorite occurs in kaolinitic sandstone with limestone lenses referred to the Parachilna Formation south and south-west of Hawker. Several further occurrences have been found in this stratigraphic unit in the north-eastern Flinders Ranges, particularly in the Frome River–Nepabunna Mission district (Callen, 1969a, 1969b). According to Daily (1976), the Parachilna Formation contains trace fossils *(Diplocraterion* sp.,*Plagiogmus arcuatus, Phycodes pedum)*, and a shelly fauna which includes the monoplacophoran genus *Bemella* permitting correlation with the upper member of the Mount Terrible Formation of the Fleurieu Peninsula, and having an inferred Tommotian age.

In the north-western Flinders Ranges, phosphorite occurs in the overlying Ajax Limestone (Callen, 1969a). More specifically, it occurs in the uppermost Wilkawillina Limestone facies of the Ajax Limestone (see Daily, 1976, Fig. 8), in hematitic and glauconitic red limestones and breccias, coinciding with beds 6–8 of Daily (1956). Numerous archaeocyathans have been described from this interval (Walter, 1967; Debrenne, 1969, 1970) and indicate to Rozanov & Debrenne (1974) an Atdabanian (Kameshkian) and Lenian (Sanashtykgolian) age.

Daily (1976, Figs 8, 9) indicates an hiatus within the Ajax Limestone which is thought to post-date his faunal assemblage 2, and it is likely that the phosphorite is associated with this break. An exact biostratigraphy of the Ajax Limestone is required to date this event and to place the shelly fossils listed by Daily (1956, 1972, 1976) and Bischoff (1976) in perspective.

(4) Yorke Peninsula Phosphorite occurrences have been long known on the Yorke Peninsula (see Johns, 1961, Fig. 1; 1976), where they are associated with the Kulpara Limestone. Phosphorite in this area is found in cavities, caves and joints, and is related to a karst topography developed on the Kulpara Limestone, a situation similar to that observed at places in the Thorntonia Limestone of the northern Georgina Basin. Daily (1976, Fig. 9) indicates that the Kulpara Limestone contains archaeocyathids, trilobites, 'camenellids', hyoliths, *'Micromitra'* and problematica, and suggests a late Tommotian to Atdabanian age (Daily 1972). More recently, Daily (1982) indicates correlation with the Wangkonda Formation of the Sellick Hill area so that the phosphatic interval, if Cambrian, may be similar in age to that between the Sellick Hill and Wangkonda Formations. It may be considerably younger, however (Fig. 23.6).

The phosphatic sediments of the Adelaide Geosyncline occur in a variety of sedimentary environments representing both transgressive and regressive phases of sedimentary cycles. They record two phosphogenic events, one in the late Riphean to Vendian and the other in the Early Cambrian. The greatest number of occurrences are of Tommotian to Atdabanian age, while a small number could be slightly younger (Lenian/Ordian).

North Australian Platform

Documented phosphorites in northern Australia occur mostly in the Georgina Basin, a segment of an epicontinental sea which covers some 325 000 km^2 of western Queensland and eastern Northern Territory, but there are also occurrences of phosphatic Late Proterozoic and Cambrian sediments in the Amadeus Basin (see below). The economically most important, and most thoroughly documented phosphorites occur in the Burke River Structural Belt, an eastern appendage of the Georgina Basin proper, but some twenty other deposits have been reported, and are shown on maps published by Howard (1971, 1972) and Cook (1976). Biochronological information is available for the Duchess, Ardmore, Lady Annie, D Tree and Thorntonia occurrences which lie along the eastern and northern basin margins. Additional information on low-grade phosphatic sediments from the southern margin of the Georgina Basin is presented here.

(1) Burke River Structural Belt The main phosphate deposit in the Burke River Structural Belt occurs in the vicinity of Phosphate Hill, some 35 km south of Duchess, where commercial grade phosphorite has been exploited. The geology of the Duchess phosphate field has been exhaustively treated elsewhere (Russell, 1967; Russell & Trueman, 1971; de Keyser & Cook, 1972; George, 1975; Fleming, 1977; Cook & Shergold, 1979; Shergold & Druce, 1980).

Phosphorite is concentrated at the top of the Beetle Creek Formation, which is divided into a basal Siltstone Member and an upper Monastery Creek Phosphorite Member. Within the Burke River Structural Belt the Beetle Creek Formation disconformably overlies Thorntonia Limestone or silicified rubble derived from it. This formation was deeply weathered, karsted and eroded prior to the deposition of the Beetle Creek Formation. The phosphatic sequence of the latter is overlain disconformably, possibly unconformably in some areas, by the Inca Formation, a dominantly carbonaceous clastic unit which is reportedly phosphatic at its base.

The lower Siltstone Member of the Beetle Creek Formation comprising some 60 m of shale, siltstone and chert, black and less siliceous below the weathering profile, contains the following taxa in the Phosphate Hill district: *Galahetes fulcrosus, Lyriaspis* sp., *Oryctocephalites gelasinus, Pagetia triaena, Peronopsis* spp., *Ptychagnostus* cf. *P. gibbus; Indota formosa*; undescribed inarticulate brachiopods; and a hyolith (Shergold, 1968; Jell, 1975; Öpik, 1975a).

By contrast, the Monastery Creek Phosphorite Member contains a rich and varied fauna derived mainly from grainstone phosphorite layers in a 10- to 12 m sequence of alternating sandy, silty or siliceous high-grade phosphorite and lower-grade phosphatic chert, siltstone and shale (see Cook & Shergold, 1979, Fig. 8). Although the phosphatic member contains a varied fauna, only one genus of eodiscid trilobites (Jell, 1970, 1975) and the bradoriid crustaceans (Fleming, 1973: in Hill, Playford & Woods, 1971; Jones & McKenzie, 1980) have been described to date. The following taxa have been isolated from acid insoluble residues: Trilobita (Oryctocephalidae undet., *Pagetia ocellata, Peronopsis* spp. Ptychopariidae undet., *Triplagnostus* sp. aff. *T. praecurrens, Xystridura* spp. *sensu lato*); Bradoriida (*Aristaluta* sp. A, *Dielymella? dubia, Flemingia duo, Indiana sipa, Monasterium dorium, Monasterium oepiki, Mononotella sestina, Svealuta* sp. A, *Zepaera rete, Zepaera* sp.); Mollusca (helcionellacean undet., *Mellopegma* sp., *Pelagiella* sp.); Hyolithida undet.; Conodontophorida? undet,; inarticulate Brachiopoda (Acrotretidae, Obolidae undet.); Pogonophora? (*Coleoloides*? sp.); sponge spicules and associated poriferan? problematica; undifferentiated echinodermal and crustacean debris.

The Beetle Creek Formation contains no species of *Redlichia* in association with its Xystriduridae, and this fact, together with the rare occurrence of *Ptychagnostus*, indicates a post-Ordian (Öpik, 1968, 1975b), Templetonian Middle Cambrian age. The phosphatic interval is overlain by Inca Formation which contains *Ptychagnostus gibbus* and younger agnostids (Öpik,1979), which suggest a mid to late Templetonian age.

Elsewhere in the Burke River Structural Belt, low grade phosphorites are associated with the transgressive base of the Inca Formation (Russell, 1967; Russell & Trueman, 1971) late in the Zone of *Triplagnostus gibbus* (latest Templetonian), and with a second trangression at the beginning of the *Ptychagnostus punctuosus* Zone (see Henderson & Southgate 1978, Fig. 3). De Keyser (1968) and de Keyser & Cook (1972) recorded phosphorite of no commercial importance at the base of the Roaring Siltstone east of Duchess, and where the Devoncourt Limestone

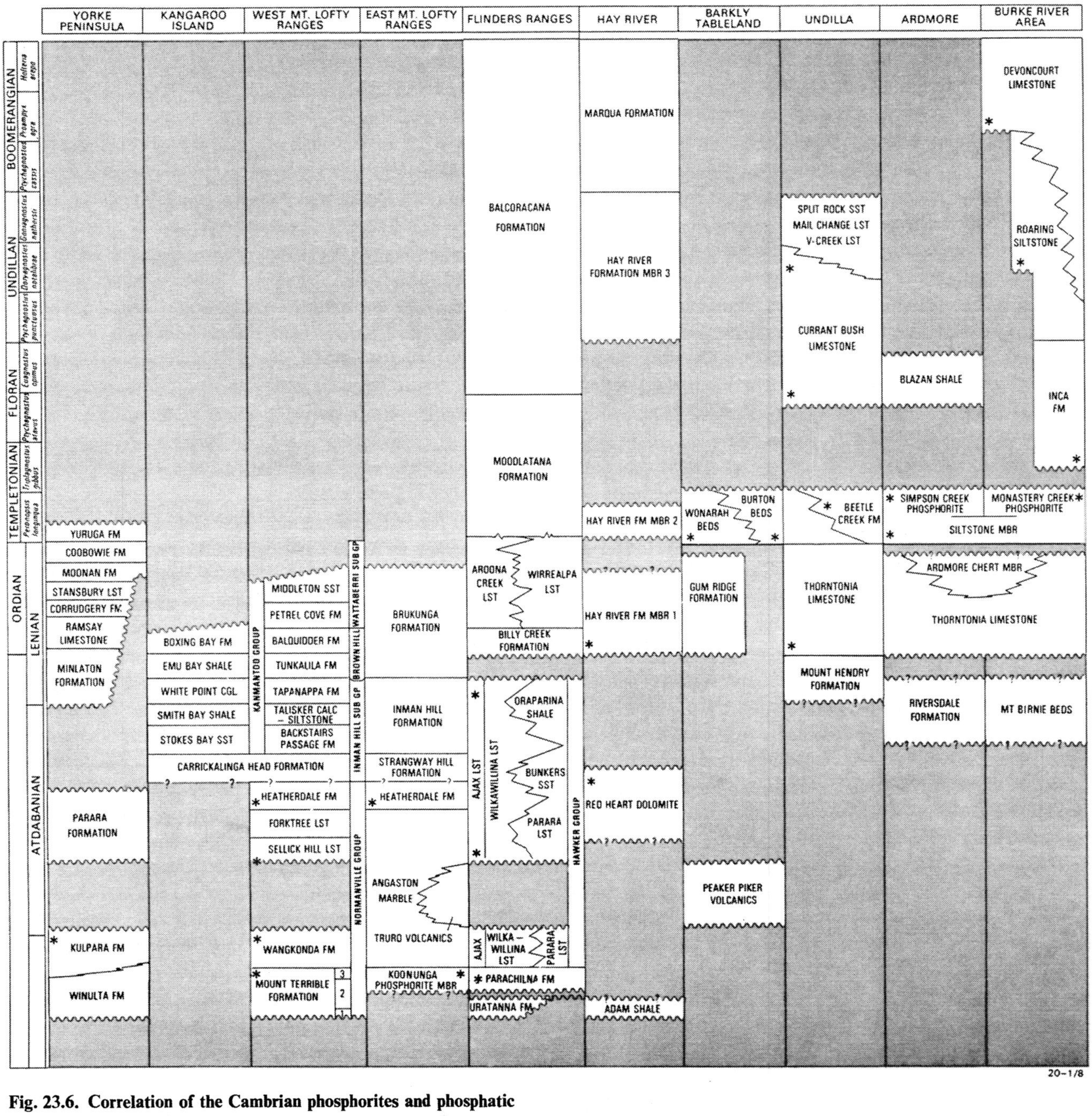

Fig. 23.6. Correlation of the Cambrian phosphorites and phosphatic sediments of Australia. Information is drawn from Abele & MacGowran (1959), Blissett & Callen (1967), Callen (1969a, 1969b), Daily (1956, 1967a, 1967b, 1972, 1976), Daily & Milnes (1971), Johns (1976), Shergold & Druce (1980), Walter (1967), and Walter *et al.* (1979).

rests disconformably on Thorntonia Limestone at the northern end of the Structural Belt. From near the former locality, Shergold (1973) has recorded *Centropleura* sp., *Hypagnostus* cf. *brevifrons, Meneviella viatrix, Ptychagnostus cassis* and *Ptychagnostus* ex. gr. *punctuosus*, considered to represent the *Pt. punctuosus*/*Goniagnostus nathorsti* interval (*Doryagnostus notalibrae* Zone of Öpik, 1979) of late Undillan age. Shergold (in de Keyser, 1968, Appendix 1) recorded *Diplagnostus* sp., *Goniagnostus fumicola, Leiopyge laevigata*, a mapaniid, and *Pseudophalacroma dubium* from the phosphatised base of the Devoncourt Limestone, indicative of the Zone of *Proampyx agra*, mid-Boomerangian (Öpik, 1961, 1979).

(2) Eastern Georgina Basin Little published biostratigraphic information is directly available to date phosphorite occurrences between Quita Creek and Riversleigh, along the eastern margin of the Georgina Basin (Cook & Shergold, 1979). At several localities, phosphatic rocks of secondary origin, phoscretes, are associated with eroded Cambrian carbonate formations. Some of these phoscretes seem to be associated with

groundwater movements and could be of Cambrian or post-Cambrian age (Cook, 1972) and are not included in this present review.

At Ardmore a similar sequence to that recorded south of Duchess has been reported, with the main phosphorite development occurring in the upper part of the Beetle Creek Formation, called the Simpson Creek Phosphorite Member (Cook & Shergold, 1979, Fig. 4). This has a similar stratigraphic position and P_2O_5 content to the Monastery Creek Phosphorite at Phosphate Hill. On superpositional evidence a mid–late Templetonian age must also be inferred. The phosphatic sequence is overstepped by the Blazan Shale whose base lies within the *Ptychagnostus atavus*/*Euagnostus opimus* interval, according to Öpik (1967). It is therefore, a little younger than the Inca Formation in the Duchess area.

Similar lithofacies associations extend to the Lady Annie/Lady Jane area, 220 km to the north of Ardmore (Rogers & Keevers, 1976; and Cook & Elgueta, this volume). Here, the Thorntonia Limestone is weakly phosphatic, the phosphate occurring as dispersed ovules and phosphatised fossil fragments. The Thorntonia Limestone has yielded species of *Redlichia*, inarticulate brachiopods, hyoliths and molluscs. According to Rogers & Keevers (1976), the formation was karsted before deposition of the Beetle Creek Formation. Phosphorite occurs most commonly, however, in the Beetle Creek Formation. Lithofacies relationships within the Beetle Creek Formation are rapid and complicated. Siliceous and phosphatic siltstones and coquinites are interbedded with phosphatic packstones, grainstones, and mudstones. The upper part of this sequence, at least, contains species of *Xystridura*, inarticulate brachiopods, hyoliths, molluscs and sponge spicules, and may be of Templetonian age. It is conceivable, however, that the whole sequence is of Ordian age. The Beetle Creek Formation is conformably overlain by Inca Formation, a finely laminated non-phosphatic chert/shale lithofacies containing ptychagnostids of the *gibbus* group and sponge spicules, as at Duchess.

The Thorntonia Limestone is also phosphatic throughout at D Tree, according to Howard & Cooney (1976). In this area, only 25 km west of Lady Annie, carbonates of early Templetonian (*Xystridura templetonensis* Zone) age have been included in the Thorntonia Limestone. Several hiati have been recognised in the Thorntonia sequence (see Southgate, this volume) and the distribution of the formation seems to have been controlled by basement topography (Howard & Cooney, 1976; Howard, this volume). The main phosphatic development is associated with a chert/siltstone lithosome referred to the Beetle Creek Formation which has yielded Xystriduridae and hyoliths. Jell (1975) has published faunal evidence which suggests that this Beetle Creek lithosome is laterally equivalent to the top of the Thorntonia Limestone. A widespread erosion surface is recognised by Southgate (this volume) overlying both facies, so that only the lower Beetle Creek Formation may be preserved. By comparison with Duchess, the phosphogenic episode must be slightly earlier at D Tree, i.e. has an early Templetonian age (Zone of *Xystridura templetonensis*).

A third phosphatic interval occurs between D Tree and Thorntonia at the base of the Currant Bush Limestone sequence in which shelly grainstone phosphorite is associated with a series of hard grounds. A well preserved molluscan fauna which includes *Tannuella* sp., *Eotebenna pontifex*, *Protowenella flemingi*, *Pelagiella deltoides*, *P. corinthiana*, *Myona? queenslandica* and *Yochelcionella cyrano* has been described (Runnegar & Jell, 1976). Additionally sponge spicules, *Chancelloria*, undetermined bradoriids, pelmatozoan debris, inarticulate brachiopods, and species of the agnostid arthropod genera *Goniagnostus (Criotypus)* and *Rhodotypiscus* also occur. Since Jell (1975, in Runnegar & Jell, 1976) has recorded the presence of *Ptychagnostus atavus* as well, an age within the *P. atavus*/*Euagnostus opimus* Zone interval is suggested for this occurrence (see Opik, 1979, Table 3).

(3) North and north-west Georgina Basin A series of small phosphorite deposits has been reported along the northern margin of the Georgina Basin, and between Riversleigh and Highland Plains (see Howard, 1971, Fig. 1; Cook, 1976, Fig. 41). The little information on them is provided by de Keyser (1969), Thomson & Russell (1971) and de Keyser & Cook (1972). The phosphorites of this area are dominantly mudstone phosphorites and siliceous siltstones, often preserved as erosional remnants, which are associated with the Border Waterhole Formation, a lateral equivalent of the Beetle Creek Formation. As in the case of the latter, the Border Waterhole Formation either overlies Thorntonia Limestone, Colless Volcanics, or Proterozoic Basement. Close to basement features, the Border Waterhole Formation is laterally equivalent to a carbonate development which is included in the Thorntonia Limestone. According to Öpik (in Carter & Öpik, 1961), the Border Waterhole Formation contains species of *Xystridura*, *Lyriaspis*, *Peronopsis* and *Helcionella*, and these probably indicate a Templetonian age.

Near Riversleigh, some 60 km north of Thorntonia, phosphatised stromatolites, referred to *Ilicta composita* (Sidorov) have been reported from the lower part of the Thorntonia Limestone (Southgate, 1980; Schmitt & Southgate, 1982). Elsewhere in the Riversleigh area, *Redlichia* spp. occur in the lower Thorntonia Limestone, indicating a probable Ordian age for this phosphatic occurrence.

In the north-west Georgina Basin, phosphorites at Wonarah, Alroy and Alexandria, are known only in the subsurface (Howard 1971, 1972; Howard & Perrino 1976). At Wonarah, mudstone phosphorites lap on to and drape earlier volcanic and carbonate sequences. They are associated with the Wonarah Beds, a chert/siltstone/limestone/phosphorite lithosome, which, on a regional basis, overlies the Gum Ridge Formation, an Ordian evaporitic unit containing *Redlichia amadeana* and *Xystridura (Polydinotes) verticosa* in association (Öpik, 1975a). The Wonarah Beds contain *Xystridura (X.) altera*, *X. (Inosacotes)* aff. *browni*, species of *Peronopsis* and *Lyriaspis*, and an oryctocephalid, indicating an early Templetonian age (Öpik, 1975a, Fig. 3).

At Alexandria (Howard, 1972), phosphatic siltstone, phosphatic chert and phosphorite, are associated with the Burton Beds whose faunas are given by Öpik (1975a) and Jell (1975). They include *Xystridura (Inosacotes) browni*, *X. (Xystridura) lauta*, *X. altera*, *X. altera obtusa*, *Lyriaspis alroiensis*, *Pagetia edura, P. prolata, P. significans, Peronopsis (Itagnostus) elkedraensis, 'Acadagnostus' scutalis, Orytocephalus reynoldsi, O. alexandrensis, Kootenia* sp.; *Oelandia* sp., *?Bradorina* sp.; *Stenotheca* sp.; eocrinoids, and inarticulate brachiopods, a fauna considered to have a probable early Templetonian age.

(4) Southern Georgina Basin Weakly phosphatic sediments occur in the Red Heart Dolomite and overlying Hay River Formation in the Hay River area, along the southern margin of the Georgina Basin (Walter *et al.*, 1979). The former formation contains phosphatic shelly problematica, inarticulate brachiopods, a hyolith, *Chancelloria*, and a poorly preserved archaeocyathan fauna: *'Cosciniptycta' bilateralis, ?Dictyocyathus* sp., Metacyathidae and *Radiocyathus minor* (Kruse & West, 1980) which suggest a late Atdabanian to early Lenian (early Ordian) age. The base of the lower member of the Hay River Formation contains poorly preserved phosphatic moulds of Monoplacophora or Gastropoda, and the upper part contains species of *Redlichia* and *Pagetia* considered to have a late Ordian age.

(5) Amadeus Basin Phosphorite occurs as thin lenses interbedded with chert, and within the quartzitic matrix of the basal conglomerates of the Areyonga Formation in the eastern Amadeus Basin (Wells *et al.*, 1967; Wells *et al.*, 1970). This is a variable conglomeratic arkosic siltstone, sandstone or dolomite unit regarded as the lower tillitic horizon of the Amadeus Basin (Preiss *et al.*, 1978). Stromatolites (not identified, see Walter, 1972) have been reported from the dolomitic intervals, but no other biota have been confirmed. The Areyonga Formation rests disconformably on Bitter Springs Formation, also reportedly phosphatic (Wells *et al.*, 1970), and this contains a stromatolite assemblage, *Boxonia, Minjaria, Inzeria, Jurusania* and *Kotuikania* having a late Riphean age in the USSR, a species of *Linella* having a Vendian age, and forms endemic to Australia, all of which are taken to suggest a late Riphean age (750–900 m.y.) (Preiss, 1976; Walter, Krylov & Preiss, 1979). The Areyonga Formation, although slightly younger, is also considered to be late Riphean because of its stratigraphic position. Phosphatic grains have also been reported from the Pertatataka Formation which overlies the Areyonga Formation. Stromatolites immediately below and above the Pertatataka Formation, described by Walter, Krylov and Preiss (1979), again suggest a late Riphean to Vendian age.

Howard (this volume) records the presence of phosphate pellets in the Todd River Dolomite in the eastern Amadeus Basin, a unit containing Daily's (1956) 'faunal assemblage 2'. Archaeocyathans described by Kruse & West (1980), associated with phosphatised and phosphatic shelly fauna, suggest a direct comparison with the assemblage recorded from the Red Heart Dolomite of the southern Georgina Basin (see above) considered to have a late Atdabanian to Lenian age.

According to Cook (in Wells *et al.*, 1970) thin low-grade grainstone phosphorites occur in the Tempe Formation which contains in its upper part trilobites, hyoliths, gastropods, and phosphatic brachiopods of possible early Middle Cambrian age (Wells, Forman & Ranford, 1965; Ranford, Cook & Wells, 1965; Wells *et al.*, 1970).

The oldest phosphatic sediments on the North Australian Platform occur among variable clastic and carbonate sediments of the Amadeus Basin. These are dated on the basis of stromatolites as late Riphean to Vendian. The earliest Cambrian phosphorites on the Platform are associated with archaeocyathan carbonates of late Atdabanian or earliest Lenian age in both the Georgina and Amadeus Basins. Carbonates of late Lenian (Ordian) age are locally phosphatic in the Georgina Basin. However, the main development of phosphorite, associated with carbonate, siltstone and chert, occurred throughout much of the Georgina Basin during Templetonian Middle Cambrian time. In this basin, younger phosphorite and phosphatic sediments are associated with local disconformities and hard grounds of late Templetonian, mid Undillan and early Boomerangian ages (Fig. 23.6).

Other areas

Apart from the main phosphogenic regions described above, there are others which have either been little explored for phosphate, contain phosphatic sediments of little economic importance, or for which little palaeontological information is available for accurate dating. Several of these areas are described by other contributions to this book. They are briefly mentioned here to complete this review.

North America

No palaeontological information is available to us for the Aphebian (pre-Riphean), Helikian (Riphean) or Hadrynian (Vendian) phosphorites described herein by Christie & Sheldon, although the Hadrynian phosphorites of the Appalachian Geosyncline are reported by them to contain acritarchs. There is, however, data for Cambrian phosphatic formations in both the Cordilleran and Appalachian regions.

The oldest Cambrian phosphatic sediments occur in the Shady Formation, mainly dolomite, which locally contains archaeocyathans and olenellid trilobites indicative of the *Bonnia–Olenellus* Zone (Fig. 23.1), of late Early Cambrian age. The overlying Rome Formation, predominantly clastic, also contains phosphorite, and the trilobite *Olenellus* indicating a similar Early Cambrian age for the Shady Formation (Resser, 1938). Early Middle Cambrian phosphatic sandstone is also reported by Wedow & Stansfield (1968) but its biochronological position is uncertain.

Rogers *et al.*, (1970) have summarised the occurrences of phosphatic Cambrian sediments in Nevada. The earliest of these is the Secret Canyon Shale in the Ruby Range, which contains trilobites of the late Middle Cambrian *Bolaspidella* Zone (Palmer, 1954, 1971). In the Yucca Flat region the upppermost Bonanza King Formation contains pelletal phosphorite and because the upper 100 m of this formation was deposited during the successive Late Cambrian trilobite zones of *Crepicephalus, Aphelaspis* and *Dicanthopyge*, they must be late Dresbachian in age. The overlying Dunderberg Shale is weakly phosphatic and was deposited during the zones of *Dunderbergia* and *Elvinia*, and thus has a latest Dresbachian and early Franconian Late Cambrian age (Palmer, 1965, 1971). Phosphate nodules also occur in the allochthonous Hales Limestone in the Hot Creek Range which Cook & Taylor (1975) regard as being deposited in a deep-water slope environment. The age of this formation ranges from Franconian into the Early Ordovician (see Taylor, 1976, 1977).

Phosphorites occurring in maritime Canada are described here in association with the phosphorites of the Avalon Platform. In addition, low-grade phosphorite also occurs in Cordilleran Canada as black nodules in shales below the Kechika Formation in north-east British Columbia. A Late Cambrian age is assigned on the basis of conodonts (Cecile & Norford, 1979). There is a second occurrence of phosphorite in the Yukon Territory, associated with the carbonates of the Road River Formation which extends across the Cambrian–Ordovician boundary (Christie, 1979).

South America

Phosphorite occurs in the Bambuí Group over a wide area of Minas Gerais, Goiás and Bahia States, Brazil (Cathcart, 1979), and is described in detail by Dardenne *et al.* (this volume). At Patos de Minas, mostly grainstone phosphorite occurs in the upper Indaia Member of the Rio Paraopeba Formation associated with sandstone and conglomerate, and overlying phosphatic carbonates which are stromatolitic near the base (Sete Lagaos Member). According to Borrello (1971), the 'Bambuí Formation' contains sponge spicules and annelid tubes, but the relationship of this fauna to the phosphorites is currently not known to us. A Late Proterozoic or Early Cambrian age has been suggested by Cathcart (1979), but the phosphorites described from Brazil in this volume indicate a late Riphean to Vendian age.

Dateable Late Proterozoic or Cambrian phosphorite is not yet reported from other areas of South America.

North and West Africa

Occurrences of phosphorite, of probable latest Proterozoic age, in West Africa, are described by Slansky (this volume) and Lucas *et al.* (this volume).

Phosphorites of probable Cambrian age have also been reported from Zemmour in the northern part of the Islamic Republic of Mauretania (Poulsen, 1960). Phosphorite nodules occur at two levels in the Lower Palaeozoic sequence: the oldest is associated with sandstone and conglomerate in a sequence overlying dated Cambrian rocks which contains five species of inarticulate brachiopods and an undetermined species of the trilobite *Redlichia*. This fauna appears to be related to that of the Kussak Formation (*Neobolus* Beds) of the eastern Salt Range of Pakistan. It suggests that the overlying phosphorite nodules may have an Ordian/Lenian or younger Cambrian age. An erosion surface separates this older phosphatic sequence from a younger one in which nodules are again associated with conglomerate and quartzite, and are overlain by graptolitic shales containing species of *Monograptus* of presumed Late Ordovician to Silurian age.

Viland (1977) has described phosphorite lenses associated with dolomite in the basal Adoudounian sequence (Calcaires inférieurs) near Taguedit, south of the Haut Atlas de Marrakech, Morocco, and a similar Adoudounian occurrence associated with dolomite, trachyte, and andesite, and a syenite intrusion, has been described at Jbel Boho, south-south-west of Marrakech, by Ducrot & Lancelot (1977). Traditionally, the Adoudounian sequence has been regarded as Riphean to Vendian in age on the basis of stromatolites such as *Linella avis, Tungussia* cf. *inna*, and *Tifounkeia* sp., and oncolites (Sdzuy, 1978; Schmitt, 1978, 1979; Boudda, Choubert & Faure-Mauret, 1979) which are found in the Série Lie de Vin and overlying Calcaires supérieurs. The first archaeocyathans and trilobites occur in the upper part of the Calcaire supérieur (Sdzuy, 1978). However, the syenite which intrudes the early Adoudounian Calcaires inférieurs has a U/Pb age based on zircons of 534 $\pm$ 10 m.y. (Ducrot & Lancelot 1977). Since this sequence predates the stromatolitic and younger Adoudounian and Cambrian sediments, a reassessment of the age of these Moroccan phosphorites is required. Indeed, Bertrand-Sarfati (1981), reinterpreting the algal data, has recognised thrombolites containing *Renalcis* and *Epiphyton*, and small, mainly non-columnar stromatolite forms in the Adoudounian Series which indicate Cambrian affinities, suggesting the Moroccan phosphorites may have a latest Proterozoic to Early Cambrian age.

South-east Asia

Phosphorite has long been known from the Socialist Republic of Vietnam, at Lao Cai, on the Red River, at the Vietnamese–Chinese border, and considered to have a late Sinian or Early Cambrian age (e.g. Bushinsky, 1966). The microflora listed by Tran Quoc An & Nguyen Dang Khoa (this volume) suggests a Vendian age, so that the Vietnamese phosphogenetic episode was probably synchronous with that of the Doushantuo Formation of neighbouring Yunnan and Guizhou Provinces of China.

Conclusions

There is insufficient biological data presently available to prove the synchroneity, or otherwise, of the Proterozoic phosphorites described herein. The majority, however, can be bracketed within the late Riphean to Vendian interval, and most are associated with, or immediately post-date Late Proterozoic glacial events and are less than 700 m.y. old. Detailed analysis of the stromatolites and microflora which may occur in association with, or close proximity to, Proterozoic phosphorites may eventually produce a more accurate age determination.

Our information on Cambrian phosphorites is most detailed for European and maritime eastern American sequences where there is some evidence for the cyclicity of phosphogenesis, derived not only from the mode of occurrence of the phosphorites and phosphatic sediments themselves, but from other sedimentological evidence. This review, and that of Brasier & Notholt (this volume) permits us to suggest that there were probably six major sedimentary and phosphogenic cycles related to transgressive/regressive events on the Avalon and Baltic Platforms during the Cambrian. Of these, three Early Cambrian cycles have been named by Mens & Pirrus (1981): Baltian, Liivian, and Aisciaian. In our conclusion we use these to calibrate and correlate Cambrian phosphogenic events elsewhere within the limits of the data available to us.

Cycle 1. In Europe, the initial Cambrian, Baltian, transgression, during the *Aldanella attleborensis–Sunnaginia neoim-*

bricata Zone, deposited glauconitic sandstone, clay and shale, and in places coarser clastic sediments, and with the exception of Nuneaton, is generally not associated with phosphorite (Fig. 23.1 to 23.3). If our correlation of this zone with the Tommotian Zones of Siberia is correct, then the Tommotian phosphorites and phosphatic sediments of Kazakhstan, southern and eastern Siberia, and possibly the Mongolian People's Republic are comparable in age with the first Cambrian phosphorites of Nuneaton. Similarly, by correlation, the phosphorites of the upper Dengying Formation of south-west China and those of the Mount Terrible Formation of South Australia, which are considered to have a Tommotian age, are comparable.

Cycle 2. In Europe, the second Cambrian phosphogenic event coincides with the Liivian sedimentary cycle, which also deposited predominantly clastic sediments during the *Rhombocorniculum insolutum– Schmidtiellus mickwitzi* Zones (Fig. 23.1 to 23.3), and is dated by the relatively widespread occurrence of the *Mobergella, Camenella* and *Coleoloides* Fauna. In Kazakhstan, we consider that the Iron–Manganese Carbonate Horizon, which is stromatolitic and contains the faunas of the *Uniformitheca/Bercutia cristata* Zones, was deposited during this cycle (Fig. 23.4) as were the early Atdabanian phosphorites of South Australia (Fig. 23.6).

Cycle 3. Cycle 3, the Aisciaian, commences with phosphorites in Europe following a widespread regression, depositing phosphatic conglomerate, grit, sandstone, carbonate and shale according to location on the Avalon and Baltic Platforms, during the *Callavia/Holmia* n.sp. and *Holmia kjerulfi* Zones which are correlated by the pandemic *Dipharus attleborensis* fauna. Phosphatic sediments are particularly widespread during the Protolenid–strenuellid/*Proampyx linnarssoni* Zones, occurring in the upper part of this third cycle, and dated by elements of the *Bathydiscus–Serrodiscus* fauna. The late Atdabanian phosphorites of central Australia (Todd River and Red Heart Dolomites) and South Australia (upper Hawker Group) appear to correlate with the early part of this cycle; and the early Lenian phosphorites of Kazakhstan, and the Yenesei River, those of the Canglangpuan in south Henan, central China, and the Ordian of northern Australia (lower Hay River Formation and Thorntonia Limestone), seem to correlate with the later part of the cycle.

Cycle 4. Extensive hiati occur between the Middle Cambrian *Oelandicus* Zones and the *Proampyx linnarssoni* Zone on the Baltic Platform, and elsewhere, and are responsible in part for the different definitions globally of the Lower/Middle Cambrian boundary. During this cycle, transgressive sediments, including grits, sandstones and often phosphatic shale were deposited during the *Paradoxides bennetti* or *P. oelandicus* Zones (Figs. 23.2 to 23.3). Phosphatic conglomerates at the top of the cycle mark a widespread period of regression. This is the time of maximum phosphogenesis in northern Australia, in the regressive upper members of the Beetle Creek Formation; and phosphorite of equivalent age is also known from Hainan Island, Guangdong Province, China.

Cycle 5. Cycle 5 commences with the widespread transgressive event which occurs on a virtual global basis at the beginning of the *Ptychagnostus gibbus* Zone and continues through the *Paradoxides paradoxissimus* Zones at least to the *Ptychagnostus punctuosus* Zone on the Avalon Platform in particular. Black shales with phosphatic nodules were commonly deposited during this interval. Regression is marked by phosphatic conglomerates during the succeeding *Goniagnostus nathorsti* and *Solenopleura brachymetopa* Zones. In northern Australia, phosphorites are associated with a local late Floran (*Euagnostus opimus* Zone) transgression, during the early part of this cycle, and it is suspected that the younger phosphorites of Hainan and of Gansu Province, south China, may be of similar age, as are the earliest Middle Cambrian phosphorites of Cordilleran North America.

Cycle 6. The sixth cycle in Europe also commences with predominantly shale deposition, often phosphatic, and containing phosphatic nodules. Sedimentation commenced during the *Leiopyge laevigata* Zone, and continued, with minor interruptions through the Late Cambrian, especially on the Avalon Platform where shales deposited during all zones, with the exception of *Parabolina spinulosa*, are weakly phosphatic. Only the lower part of the Upper Cambrian Alum Shale sequence of the Baltic Platform is phosphatic. In Cordilleran North America phosphogenesis commences almost contemporaneously in the *Bolaspidella* Zone of Nevada, and in the Great Basin phosphatic sediments occur in all zones, and continue, both there and in western Canada, into the Early Ordovician.

A seventh cycle, beyond the scope of this book, can be documented following global regression at the Cambrian–Ordovician transition, in the Yukon Territory, Canada, and elsewhere. Our findings, in general, agree with those of Eganov (1979) who considers that cyclical phosphogenetic events reflect transgressive or regressive tendencies in basin development. During the Early Cambrian, and paradoxididian Middle Cambrian, some of these events are of very wide geographical distribution. They are probably related to eustatic sea-level fluctuation and periods of intensive global lithospheric plate motion.

References

Abele, C. & McGowran, B. (1959). The geology of the Cambrian south of Adelaide (Sellick Hill to Yankalilla). *Transactions of the Royal Society of South Australia*, **82**, 301–20.

Ahlberg, P. (1980). Early Cambrian trilobites from northern Scandinavia. *Norsk Geologisk Tidsskrift*, **60**, 153–9.

Ahlberg, P. (1981). Ptychopariid trilobites in the Lower Cambrian of Scandinavia. *United States, Department of the Interior, Geological Survey, Open-File Report*, **81–743**, 5–7.

Ahlberg, P. & Bergström, J. (1978). Lower Cambrian ptychopariid trilobites from Scandinavia. *Sveriges Geologiska Undersökning*, Ca, **49**, 1–41.

Åhman, E. & Martinsson, A. (1965). Fossiliferous Lower Cambrian at Aspelund on the Skäggenäs Peninsula. *Geologiska Föreningens i Stockholm Förhandlingar*, **87**, 139–51.

Allen, P.M. & Jackson, A.A. (1978). Bryn-teg Borehole, North Wales. *Bulletin of the Geological Survey of Great Britain*, **61**, 1–48.

Allen, P.M., Jackson, A.A. & Rushton, A.W.A. (1981). The stratigraphy of the Mawddach Group in the Cambrian succession of North Wales. *Proceedings of the Yorkshire Geological Society*, **43**, 295–330.

Ambrose, G.J., Flint, R.B. & Webb, A.W. (1981). Precambrian and Palaeozoic geology of the Peake and Denison Ranges. *Geological Survey of South Australia, Bulletin*, **50**, 71pp.

Anderson, M.M. (1981). The Random Formation of southeastern Newfoundland: a discussion aimed at establishing its age and relationship to bounding formations. *American Journal of Science*, **281**, 807–30.

Aren, B., Jaworowski, K., Juskowiakowa, M., Lendzion, K. & Wichrowska, M. (1979). The Vendian and Lower Cambrian in the Polish part of the East-European Platform. *Biuletyn Instytutu geologicznego*, **318**, 43–57.

Arén, B. & Lendzion, K. (1974). Organic remains at the Vendian-Cambrian Boundary in the platform sediments in Poland. *Bulletin de l'Académie Polonaise des Sciences, Série des Sciences de la Terre*, **22**, 49–53.

Azmi, R.J., Joshi, M.N. & Juyal, K.P. (1981). Discovery of Cambro-Ordovician conodonts from the Mussoorie Tal phosphorite: its significance in correlation of the Lesser Himalaya. In *Contemporary geoscientific researches in Himalaya*, 1, ed. A.K. Sinha, pp. 245–50.

Banerjee, D.M. (1971a). Precambrian stromatolitic phosphorites of Udaipur, Rajasthan, India. *Bulletin of the Geological Society of America*, **82**, 2319–30.

Banerjee, D.M. (1971b). Aravallian stromatolites from Udaipur, Rajasthan. *Journal of the Geological Society of India*, **12(4)**, 349–55.

Bassett, D.A. & Walton, E.K. (1960). The Hell's Mouth Grits: Cambrian greywackes in St Tudwal's Peninsula, North Wales. *Quarterly Journal of the Geological Society of London*, **116**, 85–110.

Bassett, M.G., Owens, R.M. & Rushton, A.W.A. (1976). Lower Cambrian fossils from the Hell's Mouth Grits, St Tudwal's Peninsula, North Wales. *Journal of the Geological Society*, **132**, 623–44.

Bengtson, S. (1968). The problematic genus *Mobergella* from the Lower Cambrian of the Baltic area. *Lethaia*, **1**, 325–51.

Bengtson, S. (1970). The Lower Cambrian fossil *Tommotia*. *Lethaia*, **3**, 363–92.

Bengtson, S. (1976). The structure of some Middle Cambrian conodonts, and the early evolution of conodont structure and function. *Lethaia*, **9**, 185–206.

Bengtson, S. (1977). Aspects of problematic fossils in the early Palaeozoic. *Acta Universitatis Upsaliensis*, **415**, 1–71.

Bengtson, S. (1980). Redescription of the Lower Cambrian *Lapworthella cornu*. *Geologiska Föreningens i Stockholm Förhandlingar*, **102**, 53–5.

Bengtson, S. & Fletcher, T.P. (1981). The succession of skeletal fossils in the basal Lower Cambrian of southeastern Newfoundland. *United States Department of the Interior, Geological Survey, Open-File Report*, **81–743**, 18.

Bergström, J. (1973). Classification of olenellid trilobites and some Balto-Scandian species. *Norsk Geologisk Tidsskrift*, **53**, 283–314.

Bergström, J. (1980). The Caledonian margin of the Fennoscandian Shield during the Cambrian. *Memoir of the Department of Geological Sciences, Virginia Polytechnic Institute and State University*, **2**, 9–13.

Bergström, J. (1981). Lower Cambrian shelly faunas and biostratigraphy in Scandinavia. *United States Department of the Interior, Geological Survey, Open-File Report*, **81–743**, 22–5.

Bergström, J. & Ahlberg, P. (1981). Uppermost Lower Cambrian biostratigraphy in Scania, Sweden. *Geologiska Föreningens i Stockholm Förhandlingar*, **103**, 193–214.

Bertrand-Sarfati, J. (1981). Problème de la limite Précambrien–Cambrien: la section de Tiout (Maroc); les stromatolites et leur biostratigraphie (Schmitt 1979): critiques et observations. *Newsletters in Stratigraphy*, **10(1)**, 20–6.

Bischoff, G.C.O. (1976). *Dailyatia*, a new genus of the Tommotiidae from Cambrian strata of SE Australia (Crustacea, Cirripedia). *Senckenbergiana Lethaea*, **57(1)**, 1–33.

Bjamba, Z. (1973). Stratigraphy of the Riphean–Cambrian formations in north Mongolia. *Kwartalnik Geologiczny*, **17(4)**, 853–75 (in Polish), 876 (in Russian), 877 (In English).

Bjørlykke, K., Elvsborg, A. & Høy, T. (1976). Late Precambrian sedimentation in the central sparagmite basin of south Norway. *Norsk Geologisk Tidsskrift*, **56**, 233–90.

Blissett, A.H. & Callen, R.A. (1967). Myponga phosphate deposit. *Mineral Resources Review of South Australia*, **127**, 93–102.

Borisov, V.A. & Rozanov, A.Yu. (1964). New data on the biostratigraphy of the old strata of the Batenev Ridge. *Doklady Akademiya Nauk SSSR*, **158(2)**, 342–4.

Borovikov, L.I. & Kryskov, L.N. (1963). Cambrian deposits of the Kendyktas Mountains (southern Kazakhstan). *Trudy Vsesoyuznogo Nauchno-issledovatelyskogo Geologicheskogo Instituta (VSEGEI)*, **94**, 266–80, pl. 1. Gosudarstvenny, Leningrad (in Russian).

Borrello, A.V. (1971). The Cambrian of South America. In *Lower Palaeozoic Rocks of the World. 1, Cambrian of the New World*. ed. C.H. Holland, pp. 385–438. John Wiley, London.

Boudda, A., Choubert, G. & Faure-Muret, A. (1979). Essai de Stratigraphie de la Couverture sédimentaire de l'Anti-Atlas: Adoudounien–Cambrien inférieur. *Notes et Mémoires du Service géologique du Maroc*, **271**, 96pp.

Brangulis, A.P., Kirjyanov, V.V., Mens, K.A. & Rozanov, A.Yu. (1981). Vendian and Cambrian paleogeography of the East European Platform. *United States Department of the Interior, Geological Survey, Open-File Report*, **81–743**, 26–8.

Brasier, M.D. (1980). The Lower Cambrian transgression and glauconite-phosphate facies in western Europe. *Journal of the Geological Society*, **137**, 695–703.

Brasier, M.D. (1984). Microfossils and small shelly fossils from the Lower Cambrian *Hyolithes* Limestone at Nuneaton, English Midlands. *Geological Magazine*, **121(3)**, 229–53.

Brasier, M.D. & Hewitt, R.A. (1979). Environmental setting of fossiliferous rocks from the uppermost Proterozoic–Lower Cambrian of central England. *Palaeogeography, Palaeoclimatology, Palaeoecology*, **27**, 35–57.

Brasier, N.D. & Hewitt, R.A. (1981). Faunal sequence within the Lower Cambrian 'Non-trilobite Zone' (SL) of central England and correlated regions. *United States Department of the Interior, Geological Survey, Open-File Report*, **81–743**, 26–8.

Brasier, M.D., Hewitt, R.A. & Brasier, C.J. (1978). On the late Precambrian–Early Cambrian Hartshill Formation of Warwickshire. *Geological Magazine*, **115**, 21–36.

British Sulphur Corporation Ltd (1971). *World Survey of Phosphate Deposits*, 3rd edn, 180pp. British Sulphur Corporation Ltd, London.

Bushinsky, G.I. (1966). Old phosphorites of Asia and their genesis. *Akademiya Nauk SSSR, Trudy Geologichesky Institut*, **149** (In Translation by Israel Programme for Scientific Translations, 266pp. Jerusalem, 1969).

Callen, R.A. (1968). Oliver's phosphate deposits. *Mineral Resources Review of South Australia*, **128**, 127–42.

Callen, R.A. (1969a). Sedimentary phosphate exploration, part 1. The Cambrian of the Flinders Ranges. *Mineral Resources Review of South Australia*, **130**, 9–34.

Callen, R.A. (1969b). Sedimentary phosphate exploration, part 2. The Cambrian and Proterozoic of Fleurieu Peninsula, the Mount Lofty Ranges, and Yorke Peninsula. *Mineral Resources Review of South Australia*, **130**, 80–94.

Callen, R.A. (1969c). Wait's Noarlunga phosphate deposits. *Mineral Resources Review of South Australia*, **130**, 110–20.

Campana, B. & Wilson, B. (1954). *Jervis map sheet, geological atlas of South Australia, 1:63 360 series*. Geological Survey of South Australia, Adelaide.

Cao Ruiji & Zhao Wenjie (1978). The algal flora of the Tongying Formation (Upper Sinian System) in southwestern China. *Memoirs of the Nanjing Institute of Geology and Palaeontology, Academia Sinica*, **1978(10)**, 1–27 (in Chinese), 28 (in English).

Carter, E.K. & Öpik, A.A. (1961). Lawn Hill–4 mile geological series. Sheet E/54–9. *Explanatory Notes, Bureau of Mineral Resources Geology and Geophysics of Australia*, **21**, 17pp.

Cathcart, J.B. (1979). Phosphorite deposits near Patos de Minas, Minas Gerais, Brazil. *Transactions of the Society of Mining Engineers of AIME*, **264**, 1703–8.

Cecile, M.P. & Norford, B.S. (1979). Basin to platform transition,

Lower Paleozoic strata of Ware and Trutch map areas, northeastern British Columbia. *Current Research, part A, Geological Survey of Canada, Paper*, **79-1A**, 219–26.

Chang, W.T. (1980). A review of the Cambrian of China. *Journal of the Geological Society of Australia*, **27**, 137–50.

Chauhan, D.S. (1973). Stromatolites from the Precambrian phosphorite bearing stratum of the Aravalli region of Udaipur, Rajasthan (India). *Doklady Akademiya Nauk SSSR*, **208**, 1429–31.

Chauhan, D.S. (1979). Phosphate-bearing stromatolites of the Precambrian Aravalli phosphorite deposits of the Udaipur region, their environmental significance and genesis of phosphorite. *Precambrian Research*, **8(1/2)**, 95–126.

Chen Meng'e, Chen Yiyuan & Zhang Shusen (1981). Small shell fossil assemblage in the limestone of the uppermost part of the Dengying Formation at Songlingpo, Yichang. *Earth Sciences*, **1981(1)**, 32–9 (in Chinese), 40–1 (in English).

Chiang Nen-jen, Wang Tsun-chou & Chen Yung-kuang (1964). Cambrian stratigraphy of eastern Yunnan. *Acta Geologica Sinica*, **44(2)**, 137–55 (Translation of *International Geology Review*, **10(10)**, 1153–72).

Christie, R.L. (1979). Phosphorite in sedimentary basins of western Canada. *Current Research, part B, Geological Survey of Canada, Paper* **79-1B**, 253–8.

Chu Chao-ling (1960). Middle Cambrian trilobites from Datong, Qinghai Province. *Geological Gazetteer of the Chi-lien Mountains*, **4(1)**, 69–81 (in Chinese).

Cloud, P.E. & Semikhatov, M.A. (1969). Proterozoic stromatolite zonation. *American Journal of Science*, **267(9)**, 1017–61.

Cobbold, E.S. (1911). Trilobites from the Paradoxides beds of Comley (Shropshire). *Quarterly Journal of the Geological Society of London*, **67**, 282–311.

Cobbold, E.S. (1927). The stratigraphy and geological structure of the Cambrian area of Comley (Shropshire). *Quarterly Journal of the Geological Society of London*, **83**, 551–73.

Cobbold, E.S. & Pocock, R.W. (1934). The Cambrian area of Rushton (Shropshire). *Philosophical Transactions of the Royal Society*, Ser. B, **223**, 305–409.

Cook, H.G. & Taylor, M.E. (1975). Early Paleozoic continental margin sedimentation, trilobite biofacies, and the thermocline, western United States. *Geology*, **3(10)**, 559–62.

Cook, P.J. (1972). Petrology and geochemistry of the phosphate deposits of northwest Queensland, Australia. *Economic Geology*, **67**, 1193–213.

Cook, P.J. (1976). Georgina Basin phosphatic province, Queensland and Northern Territory – regional geology. In *Economic geology of Australia and Papua New Guinea, Vol. 4, Industrial Minerals and rocks*. ed. C.L. Knight, pp. 45–50. Australasian Institute of Mining and Metallurgy, Melbourne.

Cook, P.J. & Shergold, J.H. (Eds.) (1979). *Proterozoic–Cambrian Phosphorites*. 106pp. ANU Press Canberra.

Cowie, J.W. (1971). The Cambrian of the North American Arctic regions. In *Cambrian of the New World*, ed. C.H. Holland, pp. 325–84. John Wiley, London.

Cowie, J.W. (1974). The Cambrian of Spitsbergen and Scotland. In *Cambrian of the British Isles, Norden and Spitsbergen*, ed. C.H. Holland, pp. 123–56. John Wiley, London.

Cowie, J.W. (1978). IUGS/IGCP Project 29. Precambrian–Cambrian Boundary Working Group in Cambridge, 1978. *Geological Magazine*, **115(2)**, 151–2.

Cowie, J.W. & Adams, P.J. (1957). The geology of the Cambro-Ordovician rocks of central East Greenland. Part I. Stratigraphy and structure. *Meddelelser φm Grφnland*, **153(1)**, 1–150.

Cowie, J.W. & Rozanov, A.Yu. (1973). Report of the international working group on a symposium on the problems of the Cambrian–Precambrian boundary. *Izvestiya Akademiya Nauk Geologichesky Seriya*, **1973(12)**, 72–82 (in Russian).

Cowie, J.W., Rushton, A.W.A. & Stubblefield, C.J. (1972). Cambrian. *Special Report of the Geological Society*, **2**, 42pp.

Cowie, J.W. & Spencer, A.M. (1970). Trace fossils from the late Precambrian–Lower Cambrian of East Greenland. *Special Issue of the Geological Journal*, **3**, 91–100.

Crawford, A.R. (1970). The Precambrian geochronology of Rajasthan and Bundelkhand, northern India. *Canadian Journal of Earth Science*, **7**, 91–110.

Daily, B. (1956). The Cambrian of South Australia. In *El Sistema Cámbrico, su Paleogeografia y el Problema de su Base, 20th Session International Geological Congress Mexico*, **2**, 91–147.

Daily, B. (1963). The fossiliferous Cambrian succession on Fleurieu Peninsula, South Australia. *Records of the South Australian Museum*, **14(3)**, 579–601.

Daily, B. (1967a). Stansbury West No 1 and Edithburgh No 1 wells – subsurface stratigraphy and palaeontology of the Cambrian sequence. Beach Petroleum N.L. (unpublished report).

Daily, B. (1967b). Appendix in Delhi-Santos Kalladeina No 1 well completion report. South Australian Department of Mines, Open-file Envelope 813 (unpublished).

Daily, B. (1972). The base of the Cambrian and the first Cambrian faunas. *University of Adelaide Centre for Precambrian Research, Special Paper* **1**, 13–40.

Daily, B. (1976). The Cambrian of the Flinders Ranges. In *Late Precambrian and Cambrian geology of the Adelaide 'Geosyncline' and Stuart Shelf, South Australia, 25th Session of the International Geological Congress*. ed. B.P. Thomson, B. Daily, R.P. Coats & B.G. Forbes, Excursion Guide 33A, pp. 15–8.

Daily, B. (1982). Cambrian of the Flinders Ranges. In *4th International Symposium on Antarctic Earth Sciences*, Excursion Guide A, pp. 41–50, University of Adelaide.

Daily, B., Jago, J.B. & James, P.R. (1982). Lower Cambrian sediments, Precambrian-Cambrian Boundary and Delamerian tectonics of southern Fleurieu Peninsula. In *Geology of the Adelaide region. 4th International Symposium on Antarctic Earth Sciences*, ed. R.L. Oliver & C.G. Gatehouse, Excursion Guide B3, pp. 30–41. University of Adelaide.

Daily, B. & Milnes, A.R. (1971). Stratigraphic notes on Lower Cambrian fossiliferous metasediments between Campbell Creek and Tunkalilla Beach in the type section of the Kanmantoo Group, Fleurieu Peninsula, South Australia. *Transactions of the Royal Society of South Australia*, **95**, 199–214.

Daily, B. & Milnes, A.R. (1972). Revision of the stratigraphic nomenclature of the Cambrian Kanmantoo Group, South Australia. *Journal of the Geological Society of Australia*, **19(2)**, 197–202.

Debrenne, F. (1964). Archaeocyatha. Contribution à l'étude des faunes cambriennes du Maroc de Sardaigne et de France. *Notes et Mémoires du Service géologique du Maroc*, **179**, 1–371.

Debrenne, F. (1969). Lower Cambrian Archaeocyatha from the Ajax Mine, Beltana, South Australia. *Bulletin, British Museum of Natural History (Geology)*, **17**, 297–376.

Debrenne, F. (1970). A revision of Australian genera of Archaeocyatha. *Transactions of the Royal Society of South Australia*, **94**, 21–48,.

Debrenne, F., Orgeval, J.-J & Verraes, G. (1976). Présence d'Archéocyathes dans le substratum carbonate de lamine des Malines (Gard, France). *Compte rendu Société géologique de France*, **1976, 6**, 259–61.

de Keyser, F. (1968). The Cambrian of the Burke River outlier. *Record of the Bureau of Mineral Resources, Geology and Geophysics of Australia*, **1968/67.**

de Keyser, F. (1969). The phosphate-bearing Cambrian formations in the Lawn Hill and Lady Annie districts, northwestern Queensland. *Record of the Bureau of Mineral Resources, Geology and Geophysics of Australia*, **1969/147**.

de Keyser, F. & Cook, P.J. (1972). Geology of the Middle Cambrian phosphorites and associated sediments in northwestern Queensland. *Bulletin of the Bureau of Mineral Resources, Geology and Geophysics of Australia*, **138**, 79pp.

Doré, F. (1972). La transgression majeure du Paléozoïque inférieur dans le Nord-Est du massif Armoricain. *Bulletin de la Société géologique de France*, 7th series, **14**, 79–93.

Downie, C. (1974). Acritarchs from near the Precambrian-Cambrian boundary–a preliminary report. *Review of Palaeobotany and Palynology*, **18**, 57–60.

Downie, C. (1975). Precambrian of the British Isles – Palaeontology. *Precambrian Special Report of the Geological Society*, **6**, 113–15.

Dragunov, V.I., Librovich, V.L., Matukhina, V.G. & Nemtsov, V.D. (1966). Phosphatic sediments in the Lower Palaeozoic deposits of the northwestern part of the middle Siberian Platform. *Trudy Vsesoyuznogo Nauchno-issledovatelyskogo Geologicheskogo Instituta (VSEGEI)*, **146**, 95–102. Gosudarstvenny, Leningrad (in Russian).

Ducrôt, J. & Lancelot, J.R. (1977). Problème de la limite Précambrien–Cambrien: étude radiochronologique par la méthode U-Pb sur zircons du volcan du Jbel Boho (Anti-Atlas marocain). *Canadian Journal of Earth Sciences*, **14(12)**, 2771–7.

Edhorn, A.-St & Anderson, M.M. (1977). Algal remains in the Lower Cambrian Bonavista Formation, Conception Bay, southeastern Newfoundland. In *Fossil Algae*, ed. E. Flügel, pp. 113–23. Springer Verlag, Berlin.

Eganov, E.A. (1979). The role of cyclic sedimentation in the formation of phosphorite deposits. In *Proterozoic-Cambrian Phosphorites*, ed. P.J. Cook & J.H. Shergold, pp. 22–25. Australian National University Press, Canberra.

Eganov, E.A. & Sovetov, Yu.K. (1979). Karatau – a model for phosphorite deposition. *Akademiya Nauk SSSR, Sibirskoe Otdelenie, Trudy Instituta Geologii i Geofiziki*, **427**, 192pp. Izdatelystvo 'Nauka', Novosibirsk (in Russian).

Ergaliev, G.K. (1974). Stratigraphic scheme for the Cambrian deposits of southern and western central Kazakhstan. In *Pre-palaeozoic and Palaeozoic of Kazakhstan, 1. Stratigraphy of the pre-Palaeozoic, Cambrian, Ordovician and Silurian of Kazakhstan*, pp. 112–15. Akademiya Nauk Kazakhskoy SSR, Ordena Trudovogo Krasnogo Zhameni Institut Geologicheskikh Nauk K.I. Satpaeva. Izdatelystvo 'Nauka' Kazakhskoy SSR, Alma-Ata (in Russian).

Ergaliev, G.K. & Pokrovskaya, N.V. (1977). *Lower Cambrian trilobites of Lesser Karatau (southern Kazakhstan)*. Akademiya Nauk Kazakhskoy SSR, Ordena Trudovogo Krasnogo Zhameni Institut Geologicheskikh Nauk K.I. Satpaeva, 98pp. Izdatelystvo 'Nauka' Kazakhskoy SSR, Alma-Ata (in Russian).

Fedonkin, M. (1977). Precambrian-Cambrian ichnocoenoses of the east European platform. *Special Issue of the Geological Journal*, **9**, 183–94.

Fleming, P.J.G. (1973). Bradoriids from the *Xystridura* Zone of the Georgina Basin, Queensland. *Publications of the Geological Survey of Queensland, Palaeontological Paper*, **31**, 9pp.

Fleming, P.J.G. (1977). Faunas, lithologies, and the origin of phosphorites in parts of the Middle Cambrian Beetle Creek Formation of northwest Queensland. *Publications of the Geological Survey of Queensland*, **364**, 21pp.

Føyn, S. & Glaessner, M.F. (1979). *Platysolenites*, other animal fossils, and the Precambrian–Cambrian transition in Norway. *Norsk Geologisk Tidsskrift*, **59**, 25–46.

Fritz, W.H. (1972). Lower Cambrian trilobites from the Sekwi Formation type section, Mackenzie Mountains, Northwestern Canada. *Bulletin of the Geological Survey of Canada*, **212**, 1–58.

Fuchs, G. & Mostler, H. (1972). Der erste Nachweis von Fossilien (Kambrischen Alters) in der Hazira Formation, Hazara, Pakistan. *Mitteilungen der Geologischen Paläontologischen Gesellschaft Innsbruck*, **2**, 1–12.

George, R.J. (1975). Notes on the geology of the Duchess Embayment. In *Field Conference Mt Isa Region 1975*, pp. 16–9. Geological Society of Australia (Queensland Division). Brisbane.

Goldschmidt, V.M. (1925). Über fossilfürende untercambrische basalblagerungen dei Ustaoset. *Fennia*, **45**, 3–11.

Grönwall, K.A. (1902). Bornholms Paradoxideslag og deres Fauna. *Danmarks Geologiske Undersøgelse*, 2nd series, **13**, 231pp.

Hadding, A. (1927). The Pre-Quaternary sedimentary rocks of Sweden. I. A survey of the Pre-Quaternary sedimentary rocks of Sweden. II. The Paleozoic and Mesozoic conglomerates of Sweden. *Lunds Universitets Årsskrift. N.F. Avd. 2*, **23(5)**; *Kungliga Fysiografiska Sällskapets Handlingar. N.F.*, **38(5)**, 171pp.

Hadding, A. (1929). The Pre-Quaternary sedimentary rocks of Sweden. III. The Paleozoic and Mesozoic sandstones of Sweden. *Lunds Universitets Årsskrift. N.F. Avd. 2*, **25(3)**; *Kungliga Fysiografiska Sällskapets Handlingar. N.F.*, **40(3)**, 287pp.

Hadding, A. (1932). The Pre-Quaternary sedimentary rocks of Sweden. IV. Glauconite and glauconitic rocks. *Lunds Universitets Årsskrift. N.F. Avd 2*, **28(2)**: *Kungliga Fysiografiska Sällskapets Handlingar. N.F.*, **43(2)**, 175pp.

Hadding, A. (1958). The Pre-Quaternary sedimentary rocks of Sweden. VII. Cambrian and Ordovician limestones. *Lunds Universitets Årsskrift. N.F. Avd 2*, **54(5)**; *Kungliga Fysiografiska Sällskapets Handlingar. N.F.*, **63(5)**, 262pp.

Haller, J. (1971). *Geology of the East Greenland Caledonides*, 381pp. John Wiley, London.

Henderson, R.A. & Southgate, P.N. (1978). Middle Cambrian stratigraphy and sedimentology of the eastern Georgina Basin. In *Excursions Handbook, 3rd Australian Geological Convention, Townsville*, pp. 85–110.

Henningsmoen, G. (1957). The trilobite family Olenidae with description of Norwegian material and remarks of the Olenid and Tremadocian Series. *Skrifter utgilt av Det Norske Videnskaps-Akademie Oslo, 1. Mat-naturv. Klasse*, **1**, 303pp.

Henriksen, N. & Higgins, A.K. (1976). East Greenland Caledonian Fold Belt. In *Geology of Greenland*, ed. A. Escher & W. Watt, pp. 182–246 Geological Survey of Greenland Copenhagen.

Hicks, H. (1875). On the occurrence of phosphates in the Cambrian rocks. *Quarterly Journal of the Geological Society of London*, **31**, 368–83.

Hill, D., Playford, G. & Woods, J.T. (Eds) (1971). *Cambrian fossils of Queensland.* 32pp. Queensland Palaeontographical Society, Brisbane.

Holm, G. (1893). Sveriges kambrisk-siluriska Hyolithidae och Conularidae. *Sveriges Geologiska Undersökning*, C, **112**, 172pp.

Horwitz, R., Thomson, B.P. & Webb, B.P. (1959). The Cambrian–Precambrian boundary in the eastern Mount Lofty Ranges region, South Australia. *Transactions of the Royal Society of South Australia*, **82**, 205–18.

Howard, P.F. (1971). Discovery of phosphorite in northern Australia. *Transactions of the Society of Mining Engineers of AIME*, **250**, 269–73.

Howard, P.F. (1972). Explanation for phosphorite in Australia – a case history. *Economic Geology*, **67**, 1180–92.

Howard, P.F. & Cooney, A.M. (1976). D Tree phosphate deposit, Georgina Basin, Queensland. In *Economic geology of Australia and Papua New Guinea, 4. Industrial minerals and rocks*, ed. C.L. Knight, pp. 265–73. Australasian Institute of Mining and Metallurgy, Melbourne.

Howard, P.F. & Perrino, F.A. (1976). Wonarah phosphate deposit, Georgina Basin, Northern Territory. In *Economic geology of Australia and Papua New Guinea, 4. Industrial minerals and rocks.* ed. C.L. Knight pp. 273–7. Australasian Institute of Mining and Metallurgy, Melbourne.

Howell, B.F. (1925). The faunas of the Cambrian Paradoxides beds at Manuels, Newfoundland. *Bulletin of American Paleontology*, **11(43)**, 9–140.

Hutchinson, R.D. (1952). The stratigraphy and trilobite faunas of the Cambrian sedimentary rocks of Cape Breton Island, Nova Scotia. *Memoir of the Geological Survey of Canada*, **263**, 124pp.

Hutchinson, R.D. (1962). Cambrian stratigraphy and trilobite faunas of southeastern Newfoundland. *Bulletin of the Geological Survey of Canada*, **88**, 148pp.

Illing, V.C. (1916). The Paradoxidian fauna of a part of the Stockingford Shales. *Quarterly Journal of the Geological Society of London*, **71**, 386–450.

Ilyin, A.V. (1973). Khubsugul phosphate-bearing basin. *Sovmestnaya Sovetsko – Mongoliskaya Nauchno-issledovateliskaya Geologicheskaya Ekspeditisya, Trudy* **6**, 167pp. Izdatelystvo 'Nauka' Moscow (in Russian).

Ilyin, A.V. (1979). Late Precambrian-Cambrian phosphorites. In *Interregional Seminar on the development potential of Precambrian mineral deposits.* UN Centre for Natural Resources, Energy and Transport, 24pp.

Ilyin, A.V., & Bjamba, Z. (1980). *Guidebook for excursion to the phosphorites of the Khubsugul Basin, Mongolian People's Republic.* Akademiya Nauk SSSR, Sovetsky Komitet po MPGK Proekt 156. Fosfority. 135pp. Geologichesky Institut, Akademiya Nauk SSSR, Moscow (Russian/English).

Ilyin, A.V. & Ratnikova, G.I. (1981). Primary bedded, structureless phosphorite of the Khubsugul Basin, Mongolia. *Journal of Sedimentary Petrology*, **51(4)**, 1215–22.

Jell, P.A. (1970). *Pagetia ocellata*, a new Cambrian trilobite from northwestern Queensland. *Memoirs of the Queensland Museum*, **15**, 303–13.

Jell, P.A. (1975). Australian Middle Cambrian Eodiscoids with a review of the Superfamily. *Palaeontographica* **(A)150**, 97pp.

Johns, R.K. (1961). South Australian rock phosphate deposits. *Mineral Resources Review of South Australia*, **114**, 22–30.

Johns, R.K. (1976). Phosphate-South Australia. In *Economic geology of Australia and Papua New Guinea, 4. Industrial Minerals and rocks.* ed. C.L. Knight, pp. 282–285. Australasian Institute of Mining and Metallurgy, Melbourne.

Jones, P.J. & McKenzie, K.G. (1980). Queensland Middle Cambrian Bradoriida (Crustacean): new taxa, palaeobiogeography and biological affinities. *Alcheringa*, **4(3)**, 203–25.

Kanwar, S.S. & Ahluwalia, A.D. (1980). A note on the occurrence of stromatolites in Member A of Krol Formation, Narag area, district Sirmur, Himachal Pradesh. *Geological Survey of India Miscellaneous Publications*, **44**, 75–9.

Kazarinov, V.P. & Krasilnikova, N.A. (Eds) (1980). *Phosphorites and apatites of Siberia.* Akademiya Nauk SSSR, Sibirskoe Otdelenie, Institut Geologii i Geofiziki, 232pp. Izdatelistvo 'Nauka', Sibirskoe Otdelenie Novosibirsk (in Russian).

Khomentovsky, V.V. & Repina, L.N. (1965). *Lower Cambrian stratotype sections of Siberia.* Akademiya Nauk SSSR, Sibirskoe Otdelenie. Trudy Instituta Geologii i Geofiziki, 199pp. Izdatelistvo 'Nauka', Moscow (in Russian).

Kiaer, J. (1916). The Lower Cambrian Holmia fauna at Tφmten in Norway. *Skrifter utgilt av Det Norske Videnskaps-Akademie Oslo, 1. Mat-Naturv. Klasse*, **10**, 140pp.

King, W.B.R. (1940). The Cambrian fauna of the Salt Range of India. *Records of the Geological Survey of India*, **75**, *Professional Paper*, **9**, 1–15.

Knoll, A.H. & & Vidal, G. (1980). Late Proterozoic vase-shaped microfossils from the Visingsö Beds, Sweden. *Geologiska Föreningens i Stockholm Förhandlingar*, **102**, 207–11.

Korobov, M.N. (1980a). Biostratigraphy and Miomeran trilobites of the Lower Cambrian of Mongolia. *Sovmestnaya Sovetsko-Mongoliskaya Nauchno – issledovateliskaya Geologicheskaya Expeditsiya, Trudy* **26**, 108pp. Izdatelystvo 'Nauka' Moscow (in Russian).

Korobov, M.N. (1980b). Lower Cambrian trilobites and the age of the phosphorites of western Prekhubsugul (Mongolia). *Izvestiya Akademiya Nauk SSSR, Seriya Geologicheskaya*, **1980(9)**, 97–107 (in Russian).

Korobov, M.N. & Missarzhevsky, V.V. (1977). On the Cambrian and Precambrian boundary beds in western Mongolia (Khasaght-Khairkhan Ridge). In *Palaeozoic Invertebrates of Mongolia*, ed. L.P. Tamarinov, pp. 7–9. Izdatelystvo 'Nauka' Moscow (in Russian).

Kozlowski, R. (1931). Fosforyty w utworach kambryjskich Sandomierza. *Sprawozdiana Polskiego Instytutu Geologicznego*, **6**, 752–6.

Kruse, P.D. & West, P.W. (1980). Archaeocyatha of the Amadeus and Georgina Basins. *Bureau of Mineral Resources Journal of Australian Geology & Geophysics*, **5(3)**, 165–81.

Lapworth, C. (1898). A sketch of the geology of the Birmingham district. *Proceedings of the Geologists Association*, **15**, 313–416.

Lendzion, K. (1968). The Cambrian of the East-European Platform in Poland. *23rd Session International Geological Congress, Prague*, **9**, 109–17.

Lendzion, K. (1972). Kambr subholmiowy w połnocno-wschodniej Polsce. *Kwartalnik Geologiczny*, **16**, 557–68.

Liang, T.Y. & Chang, A.C. (1981). *On the characteristics and genesis of late Pre-Cambrian phosphorites associated with Gondwanaland.* Geological Institute for Mining and Chemical Industrial Minerals, Zhuo Xian, Hebei. 68pp.

Lindström, M. (1977). White sand crosses in turbated siltstone bed, basal Cambrian, Kinnekulle, Sweden. *Geologica et Palaeontologica*, **11**, 1–8.

Liu Di-yong (1979). Earliest Cambrian brachiopods from southwest China. *Acta palaeontologia Sinica*, **18(5)**, 505–11 (in Chinese), 511 (in English).

Lu Yan-hao (1979). *Cambrian mineral deposits in China and the bioenvironmental control theory.* 77pp. Geological Publishing House, Peking (in Chinese).

Ludwig, V. (1969). Lithologische Untersuchung des Kambriums im Frankenwald (Bayern). *Geologisches Jahrbuch*, **87**, 89–156.

Luo Huilin & Jiang Zhiwen (1982). On the Sinian-Cambrian boundary in the Sichuan-Yunnan region. *Scientia Geologica Sinica*, **1982(2)**, 215–9 (in Chinese), 219 (in English).

Luo Huilin, Jiang Zhiwen, Wu Xiche, Song Xueliang & Ouyang Lin, (1982). *The Sinian-Cambrian boundary in Eastern Yunnan, China.* Institute of Geological Sciences, Yunnan. 265pp. (in Chinese).

Luo Huilin, Jiang Zhiwen, Xu Zhonjiu, Song Xuiliang, & Xue Xiaofeng (1980). On the Sinian-Cambrian boundary of Meishucun and Wangjiawan. Jinning County, Yunnan. *Acta Geologica Sinica*, **1980(2)**, 95–109 (in Chinese), 110 (English).

Mambetov, A.M. & Missarzhevsky, V.V. (1971). New data on the significance of the phosphorite-bearing thickness of Lesser Karatau. In *Stratigraphy of the Precambrian of Kazakhstan and Tien-shan*, pp. 217–21. Izdatelystvo Moskovskogo Gosudarstvennogo Universiteta, Moscow.

Mambetov, A.M. & Repina, L.N. (1979). The Lower Cambrian of Talass Ala-Too and its correlation with the Lesser Karatau region and the Siberian Platform. *Akademiya Nauk SSSR, ibirskoe Otdelenie, Trudy Instituta Geologii i Geofiziki*, **406**, 98–138. Izdatelystvo 'Nauka' Sibirskoe Otdelenie Novosibirsk.

Manum, S. (1967). Microfossils from the late Precambrian sediments around Lake Mjφsa. *Norges Geologiske Undersokelse*, **251**, 45–52.

Mardla, A.K., Mens, K.A., Kala, Z.A., Kayak, K.F. & Zricalu, Z. (1968). To the stratigraphy of the Cambrian deposits of Estonia. In *Stratigrafiya nizhnego paleozoya Pribaltiki i korrelyatsiya s drugimi regionami*, pp. 22–32. Mintis (in Russian).

Marino, A., de (1980). The upper Lower Cambrian strata south of Simrishamn, Scania, Sweden. *Sveriges Geologiska Undersökning*, C, **771**, 1–22.

Martinsson, A. (1974). The Cambrian of Norden. In *Cambrian of the British Isles. Norden and Spitsbergen*, ed. C.H. Holland, C.H. pp. 185–283. John Wiley, London.

Matthew, G.F. (1889). On Cambrian organisms in Acadia. *Transactions of the Royal Society of Canada, Series 2*, **7**, 139.

Matthews, S.C. & Missarzhevsky, V.V. (1975). Small shelly fossils of late Precambrian and early Cambrian age: a review of recent work. *Journal of the Geological Society of London*, **131**, 289–304.

Mens, K.A. (1981). The Lower and Middle Cambrian boundary in the eastern Baltic area. *United States Department of the Interior, Geological Survey, Open File Report*, **81–743**, 130–1.

Mens, K.A. & Pirrus, E.A. (1975). The basal conglomerate of the Lower Cambrian Uukati Beds in Estonia. In *Geologiya kristallicheskogo fundamenta i osadochnogo chekhla pribaltiki*, pp. 132–43. Zinatne (in Russian).

Mens, K.A. & Pirrus, E.A. (1981). Sedimentological characteristics of the boundaries between the Lower Cambrian stages in the eastern Baltic area. *United States Department of the Interior, Geological Survey, Open File Report*, **81–743**, 132–3.

Missarzhevsky, V.V. & Mambetov, A.M. (1981). Stratigraphy and fauna of the boundary beds between the Cambrian and Precambrian of Lesser Karatau. *Akademiya Nauk SSSR, Trudy Geologicheskiy Institut*, **326**, 90pp. Izdatelystvo 'Nauka' Moscow (in Russian).

Müller, K.J. (1979). Phosphatocopine ostracodes with preserved appendages from the Upper Cambrian of Sweden. *Lethaia*, **12**, 1–27.

Nicholas, T.C. (1915). The Geology of the St Tudwal's Peninsula. *Quarterly Journal of the Geological Society of London*, **71**, 83–143.

Notholt, A.J.G. (1979). Resources of Precambrian and Cambrian sedimentary phosphate rock. In *Proterozoic Cambrian Phosphorites*, ed. P.J. Cook & J.H. Shergold, pp. 71–86. ANU Press, Canbera.

Öpik, A.A. (1956). Cambrian (Lower Cambrian) of Estonia. *20th Session International Geological Congress, Mexico*, **1**, 97–126.

Öpik, A.A. (1961). The geology and palaeontology of the headwaters of the Burke River, Queensland. *Bulletin of the Bureau of Mineral Resources, Geology and Geophysics of Australia*, **53**, 249pp.

Öpik, A.A. (1967). The Mindyallan fauna of northwestern Queensland. *Bulletin of the Bureau of Mineral Resources, Geology and Geophysics of Australia*, **74**, vol. 1, 404pp., vol. 2, 167pp.

Öpik, A.A. (1968). The Ordian Stage of the Cambrian and its Australian Metadoxididae. *Bulletin of the Bureau of Mineral Resources, Geology and Geophysics of Australia*, **92**, 133–68.

Öpik, A.A. (1975a). Templetonian and Ordian xystridurid trilobites of Australia. *Bulletin of the Bureau of Mineral Resources, Geology and Geophysics of Australia*, **121**, 84pp.

Öpik, A.A. (1975b). Cymbric Vale fauna of New South Wales and Early Cambrian biostratigraphy. *Bulletin of the Bureau of Mineral Resources, Geology and Geophysics of Australia*, **159**, 78pp.

Öpik, A.A. (1979). Middle Cambrian agnostids: systematics and biostratigraphy. *Bulletin of the Bureau of Mineral Resources, Geology and Geophysics of Australia*, **172**, 188pp.

Orgeval, J.-J. & Capus, G. (1978). Existence d'un horizon phosphaté uranifère et thorifère à la base des formations carbonatées cambriennes de la région du Vigan (Gard, France): son remaniement karstique. *Compte rendu Société géologique de France*, **1978, 3** , 115–7.

Palmer, A.R. (1954). An appraisal of the Great Basin Middle Cambrian trilobites described before 1900. *US Geological Survey, Professional Paper*, **264D**, 55–86.

Palmer, A.R. (1965). Trilobites of the Late Cambrian pterocephalid biomere in the Great Basin, United States. *US Geological Survey, Professional Paper*, **493**, 105pp.

Palmer, A.R. (1971). The Cambrian of the Great Basin. In *Lower Palaeozoic Rocks of the World*, vol. 1: *Cambrian of the New World*. ed. C.H. Holland, pp. 1–78. Wiley-Interscience, London.

Patel, I.M. (1976). Lower Cambrian of southern New Brunswick and its correlation with successions in northeastern Appalachians and parts of Europe. *Abstracts with programs, Geological Society of America, N.E. 11th Annual Meeting, 1976*, **8**, 243.

Patwardhan, A. (1978). First Moravamminids from the Himalaya. *National Academy of Science Letters*, **1**, 7–8.

Patwardhan, A. (1979). New data on the Mussoorie phosphorites, Lower Himalaya. In *Proterozoic-Cambrian Phosphorites*, ed. P.J. Cook & J.H. Shergold, pp. 43–4. ANU Press, Canbera.

Patwardhan, A.M. (1980). Phosphate-pyrite association and the genesis of stromatolitic and pelletal phosphorites. *Proceedings of the third Indian Geological Congress. Poona, 1980*, 347–65.

Peach, B.N., Horne, J., Gunn, W., Clough, C.T., Hinxman, L.W. & Teall, J.J.H. (1907). The geological structure of the North-West Highlands of Scotland. *Memoir of the Geological Survey of Great Britain*, xviii, 668pp.

Poulsen, C. (1932). The Lower Cambrian faunas of East Greenland. *Meddelelser ϕm Grϕnland*, **74**, 297–316.

Poulsen, C. (1960). Notes on some Lower Cambrian fossils from French West Africa. *Matematisk-fysiske Meddelelser det Kongelige Danske Videnskabernes Selskab*, **32(7)**, 3–12.

Poulsen, C. (1967). Fossils from the Lower Cambrian of Bornholm. *Det Kongelige Danske Videnskabernes Selskab, Matematisk-fysiske Meddelelser*, **36**, 1–48.

Poulsen, V. (1963). The Lower Middle Cambrian Kalby-Ler (Kalby Clay) on the Island of Bornholm. *Det Kongelige Danske Videnskabernes Selskab, Biologiske Meddelelser*, **23**, 1–13.

Poulsen, V. (1966). Cambro-Silurian stratigraphy of Bornholm. *Meddelelser fra Dansk Geologisk Forening*, **16**, 117–37.

Poulsen, V., (1978). The Precambrian–Cambrian boundary in parts of Scandinavia and Greenland. *Geological Magazine*, **115**, 131–6.

Poulsen, V. & Anderson, M.M. (1975). The Middle–Upper Cambrian transition in southeastern Newfoundland, Canada. *Canadian Journal of Earth Sciences*, **12**, 2065–79.

Prasad, K.N. (1980). Review of Precambrian microfossils from peninsular India. *Geological Survey of India, Miscellaneous Publications*, **44**, 11–5.

Preiss, W.V. (1976). Intercontinental correlations. In *Stromatolites*, ed. M.R. Walter, pp. 359–70. Elsevier, Amsterdam.

Preiss, W.V. & Forbes, B.G. (1981). Stratigraphy, correlation and sedimentary history of Adelaidean (Late Proterozoic) basins in Australia. *Precambrian Research*, **15(3/4)**, 255–304.

Preiss, W.V. & Kinsman, J.E. (1978). Stratigraphy and palaeoenvironmental interpretation of the Brighton Limestone, south of Adelaide, and its equivalents in the Orroroo region. *Geological Survey of South Australia, Report of Investigations*, **49**, 34pp.

Preiss, W.V., Walter, M.R., Coats, R.P. & Wells, A.T. (1978). Lithological correlations of Adelaidean glaciogenic rocks in parts of the Amadeus, Ngalia, and Georgina Basins. *Bureau of Mineral Resources Journal of Australian Geology and Geophysics*, **3**, 43–53.

Prian, J.-P. (1979). Charactéristiques des paléonvironnements des phosphorites cambriennes du versant septentrional de la Montagne Noire (Sud du Massif Central, France). In *Géologie Comparée des Gisements de Phosphates et de Pétrole*. Colloque International, Orléans 6–7 November, 1979; Documents du BRGM, **24**, 93–111.

Qian Yi (1977). Hyolitha and some Problematica from the Lower Cambrian Meishucun Stage in central and S.W. China. *Acta Palaeontologia Sinica*, **16(2)**, 255–74 (in Chinese), 275 (in English).

Qian Yi, Chen Meng-e & Chen Yi-yuan (1979). Hyolithids and other small shelly fossils from the Lower Cambrian Huangshandong Formation in the eastern part of the Yangtze Gorge. *Acta Palaeontologia Sinica*, **18(3)**, 207–27 (in Chinese), 228–30 (in English).

Ranford, L.C., Cook, P.J. & Wells, A.T. (1965). The geology of the central part of the Amadeus Basin, Northern Territory. *Report of the Bureau of Mineral Resources, Geology and Geophysics of Australia*, **86**, 48pp.

Reed, F.R.C. (1910). Cambrian fossils of Spiti. *Palaeontologia Indica, geological Survey of India, Series 15*, **7(1)**, 1–77.

Repina, L.N. (1981). Trilobite biostratigraphy of the Lower Cambrian Stages in Siberia. *United States Department of the Interior. Geological Survey, Open File Report* **81–743**, 173–80.

Repina, L.N., Khomentovsky, V.V. Zhuravleva, I.T. & Rozanov, A.Yu. (1964). Lower Cambrian biostratigraphy of the Sayan-Altai region. *Akademiya Nauk SSSR, Sibirskoe Otdelenie, Institut Geologii i Geofiziki*, 364pp. Izdatelystvo 'Nauka' Moscow (in Russian).

Repina, L.N. & Romanenko, E.V. (1978). Trilobites and stratigraphy of the Lower Cambrian of the Altai. *Akademiya Nauk SSSR, Sibirskoe Otdelenie, Trudy Institut Geologii i Geofiziki*, **382**, 304pp. Izdatelystvo 'Nauka' Moscow (in Russian).

Resser, C.E. (1938). Cambrian System (restricted) of the Southern Appalachians. *Geological Society of America. Special Paper*, **15**, 140pp.

Robison, R.A., Rosova, A.V., Rowell, A.J. & Fletcher, T.P. (1977). Cambrian boundaries and divisions. *Lethaia*, **10**, 257–62.

Rogers, C.L., Kleinhampl, F.J., Ziony, J.I., & Danilchik, W. (1970). Phosphate occurrences in Nye County and adjacent areas, Nevada. *US Geological Survey, Professional Paper*, **700C**, 49–60.

Rogers, J.K. & Keevers, R.E. (1976). Lady Annie–Lady Jane phosphate deposits, Georgina Basin, Queensland. In *Economic geology of*

Australia and Papua New Guinea, 4. Industrial minerals and rocks. ed. C.L. Knight. pp. 251–65. Australasian Institute of Mining and Metallurgy, Melbourne.

Rozanov, A.Yu. (1969). Altai-Sayan Folded Region. In Rozanov *et al.* (1969).

Rozanov, A.Yu. (1976). The Precambrian-Cambrian boundary. In *Boundaries of Geological Systems,* ed. A.V. Pieve, pp. 31–53. Izdatelystvo 'Nauka' Moscow (in Russian).

Rozanov, A.Yu. & Debrenne, F. (1974). Age of archaeocyathid assemblages. *American Journal of Science*, **274**, 833–48.

Rozanov, A.Yu. & Missarzhevsky, V.V. (1966). Biostratigraphy and fauna of the lower horizons of the Cambrian. *Akademiya Nauk SSSR, Trudy Geologicheskiy Institut*, **148**, 120pp. Izdatelystvo 'Nauka' Moscow (In Russian).

Rozanov, A.Yu., Missarzhevsky, V.V., Volkova, N.A., Voronova, L.C., Krylov, I.N., Keller, B.M., Korolyuk, I.K., Lendzion, K., Michniak, R., Pychova, N.G. & Sidorov, A.D. (1969). Tommotian Stage and the Cambrian lower boundary problem. *Akademiya Nauk SSSR, Trudy Geologicheskiy Institut*, **206**, 380pp. Izdatelystvo 'Nauka' Moscow (in Russian). (In translation by US Department of the Interior and National Science Foundation, Washington, DC, published by Amerind Publishing Co. Pvt. Ltd, New Delhi, 1981).

Runnegar, B. & Jell, P.A. (1976). Australian Middle Cambrian molluscs and their bearing on early molluscan evolution. *Alcheringa*, **1**, 109–38.

Rushton, A.W.A. (1973). Cambrian fossils from the Hazira Shale, Pakistan. *Nature Physical Science*, **243(130)**, 142.

Rushton, A.W.A. (1974). The Cambrian of Wales and England. In *The Cambrian of the British Isles, Norden and Spitsbergen*, ed. C.H. Holland, pp. 43–121. John Wiley, London.

Rushton, A.W.A. (1978). Fossils from the middle–upper Cambrian transition in the Nuneaton district. *Palaeontology*, **21**, 245–83.

Russell, R.T. (1967). Discovery of major phosphate deposits in northwest Queensland. *Queensland Government Mining Journal*, **58**, 153–7.

Russell, R.T. & Trueman, N.A. (1971). The geology of the Duchess phosphate deposits, northwestern Queensland, Australia. *Economic Geology*, **66**, 1186–214.

Samsonowicz, J. (1956). Cambrian paleogeography and the base of the Cambrian System in Poland. *20th Session of the International Geological Congress, Mexico*, **1**, 127–60.

Samsonowicz, J. (1960). The Lower Cambrian of the Klimontów anticlinorium. *21st Session of the International Geological Congress, Copenhagen*, **8**, 86–92.

Schmitt, M. (1978). Stromatolites from the Tiout section, Precambrian-Cambrian boundary beds, Anti-Atlas, Morocco. *Geological Magazine*, **115(2)**, 95–100.

Schmitt, M. (1979). New stromatolites from the late Precambrian of the Anti-Atlas and from the Lower Cambrian of the High Atlas, Morocco. *Senckenbergiana Lethaea*, **60(1/3)**, 39–49.

Schmitt, M. & Southgate, P.N. (1982). A phosphatic stromatolite (*Ilicta* cf. *composita* Sidorov) from the Middle Cambrian, northern Australia. *Alcheringa*, **6**, 175–83.

Sdzuy, K. (1964). Das Kambrium des Frankenwaldes. 1: Erforschungsgeschichte, Vorkommen und Stratigraphie. *Senckenbergiana Lethaea*, **45**, 201–21.

Sdzuy, K. (1972). Das Kambrium der acadobaltischen Faunenprovinz. *Zentralblatt für Geologie und Paläontologie*, **II**, 1–10.

Sdzuy, K. (1978). The Precambrian-Cambrian boundary beds in Morocco (Preliminary Report). *Geological Magazine*, **115(2)**, 83–94.

Shah, S.K., Raina, B.K. & Razdan, M.L. (1980). Redlichid fauna from the Cambrian of Kashmir. *Journal of the Geological Society of India*, **21**, 511–7.

Shah, S.M. Ibrahim (Ed.) (1977). Stratigraphy of Pakistan. *Memoirs of the Geological Survey of Pakistan*, **12**, 138pp.

Shankar, R. (1970). Stratigraphic analysis of the Chert Member, Lower Tal Formation in Dehradun and Tehri districts, U.P. *Records of the Geological Survey of India*, **106(2)**, 54–74.

Shankar, R. (1971). Stratigraphy and sedimentation of the Tal Formation, Mussoorie Syncline, Uttar Pradesh. *Journal of the Palaeontological Society of India*, **16**, 1–15.

Shankar, R. (1976). Maldeota phosphorite prospects, Dehra Dun district, Uttar Pradesh. *Geological Survey of India Miscellaneous Publications*, **24**, 465–74.

Shergold, J.H. (1968). Cambrian palaeontology. Appendix. In The Cambrian of the Burke River outlier, ed. F. de Keyser. *Records of the Bureau of Mineral Resources, Geology and Geophysics of Australia*, **1968/67**.

Shergold, J.H. (1973). *Meneviella viatrix* sp. nov., a new conocoryphid trilobite from the Middle Cambrian of western Queensland. *Bulletin of the Bureau of Mineral Resources, Geology and Geophysics of Australia*, **126**, 19–26.

Shergold, J.H. & Druce, E.C. (1980). Upper Proterozoic and Lower Palaeozoic rocks of the Georgina Basin. In *The geology and geophysics of northeastern Australia.* ed. P.J. Stephenson & R.A. Henderson, pp. 149–74. Geological Society of Australia, Queensland Division, Brisbane.

Shrivastava, R.N., Pant, A., Ahmad, A., Anil Saxena, Kothiyal, D.L. & Rao, B.V.M. (1981). *Excursion guidebook on Dehradun-Mussoorie area*, 53pp. Geological Survey of India, Bangalore.

Sin Yu-sheng & Liu Kui-chih (1976). *Microflora from the Sinian Subera of W. Hupeh and its stratigraphic significance*, 10pp. Institute of the Geology and Mineral Resources, Academy of Geological Sciences, Peking.

Singh, P. & Shukla, S.D. (1981). Fossils from the Lower Tal: their age and its bearing on the stratigraphy of the Lesser Himalaya. *Geoscience Journal*, **2(2)**, 157–76, 2 pls.

Skjeseth, S. (1963). Contributions to the geology of the Mjøsa districts and the classical sparagmite area in southern Norway. *Norges Geologiske Undersøkelse*, **220**, 126pp.

Southgate, P.N. (1980). Cambrian stromatolitic phosphorites from the Georgina Basin, Australia. *Nature*, **285(5764)**, 395–7.

Spjeldnaes, N. (1967). Studies on the latest Precambrian and Eocambrian rocks in Norway, 6. Fossils from pebbles in the Biskopasen Formation in southern Norway. *Norges Geologiske Undersøkelse*, **251**, 53–82.

Størmer, L. (1925). On a Lower Cambrian fauna at Ustaoset in Norway. *Fennia*, **45**, 12–22.

Stubblefield, C.J. (1930). A new Upper Cambrian section in south Shropshire. *Summary of progress, Geological Survey of Great Britain, for 1929*, 54–62.

Stubblefield, C.J. (1956). Cambrian palaeogeography in Britain. *20th Session International Geological Congress, Mexico*, **1**, 1–43.

Sudan, Charan Singh (1981). Stratigraphy and palaeontology of the Lower Palaeozoic sequence of Kupwara-Trahagam Belt of northwestern Kashmir. Contents and Abstract of Ph.D. Thesis, University of Jammu 6pp.

Suvorova, N.P. (1980). The boundary between the Lower and Middle Cambrian in eastern Siberia. *Byulleteny Geologii*, **55(2)**, 48–61 (in Russian).

Taylor, K. & Rushton, A.W.A. (1972). The pre-Westphalian geology of the Warwickshire Coalfield, with an account of three boreholes in the Merevale Area. *Bulletin of the Geological Survey of Great Britain*, **35**, 150pp.

Taylor, M.E. (1976). Indigenous and redeposited trilobites from Late Cambrian basinal environments of central Nevada. *Journal of Paleontology*, **50(4)**, 668–700.

Taylor, M.E. (1977). Late Cambrian of western North America: trilobite biofacies, environmental significance, and biostratigraphic implications. In *Concepts and methods of biostratigraphy*, ed. E.G. Kauffeman & J.E. Hazel, pp. 397–425. Dowden, Hutchinson & Ross Stroudsburg, Pa.

Taylor, T.G. (1910). The Archaeocyathinae from the Cambrian of South Australia with an account of the morphology and affinities of the whole class. *Memoirs of the Royal Society of South Australia*, **2**, 35–188.

Tegengren, F.R. (1962). Vassbo blymalmsfyndighet i Idre och dess geologiska inramning. *Sveriges Geologiska Undersökning*, C, **586**, 61pp.

Thomson, L.D. & Russell, R.T. (1971). Discovery, exploration and investigation of phosphate deposits in Queensland. *Australasian Institute of Mining and Metallurgy Annual Field Conference*, **1971**, 1–19.

Thorslund, P. & Westergård, A.H. (1938). Deep boring through the Cambro-Silurian at File Haidar, Gotland. *Sveriges Geologiska Undersökning, C*, **415**, 56pp.

Valdiya, K.S. (1972). Origin of phosphorite of the late Precambrian Gangolihat Dolomites of Pithoragarh, Kumaun Himalaya, India. *Sedimentology*, **19(1/2)**, 115–28.

Valdiya, K.S. (1980). Lesser Himalayan stromatolites – their biostratigraphic implications. *Geological Survey of India Miscellaneous Publications*, **44**, 117–27.

Vidal, G. (1976). Late Precambrian microfossils from the Visingsö Beds in southern Sweden. *Fossils & Strata*, **9**, 57pp.

Vidal, G. (1979). Acritarchs from the Upper Proterozoic and Lower Cambrian of East Greenland. *Grønlands Geologiske Undersøgelse, Bulletin*, **134**, 37pp.

Vidal, G. (1981a). Micropalaeontology and Biostratigraphy of the Upper Proterozoic and Lower Cambrian Sequence in East Finnmark, Northern Norway. *Norges Geologiske Undersøkelse*, **362**, 1–53.

Vidal, G. (1981b). Lower Cambrian acritarch stratigraphy in Scandinavia. *Geologiska Föreningens i Stockholm Förhandlingar*, **103**, 183–92.

Vidal, G. (1981c). Micropalaeontology and biostratigraphy of the Lower Cambrian sequence in Scandinavia. *United States Department of the Interior, Geological Survey, Open-File Report*, **81–743**, 232–5.

Viland, J.-C. (1977). Presénce d'horizons phosphatés à la base de l'Infracambrien supérieur du Haut Atlas de Marrakech (Maroc). *Notes et Mémoires de la Service géologique du Maroc*, **38(268)**, 13–22.

Vögt, T. (1924). Forholdet mellem sparagmitsystem og det marine underkambrium ved Mjøsen. *Norsk Geologisk Tidsskrift*, **7**, 281–385.

Volkova, N.A., Gnilovskaya, M.B., Lendzion, K., Kiryanov, V.V., Palij, V.M., Pachkyavichene, L.T., Piskun, L.V. Posti, E., Rozanov, A.Yu., Urbanex, A., Fedonkin, M.A. & Yankauskas, T.V. (1979). *Upper Precambrian and Cambrian paleontology of East-European Platform*. Akademiya Nauk SSSR, Geologischeskiy Institut, 210pp. Izdatelystvo 'Nauka', Moscow (in Russian).

Walcott, C.D. (1900). Lower Cambrian terrane in the Atlantic Province. *Proceedings of the Washington Academy of Sciences*, **1**, 301–39.

Wallerius, I. (1895). *Undersökningar öfver zonen med Agnostus laevigatus i Vestergötland, jamte en inledande öfversikt af Vestergötlands samtliga Paradoxideslager*, 72pp. Gleerupska Universitets-Bokhandeln, Lund.

Wallerius, I. (1930). Fran Västergötlands mellankambrium. *Geologiska Föreningens i Stockholm Fördhandlingar*, **52**, 47–62.

Walter, M.R. (1967). Archaeocyatha and biostratigraphy of the Lower Cambrian Hawker Group, South Australia. *Journal of the Geological Society of Australia*, **14**, 139–52.

Walter, M.R. (1972). Stromatolites and the biostratigraphy of the Australian Precambrian and Cambrian. *Palaeontological Association of London, Special Paper*, **11**, 190pp.

Walter, M.R. Krylov, I.N. & Preiss, W.V. (1979). Stromatolites from Adelaidean (Late proterozoic) sequences in central and South Australia. *Alcheringa*, **3**, 287–305.

Walter, M.R., Shergold, J.H., Muir, M.D. & Kruse, P.H. (1979). Early Cambrian and latest Proterozoic stratigraphy, Desert Syncline, southern Georgina Basin. *Journal of the Geological Society of Australia*, **26**, 305–12.

Wawryk, W. (1932). Ergebnisse der mikroskopisch-chemischen Analyse der kambrischen Phosphorite aus der Umgebung von Sandomierzs. *Annales de la Société Géologique de Pologne*, **VIII**, 266–72.

Wedow, H., & Stansfield, R.G. (1968). Fertilizer Raw Materials. *US Geological Survey, Professional Paper*, **580**, 281–3.

Wells, A.T., Forman, D.J. & Ranford, L.C. (1965). Geological reconnaissance of the northwestern part of the Amadeus Basin, Northern Territory. *Report of the Bureau of Mineral Resources, Geology and Geophysics of Australia*, **85**, 45pp.

Wells, A.T., Forman, D.J., Ranford, L.C. & Cook, P.J. (1970). Geology of the Amadeus Basin, central Australia. *Bulletin of the Bureau of Mineral Resources, Geology and Geophysics of Australia*, **100**, 222pp.

Wells, A.T., Ranford, L.C., Stewart, A.J., Cook, P.J. & Shaw, R.D. (1967). The geology of the northeastern part of the Amadeus Basin, Northern Territory. *Report of the Bureau of Mineral Resources, Geology and Geophysics of Australia*, **113**, 93pp.

Westergård, A.H. (1922). Sveriges Olenidskiffer. *Sveriges Geologiska Undersökning*, Ca, **18**, 1–188 (in Swedish), 189–205 (English).

Westergård, A.H. (1929). A deep boring through middle and lower Cambrian strata at Borgholm, Isle of Öland. *Svergiges Geologiska Undersökning*, C, **355**, 19pp.

Westergård, A.H. (1940). Nya djupborrningar genom äldsta ordovicium och kambrium i Östergötland och Närke. *Sveriges Geologiska Undersökning*, C, **437**, 68pp. (Swedish, English summary).

Westergård, A.H. (1944). Borrningar genom alunskifferlagret pa Östergötland. *Sveriges Geologiska Undersökning*, C, **463**, 22pp. (Swedish, English summary).

Westergård, A.H. (1946). Agnostidea of the Middle Cambrian of Sweden. *Sveriges Geologiska Undersökning*, C, **477**, 140pp.

Wucher, K. (1967). Ergebrisse der Kartierungsbohrungen Heinersdorf 1/60 und 2/62 (Thuringisches Schiefergebirge). *Jahrbuch für Geologie*, **1**, 297.

Würm, A. (1925). Uber ein Vorkommen von Mittelkambrium (Paradoxidesschichten) im bayrischen Frankenwald bei Wildenstein sülich Presseck. *Neues Jahrbuch für Mineralogie, Beilagebände*, **52(A)**, 71–93.

Xing Yusheng, Ding Qixiu & Luo Huilin (1982). Biotic characteristics of the Sinian-Cambrian boundary beds in China, and the boundary problems. *Precambrian Research*, **17**, 77–85.

Yankauskas, T.V. & Posti, E. (1973). Micropalaeontological characteristics of the Estonian Lower Cambrian. *Izvestiya Akademiya Nauk Estonskoy SSR*, **25**, *Kimiya Geologicheskiya*, 1973, 143–8 (in Russian).

Yanshin, A.L. (1979). Upper Precambrian and Lower Palaeozoic phosphate deposits in the USSR and Mongolian People's Republic. In *Proterozoic-Cambrian Phosphorites*, ed. P.J. Cook, & J.H. Shergold pp. 45–47. Canberra Publishing & Printing Co., Canberra.

Yin Gongzheng, Wang Yangeng, & Qian Yi (1983). A preliminary research into the boundary between the Sinian and the Cambrian of Guizhou: *Journal of Stratigraphy*, **6(4)**, 286–93 (in Chinese).

Yochelson, E.L., Henningsmoen, G. & Griffin, W.L. (1977). The early Cambrian genus *Volborthella* in southern Norway. *Norsk Geologisk Tidsskrift*, **57**, 133–51.

Yü Wen (1979). Earliest Cambrian monoplacophorans and gastropods from western Hubei with their biostratigraphical significance. *Acta Palaeontologia Sinica*, **18(3)**, 233–60 (in Chinese), 261–6 (in English).

Zhang Wen-tang & Zhu Zhao-ling (1979). Notes on some trilobites from the Lower Cambrian Houjiashan Formation in southern and southwestern parts of North China. *Acta Palaeontologia Sinica*, **18(6)**, 513–23 (in Chinese), 524–5 (in English).

Zhang Wen-tang, Zhu Zhao-ling, Yuan Ke-xing, Lin Huan-ling, Qian Yi, Wu Hong-ji & Yuan Jin-liang (1979). The boundary between the Cambrian and latest Precambrian in the southern and southwestern parts of North China. *Acta Stratigraphica Sinica*, **3(1)**, 51–6 (in Chinese).

Zhou Zhi-yi & Yuan Jin-liang (1980). Lower Cambrian trilobite succession in southwest China. *Acta Palaeontologia Sinica*, **19(4)**, 331–8 (in Chinese) p. 339 (in English).

Zhu Zhao-ling & Lin Tian-rui (1978). Some Middle Cambrian trilobites from Yaxian, Hainan Island. *Acta Palaeontologia Sinica*, **17(4)**, 439–43 (in Chinese), p. 443 (in English).

Zhuravleva, I.T., Repina, L.N. & Khomentovsky, V.V. (1959). The Lower Cambrian formations of Gornaya Schoriya. *Doklady Akademiya Nauk SSSR*, **128(5)**, 1030–3 (in Russian).

Zou Caiyu & Li Jiliang (1965). The first discovery of algal phosphorite in China. *Scientia Geologia Sinica*, **1(1)**, 80–5.

24

Proterozoic and Cambrian phosphorites – specialist studies: Middle Cambrian phosphatic hardgrounds, phoscrete profiles and stromatolites and their implications for phosphogenesis

P.N. SOUTHGATE

Introduction

As phosphatic sediments of marine origin occur in a specific stratigraphic context, they offer geologists the opportunity to formulate economically significant depositional models using sedimentological skills. Any study of sedimentary phosphorites should commence with an understanding of the lateral and vertical facies relationships within the deposit, and of the adjoining rock units of equivalent age. Once the facies relationships are understood, petrological, palaeontological, mineralogical and geochemical information can be integrated into an overall stratigraphic or lithofacies framework. This framework enables later observations to be placed in their context affording a means whereby original, diagenetic and much later weathering features can be separated. Even though sedimentary phosphorites should be considered as an integral part of a lithofacies mosaic, much of the published work on ancient phosphorites has concentrated on the actual deposits without placing these observations in their environmental context. In these studies the surrounding sedimentary rocks are typically referred to as dolostones, phosphatic dolostones, cherts and shales (e.g. Russell & Trueman, 1971; Howard & Cooney, 1976; Ilyin & Ratnikova, 1981).

The principal reason for this emphasis on the economically important phosphate deposits is readily appreciated. Once a potential deposit has been located, the initial aim of most companies should be to define the nature, geometry and size of the actual ore-body. However, with the growing recognition that phosphate deposits represent sites of phosphate particle accumulation, rather than phosphate production (de Keyser & Cook, 1972; Elgueta, 1981; Cook, 1984) it is clear that the genesis of most phosphate deposits will not be understood solely by studying the actual deposits. Sedimentological examination of a phosphorite deposit composed of transported phosphate particles provides information on the environmental conditions at the site of particle accumulation. It does not indicate the conditions under which phosphate was first fixed within the sediments. Phosphate grains, whether they be angular fragments, ovules or pellets, generally consist of structureless carbonate fluorapatite with variable amounts of included material (Riggs, 1979). In descriptions of the Georgina Basin phosphorites these grains have invariably been described as having a problematic origin (Russell & Trueman, 1971; de Keyser & Cook, 1972; Cook, 1972; Fleming, 1977). Phosphate grains with identifiable precursors are restricted to a small number of originally phosphatic skeletal particles. Where then do the majority of grains originate? If they are locally derived, as is suggested by the presence of angular clasts of phosphate, then evidence of *in situ* phosphatisation should occur within the laterally equivalent, dominantly calcareous rocks which in general constitute the lateral facies equivalent of the phosphate deposit. The identification within these rocks of partially-eroded surfaces or lithologies, upon or within which phosphate was fixed, should therefore provide information on the physico-chemical conditions that existed during phosphogenesis.

This chapter is therefore concerned with both the description and interpretation of a series of *in situ* phosphatic nodules, hardgrounds, crusts, stromatolites and profiles from Middle Cambrian rocks of the Georgina Basin, Australia. Similar phosphate-cemented structures in rocks of latest Proterozoic to Cambrian age in Europe and China are briefly reviewed. Their widespread occurrence suggests that the processes interpreted as having occurred in rocks of the Georgina Basin may have also taken place in many other parts of the world at this time.

Stratigraphy and regional setting of the Georgina Basin

The Georgina Basin comprises a Late Proterozoic to Early Palaeozoic, predominantly calcareous, sedimentary province covering some 325 000 square kilometres of northern Australia (Fig. 24.1). The Precambrian Mt Isa block forms the present-day eastern boundary of the Basin, while to the south the Basin is lapped over by Cretaceous strata of the Great Artesian Basin. To the north and west the carbonate succession is either bounded by Precambrian rocks or merges with Cambrian strata of the Daly River and Wiso Basins (Smith, 1972; Shergold & Druce, 1980). As shown by Cook (1982), the Georgina Basin probably represents a geomorphic remnant of a once extensive epeiric terrain that originally covered much of central Australia. The presence of Middle Cambrian sediments at Kajabbi (Wilson, Gunn & Smit, 1979) and the occurrence of early Middle Cambrian outliers within the Mt Isa Block, coupled with

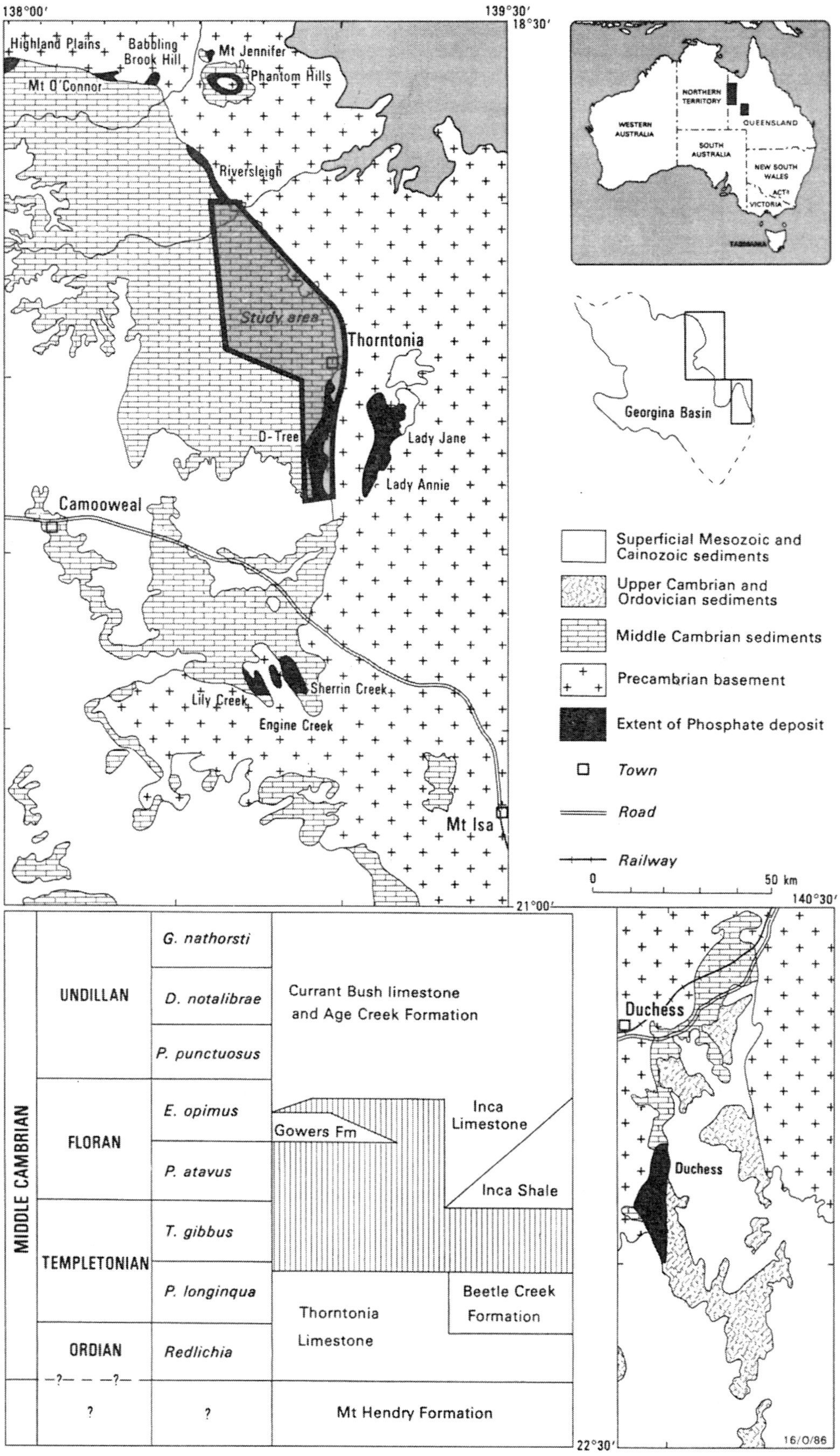

Fig. 24.1. Location map of the Georgina Basin, Australia and a geological map of the present-day north-easterly margin of the basin showing location of phosphate deposits as well as the area covered by this study. The chart in the lower left hand corner indicates the spatial and temporal distribution of the stratigraphic units found in the north-eastern portion of the Georgina Basin.

the displacement of a geomorphic surface defined by mid-Cenozoic laterite residuals (de Keyser & Cook, 1972), suggests that part of the present eastern margin may owe its presence to later epeirogenic doming and faulting. The relative position of the Mt Isa Block during the Middle Cambrian is still a matter of conjecture. This uncertainty is unfortunate as knowledge of the relative position of this Block could provide several constraints on possible models of phosphogenesis.

Although Lower Cambrian carbonates and shales were deposited in the southern parts of the Basin (Walter *et al.*, 1979), it was not until the early Middle Cambrian that marine-associated depositional environments extended throughout the Basin (Shergold & Druce, 1980). Deposition commenced in response to a marine transgression that resulted in the invasion of much of the craton by shallow submergent or epeiric seas (Cook, 1982). This Ordian Stage event (Öpik, 1967) is marked by the occurrence of *Redlichia* faunas throughout northern Australia (Öpik, 1970). Shallow submergent, semi-emergent and emergent chemical and biochemical sedimentation ensued and produced a complex facies mosaic that was dominated by calcareous sediments. Phosphorites and organic-rich sediments were also deposited during the Middle Cambrian, adding to the complexity of the mosaic. Evaporites accumulated in hypersaline semi-emergent and emergent environments. During the Late Cambrian and Early Ordovician, similar sedimentary regimes prevailed (Radke, 1980; Kennard, 1981) but the focus of sedimentation shifted to the south so that rocks of Late Cambrian and Ordovician age are restricted in distribution and occur south of latitude 21° S. The youngest rocks, of probable Middle Ordovician age (Shergold *et al.*, 1976), are located in the Toko Ranges on the southern margin of the Basin. Apart from the scattered occurrence of phosphate nodules in the lower Ordovician Nora Formation (Smith, 1972), phosphatic rocks in the Georgina Basin are most abundant in Middle Cambrian strata and particularly in the early Middle Cambrian.

Stratigraphic subdivision of the Georgina Basin has proved a difficult task. The initial basin-wide subdivision proposed by Öpik (1956; 1960) was based on a mixture of biostratigraphic and lithostratigraphic criteria. A résumé of this early stratigraphic work was compiled by Smith (1972), while problems associated with the initial subdivisions of Öpik have been outlined by de Keyser (1973). More recently, a compilation by Shergold & Druce (1980) resulted in the grouping of formations into three tectotopes. These tectonostratigraphic units grouped similar types of sediments separated by either unconformities or disconformities. The phosphorite deposits of the Georgina Basin occur within the lower part of the second tectotope. Although the actual deposits are restricted to the Beetle Creek Formation (e.g. Cook, 1976), phosphatic rocks and phosphorites are also found in the laterally equivalent and underlying calcareous rocks of the Thorntonia Limestone (Southgate, 1983). Additionally, in the Undilla region phosphatic limestones and dolostones of Floran age locally overlie phosphatic dolostones of the Thorntonia Limestone (Fig. 24.1). This overlying phosphatic interval is referred to as the Gowers Formation (Southgate, 1983). As lithofacies found in the Gowers Formation are identical to some of the lithofacies in the Thorntonia Limestone, the lithofacies descriptions of both formations are combined in the following section.

Lithofacies and mineralogical suites of rocks

The phosphorites, phosphatic carbonates and calcareous rocks of the Thorntonia Limestone, Beetle Creek Formation and Gowers Formation have been grouped into seven lithofacies (Lithofacies 1–7) (Table 24.1). These in turn can be combined into three mineralogical suites of rocks (Suites A–C). In order to place the *in situ* phosphatic pavements and lithologies found in Suite A into a regional context, the following section briefly describes and interprets the broad-scale depositional environments represented by the three suites of rocks. Detailed descriptions and interpretations of the lithofacies are given in Southgate (1983).

Suite A: The mixed-mineralogy lithofacies (Lithofacies 1, 2 and 3)

Rocks comprising Suite A are characterised by a mixture of carbonate, phosphate and glauconite particles as well as carbonate and phosphate mud. Phosphate particles are subordinate to carbonate grains; glauconite particles, although locally abundant, comprise a minor particle component usually found in association with phosphate grains. Rocks comprising the mixed-mineralogy suite are characterised by two shallowing-upward cycles, mesocycles mP_1 and gP_1 where 'm' stands for mud-dominant, 'g' for grain-dominant and P for phosphate (Fig. 24.2). The phosphate particle component of both cycles increases upwards in the cycle.

In situ phosphate-cemented nodules, hardgrounds and phoscrete profiles as well as primary phosphatic stratiform and columnar stromatolites, laminoid fenestral phos-

Table 24.1. *Lithofacies and mineralogical suites of rocks of the Thorntonia Limestone, Beetle Creek Formation and Gowers Formation, Australia*

Mineralogical suite		Lithofacies
Suite A The Mixed-Mineralogy Suite	Lithofacies 1:	Mixed-mineralogy mud dominant peloid and bioclast lithofacies
	Lithofacies 2:	Mixed-mineralogy coated peloid phosphate pavement lithofacies
	Lithofacies 3:	Mixed-mineralogy skeletal particle dominant lithofacies
Suite B The Calcareous Suite	Lithofacies 4:	Calcareous peloid and oncolite grainstone dominant lithofacies
	Lithofacies 5:	Calcareous peloid boundstone dominant lithofacies
	Lithofacies 6:	Calcareous thin-bedded and fenestrate mudstone lithofacies
Suite C The Phosphorite Suite	Lithofacies 7:	Peloid phosphorite lithofacies

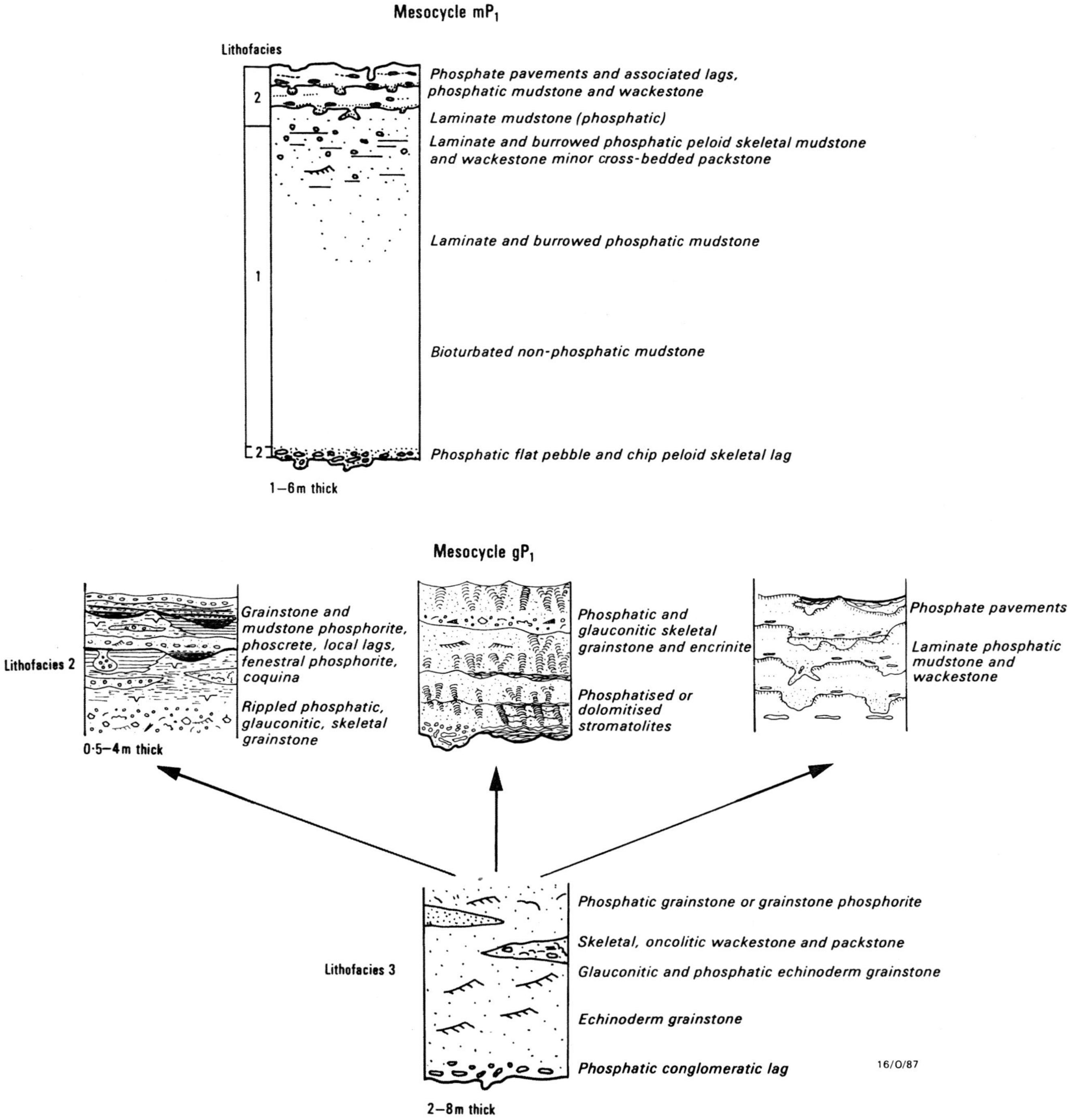

Fig. 24.2. Summary diagrams for mixed-mineralogy mesocycles mP_1 and gP_1. Included on the left hand side of these diagrams are the cyclic positions of each of the lithofacies. The three columns shown in the upper part of mesocycle gP_1 depict the different types of semi-emergent flats that develop above the echinoderm grainstones. The column shown on the uppermost, far right hand side of mesocycle gP_1 contains the same types of sediments as those found in the uppermost part of mesocycle mP_1.

phorites and mudstone phosphorites occur in the uppermost parts of mesocycles mP_1 and gP_1. In mesocycle mP_1, phosphatic lithologies are dominated by laminate bedforms; carbonate mudstones are bioturbated. Carbonate and phosphate particles coarsen upwards. Phosphate grains are usually absent in the lowermost parts of a cycle. Skeletal particles in the cycle increase upwards in diversity and abundance. Cauliflower chert nodules (Chowns & Elkins, 1974) occur in the uppermost parts of some cycles in the Riversleigh area. Sediments found in mesocycle gP_1 are dominated by coarsely crystalline and cross-bedded echinoderm and phosphatic glauconitic echinoderm grainstone. Some encrinites are characterised by herringbone cross-laminations. Coquinas composed of either platy skeletal fragments e.g. trilobite and brachiopod debris or cone-shaped hyolith shells occur in the uppermost parts of both cycles.

Mixed-mineralogy sediments belonging to the lower parts of mesocycles mP_1 and gP_1 (Lithofacies 1 & 3) were deposited in shallow submergent normal marine to metahaline salinity waters. Echinoderm grainstones accumulated in shoal complexes subjected to conditions of storm, wave and tidal current action (Fig. 24.3). Mud-supported carbonates and phosphatic carbonates of mesocycle mP_1 accumulated in the lee of echinoderm grainstone shoals and in sheltered areas between, as well as embayed areas within, emergent Precambrian and karsted early Middle Cambrian land masses and islands (Fig. 24.3). As shallowing occurred, increasing quantities of phosphate and glauconite particles, as well as skeletal fragments, were mixed with the echinoderm grains and calcareous muds. The majority of these phosphate particles were probably derived from the *in situ* phosphatic pavements, profiles and lithologies found in the uppermost parts of mesocycles mP_1 and gP_1.

Rocks comprising the uppermost parts of mesocycles mP_1 and gP_1 (Lithofacies 2) accumulated on broad semi-emergent flats (Fig. 24.3). These flats prograded over the underlying mud- or grain-supported sediments. A prolific fauna flourished on these flats in areas of near-normal marine to metahaline salinity waters. In the landward emergent portion of some shoals, hypersaline conditions developed locally and resulted in the growth of anhydrite nodules now present as cauliflower cherts (Chowns & Elkins 1974). Angular enlarged fenestrae in some phosphatic boundstones suggest the development of ephemeral evaporites, possibly halite, during periods of desiccation.

Suite B: The calcareous lithofacies (Lithofacies 4, 5 and 6).

Rocks comprising Lithofacies 4, 5 & 6 consist almost exclusively of carbonate particles or carbonate mud. Skeletal particles are rare to absent. Carbonate peloids, algal-bound particles, coated grains and intraclasts are common, as are algal-bound rocks. Rare phosphate particles occur in some lag deposits. Glauconite grains are absent.

Rocks comprising Suite B, the calcareous lithofacies, are characterised by a shallowing-upward cycle, here termed gC_1 where 'g' stands for grain-dominant and 'C' for carbonate (Fig. 24.4). Rocks found in the lower parts of this cycle are dominated by trough cross-stratified and ripple cross-laminated grainstones. Oncolites are locally dominant and increase in frequency as a cycle is ascended. Laminoid fenestral and pustular boundstones (Lithofacies 5) overlie the cross-bedded grainstones and are separated from them by a gradational boundary. Subordinate ripple cross-laminated grainstones occur in association with the boundstones. A distinctive suite of sedimentary structures and diagenetic lithologies occurs in this part of the cycle. These features include desiccation polygons, scalloped dissolution surfaces, laminar calcrete crusts, calcrete pisolites, teepee structures and sinkholes. Although fenestral boundstones often form the uppermost lithologies found in mesocycle gC_1, in some areas the cycle is capped by angular fenestral, thin-bedded to laminate, sheet-cracked and desiccated mudstone and mudstone breccia. Teepee structures are locally common and skeletal grains are absent.

The abundance of non-skeletal carbonate grains as well as algal-bound sediments and particles suggests that rocks comprising these lithofacies were deposited in hypersaline waters (Logan, 1974; Lees, 1975). Lithofacies 4 carbonates accumulated as a series of broad, low-relief shoals on a platform of subdued relief (Fig. 24.3). Circulation was gradually restricted by shallowing which eventually resulted in the development of broad semi-emergent flats. These flats were colonised by stratiform stromatolites now present as laminoid fenestral boundstones. Hypersaline conditions and periodic desiccation led to the production of indurated pavements which were subsequently ripped-up to produce flat pebble intraclasts. Calcrete profiles formed within the high semi-emergent and emergent zones, where seasonally wet periods alternated with times of aridity to produce intense vadose diagenesis (James, 1972; Harrison, 1977). Subaerial exposure also resulted in the dissolution of some carbonates, with the concomitant formation of scalloped surfaces (Read & Grover, 1977) and stratiform caverns. Teepee structures developed in areas of groundwater resurgence (Muir, Lock & Von der Borch, 1980; Warren, 1982). Angular fenestral mudstones formed by the growth of ephemeral evaporite minerals, probably halite, in hypersaline ponded waters found in the central emergent parts of large shoals (Southgate, 1983).

Suite C: The phosphorite lithofacies (Lithofacies 7).

In the north-east part of the Georgina Basin rocks comprising this mineralogical suite are represented by the Riversleigh, D Tree and Lady Annie phosphate deposits. These deposits have been studied by de Keyser & Cook (1972), Howard & Cooney (1976), Rogers & Keevers (1976), Fleming (1977) and Elgueta (1981). As a detailed discussion of the Lady Annie phosphate deposits is given by Cook & Elgueta (this volume), rocks that comprise the phosphorite lithofacies are only briefly discussed.

Elgueta (1981) described the phosphorite lithofacies as consisting of well-bedded to laminated and occasionally cross-bedded grainstone and packstone phosphorite with minor mudstone phosphorite. Grain size varies from very coarse to very fine sand. Constituent phosphate particles include originally phosphatic skeletal particles, peloids, coated grains and intraclasts. Articulated or whole skeletal particles are rare, but comminuted skeletal debris is common.

Poor outcrop precludes a detailed interpretation for rocks comprising this lithofacies. However, the presence of coquinas and well- to poorly-sorted grainstone and packstone phosphorites within a mixed-mineralogy suite of sediments suggests that sites where phosphate particles accumulated were subjected to periods of wave and/or current action. This suite of sediments is therefore interpreted as forming in a series of low-energy shoals subjected to processes of winnowing, reworking and sorting (Fig. 24.3).

Regional synthesis

The mineralogical suites of rocks described and interpreted in the preceding sections accumulated under con-

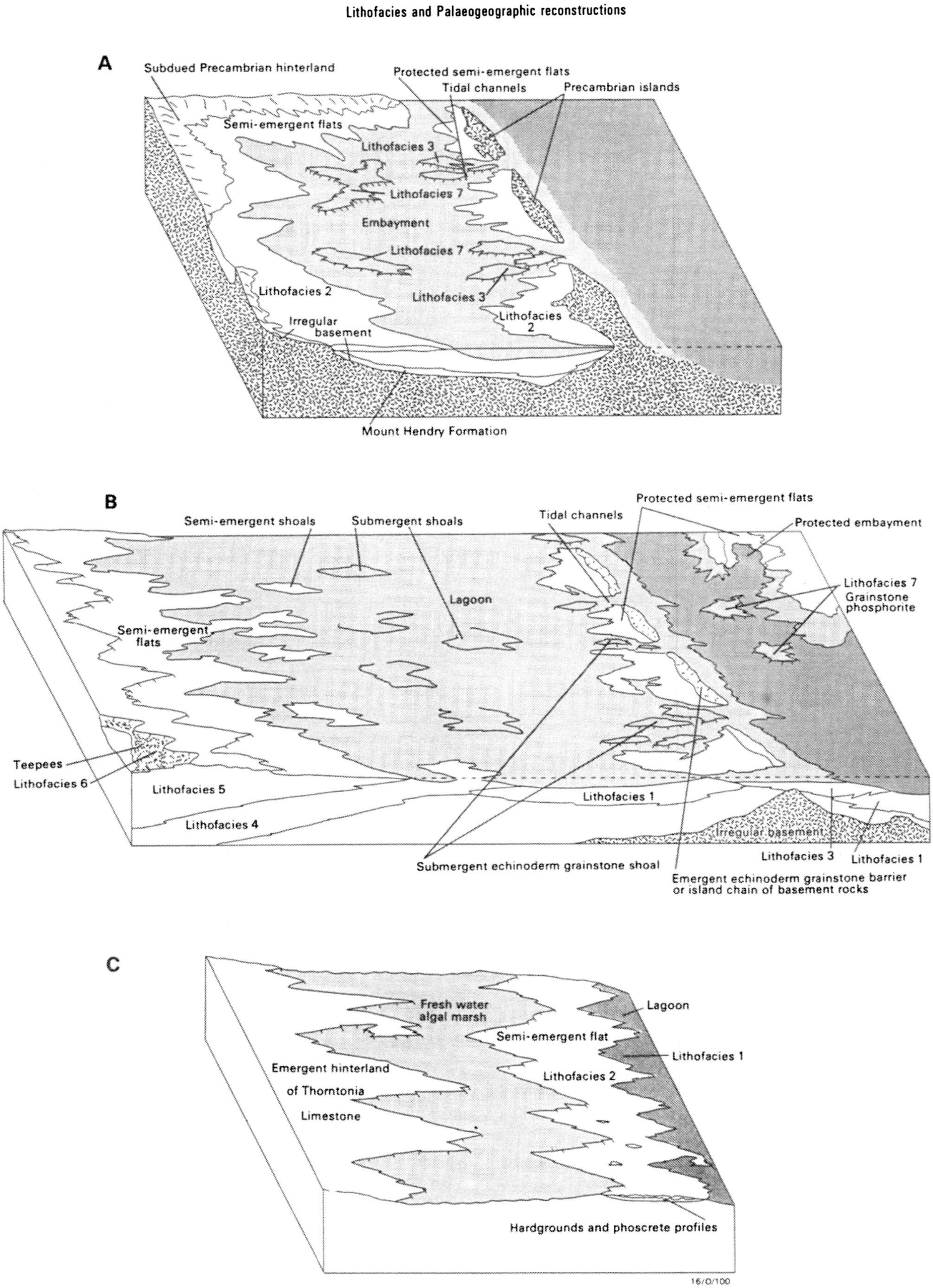
Lithofacies and Palaeogeographic reconstructions
A
Subdued Precambrian hinterland
Protected semi-emergent flats
Tidal channels
Precambrian islands
Semi-emergent flats
Lithofacies 3
Lithofacies 7
Embayment
Lithofacies 7
Lithofacies 2
Lithofacies 3
Lithofacies 2
Irregular basement
Mount Hendry Formation
B
Semi-emergent shoals
Submergent shoals
Tidal channels
Protected semi-emergent flats
Protected embayment
Lithofacies 7
Grainstone phosphorite
Lagoon
Semi-emergent flats
Teepees
Lithofacies 6
Lithofacies 5
Lithofacies 4
Lithofacies 1
Irregular basement
Lithofacies 3
Lithofacies 1
Submergent echinoderm grainstone shoal
Emergent echinoderm grainstone barrier or island chain of basement rocks
C
Fresh water algal marsh
Lagoon
Semi-emergent flat
Lithofacies 1
Emergent hinterland of Thorntonia Limestone
Lithofacies 2
Hardgrounds and phoscrete profiles
16/O/100

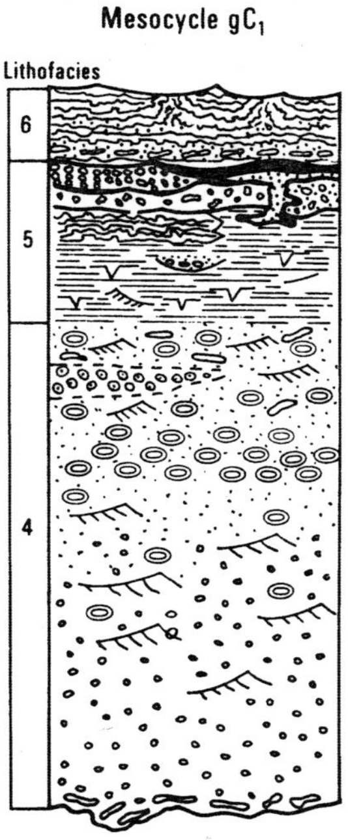

Fig. 24.4. Summary diagram for calcareous mesocycle gC_1. The cyclic positions of the lithofacies are shown in the column on the left hand side of the diagram. Sediments attributed to Lithofacies 6 are often missing.

ditions of peritidal sedimentation over an irregular, early Middle Cambrian landscape that was progressively drowned and subsequently infilled by sediment. The Ordian marine transgression invaded a hilly landscape locally characterised by a series of crudely parallel ridges and valleys. As the initial transgressive event did not result in the complete drowning of this landscape a series of embayed environments or lagoons developed between adjacent island chains (former ridges) and larger emergent land masses (Fig. 24.3A). Tidal passes occurred between adjacent islands. These passes linked embayed areas with deeper channels (sites of former valleys) where open circulatory conditions prevailed (Fig. 24.3A). Muds found in Lithofacies 1 accumulated in the protected embayed areas. Lithofacies 2 phosphatic sediments were deposited on semi-emergent flats that developed in the protected lee of island chains and as prograding sediment sheets adjacent to larger emergent areas (Fig. 24.3A). Mixed-mineralogy echinoderm grainstones accumulated in raised energy environments adjacent to the tidal passes (Fig. 24.3A). Where phosphate particles were locally sorted, they subsequently accumulated as low-relief shoals composed of grainstone phosphorite (Lithofacies 7). Some of the phosphate particles were derived from eroded phosphatic pavements and lithologies previously deposited in Lithofacies 2.

Continued basin subsidence and sediment accumulation resulted in the drowning of Precambrian basement rocks and the development of a broad, shallow epicontinental sea on a shelf with extremely low relief (Fig. 24.B). Epeiric conditions predominated in the shallow seas and produced a complex mosaic of carbonate depositional environments in which sediments belonging to Lithofacies 4, 5 and 6 were deposited (Fig. 24.3B). However, where shelf slopes increased and more open circulatory conditions occurred, echinoderm detritus accumulated to form a series of barriers (Fig. 24.3B). These barriers developed in areas of rapid shoaling (increased shelf slopes) which were subjected to tidal influence. Broad semi-emergent flats (Lithofacies 2 sediments) developed in the protected lee of the emergent encrinite shoals. Phosphate particles derived from Lithofacies 2 rocks and other possible offshore sources were eroded, transported, preferentially sorted and subsequently accumulated in higher energy environments associated with the tidal passes and lagoons (Lithofacies 7).

The onset of epeiric conditions marked the beginning of a regional regression that culminated in emergence, karstification and the eventual cessation of Thorntonia Limestone sedimentation. As this regressive event was basin-wide, a regional disconformity occurs throughout the basin at the end of Templetonian time (Shergold & Druce, 1980; Cook, 1982), and probably represents a regional lowering of sea level.

A series of minor marine inundations occurred during the interval of time represented by the disconformity surface on top of the Thorntonia Limestone. Phosphorites were deposited during some of these inundations. The next major transgressive event resulted in the deposition of phosphatic rocks of the Gowers Formation which are succeeded by carbonates of the overlying Middle Cambrian succession (Fig. 24.1).

Synsedimentary, phosphatic pavements, profiles and stromatolites

The preceding parts of this Chapter have outlined the broad-scale palaeoenvironmental associations represented by rocks that comprise the Thorntonia Limestone and Gowers Formation. The remaining sections are concerned with describing and interpreting a series of *in situ* phosphatic nodules, pavements, profiles and stromatolites which occur within the mixed-mineralogy suite of rocks. Included in this section is an outline of the types of particles that are associated with these phosphatic sedimentary structures and lithologies.

Fig. 24.3. Regional patterns of Thorntonia Limestone sedimentation through time. (A) Initial flooding of an irregular hilly landscape composed of folded Precambrian rocks resulted in the formation of protected embayments and lagoons which were bordered by semi-emergent flats. (B) Progressive drowning of this landscape and continued sediment accumulation produced a shelf of low relief on which a complex mosaic of carbonate depositional environments developed. However in areas where remnant basement highs occurred, increased shelf slopes resulted in more open circulatory conditions. Encrinite shoals developed in these areas and phosphorites both formed and accumulated in zones adjacent to the emergent regions. (C) The final phase of Thorntonia Limestone sedimentation resulted in the development of a freshwater algal marsh. Dominated by flat to domal laminated stromatolites with a palisade structure, these sediments accumulated in low ponded areas that developed above emergent rocks of the Thorntonia Limestone. Sediments of the freshwater algal marsh are separated from the underlying semi-emergent phosphatic carbonates by microkarsted and scalloped dissolution surfaces.

Phosphatic nodules and platy-laminae

Phosphate-cemented nodules and sheet-like laminae underlie some phosphatic hardgrounds (Figs 24.5, 24.6). Nodules of phosphatic mudstone form elongate bodies 0.4–4 cm in length that are no more than 3–4 mm thick. Plate-like phosphate laminae form thin discontinuous sheet-like structures that vary from 0.5 mm to 1.0 mm in thickness and are up to 5 cm in length. Carbonate mudstone occurs between laterally adjacent plates and nodules. Contacts between the phosphate-cemented structures and their host mudstones are gradational. Carbonate mudstones are extensively burrowed with burrows infilled by either carbonate mud or a mixture of carbonate mud and phosphate grains. Irregularities on the upper surfaces of nodules and platy-laminae are either moulds or casts of burrows.

Hardgrounds

In a recent review of the term hardground Bromley (1978) notes that according to present usage this term represents a discontinuity surface at which early diagenetic cementation has produced a lithified sea floor. However, in the same article Bromley notes that it is often impossible to distinguish purely submarine hardgrounds from those horizons that were cemented in a vadose environment (i.e. a semi-emergent setting). In this paper usage of the term hardground is restricted to surfaces of synsedimentary lithification but includes pavements of submergent or semi-emergent origin. Synsedimentary lithification is evidenced by a number of features which include:

(1) borings
(2) an irregular upper surface characterised by an overhanging

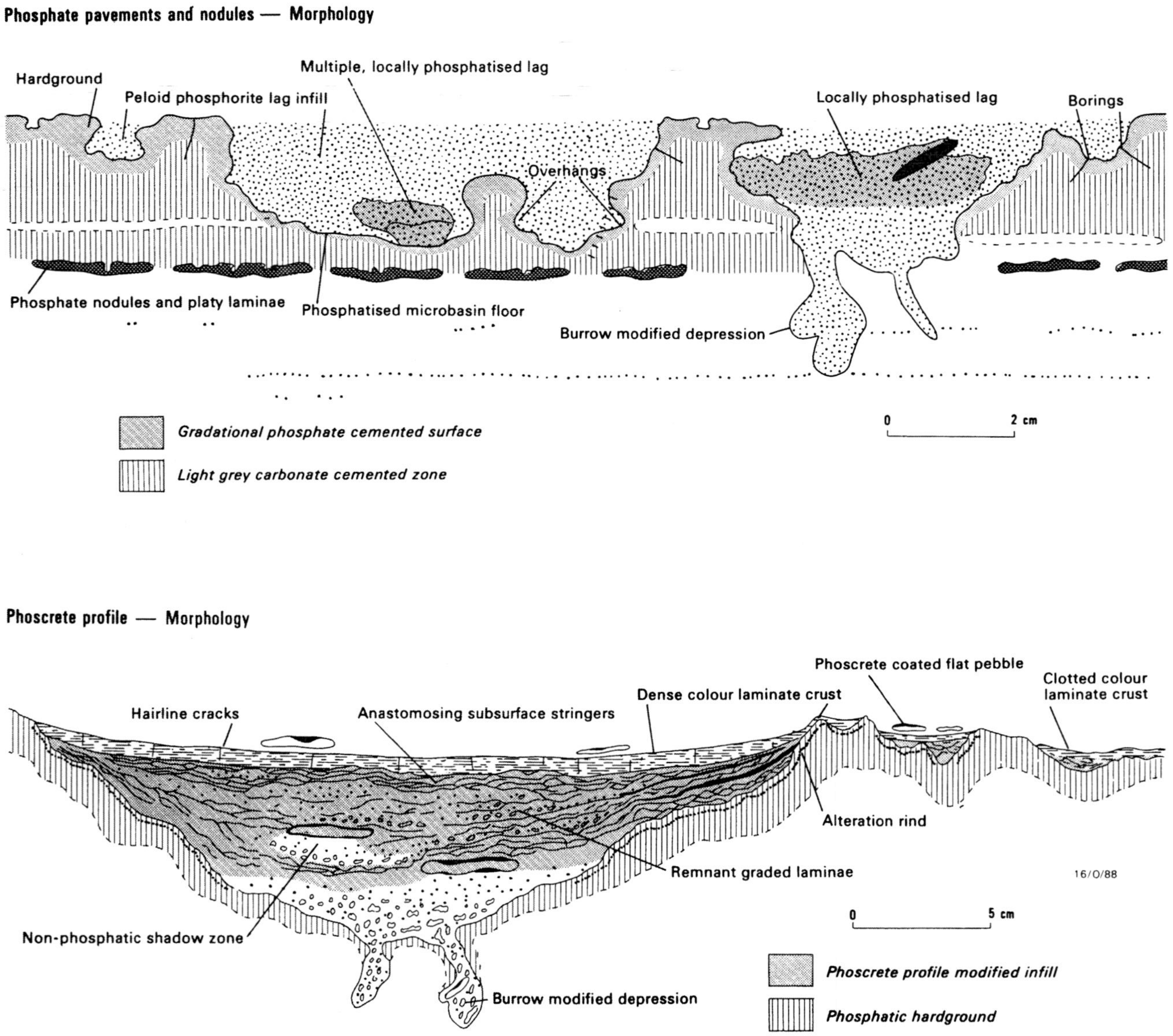

Fig. 24.5. Idealised sequence of phosphatic nodules and hardgrounds as well as a phoscrete profile, here developed above an impermeable substratum formed by a phosphatic hardground. Included in these diagrams are the principal morphological terms used in the text to describe these sedimentary structures.

Fig. 24.6. (A) Phosphatic nodules, hardgrounds and locally phosphatised lags. Basal-laminated and locally bioturbated phosphatic mudstones and wackestones are truncated by an erosional surface above which occurs a fining-upward sequence dominated by mudstones. The fining-upward sequence terminates with the development of an erosion-modified hardground (irregular white surface). Depressions on the surface are infilled by lag deposits composed of phosphate particles in a matrix of carbonate mud. Where these deposits are cemented by phosphate, locally phosphatised lags form (centre left). The basal erosion-modified hardground is overlain by two fining-upward cycles which are capped by phosphate nodules. The uppermost layer of phosphate nodules is separated from the top erosion-modified hardground by several incipient hardgrounds. Basal slab length is 26 cm. (B) Phosphatic hardground and phoscrete profile. Laminate and burrowed phosphatic mudstones are overlain by an irregular erosion-modified hardground beneath which occur several burrow-modified depressions. The impermeable hardground is overlain by a phoscrete profile here divisible into an uppermost dense colour-laminate crust and an underlying phoscrete-modified substratum. Note how the phoscrete profiles are restricted to depressions and are absent from high areas on the right hand side of the slab. Grey coloured, anastomosing subsurface micritic stringers as well as lens- and pod-shaped bodies of cryptocrystalline phosphate occur within the modified substratum which is characterised by a clotted to vaguely pelleted texture (cf. James, 1972). Except where elongate, impermeable flat pebbles occur within the modified substratum (centre), gradational contacts form between the modified profile and the underlying calcareous sediments. Bar scale divisions are 1 cm long.

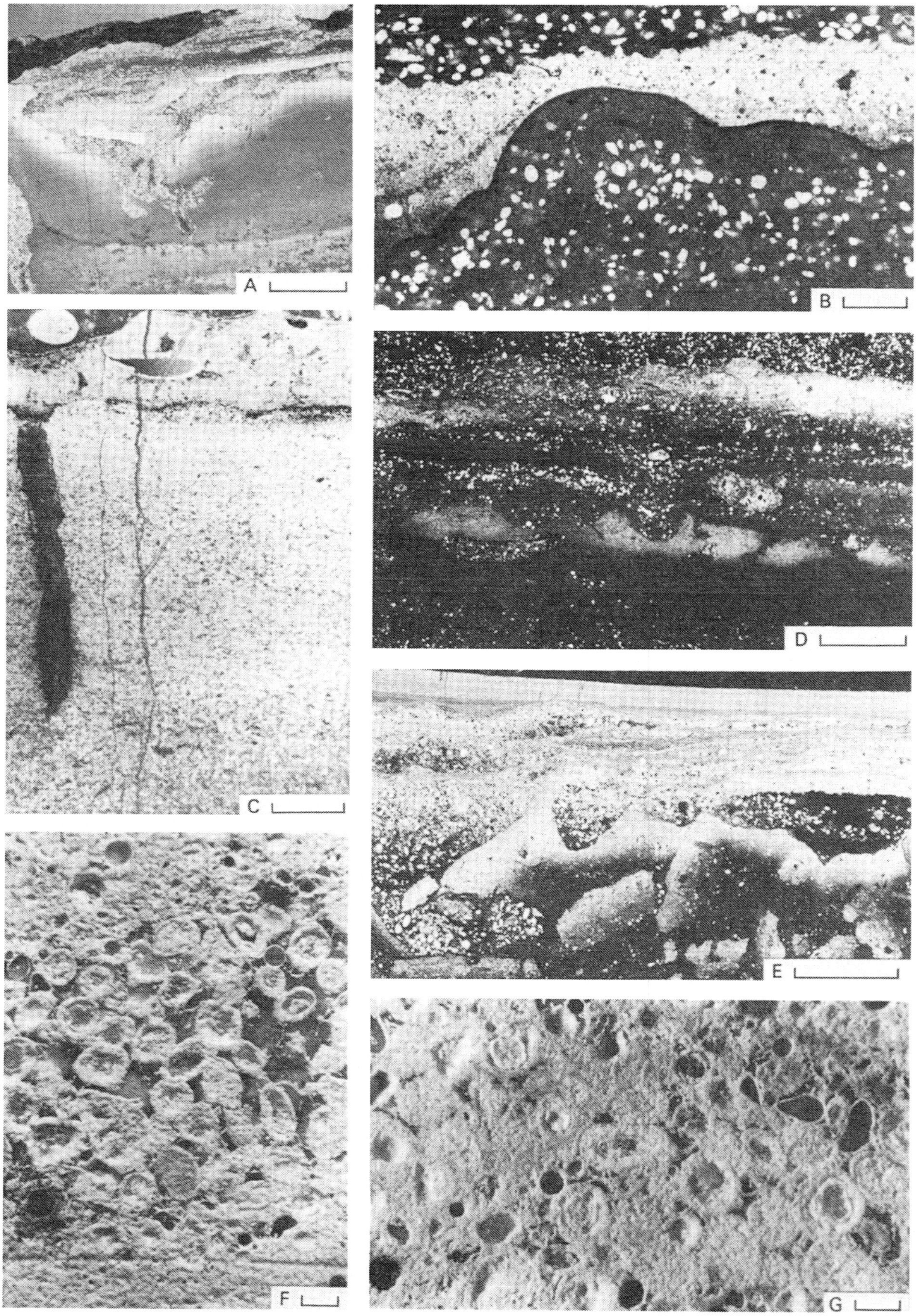
A
B
C
D
E
F
G

sculpture which could not have supported itself if unlithified
(3) a lag deposit composed of angular intraclasts derived from subjacent hardgrounds
(4) a lack of compaction

Phosphatic hardgrounds in the Thorntonia Limestone and Gowers Formation are planar to irregular pavements composed of an uppermost gradationally phosphatic rind 0.5–6.0 mm thick and an underlying zone of mudstone or wackestone 0.5–2.5 cm thick (Figs 24.5, 24.6). Hardgrounds are punctured by vertical borings up to 1 mm in diameter (Fig. 24.6). Pre-lithification soft sediment burrows are incorporated within the hardgrounds. Infills of these burrows contain phosphate particles in an otherwise non-phosphatic host mudstone. Phosphate and carbonate particles form lag deposits that either infill small surface depressions or occur as scattered grains on planar surfaces.

Hardgrounds are laterally discontinuous structures. Whereas some surfaces can be traced over distances of tens of metres others disappear or merge with adjacent hardgrounds over intervals of tens of centimetres or less. The most continuous hardgrounds have horizontal upper surfaces that parallel the bedding and define a datum level to which eroded depressions are related. Depressions on hardground surfaces can be subdivided into two groups, (1) erosion-modified pits and microbasins, and (2) burrow-modified depressions (Fig. 24.5). Erosion-modified pits and microbasins vary from less than 1 cm to tens of centimetres across. They are outlined by gradationally phosphatised walls and floors and have a local relief that varies from 1 mm to 3 cm (Figs 24.5, Fig. 24.6). Where closely spaced hardgrounds occur the local relief on some surfaces can extend to depths of 8 cm. The phosphatic rind which marks the walls and floors of the microbasin usually decreases in thickness with an increase in depth (Fig. 24.5). Overhang structures are common, and usually occur in the well-cemented upper portions of hardgrounds where they form erosional remnants. Areas of shelter porosity are often found beneath the overhangs. Burrow-modified depressions occur where microbasins extend to depths exceeding 2–3 cm (Figs. 24.5 to 24.7). These depressions are characterised by either branching, finger-like projections or bulb-shaped chambers that are outlined by successive cusp-shaped arcs. Burrows vary from 0.3 cm to 1.5 cm across and often coalesce to form semi-circular shaped chambers 2–5 cm across. Hardgrounds are undermined by the burrows and enlarged chambers.

In both phosphatic hardgrounds and nodules phosphate occurs as a series of discontinuous crescent-shaped, or less commonly continuous rims of micritic phosphate. The rims, which are up to 7 μm thick, coat a variety of substrata including carbonate peloids, skeletal fragments and vaguely defined patches of micritic carbonate. Carbonate inclusions occur within the phosphatic rims which, together with the vaguely defined patches of micrite, define a texture characterised by 'structure grumeleuse' (*sensu* Bathurst, 1975). Where elongate horizontally orientated skeletal fragments (especially those of trilobites) occur within a phosphatic hardground, a phosphatic shadow zone forms beneath them (Fig. 24.7).

Phosphatic nodules, platy-laminae and hardgrounds are interpreted as representing a synsedimentary, genetically-related suite of phosphate-cemented sedimentary structures (Southgate, 1983). Initial cementation, at discrete centres of nucleation, resulted in the formation of phosphatic nodules and platy-laminae (Fig. 24.8). As cement precipitation continued, the nodules and platy-laminae accreted laterally and merged, forming a pavement. Subsequent erosion of the overlying carbonate muds and parts of either the cemented pavement or the still soft internodule areas resulted in the formation of an irregular surface which was subjected to alternating phases of cement precipitation, and erosion (Fig. 24.8). These alternating cycles resulted in the sculpturing and eventual formation of an irregular hardground.

While the sequence of events outlined above presents information on the stages of formation of nodules, platy-laminae and hardgrounds it does not consider any of the processes by which cementation may have occurred. In order to understand the rate

Fig. 24.7. (A) Irregular, erosion-modified phosphatic hardground overlying a sequence of carbonate mudstone. The central depression has been infilled by phosphate particles subsequently cemented to form a locally phosphatised lag. Elsewhere the hardground is overlain by flat pebbles of laminoid fenestral phosphorite. The finger-like projections (centre) are infills of burrow-modified depressions. Scale bar length 2 cm. (B) Phosphatic hardground (upper white surface) underlain by a fragment of an agnostid trilobite. Note how the skeletal particle has acted as a permeability barrier to the downward movement of phosphatising solutions. This has resulted in a gradational zone of phosphate cement on the left hand side of the plate and a sharply defined phosphatic shadow zone beneath the raised central part of the trilobite Scale bar length 2 mm. (C) Trypanites boring in a phosphatic hardground. Both the hardground and sediment infilling the boring are overlain by a thin phoscrete profile. Scale bar length 2 mm. (D) Enlargement of part of the upper left hand side of the slab shown in Fig. 24.6 A. The figure shows a layer of phosphatic nodules with cusp-shaped smoothly curved upper surfaces which are locally punctured by the shafts of former burrows, now preserved within these synsedimentary cemented nodules. The U-shaped depression on the upper surface of the central nodule was caused by the shaft of an overlying *Monocraterion* burrow. The cone-shaped top of this burrow occurs on the same horizon as an incipient hardground (centre left hand side). The uppermost white surface is part of an irregular erosion-modified hardground. Scale bar length 1 cm. (E) Enlargement of part of the phoscrete profile shown in Fig. 24.6. Note the dense colour-laminate crust (uppermost white surface) with an underlying phoscrete-modified substratum consisting of grey micritic stringers, micritic lenses and pods of phosphate, and progressively ghosted phosphate particles. A phosphatic shadow zone (dark area) occurs beneath the flat pebble of phosphate in centre right. Dark spots within the phoscrete-modified substratum are carbonate particles. The phoscrete profile overlies a disrupted phosphatic hardground. Scale bar length 1 cm. (F), (G) Coated grains and a clotted to vaguely pelleted texture are typical of the modified substratum portion of a phoscrete profile. Note that whereas phosphatic nuclei merge with their micritic coats of phosphate, carbonate grains remain as distinct nucleii, e.g. calcisphere upper right in (F). Where pore spaces are finally occluded by carbonate the phosphate coats are easily distinguished. However, where rims of micritic phosphate from adjacent grains merge and eventually occlude a pore, the former grains are ghosted to form the vaguely pelleted to clotted texture. Scale bar lengths 2 mm.

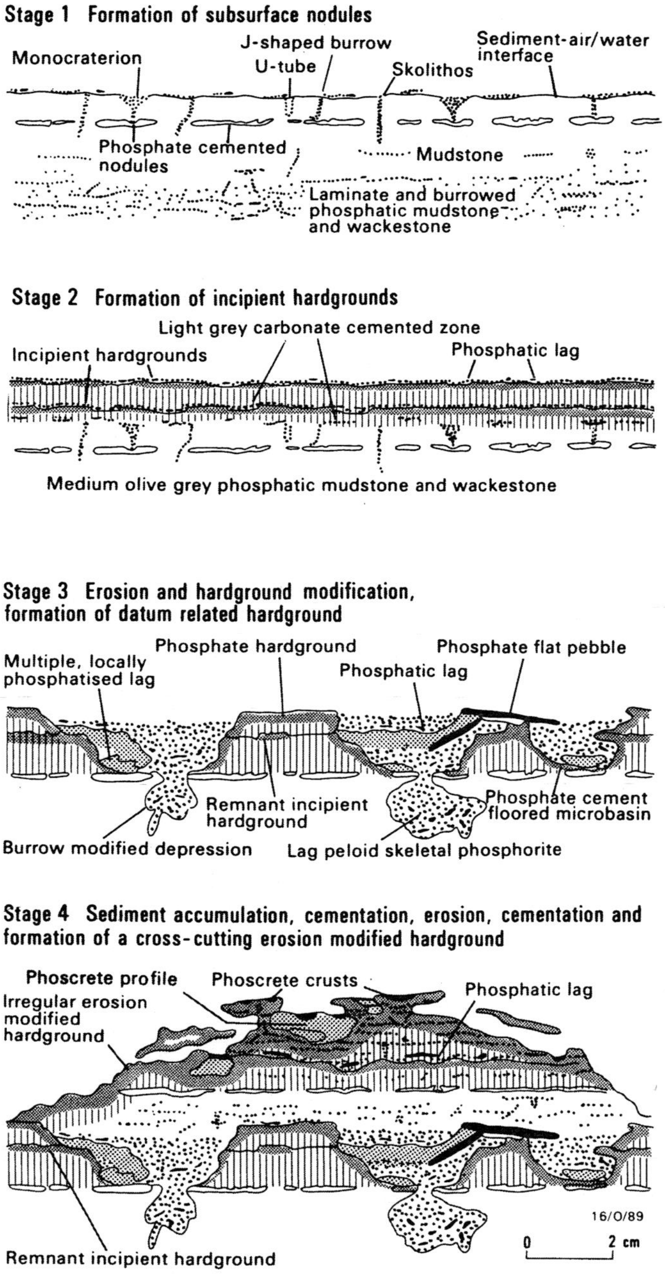

Fig. 24.8. The sequence of events, including sediment deposition, lithification and erosion, which eventually resulted in the formation of a cross-cutting erosion-modified hardground.

and depth of phosphate cementation, as well as the direction of movement of the phosphate-enriched solutions, it is necessary to examine the processes of cementation.

The bounding surfaces of laterally discontinuous phosphatic nodules and platy-laminae vary in shape. Some upper surfaces are arcuate and smoothly curved while others are straight or irregular (Figs 24.6, 24.7). Lower surfaces are generally linear and pass gradationally into the underlying mudstones. Irregularities on the upper surfaces of nodules are caused by the shafts and tunnels of burrowing organisms which either impinged upon, or penetrated, the area now occupied by the phosphatic nodules. As these irregularities are caused by burrows, and burrowing organisms are unable to penetrate lithified sediment, cementation must have occurred after burrowing had taken place. Where the upper surfaces of nodules are characterised by semi-circular shaped depressions, the overlying sediments outline where a stunted *Skolithos* or *Monocraterion* shaft impinged upon the upper surface of the nodule (Fig. 24.7) Stunted shafts vary in length from 5 mm to 10 mm and are infilled by a mixture of phosphate grains and carbonate mud.

The burrow-modified upper surfaces of both platy-laminae and nodules can indicate both the depth and rate at which the phosphate cements precipitated. For the upper surfaces of nodules to be modified by infaunal burrowing organisms, phosphate cements must have precipitated within the sediments from the interstitial waters. As the funnel-shaped top of a *Monocraterion* burrow marks the palaeosurface (Crimes *et al.*, 1977), the distance between the top of the funnel and the U-shaped top of the nodule indicates the depth at which cementation took place. In the case of the burrow shown in Fig. 24.7 this depth is 1 cm. A further consequence of the moulding of the upper surfaces of nodules by burrows is that the rate at which phosphate cements precipitated must have been rapid and coincident with either burrow occupancy or preservation of the lining of the burrow.

Systematic variations in the thickness of the phosphatic rind found on hardground surfaces also provides information relevant to processes of cementation. Where hardgrounds are partially eroded, the phosphatic rind that lines the walls and floor of the microbasin thins gradually with depth. This relationship indicates that periods of erosion and cementation alternated and that conditions conducive to cement precipitation were repeated. It also supports the previous conclusion that phosphate cementation was a near-surface process, in this example only occurring after erosion had taken place. The direction of movement of the phosphate-enriched solutions can be deduced from the phosphatic shadow zone found beneath impermeable particles. Where elongate impermeable skeletal grains, especially trilobites, occur within the phosphatic rind the typically gradational distribution of phosphate cement is locally missing (Fig. 24.7). Instead, the area above the calcareous skeletal grain is phosphate-rich, and a non-phosphatic shadow zone forms beneath the skeletal particle. Such a distribution of phosphate cement suggests that the phosphate-enriched solutions were derived from an overlying source and moved downward through the near-surface layers.

Locally-phosphatised lags

Microbasins and burrow-modified depressions are often infilled with calcareous grainstone or packstone phosphorites. Where these lag deposits are locally cemented by phosphate they form secondary pavements within the sheltered parts of microbasins (Figs 24.6, 24.7). Secondary pavements have a local relief of 1–2 mm. Their thickness is variable, but is always less than the depth of the microbasin. Some pavements form thin, irregular, often discontinuous veneers 1–2 mm thick that line the floors of microbasins, while others occur as distinct surfaces in the upper parts of some depressions. In locally phosphatised lags the distribution of phosphate cement is variable and depends upon the texture of the host material being cemented. Phosphatised wackestones have a dense upper rind with a gradational lower part. Although these surfaces are texturally identical to the underlying hardgrounds, they occur on a more restricted, localised scale. Lags of packstone and grainstone texture also have a cemented upper surface, but the cement is more evenly distributed throughout the lag and does not form a dense upper rind. Instead the cement gradually reduces with depth until the phosphatised lag eventually grades into the underlying grainstone and packstone phosphorites (Fig. 24.7).

Phoscrete profiles

The term phoscrete was initially used by Cook (1972) to describe both surface encrustations and locally thicker deposits of cryptocrystalline phosphate which intermittently blanket the present-day surface in the vicinity of the Australian phosphorite deposits. In this Chapter phoscrete or phoscrete profile is comparable to the term caliche as defined by the Glossary of Geology 1974 edition and used by James (1972) and Harrison (1977). Terms used in the description of phoscretes are essentially those used for carbonates by James (1972) and Harrison (1977).

Phoscrete profiles form an interrelated sequence of laminated to massive rock types that partially to completely infill depressions on some hardground surfaces. The profiles are divisible into two parts: an uppermost, colour-laminated phoscrete crust and an underlying zone of altered calcareous phosphorite.

(1) Laminar phoscrete, crusts. Laminar phoscrete crusts form surficial sheet-like deposits that vary from a few centimetres to several metres across. The crusts coat impermeable lithified substrata and vary in thickness from surficial rinds of less than 1 mm to dense colour-banded layers up to 2 cm thick (Figs 24.5 to 24.7). Examination of slabbed rock specimens indicate that crusts only occur within depressions. In broad shallow depressions the crusts are generally saucer-shaped with a gently inclined outer rim, while those present in narrow steep-sided depressions are U-shaped with more steeply inclined sides (Fig. 24.6). Only where surface irregularities are infilled do the dense colour-laminate crusts form sheets or rinds that extend across the high areas between adjacent microbasins.

Colour laminations are a distinctive feature of the dense colour-laminate crusts (Fig. 24.7). Laminations vary in thickness from layers less than 1 mm thick to more prominent bands several millimetres across. Laminae are defined by subtle changes in colour, with white, pink, red and grey tones present. Boundaries between adjacent colour laminations vary from sharp to diffuse.

In thin-section laminar phoscrete crusts are characterised by alternating layers of clear anastomosing microlaminae and dark, peloid or inclusion-rich microlaminae. Dark laminae contain carbonate peloids, silt-sized quartz grains, dolomite rhombs and pigment. This material is surrounded by a matrix of cryptocrystalline and microcrystalline phosphate which, where dominant, forms clear microlaminae 1–7 μm thick. These clear microlaminae accrete to form bands up to 20 μm thick.

(2) Phoscrete-modified substrata. Where laminar phoscrete crusts are underlain by calcareous phosphorites a texturally modified substratum develops between the basal phosphatic hardground and the uppermost crust (Figs 24.5 to 24.7). Primary sedimentary structures are progressively replaced by a matrix of cryptocrystalline phosphate and are varyingly overprinted and eventually dominated by a series of colour-defined micritic stringers or laminations (Fig. 24.7). The colour-defined laminations are characterised by an anastomosing network of light-grey coloured stringers which pinch and swell. Stringers vary from 0.1 mm to 1 mm thick and in a general way parallel the uppermost surface of the overlying crust.

Lens and pod-shaped bodies of cryptocrystalline phosphate, up to 2 cm long and 1–3 mm wide, occur between the anastomosing subsurface stringers (Fig. 24.7). In some profiles, areas of cryptocrystalline phosphate form vaguely pelleted units up to 2 cm thick (Fig. 24.7). In these units phosphate grains are ghosted, but carbonate particles remain as discrete, readily identifiable grains. Where primary thin beds and laminae are preserved, the matrix of cryptocrystalline phosphate is gradually replaced by the underlying micritic carbonate matrix. Areas of intense diagenetic alteration and substratum modification immediately underlie the phoscrete crusts. Beneath these intensely altered zones a gradational contact exists with the unaltered host substratum. Where parts of the underlying phosphatic hardground protrude into the phoscrete profile a grey-pigmented band develops within the phosphatic hardground (Fig. 24.5). This band is identical to the grey micritic stringers previously described. Similar grey bands of cryptocrystalline phosphate develop within the outer margins of flat pebbles and other intraclasts incorporated within the phoscrete profile.

The characteristic petrographic feature of both locally phosphatised lags and phoscrete substrata is the diverse variety of cements which combine to produce a matrix with a vaguely pelleted to clotted texture (cf. James, 1972). Although micritic cements are the dominant component of the matrix, other interparticle cements include drusy fringes of acicular crystals and colloform bands. Cements vary greatly in both thickness and their spatial relationships. This results in irregular, pendant, isopachous and radiating crystal fringes often occurring in adjacent pore spaces. Some incompletely phosphatised pore spaces are criss-crossed by micritic thread-like structures. These phosphatic fibres are interpreted as phosphate-coated chasmolithic algal filaments (Southgate, 1983).

Multiple generations of both similar and dissimilar cements accrete around individual nuclei to produce coated grains. Coats consist of successive layers of micritic phosphate between 2 μm and 30 μm thick. The coats thicken and thin around individual nuclei which can consist of skeletal fragments, carbonate peloids or phosphate particles (Fig. 24.7).

Phosphate-cemented rocks comprising both locally phosphatised lags and phoscrete profiles are interpreted as forming in a subaerially exposed vadose environment subject to alternating periods of wetting and drying. The presence of an uppermost colour-laminate crust and an underlying diagenetically altered substratum renders phoscretes very similar to the calcrete deposits described by James (1972), Read (1974), Scholle & Kinsman (1974) and Harrison (1977).

Crusts are a characteristic feature of calcretes. They are defined by colour laminations and either overlie impermeable

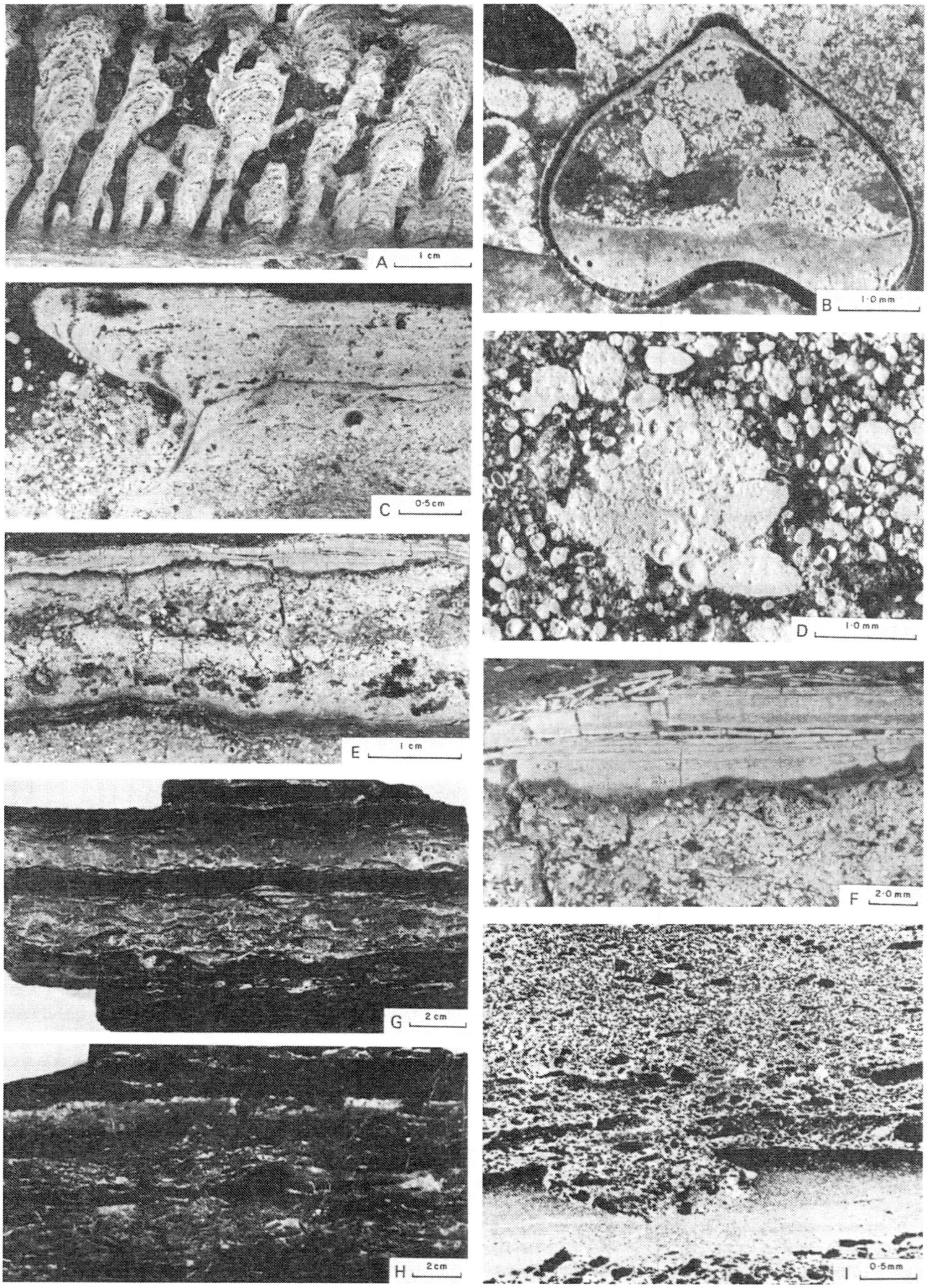
A 1 cm
B 1·0 mm
C 0·5 cm
D 1·0 mm
E 1 cm
F 2·0 mm
G 2 cm
H 2 cm
I 0·5 mm

bedrock or diagenetically altered substrata (James, 1972; Harrison, 1977). In the Gowers Formation and Thorntonia Limestone phoscrete crusts coated a variety of impermeable substrata including phosphatic hardgrounds, scalloped dissolution surfaces, phoscrete-modified substrata, flat pebbles and erosional surfaces developed on desiccated stratiform stromatolites and mudstone phosphorites. Additionally, some crusts formed sheet-like deposits overlying dissolution surfaces developed on coquinas composed of vertically stacked trilobite fragments which are indicative of deposition in a swash zone (Sanderson & Donovan, 1974). A zone where periodical subaerial exposure resulting in vadose diagenesis, therefore would have occurred. Further evidence of vadose diagenesis is provided by the diverse variety of cements found in the phoscrete modified substrata and locally phosphatised lags. The occurrence in adjacent pore spaces of meniscus, pendant and isopachous cements, coupled with the dominance of micritic cements, suggests that cementation took place in the vadose zone (Harrison, 1977). In this zone each pore space behaves differently. Blocked pore throats facilitate the retention of interstitial waters from which isopachous cements precipitate, whereas pore throats that remain open allow interstitial solutions to drain and pendant and meniscus cements to form.

Where calcrete profiles develop within a permeable substratum, the surface crust is underlain by a diagenetically altered zone on which a series of micritic subsurface stringers and lenses are superimposed (Harrison, 1977). Vadose diagenesis within this zone results in the formation of a micritic matrix and a characteristic series of complex textures and fabrics that progressively obscure the outlines of former carbonate particles. Grains are micritised and frequently become coated by successive laminations of micrite. These coated grains (James, 1972; Harrison, 1977), or calcrete ooids (Read, 1976), are a characteristic feature of calcretes. Within a calcrete profile there is a characteristic upward transition from weakly cemented grains, to grains coated by layers of micrite, to horizons of coated grains with partly shared laminations, and finally to dense predominantly micritic horizons where grains are ghosted and interparticle pores are plugged by a micritic cement (Harrison, 1977). The progressive micritisation of grains, coating of grains by micrite, and the infilling of interparticle pores by a micritic cement, results in the formation of a dense to vaguely pelleted micritic matrix (James, 1972; Harrison, 1977). It is only after these pores are plugged that impermeable substrata suitable for the development of laminated crusts are formed (Harrison, 1977).

In phoscrete profiles diagenetically modified substrata only developed within depressions infilled by calcareous packstone or grainstone phosphorites. Where colour-laminate crusts overlie impermeable hardgrounds the diagenetically modified substrates are absent. Phoscrete-modified substrata have a coated to vaguely pelleted matrix. Discrete carbonate and phosphate particles characterise the lower parts of infills. A dense, vaguely pelleted matrix with ghosted phosphate particles but distinctly outlined carbonate peloids and quartz silt, occurs in the upper parts of infills. The dense pelleted matrix occurs immediately beneath the colour-laminate crust where numerous grey subsurface stringers and lens-shaped cryptocrystalline pods are also found. Where either a colour-laminated crust or part of a diagenetically altered profile directly impinges on the underlying hardground an alteration zone outlined by grey pigment forms within the phosphatic hardground. As this alteration zone is similar in both colour and texture to the subsurface stringers and is only found where phoscrete profiles are in direct contact with a hardground, it suggests that hardgrounds formed under a different set of physico-chemical conditions to those represented by phoscrete profiles. However, the gradational distribution of the phosphate matrix, coupled with the occurrence of phosphatic shadow zones beneath elongate impermeable particles incorporated within the

Fig. 24.9. (A) Stratiform and columnar, laminate phosphatic stromatolites, form *Ilicta* cf. *composita*. Note how the intercolumnar areas are infilled by dolomite, so that apart from scattered phosphate grains they are non-phosphatic. (B) Multi-stage phosphatic infill of a hyolith. The outermost part of the calcareous shell, e.g. top and middle left-hand side, has a weakly phosphatic gradational white fringe representing a phosphatic micritic envelope. Phosphate peloids occur within the area enclosed by the shell. (C) A phosphatic stratiform stromatolite with fine laminoid fenestrae overlies a phosphatic hardground. The dark-coloured fenestrae are infilled by carbonate rather than phosphate. (D) Composite phosphate particle (aggregate) containing phosphate-coated grains and peloids. The ragged outer boundary suggests that this grain was weakly cemented and therefore friable at the time of deposition. Other phosphatic particles include lumps (top left), peloids, coated grains and skeletal fragments. (E) Extensively desiccated and disrupted phosphorite with an irregular fenestral fabric. Note how the desiccation cracks, e.g., centre left, originate from erosion surfaces. Fenestrae are infilled by carbonate and silica. A tan-coloured irregular erosion surface caps the phosphatic boundstone. This surface is overlain by a thin sequence of sheet-cracked, desiccated and partly eroded mudstone phosphorite. The dark band in the lower part of the plate is a seam of anastomosing microstylolites. (F) Enlarged view of top right of (C), showing sheet-cracked and desiccated laminae of mudstone phosphorite. Note the locally derived intraclasts of mudstone phosphorite that both overlie and occur sandwiched between some laminae. (G) Alternating thin bands of mudstone phosphorite (black) and phosphatic dolostone (grey) from the Xianfeng mine in China. Although the phosphatic hardgrounds are clearly visible in the grey phosphatic dolostones, they merge and become difficult to distinguish in the beds of mudstone phosphorite. (H) Intraclast and mudstone phosphorite from the phosphate deposits near Mussoorie in northern India. Note how the thin bed of mudstone phosphorite, upper right, is composed of multiple thin laminae of mudstone phosphorite. These laminae are similar in texture to the closely spaced phosphatic hardgrounds which comprise the thin beds of mudstone phosphorite in (G). (I) Photomicrograph of a partly eroded phosphatic hardground (black surface) in Lower Cambrian strata from the Montagne Noire in France. The hardground is overlain by a phosphatic wackestone and immediately underlain by a non-phosphatic carbonate mud. Locally derived intraclasts of phosphate (black grains) occur within the wackestone. Like the phosphatic hardgrounds from Australia and China this example has a gradationally phosphatic rind which, according to Prian (1980), formed by the phosphatisation of a carbonate mud. Photograph and details of plate description are from Prian (1980).

infills of depressions, suggests, that like hardgrounds the phosphorus-enriched solutions which resulted in phosphogenesis were derived from an overlying source.

Laminoid fenestral phosphorites

Phosphatic boundstones with a fine to medium laminoid and irregular fenestral fabric are associated with hardgrounds, phoscrete profiles and laminated mudstone phosphorites (Fig. 24.9). Laminoid fenestral phosphorites form laterally discontinuous beds 2–5 cm thick that are characterised by horizontal sheet-cracks up to 2 cm long and vertical V-shaped desiccation cracks up to 1 cm deep. Numerous erosion surfaces occur within these stromatolites. Some erosion surfaces are outlined by dense rinds of cryptocrystalline phosphate, whereas others are irregular and overlain by a lag deposit. Where rinds occur, both the fenestral and interparticle spaces are occluded by a cryptocrystalline phosphate matrix. Both the sheet cracks and fenestrae have unsupported roofs that are outlined by either bulbous protruberances of phosphate peloids or the locally arcuate rims of larger peloids, coated grains or skeletal fragments.

Laminated phosphatic stromatolites

Laminated phosphatic stromatolites from the Thorntonia Limestone have been described by Southgate (1980) and Schmitt & Southgate (1982). Phosphatic forms are restricted to the lowermost parts of stromatolitic beds where they overlie a variety of irregular impermeable substrata which include phosphatic hardgrounds, scalloped dissolution surfaces and crusts. The phosphatic stromatolites are composed of stratiform sheets up to 5 cm thick from which rise elongate, often inclined, columns and domes up to 4 cm in height (Fig. 24.9). The phosphatic stromatolites consist of alternating dark and light microlaminae.

Dark laminae are phosphatic and preserve the outlines of former algal filaments (Schmitt & Southgate, 1982). Light-coloured laminae are dolomitic and recrystallised. These laminae lack any preserved filamentous structure. In any given stromatolitic bed the phosphatic columns are overlain and pass upwards into recrystallised, dolomitic and partly silicified stromatolitic columns. This transition is accompanied by a gradual dominance of light-coloured recrystallised laminae over the darker-coloured phosphatic ones. Intercolumnar areas are infilled by recrystallised and dolomitised phosphatic packstone. Ghosted carbonate particles found within the intercolumnar areas include ooids, peloids and skeletal fragments. Phosphate particles also occur in the intercolumnar areas.

Laminated phosphatic columnar and stratiform stromatolites as well as laminoid fenestral phosphorites are interpreted as forming in semi-emergent depositional environments. In both the Thorntonia Limestone and Gowers Formations the stromatolites are associated with phosphatic hardgrounds, phoscrete crusts and flat pebbles which collectively occur in the upper semi-emergent part of mesocycles mP_1 and gP_1. Vertical desiccation cracks and horizontal sheet-cracks in the laminoid fenestral phosphorites provide evidence of subaerial exposure. The association of ridge-rill structures, scalloped dissolution surfaces, inclined columns and sulfate pseudomorphs, with laminated phosphatic stromatolites is also consistent with subaerial exposure (Southgate, 1980). Further evidence of a semi-emergent depositional environment is provided by the microstructural characteristics of the laminated stromatolites (Schmitt & Southgate, 1982).

The close spatial association between phosphatic stromatolites and lithified impermeable substrates suggests that some similarities may exist between phoscrete profiles and stratiform stromatolites in their respective processes of phosphogenesis. These processes are discussed in a following section.

Mudstone phosphorites

Thin beds of laminated mudstone phosphorite overlie both laminoid fenestral phosphorites and phoscrete crusts. The thin beds occur as infills of depressions on eroded surfaces and thin towards the high areas from which they are usually absent (Fig. 24.9). Laminae are 0.5–2 mm thick, and are separated by horizontal sheet-cracks and fragmented by vertically orientated desiccation cracks. Where mudstone phosphorites are extensively desiccated, flat pebble chips form discontinuous stratiform breccias within the mudstones (Fig. 24.9). The chips are 0.5–2 mm thick and 1–5 mm long. Phosphate chips with a similar size distribution are found in adjacent laminoid fenestral phosphorites, phoscrete crusts and packstone phosphorites.

Desiccated thin beds of mudstone phosphorite are also interpreted as forming in semi-emergent depositional environments. The restriction of these phosphorites to infills of depressions, where they occur as sheet-like draped laminae, suggests that these thin beds result from the settling of phosphate mud from shallow pools of standing water. Desiccation features indicate that these pools periodically evaporated to dryness, producing sheet and mud cracks, which led to the formation of flat pebble chips composed of mudstone phosphorite. The relative rarity of preserved *in situ* laminae and thin beds of mudstone phosphorite, coupled with the local abundance in some packstone and grainstone phosphorites of flat pebble chips of mudstone phosphorite, suggests that these thin beds have a low preservation potential. Even though the flat pebble chips found in the *in situ* brecciated laminae were locally produced, many had rounded corners and sides which suggests that these muds were firm, but not completely lithified when desiccation occurred. Consequently, their rapid abrasion during transport is likely to have produced round structureless grains, perhaps resembling peloids.

Phosphate particles

A commonly accepted and well defined system for the nomenclature of phosphate particles has yet to be agreed upon. Consequently, an unnecessarily large number of synonymous and at times contradictory terms are used for the description of these particles (e.g. Mabie & Hess, 1964; Cook, 1976; Riggs, 1979; Birch, 1979; Soudry & Nathan, 1980). The loose definition of terms not only results in descriptive ambiguities, but also renders the comparison of phosphate particles with other chemically precipitated and subsequently reworked sedimentary grains impossible. In order to simplify this situation the commonly accepted terms used for the description of carbonate grains have been applied to the phosphate particles present in the Gowers Formation and Thorntonia Limestone. A variety of phosphate particles are associated with the phosphatic hardgrounds and profiles. Some of these particles are characterised by textures and fabrics which suggest that they were locally derived; others have an equivocal origin and may have been transported from another source.

Phosphatic peloids (McKee & Gutschick, 1969; Soudry & Nathan, 1980) are rounded to sub-rounded, circular to elongate particles that contain variable amounts of included material (Fig. 24.9). Phosphate peloids vary from 100 μm to 400 μm across and contain inclusions of micritic carbonate patches and carbonate peloids, dolomite rhombs, silt-sized quartz grains and clay. These particles were probably derived from a number of sources, some of which may have included hardgrounds, laminae of mudstone phosphorites or phosphatic stromatolites.

Peloids, skeletal grains and intraclasts of both phosphate and carbonate composition are often coated by single or multiple layers of cryptocrystalline phosphate to produce coated phosphate grains (Figs 24.7, 24.9). Successive accretionary layers form multi-layered coatings of micrite up to 200 μm across. The basal coats typically thicken in depressions on the nucleus so that although the general shape of the nucleus is maintained the surface of the particle is progressively rounded.

Phosphatic ooids (Milliman, 1974) are characterised by alternating layers of micritic carbonate and phosphate. They have a concentric structure, a ghosted radial fabric and come in simple, superficial and compound forms.

Phosphatic skeletal grains and skeletal-shaped grains consist of phosphatic shell debris of primary origin, e.g. inarticulate brachiopods; phosphatic moulds of shelly material (Fig. 24.9), e.g. hyoliths and trilobites; and phosphate-infilled pore spaces, e.g. echinoderm particles. Phosphate replacement of originally calcareous shelly material was only rarely observed and where present resulted in the formation of a phosphatic micritic envelope (Bathurst, 1966; Fig. 24.9).

Composite phosphate particles similar in fabric and texture to the aggregates and lumps of Milliman (1974) have a cryptocrystalline phosphate matrix, a clotted texture and contain phosphate-coated grains (Fig. 24.9). Phosphatic aggregates and lumps usually have ragged outer surfaces which suggests that they represent friable particles, transported only a short distance from their site of origin. These particles, like phosphate-coated grains, were probably derived by the erosion of locally phosphatised lags and phoscrete substrates.

Locally derived flat pebbles and round pebble chips of phosphate composition consist of phosphate-coated grains and peloids with a carbonate or phosphate mineralogy in a matrix of cryptocrystalline phosphate (Figs 24.7, 24.9). Both flat pebbles and round pebble chips are often coated by multi-layered rims of cryptocrystalline phosphate. Some intraclasts have a laminoid fenestral fabric, others have a pigment-defined laminate texture identical to that found in colour-laminate phoscrete crusts.

Other occurrences of phosphatic hardgrounds, nodules and stromatolites of Proterozoic–Cambrian age

The significance of phosphatic hardgrounds, nodules, stromatolites, phoscrete profiles and mudstone phosphorites in processes of phosphogenesis in rocks of Proterozoic–Cambrian age is difficult to assess. While it has been shown that a variety of phosphate-coated particles, peloids, composite grains and intraclasts can be produced by the erosion of *in situ* pavements and lithologies, it is difficult to determine their quantitative importance in the production of phosphate grains. Indeed, examination of the literature on Proterozoic–Cambrian phosphorites would suggest that similar synsedimentary cemented structures and lithologies are comparatively rare. However, as noted in the introduction, their apparent absence may be due to the lack of work undertaken on the predominantly calcareous rocks that surround most deposits. Where then have similar synsedimentary, phosphate-cemented sedimentary structures and lithologies been recognised? To the author's knowledge the only publications on similar phosphatic hardgrounds of Proterozoic–Cambrian age are by Brasier & Hewitt (1979) and Prian (1980).

Phosphatic hardgrounds of Mesozoic and Cainozoic age are well known from Europe, the United States and South Africa (e.g. Kennedy & Garrison, 1975; Riggs, 1979; Jarvis, 1980; Birch, 1980). These geologically younger surfaces have some features in common with the older Proterozoic–Cambrian hardgrounds described in this paper. However, as they are younger in age and therefore influenced by a later, more diverse, burrowing, boring and encrusting fauna (e.g. Bromley, 1975) they are not considered in any detail in this paper.

Brasier & Hewitt (1979) have briefly described a suite of phosphatic hardgrounds, phosphate-encrusted intraclasts and a variety of phosphatic internal moulds of tubular fossils from the Lower Cambrian Hartshill Formation of Warwickshire in England. The line diagrams of Brasier & Hewitt (1979), suggest many similarities with the Georgina Basin hardgrounds in that they are irregular, partly-eroded, multi-developed surfaces marked by phosphatic rinds approximately 1 mm thick. Brasier & Hewitt (1979) interpret this sequence as forming in shallow submergent settings.

Phosphate-cemented pavements of Lower Cambrian age have also been described by Prian (1980) from the Montagne-Noire in Southern France (Fig. 24.9). These pavements have a gradationally phosphatised upper rind 1–2 mm thick, are black in colour and have been locally eroded. They are associated with flat and round pebbles as well as angular phosphatic intraclasts, phosphate-coated grains and composite particles. In some areas phosphate occurs as phosphatic algal filaments. On the basis of their variable colour (black to beige), association with algae and possible birdseye structures, as well as the associated suite of phosphate particles, Prian concluded that these surfaces, and therefore phosphogenesis, took place in shallow submergent to semi-emergent environments. Prian (1980) also concluded that periods of synsedimentary diagenetic phosphogenesis were associated with times of decreased rates of sedimentation in quiet water environments.

The opportunity provided by IGCP 156 to visit deposits of Proterozoic to Cambrian age in Mongolia, India and China has led to the recognition of phosphatic hardgrounds elsewhere. Phosphatic hardgrounds from the Xianfeng mine in China (Fig. 24.9) occur in an interbedded sequence of black mudstone and wackestone phosphorite and light-grey phosphatic dolostone. Although hardgrounds occur in both the light-grey dolomitic units and the black phosphate-rich beds (Fig. 24.9), they are most easily recognised in the phosphatic dolostones, where they form a series of closely spaced, but laterally discontinuous phosphatic surfaces 0.5–1.5 mm thick. The surfaces usually extend for distances of 30 cm and less and have a local relief of 1–5 mm (Fig. 24.9). Surfaces are generally overlain by a phosphatic lag and are characterised by a dark-grey to black upper rind that gradually merges with the underlying weakly phosphatic and lighter-coloured dolomite. Boundaries between the black phosphate-rich beds and the phosphatic dolostones vary from sharp to gradational. However, as a general rule the upper surfaces of most mudstone and wackestone phosphorite beds are sharp erosional surfaces whereas the lower boundaries are gradational. Gradational boundaries are marked by the gradual merging of phosphatic hardgrounds which increase in frequency toward the overlying phosphate-rich bed. Phosphate-rich beds usually have a mudstone or wackestone texture and consist of a variety of phosphate particles including peloids, coated grains, skeletal fragments and intraclasts which are surrounded by a cryptocrystalline phosphate matrix. Peloids are the most common type of carbonate particle present. These grains are frequently coated by rims of micritic phosphate but show no evidence of replacement by phosphate. Black phosphate-rich beds consist of either a number of closely spaced phosphatic hardgrounds 0.5–2 mm thick or a comparatively uniform sequence of mudstone phosphorite. Both the closely-spaced hardgrounds and mudstone phosphorites are normally graded sequences floored by phosphatic wackestones or packstones that are overlain by, and grade into, mudstone phosphorites.

Alternating thin beds of mudstone phosphorite and phosphatic dolostone also occur in other deposits, e.g. Yunnan Province, China, as well as in Mongolia and northern India. The Mongolian occurrences, from the Khubsugul deposit, have been described by Ilyin & Ratnikova (1981) and Ilyin *et al.* (this volume). Like the thin beds of mudstone phosphorite found in China, the Indian (Fig. 24.9) and Mongolian occurrences usually have upper erosion surfaces overlain by a phosphatic lag composed of phosphate peloids and intraclasts in a dolomite matrix. Locally derived flat and round pebbles of mudstone phosphorite are also found in these lags. Phosphatic hardgrounds were not observed during the field examination of these interbedded mudstone phosphorite and phosphatic dolostone sequences. However a sample supplied by Pyrites, Phosphates and Chemicals Ltd from the phosphate deposits near Mussoorie in northern India contains a mudstone phosphorite bed composed of a number of closely spaced phosphorite laminae (Fig. 24.9). The internal structure of this sample is very similar to the beds of mudstone phosphorite from the Xianfeng mine in China.

Possible processes by which phosphatic hardgrounds and phoscrete profiles may have formed

Some sedimentological constraints

The discovery of texturally similar phosphatic hardgrounds in Australia, Europe and Asia suggests that similar processes of phosphogenesis occurred on a wide scale during the Proterozoic–Cambrian interval. Furthermore, the close association between mudstone phosphorites, phosphatic hardgrounds and phosphatic dolostones in the Xianfeng mine suggests that the physico-chemical conditions responsible for the formation of phosphatic hardgrounds may have led elsewhere to the development of mudstone phosphorites. These phosphorites could then have accumulated in a quiet-water protected environment where phosphate muds dominated over the periodical influx of mixed carbonate and phosphate particles.

Apart from the textural similarities between the phosphatic hardgrounds found in Australia, Europe and Asia they are all characterised by relatively rapid changes in the mineralogy of the sediments. For example phosphatic hardgrounds in the Georgina Basin are most common in the uppermost part of a calcareous shallowing-upward cycle dominated by the accumulation of carbonate mud. The hardgrounds described by Prian (1980) and those from China, formed by phosphate cements precipitating in a calcareous sediment, now occur as a dolomitic mudstone or wackestone. Alternating beds of mudstone phosphorite and phosphatic dolostone also represent similar sudden switches in the dominant mineralogy of the sediments. What variables could therefore cause these switches in mineralogy and why should they be restricted to such a relatively small stratigraphic interval? Some of the answers to these questions can be gained by studying the phoscrete profiles that overlie some of the Georgina Basin hardgrounds as well as by examining the upper and lower boundaries of the phosphate-rich beds and hardgrounds. As phoscrete profiles are the phosphate equivalents of calcretes (Southgate, 1983) it should be possible to apply processes known to occur in the formation of modern calcretes to assist in the understanding of these ancient phoscretes. Harrison (1977) has proposed a four-stage model to explain the formation of a calcrete profile. Critical to this model is the development of a friable, uncemented soil horizon above the zone where carbonate reprecipitation and calcrete diagenesis takes place. In essence Harrison's model involves the dissolution of calcium carbonate from the near-surface friable soil zone and its subsequent reprecipitation within the underlying calcrete profile during periods of aridity and pore water evaporation. Colour-laminate crusts formed where the substratum formed a barrier locally impermeable to the downward moving carbonate-enriched solutions. Crusts formed by processes of aggradation.

A similar model can be used to explain the characteristic textures and fabrics found in Cambrian phoscrete profiles and locally phosphatised lags. Colour-laminate crusts only formed above impermeable substrata which included scalloped dissolution surfaces and the uppermost 'plugged' portion of phoscrete profiles. Locally phosphatised lags and phoscrete layers followed a similar distribution, so that like the phosphate crusts they only formed above impermeable layers. As both locally phosphatised lags and phoscrete profiles overlie an impermeable substratum an underlying source of phosphorus can be dismissed. Therefore, like phosphatic hardgrounds and nodules the phosphorus now present in these Cambrian phoscrete profiles and locally phosphatised lags was most probably derived from an overlying as-yet-unidentified source. In calcretes an overlying friable, calcareous soil horizon provided the source for the cement and matrix material (Harrison, 1977). As processes of calcrete formation necessitate the removal of this material it is unlikely that it will be preserved. Similarly in phoscretes, the phosphate source is also unlikely to be preserved. However, in phoscretes there is a slightly different problem. Whereas calcrete profiles formed in a calcareous depositional environment by the dissolution and reprecipitation of the dominant mineral component, phoscrete profiles and locally phosphatised lags record the perturbation of a depositional environment otherwise dominated by carbonates. This switch in mineralogy suggests that the phoscrete equivalent of the calcareous 'soil profile' and therefore the phosphate now found as cement and matrix material, must have been introduced into the depositional environment. This conclusion enables several constraints to be placed on the source of phosphate and possible mechanisms of phosphogenesis. Whereas hardgrounds and phoscrete profiles found in the Georgina Basin have been interpreted as forming in semi-emergent environments, hardgrounds found in Europe have been assigned a submergent origin (Brasier & Hewitt, 1979; Prian, 1980). The textural similarities between the Australian hardgrounds and those found in Europe and China suggest that they may have all formed by similar processes. Their rapid vertical repetition, the lateral discontinuity of each surface, thinness of the phosphatic rind and incorporation of the moulds of burrows within the Australian hardgrounds and nodules suggest that these structures formed over relatively short intervals of time. Additionally, the occurrence of graded laminae and lag deposits in association with hardgrounds suggests that these depositional environments were subjected to sediment winnowing and storms. Where both hardgrounds and phoscretes formed from an introduced source of phosphate over relatively short intervals of time by repeated processes of cementation (see earlier discussions), a transient source of phosphate and/or transient conditions of phosphogenesis are implied.

Although the marine geochemical cycle of phosphorus is poorly known (Bentor, 1980), phosphorus present in near-surface waters and therefore shallow submergent and semi-emergent depositional environments is rapidly incorporated into and recycled by the biomass (Gulbrandsen, 1969). Consequently if phosphate is to be supplied to sediments in a peritidal setting the most likely source will be via organic matter. Most modern phosphorites occur in areas of upwelling where nutrient-rich surface waters result in areas of high productivity and the accumulation of organic-rich sediments on the seafloor (e.g. Burnett, 1977; Bremner, 1980a, b). Similarly, ancient phosphorites, including those of the Georgina Basin, are generally spatially related to organic-rich facies (Gulbrandsen, 1969; Bentor, 1980; Sandstrom, this volume). If organic matter is accepted as the medium by which phosphate is delivered to the sediments it is apparent that this introduced particulate source of phosphate will have the properties of sediment and will therefore be subject to repeated cycles of deposition, erosion and or degradation. This suggests that some controls on phosphogenesis may be related to sites where organic matter is deposited and later degraded releasing phosphate to the interstitial waters.

It is beyond the scope of this chapter to outline the various steps by which organic matter is degraded and phosphate either returned to seawater or incorporated within interstitial waters. These subjects have been discussed by Gulbrandsen (1969), Stumm & Leckie (1970), Burnett (1977) and Sandstrom (1982). However, it is possible to suggest some processes by which the organic material may have been deposited as well as providing possible reasons why shallow submergent and semi-emergent depositional environments seem to have been preferred sites of phosphogenesis.

Phosphatic nodules, hardgrounds and phoscrete profiles—a proposed depositional model

Evidence indicative of Middle Cambrian zones of high productivity in the Georgina Basin are found in the Beetle Creek and Inca Formations. Whereas phosphorites, siltstones and shales of the Beetle Creek and Inca Formations are white and highly weathered in surface outcrops, fresh drill core from the Duchess, Ardmore and Lady Annie areas shows that these sediments are black and organic-rich when fresh (de Keyser & Cook, 1972; Fleming, 1977; McKirdy, 1977; Sandstrom, 1982). The abundance of organic matter in these rocks has resulted in the shales and siltstones being considered as possible source rocks for petroleum (Powell, Cook & McKirdy, 1975; McKirdy, 1977, Sandstrom, 1982). Even though the origin of this organic material is uncertain, two sources have been proposed. McKirdy (1977) suggested that some of the organic matter may have been derived from phytoplankton or cold water zooplankton, while Sandstrom (1982) and Southgate (1983) have suggested that some of this material may have been of non-planktonic algal origin. However, whatever the origin it is apparent that storms could result in the periodic dumping of large quantities of organic matter on the semi-emergent flats (Fig. 24.10). Algal material growing in both submergent and semi-emergent settings could have been ripped-up and placed in suspension during storms, while phytoplankton present as possible red sea blooms may have been transported and deposited onshore during these storms. In many instances any organic material deposited on the semi-emergent flats was probably oxidised and its phosphorus content either returned to the biomass or trapped within the interstitial waters. However, where the organic matter was deposited in shallow ponded environments, in depressions floored by impermeable substrata or in relatively thick piles (Fig. 24.10), anoxic conditions may have developed within the underlying water-saturated, predominantly carbonate muds. Studies aimed at understanding the distribution and fate of organic matter found on intertidal flats were recently undertaken by Ferguson & Burne (1981). They concluded that even though sulfate-reducing bacteria produced a zone of anoxic conditions 1–5 cm thick just below the surface, subsequent groundwater seepage and tidal flushing of the sediments by underlying oxygenated waters rapidly oxidised the sediments. Consequently, in a typical profile through the top few centimetres of the intertidal flats studied by Ferguson & Burne (1981) the black anoxic sediments found just below the surface became progressively tan-coloured and oxidised with depth. This relationship may partly explain the distribution of phosphatic hardgrounds and phoscrete profiles found in the Gowers

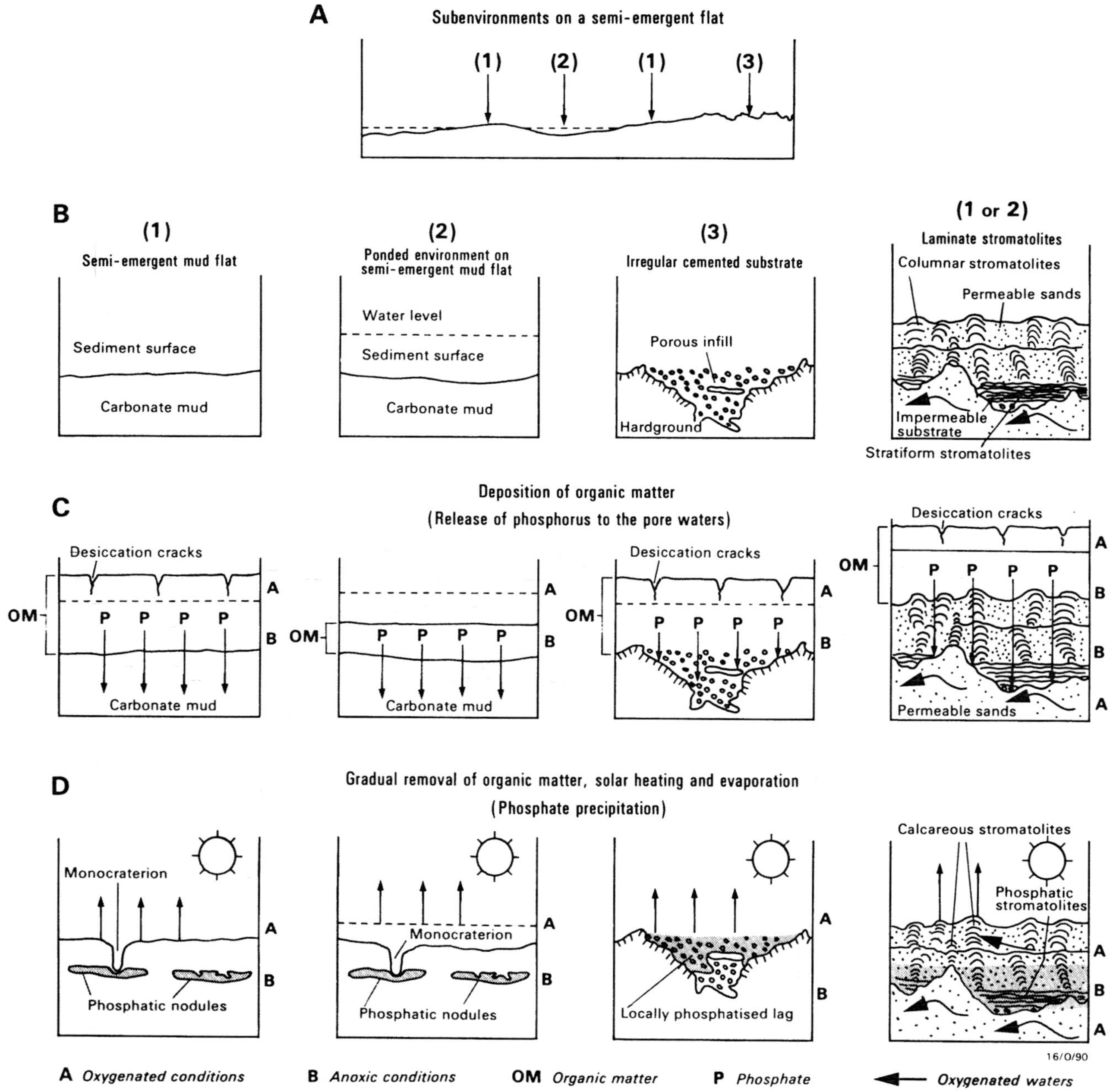

Fig. 24.10. Generalised diagram depicting the major variables in the repetitive 3-stage cycle proposed for the precipitation of phosphate cements in phosphatic nodules and hardgrounds, phoscrete profiles, locally phosphatised lags and phosphatic stromatolites. (A) The three principal subenvironments found on the semi-emergent flats. (B) Normal conditions of sedimentation with the sediment surface either subaerially exposed or locally covered by ponded waters. Surface conditions are oxygenated. Note how in the laminated stromatolites the impermeable substratum shields the immediately overlying stratiform and columnar stromatolites from the oxidising groundwaters. (C) Following a storm and flooding of the semi-emergent flat, organic matter is deposited in subenvironments (1, 2 & 3). This organic matter undergoes bacterial degradation releasing phosphorus to the interstitial waters. Anoxic conditions develop beneath the organic layer. In 1 & 3 the upper surface is progressively desiccated and oxidised. In 2 the organic matter is preserved for a longer period of time. (D) Progressive oxidation and desiccation removes the blanket of organic matter. Solar heating results in the elevation of pore water temperatures and pH. Concomitant evaporation causes pore waters to become more concentrated. Under these conditions phosphate precipitates on suitable substrata within the host sediments forming phosphatic nodules, hardgrounds, stromatolites or phoscrete profiles.

Formation and Thorntonia Limestone. Where phosphatic hardgrounds occur within sediments primarily composed of carbonate mud the early oxidation of any buried or overlying organic matter by groundwater or tidal-water flushing may have been inhibited by the low permeability of the muddy sediments. Similarly, where phoscrete profiles formed within porous, permeable sediments the underlying impermeable substratum may have prevented any oxidising groundwaters from flushing the sediments, leading to the possible local preservation of organic matter and the maintenance of anoxic conditions (Fig. 24.10). This sequence of storm-initiated depositional events would have resulted in a readily renewable supply of organic matter and associated phosphorus to the semi-emergent flats. Furthermore the initial oxidation of this organic material followed by the onset of sulfate reduction as free oxygen became depleted, would have resulted in the release of phosphate to the interstitial waters and the subsequent formation of anaerobic conditions (Trudinger, 1979).

Once phosphate is present in the interstitial waters certain conditions are necessary to cause its precipitation. Although the chemical circumstances which determine whether or not phosphate is fixed within anaerobic sediments are poorly known (Trudinger, 1979) several constraints on the precipitation of phosphate have been identified. Two factors known to favour the precipitation of phosphate, especially in shallow submergent and semi-emergent settings, are: (1) increases in temperatures (Gulbrandsen, 1969); and (2) increases in the pH of the interstitial or lagoonal waters (Roberson, 1966; Pytkowicz & Kester, 1969; Gulbrandsen, 1969).

Where the temperatures of shallow lagoonal, ponded and interstitial semi-emergent waters are raised by solar heating, increased rates of photosynthesis follow and result in an increase in pH which in turn should induce phosphate cements to precipitate. In anoxic sediments raised levels of pH may also result from the decomposition of organic matter (Berner, 1969).

While raised temperatures and pH levels are conducive to phosphate precipitation, magnesium ions are regarded as an inhibitor (Martens & Harriss, 1970). However, work by Baker & Kastner (1981) has shown that anaerobic environments can lead to pore waters becoming depleted in Mg^{2+} by the formation of dolomite. This occurs by the removal of SO_4^{2-} ions which according to Baker & Kastner (1981) inhibit the formation of dolomite. Other ways of depleting the Mg^{2+} ion concentration of interstitial waters are by the formation of diagenetic Mg^{2+}-rich clay minerals such as sepiolite and chlorite (Burnett, 1977).

From the preceding discussion it is evident that a series of physico-chemical factors which are conducive to phosphate precipitation occur in shallow submergent and semi-emergent settings. From a sedimentological viewpoint these settings allow for the rapid repeated deposition of organic matter on semi-emergent flats as well as within ponds and lagoons. Oxidation of the organic matter leads to the repeated release of phosphate to the interstitial waters and the eventual formation of anaerobic conditions as sulfate reduction takes place. This assists in raising the alkalinity of pore waters (Baker & Kastner, 1981), which along with increases in pore water temperature caused by solar heating, results in the precipitation of phosphate cements. Anaerobic conditions inhibit the recycling of phosphorus to the biomass and at the same time promote dolomite formation, which in turn enhances the likelihood of phosphate precipitation. Therefore, even though phosphate is added to the pore waters under both aerobic and anaerobic conditions, the phosphate is preferentially precipitated under anoxic conditions which develop locally beneath the layer (blanket) of organic matter (Fig. 24.10).

Where the upper surfaces of some phosphatic nodules are moulded by overlying burrows it is also evident that phosphate precipitation continues, even after the organic layer has been removed. As *Monocraterion* and *Skolithos* burrows represent former filter feeders (Crimes *et al.*, 1977) they must have had access to the sediment–air–water interface. Consequently, where the upper surfaces of phosphate nodules are moulded by these burrows phosphate cementation would have occurred after the organic layer was removed. Removal of this layer would therefore subject the muddy sediments to the direct effects of solar heating, thus facilitating the rapid rise of pH which should have been accompanied by phosphate cement precipitation (Fig. 24.10).

Whereas periods of phosphate cement precipitation were probably associated with the deposition of an organic-rich layer resulting in anaerobic conditions, sediment erosion and hardground exposure at the sediment–air–water interface would have been accompanied by oxygenated conditions. During these periods, hardgrounds were bored and phosphatic pavements and profiles altered by sediment erosion, and the undermining effects of burrowing organisms. However, during later periods of burial and blanketing of the surface by organic matter, renewed cycles of phosphogenesis probably took place.

A similar model can be used to explain the genesis of phosphatic hardgrounds as well as interbedded sequences of phosphatic dolostone and mudstone–wackestone phosphorites of India, China and Mongolia. Whereas in semi-emergent environments the storm-deposited organic matter would have formed a distinct yet gradually desiccating layer, in submergent environments the same material may have formed a layer of partly suspended material that gradually increased in density with an increase in thickness. As a dense blanket of organic matter would affect sediment dynamics, it is pertinent to consider the physical properties of such a layer. The closest modern analogue available for determining the effects on sedimentation of a layer of suspended organic matter is provided by Bremner (1980a) who has described the physical properties of the diatomaceous mud belt that mantles the sea-floor on the inner continental shelf off south-west Africa. This mud belt occurs in water depths of 1–222 m and has formed in response to high levels of organic productivity in upwelling waters associated with the Benguela Current. Pre-Holocene as well as Recent phosphate concretions with a diagenetic origin occur within the mud belt (Bremner, 1980a). On the basis of theoretical considerations, Bremner (1980a) concluded that wave-induced bottom currents should have inhibited the formation of this layer. However there are several physical properties inherent in these muds which add to their stability:

(1) The sediment–water interface is characterised by an ill-defined turbid boundary. Water turbidity is caused by large quantities of suspended organic matter and mud.

(2) The non-cohesive bottom muds have a damping effect on incoming waves. Consequently much of the energy associated with these waves is absorbed by the mud layer, thus reducing the potential for sea-floor erosion.

(3) The bottom mud layer has a density in excess of 100 mg/l; above this density-level muddy sediments will accumulate irrespective of wave activity at the bed (McCave, 1971).

From the above discussion it is apparent that the loss of wave activity at the bed will result in typically quiet-water fine-grained sediments accumulating in an environment of moderate energy where swell or wind waves would normally feel the bottom. Under these conditions sediments regarded as normally being deposited below wave base would have accumulated in much shallower depths. Furthermore, in such a sedimentary system, small-scale sedimentary cycles such as alternating sequences of mudstone phosphorite and phosphatic dolostone could be related to the expansion and/or contraction of an organic layer rather than storms, as is normally the case in shallow seas (Hardie & Ginsburg, 1977). Under such conditions of sedimentation, the phosphatic hardgrounds found in the Xianfeng mine would have formed beneath a thin transient layer of organic matter. Therefore, the hardgrounds are assumed to mark times of increased organic input and expansion of the organic layer. Thin beds of mudstone phosphorite would then represent periods when the organic layer had expanded, promoting the onset of anaerobic conditions, and supply of phosphate to the pore waters resulting in phosphogenesis. Phosphatic dolostones would then record contraction of the organic layer, and feeling of the bottom by storm and wind waves, followed by erosion of the previously deposited mudstone phosphorites to produce the phosphate particles now found in the phosphatic dolostones. Contraction of the organic layer would result from the return to predominantly oxygenated conditions which coincided with the dominance of carbonate rather than phosphate.

In this model, conditions of raised energy at the bed are not related to shallowing, but to the relative thickness and consequent damping effects of the organic layer. This relationship provides a possible explanation for the otherwise perplexing problem of sediments apparently displaying features of both deep and shallow water depositional environments. For example, the close spatial association between beds of quiet-water mudstone phosphorites characterised by an uppermost erosional surface and overlain by higher-energy lag deposits, which contain flat and round pebbles of mudstone phosphorite, can be explained in the following way: as the organic layer gradually declines in thickness and contracts, phosphogenesis is halted and waves begin to feel the bottom, eroding the previously formed beds of mudstone phosphorite. This results in the formation of an upper erosion surface which is overlain by a phosphatic lag containing locally derived phosphate intraclasts up to pebble size. The same model can also be used to explain the sharp to gradational boundaries that occur between the upper parts of the dolomitic beds and the lower part of the mudstone–wackestone phosphorite beds. Gradational boundaries are marked by: (1) a fining-upward sequence, and (2) a gradual increase in the number of phosphatic hardgrounds which become progressively more closely spaced until they eventually merge to form a phosphate bed. This sequence would represent the gradual expansion of the organic layer causing a reduction of energy levels at the bed, and the formation of phosphatic hardgrounds. If this model is correct then sharp boundaries would represent the rapid expansion of the organic layer which resulted in a sudden switch from a system dominated by carbonates to one in which phosphogenesis took place.

Phosphatic stromatolites

The close spatial relationship between phosphatic stromatolites and cemented impermeable substrata suggests that a chain of synsedimentary–diagenetic events similar to those previously described for phoscrete profiles may have elsewhere led to the formation of phosphatic stromatolites. These stromatolites accreted in semi-emergent settings (Southgate, 1980) and by definition originally consisted of an intertwined array of algal filaments. Even though these filaments consisted of organic matter, insufficient quantitites of phosphorus would have been present in this material to account for the present phosphate content of the stromatolite. Consequently, it is necessary to invoke an extraneous source of phosphate. As the phosphatic stromatolites overlie impermeable substrata, this extraneous source was most likely to be an overlying one. In some cases the outlines of former algal filaments are preserved within the phosphatic stromatolites (Schmitt & Southgate, 1982). This suggests that phosphorus, once present within the interstitial waters, precipitated on these filaments as a post-mortem diagenetic cement. However, where phosphatic stromatolites are surrounded by porous and permeable phosphatic echinoderm sands, rather than impermeable muds, the role played by the impermeable substratum assumes a greater degree of importance, shielding the stromatolitic organic matter from the oxidising effects of groundwater flow and tidal flushing (Fig. 24.10). This resulted in the maintenance of locally anoxic conditions above the impermeable substratum and the slow bacterial degradation of the stromatolitic organic matter found within this protected zone. Elsewhere, the organic material was oxidised so that calcareous, but now dolomitic, stromatolites formed. Storms resulted in the deposition of organic matter on the intertidal flats. Phosphate present within this organic matter was released to the underlying interstitial waters. This phosphate subsequently precipitated on organic templates preserved in areas where locally anaerobic conditions were maintained above the impermeable substrata. The sequence of events outlined above and the physical constraints placed on this model by the impermeable substratum explain why only the basal parts of these stromatolitic beds were phosphatised. The model also explains why the intercolumnar areas were not phosphatised, the lack of suitable organic substrata and microenvironments having precluded the precipitation of phosphate in these areas. In this model organic matter is involved in two separate but interrelated processes of phosphogenesis. While the storm-deposited organic matter acts as a source of phosphate and results in generally anoxic conditions in the underlying sediments, organic matter present within the sediments acts as a template for phosphate cements to precipitate on. This dual role of organic matter is discussed further by Southgate (in prep.).

Although phosphatic stromatolites occur only rarely in the Georgina Basin deposits they are particularly common in phosphorites found on the Indian subcontinent (Banerjee, this volume) and have also been recorded from China and the USSR (e.g. Bushinsky; 1964, 1969). Whereas very little information is available on the phosphatic stromatolites from the USSR and China, the Indian occurrences have been studied by several workers including Banerjee (1971), Chauhan (1979) and Choudhuri & Roy (this volume).

Phosphatic stromatolites from Rajasthan, India have been interpreted as accreting in submergent, semi-emergent and emergent depositional environments (e.g. Banerjee, 1971, Chauhan 1979; Sant, 1979; Choudhuri & Roy, this volume). Phosphatic stromatolites with an interpreted semi-emergent depositional environment are well developed in the Nimach Mata area, where they are characterised by a number of forms including stratiform sheets as well as single and laterally linked columns (Chauhan, 1979). Both forms are incompletely phosphatised and are frequently partly eroded. Phosphate particles are common in the intercolumnar areas and are of stromatolitic derivation (Chauhan, 1979). These intercolumnar areas have a grainstone or possible packstone texture (intrasparite or intrasparudite texture of Chauhan, 1979).

Unlike stromatolites from the Nimach Mata area, those from the Maton and Jhamarkotra deposits are similar in form consisting of branched and single phosphatic columns up to 40 cm in length (Chauhan, 1979). In both these places intercolumnar areas are non-phosphatic and generally infilled by a micritic dolomite. Sedimentary structures indicative of subaerial exposure are lacking, so that the stromatolites are interpreted as accreting in shallow submergent depositional environments (Chauhan, 1979).

Little agreement exists as to the precise role played by these algal structures in phosphogenesis. However, Choudhuri & Roy (this volume) have summarised the main arguments which propose either the direct precipitation of phosphate from seawater as a metabolic function of the algae (Verma & Barman, 1980), or the action of algae as catalysts in causing the precipitation of phosphate from seawater, by changing the chemical environment of the surrounding waters (Banerjee, 1978; Chauhan, 1979). In refuting the ideas of Verma & Barman (1980), Banerjee claimed that the uptake of phosphate by algae could not alone explain the high concentration of phosphate in these stromatolites. Furthermore the change in mineralogy from phosphate to carbonate on both a cyclic laminar scale (Banerjee, 1978; Chauhan 1979) and in some stromatolites on a columnar scale (Chauhan, 1979) suggests that phosphogenesis was not a primary metabolic function.

A phosphogenic model involving the precipitation of phosphorus from seawater due to chemical changes in the surrounding waters is also difficult to envisage. Whether these processes were possibly aided by algal photosynthesis (Chauhan, 1979) or a combination of photosynthesis and variations in diurnal effects and tidal fluctuations (Banerjee, 1978), the problem remains as to how to prevent mixing within the water column and preserve the assumed microenvironments long enough for phosphate to precipitate. However, an alternative model similar to the one invoked by Southgate (1980) and Schmitt & Southgate (1982) to explain the Australian phosphatic stromatolites may apply to the Indian examples. If phosphogenesis took place as a synsedimentary diagenetic process, with phosphate being supplied from the interstitial waters, microenvironments conducive to phosphate precipitation would not be lost by seawater-mixing. Instead conditions of both raised pH and magnesium depletion could be a consequence of the slow bacterial degradation of organic matter associated with the stromatolites (see earlier discussions). Furthermore as these processes would have taken place in the interstitial waters of semi-emergent and shallow submergent environments they would have been subject to solar heating which would have assisted in phosphate precipitation.

The interpreted relationship between phosphatic stromatolites and areas where anaerobic conditions were locally maintained may also assist in explaining the distribution of phosphate and carbonate in the Indian stromatolites. Stromatolites from Nimach Mata vary in mineralogy from dominantly phosphatic to calcareous and have packstone- and grainstone-textured intercolumnar areas. As these stromatolites accreted in semi-emergent environments (Chauhan, 1979) oxidising groundwaters moving through the permeable intercolumnar sediments may have resulted in the partial destruction of the stromatolitic organic matter and the formation of calcareous rather than phosphatic stromatolites. However, in quiet-water lagoonal settings where the intercolumnar areas were infilled by relatively impermeable micritic carbonate, the stromatolitic organic material may have been preserved within dominantly anaerobic sediments. If this was the case any phosphate added to the interstitial waters from an overlying source could have then resulted in precipitation of phosphate under microenvironmental conditions associated with the slowly degrading stromatolitic organic material. If this chain of events is correct then the phosphogenic model developed here for the Australian stromatolitic phosphorites may also explain these Indian occurrences.

Summary

Early Middle Cambrian phosphorites of the Georgina Basin accumulated in shallow submergent and semi-emergent depositional environments within a lithofacies mosaic dominated by calcareous sediments. Rocks comprising the mixed-mineralogy suite were deposited under the influence of open circulatory conditions in both locally protected and high-energy environments. Normal to metahaline salinities prevailed in areas of tidal influence. The calcareous suite of rocks accumulated on a stable, shallow-water shelf characterised by negligible relief. Water circulation was restricted so that hypersaline conditions developed and non-skeletal particles dominated. The phosphorite suite of sediments were locally winnowed, reworked and sorted so that they accumulated in raised-energy environments found within the lagoons and embayments.

On semi-emergent flats phosphogenesis resulted in the formation of phosphatic nodules, hardgrounds, phoscrete profiles, phosphatic stromatolites and mudstone phosphorites. Phosphatic pavements and profiles formed during early diagenesis by the precipitation of phosphate cements and matrix material from interstitial solutions. Sheet-cracked laminae and beds of mudstone phosphorite formed by the settling of mud-sized particles from suspension. The erosion of these synsedimentary surfaces and lithologies produced a variety of phosphate particles including peloids, coated grains, aggregates, lumps and flat pebbles. Phosphatic moulds of skeletal particles formed in either semi-emergent or shallow submergent settings. In phosphatic hardgrounds and phoscrete profiles the gradational distribution of phosphate cements and shadow zones indicates that solutions from which phosphate precipitated moved down through the sediments and were derived from an overlying source. Rates of phosphate precipitation were rapid, so that in phosphatic nodules they coincided with either burrow occupancy or the preservation of the lining of the burrow. Shallow submergent sites of phosphogenesis have not been identified in the deposits of the Georgina Basin. However, in Europe, Asia and India they occur in association with both economic phosphate deposits and non-economic occurrences of phosphorites. The textural similarity between phosphatic hardgrounds found in Australia and those of Asia and Europe suggest that the processes of phosphogenesis interpreted from a detailed study of the Georgina Basin deposits may be widespread.

References

Baker, P.A. & Kastner, M. (1981). Constraints on the formation of sedimentary dolomite. *Science*, **213**, 1214–6.

Banerjee, D.M. (1978). Chemical rhythmicity in the Precambrian laminated phosphatic stromatolites and its bearing on the origin of algal phosphorite. *Indian Journal of Earth Sciences*, **5**, 102–10.

Banerjee, D.M. (1971). Aravallian stromatolites from Udaipur Rajasthan. *Journal Geological Society of India*, **12**, 529–42.

Bathurst, R.G.C. (1966). Boring algae, micritic envelopes and lithifiction of molluscan biosparites. *Geology Journal*, **5**, 15–32.

Bathurst, R.G.C. (1975). *Carbonate sediments and their diagenesis*. Developments in Sedimentology 12, 658pp. Elsevier, Amsterdam.

Bentor, Y.K. (1980). Phosphorites – the unsolved problems. In *Marine Phosphorites*, ed. Y.K. Bentor, pp. 3–15. Society of Economic Palaeontologists and Mineralogists Special Publication 29.

Berner, R.A. (1969). Chemical changes affecting dissolved calcium during the bacterial decomposition of fish and clams in seawater. *Marine Geology*, **7**, 253–74.

Birch, G.F. (1979). Phosphorite pellets and rocks from the western continental margin and adjacent coastal terrace of South Africa. *Marine Geology*, **33**, 91–116.

Birch, G.F. (1980). A model of penecontemporaneous phosphatization by diagenetic and authigenic mechanisms from the western margin of southern Africa. In *Marine Phosphorites*, ed. Y.K. Bentor, pp. 79–100. Society of Economic Palaeontologists and Mineralogists Special Publication No. 29.

Brasier, M.D. & Hewitt, R.A. (1979). Environmental setting of fossiliferous rocks from the uppermost Proterozoic–Lower Cambrian of central England. *Palaeogeography, Palaeoclimatology, Palaeoecology*, **29**, 35–57.

Bremner, J.M. (1980a). Physical properties of the diatomaceous mud belt of South West Africa. *Marine Geology*, **34**, 1767–76.

Bremner, J.M. (1980b). Concretionary phosphorite from S.W. Africa. *Journal of the Geological Society* **137**, 733–86.

Bromley, R.G. (1978). Hardground diagenesis. *In The Encyclopedia of Sedimentology*, ed. R.W. Fairbridge & J. Bourgeois, pp. 397–9. Dowden, Hutchinson & Ross, Stroudsburg, Pa.

Bromley, R.G. (1975). Trace fossils at omission surfaces. In *The Study of Trace Fossils*, ed. R.W. Frey, pp. 399–428. Springer-Verlag, Berlin.

Burnett, W.C. (1977). Geochemistry and origin of phosphorite deposits from off Peru and Chile. *Bulletin, Geological Society of America*, **88**, 813–23.

Bushinsky, G.I. (1964). On the shallow water origin of phosphorite sediments. In *Deltaic and Shallow Marine Deposits*, ed. M.L.J.U. Van Straaten, pp. 62–70. Elsevier, Amsterdam.

Bushinsky, G.I. (1969). *Old phosphorites of Asia and their genesis*. Translated from Russian by Israel Program for Scientific Translations, Jerusalem.

Chauhan, D.S. (1979). Phosphate-bearing stromatolites of the Precambrian Aravalli phosphorite deposits of the Udaipur region, their environmental significance and genesis of phosphorite. *Precambrian Research*, **8**, 95, 126.

Chowns, T.M. & Elkins, J.F. (1974). The origin of quartz geodes and cauliflower cherts through the silicification of anhydrite nodules. *Journal of Sedimentary Petrology*, **44**, 885–903.

Cook, P.J. (1972). Petrology and geochemistry of the phosphate deposits of northwest Queensland, Australia. *Economic Geology*, **67**, 1193–213.

Cook, P.J. (1976). Sedimentary phosphate deposits, vol. 7, In *Handbook of Stratabound and Stratiform Ore Deposits*, ed. K.H. Wolf, pp. 505–35. Elsevier, Amsterdam.

Cook, P.J. (1982). The Cambrian palaeogeography of Australia and opportunities for petroleum exploration. *Australian Petroleum Exploration Association Journal*, **22**, 42–64.

Cook, P.J. (1984). Spatial and temporal controls on the formation of phosphate deposits – a review. In *Phosphate Minerals* ed. J.O. Nriagu & P.B. Moore, pp. 242–74. Springer-Verlag, Berlin.

Crimes, T.P., Legg, I., Marcos, A. & Arboleya, M. (1977). Late Precambrian–Lower Cambrian trace fossils from Spain. pp. 91–138, In *Trace Fossils 2*. ed. T.P. Crimes & J.C. Harper, pp. 91–138. Special Issue 9, Geological Journal, Chichester.

de Keyser, F. (1973). A review of the Middle Cambrian stratigraphy in the Queensland portion of the Georgina Basin. *Bulletin, Bureau of Mineral Resources, Geology and Geophysics, Australia*, **139**, 13–27.

de Keyser, F. & Cook, P.J. (1972). Geology of the Middle Cambrian phosphorites and associated sediments in northwestern Queensland. *Bulletin, Bureau of Mineral Resources, Geology and Geophysics, Australia*, **138**, 79pp.

Elgueta, S. (1981). Sedimentological study of the Western Zone of the Lady Annie phosphate deposit, Queensland, Australia, M.Sc. Thesis, Australian National University, Canberra (unpublished).

Ferguson, J. & Burne, R.V. (1981). Interactions between saline red bed groundwater and peritidal carbonates, Spencer Gulf, South Australia: significance for models of stratiform copper ore genesis. *Bureau of Mineral Resources Journal of Australian Geology and Geophysics*, **6**, 319–25.

Fleming, P.J.G. (1977). Faunas, lithologies and the origin of phosphorites in parts of the Middle Cambrian Beetle Creek Formation of northwest Queensland. *Geological Survey of Queensland Special Publication*, **364**, 21pp.

Glossary of Geology (1974). American Geological Institute, Washington DC.

Gulbrandsen, R.A. (1969). Physical and chemical factors in the formation of marine apatite. *Economic Geology*, **64**, 365–82.

Hardie, L.A. & Ginsburg, R.N. (1977). Layering, the origin and environmental significance of lamination and thin bedding. In *Sedimentation on the Modern Carbonate Tidal Flats of Andros Island, Bahamas*. ed. L.A. Hardie, pp. 50–123. Studies in Geology 22. Johns Hopkins University, Baltimore.

Harrison, R.S. (1977). Caliche profiles: indicators of nearsurface subaerial diagenesis, Barbados, West Indies. *Bulletin Canadian Petroleum Geology*, **25**, 123–73.

Howard, P.F. & Cooney, A.M. (1976). D-Tree phosphate deposit, Georgina Basin Queensland. In *Mineral Deposits of Australia and Papua New Guinea*, ed. C.L. Knight, C.L., pp. 265–73. Australasian Institute of Mining and Metallurgy, Melbourne.

Ilyin, A.V. & Ratnikova, G.I. (1981). Primary, bedded structureless phosphorites of the Khubsugul Basin, Mongolia, *Journal of Sedimentary Petrology*, **51**, 1215–22.

James, N.P. (1972). Holocene and Pleistocene calcareous crust (caliche) profiles: criteria for subaerial exposure. *Journal of Sedimentary Petrology*, **42**, 817–36.

Jarvis, I. (1980). The initiation of phosphatic chalk sedimentation – the Senonian (Cretaceous) of the Anglo-Paris Basin. In *Marine Phosphorites*, ed. Y.K. Bentor, Y.K. pp. 167–197. Special Publication 29, Society of Economic Palaeontologists and Mineralogists.

Kennard, J. (1981). The Arrinthrunga Formation: Upper Cambrian epeiric carbonates in the Georgina Basin, Central Australia. *Bulletin, Bureau of Mineral Resources Geology and Geophysics, Australia*, **211**, 61pp.

Kennedy, W.J. & Garrison, R.E. (1975). Morphology and genesis of nodular chalks and hardgrounds in the Upper Cretaceous of southern England. *Sedimentology*, **22**, 311–86.

Lees, A. (1975). Possible influences of salinity and temperature on modern shelf carbonate sedimentation. *Marine Geology*, **19**, 159–98.

Logan, B.W. (1974). Inventory of diagenesis in Holocene—Recent carbonate sediments, Shark Bay, Western Australia. In *Evolution and diagenesis of Quaternary sequences, Shark Bay, Western Australia*, ed. B.W. Logan, pp. 195–249. Memoir 22, American Association of Petroleum Geologists, Tulsa.

Mabie, C.P. & Hess, H.D. (1964). Petrographic study and classification of western phosphate ores. *United States Bureau of Mines, Report* **6468**, 95pp.

McCave, I.N. (1971). Wave effectiveness of the seabed and its relationship to bedforms and deposition of mud. *Journal of Sedimentary Petrology*, **41**, 89–96.

McKee, E.D. & Gutschick, R.C. (1969). History of Redwall Limestone of northern Arizona. *Geological Society of America, Memoir*, **114**, 726pp.

McKirdy, D.M. (1977). Diagenesis of microbial organic matter: a geochemical classification and its use in evaluating hydrocarbon-generating potential of Proterozoic and Lower Palaeozoic sediments Amadeus Basin, Central Australia. Ph.D. Thesis, Australian National University, Canberra (unpublished).

Martens, C.S., Harriss R.C. (1970). Inhibition of apatite precipitation in the marine environment by magnesium ions. *Geochimica et Cosmochimica Acta*, **34**, 621–5.

Milliman, J.D. (1974). *Marine Carbonates, Recent sedimentary carbonates*, part 1. 375pp. Springer-Verlag, Berlin.

Muir, M., Lock, D. & von der Borch, C.C. (1980). The Coorong model for penecontemporaneous dolomite formation in the middle Proterozoic McArthur Group, Northern Territory, Australia. In *Concepts and Models of Dolomitization*, ed. D.H. Zenger, J.B. Dunham & R.L. Ethington, pp. 51–67. Special Publication 28, Society of Economic Palaeontologists and Mineralogists.

Öpik, A.A. (1956). Cambrian Geology of Queensland. *Bulletin, Bureau of Mineral Resources, Geology and Geophysics, Australia*, **49**, 1–24.

Öpik, A.A. (1960). Cambrian and Ordovician geology of Queensland. *Journal of the Geological Society of Australia*, **7**, 89–109.

Öpik, A.A. (1967). The Ordian stage of the Cambrian and its Australian Metadoxididae. *Bulletin, Bureau of Mineral Resources, Geology and Geophysics, Australia*, **92**, 133–69.

Öpik, A.A. (1970). *Redlichia* of the Ordian (Cambrian) of northern Australia and New South Wales. *Bulletin, Bureau of Mineral Resources, Geology and Geophysics, Australia*, **114**, 66pp.

Powell, T.G., Cook, P.J. & McKirdy, D.M. (1975). Organic geochemistry of phosphorites: relevance to petroleum geochemistry. *American Association of Petroleum Geologists, Bulletin*, **59**, 618–32.

Prian, J.P. (1980). Les Phosphorites Cambriennes du Versant Septentrional de la Montagne-Noire, au sud du Bassin Permien de Camares (Avegron). Ph.D. Thesis, L'Université P. et M. Curie, Paris (unpublished).

Pytkowicz, R.M. & Kester, D.R., (1969). Relative calcium phosphate saturation in two regions of the North Pacific. *Limnology and Oceanography*, **12**, 714–8.

Radke, B.M. (1980). Epeiric carbonate sedimentation of the Ninmaroo Formation (Upper Cambrian–Lower Ordovician), Georgina Basin. *Bureau of Mineral Resources Journal of Australian Geology and Geophysics*, **5**, 183–200.

Read, J.F. (1974). Calcrete deposits and Quaternary sediments, Edel Province, Shark Bay, Western Australia. In *Quarternary Carbonate Sequences, Shark Bay, Western Australia*, ed. B.W. Logan, pp. 250–82. Memoir 22, American Association of Petroleum Geologists, Tulsa.

Read, J.F. (1976). Calcretes and their distinction from stromatolites. In *Stromatolites*, ed. M.R. Walter, M.R., pp. 55–72. Developments in Sedimentology, 20. Elsevier, Amsterdam.

Read, J.F. & Grover, G.A. (1977). Scalloped and planar erosion surfaces, Middle Cambrian limestones, Virginia: analogues of Holocene exposed karst or tidal rock platforms. *Journal of Sedimentary Petrology*, **47**, 956–72.

Riggs, S.R. (1979). Phosphorite sedimentation in Florida – a model phosphogenic system. *Economic Geology*, **74**, 285–314.

Roberson, C.E. (1966). Solubility implications of apatite in seawater. *United States Geological Survey Professional Paper* **550-D**, D178–85.

Rogers, J.K. & Keevers, R.E. (1976). Lady Annie–Lady Jane phosphate deposits, Georgina Basin. Queensland. In *Mineral Deposits of Australia and Papua New Guinea*, ed. C.L. Knight, pp. 251–64. Australasian Institute of Mining and Metallurgy, Melbourne.

Russell, R.T. & Trueman, N.A. (1971). The geology of the Duchess phosphate deposits, Northwest Queensland, Australia. *Economic Geology*, **66**, 1186–214.

Sanderson, D.J. & Donovan, R.N. (1974). The vertical packing of shells on recent beaches. *Journal of Sedimentary Petrology*, **44**, 680–8.

Sandstrom, M.W. (1982). Organic geochemistry of phosphorites. Ph.D. Thesis, Australian National University, Canberra. 346pp. (unpublished).

Sant, V.N. (1979). Precambrian phosphorite of Rajasthan, India, a case history (abstract). In *Proterozoic and Cambrian Phosphorites*, ed. P.J. Cook & J.H. Shergold, pp. 39–40. ANU Press, Canberra.

Schmitt, M. & Southgate, P.N. (1982). A phosphatic stromatolite, (*Ilicta* cf. *composita* Sidorov) from the Middle Cambrian, northern Australia. *Alcheringa*, **6**, 175–83.

Scholle, P.A. & Kinsman, D.J.J. (1974). Aragonite and high Mg calcite from the Persian Gulf, a modern analogue for the Permian of Texas and New Mexico. *Journal of Sedimentary Petrology*, **44**, 904–16.

Shergold, J.H. Druce, E.C., Radke, B.M. & Draper, J.J. (1976). *Cambrian and Ordovician stratigraphy of the eastern portion of the Georgina Basin, Queensland and the eastern Northern Territory.* Excursion Guidebook 4C, 25th Session International Geological Congress, Sydney. 54pp.

Shergold, J.H. & Druce, E.C. (1980). Upper Proterozoic and Lower Palaeozoic rocks of the Georgina Basin. In *The Geology and Geophysics of north-eastern Australia*. ed. R.A. Henderson & P.J. Stephenson, pp. 149–74. Geological Society of Australia, Queensland Division.

Smith, K.G. (1972). Stratigraphy of the Georgina Basin. *Bulletin, Bureau of Mineral Resources, Geology and Geophysics, Australia*, **111**, 156pp.

Soudry, D. & Nathan, Y. (1980). Phosphate peloids from the Negev phosphorites. *Journal of the Geological Society* **137**, 749–55.

Southgate, P.N. (1980). Cambrian stromatolitic phosphorites from the Georgina Basin. *Nature*, **285**, 395–7.

Southgate, P.N. (1983). Middle Cambrian phosphatic and calcareous depositional environments, the Undilla region of the Georgina Basin, Australia. Ph.D. Thesis, Australian National University, Canberra. (unpublished).

Stumm, W. & Leckie, J.O. (1970). Phosphate exchange with sediments; its role in the productivity of surface waters. In *Advances in water pollution research*, vol. 2, pp. II 26/16. Pergamon Press. Oxford.

Trudinger, P.A. (1979). Microbial controls on phosphate accumulation. in *Proterozoic and Cambrian phosphorites*, ed. P.J. Cook & J.H. Shergold, pp. 87–92. ANU Press, Canberra.

Verma, K.K., & Barman, G. (1980). Relationship between stromatolites and phosphate localisation within the carbonate sequence of the Aravalli Supergroup in Udaipur and Banswara districts, Rajasthan. In *Stromatolites*, ed. M.V.N. Murthy, *Geological Survey of India Miscellaneous Publication*, **44**, 355–66.

Walter, M.R., Shergold, J.H., Muir M.D. & Kruse, P.D. (1979). Earliest Cambrian and latest Proterozoic stratigraphy, Desert Syncline, southern Georgina Basin. *Journal Geological Society of Australia*, **26**, 306–12.

Warren, J. (1982). The hydrological significance of Holocene teepees, stromatolites and boxwork limestones in coastal salinas in South Australia. *Journal of Sedimentary Petrology*, **52**, 1171–202.

Wilson, I.H., Gunn, M.J. & Smit, J.A.J. (1979). The Kajabbi Formation. A Middle Cambrian unit northwest of Cloncurry. *Queensland Government Mining Journal*, **80(938)**, 629–33.

25

Proterozoic and Cambrian phosphorites – specialist studies: phosphogenesis and its relationship to exploration for Proterozoic and Cambrian phosphorites

S.R. RIGGS

Introduction

Most sedimentary rocks in the geological column and sediments on the seafloor contain very minor concentrations of phosphatic components (less then 0.3% P_2O_5). However, periodically through geological time, phosphorites (taken here as sedimentary rocks with 5% P_2O_5 or greater) have formed on the seafloor in response to specialised oceanic conditions and accumulated in sufficient concentrations to form major stratigraphic units of regional extent. The poorly understood prerequisites and processes that led to the origin of these anomalous phosphorite units include a multitude of complex variables which interact through time.

The formation and subsequent accumulation of phosphorites sufficient to form a major stratigraphic unit with the potential of becoming an economic deposit requires a specialised set of conditions in specific geological settings. The prerequisites for such phosphorite sedimentation define the phosphogenic system which is largely controlled by the tectonic setting. Since sediments are direct products of the regional tectonic framework and geological history, stratigraphic patterns represent the recorded response of the type, magnitude, duration, and spatial location of regional tectonic events or lack of such events. The tectonic framework largely determines the geographical latitude; the physical geometry of the deep-ocean plateaux, continental margins, and epicontinental seas; global sea level position and pattern of fluctuations; and the associated oceanographic current regimes which determine the physical, chemical and biological conditions of the seafloor environments. The regional structural framework then determines the detailed sedimentary environments, their geometries, and the physical, chemical, and biological processes producing the phosphorites and associated sediment assemblages.

Phosphorites, a product of anomalous marine conditions, occur simultaneously with and adjacent to a group of additional anomalous sediments. These associated sediments become mixed and interbedded with the phosphates as the chemical systems pulse and oscillate with changing depositional conditions and fluctuating sea levels. This complex of sediments constitutes the phosphogenic assemblage and is outlined in Table 25.1. No single phosphatic sequence contains all of these associated minerals; each deposit is associated with some of them either as precursor or postcursor events, interbedded facies, or as contemporaneous and laterally equivalent depositional sequences. Different combinations of the associated sediments constitute the various phosphate assemblages, some of which are listed in Table 25.2. Each assemblage is a product of specific environmental conditions of deposition and reflects a specific genesis. If the complexities of phosphate genesis are to be unravelled, these associated sediments must be considered since they are integral pieces of the phosphate puzzle.

Although phosphogenic sediment systems represent abnormal conditions of sedimentation, these conditions have occurred often enough in the geological past to produce deposits which are not uncommon on a worldwide basis. The future economics of developing and recovering adequate supplies of phosphorus for the fertilizers necessary to feed the world are not going to be a function of whether phosphorus exists or not, but rather of its availability with respect to international political situations and of its minability. The world has abundant major phosphate deposits; it will be a long time before this critical commodity runs out. However, many individual countries do not have an adequate internal supply of fertilizer raw materials to meet their rapidly expanding needs. This is particularly true of many developing third world nations who, in addition, cannot afford to buy the raw materials on the open international market. Thus, it becomes increasingly imperative to refine and improve our understanding of the basic models for the origin and occurrence of phosphorite. This sophistication can only increase the probability of discovering 'new' phosphate deposits in those developing areas.

Phosphate models

Models for the genesis of phosphorites are nothing more than patterns or sets of plans capable of leading us to an understanding of the formation and development of a complex sedimentological system. The value of having such models is obvious; they are important vehicles for organising the multitude of complex variables interacting through time to produce this sequence of sediments. Of course, as models evolve towards reality they become increasingly more important and valuable

Table 25.1. *Sediments associated with phosphorites*

Associated sediments	Most common components	Environmental significance and common interpretation
1. Phosphate	Carbonate fluorapatite or carbonate hydroxyapatite	See text and Table 25.5
2. Carbonaceous mudstone and black shales	Organic matter	High organic productivity in the bathyl realm, low oxygen and reducing conditions in the benthic realm
3. Chert, diatomites, and porcellanites	Opal, chalcedony, and microcrystalline quartz	Cold-water, nutrient-rich upwelling currents supplying Si to areas of high organic productivity; diagenetic alteration of volcanic ash
4. Carbonates	Dolomite	Refluxing of Mg brines in (1) shallow-marine evaporite lagoons and supratidal flats in warm arid climates, or (2) interstitial diagenetic alteration of calcite.
	Calcite	Less common than dolomite; usually occurs as interbedded sequences
5. Mg-rich clays	Attapulgite, palygorskite, and sepiolite	Products of the alteration of volcanic ash; diagenetic formation from interstitial waters
6. Iron-rich sediments	Oxides (hematite, limonite, and goethite)	Dominant in the Proterozoic and less common in Palaeozoic sediments; occurs as lateral facies with minor interbeds
	Silicates (glauconite)	Dominant in the Mesozoic and Cenozoic; occurs as lateral facies
7. Manganese and ferromanganese	Mn oxides and carbonates	Occurs as lateral facies
8. Bedded sulfides	Pb, Zn, Fe, and Cu sulfides	Hydrothermal solutions associated with rifting, transform faulting, or submarine volcanism; occurs as lateral facies
9. Volcanic sediments	Tuff, tuffaceous mud, and bentonite	Derived from associated volcanic activity; clays, silica, and zeolites derived from altered volcanic ash
10. Evaporites	Halite, gypsum, and anhydrite	Occurs as lateral facies; forms in shallow coastal lagoons in warm and arid climatic zones
11. Glacial deposits	Tillites	Occurs as basal facies below the marine transgression
12. Other associated authigenic sediments	Barite, celestite, fluorite, etc.	
13. Anomalous trace elements	Ag, As, Cd, Cr, La, Mn, Mo, Pb, Se, Sn, Sr, U, Y, Yb, Zn	

as tools to guide us in our exploration efforts for less obvious and 'new' types of deposits.

The literature on phosphorites is voluminous; most of it deals only with facets of an extremely complex sediment system and inevitably focuses on a model which often becomes the definitive model for the origin of phosphorites. This has resulted in the development of many models for the genesis of phosphorite (Table 25.3). In 1829 William Buckland first theorised that phosphorites were coprolites or fossil faeces. Since then organisms and organic productivity have been considered extremely important components in most models. In 1937 and 1938 Kazakov introduced the idea of upwelling currents providing a continuous supply of phosphorus from the deeper and colder waters. This concept, in conjunction with the high

Table 25.2. *Common phosphate assemblages and their relationship to tectonism*

Degree of tectonism	Sediment assemblage	Examples
High tectonism proximal to the environments of deposition	1. Phosphate-bedded sulfides–banded iron–volcanic sediment	Broken Hill, Australia; Vihanti, Finland; Carlton, Minn., USA
	2. Phosphate–chert–tuffaceous sediment	Volta Basin, Niger; Sechura, Peru; Monterrey Fm, Mexico
	3. Phosphate–banded iron–chert	Baruga Basin, Mich., USA; Doushantuo, China
	4. Phosphate–black shale–chert	Phosphoria Fm, USA; Mussoorie, India
	5. Phosphate–manganese oxide–ferromanganese oxide	Blake Plateau, USA
Low tectonism in the environments of deposition	6. Phosphate–dolomite–Mg clays–glauconite–chert	Hawthorn Fm, SE USA; Pungo River Fm, NC USA
	7. Stromatolite phosphate–dolomite–chert	Jhabua, India; Udaipur, India

Table 25.3. *Models of phosphorite genesis*

1. Buckland (1829)	Coprolites and fossil faeces.
2. Seeley (1866)	Gelatinous phosphate of lime with the phosphorus derived from the decomposition of marine plants and animals plus skeletal material.
3. Fisher (1873)	Nodular phosphates representing condensed sections.
4. Cornet (1886), Keyserling (*in* Penrose, 1888)	Mass mortality of fish as a major source of phosphorus which was dissolved and subsequently redeposited.
5. Shaler (*in* Penrose, 1888)	Formed at the bottom of swamps associated with peat.
6. Kazakov (1937, 1938)	Upwelling of deep phosphorus-rich waters into geosynclinal basins and shelf platforms leading to an increase in T and decrease of P and CO_2 in solution causing apatite to precipitate below euphotic zone (between 50 m and 200 m)
7. Mansfield (1940)	Time and space association of volcanism with phosphate deposition; the rapid dissolution of volcanic glass supplies increased concentrations of P, F, and Si resulting in an increased organic productivity including silica-utilising organisms.
8. Brongersma-Sanders (1957)	Marine mass mortalities occur in regions of phosphorus-rich upwelling currents and high organic productivity. These are often areas of bituminous-rich sediments, diatomaceous oozes, and occasionally phosphatic sediments.
9. Gimmelfarb (1958)	Increased input of phosphorus due to erosion of adjacent elevated land masses.
10. Ames (1959)	Physico-chemical replacement of carbonate sediments.
11. Bushinsky (1966)	Dead organisms are the immediate source of P. The rate of supply of organisms must be high and in a warm and shallow environment to facilitate rapid deposition and decomposition of the organism.
12. Pevear (1966)	P-rich waters supplied from estuarine marshes.
13. Rooney & Kerr (1967)	Initiation of mass mortalities as a result of massive ash falls.
14. Baturin (1971) Bremmer & Willis (1975) Cook (1967, 1976)	Diagenetic formation of a low-grade phosphorite immediately below the sediment–water interface, followed by a period of mechanical reworking and concentration to form a high-grade deposit.
15. Doyle *et al.* (1978)	Mollusc kidney stones.
16. Hite (1978)	Phosphate precipitated as phosphorus-rich continental brines mix with cold ocean waters.
17. Influence of IGCP 156-Phosphorites upon scientific community (1978 to present)	Recognition of the inter-relationships of global tectonism upon palaeoceanographic and palaeoclimatic conditions has led to the understanding that there are many variables interacting to produce different types of phosphate deposits episodically in variable oceanographic settings and environments.

organic productivity often associated with such upwellings, has dominated most hypotheses since then. Periodically, a few other models receive local support, such as the input of phosphorus with volcanogenic material (Mansfield, 1940) or the physicochemical replacement of limestone (Ames, 1959).

Most workers of the last three decades have generally interpreted phosphate deposition to represent periods of low rates of sedimentation in combination with a large supply of nutrient phosphorus derived through some form of upwelling on west coast, shallow, marine shelf environments in low- to mid-latitudes. Cold, nutrient-rich upwelling currents have generally been considered prerequisite mechanisms supplying essential elements for production of large volumes of organic matter to the sediments. Nutrient recycling by bacteria at either the sediment–water interface or within interstitial waters then leads to the formation and growth of phosphate grains.

Since 1978, a new school of thought about the genesis of phosphates has been gradually evolving, largely as the result of the interaction of world scientists through the International Geological Correlation Program 156 on Phosphorites. This new direction has resulted in recognising the importance of global tectonism through time and its influence upon a complex variety of palaeoceanographic processes involved in phosphogenesis (i.e. sea level fluctuations, deep-ocean sinks and vertical mixing, major oceanic chemical cycles, geometry and orientation of oceanic basins, palaeoclimatic conditions, location and movements of crustal plates and development of oceanic basins, etc.). This has led to the following realisations: (1) there are many different types of phosphorites which form in response to different sets of oceanographic conditions and processes and in different environments of deposition; (2) phosphogenesis requires more than just upwelling or organisms alone; and (3) there are many types of upwelling, each of which results in phosphogenesis if all the other pre-requisite variables are correct. The rest of this paper will develop these ideas further.

It must be recognised that most of the diverse ideas and theories of phosphogenesis which have been presented in the past as the definitive model are not adequate by themselves to explain the total complex system. However, they each appear to be a valid component of a puzzle which has many pieces. The development of models which can be used in the exploration for new deposits must take into consideration the entire spectrum of types of phosphate deposits as well as the complex of associated sediments that together are products of the phosphogenic system. The diverse conflict of ideas between many workers is a natural product of looking only at individual pieces of the puzzle.

Relationship of phosphogenesis to palaeoceanography

Episodicity of phosphogenesis

Many workers have demonstrated a strong episodicity to periods of phosphogenesis, including Cook & McElhinny (1979), Bentor (1980), and Sheldon (1980, 1981). Cook & McElhinny present evidence that major episodes of phosphogenesis occurred during the Miocene to Pliocene, Late Cretaceous to Eocene, Jurassic, Permian, Ordovician, Cambrian, and during four periods of the Proterozoic including the Late Proterozoic (at 620 and between 700 and 800 m.y.); Middle Proterozoic (between 1200 and 1600 m.y.); and Early Proterozoic (between 1800 and 2200 m.y.).

Phosphorite formation is not only episodic, but is generally of global extent when it occurs. This has been well documented for the Cambrian and various Proterozoic sequences by Notholt (1979), as well as within the associated chapters of this book. Notholt recorded at least 92 occurrences of Cambrian and Precambrian sedimentary phosphate world-wide, 54 of which are Cambrian in age. Many additional deposits have been discovered or uncovered from the literature since Notholt's paper. The Cambrian deposits include a major phosphogenic belt which extends from South Australia, through the Georgina Basin of northern Queensland; Vietnam; Hazara, Pakistan; Hainan and Yunnan, China; Khubsugul, Mongolia, and the major deposits of Karatau, USSR. In addition, Cambrian phosphate is known from numerous localities in Europe (Notholt, 1979) and from a few localities in Canada and the USA. Age assignments for Proterozoic phosphate deposits are often of poor quality and imprecise. Nevertheless, major deposits appear to occur with a world-wide distribution during each of the four proposed phosphogenic episodes as indicated by Cook & McElhinny (1979) and Cook & Shergold (1980).

The global nature of the phosphogenic episodes is apparent with each episode, regardless of age. For example, during the youngest and most recent episode, Miocene phosphorites formed throughout the south-eastern USA continental margin (Riggs, 1984), the Blake Plateau (Manheim, Pratt & McFarlin, 1980) and Miami Terrace (Mullins & Neumann, 1979); off north-west Africa (Summerhayes, Nutter & Tooms, 1972), south-west Africa (Birch, 1979), south-west California (Pisciotto, 1978), Baja California (Salas, 1978), north-west Peru (Cheney, McClellan & Montgomery, 1979), and on the Chatham Rise east of New Zealand (Cullen, 1980), to name a few. Similar global distributions are characteristic of the Cretaceous to Eocene deposits. Thus, each major phosphogenic episode does appear to be global in extent, whereas all other geological periods may be characterised by only local and random occurrences of phosphate and associated lithologies.

Sea level fluctuations

That the stratigraphic section represents a very sensitive record of global, regional, and local fluctuations in relative sea level is a well-established and self-evident truth in geology. However, sedimentological patterns of deposition resulting from these fluctuations, as well as the causes and nature of such cyclical events, have been subjected to much controversy and debate. Vail *et al.* (1977) defined three orders of relative sea level change. First-order or 'global' cycles are records of world-wide sea level responses to major, large-scale geotectonic processes with durations of 100–300 m.y. Second- and third-order cycles have durations of 10–80 and 1–10 m.y., respectively. Vail *et al.* (1977) believe that all the first-order cycles are responses to geotectonic mechanisms and some second-order cycles may be of sufficient duration and magnitude to be similar responses. They believe that glaciation and deglaciation may account for some second-order and many third-order cycles. They also recognise that there may be other causes working 'in combination with geotectonics or glaciation to accentuate or diminish the changes'. For example, Pitman (1978, 1979) proposed that abrupt changes in rates of sea floor spreading would cause sea level fluctuations, affecting the sediment flux and the resulting sediment patterns along continental shelves.

Vail *et al.* (1977) and Vail & Mitchum (1979) have delineated a series of global highstands and lowstands of the sea. Global highstands occur when sea level is above the shelf edge in most regions of the world and consequently are characterised by widespread marine sediments on continental shelves and adjacent coastal plains. Global lowstands occur when sea level is below the shelf edge and are characterised by erosion and non-deposition, producing major inter-regional unconformities.

Cook & McElhinny (1979) plotted estimated volume of phosphate formed and specific number of deposits versus geological time during the Phanerozoic. Sheldon (1980) superimposed this phosphate data on the Phanerozoic sea level curve of Tissot (1979) displaying a remarkable relationship. Plotting the same data on the more detailed curves of relative sea level change of Vail *et al.* (1977) and Vail & Mitchum (1979) demonstrates a strong correlation between the major phosphogenic episodes and first-order global highstands. The Cambrian and Jurassic–Cretaceous phosphogenic global episodes occur in conjunction with the higher sea levels on the rising side of the first-order highstands of the sea. The Ordovician and Tertiary episodes occur in conjunction with the higher sea levels on the falling side of the first-order highstands. Other geological periods either do not contain major phosphate deposits or contain local or regional deposits which are not of global extent. For example, the Phosphoria Formation, which is itself very large, appears to be an isolated deposit in the Permian which formed in response to second-order perturbations in sea level when conditions were temporarily adequate to produce a phosphogenic sequence on a regional basis. Other small-scale events which formed in response to regional and ephemeral conditions produced local, scattered deposits through the geological column. This depositional pattern is distinctly different from the major episodes of global extent.

Global sea level curves do not exist for the Proterozoic. However, if the relationship of phosphogenic episodes to first-order highstands is real, then there is a mechanism for creating a preliminary curve. Each of the four Proterozoic phosphogenic episodes defined by Cook & McElhinny (1979) and Cook & Shergold (1980) – Upper Proterozoic I (620 m.y. mean age); Upper Proterozoic II (700–800 m.y.); Middle Proterozoic? (1200–1600 m.y.); Lower Proterozoic (1800–2200 m.y.) – is characterised by extensive sequences of anomalous sediments associated with various phosphate assemblages. They probably formed in response to processes which were global in extent, assuming that each episode represents changing sea level associated with first-order highstands of the sea. If so, then each of these periods and each sediment sequence containing the associated sediment assemblages are prime stratigraphic targets for further exploration and potential expansion of the world's phosphate resources.

Vail *et al.* (1977) identified a major highstand in the Middle Miocene, a moderate highstand in the Early to Middle Pliocene, and three brief, lower highstands during the Pleistocene. Major lowstands occurred in the Late Miocene, in the Late Pliocene to Early Pleistocene, and three brief lows during the Pleistocene. Arthur & Jenkyns (1981) have demonstrated that each of the three major world phosphogenic events of the Mesozoic–Cenozoic, including the Miocene, occurred during periods of relative highstands of the sea. The generally phosphate-poor Pliocene and Pleistocene are characterised by brief periods of time and rapidly fluctuating sea levels. The Neogene sediments of North Carolina show a strong correlation between patterns of sedimentation, phosphorite formation, and relative position and movement of eustatic sea level (Riggs, 1984; Riggs, Snyder *et al.*, 1982). They demonstrate that the major Miocene phosphogenic episode occurred on the Miocene transgression and that phosphogenesis increased with rising sea level. A minor phosphogenic episode also occurred during the shorter Pliocene transgression, with possibilities of occasional and local phosphate formation on the North Carolina continental shelf during very brief Pleistocene highstands (Riggs, Snyder *et al.*, 1982). Burnett (1980) has shown that the small amounts of phosphate on the Peru–Chile continental margin were formed during Pleistocene interglacial highstands of the sea. Thus, in addition to being related to transgressive sea level events and associated highstands, it appears that the duration of highstands in an important factor in providing adequate time for phosphogenesis to accumulate significant concentrations of phosphate within the depositional environment.

Climatic fluctuations and the influence of glaciation

A vast array of oxygen isotope, sedimentological, and palaeontological evidence supports the interpretation that many third-order and some second-order cycles of global sea level fluctuation are direct results of glaciation and deglaciation (Vail *et al.*, 1977). This is particularly true for the Neogene where a general cooling trend throughout the Paleogene culminated in a series of well-documented glacial episodes (Frakes, 1979). Frakes believes that sea level plays a major role in controlling world climate due to the greater heat capacity of water versus land. Thus, during periods of high sea level, more incoming radiation is absorbed to produce a warmer and more humid climate in which there is increased chemical weathering. Arthur & Jenkyns (1981) believe that such periods of warm and humid climates could result in higher alkalinities and nutrient supplies, allowing for greater rates of sedimentation of organic carbon, phosphate, silica, and carbonate in the world oceans. During lowstands of the sea, less incoming radiation is absorbed, resulting in wider temperature fluctuations over large landmasses with low evaporation and precipitation rates due to decreased oceanic area. These conditions could generate or accelerate the growth of glaciers and lead to an expansion of mid-latitude deserts. The extreme palaeoclimatic effect of glaciation leads to the extensive development of continental desert conditions (Sheldon & Burnett, 1981) and a lack of major phosphate deposits (Arthur & Jenkyns, 1981). Thus, there seems to be adequate evidence to demonstrate that glacial episodes can be responsible for third-order sea level fluctuations, that glacial lowstands are generally characterised by cooler and more arid climatic conditions, while highstands are characterised by warmer and wetter conditions with increased authigenic sedimentation.

Cook & McElhinny (1979) believe there is a strong association between phosphogenesis and glaciation as demonstrated by the numerous phosphate deposits which display a close stratigraphic association between tillites and phosphorites (i.e. Georgina Basin in Australia, Volta Basin in Africa, Laisvall deposit in Sweden, Karatau phosphorites in USSR, and eastern Yunnan deposits in China). The major glacial episodes which they believe show this close correspondence include those occurring at about 770 m.y., 615 m.y., 560 m.y., 445 m.y., 295 m.y., and from 26 m.y. to the present. Sheldon (1980) related the episodicity of phosphate deposition during the Cambrian, Ordovician, Permian, and Miocene to the transgression from low and cold oceans associated with glacial periods when a deep-ocean phosphorus sink developed, to the high and warm oceans associated with interglacial episodes and characterised by upwelling in the trade-wind belt. Riggs's (1984) model of sedimentation for the Neogene phosphorites of south-eastern United States was dependent upon fluctuations of sea level which resulted from Neogene glaciation and associated climatic changes.

Relationship of upwelling to phosphogenesis

Since Kazakov suggested that upwelling currents might play an important role in the genesis of phosphate (1937, 1938), there has been an overwhelming interest by phosphate geologists and geochemists in the oceanographic processes of upwelling. However, this interest has been largely directed towards only one or two types of upwelling, in particular those areas characterised by west coast trade-wind belt and equatorial upwelling.

In papers published by Cook & McElhinny (1979) and Sheldon (1980), hypotheses dealing with upwelling and phosphogenesis were summarised. Both papers presented two oceanographic models to explain the bulk of the world's phosphate deposits. These models were: (1) equatorial upwelling associated with east–west seaways in low latitudes; and (2) trade-wind belt upwelling associated with oceanic gyres of north–south seaways in which coastal upwelling takes place along western continental boundaries. Cook & McElhinny (1979) believe there is an interrelationship between phosphorites and periods of glaciation. They believe that because colder waters contain a higher concentration of dissolved phosphorus, at the onset of glaciation and a greater temperature differential, upwelling would be intensified and oceanic overturn would be initiated.

Sheldon (1980) believes that sea level is a very important consideration in phosphogenesis and that there is long-term phosphorus build-up in oceanic sinks during stable periods associated with low sea level stands. Phosphogenic episodes occur during periods of active vertical mixing (upwelling) and phosphorus withdrawal from deep-ocean sinks. Equatorial upwelling takes place during highstands of warm seas while trade-wind belt upwelling takes place during transitions from high to low sea levels with its cold oceans resulting from glaciation.

Both Cook & McElhinny (1979) and Sheldon (1980, 1981) acknowledge that there are many phosphorites that do not flt these two models, including those formed in shallow intracontinental epeiric seas and those occurring along eastern continental margins. There are also many major phosphate deposits which do not fit into these two models of upwelling, including the south-eastern US continental margin (Riggs, 1984); the Blake Plateau (Manheim *et al.*, 1980); the Miami Terrace (Mullins & Neumann, 1979); East Australian continental shelf (Marshall & Cook, 1980; Cook & Marshall, 1981; O'Brien & Veeh, 1980), as well as other deposits which fit the existing upwelling models with difficulty, such as the Georgina Basin of Australia (Cook & Shergold, 1980), and Chatham Rise of New Zealand (Cullen, 1980). Other models have not been developed largely because of the void in scientific understanding of other types of upwelling systems and the lack of an adequate synthesis.

Parrish (1982) identified two causes for upwelling systems. First, the primary driving force for the most extensive forms of upwelling is atmospheric circulation which produces coastal upwelling, open-ocean divergence, and the less-common channelled-flow upwelling. Second, the interaction between oceanic current regimes, bathymetry and the structure of coastal margins produces dynamic or topographic upwelling. She defined the following types of upwelling operating within modern oceanic systems.

(1) *Coastal Upwelling.* Coastal upwelling requires steady winds along continental margins that prevail from the proper direction as determined by the Coriolis effect. Such winds, associated with permanent subtropical high-pressure cells, blow parallel to the coast causing a net offshore water transport which is replaced by cold, nutrient-rich upwelled waters from below. Three subvarieties are based upon geographic setting:

1. Meridional: west-facing coastlines which are longitudinally oriented (i.e. western coasts of North and South America; north-west and south-west Africa).
2. Zonal: north- or south-facing coastlines that are situated at the proper latitude relative to major zonal wind systems (i.e. northern South America and Yucatan).
3. Monsoonal: when monsoonal winds blow parallel to a coastline oriented in the proper direction (i.e. Somalia and western India).

(2) *Open-ocean divergence.* This form of upwelling occurs under atmospheric low-pressure systems such as the equatorial and Antarctic low-pressure zones and includes cellular or radial, zonal, and equatorial upwellings depending upon the configuration of the winds that create them. The resulting wind patterns cause the ocean surface waters to be driven directly away from the centre of the low-pressure system producing upwelling from below.

(3) *Channelled-flow upwelling.* This upwelling occurs when persistent winds blow down an elongated narrow seaway causing upwelling along one side and downwelling along the opposite side. The Gulfs of California and Thailand are modern examples of this variety of upwelling system.

(4) *Dynamic or topographic upwelling.* Along the eastern continental margins, upwelling is primarily associated with the interaction between the oceanic gyres, coastline configuration,

and bathymetry with some augmentation from the wind patterns. Pietrafesa (1983) explained this process as follows: if strongly sheared currents flow over the shelfbreak and over bottom topography with diverging isobaths in the downstream direction, and if the current maximum is deflected seaward of the shelfbreak, upwelling and onshore flow of cold bottom waters will result. Modern examples where this process is important include the Gulf Stream along the south-eastern United States, the Kuroshiro Current along the east coast of Japan, and the East Australian Current of Australia.

Relationship of phosphogenesis to tectonism

Tectonic setting

Cook & McElhinny (1979) proposed an indirect genetic link between periods of volcanism, orogenesis, formation of evaporites, and episodes of phosphogenesis. They related each of these periods to specific phases in the rifting and separation of continental and oceanic components of the plates. In the case of diverging plates, they believe that volcanism and tectonism are associated with the initiation and early-stage rifting features, evaporites occur in the intermediate rift-valley stage, and phosphorites form in the late stages after the rift-valley is breached and a seaway established. Converging plate models are essentially mirror images with the stages of continental sedimentation and rifting being replaced by a phase of mountain building.

However, if the most recent global phosphogenic event is considered – the Miocene – it becomes quite apparent that the model of Cook & McElhinny (1979) is far too simplistic. In fact, phosphate forms in spite of the proximity or degree of regional tectonic activity within an area. Conditions favourable for the development of major phosphogenic events did exist along the trailing plate margins as suggested by Cook & McElhinny. Favourable areas included the Miocene deposits of the east coast of North America and west coast of Africa (the Pungo River and Hawthorn Formations, the Blake Plateau deposits, and the Miami–Pourtales Terrace deposits of south-eastern USA; the Varswater Formation and Agulhas Bank of South Africa; and the continental shelf deposits of south-west and north-west Africa). However, suitable conditions also existed contemporaneously proximal to major plate boundaries associated with severe tectonism. This was the case for the Miocene deposits along the west coast of North and South America including the extensive Monterey Formation of California, USA and Baja California, Mexico, and the Sechura deposits of Peru. Thus, phosphate deposits do occur in depositional environments characterised by both end members of tectonic activity.

Areas of severe tectonism are generally characterised by: (1) a flood of terrigenous and volcanic sediments forming great thicknesses of stratigraphic section; (2) a severe relief dominated by abundant deep basins with narrow shallow-water environments; and (3) a stratigraphic record which may be structurally deformed, metamorphosed, and even lost from the record with continued tectonism. Conversely, areas of stable continental margins with low tectonism are characterised by: (1) normal sequences of sedimentation of either authigenic carbonates or fine-grained terrigenous sediments; (2) broad, shallow-water depositional environments; and (3) a stratigraphic record which is thin to moderately thick with extreme lateral extents.

Phosphate sequences do occur within the tectonic framework of either of the extremes described above, or anywhere in between; but they do not form contemporaneously with either of the dominant sediment sequences in either end member. Rather, the phosphate sequences form during lulls or changes in tectonism which appear to reflect major fluctuations in sea level and are characterised by major changes in the dominant sediment type. Consequently, the phosphate sequence usually occurs between two distinctly different lithological groups of sediments. The phosphorites associated with each portion of the tectonic continuum are distinctive, but variable, depending upon the different environmental conditions of deposition.

Cook & Shergold (1980) demonstrated a striking similarity of vertical lithological patterns between the Cambrian deposits of the Georgina Basin, Australia; eastern Yunnan, China; Hazara, Pakistan; and Karatau deposits in the USSR. The section at Lake Khubsugul in the Mongolian People's Republic also has a similar lithological pattern. The basic pattern from the base upward is as follows: (1) a basal sequence of conglomerates, tillites, and sandstones; (2) dolomites which associated cherts and siltstones; (3) a phosphorite sequence which grades laterally and vertically into (4) interbedded facies of argillites, cherts, and dolomites; and (5) a thick and extensive calcite limestone sequence. The Cambrian deposits of northern Europe, including Laisvall, Sweden (Rickard *et al.*, 1979), Vassbo, Sweden (Christofferson *et al.*, 1979), Bornholm, Denmark (Poulsen, 1960), and Newfoundland, Canada (Dale, 1915) also reflect similar vertical patterns of deposition, with two exceptions: total concentration of phosphate is considerably less, and phosphorites grade upward into organic and iron-rich shales instead of limestones at the end of the depositional sequence. Minor lithological variations do exist from area to area; these are interpreted to be responses to third- and fourth-order sea level fluctuations, local tectonic processes and, as pointed out by Brasier (1980), the palaeogeographic location within the Cambrian depositional environments (this determines the resulting facies). Thus, the general lithological patterns are distinctive and similar, and they appear to reflect changing depositional conditions in response to global oceanographic processes.

When the Early to Middle Cambrian phosphate deposits are plotted on the global sea level curves of Vail *et al.* (1977), they occur on the steep side of a first-order sea level transgression. Thus, the tillites and coarse clastics dominated the depositional regime during low sea level, carbonates and fine clastics dominated the regime during high sea level, whereas phosphate and associated sediments formed in the interim during changing depositional regimes.

Similar changes in depositional patterns are reflected in other phosphogenic episodes. For example, in continental margin deposits of south-eastern USA, Riggs (1984) described two orders of changing lithology. On the regional scale, the Eocene and Oligocene sediments are dominated by limestones in the south and sandy limestones and calcareous sandstones in the north. This was followed by the Miocene phosphogenic episode which is characterised by phosphate, a palygorskite–attapulgite and sepiolite clay assemblage, terrigenous sand and clay, carbonates which are dominantly dolomite, glauconites in the northern section, and diatomites locally. Post-phosphogenic sediments are dominantly terrigenous with minor and local carbonates, particularly in the southern areas. Variations in this general pattern are direct responses to variations within the regional tectonic framework. There is a general northward increase in terrigenous sediments in response to latitudinal changes in climate and the increasing proximity of the Appalachian terrigenous source area.

On the smaller and more local scale, Riggs, Lewis *et al.* (1982) described numerous cycles of phosphate deposition. Each cycle is characterised by a vertical sequence in which the dominant sediments grade upward from terrigenous to phosphates to carbonates. Each cycle is capped by an unconformity surface and followed by another sedimentary sequence.

Large-scale or regional lithological variations in south-eastern USA are interpreted to have resulted from the falling or regression of global sea level associated with the first-order sea level cycle of Vail *et al.* (1977). The phosphogenic episode occurs above the extensive carbonates deposited during the first-order sea level high and below the terrigenous sediments deposited during the first-order sea level low of the Upper Miocene to Pleistocene. This pattern is the reverse of that displayed by the Cambrian phosphorites and represents the opposite side, or regressive phase of a first-order sea level event. However, phosphate occupies the same relative position with respect to changing depositional regimes. Local lithological variations within the phosphogenic episode reflect deposition on the transgressive phase associated with third-order or possibly local fourth- and fifth-order sea level fluctuations.

The Miocene phosphate deposits in Baja California, Mexico, are believed to be contemporaneous with those in south-eastern USA. However, the Baja California deposits formed proximal to a major plate boundary and thus changing patterns of authigenic, terrigenous, and volcanic sediments are direct responses to the type, intensity, and duration of each tectonic event and the resulting sediment pulses.

The lithological patterns within the stratigraphic section of Baja California have been summarised as follows by Riggs (1980). The phosphatic sediments were deposited unconformably on the Eocene non-phosphatic sandstones of the Tepetates Formation. The Monterey and San Isidro Formations are characterised by alternating beds of phosphorite silts and sands, diatomites, calcareous sediments, and tuffaceous silts and sands. Vertically upward, the authigenic sediments decrease and grade into a volcanic sedimentary sequence of interbedded tuffaceous sands and bentonite clays in the upper portion of the San Isidro Formation. The overlying Comondu Formation, which consists of over 2000 m of volcanic tuffaceous sandstones and conglomerates, completely overwhelmed all authigenic sedimentation. As volcanism subsided, phosphorites began to form and accumulate within the sands of the lower portion of the Salada Formation, which grades upward into a non-phosphatic carbonate in the upper portion of the formation. Phosphorites of the Salada are characterised by smaller-scale cyclic patterns similar to those described in North Carolina by Riggs, Lewis *et al.* (1982) and Snyder, Hine & Riggs (1982). Subsequent reworking and brief new episodes of phosphogenesis produced the extensive phosphate sequence occurring within the Pleistocene sediments along the coastal zone and on the continental shelf. Sediments occurring with the phosphorites can be either volcanic, terrigenous, or authigenic carbonates and diatomites: which sediment actually forms is the result of the complex interaction of the tectonic activity at any given period of time, the pattern of sea level fluctuation and associated palaeoceanographic processes during that same period, and the climatic conditions which affect vegetative cover, type of weathering and terrigenous sediment influx.

The most important aspect of this discussion is that phosphogenesis appears to occur at times of changing sediment regimes during global phosphogenic episodes, regardless of location with respect to the classic west-coast upwelling model or its proximity to active tectonism. Rather, it appears that a change in global tectonism causes major fluctuations in global sea level, oceanographic current regimes, and world climatic conditions. The result of these changes is reflected in the patterns of associated sedimentary units of the stratigraphic column. Most phosphorite sequences appear to have formed at times when the patterns of sedimentation were changing from one type to another. This is apparent in both the very large scale involving geological epochs, regions and formations or in the smaller time framework associated with local facies and unconformities.

A more direct relationship exists between tectonism and specific types of phosphogenesis. Certain phosphate assemblages (e.g. those characterised by volcanic sediments and bedded sulfides) form proximal to and are directly related to active tectonism associated with plate boundary processes. The resulting stratiform deposits often contain major concentrations of sedimentary phosphate and are not uncommon in the geological column. Several examples of such deposits include the Baruga deposits in Michigan, the Broken Hill deposits of Australia, and the Vihanti area of Finland.

Palaeolatitude

Consideration of the broad architecture of the Earth's surface upon phosphogenesis was first addressed by Sheldon (1964). He reconstructed the continental positions at the time of formation of major phosphorites, demonstrating that the origin of phosphorite had definite palaeolatitudinal controls. He found that phosphorites formed between 5° and 42° palaeolatitude with a mean latitude of 23°. The positions of the continents themselves, a direct result of major tectonic processes operating on the Earth's surface, determine the resulting size and shape of ocean basins, the geographical or latitudinal position, and the climatic and current patterns at the time of formation. Freas & Eckstrom (1968) tested this control with good results by reconstructing the continents from the Permian to the Neogene with respect to the occurrence and distribution of known phosphate deposits. They also defined important areas of potential phosphate deposits where phosphorites were unknown. Cook & McElhinny (1979) confirmed the earlier work of Sheldon and demonstrated that the majority of phosphate deposits formed at low-latitude locations.

Regional structural control of phosphate sedimentation

Studies in generally undeformed phosphate deposits of the major Miocene phosphogenic episode of south-eastern USA (Riggs, 1984) have demonstrated that the patterns of sedimentation throughout the depositional system were controlled and defined by pre-depositional structural framework and erosional topography. Two scales or orders of structures and topographic features controlled the formation and subsequent deposition of phosphorites. First-order or large-scale structures defined the regional setting and determined location, size, and geometry of the phosphogenic provinces. Maximum concentrations of phosphate formed and accumulated in the shelf and nearshore environments around the nose and flanks of these first-order structural features and decreased away from the structural noses to minimums in adjacent first-order embayments.

Superimposed upon this regional pattern of phosphate distribution were a series of second-order structural or topographic highs and adjacent entrapment basins which defined the various phosphate districts. The geographic location of each second-order feature, within the first-order structural framework, determined the magnitude of phosphogenesis and type of associated sedimentary components. In the deposits which occurred on the flanks of first-order structures, the loci of phosphate formation were in mid-slope areas of entrapment basins adjacent to second-order structural or topographic highs. Away from these loci, the phosphate became finer-grained and was generally more disseminated in thicker sequences; this probably reflects lower production levels and increasing dilution by terrigenous and carbonate sediments. Specific size and geometry of each district was dictated by the genesis of the basin where the phosphorites accumulated (e.g. structural deformation, primary depositional processes, subaerial or submarine erosion, etc.) and the post-depositional erosional history of the basin.

Riggs (1984) attributed this strong relationship between phosphate deposition and structural framework to the direct interaction between sea level fluctuation, the palaeoceanographic current regime, and the geometry of the continental margin. On a rising sea level, structural highs deflected the currents of the oceanic gyre system, causing increased magnitude and duration of topographic upwelling currents to interact farther across the continental margin.

There is little reason to believe that structure has not produced a similar control over phosphate deposition during all periods of geological time. However, it becomes essential to incorporate good sedimentation and petrology studies in order to form an accurate reconstruction of the palaeogeography at the time of sediment deposition and remove post-depositional deformation and erosion. If this can be done, structure becomes a critical component in the exploration for phosphate deposits, particularly in a known phosphogenic system.

Exploration for phosphate deposits

Types of phosphate deposits

Sediments constituting the stratigraphic column are the recorded response to the regional tectonic history and its interaction with the sedimentary environments of deposition, regardless of whether tectonism is passive or active. Three factors are of prime importance in determining what the resultant sediment sequence will be. First is the variation in the type, magnitude, duration, and proximity of active tectonism to the depositional system. Second is the nature of the palaeoceanographic system including the level of the sea, regional climatic conditions, and type of oceanic current regime and associated upwelling systems. Third is the geometry, structural setting, and specific environmental conditions of the depositional basin. Consequently, various sediment sequences result as the complex variables of these three major factors interact. The sedimentary components (Table 25.1) combine to produce various sedimentary assemblages (Table 25.2) which represent a continuum of sedimentation. Even though much work needs to be done to understand fully the processes leading to the formation of each of these assemblages, these anomalous sedimentary sequences are real. Table 25.4 summarises the necessary ingredients of the phosphogenic system–prerequisites for the formation of a major phosphate deposit.

When all variables are considered, there are many different types of phosphate deposits which demand multiple models to explain their genesis. Table 25.5. presents a summary of five of the more important types, each type having distinct parameters which determine its location within the stratigraphic column and its resource potential. These characteristics are critical factors when considering an exploration programme for phosphate resources. In considering the exploration for Proterozoic–Cambrian phosphate deposits, two of the five types can be totally disregarded because they do not occur in rocks of this age. Two other types have low probabilities of containing enough phosphate resources to warrant anything other than scientific investigation. It is extremely important to be able to recognise these latter two types of phosphate, which rarely amount to much more than mineral occurrences, in order to prevent expending undue energy and money on low-probability exploration targets.

If insular and seamount phosphorites (Type 4 in Table 25.5) are indeed the result of avian guano deposits, this type of phosphate and the fossil bone beds (Type 5 in Table 25.5) will not be represented in the pre-vertebrate geological record. All the known examples of plateau phosphorite (Type 3 in Table 25.5) occur in intermediate-water depths in the modern ocean basins and are of Mesozoic–Cenozoic age. However, if plateau phosphorites did form in comparable environments in the Proterozoic–Cambrian seas, the probability is high that they have been greatly modified by tectonism and metamorphism. Because of their occurrence as thin and extensive beds, major deformational processes would probably render such a sequence very difficult to explore for and uneconomic except on a very local scale. Unconformity phosphate (Type 2 in Table 25.5) constitutes a common mineral occurrence in the Proterozoic–Cambrian sediments: however, due to the negligible accumulation of phosphate on such surfaces, it represents an exploration target with a very low probability and, at best, of very local, small-scale potential. This leaves the phosphate deposits associated with shallow-water continental margins as potentially the most important type and the prime exploration target for new Proterozoic–Cambrian phosphate deposits.

Table 25.4 *Essential components of a continental margin phosphogenic system*

1. Regional tectonism	A. Palaeo-reconstruction of continents and oceans B. Latitude: tropical to temperate C. Regional development of a continental margin basin system D. Structural development of highs and associated embayments E. Both low and high degrees of tectonism F. Changing tectonism G. Periods of quiescence during active tectonism
2. Source of phosphorus	A. Deep, cold, nutrient-rich waters associated with currents B. Pericontinental coastal waters C. Subaerial weathering and fluvial sources D. Submarine hydrothermal and/or volcanogenic solutions
3. Mechanisms of moving and concentrating phosphorus solutions	A. Upwelling current systems (see text for different types) B. Fluvial density plumes C. Density plumes off vents and fracture systems D. Life cycle blooms in the euphotic zone E. Faunal and floral mass-kills
4. Mechanisms for precipitation and formation of sedimentary particles	A. Organic matter and skeletal material B. Biochemical processes of bacteria, algae, etc. C. Physico-chemical precipitation as colloids, clays, etc. a. At or above the sediment–water interface b. Below the sediment–water interface D. Replacement of pre-existing carbonates etc. E. Diagenetic alteration of pre-existing mineral forms
5. Sedimentary environments of deposition	A. Intermediate-water, open marine plateaux and seamounts B. Shallow-water, open marine shelf, platform, or miogeocline C. Unconformity marine hardgrounds with condensed sections D. Shallow-water, semi-restricted, marine embayments E. Nonmarine lacustrine environments on continental margins F. Insular guano deposits G. Secondary replacement or weathered residuum of guano deposits
6. Structural framework	A. Regional structural highs or headlands and embayments B. Local structural highs and adjacent entrapment basins
7. Phosphate sedimentation	A. Contemporaneous with, at or above sediment–water interface a. Mudstone phosphorite-intraclastic grains b. Ovular grains c. Oolitic grains d. Skeletal grains B. Diagenetic and interstitial, below sediment–water interface a. Apatite crystallites b. Replacement of pre-existing mineral phases
8. Contemporaneous associated sedimentation	A. Diluent sedimentation in phosphate facies B. Laterally adjacent facies C. Interbedded cyclical or vertical facies
9. Secondary modification	A. Burial and induration B. Erosion, truncation, and reworking C. Structural deformation D. Diagenetic alteration of authigenic mineral components
10. RESULTANT PHOSPHORITE SEDIMENT AND STRATIGRAPHIC UNIT.	

Depositional model of continental margin phosphogenesis

Neogene phosphorites of south-eastern USA are among the largest and economically most important phosphate deposits in the world. Due to their young age and location on a passive continental margin, these deposits are essentially undeformed records of the world's most recent phosphogenic episode. In addition, recent studies have demonstrated that they are very similar to most of the major continental shelf phosphate deposits of the world. Since these deposits are now among the best known geologically, it is justified to inlude them as a model to help understand and explore for the similar, but more highly modified and less well known Proterozoic and Cambrian phosphorites.

Complex depositional patterns within the Neogene phos-

Table 25.5. *Types of phosphate deposits*

1. *CONTINENTAL MARGIN PHOSPHATE*

A. Associated with coastal upwelling

a. Occurrence:

Meridional or Trade-Wind-Belt: western continental margins associated with oceanic gyres in north–south seaways.

Equatorial or Zonal: northern or southern continental margins associated with equatorial upwelling in low latitude east–west seaways.

Monsoonal: Seasonal monsoonal winds blow parallel to a continental margin and in the appropriate direction to develop upwelling.

b. Geometry:

Very thick and laterally extensive, shallow-water continental shelf deposits.

c. Phosphorite type:

Dominantly sand- and silt-size grains; ovules. intraclasts, ooliths, and skeletal grains, with local concentrations of mudstone phosphorite (microsphorite) and associated nodules and plates.

d. Sediment assemblage:

Black shale, chert or diatomite, carbonates (dolomite or calcite).

e. Examples:

Permian Phosphoria Formation, USA; Cambro-Ordovician (?) Tal Formation, Mussoorie, Uttar Pradesh, India; Cambrian Khubsugul deposits, Mongolian People's Republic.

f. Resource potential:

Vast; these deposits are among the most extensive accumulations of sedimentary phosphate in the world with a very high resource potential.

B. Associated with dynamic or topographic upwelling

a. Occurrence:

Eastern continental margins associated with the boundary currents produced by oceanic gyres in north–south seaways; however, the interaction of any major oceanic surface current system with major structural or bathymetric features on adjacent landmasses will cause upwelling.

b. Geometry:

Thin to very thick interbedded sequences which occur as laterally extensive deposits associated with first- and second-order structures in shallow- to intermediate-water continental shelf and platform deposits.

c. Phosphorite type:

Dominantly sand- and silt-size grains; ovules, intraclasts, ooliths, and skeletal grains with local concentrations of mudstone phosphorite (microsphorite) and associated nodules and plates.

d. Sediment assemblage:

Dolomite, Mg-rich clays, and glauconite.

e. Examples:

Miocene Hawthorn and Pungo River Formations, south-eastern USA; eastern Australian continental shelf; Chatham Rise, New Zealand.

f. Resource potential:

Vast; these deposits are among the most extensive accumulations of sedimentary phosphate in the world with a very high resource potential.

C. Associated with channelled-flow upwelling

a. Occurrence:

One or both sides of elongated narrow seaways in which seasonal winds blow parallel to the seaway.

b. Geometry:

A series of thin to intermediate thickness beds of moderate extent occurring as a series of very shallow-water deposits around the perimeter of an elongated bay or basin.

c. Phosphorite type:

Dominantly sand- and silt-size grains; ovules, intraclasts, ooliths, and skeletal grains with local concentrations of mudstone phosphorite (microsphorite) and associated nodules and plates.

d. Sediment assemblage:

Carbonates (dolomite and calcite), organic-rich shales, diatomites and cherts, stromatolites, and minor evaporites.

e. Examples:

Cambrian Georgina Basin, Australia; Miocene Monterey Formation, south-western California, USA.

f. Resource potential:

Vast accumulations, very high potential.

D. Associated with tectonically active continental margins

a. Occurrence:

Any of the above upwelling occurrences in which the landmass or continental margin is proximal to an active plate boundary.

b. Geometry:

Thin to thick interbedded sequences in deep- to shallow-water marine and lacustrine, small- to medium-sized basins with steep and irregular topography.

c. Phosphorite type:

Dominatly sand- and silt-size ovules, intraclasts, ooliths, and skeletal grains with local concentrations of mudstone phosphorite (microphorite).

d. Sediment assemblage:

Dolomite, diatomites and cherts, volcanic tuffs and clays, and some deposits may be interbedded with, vertically adjacent to, or laterally equivalent to sequences of bedded iron oxides, sulfides, manganese, barite, etc.

e. Examples:

Proterozoic deposits in Baruga Basin, Mich. and Thomsen Formation in Carlton Co., Minn., USA; Grenville Series in Olden Bedford, Ontario, Canada; Miocene Monterey Formation in south-eastern Baja California, Mexico; Miocene Sechura deposits in Peru; Proterozoic deposits at Broken Hill, Australia; the Doushantuo deposits in the People's Republic of China; Proterozoic Vihanti deposit, Finland; Cambrian deposits at Vassbo and Laisvall, Sweden.

f. Resource potential:

Vast accumulations, very high potential.

E. Stromatolitic phosphate

a. Occurrence:

Dominantly Proterozoic in age and occurring in shield areas of continental landmasses; formed in association with any of the above upwelling systems; deposited in very shallow-water, marginal environments on continental shelves or in epicontinental seas.

Table 25.5 *Contd.*

b. Geometry:
Laterally discontinuous, biohermal deposits with highly variable thicknesses.
c. Phosphorite type:
In stromatolitic structures either as columnar forms, fragmental pieces of the columns, or as bedded algal mats.
d. Sediment assemblage:
Dolomites, cherts, phyllites, and quartzites.
e. Examples:
Aravalli phosphogenic province between Udaipur, Rajasthan and Jhabua, Madhya Pradesh, India.
f. Resource potential:
Moderate to vast accumulations in many small deposits; very high regional potential.

2. *UNCONFORMITY PHOSPHATE*
a. Occurrence:
Occurs in sediments and rock units of all ages; formed in shallow-water zones of sediment by-pass and non-deposition on continental shelves and epicontinental or epeiric seas.
b. Geometry:
Very extensive regional to very local, thin surfaces or pavements.
c. Phosphate type:
Highly burrowed and bored, thin phosphate pavements occurring on corroded hardgrounds with abundant intraclastic pebbles.
d. Sediment assemblage:
Glauconite, manganese and pyrite; often occurring on a carbonate hardground.
e. Examples:
Surfaces of the Eocene Castle Hayne Limestone and Miocene Pungo River Formation, North Carolina; Cretaceous of the Anglo-Paris Basin, France; Tertiary sediments of eastern England.
f. Resource potential:
Negligible accumulation; local, small-scale potential only.

3. *PLATEAU PHOSPHATE*
a. Occurrence:
Dominantly Cenozoic in age; occurring on intermediate to deep-water marine plateaux, continental slopes, and seamounts in the oceanic environment.
b. Geometry:
Laterally extensive, thin pavements.
c. Phosphorite type:
Mudstone phosphorite (microsphorite) pavements and coarse plates and nodules.
d. Sediment assemblage:
Manganese and ferromanganese oxides, glauconite, diatomite.
e. Examples:
Blake Plateau and Miami–Pourtales Terrace, south-eastern USA; Agulhas Bank, South Afrcia.
f. Resource potential:
Vast accumulations, long-term future potential.

4. *INSULAR AND SEAMOUNT PHOSPHATE*
a. Occurrence:
Dominantly Cenozoic in age; formed on oceanic islands in the equatorial climatic zone; seamounts are generally considered to be submerged insular deposits.
b. Geometry:
Thin to moderately thick deposits with great irregularity.
c. Phosphorite type:
Replacement of underlying limestones or volcanic weathered materials by aerobic decay solutions of avian guano.
d. Sediment assemblage:
Calcium phosphates in limestones; aluminium phosphates in weathered volcanic clay and debris.
e. Examples:
Insular deposits: Nauru, Ocean, Makatea, and Christmas Islands; Seamount deposits: mid-Pacific Mountains, Milwaukee Bank, Sylvania Seamount, Tasman Sea Guyots, and the New England Seamounts.
f. Resource potential:
Diminishing; local and short-term potential.

5. *FOSSIL BONE BEDS*
a. Occurrence:
Dominantly Mesozoic and Cenozoic in age; deposited in highly stratified lacustrine, epicontinental or epeiric seas, or semirestricted to restricted marine bays with anoxic benthic conditions; often resulting from mass-kills.
b. Geometry:
Thin, irregular, and generally small deposits.
c. Phosphorite type:
Disseminated, high-grade vertebrate bone material.
d. Sediment assemblage:
Organic-rich shales.
e. Examples:
Morrison Shale, western USA; Triassic of southern Germany; Maikop Formation, USSR
f. Resource potential:
Very small, local, and short-term potential

phorites have been delineated by many workers. Regional studies and detailed lithofacies analyses have been carried out on the Miocene Pungo River and Pliocene Yorktown Formations in the North Carolina Phosphate District by Gibson (1967), Miller (1971), Riggs, Lewis *et al.* (1982), Scarborough, Riggs & Snyder (1982), and Snyder *et al.* (1982), and on the Miocene–Pliocene Hawthorn Formation in Florida by Altschuler, Cathcart & Young (1964), Cathcart (1968), and Riggs (1979a, 1979b). Extensive high-resolution seismic profiling and vibracoring have defined the Miocene Pungo River Formation and Pleistocene phosphorites in Onslow Bay, North Carolina (Lewis, 1981; Riggs, Snyder *et al.*, 1982, 1985; and Snyder *et al.*, 1982).

Riggs (1984) demonstrated that changing patterns of terrigenous, carbonate, and phosphate sediments in the Neogene section of the Atlantic continental margin were direct responses to the type, intensity, and duration of various palaeoceanographic and palaeoclimatic events. He presented a model explaining how the changing patterns of sea level, oceanographic current regimes, and associated climatic fluctuations interacted with regional structures to affect shelf sedimentation and produce phosphogenesis. The main processes included: (1) the wind-driven clockwise Gulf Stream gyre system of the northwest Atlantic Ocean which interacts with the continental margin as it moves west and north; and (2) major fluctuations in global sea level which cause the transgressive–regressive pulses of oceanic water on to the continental margin allowing for the formation of a marine sediment wedge. Fluctuations in sea level during the Neogene, largely the result of continental glaciation and deglaciation (Vail *et al.*, 1977), were accompanied by three effects: (1) significant modification of the path and dynamics of the Gulf Stream causing different oceanographic water masses to dominate the continental margin, including periods of extensive topographic upwelling; (2) extensive climatic changes resulting in modification of vegetative cover, rates and types of erosion, fluvial processes, and resultant terrigenous sediment and nutrient supplies to the ocean; and (3) alternating conditions between sediment deposition and accumulation and subsequent erosion and truncation.

Riggs (1984) concluded that the Pungo River and Hawthorn Formations represent a major phosphogenic episode which was coincident with the extensive global rise in sea level of the second-order Miocene transgression. The Pliocene sediment sequence, with its much smaller concentration of associated primary phosphorites, was deposited on the transgression of the subsequent second-order sea level cycle which was much shorter than the first. Preliminary studies suggest that the very local concentration of Pleistocene phosphorites on the North Carolina shelf may not be totally reworked from the Pungo River. Some of it may in part be primary phosphate that is the product of one or more very brief Pleistocene phosphogenic episodes which coincided with the Pleistocene transgressive cycles (Riggs *et al.*, 1985).

The present Holocene transgression is approximately at a midshelf stage with the shoreline about 50 m above the shelf-break in Onslow Bay, North Carolina. According to Parrish (1982), 50 m is the minimum water depth required on a shallow shelf to initiate active upwelling and maintain good bottom-water intrusion layers across the shelf. This sea level position has recently placed the axis of the Gulf Stream against the Florida–Hatteras Slope and at the edge of the continental shelf. This is the approximate position occupied by the Gulf Stream during the early stages of each Cenozoic second-order high stand of the sea described by Pinet, Popenoe & Nelligan (1981), Pinet & Popenoe (1982) and Popenoe *et al* (in press). Sea level has now risen high enough on to the shelf to cause the Gulf Stream to interact with the shelf's geometry initiating the early stages of dynamic or topographically induced upwelling. These upwelling processes have been recently described from the modern system (Bane & Brooks, 1979; Blanton *et al.*, 1981; Brooks & Bane, 1978; Janowitz & Pietrafesa, 1980; Lee, Atkinson & Legeckis, 1981; and Pietrafesa, 1983) and presently cause abundant short-term pulses of nutrient-enriched bottom waters to flow across the continental shelf (Atkinson, Pietrafesa & Hofmann, 1982; Hofmann *et al.*, 1980; and Hofmann, Pietrafesa & Atkinson, 1981). Assuming that the present transgression continues, along with the warming climate and decreased terrigenous sediment influx, it is logical to conclude that the dynamic upwelling processes associated with the Gulf Stream would increase. This could lead in time from the past and present terrigenous sediment regime on the shelf to an increased dominance of authigenic sediment deposition including the formation and deposition of phosphates and carbonates.

Smaller-scale cyclical patterns of deposition occur within each of the three Neogene phosphorite units, which Riggs (1984) interpreted to reflect sea level fluctuations similar to that of the Holocene. Each depositional unit and associated unconformity surfaces are believed to correlate with established third-order, or higher frequency, eustatic sea level cycles. During each transgressive cycle, increased phosphate and decreased terrigenous sedimentation reflect the growing influence of cold, nutrient-rich upwelled Gulf Stream waters and moderating climatic conditions. Each transgression culminated in a brief period of subtropical conditions and carbonate deposition during the highest sea level stands. The regressive phase is marked by the unconformity surfaces which reflect either erosion or the development of hardgrounds with exposure and non-deposition.

First-order structural highs produced extensive shallow-shelf platform environments which projected seaward across the continental margin delineating the regional embayments, and interacted with the major oceanic current system causing topographic upwelling. The location, size, and origin of second-order structural or topographic features, superimposed upon the regional structural framework, determined the nature of the interaction with the oceanic system. The outer nose and adjacent flanks of the first-order structures became sites of major phosphorite sedimentation when the palaeoceanographic conditions were right to produce a phosphogenic episode. These included shoaling and large-scale upwelling of nutrient-rich waters in response to quasi-permanent deflection of the oceanic boundary current in conjunction with a slowly transgressing sea level. Phosphorite sedimentation generally decreased into the regional first-order embayments as the shelf environments became more remote from the oceanic system. Away from the nose of the first-order structures, frontal eddies and small-scale topographically induced upwellings became important and the resultant phosphate formed along the mid-slope portion of second-order embayments. The phosphate depocentres migrated farther on to the inner shelf of the first-order embayments through the transgression as the oceanic current system increasingly impinged upon the shelf system.

Passive continental margins

Phosphate deposits associated with passive continental margins are not only the most common type of deposit, but are also the most extensive. They occur throughout geological time from the lower Proterozoic to the Cenozoic. Continental margin deposits can result from either coastal, topographic or channelled-flow upwelling currents (Table 25.5), all of which are capable of delivering deeper, colder, and nutrient-enriched waters into shallow-water environments of continental margins for extended periods of geological time. Consequently, most continental margins occurring within tropical to temperate latitudes become potential sites for phosphogenesis.

Stromatolite sequences are very common in Proterozoic carbonates, but rarely are they associated with phosphogenesis. Obviously, at certain times or in specific environmental settings, conditions were suitable for phosphogenesis to take place. Consequently, it is not enough just to identify areas of Proterozoic stromatolitic carbonates; this is only the first step in exploration. The next critical step is to sort out the patterns of deposition within the carbonate sequence, and in particular, identify periods of changing lithologies, both into and out of periods of major terrigenous, volcanic, or carbonate deposition. These patterns reflect changes in the tectonic regime and fluctuations in sea level and climatic conditions, all of which modify sediment supplies, water chemistry, and oceanographic current regimes. For example, a prime target zone would be an interbedded sequence of marine shales and dolomites occurring between major conglomerate–sandstone and carbonate sequences, either on the transgressive or regressive side of the depositional event. This target would be particularly intriguing if there were any known occurrences of chert, banded iron, manganese, barite, metallic sulfide etc., in these transitional sediments. Such a Proterozoic continental margin sediment sequence has a high probability of containing the record of a phosphogenic event regardless of whether stromatolites are present or not. However, all stromatolitic sediments occurring within such a target, should be investigated for potential phosphate.

Phosphate exploration within the Cambrian is much more direct than in the complex and modified Proterozoic sediments. In general, initiation of Cambrian sedimentation was characterised by the cessation of glacial activity followed by moderating climatic conditions, and a global sea level transgression across the continental margins (Sloss & Speed, 1974; Vail *et al.*, 1977). This global event produced similar patterns of change in Cambrian sediments of most basinal systems; the resulting sedimentary sequence generally grades upward from tillites or conglomerates at the base, to sands and silts, to dolomites and cherts, and finally to extensive sequences of carbonates or black shales.

Cook & McElhinny (1979) and Cook & Shergold (1980) point out that the Cambrian represents a major phosphogenic episode, and that many of the phosphate deposits occur in the dolomite–chert facies of the rapidly changing lithofacies sequence produced by the global Cambrian transgression. Slight variations in the age as well as in detailed lithologies of the many Cambrian deposits are expected because of regional variations in both the tectonic setting and activity (see Shergold & Brasier this volume). Thus, the transitional facies which formed during this stage of the Cambrian sea level rise is a prime exploration target zone along any Cambrian continental margin which had the proper palaeogeographic location and palaeoceanographic conditions to generate the phosphate facies. Similar phosphorite-associated sediment relationships could also exist on any higher frequency sea level fluctuations superimposed upon the first-order global curves of the Cambrian. Such events, recognised by major changes in the patterns of sedimentation, appear to be the time of phosphogenesis. In other words it is not worth looking for phosphorite in massive sections of uniform carbonates or terrigenous sands and clays in the depositional sequence of any passive continental margin system.

Tectonically active continental margins

The geological column contains many known examples of phosphatic sediments which are associated with stratiform sulfides, banded iron formations, and sequences of volcanogenic sediments, particularly during the Proterozoic (Table 25.5). Cook & McElhinny (1979) recognised a strong correlation between periods of phosphogenesis and banded iron formation, stating that even though the banded iron formations are not phosphate deposits, they do contain extremely large volumes of phosphorus. For example, they calculated that the Hamersley Group Brockman Iron Formation in Australia contains an estimated 5×10^{11} tonnes of P_2O_5, as much phosphorus as occurs in the world's largest phosphate deposits. Most of the phosphorus in iron formations is absorbed on to iron oxides or occurs in a poorly crystalline iron phosphate phase, but in some deposits phosphate occurs as discrete laminae and beds of phosphate grains. Most of these anomalous sedimentary sequences have close associations with plate boundary processes in an active tectonic regime.

Assuming that phosphatic sediments do form in association with active as well as passive continental margin environments, which previous discussions document fairly convincingly, this phosphate association presents many new exploration opportunities. For example, the world's classic banded iron formations and stratiform base metal sulfide deposits merit serious exploration efforts. The recently discovered Tynagh deposit in Ireland, along with three other new deposits which occur along strike, are Lower Carboniferous bedded sulfides of pyrite, galena, sphalerite, and some chalcopyrite and barite (Russell, 1968). At the same stratigraphic horizon to the north, a banded iron formation occurs interbedded with tuffs. Could this sequence of sediments grade up the depositional slope into a phosphatic sequence? Kanasewich (1968) proposed that the stratiform sulfides which occur within the Precambrian Purcell Belt Series (the Sullivan deposit at Kimberly, British Columbia; the Coeur d'Alene deposits of Idaho; and the Spar Lake deposit in Montana) had their origin in a Precambrian rift/transform fault complex that crosses western Canada and the United States. What is the relationship of the poorly known phosphates which occur within the Spokane Formation of the Precambrian Belt Series (Gulbrandsen, 1966) and which also contain stratabound copper and silver (Lange & Eby, 1981), to the stratiform sulfide deposits? Establishing the lateral and vertical stratigraphic relationships along with the palaeogeographic and tectonic reconstructions might lead to the discovery of more extensive and richer phosphate deposits. What about other similar major deposits of the world such as Mount Isa in Australia? The tectonic setting that produced many of these anomalous sedimentary deposits also could have produced a lateral upslope or shallower-water facies of phosphorus-rich sediments, assuming an adequate basin geometry with the necessary physical-chemical conditions.

If the tectonic hypothesis is to be tested, several major points must be kept in mind: (1) not all tectonic processes leading to bedded sulfide deposits will have a lateral phosphate facies associated with them; (2) the physical geometry of the associated depositional environments or associated sediment deposition may not have been conducive to the ultimate formation and preservation of phosphatic sediments; (3) the lateral phosphatic sediment sequence, if formed, could have been subsequently destroyed by continued tectonism and erosion. Thus, the associations tend to be interesting, but the complexity and extreme variability of the different types of phosphogenic systems must be remembered.

Summary and recommendations

Not all continental margins or periods of geological time are characterised by the formation and occurrence of phosphorites. There are many types of upwelling systems which occur along most types of continental margins. Each is theoretically capable of supplying the necessary phosphorus to produce the associated organic productivity and resultant phosphatic sediments. Consequently the ocean basin geometry and orientation is not all that important. What is important is for the situation which allows optimum upwelling to persist for extended periods of geological time. Therefore, other prerequisites for phosphogenesis must be considered in defining targets and exploring for depositional systems in which phosphorites would have formed. The parameters with the greatest probability of leading to the formation of a major phosphate deposit during the Proterozoic and Cambrian include the following:

1. a rock sequence representing one or more of the five recognised phosphogenic episodes (mean ages of 2200 to 1800 m.y., 1600 to 1200 m.y., 800 to 700 m.y., 650 m.y. and 540 m.y.);
2. a palaeogeographic location within low to intermediate latitudes (between 5° to 42° latitude);
3. an intermediate stage of sea level on either the transgressive or regressive phase of a first-or second-order global sea level cycle; higher frequency sea level cycles produce the interbedded phosphorites and associated anomalous sediments;
4. a changing pattern of tectonic activity or a lull in a section reflecting highly active tectonism; this is characterised by a zone of variable and changing lithologies;
5. a stratigraphic section containing a significant accumulation of shallow-water, marine, continental margin sediments;
6. a zone of variable and changing lithologies within the stratigraphic pile which separates thick sections of uniform sediments of either terrigenous sands and clays, authigenic carbonates, or volcanic sediments;
7. a lithologic zone characterised by one or more combinations of the authigenic sediments commonly associated with phosphate assemblages (i.e. cherts, dolomites, black shales, Mg-rich clays, iron-rich sediments, manganese, bedded sulfides, etc);
8. a regional structural setting which provides: (a) a shallow-water continental shelf environment; (b) major headlands or bathymetric highs extending across the continental margin to interact with the oceanic current regime and produce maximum upwelling; and (c) adjacent embayments or entrapment basins for the accumulations of the resulting sediments.

A simple, first-phase exploration programme can be developed for continental margin-type phosphate deposits. The following outline is based upon utilising existing geological information of a given country or region, not the very costly exercise of drilling and developing a new data base or resource evaluation, which are phases II and III of an exploration programme. Depending upon the size of the area of interest and the amount of geological information available, a small team of trained geologists (from 2 to 6 people) with a good geological library and several months, can carry out a phase I level exploration programme for continental margin type phosphate deposits. Such a programme would include the following activities:

1. Collect and collate all existing information on global tectonic events, sea level curves, and phosphogenic episodes.
2. Utilise existing global palaeogeographic reconstructions to develop one or more general working models concerning the palaeogeographic locations and historical development of the continental margins through geological time.
3. Delineate the sedimentary basins of the continental margins and define the periods of geological time during which each was active.
4. Identify all periods of changing depositional patterns which contain lithologies commonly associated with phosphate sedimentation within each of the basinal systems. If the data are available, generate crude facies maps for each period of changing depositional patterns or associated sediment assemblages.
5. Summarise all the above information; then, utilising criteria established in the earlier portion of this paper (latitudinal controls, global sea level and climatic controls, associated sediment assemblages, etc.), identify all potential target zones within the geological column and target areas within each basin for field examination. Remember, each region or basin may have a regional tectonic imprint which would cause major deviations from global patterns.
6. Examine all potential target areas in the field to identify those targets with the greatest potential for proceeding into phase II of the exploration programme.
7. Begin phase II, the detailed geological field mapping of those sedimentary units and areas which still appear favourable after the preliminary field examination.

Continental margin type phosphorites, regardless of age, are similar in their general features, but do vary in the net volume and concentration of phosphate, the phosphate grain types, the associated carbonate, terrigenous and other sediment components, and the stratigraphic cycles including deposition/erosion and changing lithological patterns. These variations in the sedimentological patterns of an abnormal stratigraphic sequence can be explained by the variable interaction of at least four environmental parameters operating through geological time; (1) size, geometry, and structural framework of the continental margin depositional system; (2) intensity, stability, and duration of the oceanographic conditions controlling the physics/chemistry/biology of the shelf environment (i.e. upwellings, boundary currents, eustatic sea level oscillations, etc.); (3) amount and rate of diluent sedimentation resulting from contemporaneous terrigenous and authigenic deposition; and (4) additional secondary factors (including subsequent history of deposition, structural deformation, erosion, burial, diagenetic history, and weathering of the sediments) which can severely modify the stratigraphic unit if the phosphorites were formed. These variables, which produce the differences between deposits, are the factors determining the specific economics of a deposit once it has been discovered and evaluated.

References

Altschuler, Z.S., Cathcart, J.B. & Young, E.J. (1964). *The geology and geochemistry of the Bone Valley Formation and its phosphate deposits, west central Florida.* Guidebook to Field Trip No. 6, 68pp. Annual meeting, Geological Society of America, Boulder.

Ames, L. L. Jr (1959). The genesis of carbonate apatites. *Economic Geology*, **54**, 829–41.

Arthur, M.A. & Jenkyns, H.C. (1981). Phosphorites and paleoceanography. In *Proceedings 26th Session International Geological Congress, Paris*, pp. 83–96. Oceanologica Acta, Montreuil.

Atkinson, L.P., Pietrafesa, L.J. & Hofmann, E.E. (1982). Nutrient sources to Onslow Bay, North Carolina. *Journal of Marine Research*, **40**, 679.

Bane, J.M. & Brooks, D.A. (1979). Gulf Stream meanders along the continental margin from the Florida Straits to Cape Hatteras. *Geophysical Research Letters*, **6**, 280–2.

Baturin, G.N. (1971). Stages of phosphorite formation on the sea floor. *Nature (Phys. Sci.)*, **232**, 61–2.

Bentor, Y.K. (1980). Phosphorites–the unsolved problems. *Society of Economic Paleontologists and Mineralogists, Special Publication*, **29**, 3–18.

Birch, G.B. (1979). Phosphatic rocks on the western margin of South Africa. *Journal of Sedimentary Petrology*, **49**, 93–110.

Blanton, J.O., Atkinson, L.P., Pietrafesa, L.J. & Lee, T.N. (1981). The intrusion of Gulf Stream water across the continental shelf due to topographically-induced upwelling. *Deep-sea Research*, **28A**. 393–405.

Brasier, M.D. (1980). The lower Cambrian transgression and glauconite-phosphate facies of western Europe. *Journal of the Geological Society*, **137**, 695–704.

Bremner, J.M. & Willis, J.P. (1975). Glauconite and phosphorites from Southwest Africa. *Proceedings of the Electron Microscopy Society of Southern Africa*, **5**, 127–8.

Brongersma-Sanders, M. (1957). Mass mortality in the sea. In *Treatise on Marine Ecology and Paleoecology*, ed. J.W. Hedgepeth, pp. 941–1010. Memoir 67(1), Geological Society of America, Boulder.

Brooks, D.A. & Bane, J.M. (1978). Gulf Stream deflection by a bottom feature off Charleston, South Carolina. *Science*, **201**, 1225–6.

Buckland, W. (1829). On the discovery of coprolites, or fossil faeces, in the Lias at Lyme Regis, and in other formations. *Geological Society of London, Transactions*, 2nd. ser., **3**, 223–38.

Burnett, W.C. (1980). Apatite-glauconite associations off Peru and Chile: paleoceanographic implications. *Journal of the Geological Society*, **137**, 757–64.

Bushinsky, G.I. (1966). Ancient phosphorites of Asia and their genesis. *Trudy Geologicheskiy Institut, Akademiya Nauk SSSR*, **149**, 192pp.

Cathcart, J.B. (1968). Florida-type phosphate deposits of the United States–origin and techniques for prospecting. In *Proceedings of the Seminar on Sources of Mineral Raw Materials for the Fertilizer Industry in Asia and the Far East*, pp. 178–86. Mineral Resources Development Series 32. United Nations ECAFE.

Cheney, T.M., McClellan, G.H. & Montgomery, E.S. (1979). Sechura phosphate deposits, their stratigraphy, origin and composition. *Economic Geology*, 74, 232–59.

Christofferson, H.C., Wallin, B., Selkman, S. & Rickard, D.T. (1979). Mineralization controls in the sandstone lead-zinc deposits at Vassbo, Sweden. *Economic Geology*, 74, 1239–49.

Cook, P.J. (1967). Winnowing-an important process in the concentration of the Stairway Sandstone (Ordovician) phosphorites of central Australia. *Journal of Sedimentary Petrology*, **37**, 818–828.

Cook, P.J. (1976). Sedimentary phosphate deposits. In *Handbook of Strata-bound and Stratiform Ore Deposits*, ed. K.H. Wolf, pp. 505–35. Elsevier, New York.

Cook, P.J. & McElhinny, M.W. (1979). A re-evaluation of the spatial and temporal distribution of sedimentary phosphate deposits in the light of plate tectonics. *Economic Geology*, **74**, 315–30.

Cook, P.J. & Marshall, J.E. (1981). Geochemistry of iron and phosphorus-rich nodules from the East Australian continental shelf. *Marine Geology*, **41**, 205–21.

Cook, P.J. & Shergold, J.H. (1980). Proterozoic and Cambrian phosphorites of Asia and Australia. In *Fertilizer Mineral Potential in Asia and the Pacific*, ed. R.P. Sheldon & W.C. Burnett, pp. 207–24. East–West Resources Systems Institute, Honolulu.

Cornet, M.F.L. (1886). On the Upper Cretaceous Series and the phosphate beds in the neigbourhood of Mons (Belgium). *Quarterly Journal of the Geological Society of London*, **42**, 325–40.

Cullen, D.J. (1980). Distribution, composition and age of submarine phosphorites on Chatham Rise, east of New Zealand. *Society of Economic Paleontologists and Mineralogists, Special Publication*, **29**, 139–48.

Dale, N.C. (1915). The Cambrian manganese deposits of Conception and Trinity Bays, Newfoundland. *Proceedings of the American Philosophical Society*, **54**, 371–456.

Doyle, L.J., Blake, N.J., Woo, C.C. & Yevich, P. (1978). Recent biogenic phosphorite: concretions in mollusk kidneys. *Science*, **199**, 1431–3.

Fisher, O. (1873). On the phosphatic nodules of the Cretaceous rock of Cambridgeshire. *Quarterly Journal of the Geological Society of London*, **29**, 52–63.

Frakes, L.A. (1979). *Climates Throughout Geologic Time*, 310pp. Elsevier, Amsterdam.

Freas, D.H. & Eckstrom, C.L. (1968). Area of potential upwelling and phosphorite deposition during Tertiary, Mesozoic and late Paleozoic time. In *Proceedings of the Seminar on Sources of Mineral Raw Materials for the Fertilizer Industry in Asia and the Far East*, pp. 228–37. Mineral Resources Development Series, 32. United Nations ECAFE.

Gibson, T.G. (1967). Stratigraphy and paleoenvironment of the phosphatic Miocene strata of North Carolina. *Geological Society of America, Bulletin*, **78**, 631–50.

Gimmelfarb, B.M. (1958). *Regularity of the tectonic distribution of phosphorite deposits in the USSR.* Akademiya Nauk SSSR, Izdatelstvo, Moscow, 487–516.

Gulbrandsen, R.A. (1966). Precambrian phosphorite in the Belt Series in Montana. *United States Geological Survey professional paper*, **550-D**, D199–202.

Hite, R.J. (1978). Possible genetic relationship between evaporites, phosphorites, and iron-rich sediments. *Mountain Geologist*, **14(3)**, 97–107.

Hofmann, E.E., Pietrafesa, L.J. & Atkinson, L.P. (1981). A bottom water intrusion in Onslow Bay, North Carolina. *Deep-Sea Research*, **28A**, 329–45.

Hofmann, E.E., Pietrafesa, L.J., Klinck, J.M. & Atkinson L.P. (1980). A time-dependent model of nutrient distribution in continental shelf waters. *Ecological Modelling*, **10**, 193–214.

Janowitz, G.S. & Pietrafesa, L.J. (1980). A model and observations of time dependent upwelling over the mid-shelf and slope. *Journal of physical Oceanography*, **10**, 1574–83.

Kanasewich, E.R. (1968). Precambrian rift: genesis of strata-bound ore deposits. *Science*, **161**, 1002–5.

Kazakov, A.V. (1937). The phosphorite facies and the genesis of phosphorites. In *Geological investigations of agricultural ores*, pp. 95–113. Transactions of the Science Institute of Fertilizers and Insecto-fungicides **142** (publ. for the 17th Session International Geological Congress, Leningrad).

Kazakov, A.V. (1938). The phosphorite facies and the genesis of natural phosphates. *Sovetskaya Geologiya*, **8**, 33–47.

Lange, I.M. & Eby, D.E. (1981). Stratabound copper-silver-bearing oolitic deposits in the Belt Supergroup of Montana. *Economic Geology*, **76**, 933–6.

Lee, T.N., Atkinson, L.P. & Legeckis, R. (1981). Detailed observations of a Gulf Stream spin-off eddy on the Georgia continental shelf, April 1977. *Deep-Sea Research*, **28A**, 347–78.

Lewis, D.W. (1981). Preliminary Stratigraphic Evaluation of the Pungo River Formation in Onslow Bay, North Carolina Continental Shelf. M.Sc. Thesis, East Carolina University, Greenville, North Carolina, 100pp. (unpublished).

Manheim, F.T., Pratt, R.M. & McFarlin, P.F. (1980). Composition and origin of phosphorite deposits of the Blake Plateau. *Society of Economic Paleontologists and Mineralogists, Special Publication*, **29**, 117–37.

Mansfield, G.R. (1940). The role of fluorine in phosphate deposition. *American Journal of Science*, **238(12)**, 833–79.

Marshall, J.F. & Cook, P.J. (1980). Petrology or iron- and phosphorus-rich nodules from the E. Australian continental shelf. *Journal of the Geological Society*, **137**, 765–71.

Miller, J.A. (1971). Stratigraphic and Structural Setting of the Middle Miocene Pungo River Formation of North Carolina. Ph.D. Thesis, University of North Carolina. Chapel Hill. North Carolina, 82pp. (unpublished).

Mullins, H.T. & Neumann, A.C. (1979). Geology of the Miami Terrace and its paleo-oceanographic implications. *Marine Geology*, **30**, 205–32.

Notholt, A.J.G. (1979). Resources of Precambrian and Cambrian sedimentary phosphate rock. In *Proterozoic-Cambrian Phosphorites*, ed. P.J. Cook & J.H. Shergold, pp. 71–86. ANU Press, Canberra.

O'Brien, G.W. & Veeh, H.H. (1980). Holocene phosphorite on the east Australian continental margins. *Nature*, **288(5792)**, 690–2.

Parrish, J.T. (1982). Upwelling and petroleum source beds, with reference to Paleozoic. *American Association of Petroleum Geologists, Bulletin*, **66**, 750–74.

Penrose, R.A.F. (1888). Nature and origin of deposits of phosphate of lime, with an introduction by N.S. Shaler. *United States Geological Survey Bulletin*, **46**, 143pp.

Pevear, D.R. (1966). The estuarine formation of United States Atlantic coastal plain phosphorite. *Economic Geology*, **61(2)**, 251–6.

Pietrafesa, L.J. (1983). Shelfbreak circulation, fronts and physical oceanography: east and west coast perspectives. *Society of Economic Paleontologists and Mineralogists, Special Publication*, **33**, 233–50.

Pinet, P.R. & Popenoe, P. (1982). Blake Plateau: control of Miocene sedimentation patterns by large-scale shifts of the Gulf Stream axis. *Geology*, **10**, 257–9.

Pinet, P.R., Popenoe, P. & Nelligan, D.F. (1981). Gulf Stream: reconstruction of Cenozoic flow patterns over the Blake Plateau. *Geology*, **9**, 266–70.

Pisciotto, K.A. (1978). Basinal Sedimentary Facies and Diagenetic Aspects of the Monterey Shale, California. Ph.D. Thesis, University of California, Santa Cruz, California, 450pp. (unpublished).

Pitman, W.C. (1978). The relationship between eustacy and stratigraphic sequences of passive margins. *Geological Society of America, Bulletin*, **89**, 1389–403.

Pitman, W.C. (1979). The effect of eustatic sea level changes on stratigraphic sequences at Atlantic margins. *American Association of Petroleum Geologists, Memoir*, **29**, 453–60.

Popenoe, P. (in press). Seismic stratigraphy and Tertiary development of the North Carolina Continental margin. In *Geological Evolution of the US Atlantic Continental Margin*, ed. C.W. Poag, Hutchinson & Ross.

Poulsen, Chr. (1960). The Paleozoic of Bornholm. *21st Session International Congress, Norden.* Guide to Excursions A46 C41, 15pp.

Rickard, D.T., Willdén, M.Y., Marinder, N.E. & Donnelly, T.H. (1979). Studies on the genesis of the Laisvall sandstone lead-zinc deposit, Sweden. *Economic Geology*, **74**, 1255–85.

Riggs, S.R. (1979a). Petrology of the Tertiary phosphorite system of Florida. *Economic Geology*, **74**, 195–220.

Riggs, S.R. (1979b). Phosphorite sedimentation in Florida – a model phosphogenic system. *Economic Geology*, **74**, 285–314.

Riggs, S.R. (1980). Tectonic model of phosphate genesis. In *Fertilizer Mineral Potential in Asia and the Pacific*, ed. R.P. Sheldon & W.C. Burnett, pp. 159–90. East–West Resource Systems Institute, Honolulu.

Riggs, S.R. (1984). Paleoceanographic model of Neogene phosphorite deposition, U.S. Atlantic continental margin. *Science*, **233(4632)**, 123–31.

Riggs, S.R., Lewis, D.W., Scarborough, A.K. & Snyder, S.W. (1982). Cyclic deposition of Neogene phosphorites in the Aurora Area, North Carolina, and their possible relationship to global sea level fluctuations. *Southeastern Geology*, **23(4)**, 189–204.

Riggs, S.R., Snyder, S.W., Ellington, M.D., Burnett, W.C. & Beers, M. (1982). Pleistocene/Holocene phosphorite formation – North Carolina continental. *Geological Society of America Programs with Abstracts*, **14**, 77.

Riggs, S.R., Snyder, S.W.P., Hine, A.C. Snyder, S.W. Ellington, M.D. & Mallette, P.M. (1985). Geologic framework of phosphate resources in Onslow Bay, North Carolina Continental Shelf. *Economic Geology*, **80**, 716–38.

Rooney, T.P. & Kerr, P.F. (1967). Mineralogic nature and origin of phosphorite, Beaufort Co., North Carolina. *Geological Society of America Bulletin*, **78**, 731–48.

Russell, M.J. (1968). Structural controls of base metal mineralization in Ireland in relation to continental drift. *Transactions of the Institution of Mining and Metallurgy*, **77**, B117–28.

Salas, G.P. (1978). Sedimentary phosphate deposits in Baja California, Mexico. *American Institute of Mining Engineers, Preprint* **78-H-75**, 21pp.

Scarborough, A.K., Riggs, S.R. & Snyder, S.W. (1982). Stratigraphy and petrology of the Pungo River Formation, central coastal plain, North Carolina. *Southeastern Geology*, **23(4)**, 205–16.

Seeley, H. (1866). The rock of the Cambridge Greensand. *Geological Magazine*, **3**, 302–7.

Sheldon, R.P. (1964). Paleolatitudinal and paleogeographic distribution of phosphorite. *United States Geological Survey, Professional Paper*, **501-C**, 106–13.

Sheldon, R.P. (1980). Episodicity of phosphate deposition and deep ocean circulation – an hypothesis. *Society of Economic Paleontologists and Mineralogists, Special Publication*, **29**, 239–48.

Sheldon, R.P. (1981). Ancient marine phosphates. *Annual Review of Earth and Planetary Science*, **9**, 251–84.

Sheldon, R.P. & Burnett, W.C. (1981). Paleogeography of Phanerozoic phosphorites and extreme upwelling systems. *American Association of Petroleum Geologists, Bulletin*, **65**, p. 991.

Sloss, L.L. & Speed, R.C. (1974). Relationships of cratonic and continental margin tectonic episodes. *Society of Economic Paleontologists and Mineralogists, Special Publication*, **22**, 98–119.

Snyder, S.W., Riggs, S.R., Katrosh, M.R., Lewis, D.W. & Scarborough, A.K. (1982). Synthesis of phosphatic sediment–faunal relationships within the Pungo River Formation: paleoenvironmental implications. *Southeastern Geology*, **23(4)**, 233–46.

Snyder, S.W.P., Hine, A.C. & Riggs, S.R. (1982). Miocene seismic stratigraphy, structural framework and sea level cyclicity: North Carolina continental shelf. *Southeastern Geology*, **23(4)**, 247–66.

Summerhayes, C.P., Nutter, A.H. & Tooms, J.S. (1972). The distribution and origin of phosphate in sediments off northwest Africa. *Sedimentary Geology*, **18**, 3–28.

Tissot, B. (1979). Effects on prolific petroleum source rocks and major coal deposits caused by sea-level changes. *Nature*, **277**, 463–5.

Vail, P.R. & Mitchum, R.M. Jr (1979). Global cycles of relative changes of sea level from seismic stratigraphy. *American Association of Petroleum Geologists, Memoir*, **29**, 469–72.

Vail, P.R., Mitchum, R.M. Jr, Todd, R.G., Widmier, J.M., Thompson, S. III, Sangree, J.B., Bubb, J.N. & Hatelid, W.G. (1977). Seismic stratigraphy and global changes of sea level. *American Association of Petroleum Geologists, Memoir*, **26**, 49–212.

26

Proterozoic and Cambrian phosphorites – nature and origin

P.J. COOK AND J.H. SHERGOLD

Introduction

In the previous chapters of this book basic information on Proterozoic and Cambrian phosphorites has been presented. It is not our intention to reiterate that wealth of information. Instead, we shall highlight some of the more important features, particularly those common to a number of deposits. No doubt the reader will have already noted a number of common threads in many of these chapters and it is those threads that we shall use to test existing hypotheses for phosphogenesis as well as to provide a basis for a model. The separation of fact and speculation cannot always be as clear as we would wish and indeed the development of any geological hypothesis involves observation, the development of a model, and the identification of the additional observations needed to test that model. This process is no different in other areas of science. What is different in the development of models for phosphogenesis is the diversity of observations which need to be drawn on if the integrated model is to be credible.

The model

The model presented in Fig. 26.1 shows by means of the standard symbolism of a computer flowchart, the sequential steps involved in the formation of a phosphorite (primarily the vertical flow paths) and some of the processes of 'reinforcement' between the various sequences (primarily represented by horizontal flow paths). Just as in a computer flowchart, 'loops' represent stages in the model where several iterations may occur. Major input or output is represented by trapezoids, and the beginning or ending of the 'program' or model is represented by an oval. The structure of the model could be regarded as a matrix, with the vertical columns (A, B, C, D) representing various scales, with A the largest (global) scale and C the smallest (local) scale. This hierarchy is not completely consistent in that D represents processes acting on a range of scales, though focussing mainly on the local and with considerable interaction between columns C and D. Scale refers to both temporal and spatial magnitudes. Thus, major evolutionary changes (Column A) may perhaps occur over 10^7–10^8 years, medium term changes (Column B) over 10^5–10^6 years, and short term changes over 10^3–10^5 years. Global phenomena (Column A) may occur over spatial scales of perhaps 10^7–10^8 km^2, regional phenomena (Column B) over 10^5–10^6 km^2 and local phenomena (Columns C and D) over 10^1–10^4 km^2. Thus, there is an implicit hierarchy in the vertical structure of the matrix. The horizontal structure of the matrix (1–9) is broadly sequential, commencing with the basic driving mechanism of line 1, such as climatic change or plate tectonics and ending with what is a rather localised and relatively small scale process – the accumulation of phosphate grains (line 8), finally leading to the formation of a phosphorite deposit (line 9).

There is also a philosophical basis for the model in that the basic driving mechanisms are plate tectonics and climate. It could be argued that even climatic change is frequently a consequence of the location of the continents relative to the poles, but it could also be argued that local (?intraplate) tectonics owe very little to plate tectonics. To attempt to do justice to all the options would result in a highly complex and almost certainly incomprehensible diagram. Therefore Fig. 26.1 of necessity is a compromise. At the lower end of the columns, again an underlying philosophy is apparent in that the formation of phosphorite deposits is related to floral and faunal changes ranging up to major evolutionary changes. These are all points which have been touched on earlier in this volume but they need to be re-examined with our model in mind.

It is worth briefly examining the structure of the four columns (A–D) which make up the most obvious components in the flowchart as these are to some extent derived from existing models which address a part of the total phosphogenic model represented in Fig. 26.1.

Column A

The portion of our flowchart in Column A represents in a very simplistic form the model presented by Cook & McElhinny (1979) as spatial models A and B and temporal model C. In these models those authors saw plate tectonics leading to the development of major new seaways, which produced major changes in the patterns of global oceanic circulation and which in turn lead to a redistribution of phosphorus from the deep to the shallow ocean and ultimately to major floral and faunal changes. We need to examine the extent to which these changes can be

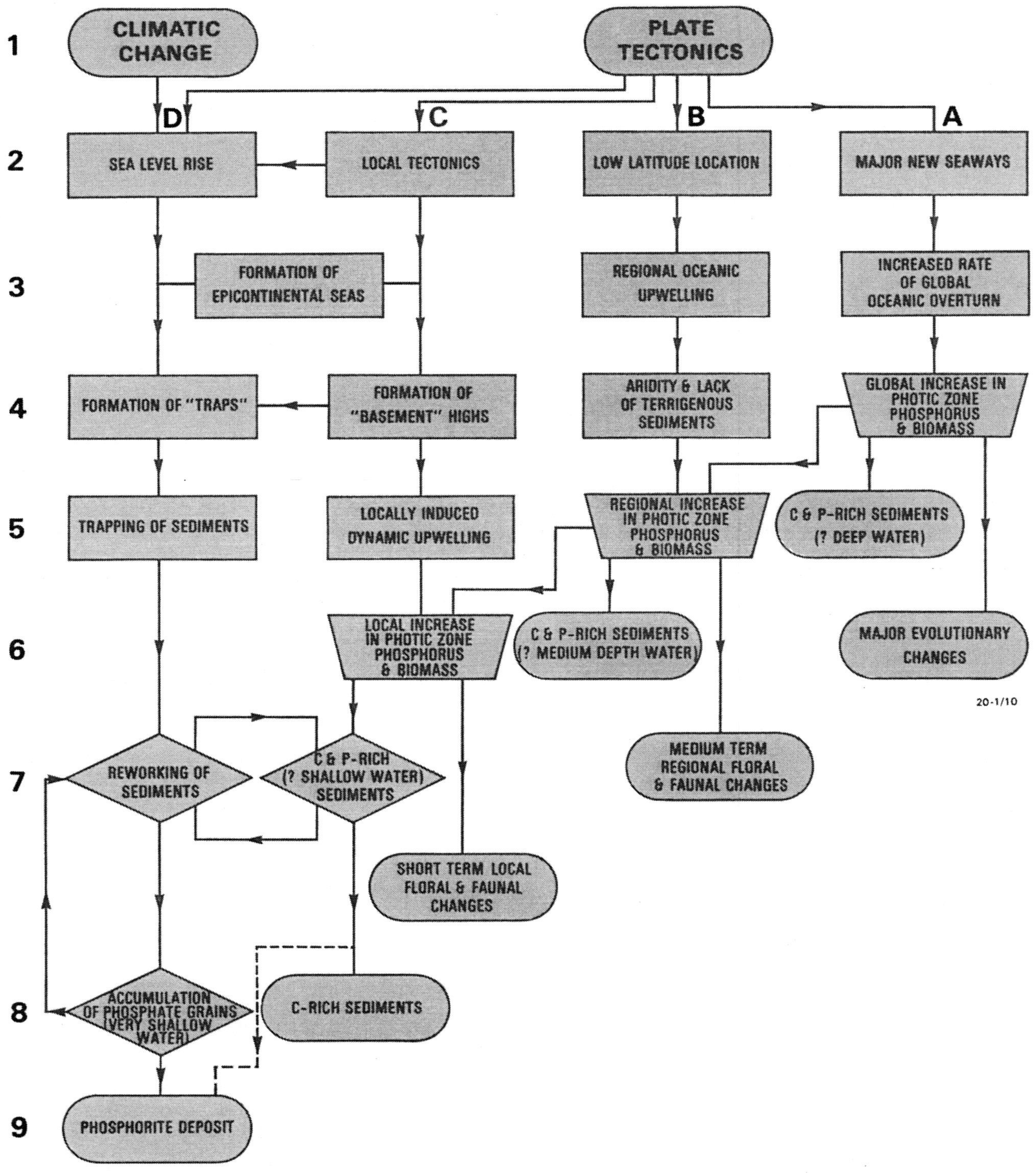

Fig. 26.1. A model for phosphogenesis.

documented for the Late Proterozoic and the Cambrian, through the organic and geochemical record for this time.

Column B

This portion of our flowchart is perhaps best seen as the upwelling model developed by Kazakov (1938) and applied by McKelvey *et al.* (1953) and many other workers to the Phosphoria Formation. As in Column A we see the starting point as plate tectonics so that in this respect it also owes something to the Cook & McElhinny (1979) model. In this instance the connection with plate tectonics is provided through the latitude: lithospheric plate movement results in a portion of a plate moving to a low latitude location where oceanic upwelling can occur on a regional scale. The mechanics involved in this form of upwelling are discussed by Parrish *et al.* (this volume). Upwelling produces aridity and a consequent lack of detrital sediment. It also produces an increase of phosphorus in the water column and an associated increase of the biomass.

Kazakov proposed that phosphate precipitated directly from the water column although there is little evidence for this. Therefore we need to examine the extent to which evidence of upwelling in the Proterozoic and Cambrian supports our model.

Column C

This part of the flowchart is concerned with the more local scale processes perhaps analogous to the scale of model presented by Riggs (1984) for parts of the southeastern United States. Again, these processes may have been largely initiated by plate tectonics though it must be acknowledged that intra-plate tectonism may in part be a consequence of other processes. The important elements in Column C include the development of an epicontinental sea or shelf (also commonly associated with sea level changes), and the development of basement structures, particularly basement highs, which can profoundly influence sedimentation. One of the important consequences of the formation of such a 'high' is that it produces dynamic upwelling and this in turn produces a local increase of phosphorus in the water column and an associated increase of the biomass. Therefore we shall briefly examine in this review the extent to which local tectonics have influenced sedimentation.

Before briefly reviewing Column D, it is important to point out that up to and including the stage of Column C we have not in fact produced a major phosphorite. We have proposed here that major phosphorites will not be produced by a local, regional or global increase in the phosphate content of the photic zone if these processes occur in isolation. To produce a phosphorite, it is necessary for these processes to act in concert – to reinforce each other. This is represented in our model by the flow path from A4 to B5 to C6. If this is really what happens, then there should at times be global synchroneity in phosphogenesis. We need to review evidence for this synchroneity. It also requires that there are major peaks in phosphogenesis, a topic broadly considered by Cook & McElhinny (1979), but which needs more detailed consideration for the Proterozoic and Cambrian if we are to test the present model. One further point regarding Column C is that we do not propose the direct formation of a major phosphorite. We have suggested that reworking of organic-rich sediment (represented by the flow paths between C7 and D7) is frequently necessary if we are to 'upgrade' the phosphate content of the sediment and have both a general association and yet a facies separation between organic rich sediments and phosphorites. There is a need to examine this relationship further.

Column D

In this final column we have had to compromise on the orderly arrangement of scale represented by Columns A, B and C. In D we are concerned in part with a representation of the model proposed by Arthur & Jenkyns (1981), with sea-level change on scales ranging from long term (global and eustatic) at D2 to short term (due to local tectonics), down to, at the extreme, the change of sea-level over a tidal cycle) at D8. It is therefore essential that we review the evidence for sea-level change–phosphorite associations in the Proterozoic–Cambrian to support this section of the model. An important feature of this part of the model is the extent to which there is clear interaction between Columns C and D. For example, in line 2, a sea-level change may result from climatic or tectonic events; 'traps' (locations where phosphate grains are physically trapped to form accumulations) may form through sea-level change (D4) but equally well they may form through tectonics producing topographic changes of the type represented by C4.

One important feature of this part of the model is the iterations indicated between D7 and D8. In most cases these are essential if the sediments (including those derived from C7) are to be 'upgraded' to form a phosphorite. Therefore if we are to test the application of this part of the model we must examine the extent to which phosphate grains accumulate and are reworked in Proterozoic and Cambrian phosphorites. Accordingly, our model finally 'produces' a phosphorite at D9. There is, however, a further implication from our model, namely that the 'products' of the columns should be temporally related. Thus, at the extreme ends of the model, the phosphorites of D9 should be contemporaneous with the major evolutionary changes of A6. Consequently, a question of fundamental importance which must be addressed is the extent to which phosphogenesis and major evolutionary changes are related in the Proterozoic and Cambrian.

Testing the model

The model has been created out of our knowledge of phosphorites of all ages in the geological record. In explaining the basis for the model in the previous section, a number of important questions have been raised. All of these have been posed, with varying degrees of clarity throughout this volume, but it seems important to pose them again in the light of the new information that has been brought together in this volume. Accordingly, we will now attempt to test the model by using data from Proterozoic and Cambrian sources, first considering broadscale questions, and then those which are more detailed.

1. Is there a Proterozoic/Cambrian global phosphogenic event?

The sheer magnitude of phosphate rock currently produced from Proterozoic and Cambrian sediments strongly suggests a major phosphogenic event. Production of Proterozoic and Cambrian phosphorites, estimated at 23.5 million tonnes in 1982 has increased significantly over the past ten years (compare our Fig. 26.2 with figure 1 of Notholt, 1980) largely due to expansion of mining activities in the Soviet Union and China. But this is small when compared to the resources of Proterozoic and Cambrian phosphorites estimated to exceed 34×10^9 tonnes with a P_2O_5 content of more than 4% (Cook, 1983). Approximately 60% of these reserves are of Cambrian age (Notholt, 1979, 1980). Proterozoic and Cambrian phosphorites have been documented on all continents, with the exception, so far, of Antarctica. Almost 350 significant occurrences or deposits have been listed by Notholt (1979), Lee (1980) and in Cook & Shergold (this volume). The distribution of the most important of these deposits is discussed by Notholt & Sheldon (this volume). There is a marked concentration of

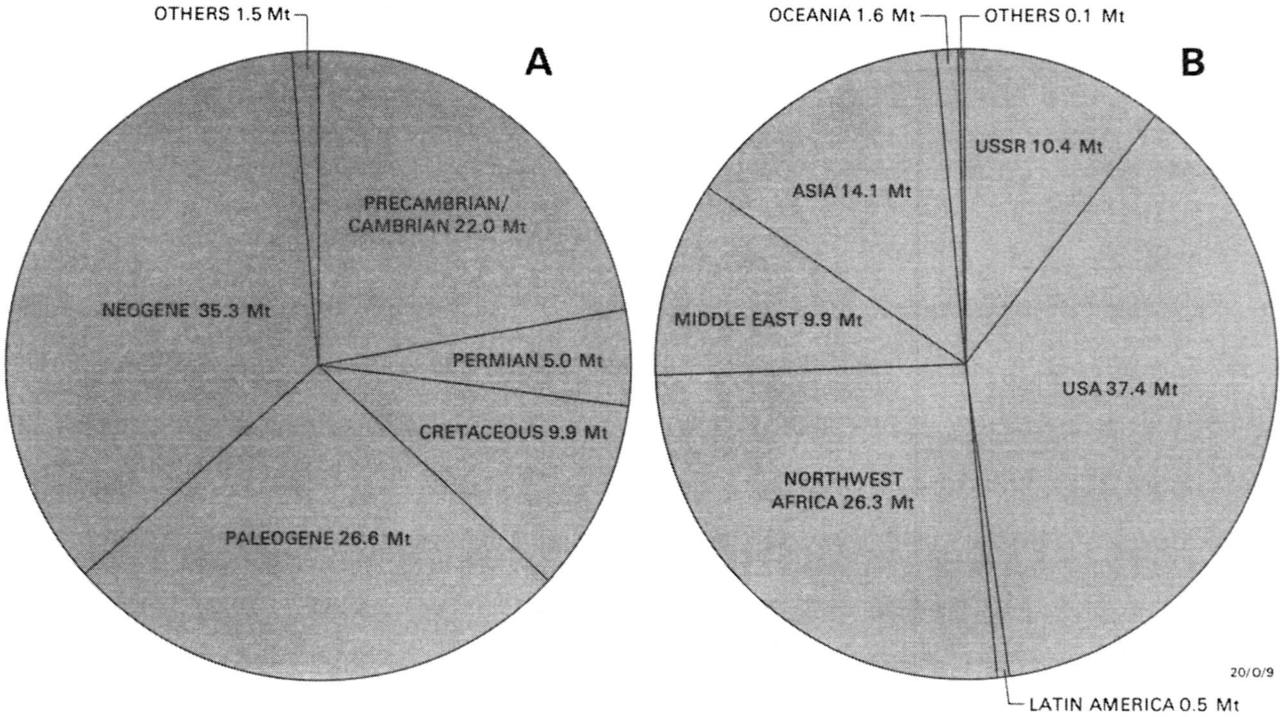

Fig.26.2. Stratigraphical (A) and geographical (B) distribution of phosphorites based on 1982 production figures.

deposits in a belt which extends from Kazakhstan through central Asia into southern China, Vietnam and Australia. Fig. 26.3A–D provides a summary of the temporal distribution of Proterozoic and Cambrian phosphorites. It is evident that phosphogenesis was most widespread during the Late Proterozoic and Early Cambrian. Major deposits formed in the Late Proterozoic (Fig. 26.3A) in S.E. Asia, central Asia, India and West Africa, with minor occurrences scattered throughout North and South America, northern Africa, Europe, India, Asia and Australia. In the Early Cambrian (Fig. 26.3B) major deposits were formed in central and eastern Asia with minor occurrences in Europe, eastern North America, Asia and Australia.

It is difficult to quantify the abundance of phosphorus at any particular time. Cook & McElhinny (1979) attempted to determine total phosphate abundance in the sedimentary column and from this derived a curve showing phosphate distribution throughout the Phanerozoic. This curve was found to be closely mirrored by a histogram based on the abundance of phosphorite deposits and occurrences. We have therefore used the more easily determined deposit plus occurrences parameter to quantify relative abundance of phosphorites in the Proterozoic and Cambrian. The distribution curve of Cook & McElhinny (1979) used an early compilation of phosphate deposits prepared by The British Sulphur Corporation (1971). A more recent compilation by Notholt (1979) detailed approximately 100 deposits and occurrences of Proterozoic and Cambrian age. Using the sources indicated in the caption to Table 26.1, we find that there are in fact almost 350 discrete deposits and occurrences of Proterozoic and Cambrian age (Table 26.1). Of these, half are of Early Cambrian age, and indeed the majority of these are found in the initial part of the Early Cambrian. The Proterozoic represents the period with the next greatest abundance of phosphate deposits and occurrences and the majority of these are of Late Proterozoic (Vendian/Sinian) age. The abundance of Middle Cambrian deposits is biased by the numerous Australian examples. Deposits and occurrences are least abundant in the Late Cambrian and indeed there are no major deposits of this age. Therefore the picture which emerges is of abundant phosphogenesis in the Late Proterozoic reaching a peak in the Early Cambrian decreasing, though with possibly a minor peak, in the early Middle Cambrian before sharply falling off during the remainder of Cambrian time. A phosphogenic event of global extent does indeed seem to have occurred therefore during the Late Proterozoic and Early Cambrian.

Table 26.1 *Deposits and occurrences of Proterozoic and Cambrian phosphorites. Figures based on data from Notholt, 1979; Lee, 1980; Ilyin & Bjamba, 1980; Eganov et al., 1984 and various contributions to this volume*

	Proterozoic	Early Cambrian	Middle Cambrian	Late Cambrian
China	33	48	3	1
Australia	46	40	18	
USSR/MPR	33	39	5	1
India/Pakistan	8	1		1
Africa	4	1		1
S. America	1			
N. America	8	4	3	6
Europe	5	18	11	5
	138	151	40	15

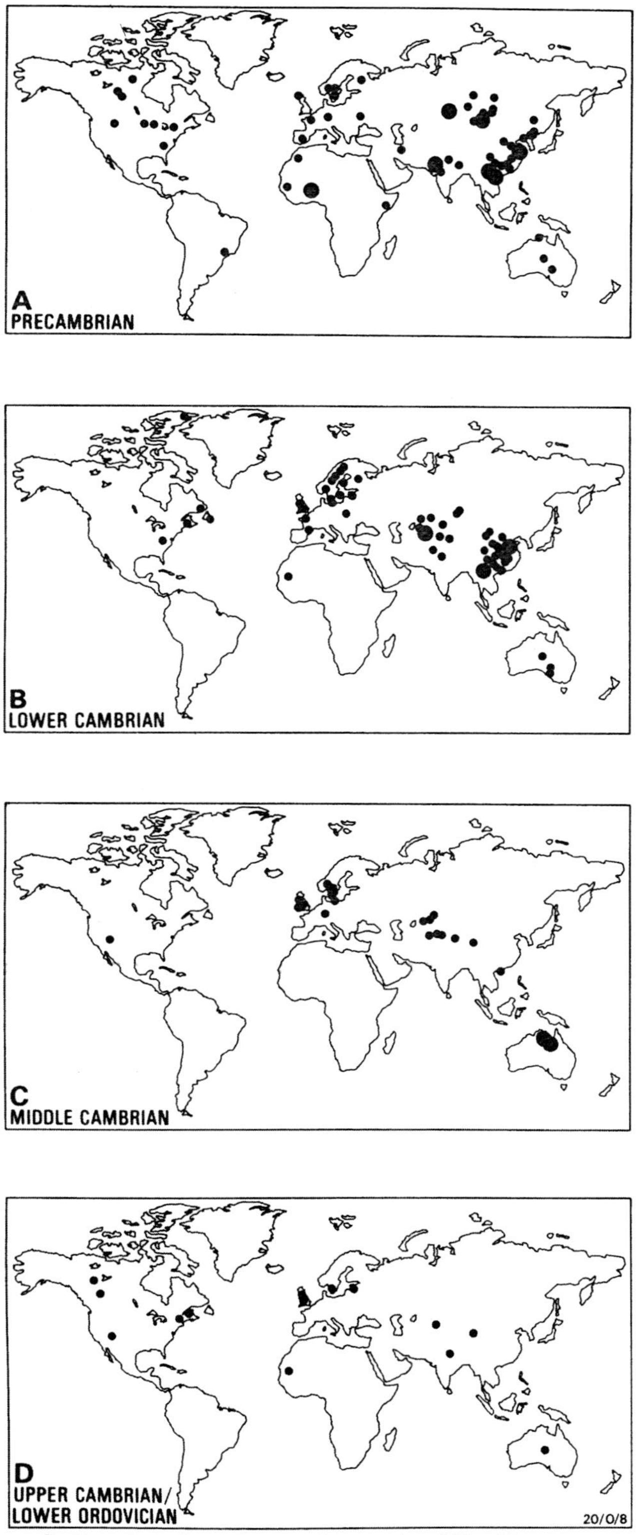

● *Deposits estimated to contain more than 100 million tonnes of phosphate rock*

• *Minor deposits and occurrences for which resources estimates are not known*

Fig. 26.3. Distribution of Late Proterozoic and Cambrian phosphorite deposits and occurrences.

2. How many phosphogenic events were there, and were they globally synchronous?

In a first attempt to delineate major phosphogenic episodes, Cook & McElhinny (1979) suggested that phosphogenesis was most common during four periods in the Proterozoic. They identified these times as Upper Proterozoic I (about 620 m.y.), Upper Proterozoic II (700–800 m.y.), Middle Proterozoic (1200–1600 m.y.) and Lower Proterozoic (1800–2200 m.y.). It is immediately apparent that resolution of the early compilation becomes less precise with increasing age and that the older events become progressively more speculative. In the absence of a highly resolved biochronological scale it is difficult to define these events accurately and to correlate them. The reason for this is that during the Riphean and earlier Precambrian, prior to the appearance of the metazoans, the biosphere was dominated by algal and bacterial life preserved in the fossil record only under very specific sedimentological conditions (for dominant components of the biota see Fig. 26.4). The organo-sedimentary stromatolites (algal mats) alone provide a potential biochronological framework (Cloud & Semikhatov, 1969; Bertrand-Sarfati & Walter, 1980) at this time, but require cautious application considering the possible repetition of stromatolite morphology with time (see Banerjee, and Shergold & Brasier, this volume for discussion on the age of stromatolitic phosphorites in India and China). During the Late Riphean, however, at about 800 m.y., organic walled microfossils (acritarchs – which first appeared at ± 1400 m.y.) became increasingly important both as contributors to the phytoplankton (and hence the phosphogenetic process), and as biochronological discriminators (Brasier, 1979; Vidal & Knoll, 1982, 1983). Thus the Upper Proterozoic I event can now be fairly well defined (Fig. 26.3A) and does indeed occur just below the Precambrian–Cambrian boundary. A number of deposits of probable Late Riphean age, grouped around the so-called Upper Proterozoic II event, are less easy to synchronise. There is now little evidence to support a Middle Proterozoic event given the degree of uncertainty surrounding the age of the Rajasthan and associated Indian deposits on which this episode was originally based. Evidence for an Early Proterozoic event is somewhat more convincing, being based on the concentration of banded iron formations (BIFs) around the Archaean–Proterozoic boundary. Even here, however, there is some uncertainty regarding the abundance of phosphorite in BIFs. They certainly occur with some, but whether they occur with all has yet to be determined. It should also be pointed out that the duration of the Early Proterozoic is equivalent to that of the whole of the Palaeozoic and may embrace more than one discrete phosphogenic episode.

To some extent consideration of Cambrian phosphogenesis is hindered by the fact that there is no universal agreement on the boundaries of the Cambrian system. Historically, the lower boundary was taken at the first unconformity above Precambrian basement rocks from which Cambrian fossils were recovered. Later, following detailed palaeontological investigations, the first trilobite faunas were considered to herald the beginning of Cambrian time, and the Early Cambrian was defined as being characterised by the evolution and radiation of olenellid trilobites, a view held until the sixties, and still maintained by some. Subsequent research, on soft-bodied metazoans and ichnofossils found stratigraphically below the first trilobites, has led to the generally accepted conclusion of a pre-trilobite interval of Cambrian time.

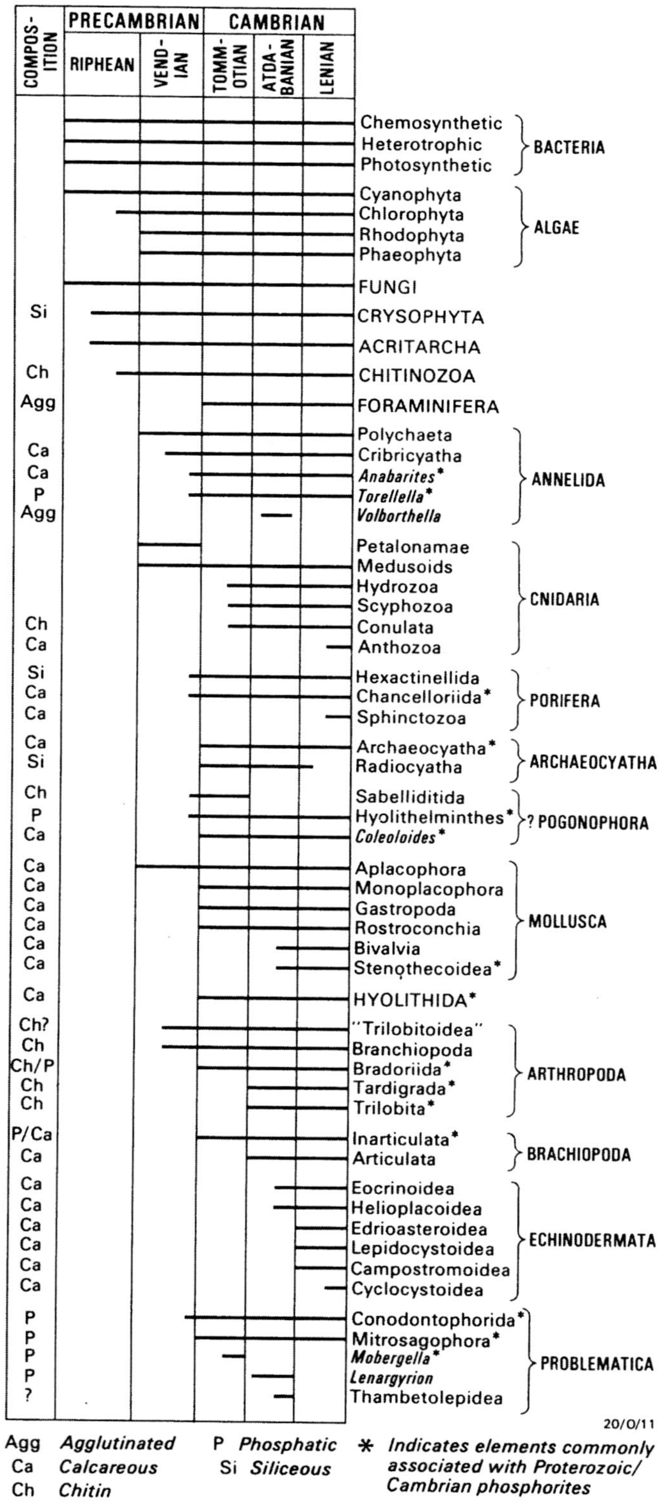

Fig. 26.4. Biota and its skeletal composition during the Late Proterozoic and Early Cambrian.

The Early Cambrian is the age of the Archaeocyatha (Zhuravleva, 1970; Rozanov & Debrenne, 1974; Rozanov, 1973) but because these organisms are strongly environmentally controlled in their distribution, they cannot be recognised universally. Their association with 'small shelly fossils' of various kinds in the earliest Cambrian and with trilobites in the later part of the Early Cambrian has imparted a biochronological significance to the archaeocyathans. Since their evolution is most completely developed and documented on the Siberian Platform and Altai–Sayan regions of the USSR, Soviet stage nomenclature is commonly adopted for the broader biochronological division of the Early Cambrian.

Faunas predating the first trilobites were initially reported from the Pestrotsvet Formation in the Aldan and Lena River areas of the southern Siberian Platform by Missarzhevsky & Rozanov (1963, 1973), Rozanov & Missarzhevsky (1966) and Rozanov *et al.* (1969), and from the Altai–Sayan region by Repina *et al.* (1964). A variety of tubes, early molluscs, brachiopods, and taxonomically and palaeobiologically problematical phosphatic and phosphatised sclerites of segmented organisms were found in association with the earliest Archaeocyatha (Rozanov, 1966), and were considered to form the palaeontological basis of the Tommotian Stage, the initial pre-trilobite interval of the Cambrian System (Rozanov *et al.* 1969). This is succeeded by the Atdabanian and Lenian Stages which retain small shelly fossils, both phosphatic and otherwise, together with Archaeocyatha and trilobites, the last becoming increasingly important for correlation purposes.

Subsequent to the discovery of the Russian terminal Proterozoic and initial Cambrian faunas, primitive mineralised Metazoa have been reported globally, but particularly from Newfoundland, South Australia and China. Currently, only a broad correlation can be effected between southwestern China (Meishucun Stage) and the Siberian Platform (Tommotian), and it is not possible to resolve correlations accurately at zonal level. The correlation of the pre-trilobite Early Cambrian into Europe is equally difficult to resolve in detail on the basis of shelly fossils in the absence of archaeocyathans because of the incompleteness of the sedimentary record in many areas. The correlation proposed by Shergold & Brasier (this volume, Ch. 23, Fig. 23.3) is simplified to form a basis for the discussion of Early Cambrian phosphogenic events.

The Middle Cambrian is characterised by the wide dispersal and rapid evolution of the agnostid trilobites which are used globally with a high degree of biochronological resolution. The Middle Cambrian is also the time of the paradoxidid trilobites which are both widely distributed and have great biochronological potential. Since the biochronology of the Middle Cambrian was essentially first established in Scandinavia (Linnarsson, 1877; Westergård, 1936, 1944, 1947), this region provides a standard – the paradoxidid Middle Cambrian – for correlation. However, only one major Middle Cambrian deposit, at Duchess, western Queensland, Australia, has been studied in palaeontological detail.

The Late Cambrian, like the Middle, is finely resolved biochronologically on the basis of agnostid and olenid trilobites, and again the standard scheme was established and refined in Scandinavia (Linnarsson, 1877; Westergård, 1947; Henningsmoen, 1957). At the end of the Late Cambrian, conodonts, also of global distribution, become increasingly important for international correlation and are particularly helpful in correlating the endemic faunas of the Late Cambrian and Early Ordovician carbonate platforms.

There are no major phosphate deposits of Late Cambrian age, but there are a number of occurrences of interest, generally associated with the development of organic rich black shales in Himalayan India, the Avalon Platform of Europe and its extension into maritime Eastern Canada, and the Cordilleran region of North America, from Nevada in the USA to the Northwestern Territory of Canada.

In summary, Early Cambrian time is initially calibrated by Archaeocyatha and associated small shelly fossils which are often of apatitic composition. Later in the Early Cambrian calcareous shelled organisms predominated and trilobites became increasingly important biochronologically, a role they maintained until the proliferation of conodonts (also composed of apatite) at the beginning of the Ordovician. As a result, Cambrian phosphogenic events can be dated with a degree of confidence which is not attainable in the Proterozoic. Accordingly, during the Cambrian, Shergold & Brasier (this volume) have been able to describe six sedimentary and phosphogenic cycles on the Baltic Platform which are related to transgressive–regressive events. Three of these, the main ones, occur in the Early Cambrian (Baltian, Liivian and Aisciaian), the first two apparently predating the appearance of trilobites and being associated with the earliest biomineralised metazoans. Two cycles occur within the Middle Cambrian, the most important of which is the early part of the Middle Cambrian; and a single cycle occurs during the Late Cambrian. These cycles can be used to correlate the phosphogenic events of other areas. Thus it is possible to infer that the Tommotian phosphorites of central Asia, southwestern China, and South Australia predate the first substantial phase of Cambrian phosphogenesis in Europe. Some cycles, like that commencing in the *Ptychagnostus gibbus* Zone of the Middle Cambrian represent global events. However, summarising the available biochronological evidence (Fig. 26.4), there is no doubt that the phosphogenic event around the Proterozoic–Cambrian boundary was the most widespread, and that within the resolution of palaeontological information available, it occurred in many parts of the world at about the same time. Therefore, the answer to our question must be that there is evidence of several phosphogenic events during the Proterozoic and Cambrian but that at the present time we are uncertain as to the global synchroneity of many of them, particularly those of Proterozoic age. Our knowledge of Cambrian events is rather better and it appears that phosphogenesis occurred over wide areas at essentially the same time. The event around the Proterozoic–Cambrian boundary appears to have been particularly important and may well have been more sharply defined than most. This then would lend credence to the flow path A4–B5–C6 in our model (Fig. 26.1) though it does not necessarily imply that A2 and A3 are precursors of A4.

3. Is there evidence of the low latitude, oceanic upwelling and aridity implied in Column B?

The most recent palaeolatitude compilation (Parrish *et al.* 1983, this volume) together with the earlier work of Cook & McElhinny (1979) indicate that the low palaeolatitude–phosphorite correlation first proposed by Sheldon (1964) for the Phanerozoic in general applies to the Cambrian, but our knowledge is too incomplete to make the same observation for Proterozoic deposits. A low latitude would be consistent with upwelling, but does not of course prove that oceanic upwelling took place. Nevertheless, there is a consistency in the pattern of palaeogeography, sedimentary assemblages and phosphorites demonstrated by Parrish *et al.* which lends credence to the upwelling hypothesis. The black shale–chert–phosphorite assemblage is generally taken to indicate a marine upwelling environment, and evaporites commonly found associated with the phosphorites of this assemblage are taken to be a manifestation of arid environments characteristic of areas of marine upwelling. The Permian Phosphoria Formation is the classic example of this type of depositional setting (McKelvey *et al.*, 1953). For many years in the English language literature as well as in the Russian literature, the upwelling model, exemplified by the Phosphoria Formation, was pre-eminent and served as an important (and successful) exploration tool. The black shale–chert–phosphorite lithological association has not been widely recognised in the Proterozoic. However, it is more common in the Cambrian, and several major Early Cambrian deposits, e.g. the Karatau and Khubsugul deposits of the USSR and MPR, and of southwestern China, are characterised by this assemblage. Evidence of aridity is less immediately obvious, but may be suspected by the lack of terrigenous sediment in many phosphorites, and by the common incidence of evaporites in associated sediments, particularly in the Early Cambrian. The last are relatively widespread and are represented by actual evaporite minerals or by pseudomorphs after a variety of sulfate and halide minerals.

At this stage we are forced to give a qualified response to our question on oceanic upwelling. Little information is available for Proterozoic phosphorites to test Column B of the model. Cambrian deposits, however, more convincingly demonstrate that phosphorites accumulated in low latitude locations and do yield evidence for arid climatic conditions. Nevertheless, it is admitted that the proposed link between phosphogenesis and oceanic upwelling implied in B3 of the model is not as convincing as we might desire and we have been forced to fall back on Neogene analogues to support the association, e.g. the black shale–chert–phosphorite assemblage which is associated with oceanic upwelling off the Namibian coast of southwest Africa. Use of this analogue implies uniformitarianism and yet our model carries with it the implication of a non-steady state and departure from uniformitarianism! We strongly suspect that regional oceanic upwelling is intimately tied to phosphogenesis, but acknowledge that this is one of the less well defined areas of our model.

4. Is there evidence for the link between tectonics, basement highs and phosphogenesis, as implied in Column C?

This part of the model deals with more localised processes occurring at the scale of a basin or an individual deposit, and is perhaps one of the better defined areas of our model. There is compelling evidence from the Late Proterozoic and Cambrian deposits of China (Li, this volume) that phosphate deposits are localised on or around basement highs. Similarly, the Cambrian deposits of the Georgina Basin (Howard, this volume) can be related to basement structures such as the Alexandria–Wonarah High. This association with 'basement' highs is commonly taken to carry the implication that there is dynamic upwelling generated by currents moving over the high, and it is this which is responsible for the formation of a phosphorite as indicated in C5 of our model. This may be so, but as in the case of oceanic upwelling we cannot provide incontrovertible supporting evidence from Proterozoic or Cambrian rocks. Until such time as a definitive indicator can be found for upwelling, such as a particular organism or assemblage of organisms or biochemical marker, we are likely to continue to rely on the presence of phosphorites and the black shale–chert–phosphorite assemblage to imply upwelling. This hypothesis, implied in section A4, B5 and C6 of our model (Fig. 26.1), is certainly acceptable, but it is important to be wary of the hazard of then taking the 'evidence' of upwelling as *proof* that upwelling and phosphogenesis are related – even though we suspect they are, from the evidence of Neogene deposits.

It appears, therefore, that the answer to our question must be that there is strong evidence to support a link between phosphorites, tectonics and basement structure as indicated in Column C. The link implied by flow path C4 to C5 is supported by Neogene examples but cannot be proved by evidence from the Cambrian or Proterozoic.

5. Can phosphorites be related to changes of climate and in turn to sea-level changes?

Despite the imperfections of the Proterozoic and Cambrian stratigraphic record, similarities with the eastern United States Neogene model (Riggs, 1979 a, b, 1984) are quite striking. Cook (1984) and Cook & Shergold (1984) have pointed to the association of Cambrian sea-level changes in Australia with the deposition of phosphorites. At times of low sea-level stands the Australian 'phosphorites' (generally little more than scattered phosphate concretions) formed in outer shelf locations. With subsequent transgression phosphogenesis became a more prominent feature, culminating in the widespread Middle Cambrian phosphorites of the northern Australian craton. Like their Neogene analogues, they appear to be formed close to, but not at, the culmination of the transgression. Eganov *et al.* (this volume) have similarly demonstrated the influence of sea-level changes on the Karatau deposits. Comparable data are not available for all Proterozoic and Cambrian deposits, but nevertheless the majority of deposits, and even many occurrences, are associated with transgressions. The extent to which the transgressions around the Precambrian–Cambrian boundary were synchronous (and therefore, by implication, the consequence of global sea-level changes) can be judged from the discussion (Shergold & Brasier, this volume) on the biochronology of the deposits (see also Brasier, 1980a). Our model implies that the relationship of phosphogenesis to eustatic rises of sea-level could be due to either plate tectonics or climatic changes. During the Late Proterozoic and Early Cambrian there is some evidence that climatic change may have been a factor.

The association of phosphorites with tilloids, or undoubted tillites, has been commented on by many previous authors since Mansfield (1927) first suggested that the occurrence of phosphorite beds in the Phosphoria Formation could be correlated with Permian glacial events. More recently Riggs (1984) has shown that there is a clear correlation between sea-level changes during the Miocene and the stratigraphic location of phosphorites. Thus, if sea-level changes are an important component in the formation of phosphorites, it is reasonable to expect that glacial epochs, times when there were major changes of sea-level, would be related to times of phosphogenesis. It is not possible to test this hypothesis in ancient sediments in the detail that Riggs (1984) has found possible for the Neogene. Cook (1982) has shown that the occurrence of Australian Cambrian phosphorites is related to sea-level changes, though there is no evidence that these eustatic sea-level changes are a result of glaciation or deglaciation, or that there ever was a Middle Cambrian glaciation. Evidence of glaciation has been found in the section below the Georgina Basin phosphatic sequence: the Mount Birnie Beds, a tillitic unit found in the eastern part of the Georgina Basin, occur a few tens of metres below the major phosphatic interval (de Keyser & Cook, 1972). This might be taken as evidence to support a phosphogenesis–glaciation correlation, but in fact the tillites are the result of a Late Proterozoic glacial event in Australia which could be as old as 620–650 m.y. whereas the Middle Cambrian Beetle Creek phosphorites could be as young as 540–520 m.y., indicating that there may be a time gap of as much as 100 m.y. between glaciation and phosphogenesis. However, given the paucity of radiometric ages to reliably date sedimentary sequences around the Precambrian–Cambrian boundary, the gap between glaciation and phosphogenesis could equally well be comparatively brief. In addition to the Australian sequence, late Proterozoic tillites are overlain by phosphorites in southwestern China, the Khubsugul Basin of the MPR, in the Hazara district of Pakistan, Himalayan India, the Volta Basin of West Africa, and in the Bambuí sequence of Brazil (Fig. 26.5). The thickness of the sedimentary sequence between the termination of the tillitic episode(s) and phosphogenesis ranges from a few metres to a few hundred metres. As in the case of Australia, there is great uncertainty in all these basins regarding the duration between glaciation and phosphogenesis. In some areas it appears to represent much of the Vendian and all or part of the Early Cambrian, and this would normally be taken to represent some 50–90 m.y. In our view there is considerable doubt that this is the case from the evidence provided by phosphatic sequences in southwestern China where major areas of Cambrian phosphorites are also the areas where there are Vendian phosphorites. It is of course possible that some of the Cambrian phosphorites are derived by reworking of Vendian phosphorites, but there is no evidence that this has happened on a large scale. A second option is that in this region there was repeated and widespread phosphogenesis extending from the Vendian into the Cambrian, over a period of perhaps 100 m.y. or more. However, when we look at Phanerozoic phosphogenic provinces such as the North African province where there is good biostratigraphic control, we see phosphogenesis occurring within a given province for 20 or even 30 m.y. but not for 100 m.y. Therefore, by analogy with the later Phanerozoic phosphogenic episodes we are forced to conclude that the Vendian and the Early Cambrian may represent a much shorter time interval than is normally taken to be the case. The point to be made is that in many areas phosphogenesis is preceded by

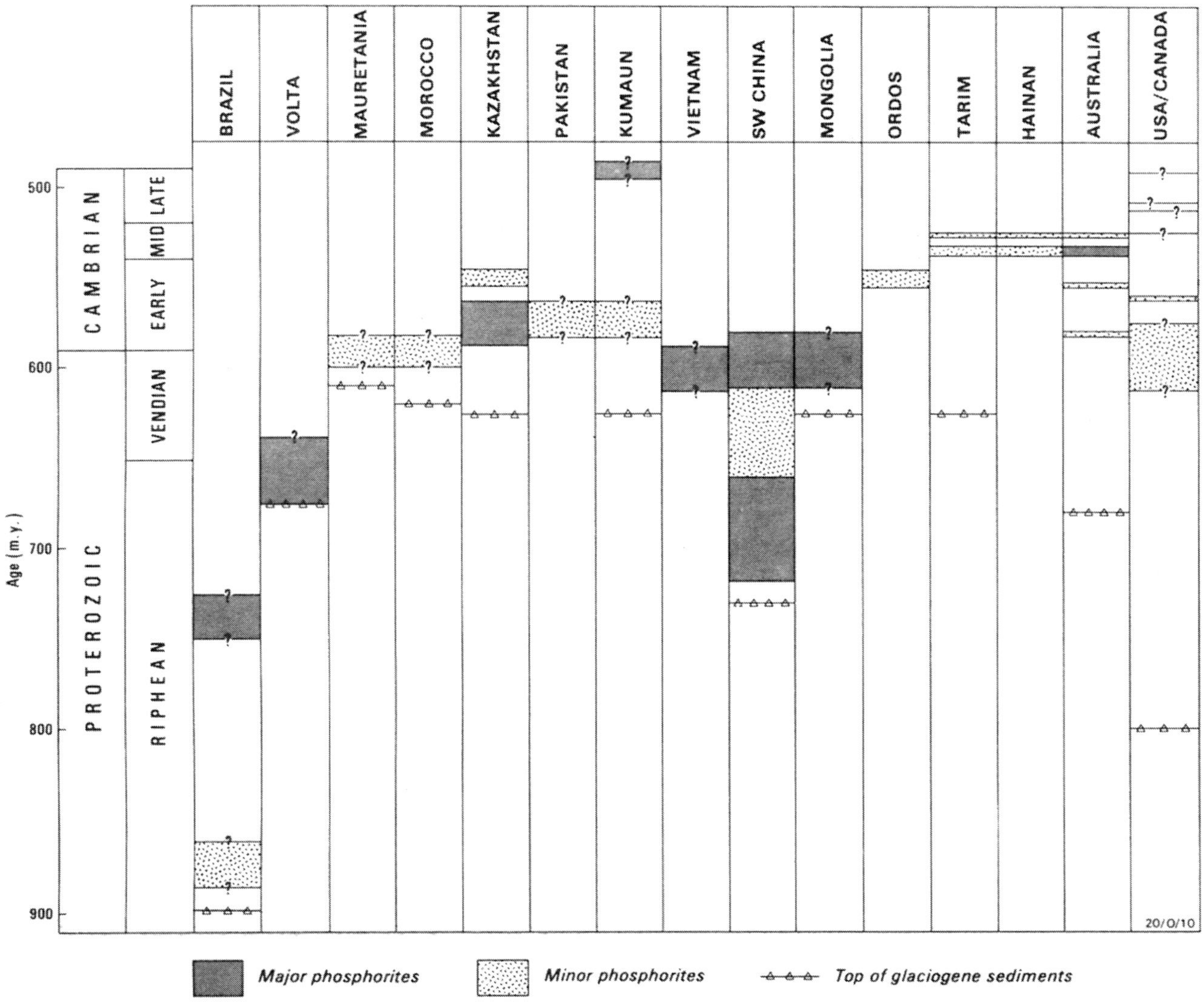

Fig. 26.5. Relationship of Late Proterozoic and Cambrian phosphorites to preceding glacial sequences. Distribution of glaciogene sediments taken largely from Hambrey & Harland (1981); stratigraphic position of phosphorites mainly follows Cook & Shergold (1984).

glaciation and that the period between the two events is more likely to be of the order of 10^7 years than 10^8 years.

A possible link between glaciation and phosphogenesis at the Proterozoic–Cambrian boundary is consistent with Column A of the model. A period of glaciation could have produced a large volume of cold, nutrient-rich water which, after a long residence time, was eventually recycled into the shallow zone where it produced a major burst of organic productivity in the photic zone. This would be analogous to the polytaxic–oligotaxic events documented in the late Phanerozoic by Fischer & Arthur (1977). In an ingenious hypothesis linking phosphogenesis and glaciation, Sheldon (1982) has argued from the presence of sedimentary 'doublets' that ice rings around the Earth produced low latitude glaciations. The hypothesis provides a possible explanation for why phosphorites, which are known to form preferentially in low latitudes (Parrish *et al.*, this volume), should be found in the same location as tillites, which are normally taken to form in high latitude locations. Alternatively, the low latitude association of both phosphorites and glacials could be taken as merely a function of an exceptionally widespread glacial period in the Late Proterozoic when glaciation did extend to low latitudes. Further models are possible, involving for example the generation of exceptionally cold bottom waters which could have produced abnormally cold climates where these waters upwelled in near equatorial locations, analogous to the comparatively cool conditions that prevail today in near equatorial locations off the west coast of South America. A problem with this model is that cold upwelling waters also normally generate exceptionally arid deserts thus precluding the development of continental ice on a scale sufficient to account for the widespread tillites. At this stage any model to explain the common spatial distribution has problems and it is difficult to escape the conclusion that a rigorous application of uniformitarianism does not explain that distribution. In addition to the geochemical model proposed by Cook & McElhinny (1979) to explain a temporal association of glaciation and phosphogenesis the compelling evidence of Riggs *et al.* (1982) of the link between sea-level changes and phosphogenesis during the Neogene provides, by analogy, a link between these same two events in the Proterozoic and Cambrian. It would be reasonable to expect that the glacial event was associated with waxing and waning of the ice caps and that associated sea-level changes would have occurred long after the main glaciation. Arthur & Jenkyns (1981) also considered that sea-level changes were of importance to the formation of major phosphorite deposits. We are forced to the view that the occurrence of phosphorites underlain by tillites (usually with intervening sediments especially dolomite) is sufficiently common that it cannot be dismissed as fortuitous. In Fig. 26.1, there probably should be a flow path from Column D to Column A, which cannot be proved, but which is there by implication. Therefore, the answer to our question must be that there is strong evidence from the Cambrian of links between phosphogenesis and sea-level changes (on a variety of scales). There is also evidence of an association with climatic change–notably from the evidence of glaciation preceding phosphogenesis close to the Proterozoic–Cambrian boundary.

6. Do phosphorites preferentially occur in epicontinental seas?

In this instance, the evidence provided by contemporary phosphorites would indicate that the answer to this question should be no. However, there is strong evidence from the Proterozoic, and particularly the Cambrian, that epicontinental seas were preferred sites for the formation of phosphate deposits. The best evidence for this is provided mainly by Chinese and Australian deposits, but those of Karatau, Khubsugul and West Africa all appear to fit better into an epicontinental model than any other. The origin of the epicontinental sea appears to matter little, hence this portion of our model is indicated on line 3 as intermediate between Columns C and D. The preferred epicontinental location carries with it the implication that oceanic upwelling, normally associated with relatively deep open oceanic locations such as the shelf off southwest Africa or off Peru and northern Chile at the present day, is unlikely to have occurred. Our model, which contains various scales of upwelling as important components, avoids this problem by permitting deep oceanic upwelling to impinge only indirectly (as indicated by the lateral flow path D5–C6–D7) perhaps through progressively shallower upwelling 'cells' into the very shallow seas where phosphate deposits actually form. Alternatively, it may be that the effects of oceanic upwelling result in a massive increase in the biomass whose remains were swept into the epicontinental sea. But whatever the mechanism the evidence from the Proterozoic and Cambrian shows quite clearly that the answer to our question is yes, phosphorites do preferentially occur in epicontinental seas.

7. Are 'traps' important as sites for the accumulation of phosphate deposits?

The English and Russian literature has tended to be dominated by the upwelling model of phosphogenesis. In French literature on the other hand, as a result of 'exposure' to the Tethyan phosphorites, a somewhat different phosphogenic model has been developed which has emphasised structural controls, recognition of the reworked nature of many phosphate grains, and the importance of a 'trap' for the accumulation of phosphate grains to form a deposit.

'Traps' may form by either sea-level change or by local tectonics (hence the link between C4 and D4) and are sites where phosphate grains are presumed to accumulate. Hydrodynamic 'traps' – sites where wave or current action may be at a minimum – may exist, but cannot be proved for the Proterozoic or Cambrian. Topographic 'traps' on the other hand are more readily apparent, one of the best examples of this type of location being provided by the Cambrian deposits of the Georgina Basin, northern Australia (Howard, this volume; Cook & Elgueta, this volume). There the Duchess, Ardmore and Lady Annie deposits all occur within grabens that have been demonstrated to have been topographic lows during the Cambrian. While the same structural and palaeogeographic detail is not available for many other deposits, commonly many are found in what appear to be embayments or depressions. Accordingly, there is some evidence for the Proterozoic, more obvious in the Cambrian, that traps' are important in the localisation of phosphate deposits.

8. Is the reworking and accumulation of sediments under very shallow water conditions important to the formation of major phosphorite deposits?

In our model, reworking of sediments is indicated through two circular flow paths C7–D7–C7 and D7–D8–D7. By implication, there could also be a larger circular flow path C7–D7–D8–D7–C7. There is evidence of mechanical reworking of sediments in a number of Proterozoic and Cambrian phosphorite deposits and occurrences. Evidence of reworking involving the biomass (i.e. C7–D7–C7) is apparent from the presence of condensed sequences, for example the phosphorite occurrences of western Europe. Condensed sequences are also frequently associated with hardgrounds, which may indicate breaks in sedimentation, thus giving the opportunity for mechanical reworking. Reworking of biogenic material is also indicated by the presence of layers composed primarily of phosphatic or phosphatised shells, many of which may be abraded and show evidence of current orientation and imbrication. In addition to mechanical reworking of shells there is also evidence of some biogenic reworking: there are marked differences in the organic geochemistry of phosphorites and of their associated sediments. Some of these differences may be a response to different diagenetic conditions but most are probably a consequence of the depositional conditions. For example, winnowing and associated aerobic degradation of the intergranular organic matter in the phosphorites would have a profound effect on the final organic geochemistry. However, most of the organic matter is in the form of very fine lamellar filaments between the phosphate grains which suggests that winnowing was relatively minor. Within the phosphate grains themselves there has been abundant pyritisation of organic matter, presumably as a consequence of sulfate reduction. There are no known Cambrian vascular plants so that it can reasonably be assumed that the organic matter was largely of marine algal origin. Much of it may well have been allochthonous, as were many of the phosphate grains themselves. However, the presence of some large organic filaments suggests that some was certainly autochthonous. In addition it is conceivable that, for example in an embayment where phosphate grains were trapped, there would be some dissolution of phosphate from the grains and movement of soluble phosphate into the overlying water column. Vigorous organic growth would occur as a result of this abundant phosphate, and a secondary high productivity system may have been generated in the nearshore zone which, rather than being the source of phosphorus for phosphorites, was in fact the consequence of the pre-existing phosphorites.

However, in our model it is the iteration represented by D7–D8–D7 which is considered to be the most important step in moving from a phosphatic sediment to a major phosphorite deposit. The textural evidence discussed below shows that reworking is important in many phosphorites.

9. Do all phosphorites undergo reworking and accumulation under very shallow marine conditions as implied in D8 of our model?

In order to answer this question we are forced to examine the question of 'which phosphorite?' and in order to do this we must briefly review the main phosphorite types.

The petrology of Cambrian grainstone phosphorites demonstrates that there is a biological component in many of them, such as fragments of pre-existing phosphatic shells or rounded fragments of phosphatised calcareous fossils. However, the great majority of phosphate grains (ovules using the nomenclature of Mabie & Hess, 1964) have no clearly identifiable origin. For recent phosphatic grains, such as those off the coast of South America, it is possible to demonstrate that some formed as a result of post-mortem phosphatisation of the body cavity or the soft parts of shelly organisms such as calcareous foraminifera (Manheim *et al.*, 1975) and subsequent removal of the exoskeleton by abrasion or dissolution leaving the internal filling to form phosphate grains. Obviously in the Proterozoic the absence of a mineralised exoskeleton would have precluded the formation of phosphate grains by this kind of process. Grainstone phosphorites do nevertheless occur in the Proterozoic. In some cases, such as the Rajasthan phosphorites, rare phosphate grains have developed by the abrasion of pre-existing boundstone phosphorites. In the West African Proterozoic phosphorites grains are abundant (Lucas *et al.*, this volume), and many are ooids. In calcareous sediments it is recognised that micro-organisms play an essential role in the formation of some forms of ooids (Bathurst, 1975), and recent work, Davies *et al.* (1978) and Ferguson *et al.* (1978) has demonstrated that ?inorganic ooids may form preferentially in the presence of abundant humic acids – a major component of the organic fraction of many phosphorites. However, as this humic component is itself derived from marine organisms the extent to which the formation of any ooid (whether calcareous or phosphatic) can be regarded as purely inorganic is open to question. One further mechanism for the formation of grains is by derivation through reworking of pre-existing phosphorites, and certainly some grains are demonstrably of intraclastic origin. There is, however, no evidence that the majority of grains are formed by this mechanism and the evidence from scanning electron microscopy that the apatite crystallites in grainstone phosphorites and mudstone phosphorites are different in morphology and size (see, for example, Cook, 1972; Ilyin *et al.*, this volume) is inconsistent with one being derived from the other.

The relative abundance of mudstone phosphorite in Proterozoic and Early Cambrian deposits may be an indication that some different mechanisms for phosphate concentration and precipitation took place in the Proterozoic. We know that algae may trap, bind or form sites for the precipitation of apatite crystallites to produce boundstone (stromatolitic) phosphorites, but offshore phosphorites (O'Brien *et al.*, 1981) and experimental work (Lucas & Prévôt, 1981, 1984) show that bacteria can also play an important role in directly precipitating apatite. Apatite crystallites in mudstone phosphorites, show no evidence of abrasion through transport but as they are very fine grained, any transport is likely to have been primarily in suspension. Alternatively, the crystallites were directly precipitated (abiologically or biologically) from a shallow water column overlying the sediment–water interface. There are some difficulties in inorganically precipitating apatite from seawater due to the inhibiting effects of magnesium (Martens & Harris, 1970; Lucas & Prévôt, 1981.), but given sufficient time and adequate concentration of phosphate, and probably also very shallow conditions, phosphate can precipitate from marine environments. Whilst it is possible that the precipitated apatite was then subsequently reworked to produce a concentration of mudstone phosphorite, any such reworking must have only been over short distances. The possibility of *in situ* phosphatisation of a pre-existing calcareous or terrigenous mud cannot be discounted but there is no evidence of this primary non-phosphatic sediment in most cases.

Intraclastic phosphorites have been described from several Late Proterozoic deposits, such as Chaoyang, China (Li, this volume), and result from the break-up of fine grained semi-lithified mudstone or grainstone phosphorite. At Jhamarkotra intraclastic phosphorites are composed of angular clasts and fragments of boundstone phosphorite. In some instances (see Southgate, this volume) angular clasts result from the partial break-up of disconformity surfaces such as hardgrounds or phoscretes due either to organic (burrowing and boring) or mechanical reworking. Other discontinuities are represented by conglomeratic phosphorites, where the clasts are composed of well-rounded phosphate pebbles and cobbles derived from the erosion of a pre-existing phosphorite. The Lower Cambrian Xianji deposit of China is partly of this type. This form of coarse reworked phosphorite characteristically is relatively low-grade due to the inclusion of abundant sand-size terrigenous grains.

Boundstone phosphorites are most common in the Proterozoic of India (Banerjee, 1971) and China, though they also occur in the Cambrian of Australia (Southgate, 1980) and France (Prian, 1980). Boundstone phosphorites are much less common in the remainder of the Phanerozoic, but this may be merely a reflection of the general decrease of stromatolite abundance. There do, however, seem to be other factors involved for the peak in stromatolite development in the Proterozoic (Walter & Heys, 1985) predating the maximum development of phosphatic stromatolites by at least 100 m.y. and possibly as much as 200 m.y.

The occurrence of totally non-phosphatised inter-columnar

areas, the incorporation of small intraclasts of phosphatic stromatolite, and the fact (in the case of the Rajasthan deposits) that the phosphatic stromatolites are a different form to the non-phosphatic (calcareous) stromatolites all support primary precipitation and/or fixation of apatite, presumably by cyanobacteria.

Most commonly nodular phosphorites occur as scattered nodules or concretions within black organic-rich clastic rock sequences. The nodules range in size from a few millimetres to tens of centimetres, and are described in this volume from the Upper Proterozoic of Scotland, the Lower Cambrian of South Australia, the Middle and Upper Cambrian of Scandinavia, and around the Cambrian–Ordovician boundary in both maritime and cordilleran Canada. They are analogous to those found off the present-day coasts of Chile and Peru (Burnett & Veeh, 1977), and Southwest Africa (Bremner, 1980) in high organic productivity zones and in relatively deep water at outer shelf or upper continental slope locations. They have undergone little or no reworking and appear to have been formed in most cases at or just below the sediment/water interface. Most recent research (O'Brien *et al.*, 1981; Froelich & Burnett, personal communication) suggest that their origin also is related to bacterial activity.

In reviewing the various common phosphorite types it becomes apparent that facies relationships can be related to detailed small scale depositional environments, and that phosphorite type can be related to changes in water depth. Therefore from our present knowledge of Proterozoic and Cambrian phosphorites it is possible to propose an idealised section from outer shelf to beach locations, along which we find progressively: (1) minor concretionary and nodular phosphorites; (2) minor mudstone phosphorite; (3) major grainstone phosphorite with some boundstone phosphorite; (4) major mudstone phosphorite with minor boundstone phosphorite; and finally in the most shoreward location (5) phoscrete, rarely with conglomeratic phosphorite, at or above the littoral zone. Major phosphorite types and their proposed origins are summarised in Table 26.2, and their likely depositional environments are illustrated in Fig. 26.6. From this we see that D7–D8–D7 cycles in our model do not account for accumulation of nodular phosphorites, and indeed this type of phosphorite probably represents a minor output from Column B (such as that represented by B6). Boundstone phosphorites undergo very little reworking and can perhaps be considered as a very minor output from Column C. Mudstone phosphorites occupy an uncertain place in our model. They may represent an additional, and at times rather more significant, output from Column C, as shown by the dotted flow path C7–D9 where there is no implied reworking to produce a mudstone phosphorite deposit in a very shallow environment. Hence, we are dealing with a primary accumulation.

The remainder of the deposit types, such as grainstone phosphorite (by far the most important deposit type, particularly in the Phanerozoic) and the very minor conglomeratic and intraclastic types, have all undergone mechanical reworking under very shallow water conditions as indicated by the associated textures (cross-bedding, ripple marks, etc.) and by their grainsize characteristics. Therefore, the answer to our question is that the majority of grainstone phosphorites, undergo mechanical reworking and accumulation under shallow water conditions. Boundstone, and perhaps most mudstone phosphorites, undergo little if any reworking but nevertheless also accumulate under very shallow marine conditions. Nodular phosphorites accumulate under somewhat deeper water conditions. Our model fairly effectively takes all the possible options for depositional settings into account.

Table 26.2. *Summary of phosphorite types and their origin*

A. Grainstone phosphorite
1. Phosphatisation of non-phosphatic rounded grains:
   (a) as an inorganic process,
   (b) in response to the action of endolithic algae.
2. Phosphatisation of body cavity followed by removal of exoskeleton by abrasion or chemical action.
3. Abrasion of pre-existing phosphatic material (of any origin but especially mudstone and boundstone phosphorite) to produce sand-size phosphate grains.
4. Accumulation of originally phosphatic bioclastic remains.

B. Packstone and wackestone phosphorite
1. Mixing of grainstone and mudstone phosphorite.
2. Formation of a diagenetic matrix-cement within a grainstone phosphorite.

C. Mudstone phosphorite
1. Precipitation of apatite and/or settling of a suspended phosphate mud.
2. Precipitation-binding-trapping by cyanobacteria.
3. Phosphatisation of pre-existing non-phosphatic fine-grained sediments:
   (a) as an early diagenetic (pre-lithification) process,
   (b) as a late diagenetic (post-lithification) process.

D. Boundstone phosphorite
1. Precipitating/binding/trapping of phosphate to produce:
   (a) laminate stromatolitic phosphorite,
   (b) laminoid phosphorite.

E. Intraclastic phosphorite
1. Break-up of partly indurated phosphorite.

F. Conglomeratic phosphorite
1. Break-up abrasion and transportation of indurated phosphorite.

G. Nodular phosphorite
1. Development of diagenetic nodules within soft sediment.

H. Phosphate hardgrounds
1. Early diagenetic precipitation of phosphate cement within soft sediment commonly followed by removal of overlying soft sediment.

I. Phoscrete
1. Precipitation (possibly associated with chasmolithic and epilithic algae) of phosphate and phosphatisation of surficial cover associated with soil profile.

10. Are phosphorites always associated with organic-rich sediments?

This association is implied by the fact that organic-rich sediments are the end products of both Columns D and C of our

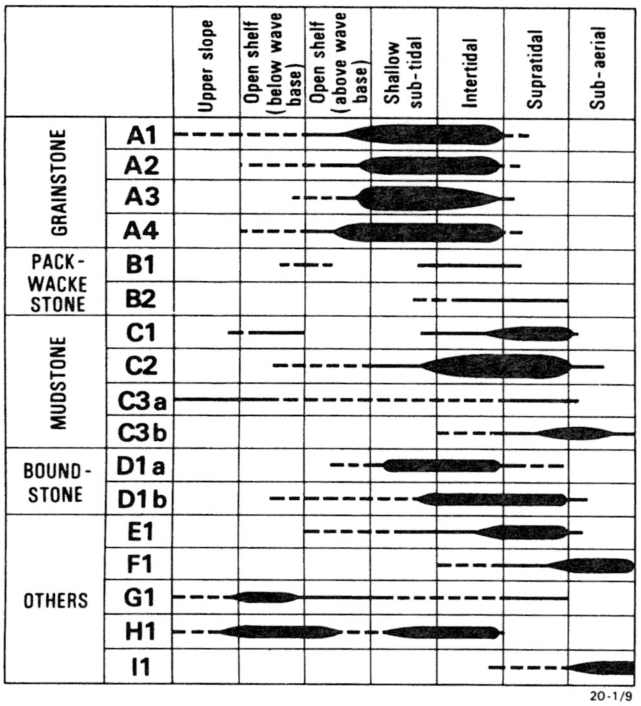

Fig. 26.6. Summary of the depositional environments of phosphorite petrological types recognised in this paper (see Table 26.2).

model. The picture which emerges from studies of the organic petrology and geochemistry of phosphorites (Sandstrom, this volume) is of abundant organic matter in the total phosphatic assemblage, but the highest organic contents are found in the associated black shales (for example, the Inca Formation of the Georgina Basin; the 'Stone Coals' of China). This same relationship applies in younger phosphatic sequences – for example, in the Phosphoria Formation it is the associated black shales which are very high in organic matter (and constitute important petroleum source rocks) rather than the actual phosphorites.

In his original upwelling model for phosphogenesis, Kazakov (1937, 1938) largely ignored the role of the organic world, considering the process of phosphate precipitation to be essentially inorganic. However, as indicated above it is apparent that organisms are vitally involved in phosphogenic processes, a relationship first recognised by Buckland (1829), who interpreted phosphate nodules as coprolites. The various hypotheses for the formation of phosphorites through the agency of organic activity have been reviewed by Gulbrandsen (1969) and Sheldon (1981). The activities involve direct contribution to the phosphorus sink, modification of the chemistry of the hydrosphere, and involvement in diagenetic processes. Both plants and animals are active in the phosphogenic process at various levels and the interactions are extremely complex. A measure of the potential importance of organic activity in the contemporary phosphogenic process is indicated by estimates of global phosphorus fluxes from terrestrial sources (land plants and animals, soils etc.) to the shallow oceans (3×10^{10} moles yr^{-1} P – Burnett & Oas, 1979), of which fully two-thirds is of biogenic origin, and from the surface waters to the deep oceans via the agency of marine plants (135×10^{10} moles yr^{-1}P – Sandstrom, 1982). In contemporary phosphogenic situations, where the biomass components are known, the most important contributions come from the phytoplankton, zooplankton, bacteria and higher plants, with a small but locally important contribution from marine arthopods and vertebrates. It is not so easy to assess the composition of the Proterozoic–Cambrian biomass. At this time, vertebrates and higher plants had not evolved, and the composition of the phytoplankton and zooplankton were markedly different (Fig. 26.4), bacteria and algae being apparently of primary importance. However, these contributions are presently impossible to quantify. Except for the organo-sedimentary stromatolites, algal remains are not readily fossilised, and bacterial remains very rarely so. The zooplanktonic components of the biomass are likely to have existed in quantity, but in non-mineralised, and thus unpreserved, form. Analyses of phosphorus content from complete organisms of a selected variety of hard tissues have been conducted on a range of animals and plants (e.g. Clarke & Wheeler, 1922; Vinogradov, 1953; Rhodes & Bloxam, 1971; Lowenstam, 1972). Those having some relevance to the Proterozoic–Cambrian biota include analyses on algae, coelenterates, hydrozoans, echinoderms, bivalve and gastropod molluscs, inarticulate brachiopods and crustaceans (see Table 26.3). Additionally, information on the P_2O_5 content of certain phytoplankton (mainly diatoms and dinoflagellates) and algae has been published by Fauconnier & Slansky (1979) and Slansky (1980).

Although the percentage of phosphorus is relatively low in many groups (e.g. the phytoplankton), the sheer numbers of individual cysts which may occur during periods of oceanic upwelling or seasonal bloom, make them a very important potential source of phosphorus. For example, seasonal fluxes in the environmental conditions of diatoms may give rise to blooms containing up to 10^9 cells/m^3 water (Brasier, 1980b) and form oozes which contain as much as 2% P_2O_5 (Baturin, 1978).

The phytoplankton are poorly represented in the Proterozoic–Cambrian biological record. Acritarchs appear to have occupied the ecological position of the later dinoflagellates, but are not common and do not occur in the concentrations recorded, for example, in the Paleocene Gafsa Basin of Tunisia where one sample contains 40 000 cysts/gram rock (Fauconnier & Slansky, 1979). At various places in this chapter we have discussed the potential importance of upwelling to our model for phosphogenesis. When it occurs at the present day it is invariably accompanied by a high level of biological activity. The primary production of organic matter in regions of upwelling may attain 27 tonnes/sq km/day dry weight (Peres & Deveze, 1964). Riley *et al.* (1949), however, estimated an annual organic carbon production from phytoplanktonic sources at 475 tonnes/sq km/day, from which value Fauconnier & Slansky (1979) calculated an annual production of phosphorus at 21 tonnes P_2O_5/sq km. At the present time, upwelling waters are mirrored by the 'red tides' – phytoplanktonic blooms composed largely of dinoflagellates with concentrations as great as 50×10^6 cells/litre water (Fauconnier & Slansky, 1979; Slansky, 1980). There is little direct evidence from the Proterozoic–Cambrian fossil record that ancient upwelling systems existed, although if they did, they would have been associated with a largely algal and bacterial biomass. However,

Table 26.3 *Phosphorus content of some invertebrates and plants. (1) Clarke & Wheeler (1922); (2) Rhodes & Bloxam (1971); (3) Slansky (1980)*

	P_2O_5 (%)	
Foraminiferans	(1) trace only	
Poriferans	(2) 0.17–0.77	
Hexacorals	(1) trace only	(2) 0.00–2.05
Octacorals	(1) 0.10–2.97	(2) 0.00–8.57
Hydroids	(1) trace only	(2) 0.18–3.23
*Annelids	(1) 0.22–20.72	(2) 6.30–21.58
Crinoids	(1) 0.10–0.43	
Echinoids	(1) 0.05–0.64	
Asteroids	(1) 0.07–0.81	
Ophiuroids	(1) 0.05–0.32	
Holothurians	(1) 0.85–3.15	
Calcareous brachiopods	(1) 0.17–0.25	
Phosphatic brachiopods	(2) 19.62–40.62	
Bryozoans	(1) 0.09–0.60	(2) 0.00–2.68
Bivalves	(1) 0.03–0.17	
Scaphopods	(1) trace only	
**Gastropods	(1) 0.04–0.38	
Cephalopods	(1) trace only	
*Crustaceans	(1) 1.87–7.75	
Phaeophyceae	(2) 0.88–7.17	
Rhodophyceae	(2) 0.50–0.93	
Corrallinaceae	(2) 0.00–0.48	
Chlorophyceae	(2) 1.21–4.82	
Dinoflagellates	(3) 0.00–0.08	
Diatoms	(3) 0.32–1.35	

*Clarke & Wheeler (1922) note that the phosphatic annelid *Onuphis tubicola* contains 21.58% P_2O_5 (57.20% in calcined ash), and that the phosphate content of different kinds of Crustacea varies considerably.
**Lowenstam (1972) records P_2O_5% values of up to 39.7% in the gizzard plates of eight gastropod species.

the organic carbon content of the phosphorites themselves, and particularly of the associated black shales clearly indicates that, at times, there was a high level of organic activity.

The work of Sandstrom (this volume) has shown that phosphorites have a somewhat different organic composition to the associated sediments: phosphorites contain abundant humic acids whereas they are fairly rare in the associated sediments. This carries the implication that the nature of the organic matter has a direct bearing on the formation of phosphate grains, or vice versa. But whichever way round it is, it suggests a strong C–P link. This is hardly surprising in view of the intimate link between these elements in all living organisms at the present day. There is no reason to believe that phosphorus was any less essential to organisms in the past.

The answer to our question must be that whilst the phosphorites themselves may be relatively low in organic carbon, all major deposits are invariably intimately associated with organic-rich sediments, if the total assemblage is preserved and has not been subjected to post-depositional weathering.

11. Are there facies relationships which the model does not presently cover?

Earlier in this volume Riggs examined the broadscale stratigraphic relationships of phosphorites to other sediments, pointing to the fact that phosphorites occur adjacent to, and form simultaneously with what he terms 'anomalous sediments'. We have already discussed the phosphorite–evaporite and phosphorite–black shale assemblages, both of which are included in the model. However, a commonly observed, as yet poorly understood, relationship exists between phosphorite and Mn or Mn–Fe facies. Examples of this association include the Proterozoic deposits of the Bawah area of India (Banerjee, this volume), and the Cambrian deposits and occurrences of the Avalon Platform (Notholt & Brasier, this volume), the Ocoee Supergroup of the eastern United States (Christie & Sheldon, this volume), the Khubsugul Basin (Ilyin *et al.*, this volume) and the Karatau Basin (Eganov *et al.*, this volume). In the case of the Khubsugul Basin, Ilyin *et al.* (this volume) note that in northern Mongolia the manganiferous interval is associated with a low grade bauxite. Manganiferous horizons in the Georgina Basin (Southgate, this volume) are also associated with phosphatic and non-phosphatic stromatolites and occur on subaerially exposed disconformity surfaces. Taken together these features are in general consistent with inter-tidal to sub-aerial locations. Enrichment of Mn and Fe commonly takes place at or above the intertidal zone where there is discharge of Fe and Mn-enriched groundwaters and precipitation of metals in response to a change of redox conditions. Alternatively, metal enrichment may occur as a consequence of sub-aerial lateritic-type weathering under tropical conditions, forming ferruginous, manganiferous or aluminous concentrations at horizons laterally equivalent to the phosphorite. Such an explanation seems to be consistent with the low palaeolatitude component of our model. A recent review of the origin of 'manganese giants' by Frakes & Bolton (1984) concludes that sea-level changes and the oxic–anoxic history of the water column have been major factors controlling the formation of manganese ore bodies. They also propose that the manganese grains are concentrated around the basin margin, for example in coastal embayments, by what they term the 'broom effect' (whereby the grains are driven shorewards by currents). Although there are some differences in detail, there are similarities in the model proposed by Frakes & Bolton (1984) for manganese deposits and our model for phosphogenesis (Fig. 26.1). The fact that manganese deposits and phosphorites do occur as lateral equivalents in the Proterozoic and Cambrian would seem to strengthen the view that these two chemical (or biochemical) sediments have much in common in their mode of origin and particularly in the importance of mechanical processes in concentrating the ore i.e. the D7–D8–D7 iterations of our model. The association between phosphorite and glauconite has been explored in a series of Geological Society of London papers (1980) which clearly showed both phosphorites and glauconites associated with shallow water low latitude marine conditions. Organic matter appears to play an important role (possibly in producing the right redox condition) and mechanical reworking is essential if the glauconite grains are to form a greensand (or the phosphate grains are to form a phosphorite). Therefore, our model will account for the phosphorite–glauconite association.

A phosphorite-banded iron formation (BIF) association has been alluded to earlier. It is important to emphasise that this association is dominated by iron. The phosphatic portions of the sequence are no more than thin intervals generally only a few millimetres thick within the BIFs (Howard, this volume). Nevertheless, these thin partings are clearly sedimentary and richly phosphatic, and therefore they can reasonably be regarded as phosphorites despite the overall minute contribution that they make to the composition of a BIF sequence. There are many theories for the origin of BIFs but the only one which can be applied to explain the BIF–phosphorite association is the upwelling model proposed for BIFs by Holland (1973) and applied by Cook & McElhinny (1979) to the phosphorite association. Simplistically, this hypothesis requires that the iron, associated silica and the phosphorus in Precambrian BIFs is derived from deep cold anoxic waters which upwelled into shallow marine areas and there precipitated out as BIFs. Our model is consistent with the occurrence of thin phosphorites in BIFs.

A suggested association of phosphorites and base metal mineralisation is less well defined. The Proterozoic base metal deposits of Broken Hill provide one of the few well-documented examples. Both the base metals and the apatite may be of syndepositional volcanogenic origin, and Stanton (1972) provided geochemical evidence to support this. This cannot be accommodated comfortably into our model. However, many base metals are hosted in organic-rich black pyritic shales (Gustafson & Williams, 1981) and are commonly associated with evaporites. Phosphorites are similarly found associated with black pyritic shales and evaporites. Therefore phosphorites and base metals may share a common lithological association, though the detailed genetic processes involved for base metals and phosphorites are in themselves quite different. The better defined association between phosphorites and uranium mineralisation is a consequence of somewhat different processes. The first of these is a consequence of the fact that uranium readily substitutes in the apatite lattice so that phosphorites can have high initial uranium contents. Weathering of phosphorites, can result in an extensive leaching of uranium followed by redeposition and concentration of the mobilised uranium in adjacent areas where the redox conditions are favourable. A second type relies on the fact that phosphate, like uranium, can be concentrated by weathering on disconformity surfaces. The phosphorite–uranium association is one which must lie outside the purview of our model. Accordingly, the answer to the question must be that there are some facies relationships which our model does not cover, but that overall it is able to accommodate all the major associations including the P–Mn and P–Fe associations.

12. Is organic evolution related to major periods of phosphogenesis?

One of the most fundamental questions raised by the model is related to the output from Column D where phosphorites are matched by the outputs of Columns A–C which involve organic changes on various scales. We have previously noted (Cook & Shergold, 1984) the correspondence of rapid organic evolution at the Proterozoic–Cambrian boundary with periods of intense phosphogenesis.

The rapid acquisition of mineralised skeletal structures during the initial Early Cambrian, and consequent sudden influx of shelly fossils has been termed the Cambrian explosion or Cambrian radiation event (Brasier, 1979). Some of the main hypotheses previously advanced to explain this sudden increase in the fossil record have been discussed by Brasier (1982), Runnegar (1982) and Cook & Shergold (1984). The last authors point out that the first appearance of the metazoans (*c.*670 m.y.) corresponds reasonably well with the onset of Late Proterozoic–Early Cambrian phosphogenesis, and that the incidence of metazoan biomineralisation at the Proterozoic–Cambrian transition coincides remarkably with the main phase of phosphogenesis, and with associated anomalous isotope geochemistry of sulfur and oxygen (Cook & Shergold, 1984, Fig. 2). It is postulated that an increase in phosphorus and other nutrients in the marine environment at this time may have led to an increase in the shallow marine biomass (phytoplankton), which in turn may have produced an increase in the content of atmospheric oxgen due to increased photosynthesis. The abundance of organisms secreting primary phosphatic skeletal structures also correlates with the peak of phosphogenesis in the Tommotian Stage of the Early Cambrian. As phosphatic organisms seem to be more abundant near phosphate deposits, there appears to be a direct link between phosphorites and phosphate biomineralisation. Maximum diversification of morphological structure also occurs at this time. Biomineralisation may therefore by associated initially with the disposal, by secretion as skeletal structures, of perhaps unwanted quantities of phosphorus.

Some terminal Proterozoic and Early Cambrian organisms, like the inarticulate brachiopods, the conodonts, the hyolithelminths, mitrosagophorans and phosphatocopine bradoriids, are composed almost entirely of calcium phosphate and would conceivably have made a contribution to phosphogenesis by the direct precipitation of skeletal apatite and its incorporation into the sediments. Those organisms which utilised calcium phosphate to secrete skeletal structures in the latest Proterozoic and Cambrian are noted in Fig. 26.4. With the waning of phosphogenesis in the later Early Cambrian, calcium carbonate became the dominant biomineralising compound, perhaps because the precipitation of calcite was no longer inhibited by the presence of excess phosphorus in the environment. At this stage, metabolism based on calcium ion regulation became increasingly refined (Lowenstam & Margulis, 1980). Therefore we do indeed observe that phosphogenesis and major evolutionary changes are related, and it is most unlikely that this association is fortuitous for phosphorus is commonly a biolimiting element. A marked increase in phosphate availability is likely to have led to a marked increase in the biomass, particularly in phytoplankton. This would in turn have affected not only the hydrosphere by modifying the phosphate content, the pH, and the partial pressure of CO_2, but also the atmosphere through a marked increase in the abundance of photosynthesisers.

13. Can phosphogenesis be related to plate tectonics?

With this question we return to the starting point of our model i.e. line 1, and it follows that to completely answer this question we would virtually have to repeat much of the prior discussion in this chapter. It is also perhaps the most difficult question to answer other than by adding yet more details to the model. It is a question which has been examined by Cook & McElhinny (1979) who found no evidence for a direct genetic relationship, but suggested that there is a temporal and spatial relationship to phases of plate tectonics. They pointed out that a tensional and rifting phase precedes the separation of continental fragments and the initial development of a seaway, in which phosphorites may have subsequently been deposited. Thus, tholeiitic basalts (perhaps associated with early rifting) could reasonably be expected to precede phosphorites. This is indeed the case in the Georgina Basin of northern Australia for example. Conversely, if a compressional phase results in the formation of a narrower seaway (in which phosphorites were then deposited) then an andesitic basalt reasonably could be expected to occur synchronously or follow the deposition of phosphorites. These models of Cook & McElhinny admittedly represent oversimplification of the relationships that exist between phosphorites and volcanics but they do serve to illustrate the point that any relationship between phosphogenesis and volcanism can be more readily related to phases of plate tectonics than to a direct genetic (i.e. geochemical) link.

Cook & McElhinny (1979) point out that phosphorites form mainly at low palaeolatitudes, and that a new phase of sea floor spreading which results in a continental plate moving to a low latitude could be considered in part responsible for placing that plate or the associated continental margin in a location where phosphogenesis might take place. In other words, whilst plate tectonics might not be responsible for the formation of a phosphate deposit it might be responsible for 'setting the scene'. The Cook & McElhinny models emphasise the importance that the orientation of seaways have to phosphogenesis: they postulate that generally phosphogenic conditions can occur relatively soon after the initiation of sea-floor spreading if the narrow seaway formed is in a near-equatorial location and has a latitudinal (i.e. east–west) orientation. Broad-scale phosphogenesis is likely only to occur in a longitudinally (north–south) oriented seaway when sea-floor spreading has proceeded to the stage where the ocean is sufficiently broad for stable oceanic gyrals to form in low latitude locations. A third phosphogenic system may be formed in the monsoonal season (Parrish *et al.* this volume). All three scenarios could give rise to phosphogenic systems in Proterozoic and Cambrian time. However, there is some uncertainty regarding the application of the theory of plate tectonics to the Precambrian and it could be argued that the Cook & McElhinny models have no relevance to that period of Earth history. For our part we see no evidence that patterns of sedimentation and the underlying causes of these patterns, including tectonics, were greatly different in the Proterozoic and the Phanerozoic and we would therefore argue that broad-scale Phanerozoic models of phosphogenesis are likely to be applicable in the Proterozoic with any differences being more a response to changes in the biota rather than changes in tectonic style. Nevertheless, it must be admitted that there is a need to acquire far more data for the Proterozoic and Cambrian to more confidently fit phosphogenesis into a temporal and spatial pattern of plate movement.

We are forced to conclude that we are unable to provide a definitive answer to the question posed. The inferred association between phosphogenesis and plate tectonics obviously is an important element in our model, but as far as the Proterozoic and Cambrian are concerned we need far more data on palaeogeography, sea-floor spreading and global tectonics before we can provide the evidence to support this aspect of our model.

Conclusions

In this final chapter we have proposed a model and posed a series of questions designed to test that model. In most instances the answers elicited from the evidence provided in the previous chapters of this volume and summarised here, give grounds for believing that the model is realistic. There is obviously a great deal of further testing to be done and no doubt there will be a need to modify the model as new information comes to light. This new information may even ultimately enable us to quantify the model.

We have seen throughout this volume evidence that no single mechanism is responsible for the formation of phosphorites and that phosphogenesis is best represented by a fairly complex model. Phosphorite deposits form in response to factors ranging in scale from the depositional conditions in an area a few square kilometres in extent to changes in oceanic conditions which may be global in extent. They are also profoundly affected by processes acting on a variety of time scales ranging from perhaps tens of years to tens of millions of years. Factors such as climatic changes, changes in the phosphate content of the world ocean, and plate tectonics create the general conditions for phosphogenesis, but they will not in themselves produce phosphate deposits. Our model indicates that this is the consequence of more localised conditions including upwelling (oceanic, shelf or estuarine), the development of shallow (commonly epicontinental) seas, the occurrence of currents and minor relative sea-level changes to rework and concentrate the phosphate grains, and the development of 'traps' in which the phosphate grains accumulate. There are clearly variations on this set of parameters. For example, upwelling may not have been a pre-requisite for deposits located well within very shallow intracratonic basins. For these it may be necessary to invoke other mechanisms for getting nutrient-rich waters into extensive very shallow seas. Alternatively, upwelling may have taken place, but perhaps hundreds of kilometres from the site where the phosphate grains finally accumulated. Mechanical reworking whilst quite clearly a pre-requisite for most grainstone phosphorites may not have been a pre-requisite for the mudstone phosphorites and certainly was not a necessary feature of boundstone phosphorites. Again, this can be accommodated within our model.

Almost 50 years ago Kazakov profoundly influenced our thinking by pointing to the importance of upwelling to phosphogenesis. Our story is not so elegantly simple and therefore perhaps not quite so palatable as that presented by Kazakov, for we are forced to conclude that upwelling is just one of the potential pre-requisites, albeit an important one, for many phosphate deposits. Our phosphogenic model is complex and involves a range ofphysicochemical processes acting on a variety of temporal and spatial scales. It may be argued that the model we have provided is specific to the Proterozoic and Cambrian periods. This may be so, and certainly the major changes in flora and fauna around the Proterozoic–Cambrian boundary clearly had a profound effect on the distribution of phosphorus in the biomass, but we suspect that our model may be broadly applicable to other periods of phosphogenesis, such as those that occurred in the late Cretaceous and early Tertiary or in the Miocene. We also suspect that just as the phosphogenic event around the Precambrian–Cambrian boundary had far-reaching effects on the biota at that time so too did the later phosphogenic events. The Cretaceous–Tertiary boundary is one where some of the most spectacular evolutionary changes occur. This was also a time when phosphogenesis occurred on a scale equivalent to that around the Precambrian–Cambrian boundary.

Our final conclusion must be that we have a new found appreciation of the importance of phosphorites not only in their own right but because of the light they throw on other processes of Earth history ranging from the geochemistry of the hydrosphere to the many processes of organic evolution which had influence on and were profoundly influenced by the distribution of phosphorus in the biosphere. Not only are phosphorites now better understood but they are also now established as sediments of great economic and scientific importance.

References

Arthur, M.A. & Jenkyns, H.C. (1981). Phosphorites and paleoceanography. *Oceanologia Acta*, SP, 83–96.

Banerjee, D.M. (1971). Precambrian stromatolitic phosphorite of Udaipur, Rajasthan, India. *Geological Society of America Bulletin*, **82**, 2319–30.

Bathurst, R.J.C. (1975). Carbonate sediments and their diagenesis. In *Developments in Sedimentology*, **12**, 620pp. Elsevier, Amsterdam.

Baturin, G.N. (1978). *Phosphorites*. P.P. Shirshov Institute of Oceanology, USSR Academy of Sciences, 232pp. Nauka, Moscow.

Bertrand-Sarfati, J. & Walter, M.R. (1980). Stromatolite biostratigraphy. *Precambrian Research*, **15**, 353–71.

Brasier, M.D. (1979). The Cambrian radiation event. In *The Origin of Major Invertebrate Groups*, ed. M.R. House. Systematics Association Special Volume, **12**, 103–59.

Brasier, M.D. (1980a). The Lower Cambrian transgression and glauconite–phosphate facies in western Europe. *Journal of the Geological Society*, **137** (6), 695–703.

Brasier. M.D. (1980b). *Microfossils*. George Allen & Unwin, London.

Brasier, M.D. (1982). Sea level changes, facies changes, and the Late Precambrian–Early Cambrian evolutionary explosion. *Precambrian Research*, **17**, 105–23.

Bremner, J.M. (1980). Concretionary phosphorite from SW Africa. *Journal of the Geological Society London*, **137**, 773–86.

Buckland, W. (1829). On the discovery of coprolites, or fossil faeces, in the Lias at Lyme Regis, and in other formations. *Transactions of the Geological Society London 2nd Series*, **3**, 223–36.

Burnett, W.C. & Oas, T.G. (1979). Environment of deposition of marine phosphate deposits off Peru and Chile. In *Proterozoic–Cambrian Phosphorites*, ed. P.J. Cook & J.H. Shergold, pp. 54–6. ANU Press, Canberra.

Burnett, W.C. & Veeh, H.H. (1977). Uranium-series disequilibrium studies on phosphorite nodules from the west coast of South America. *Geochimica et Cosmochimica Acta*, **41**, 755–64.

Clarke, F.W. & Wheeler, W.C. (1922). The inorganic constituents of marine invertebrates. *US Geological Survey Professional Paper*, **124**, 1–62.

Cloud, P.E. & Semikhatov, M.A. (1969). Proterozoic stromatolite zonation. *American Journal of Science*, **267**, 1017–61.

Cook, P.J. (1972). Petrology and geochemistry of the phosphate deposits of northwest Queensland, Australia. *Economic Geology*, **64**, 1193–213.

Cook, P.J. (1982). The Cambrian palaeogeography of Australia and opportunities for petroleum exploration. *APEA Journal*, **22**, 42–64.

Cook, P.J. (1983). World availability of phosphorus: an Australian perspective. In *Phosphorus in Australia*, ed. A.B. Costin & C.H. Williams, pp. 1–41. Centre for Resource and Environment Studies, Canberra.

Cook, P.J. (1984). Spatial and temporal controls on the formation of phosphate deposits – a review. In *Phosphate Minerals*, ed. J.O, Nriagu & P.B. Moore, pp. 242–74. Springer-Verlag, Berlin.

Cook, P.J. & McElhinny, M.W. (1979). A re-evaluation of the spatial and temporal distribution of sedimentary phosphate deposits in the light of plate tectonics. *Economic Geology*, **74**, 315–30.

Cook, P.J. & Shergold, J.H. (1984). Phosphorus, phosphorites and skeletal evolution at the Precambrian–Cambrian boundary. *Nature*, **308**, 231–6.

Davies, P.J., Bubela, B. & Ferguson, J. (1978). Formation of ooids. *Sedimentology*, **25**, 703–30.

Eganov, E.A., Ergaliev, G.H., Ilyin, A.V. & Krasnov, A.A. (1984). Karatau Phosphorite Basin, Kazakhastan. *27th Session International Geological Congress Moscow, Excursion Guidebook* 045, 3–46 (Russian), 47–69 (English).

Fauconnier, D. & Slansky, M. (1979). The possible role of dinoflagellates in phosphate sedimentation. In *Proterozoic–Cambrian Phosphorites*, ed. P.J. Cook & J.H. Shergold, pp. 93–101. ANU Press, Canberra.

Ferguson, J., Bubela, B. & Davies, P.J. (1978). Synthesis and possible mechanism of formation of radial carbonate ooids. *Chemical Geology*, **22**, 285–308.

Fischer, A.G. & Arthur, M.A. (1977). Secular variations in the pelagic realm. *Society of Economic Paleontologists and Mineralogists Special Publication*, **25**, 19–50.

Frakes, L.A. & Bolton, B.R. (1984). Origin of manganese giants; sea-level changes and anoxic–oxic history. *Geology*, **12**(2), 83–6.

Gulbrandsen, R.A. (1969). Physical and chemical factors in the formation of marine apatite. *Economic Geology*, **64**, 365–82.

Gustafson, L.B. & Williams, N. (1981). Sediment-hosted stratiform deposits of copper, lead and zinc. *Economic Geology*, **75**, 139–78.

Hambrey, M.J. & Harland, W.B. (1981). *Earth's pre-Pleistocene Glacial Record*. Cambridge University Press, Cambridge.

Henningsmoen, G. (1957). The trilobite family Olenidae with description of Norwegian material and remarks on the Olenid and Tremadocian Series. *Skrifter utgilt av Det Norske Videnskaps-Akademie Oslo*, 1, *Mat-naturv, Klasse*, **1**, 1–303.

Holland, H.D. (1973). The oceans: a possible source of iron in iron-formations. *Economic Geology*, **68**, 1169–72.

Ilyin, A.V. & Bjamba, Z. (1980). *Guidebook for Excursion Phosphorites of the Khubsugul Basin, Mongolian Peoples Republic, 17–28 June, 1980*. Soviet Committee for IGCP Project 156, Phosphorites, Geological Institute, USSR Academy of Sciences.

Kazakov, A.V. (1937). The phosphorite facies and the genesis of phosphorites. In *Geological Investigations of Agricultural Ores. Transactions Scientific Institute for Fertilizers and Insecto-fungicides*, **142**, 93–113.

Kazakov, A.V. (1938). The phosphorite facies and the genesis of phosphorites. *17th Session International Geological Congress Moscow*, pp. 95–113.

Keyser, F. de & Cook, P.J. (1972). Geology of the Middle Cambrian phosphorites and associated sediments of northwestern Queensland. *Bureau of Mineral Resources, Geology and Geophysics of Australia, Bulletin*, **138**, 79pp.

Lee, A.I.N. (ed.). (1980). *Fertilizer Mineral Occurrences in the Asia–Pacific Region*. East–West Resource Systems Institute, East–West Center, Honolulu.

Linnarsson, J.G.O. (1877). Om faunan i lagren med *Paradoxides Olandicus. Geologiska Foreningens i Stockholm Forhandlingar*, **3**, 352–75.

Lowenstam. H.A. (1972). Phosphatic hard tissues of marine invertebrates: their nature and mechanical function, and some fossil implications. *Chemical Geology*, **9**, 153–66.

Lowenstam, H.A. & Margulis, L. (1980). Evolutionary prerequisites for Early Phanerozoic calcareous skeletons. *Biosystems*, **12**, 27–41.

Lucas, J. & Prévôt, L. (1981). Synthèse d'apatite à partir de matière organique phosphorée (ARN) et de calcite par voie bactérienne. *Compte Rendus d'Academie des Sciences Paris*, **292** Serie II, 1203–8.

Lucas, J. & Prévôt, L. (1984). Synthèse de l'apatite par voie bactérienne à partir de matière organique phosphatée et de divers carbonates de calcium dans des eaux douce et marine naturelles. *Chemical Geology*, **42**, 101–18.

Mabie, C.P. & Hess, H.D. (1964). Petrographic study and classification of western phosphate ores. *US Bureau of Mines Report*, **6468**, 1–95.

McKelvey, V.E., Swanson, R.W. & Sheldon, R.P. (1953). The Permian phosphorite deposits of the western United States. *19th Session, International Geological Congress, Algiers*, **11**, 45–65.

Manheim, F., Rowe, G.T. & Jipa, D. (1975). Marine phosphorite formation off Peru. *Journal of Sedimentary Petrology*, **45**, 743–51.

Mansfield, G.R. (1927). Geography, geology and mineral resources of part of Southeast Idaho. *US Geological Survey Professional Paper*, **152**, 1–409.

Martens, C.S. & Harris, R. (1970). Inhibition of apatite precipitation in the marine environment by magnesium ions. *Geochimica et Cosmochimica Acta*, **34**, 621–5.

Missarzhevsky, V.V. & Rozanov, A.Yu. (1963). On the Cambrian lower boundary. *Izvestiya Akademiya Nauk SSSR, Geologichesky Seriya*, **2**, 60–72.

Missarzhevsky, V.V. & Rozanov, A.Yu. (1973). *Excursion guide book to the Aldan and Lena Rivers. International Excursion for the Cambrian and Precambrian boundary problem*, pp. 1–69 (Russian), 71–118 (English). USSR Academy of Science, Moscow–Irkutsk.

Notholt, A.J.G. (1979). Resources of Precambrian and Cambrian sedimentary phosphate rock. In *Proterozoic–Cambrian Phosphorites*, ed. P.J. Cook & J.H. Shergold, pp. 71–86. ANU Press, Canberra.

Notholt, A.J.G. (1980). Economic phosphatic sediments: mode of occurrence and stratigraphical distribution. *Journal of the Geological Society of London*, **137**, 793–805.

O'Brien, G.W., Harris, J.R., Milnes, A.R. & Veeh, H.H. (1981). Bacterial origin of East Australian continental margin phosphorites. *Nature*, **294**, 442–4.

Parrish, J.T., Ziegler, A.M. & Humphreville, R.G. (1983). Upwelling in the Paleozoic Era. In *Coastal Upwelling. Its Sediment Record.* Part B: *Sedimentary Records of Ancient Coastal Upwelling*, ed. J. Thiede & E. Suess, pp. 553–78. Plenum Press, New York.

Peres, J.M. & Deveze, L. (1964). *Oceanographic biologique et biologie marine.* Volume 2. *La vie pelagique.* Universités de France Press, Paris.

Prian, J.-P (1980). Les phosphorites Cambriennes du versant septentrional de La Montague Noire, en sud du Bassin Permien de Camares (Avegron). Ph.D. Thesis, L'Universite P. et M. Curie, Paris (unpublished).

Repina, L.N., Khometovsky, V.V., Zhuravela, I.T. & Rozanov, A.Yu. (1964). Lower Cambrian biostratigraphy of the Sayan–Altai region. *Akademiya nauk SSR, Sibirskoe Otdelenie, Institut Geologii i Geofiziki*, 364pp. Izdatelystvo 'Nauka', Moscow.

Rhodes, F.H.T. & Bloxam, T.W. (1971). Phosphatic organisms in the Paleozoic and their evolutionary significance. *Proceedings North American Paleontological Convention, September 1969.* Part K, 1485–513.

Riggs, S.R. (1979a). Petrology of the Tertiary phosphorite system in Florida. *Economic Geology*, **74**(2), 195–220.

Riggs, S.R. (1979b). Phosphorite sedimentation in Florida – a model phosphogenic system. *Economic Geology*, **74**, 285–314.

Riggs, S.R. (1984). Paleoceanographic model of Neogene phosphorite deposition, US Atlantic continental margin. *Science*, *223* (4632), 123–31.

Riggs, S.R., Lewis, D.W., Scarborough, A.K. & Snyder, S.W. (1982). Cyclic deposition of Neogene phosphorites in the Aurora area, North Carolina, and their possible relationship to global sea-level fluctuations. *Southeastern Geology*, **23**(4), 189–204.

Riley, G.A., Stommel, M. & Bumpus, D.F. (1949). Qualitative ecology of the plankton of the western North Atlantic. *Yale University Oceanographic Laboratory*, **12**, 1–69.

Rozanov, A.Yu. (1966). The Cambrian lower boundary problem. *Sbornik Itogi Nauki, Moskva*, 92–111.

Rozanov, A.Yu. (1973). Regulations for the morphological evolution of the Archaeocyatta and problems of the development of the Lower Cambrian Stages. *Akademiya Nauk SSSR, Trudy Geologicheskiy Institut*, **241**, 153–60.

Rozanov, A.Yu. & Debrenne, F. (1974). Age of archaeocyathid assemblages. *American Journal of Science*, *274*, 833–48.

Rozanov, A.Yu. & Missarzhevsky, V.V. (1966). Biostratigraphy and fauna of the lower horizons of the Cambrian. *Akademiya Nauk SSR, Trudy Geologicheskiy Institut*, **148**, 120pp. Izdatelystvo 'Nauka', Moscow.

Rozanov, A.Yu., Missarzhevsky, V.V., Volkova, N.A., Voronova, L.C., Krylov, I.N., Keller, B.M., Korolyuk, I.K., Lendzion, K., Michniak, R., Pychova, N.G. & Sidorov, A.D. (1969). Tommotian Stage and the Cambrian lower boundary problem. *Akademiya Nauk SSSR, Trudy Geologicheskiy Institut*, **206**, 380pp. Moscow: Izdatelystuo 'Nauka'. (In translation by US Department of the Interior and National Science Foundation, Washington, DC, published by Amerind Publishing Co Pvt Ltd, New Delhi, 1981).

Runnegar, B.N. (1982). The Cambrian explosion: animals or fossils? *Journal of the Geological Society of Australia*, **29**, 395–411.

Sandstrom, M.W. (1982). Organic geochemistry of phosphorites and associated sediments. Ph.D. Thesis, Australian National University (unpublished).

Sheldon, R.P. (1964). Paleolatitudinal and paleogeographic distribution of phosphorite. *US Geological Survey Professional Paper*, **501**-C, 106–13.

Sheldon, R.P. (1981). Ancient marine phosphorites. *Annual Reviews Earth and Planetary Sciences*, **9**, 251–84.

Sheldon, R.P. (1982). The Precambrian earth ice-ring model to account for changes in exogenic regimes from Proterozoic to Phanerozoic eras. *IGCP Project 156, International Field Workshop and Seminar on phosphorite*, **5**, 89–90 (Abstract).

Slansky, M. (1980). Géologie des phosphates sedimentaires. *Memoire Bureau de Recherches géologiques et minières*, **114**, 92pp.

Southgate, P.N. (1980). Cambrian stromatolitic phosphorites from the Georgina Basin. *Nature*, **285**, 395–7.

Stanton, R.L. (1972). A preliminary account of chemical relationships between sulfide lode and 'banded iron formation' at Broken Hill, New South Wales. *Economic Geology*, **67**, 1128–45.

The British Sulphur Corporation Ltd (1971). *World Survey of Phosphate Deposits*, 3rd edn. The British Sulphur Corporation Ltd, London.

Vidal, G. & Knoll, A.H. (1982). Radiations and extinctions of plankton in the Late Proterozoic and Early Cambrian. *Nature*, **297**, 57–60.

Vidal, G. & Knoll, A.H. (1983). Proterozoic plankton. *Geological Society of America Memoir*, **161**, 265–77.

Vinogradov, A.P. (1953). The elementary chemical composition of marine organisms. *Sears Foundation for Marine Research Memoir*, **2**, 647pp.

Walter, M.R. & Heys, G.R. (1985). Links between the rise of the Metazoa and the decline of stromatolites. *Precambrian Research*, 29, 149–174.

Westergård, A.H. (1936). Paradoxides oelandicus beds of Öland, with the account of a diamond boring through the Cambrian of Mossberga. *Sveriges Geologiska Undersökning*, C, **394**, 5–66.

Westergård, A.H. (1944). Borrningar genom alunskifferlagret pa Östergötland. *Sveriges Geologiska Undersökning*, C, **463**, 1–22.

Westergård, A.H. (1947). Supplementary notes on the Upper Cambrian trilobites of Sweden. *Sveriges Geologiska Undersökning*, C, **489**, 3–34.

Zhuravleva, I.T. (1970). Marine faunas and Lower Cambrian stratigraphy. *American Journal of Science*, **269**, 417–55.